THE TRANSPORT MANAGER'S & OPERATOR'S HANDBOOK 2006

36th edition

DAVID LOWE

IVECO

Truck Point
BRIDGESTONE
Firestone

KOGAN PAGE

The masculine pronoun has been used in this book. This stems from a desire to avoid ugly and cumbersome language, and no discrimination, prejudice or bias is intended.

Publisher's note
Every possible effort has been made to ensure that the information contained in this book is accurate at the time of going to press, and the publishers and authors cannot accept responsibility for any errors or omissions, however caused. No responsibility for loss or damage occasioned to any person acting, or refraining from action, as a result of the material in this publication can be accepted by the editor, the publisher or the author.

First published as *The Transport Manager's Handbook*

This thirty-sixth edition published in Great Britain in 2006

Kogan Page Limited
120 Pentonville Road
London N1 9JN
United Kingdom
www.kogan-page.co.uk

British Library Cataloguing in Publication Data

A CIP record for this book is available from the British Library.

ISBN 0 7494 4488 6

Typeset by Saxon Graphics Ltd, Derby
Printed and bound in Great Britain by Cambrian Printers Ltd, Aberystwyth, Wales

Contents

Biographical Note

David Lowe FCILT

David Lowe is a freelance writer and lecturer who has been actively involved with the road transport industry for many years.

He is the sole author of *The Transport Manager's and Operator's Handbook* now in its 36th year of annual publication. His many other works include two best-selling titles for CPC students: *A Study Manual of Professional Competence in Road Transport Management* (11 editions) and *1001 Typical CPC Questions and Answers* (three editions), *Goods Vehicle Costing and Pricing Handbook* (four editions), *The Tachograph Manual* (two editions), and *The European Road Freighting Handbook* (1994), all published by Kogan Page. His latest works, *The Dictionary of Transport and Logistics* and *The Pocket Guide to LGV Drivers' Hours and Tachograph Law*, were published by Kogan Page in 2002 and 2004 respectively.

He has written extensively for the transport press, including *Commercial Motor*, and also wrote a number of successful operator and driver handbooks for *Headlight* magazine. He was the author of the Chartered Institute of Purchasing and Supply (CIPS) graduate diploma study guide *Distribution* published in 2000, wrote the texts for Dunlop Tyres' *Usercare* long-running series for commercial vehicle operators and its *Facts About...* series for motorists, and was editor of Ford Motor Company's popular *Fleet Efficiency Guide* series.

With his detailed practical knowledge of UK and European transport law, David Lowe is well qualified to explain complex legal issues in layman's terms. He understands what the transport manager and small fleet operator need to know, and he tells them in no uncertain terms what they must and must not do to stay on the right side of the law.

He is a Chartered Fellow of the Institute of Transport and Logistics, and a past Champion of its Freight Transport Special Interest Group (SIG). He is a Liveryman of the Worshipful Company of Carmen and a Freeman of the City of London.

Acknowledgements

I am grateful to all the many individuals and organizations that have provided helpful information for inclusion in this *Handbook*. During the 36 years I have been writing the *Handbook* I have approached so many individuals, firms and industry-related organizations for information, guidance and assistance that it really is impossible to thank them all individually. Some are consulted regularly year after year; some have been newly sought out for guidance. There is no doubt that without such help the *Handbook* would not have achieved its prominence as the so-called 'bible' for transport managers and others. To all concerned, once again, I express my sincere thanks for their help.

Reference for information has been made to many sources – the legislation itself, both British and EU; published guides and notes; to the Department for Transport (D*f*T) and its forebears, the Traffic Commissioners and their Traffic Area Office staff, and to the enforcement authorities. Many other relevant bodies have been consulted and both the national press and the transport trade press have provided a continuous reminder of changing legislation and developments in the industry. Similarly, a constant flow of comments and questions from consultancy clients and seminar delegates has provided the source of many leads for inclusion of information and the inspiration to continue revising the contents of the *Handbook* to include more and more legislative detail. To all these publications, organizations and nameless individuals I am greatly indebted.

I would like to thank especially my wife Patricia and the many friends in the industry who are a constant source of encouragement and, as always, too, the editorial staff of Kogan Page whose help ensures yet another edition of the *Handbook*.

Every effort has been made to ensure that the material contained in this book is as correct and as up-to-date as possible and I accept full responsibility for any errors which may be found. However, I cannot stem the progress of legislation which inevitably occurs during the life of the current edition of this annual publication.

Preface to the 36th Edition

The past year, 2004–05, has not been a happy one for the road haulage industry. While costs have continued to rise – driven ever upwards by rising fuel prices (the price for crude oil was verging on $60 a barrel at the time of writing) and the high level of government duty on every litre purchased, and as a result of the reduced working hours emanating from the Road Transport Directive's imposition of a 48-hour week (estimated to cost the haulage industry up to £4 billion), which finally came into effect in April 2005 – new and proposed legislation has continued to burden operators at an alarming rate, as the many new provisions recorded in this revised and updated 2006 edition of the *Handbook* clearly indicate. There is everything here, from the farce over the on/off introduction of digital tachographs to the build-up to the Lorry Road User Charge which was finally, and very abruptly, dumped by the government just a short time before this edition of the *Handbook* went to press – another farce? In between these particular absurdities there is still the uncertainty about proposed changes to the EU hours' law. But more significantly at the moment is the reality of new legal requirements; for example, on the security of dangerous goods carriage, revised rules on the enforcement of the EU drivers' hours and rest period requirements, new rules on working at height, new rules on speed limiters for light vans – a sector of the haulage market that seems to be growing inexorably according to DfT statistics – and a host of other provisions set to cause grief to the struggling road haulier. Amid all this excitement one could be forgiven for almost overlooking the fact that in July 2004 the government admitted that its former 10-year Transport Plan was both too optimistic 'and looked at the wrong things'. In its place, however, it launched a 30-year plan, one of the keystones of which was to be the introduction of the Lorry Road User Charge, which, of course, has since been abandoned; so much for government 10-year and 30-year planning.

Keeping up to date with the latest legislation is reputed to be the biggest challenge facing the road haulage industry and, in fact, there is so much legislation coming through these days that your eyes glaze over, according to Bob Durward, director of the British Aggregates Association. On a positive note, however, readers of this *Handbook* should have no such problems of comprehending the law or recognizing their statutory responsibilities since, as always, its purpose is to help goods vehicle operators, transport managers and the many other individuals responsible for road transport operations to know and understand their legal obligations. More particularly, it should help them avoid the risks of prosecution, along with the heavy financial penalties imposed these days on conviction and, even more worrying, the imposition of punitive action against their operator's licence; these being the almost certain consequences of illegal activities that the transport trade press continues to report week by week with unfailing regularity. Hopefully, this *Handbook* will, as ever, help to steer readers clear of such problems and provide both interesting and intelligible reading as well as a useful and ready source of reference to the many and complex legal requirements relating to goods vehicle ownership and use.

READERS PLEASE NOTE

The Inland Revenue and Her Majesty's Customs and Excise

As a result of provisions contained in the *Commissioners for Revenue and Customs Act 2005*, from 18 April 2005 two new integrated departments have been established as follows:

- Her Majesty's Revenue and Customs (HMRC), and
- The Revenue and Customs Prosecutions Office (RCPO).

The new department, HMRC, is responsible for collecting the bulk of all tax revenues as well as strengthening the UK's frontiers. It manages many personal and business issues, but particularly so far as transport operators are concerned:

- income, corporation, capital gains, inheritance, insurance premium, stamp, land and petroleum revenue taxes;
- environmental taxes: climate change and aggregates levy and landfill tax;
- VAT;
- customs duties and frontier protection;
- excise duties;
- National Insurance;
- enforcement of the National Minimum Wage.

Previous Editions of *The Transport Manager's and Operator's Handbook*

Readers may be interested to note that a full set of all previous editions of the *Handbook* (ie 35 editions to date from the first in 1970) are held by the Chartered Institute of Logistics and Transport in its Corby Library where they may be accessed by anybody researching the progress of transport legislation over this period.

David Lowe

Introduction

Historically, the road transport industry in Great Britain has had to contend with a mass of legislation which has imposed considerable restriction on operations, to say nothing of the burdens of high cost and considerable worry for the owners, operators and managers of goods vehicle fleets.

The past 35 or so years in particular have seen more than their fair share of such legislation. Looking back to the days when the Road Safety Act 1967 (long defunct) and the Transport Act 1968 were about to be introduced, it would have been difficult then to foresee the mass of legal requirements which would have to be faced over the years ahead. Who could have anticipated the construction of a tunnel under the Channel and the subsequent development of combined road and rail transport networks all over Europe?

In 1970 we had changes in drivers' hours and records which were eventually understood and accepted; and in place of carriers' licensing, a new system of operators' licensing was also introduced in 1970 bringing own-account operators and professional hauliers under one common scheme for the control of goods vehicle operations. This, too, was accepted and largely appreciated by the industry which recognized the need for increased road safety measures.

By 1978, however, many changes in these items alone, as respective chapters in the book show, had taken place and in 1984 further significant changes were made, and yet again in 1996. The hours regulations have become those of Europe; the records requirements changed to meet European law late in 1976 and subsequently have been largely replaced by tachographs on a mandatory basis from 1981; and the one-for-all system of 'O' licensing was torn apart to satisfy the demands for establishing standards of professional competence. In 1986 major changes were made to both the European and the British drivers' hours rules and the European tachograph rules, with the object of making the law simpler to understand and more flexible for operators to apply.

Besides these particular items of legislation there have been many more changes which earlier editions of this book have charted year by year. In fact, each year has seen something new in legal responsibilities for the vehicle operator and transport manager to digest. If it has not always been significant in transport terms it has been significant in other terms, significant Acts being the ill-fated Industrial Relations Act, the Equal Pay Act, the Health and Safety at Work etc Act, the Employment Protection Act and the Employment Protection (Consolidation) Act 1978; then the Employment Act 1980 and the Employment Act 1982, and others since then.

In previous years this introduction has referred to the great difficulty which confronts those engaged in the industry, no matter what their capacity or title, in keeping track of what must, and must not, be done (and the penalties involved for non-compliance). The pile of transport, social and environmental legislation is constantly mounting, and as it does so the problems of keeping up-to-date with it all – and putting it into practice – become increasingly complex and time-consuming. The industry has had to take on board the effects of the Single European Market with all that this has entailed by way of changed administrative and operational practices as well as increased competition. From a legislative viewpoint, international transport operators have found the way eased with barriers removed and restrictive measures, such as the requirement for road haulage permits and certain customs procedures, now eliminated. On the other hand, the domestic road haulage industry has found such steps as the legalization of cabotage presenting them with unwelcome competition on their own doorsteps from large numbers of marauding Euro-hauliers looking to pay their running costs home after delivering international loads into Britain.

So it seems that the tide of legislation cannot, and will not, be stemmed even for a brief period to allow the industry some respite from pressure and change, from new restrictions and new burdens.

The purpose of this *Handbook*, therefore, is to gather as much of this legislative material together as can be reasonably squeezed between the covers, to explain what it is all about in lay terms which are both easy to read and to understand and apply, and thereby to provide the hard-pressed vehicle operator or transport manager with one accessible, intelligible source of information on the responsibilities laid on him by law emanating from both the British Government and the EU.

The *Handbook* is intended for the fleet operator (whatever the size of his fleet – large or very small), the transport manager, the owner-driver haulier or anybody else, whatever his title, whose responsibilities include the day-to-day control, administration or operation of goods vehicles, and the small operator who has found that there is a lot more to running a goods vehicle besides just taxing and insuring it. In addition to its function as a ready source of reference for the main legal requirements affecting goods vehicle operation and other useful information, the book may also be found, by those studying for transport examinations, to be an additional means of acquiring a detailed knowledge of the relevant legislation currently applicable to the UK road freight industry.

The major items of legislation affecting the operator and driver of goods vehicles have been covered, together with some of the other legal requirements which are not necessarily new but which it is important for him to know about. Besides transport legislation the owner or manager has been increasingly confronted with social and environmental legislation which affects him both as an employer of staff and as the occupier of premises.

However, while this *Handbook* essentially concentrates on transport legislation rather than transport policy, it is important just to mention here two key policy documents that are helping to shape the way that governments determine their strategies for legislating the operation of goods vehicles, among other things. First of these was the UK Government's 1998 White Paper, *A New Deal for Transport: Better for Everyone,* which has been followed by a number of further documents reporting progress. Then in September 2001 the EU published its White Paper, *European Transport Policy for 2010: Time to Decide.*

Apart from this there is still a great deal more information that the operator and manager should have at his fingertips. Some of it concerns particular branches of the transport industry; the carriage of dangerous goods, abnormal loads, food or livestock, for example. The person responsible for these specialized traffics should have acquired, through experience, some knowledge of the regulations concerning his particular field. The newcomer to such operations will need to obtain the appropriate regulations direct from The Stationery Office (TSO) – see list of addresses at the end of this Introduction – or from a bookseller who stocks TSO publications, and study them carefully.

The principal Acts of Parliament which form the basis of the legislation explained in this book are supported by, and their provisions brought into effect by, a much larger number of regulations, orders and amendments. These are the means by which the Secretary of State for Transport (D*f*T) puts into effect the legal requirements laid down in principle in an Act of Parliament. Legislation contained in an Act is not of itself effective until brought into force by regulations or orders made by the Secretary of State, for which purpose he is given the necessary powers in the Act. Many provisions contained in such Acts may, in fact, never be brought into use, but they do not automatically become extinct by lack of use. They remain dormant on the statute book unless repealed by a further Act.

European legislation arises as the result of draft proposals, similar to British Government White Papers, which are circulated among interested parties (the trade unions, trade associations, and so on) for comment prior to legislative action being taken, and are then enforced by means of directives and regulations of the Council of the European Union.

It is most important for transport people to be aware of the regulations affecting both their own special type of operations and transport in general. To keep up with all the individual Acts, regulations, amendments and modifications which The Stationery Office publishes on behalf of the Government in connection with transport, and the statutory publications of the European Communities, is no mean task (see accompanying list of TSO bookshops where these publications can be bought – or they can be ordered through other, TSO stockist, booksellers). A more convenient method of keeping up-to-date on all the new measures affecting the industry is to read the trade press regularly.

The principal journals covering the road goods transport field are *Motor Transport* and *Commercial Motor,* published weekly, and *Trucking,* published monthly. Contents include the latest industry news, new regulations with simple explanations and articles on subjects of special interest such as insurance, vehicle maintenance, road tests of new vehicles, costing, education and training, and management topics. In particular, these journals are of special interest, because of their extensive reporting of court and tribunal cases, which give the reader a good indication of how the law is applied, how the enforcement agencies, and subsequently the courts, interpret legal provisions and the levels of penalties imposed on offenders and those who lose out in tribunal cases.

It must be emphasized that the fleet operator and transport manager should keep abreast of what is happening in the industry if they want to be efficient, progressive and stay on the right side of the law. To concentrate on the job in hand to the exclusion of all that is happening in the industry at large is an attitude adopted by many operators and many of them have already suffered the consequences of not being prepared to meet some of the drastic changes that have taken place in recent years.

A valuable aid to the operator is membership of one or other of the trade associations. The Freight Transport Association which represents the own-account operator, and the Road Haulage Association, looking after the interests of the hire and reward professional haulier, provide a number of services to their members as well as keeping them informed about what is happening in the industry through their respective journals, *Freight Transport* and *Roadway.* Both organizations hold open meetings and training sessions, addressed by specialists on various topics of importance, at which members can hear first-hand details of current and new legislation, ask questions and air their views.

Additionally, for the operator or manager who wants to keep abreast of current legislation and practices in the industry there are commercial seminars and training sessions ranging from one day to over a week in duration. These provide an excellent means of keeping up-to-date with what is going on, and of meeting and talking to other people in the industry with similar interests and problems.

The road transport industry plays an important part in the life of Great Britain. Its safe operation is essential for the well-being of the people and its efficient operation is vital to the economy. Higher standards of management and control within the industry with a far greater awareness and understanding of the legal requirements, the operational demands and the economic considerations are necessary if these essential criteria of safety and efficient operation are to be achieved and upheld.

Special Note

It should be noted that this *Handbook* is intended purely as a practical interpretation of legal matters for the lay reader and is only a *guide* to matters current at the time of writing. It is not a definitive legal work of reference and should not be used as such. Readers are advised to check the legislation itself before committing time or expenditure to any particular course of action and any operator needing detailed legal advice is recommended to acquire it through normal legal channels. The author and publishers accept no responsibility whatsoever for decisions taken or other irrevocable actions based on the contents herein.

TSO Bookshops

London	123 Kingsway, London WC2B 6PQ. Tel: 020 7242 6393 Fax: 020 7242 6394 (telephone orders are not accepted – see below)
Belfast	16 Arthur Street, Belfast BT1 4GD. Tel: 02890 238451 Fax: 02890 235401
Birmingham	68–69 Bull Street, Birmingham B4 6AD. Tel: 0121 236 9696 Fax: 0121 236 9699
Manchester	9–21 Princess Street, Manchester M60 8AS. Tel: 0161 834 7201 Fax: 0161 833 0634
Edinburgh	71 Lothian Road, Edinburgh EH3 9AZ. Tel: 0870 606 5566 Fax: 0870 606 5588
Cardiff	18–19 High Street, Cardiff CF10 1PT. Tel: 02920 395548 Fax: 02920 384347

The Stationery Office

PO Box 29, Norwich NR3 1GN
Tel: 0870 600 5522 (telephone orders, general enquiries, subscriptions and standing orders)

E-mail: book.orders@tso.co.uk
Website: www.tso.co.uk

Parliamentary publications may be obtained from:

The Parliamentary Bookshop
12 Bridge Street
Parliament Square
London SW1A 2JX

Telephone orders: 020 7219 3890
General enquiries: 020 7219 3890
Fax orders: 020 7219 3866
E-mail orders: bookshop@parliament.uk

TRANSPORT MINISTERS 1970–2005

The author has been asked many times about the names and dates in office of successive Transport Ministers. The following list of such Ministers in post during the life of this *Handbook* from 1970 shows what a potentially vulnerable position this is, with some incumbents lasting only a matter of months. The very first Minister of Transport was Sir Eric Geddes who served from 19 May 1919 to 7 November 1921.

Minister for Transport Industries

- John Peyton (15 October 1970 to 4 March 1974)

Minister for Transport

- Fred Mulley (7 March 1974 to 12 June 1975)
- John Gilbert (12 June 1975 to 10 September 1976)

Secretary of State for Transport

- William (Bill) Rodgers (10 September 1976 to 4 May 1979)

Minister of Transport

- Norman Fowler (11 May 1979 to 5 January 1981)

Secretary of State for Transport

- Norman Fowler (5 January 1981 to 14 September 1981)
- David Howell (14 September 1981 to 11 June 1983)
- Tom King (11 June 1983 to 16 October 1983)
- Nicholas Ridley (16 October 1983 to 21 May 1986)
- John Moore (21 May 1986 to 13 June 1987)
- Paul Channon (13 June 1987 to 24 July 1989)
- Cecil Parkinson (24 July 1989 to 28 November 1990)
- Malcolm Rifkind (28 November 1990 to 10 April 1992)
- John MacGregor (10 April 1992 to 5 July 1995)
- Sir George Young, Bt. (5 July 1995 to 2 May 1997)

Secretary of State for Environment, Transport and the Regions

- John Prescott (2 May 1997 to 8 June 2001)

From 1997 to 2001, the Ministers of State with responsibility for Transport were:

- Gavin Strang (3 May 1997 to 27 July 1998)
- John Reid (27 July 1998 to 17 May 1999)
- Helen Liddell (17 May 1999 to 29 July 1999)
- Lord Macdonald of Tradeston (29 July 1999 to 8 June 2001)

NB: Gavin Strang and John Reid attended Cabinet meetings, but were not formally members of the Cabinet.

Secretary of State for Transport, Local Government and the Regions

- Stephen Byers (8 June 2001 to 29 May 2002) (Resigned)

Secretary of State for Transport

- Alistair Darling (29 May 2002 to still in office in July 2005)

Acknowledgement: www.nationmaster.com/encyclopedia

1 Goods Vehicle Operator Licensing

Operator ('O') licensing is the regulatory 'quality' control system imposed by government to ensure the safe and legal operation of most goods vehicles in Great Britain. While other individual aspects of legislation also apply to such vehicles, the 'O' licensing system provides the overriding control of road freight transport operations. Failure to observe the requirements and conditions under which 'O' licences are granted will lead to severe penalty; likewise, breach of other legislation can result in appropriate penalties as set out in respective statutes and, subsequently, will involve penalties against the operators' licence itself. Similar licensing controls apply to goods vehicle operations in Northern Ireland under the separate Road Freight Operator's Licence scheme operated in the Province (see pp 52–55).

Under the current British licensing scheme, trade or business users of most goods vehicles over 3.5 tonnes maximum permissible weight must hold an 'O' licence for such vehicles, whether they are used for carrying goods in connection with the operator's main trade or business as an own-account operator (ie a trade or business other than that of carrying goods for hire or reward) or are used for hire or reward road haulage operations. Certain goods vehicles, including those used exclusively for private purposes, are exempt from the licensing requirements. Details of the exempt vehicles to which 'O' licensing does not apply are given on pp 10–13.

The original system of operators' licensing was established by the Transport Act 1968, the relevant provisions of which are now consolidated into the Goods Vehicle (Licensing of Operators) Act 1995. This Act states that no person may use a goods vehicle on a road for hire or reward or in connection with any trade or business carried on by him except under an operator's licence.

Regulations which took effect on 1 January 1978 made substantial changes to the 'O' licensing system which had existed from 1970 to the end of 1977. These regulations, The Goods Vehicles (Operators' Licences) Regulations 1977, introduced a three-tier system of 'O' licensing in contrast to the previous single-tier system which applied similarly to both professional hauliers and own-account operators. From 1 June 1984 further significant changes to the system were introduced by The Goods Vehicles (Operators' Licences, Qualifications and Fees) Regulations 1984 (as amended) which, principally, gave Licensing Authorities (now called Traffic Commissioners (TCs)) powers to consider representations to 'O' licence applications on the grounds that environmental nuisance or discomfort would be caused to local residents by the presence of goods vehicles at or operating from particular locations.

Further changes were made from 1 January 1996 when the system of continuous 'O' licensing was introduced whereby licences, once granted, remain valid indefinitely, provided the holder complies with all relevant conditions and pays the necessary fees on time – failure to do so bringing with it risk of loss of the licence – or unless a major variation of the licence is required (eg to increase the number of authorized vehicles or to change or add a new operating centre to the licence). The requirement for five-yearly licence renewals has been abolished for cases where details remain unchanged although operating centres are subject to periodic reviews to ensure that they remain suitable.

Besides this principal change, the 1995 Act also provides new definitions for vehicle operating centres (they must be sufficiently large to accommodate all the vehicles used under the licence), where newspaper advertisements must be placed (ie in one or more local newspapers circulating in the locality), and the period of time in which objectors and environmental representors may make their case (five years for environmental representors). It also gives wider powers to TCs in respect of road safety matters (particularly in regard to situations where vehicles may cause danger to the public).

From 1 October 1999 yet more changes were introduced as a result of new UK regulations (the Goods Vehicle Operators (Qualifications) Regulations 1999) implementing the provisions of EC Council Directive 98/76/EC of 1 October 1998 which amends Directive 96/26/EC. Principal among these provisions are stricter good repute requirements, increased financial commitments for new (and later for existing) operators, a more extensive professional competence examination syllabus and a tougher examination regime. The aim of these is to ensure higher levels of professionalism and law abidance in road haulage (similar steps are being applied in passenger transport) and a harmonized approach throughout all the member states of the EU.

Proposals are being discussed to make radical, modernizing, changes to the present 'O' licensing system described here. A consultation paper is expected in 2005 which may include such ideas as making the whole of England into a single Traffic Area, doing away with paper windscreen 'O' discs and abolishing the current 28-day margin. Operators will be increasingly encouraged to manage the administration of their own fleet licences through the Operator Self-Service System (see pp 9–10) via the internet.

NB: Relevant regulations

- The Goods Vehicles (Licensing of Operators) Act 1995 (Commencement and Transitional Provisions) Order 1995 SI 2181/1995
- The Goods Vehicles (Licensing of Operators) Regulations 1995 SI 2869/1995
- The Goods Vehicles (Licensing of Operators) (Fees) Regulations 1995 SI 3000/1995
- The Goods Vehicles (Licensing of Operators)(Temporary Use in Great Britain) Regulations 1996 SI 2186/1996
- The Goods Vehicle Operators (Qualifications) Regulations 1999 SI 2430/1999
- The Transport Tribunal Rules 2000 SI 3226/2000
- The Goods Vehicles (Enforcement Powers) Regulations 2001 SI 3981/2001
- The Transport Tribunal (Amendment) Rules 2001 SI 4041/2001
- The Transport Tribunal (Amendment) Rules 2002 SI 643/2002
- The Goods Vehicle Operators (Qualifications) (Amendment) Regulations 2004 SI 3222/2004

ADMINISTRATION OF LICENSING SYSTEM

The 'O' licensing system, which is based on the concept of ensuring legal and safe operation and thus is a system of 'quality' as opposed to 'quantity' licensing, is administered on a regional (ie Traffic Area) basis throughout Great Britain. (Northern Ireland's Road Freight Operators' Licensing system is dealt with separately by the Department of the Environment in Belfast.) The Traffic Area Offices (TAOs), each with its own Traffic Commissioner (TC), administer the eight Traffic Areas that form the network (see Appendix I for a list).

These TCs are appointed by the Secretary of State for Transport and are 'independent quasi-judicial authorities', who have the statutory power to grant or refuse operators' licences, to place road safety and environmental conditions or restrictions on such licences where necessary, and subsequently to impose penalties against licences in the event of the holder being convicted for goods vehicle-related offences.

NB: The male gender is used throughout this Handbook when referring to Traffic Commissioners. No discrimination, prejudice or bias is intended by such use.

As of 31 March 2004, 407,456 goods vehicles were specified in 102,946 'O' licences in Great Britain, of which 49,463 were restricted licences, 41,111 were standard national licences and 12,372 were standard international licences. These numbers show a small increase on the previous figures, and are well below the 1989/90 figure of 132,236 valid 'O' licences in issue, covering 441,656 goods vehicles.

Source: Annual reports of the Traffic Commissioners (covering Goods Vehicle Operator Licensing) for the period 1 April 2003 to 31 March 2004.

THE OPERATOR SELF-SERVICE SYSTEM

The Traffic Area Network (TAN) computer system finally came into operation in mid-2002. It was merged with the VOSA system and re-launched on 31 July 2004 and is to be found at: www.tan.gov.uk (click on Operator Licensing Self Service). Principally, this system significantly reduces the time taken to process 'O' licence changes and particularly results in the 28-day period of grace for advising the TC of vehicle changes being abolished. Operators will be expected to transact such business online. They must use a pre-registered user ID and a case-sensitive password. 'Applications and Decisions' (see p 37) will subsequently be produced weekly instead of fortnightly. Additionally, the system will provide enforcement agencies with 'real-time' data to help them catch illegal operators.

This online Government service enables goods vehicle operators to add and remove vehicles from their Operator's licence at any time of the day or night without the need to complete an application form and without any intervention by VOSA staff. It reduces the time taken to add a vehicle to a licence from 14 days to within a couple of minutes. Operators can register online to join the self-service system; check their own licence details held on VOSA's operator licence computer system; transfer vehicles between licences they hold in different areas; track the progress of licence applications and set up access for other members of their staff, allowing greater access to records and transactions for larger companies. In due course the system will allow operators to pay their licence invoices and renew licences online.

'O' LICENCES NOT FOR SALE

Operator's licences are issued only to applicants who meet strict criteria as set out in the regulations and as described in this chapter. As such, 'O' licences are *not* transferable between operators and may *not* be borrowed, used on loan or sold, with or without the vehicles to which they relate, by any person other than the authorized holder. Advertisements in the transport press purporting to be for the sale of such licences are misleading because they encourage respondents to act illegally and should be treated with great caution.

EXEMPTIONS FROM 'O' LICENSING

There are a number of categories of vehicle which are exempt from 'O' licensing requirements as described below.

Small Vehicles

The principal exemption applies to 'small' vehicles identified as follows.

Rigid vehicles are 'small' if:

- they are plated and the gross plated (ie maximum permissible) weight is not more than 3.5 tonnes;
- they are unplated and have an unladen weight of not more than 1525 kg.

A combination of a rigid vehicle and a drawbar trailer is 'small' if:

- both the vehicle and the trailer are plated, and the total of the gross *plated* weights is not more than 3.5 tonnes;

- either the vehicle or the trailer is not plated, and the total of the *unladen* weights is not more than 1525 kg.

However, if the unladen weight of an unplated trailer is not more than 1020 kg this weight does not have to be added to the unladen weight of the drawing vehicle for the purposes of deciding if the combination is 'small' for 'O' licensing purposes.

Articulated vehicles are 'small' if:

- the semi-trailer is plated, and the total of the *unladen* weight of the tractive unit and the plated weight of the semi-trailer is not more than 3.5 tonnes;
- the semi-trailer is not plated, and the total of the *unladen* weights of the tractive unit and the semi-trailer is not more than 1525 kg.

Older Vehicles

Also included in the exemptions are pre-1 January 1977 vehicles which have an unladen weight not exceeding 1525 kg and a gross weight greater than 3.5 tonnes.

Other Exemptions

Regulations list the following further specific exemptions from 'O' licensing requirements.

1. Vehicles licensed as agricultural machines used solely for handling specified goods, and any trailer drawn by them.
2. Dual-purpose vehicles and any trailer drawn by them.
3. Vehicles used on roads only for the purpose of passing between private premises in the immediate neighbourhood and belonging to the same person (except in the case of a vehicle used only in connection with excavation or demolition) provided that the distance travelled on the road in any one week does not exceed in aggregate 9.654 km (ie six miles).
4. Motor vehicles constructed or adapted primarily for the carriage of passengers and their effects and any trailer drawn by them while being so used.
5. Vehicles being used for funerals.
6. Vehicles being used for police, fire brigade and ambulance service purposes.
7. Vehicles being used for fire fighting or rescue work at mines.
8. Vehicles on which a permanent body has not yet been built carrying goods for trial or for use in building the body.
9. Vehicles being used under a trade licence.
10. Vehicles used in the service of a visiting force or headquarters.
11. Vehicles used by or under the control of HM United Kingdom forces.
12. Trailers not constructed for the carriage of goods but which are used incidentally for that purpose in connection with the construction, maintenance or repair of roads.
13. Road rollers or any trailer drawn by them.
14. Vehicles used by HM Coastguards or the Royal National Lifeboat Institution for the carriage of lifeboats, life saving appliances or crew.
15. Vehicles fitted with permanent equipment (ie machines or appliances) so that the only goods carried are:

(a) for use in connection with the equipment;
(b) for threshing, grading, cleaning or chemically treating grain or for mixing by the equipment with other goods not carried on the vehicle to make animal fodder; or
(c) mud or other matter swept up from the road by the equipment.

16. Vehicles while being used by a local authority for the purpose of enactments relating to weights and measures or the sale of food or drugs.
17. Vehicles used by a local authority under the Civil Defence Act 1948.
18. Steam-propelled vehicles.
19. Tower wagons or any trailer drawn by them provided that any goods carried on the trailer are required for use in connection with the work on which the tower wagon is used.
20. Vehicles used on airports under the Civil Aviation Act 1982.
21. Electrically propelled vehicles.
22. Showmen's goods vehicles and any trailer drawn by such vehicles.
23. Vehicle of non-resident operator carrying out cabotage operations in the UK under EU regulation 3118/93/EEC.
24. Vehicles first registered prior to 1 January 1977 which are not over 1525 kg unladen weight and are plated for more than 3500 kg but not more than 3556.21 kg (3.5 tonnes).
25. Vehicles used by a highway authority in connection with weighbridges.
26. Vehicles used for emergency operations by the water, electricity, gas and telephone services.
27. Recovery vehicles.
28. Vehicles used for snow clearing or the distribution of grit, salt or other materials on frosted, ice-bound or snow covered roads and for any other purpose connected with such activities.
    NB: This exemption is not restricted solely to local authority-owned vehicles.
29. Vehicles going to or coming from a test station and carrying a load which is required for the test at the request of the Secretary of State for Transport (ie by the test station).

Exemption for Private Vehicles

Exemptions also apply to vehicles used privately (ie for carrying goods for solely private purposes and not in any way connected with a business activity) and by voluntary organizations.

Northern Ireland Vehicles

Northern Ireland-based operators do not need 'O' licences for vehicles running, laden or unladen, in or through the UK while on international journeys. Similarly, there is no requirement for 'O' licences to be held by hauliers established in other EU member states (and not established in the UK) for vehicles operating within the UK on international journeys.

Non-Exempt Vehicles

All other goods-carrying vehicles over 3.5 tonnes gross weight not specifically shown as exempt in the list above must be covered by an 'O' licence. This includes such vehicles that are only temporarily in the operator's possession, or are hired or borrowed on a short-term basis, if they are used in connection with a business (even a part-time business).

No Exemption for Fast Agricultural Tractors

Fastrac-type agricultural tractors capable of pulling substantial loads at speeds of up to 40 mph on public roads must be specified on an 'O' licence if used for hire or reward haulage work. The agricultural exemption mentioned above applies only when such machines are used by farmers in connection with their own agricultural business.

These vehicles provide unfair competition to licensed road hauliers by using red diesel and under-age drivers, by paying a reduced rate vehicle excise duty and by not having any requirements to observe the law on goods vehicle plating and testing, drivers' hours, breaks and rest periods, and tachographs.

THE VEHICLE USER

An 'O' licence must be obtained by the 'user' of the vehicle for all the vehicles he operates to which the regulations apply. The 'user' may be the owner of the vehicle or he may have hired it. If the vehicle was hired without a driver, the hirer is the 'user'. There is considerable importance attached to the word 'user', and its exact meaning, both for the purposes of 'O' licensing and in other regulations. It may be explained simply as follows:

- An owner-driver who uses his vehicle in connection with his own business is the 'user' of his own vehicle.
- If the owner of a vehicle employs a driver to drive it for him and he pays the driver's wages then the owner is the 'user' because he is the employer of the driver.
- If a vehicle is borrowed, leased or hired without a driver and the borrower or hirer drives it himself or pays the wages of a driver he employs to drive it then the borrower or hirer is the 'user'.

From this it can be seen that, in general, the person who pays the driver's wages is the 'user' of a vehicle, and it is this person (or company) who is responsible for holding an 'O' licence and for the safe condition of the vehicle on the road and for ensuring that it is operated in accordance with the law. However, it must be remembered that the driver himself, although an employee, is still also the user of the vehicle in the context of certain legislation (eg The Road Vehicles (Construction and Use) Regulations 1986, as amended) and he, too, is responsible for its safe condition on the road and is liable to prosecution if it is not in safe and legal condition.

A situation has arisen in recent times where owner-drivers of goods vehicles who cannot themselves obtain the professional competence qualification have had their vehicles specified on the 'O' licence of another operator but have nevertheless been paid as self-employed contractors to the other operator. This practice is illegal because if the driver owns the vehicle and uses it in connection with his business then by virtue of the regulations he is the 'user' and is therefore responsible for holding the 'O' licence for it.

Agency Drivers

Dependence on agencies for the supply of temporary drivers to provide relief manpower when regular drivers are not available has caused difficulty in interpretation of the term 'user' and in deciding who should hold the 'O' licence; the vehicle owner or the agency which employs the driver. It can be seen from item 2 above that the person who pays the driver's wages is the 'user', and is therefore the person who should hold the 'O' licence.

However, the status of the vehicle 'user' in these circumstances has been determined by the agencies getting operators to sign agreements whereby the vehicle operator technically becomes the employer of the driver rather than the agency being the employer and consequently the operator remains the legal 'user' of the vehicle. Usually the agency asks the hirer to sign an agreement whereby the agency becomes the 'agent' of the operator for these purposes in paying the driver's wages. This practice has been proved in court to be legally acceptable on the grounds that the Transport Act 1968 section 92(2) states that 'the person whose servant or agent the driver is, shall be deemed to be the person using the vehicle'. The driver is considered to be the servant of the hirer because the hirer gives instructions and directs the activities of the driver who is temporarily in his employ. The key issue here is who controls how the drivers do their job. This must be the 'user' of the vehicle and the holder of the 'O' licence.

NB: The High Court decision in the controversial 1999 case concerning Interlink Express Parcels and Night Trunkers was overturned by the Appeal Court in 2001, so interpretation of the law on the employment of agency drivers remains as stated above.

The great danger with agency drivers is that the operator has no sound means of establishing whether the driver is legally qualified to drive or whether he has already exceeded his permitted driving hours on previous days and whether he has had adequate rest periods. Reputable agencies usually go to considerable lengths to ensure that drivers provided by them for their clients are properly licensed and have complied with the driving hours rules in all respects. It is worth also remembering that the use of casually hired or temporary drivers (whose backgrounds and previous experiences may not be fully known) can result in jeopardy of the contract of insurance covering the use of vehicles and there could also be serious security risks as well as possible 'O' licence penalties for infringements of the law. For this reason the operator should confine himself to obtaining drivers from reputable agencies who are known to have vetted drivers satisfactorily.

Operators who use agency drivers should be aware that they may be held liable for negligence or driving offences committed by such drivers. They are also liable for ensuring the health and safety of hired drivers, and must apprise them of the legal requirements of vehicle operations, such as the 'hours' law, tachographs, safe loading and vehicle checks.

A Code of Practice for the employment of agency drivers is available from offices of the Freight Transport Association (FTA), the Road Haulage Association (RHA) and the Recruitment and Employment Confederation (REC). The Code sets out the respective duties and responsibilities of the haulier on the one hand and the supplying agency on the other, with a checklist for each to ensure that full and correct information is exchanged as to the requirement for the driver (eg the skills and personal attributes required) and the particular job to be done. It also contains a model set of instructions and procedures which should be given to drivers.

Copies of the Code can be obtained from local FTA and RHA offices or from the REC at 36–38 Mortimer Street, London W1W 7RG (tel: 020 7462 3260).

RESTRICTED AND STANDARD 'O' LICENCES

There are three main types of 'O' licence as described below and, in certain circumstances, a temporary licence known as an interim direction may be granted in exceptional circumstances (see p 29).

1. **Restricted licences:** available only to own-account operators who carry nothing other than goods in connection with their own trade or business, which is a business other than that of carrying goods for hire or reward. These licences cover both national and international transport operations with own-account goods. Restricted 'O' licence holders must not use their vehicles to carry goods for hire or reward or on behalf of customers' businesses – even if it is done only as a favour or is seen as being part of the service provided to a customer and even if no charges are raised (see below). Such activities are illegal and could result in penalties.

2. **Standard licences (national transport operations)**: for hire or reward (ie professional) hauliers, or own-account operators who also engage in hire or reward operations, but restricted solely to national transport operations (ie operations exclusively within the UK). Own-account holders of such licences may also carry their own goods (but not goods for hire or reward) on international journeys.

3. **Standard licences (national and international transport operations):** for hire or reward (ie professional) hauliers, or own-account operators who also engage in hire or reward carrying, on both national and international transport operations.

National transport operations in this context include journeys to and from ports with loaded trailers which were previously considered to be international journeys.

Standard Licences for Own-Account Operators

Own-account operators may voluntarily choose to hold a standard 'O' licence for national or both national and international transport operations instead of a restricted licence provided they are prepared to meet the necessary additional qualifying requirements (principally the professional competence qualification – see Chapter 2). Among the reasons which may influence them to take this step is the desire to carry goods for hire or reward to utilize spare capacity on their vehicles, especially on return trips. Such a requirement may also arise because a firm is involved in carrying goods for associate companies on a reciprocal or integrated working basis which does not come within the scope of activities which are permitted under 'O' licensing between subsidiary companies and holding companies (see p 43) or firms may find themselves in the position where they carry goods in connection with their customers', as opposed to their own, businesses.

Firms holding restricted 'O' licences may not carry on their vehicles goods on behalf of customers (ie in connection with the trade or business of the customer rather than in connection with their own business) or other firms even if such operations are described as being a 'favour' or 'part of the service' to the customer and involve no payment whatsoever. This may occur, for example, when a vehicle delivers goods to a customer and the customer then asks the driver to drop off items on his return journey because he is 'going past the door' and their own vehicle is not available. Such activities would be illegal under the terms of a restricted 'O' licence and if two convictions for such an offence are made within five years, the licence must be revoked by the Traffic Commissioner.

REQUIREMENTS FOR 'O' LICENSING

In order to obtain an 'O' licence, applicants must satisfy certain conditions specified in regulations.

Restricted Licences

Applicants must be:

- fit and proper persons;
- of appropriate financial standing.

Standard Licences (national transport operations)

Applicants must be:

- of good repute;
- of appropriate financial standing;
- professionally competent, or must employ a person who is professionally competent, in national transport operations.

Standard Licences (national and international transport operations)

Applicants must be:

- of good repute;
- of appropriate financial standing;
- professionally competent, or must employ a person who is professionally competent, in both national and international transport operations.

Other Legal Requirements

Besides the specific requirements mentioned above, licence applicants and holders have to satisfy further legal requirements relating to the suitability and environmental acceptability of their vehicle operating centres, the suitability of their vehicle maintenance facilities or arrangements and, overall, their ability and willingness to comply with the law in regard to vehicle operating as demonstrated by signing the undertakings on the 'O' licence application form. These matters are dealt with in detail in this chapter.

NB: It is also worth noting that increasingly Traffic Commissioners are urging licensed operators to check carefully that any subcontract road hauliers to whom they pass on work are fully licensed to carry out the work. See p 48.

Good Repute

For a TC to be able to grant an 'O' licence, he must determine that the applicant is of 'good repute'. Without this particular requirement being well established, the fact that the applicant may meet all other relevant criteria is of no account, no licence will be granted.

With the introduction of new provisions from EU Council Directive 98/76/EC in UK regulations (The Goods Vehicle Operators (Qualifications) Regulations 1999), much tougher good repute standards are imposed on both new and existing operators.

The term 'good repute' is defined on the basis that an individual is not of good repute if he has been convicted of more than one serious offence or of road transport offences concerning:

- pay and employment conditions in the profession (ie of road haulier); or
- drivers' hours and rest periods;
- weights and dimensions of goods vehicles;
- road and vehicle safety;
- protection of the environment; and
- rules concerning professional liability.

For the purposes of the standard 'O' licensing scheme this means that the applicant for a licence (ie an individual) must not have a past record which includes more than one conviction for serious offences or conviction for road transport offences during the previous five years (excluding convictions that are 'spent' – see below) relating to the above issues. Similarly, to be a fit and proper person in order to obtain a restricted 'O' licence means that there should not be a past record of such offences. If it is a limited liability company applying for a licence and the company has relevant convictions on its record (ie for serious offences) then the TC may use his discretion in deciding whether the firm is of good repute, a facility he does not have in the case of individuals or partnership businesses.

It should be noted that the TC will not necessarily refuse to grant a licence to an applicant who has had convictions – but he must if they are for more than one serious offence which affects the applicant's good repute – but he will consider the number and seriousness of the convictions before making a grant. He may, for example, issue a licence for a shorter period to see if the applicant has 'mended his ways', or grant a licence for fewer vehicles than the number requested. If, during the currency of a licence, an 'O' licence holder is convicted of offences related to goods vehicle operations, the TC may call the operator to a public inquiry and determine whether he is still a fit and proper person or of good repute and whether he should be allowed to continue holding an 'O' licence (see also p 49).

In the case of partnership applications for licences, if one of the partners is considered not to be of good repute, then the TC will be bound to conclude that the partnership firm is not of good repute and, on that basis, he will refuse to grant a licence.

The relevant offences for which conviction damages a person's good repute are those specified in the Goods Vehicles (Licensing of Operators) Act 1995, schedules 2(4) and (5) and include matters relating to:

- unlawful use of vehicles;
- unroadworthiness of vehicles, maintenance and maintenance records;
- plating and testing;
- speed limits;
- weight limits, safe loading and overloading of vehicles;
- licensing of drivers;
- drivers' hours and record keeping (including tachographs);
- certain traffic offences (eg parking);
- certain waste-related offences;
- illegal use of rebated fuel oil;
- forgery;
- international road haulage permits.

NB: Offences concerning protection of the environment and commercial matters such as professional liability now included in the EU Directive mentioned above will be included in UK regulations in due course.

Increasingly, convictions for vehicle excise licensing offences are featuring in consideration of 'O' licence applicants' good repute by the TCs and a number of licences have been either refused or penalized on these grounds.

Additionally, in the case of applications for standard international 'O' licences, the TC must be told of convictions for any other offence besides those concerning the matters listed above.

Serious Offences

A serious offence, by an individual or by a company or its management, as referred to above is defined as one where, if committed in the UK, a sentence of more than three months' imprisonment, or a community service order of more than 60 hours was ordered, or a fine exceeding level four on the standard scale (currently £2,500) was imposed. If committed abroad the seriousness of the offence would be determined by assessing the punishment relative to UK standards.

Criminal Offences

Since 1 April 1998 convictions for offences such as falsifying records (eg tachograph charts), forgery (eg of insurance documents or a driving licence) and fraudulent use or display of an official document (eg an 'O' licence disc or vehicle excise licence), including aiding and abetting these offences, have branded the offender a criminal. These convictions will be recorded on the Police National Computer and can be accessed at will by the police, other organizations and possibly, in due course, the Traffic Commissioners, in determining a person's good repute for 'O' licensing purposes.

Spent Convictions

Spent convictions, which are referred to above in the context of good repute, are those which *do not* have to be declared on the licence application form, because they were incurred sufficiently long ago to be considered legally invalid for determining a person's past record. Under the Rehabilitation of Offenders Act 1974 a person who was convicted of an offence for which he was fined by a court need not disclose that conviction after five years (or 2½ years if aged under 17 at the time), or if a prison sentence of six months or less was imposed it need not be disclosed after seven years (or 3½ years if aged under 17 at the time). In the case of a prison sentence of between six months and 2½ years the rehabilitation period (after which no disclosure is necessary) is 10 years, or 5 years if aged under 17 at the time. A table showing the full range of rehabilitation periods is given below.

In Scotland, supervision requirements made by children's hearings attract similar rehabilitation periods to those for care or supervision orders.

(i) Fixed rehabilitation periods:

Sentence	Rehabilitation period
• For a prison sentence between 6 months and 2½ years	10 years (5 years if aged under 17 years at the time)
• For a prison sentence of 6 months or less	7 years (3½ years if aged under 17 years at the time)
• For a fine or a community service order	5 years (2½ years if aged under 17 years at the time)
• For an absolute discharge	6 months
• For Borstal*	7 years
• For a detention centre	3 years

(ii) Variable rehabilitation periods:

Sentence	Rehabilitation period
• A probation order, conditional discharge or bind over	1 year or until the order expires (whichever is longer)
• A care order or supervision order	1 year or until the order expires (whichever is longer)
• An order for custody in a remand home, an approved school* or an attendance centre order	A period ending 1 year after the order expires
• A hospital order (with or without a restriction order)	A period ending 1 year after the order expires 5 years or a period ending 2 years after the order expires (whichever is longer)

* *NB: These are more commonly known as 'young offenders institutions'*

Financial Standing

The requirement for financial standing means the applicant being able to prove to, or assure, the TC that sufficient funds (ie money) are readily available to maintain the vehicles to be covered by the licence to the standards of fitness and safety required by law.

New Requirements

Under new rules brought in by the UK Government in response to the EU's Council Directive 98/76/EC of 1 October 1998 (ie The Goods Vehicle Operators (Qualifications) Regulations 1999), significantly increased levels of financial standing have had to be shown both by new 'O' licence applicants and by existing licence holders since the following dates:

- New applicants from 1 October 1999.
- Existing 'O' licence holders from 1 October 2001.

The Directive requires that road haulage operators must have available capital and reserves of at least £6,200 for their first vehicles and at least £3,400 for each additional vehicle; thus for a five-vehicle fleet the operator will have to prove available finances of at least £19,800 and for a 10-vehicle fleet the figure would be £36,800. These financial resources will need to be reviewed on a five-yearly basis.

These values are converted from euros which vary daily in line with exchange rate fluctuations (in mid-2005 the euro was worth about 70 pence sterling), for the purposes of the Directive its value against national currencies is to be fixed annually based on its value on the first working day of October each year to take effect from 1 January of the following calendar year (the rate is published in the *Official Journal of the European Communities*).

The former lower figures applicable to both national operators and restricted licence holders have been abolished. All operators must now meet the standards described above.

It is important to realize that the requirement for minimum levels of capital and reserves applies to the total number of vehicles authorised on an 'O' licence, not just to the vehicles currently specified. In the case where an operator has a significant margin between the number of authorized vehicles and those actually specified, he may wish to consider decreasing this margin to reduce the amount of financial resources he has to prove to the Traffic Commissioner. The number of authorized vehicles on an 'O' licence can be reduced by application to the TC using form GV81 (see p 42).

The EU Directive allows for the financial standards described above to be established by means of confirmation or assurance from a bank or from other properly qualified institutions that such funds are available, in the form of a bank guarantee, pledge or security, or by similar means.

A key point that has been reiterated recently is that proof of sufficient finances to meet the legal requirement should not relate to a single day when a bank balance may have been artificially boosted by a temporary injection of funds that are, in effect, moved away again the next day, but rather should relate to an average balance over the life of the licence. The TC has to be certain that the licence applicant or operator has sufficient funds 'available'* to establish and properly administer the business on a daily basis, not just to meet a target balance on a particular day.

**NB: 'Available' in this context means, according to the Transport Tribunal: 'capable of being used'; 'at one's disposal'; 'within one's reach'; 'easy to get at'.*

Failure to pay fines, other penalties and business debts as well as nonpayment of vehicle excise duties will suggest to the TC that an operator has cash-flow problems and is therefore unlikely to meet the legal requirement for financial standing. This could lead to loss of the 'O' licence.

Assessment of Financial Standing

On the fifth anniversary of the grant of a licence, the TC will carry out a 'wealth' check to determine whether the required reserves have been available during the previous five years. If they have *not*, the licence holder will be deemed to be no longer of the required financial standing and the licence will be revoked.

The TC has considerable powers to inquire into the finances of applicants, including the right to ask for the production of proof of financial standing by means of audited accounts and bank statements or bank references, savings or deposit account books or other evidence of funds stated to be available for the maintenance of vehicles. He will particularly look at the firm's balance sheet within the audited accounts and determine its liquidity (ie its capability of paying its debts as they fall due). He will examine the relevant financial ratios such as current assets to current liabilities, which ideally should not be less than 2:1, and the so-called quick ratio of quickly realizable assets (ie items which can quickly be turned into cash) to current liabilities which, in this case, should not be less than 1:1. It is believed that where a firm's accounts show a current ratio of less than 0.5:1 the TC should consult with the financial assessors he is empowered to call on.

NB: Road hauliers who make a practice of factoring their debts (ie selling their unpaid invoices at less than face value to secure a quick cash return) may find the TC questioning their financial standing on the grounds that such practices are an indicator that all is not well within a firm and that it might not meet the required financial standards for holding an 'O' licence. However, new invoice factoring arrangements established by the Royal Bank of Scotland Commercial Services (RBSCS) may satisfy the Traffic Commissioners on the financial standing of an operator.

In complex cases, usually involving companies where, perhaps, funds are moved between one subsidiary and another, and there is cross-accounting and such like, the TCs can call on financial experts (assessors) to help determine the true position of an applicant.

Professional Competence

Details of the professional competence requirements for standard 'O' licence holders are given in Chapter 2.

Wrong Licences

It is illegal to operate on the wrong type of 'O' licence. Applicants are required to specify which type of licence they require and those who specify restricted licences will be subject to severe penalties if they subsequently carry goods for hire or reward. Operators who specify standard licences covering only national operations and who engage in international operations will be similarly penalized.

Operating Centres

The vehicle operating centre is defined as the place where the vehicle is 'normally kept'. This is commonly taken to mean the place where the vehicle is regularly parked when it is not in use. However, places where vehicles are parked occasionally, even if on a regular basis, in circumstances that are exceptional to the normal conduct of the business, are not considered to be operating centres.

Where operators regularly permit drivers to take vehicles home with them at night and at weekends in circumstances that are not 'exceptional' to the normal conduct of the business, then the place where the drivers park vehicles near to their home becomes the vehicle operating centre. This place then must be declared on the 'O' licence application form.

In these circumstances an operator could have to declare a number of separate operating centres in addition to his normal depot or base and he could face environmental representation against each of these places and have restrictive environmental conditions placed on his licence in respect of their use. Alternatively, he could lose his licence if he fails to declare such places as operating centres. Failure to notify the TC of new or additional operating centres is an offence. Similarly, the practice of allowing drivers to take vehicles home regularly or, for other reasons, park away from the operating centre regularly (except when on genuine journeys away from base) puts the 'O' licence in jeopardy, as well as risking prosecution.

What is abundantly clear is that Traffic Commissioners expect licence applicants to be able to show that their proposed operating centre is both suitable in environmental terms and sufficiently large to accommodate all the vehicles authorized to be based there. Should this not be the case a licence may be refused or a grant made authorizing fewer vehicles (ie only as many, possibly, as can be parked at the centre).

It is useful to stress again that the use of an unauthorized operating centre or failure to notify the TC of a change of operating centre is an offence which can result in a fine of up to £2,500 on conviction and penalty against the 'O' licence.

Operating centres are subject to review as to their continued suitability at five-yearly intervals – at the TC's discretion. In other words the TC may call an operator for review once every five years, but should he decide not to do so the centre remains 'suitable' for another five years.

LICENCE APPLICATION

Applications for 'O' licences must be made to the TC for each Traffic Area in which the operator has vehicles based. These bases will be the operating centres (see above for definition) for the vehicles. One 'O' licence will be sufficient to cover any number of vehicles operating at one centre and any number of operating centres in any one Traffic Area. If operating centres are in different Traffic Areas then separate 'O' licences will be required for each Traffic Area. (A list of Traffic Area Office addresses is to be found in Appendix I.)

Form GV79

Application for a licence has to be made on the appropriate form – form GV79 obtainable from the Traffic Area Office – which is straightforward and simple to answer. This form incorporates questions relating to vehicle operating centres, the previous history of licence applicants during the past five years and the type of licence required.

There are questions to be answered on the form relating to the name and address of the business, its partners or directors; information regarding vehicles currently owned and those which it is planned to acquire is also required and the address of their respective operating centres. Questions ask if the applicant company or individual or any partners of the business have convictions which are not 'spent' (under the Rehabilitation of Offenders Act 1974 a person is relieved of the obligation to disclose information about a conviction which is 'spent' – see p 19). Details are required of any such convictions including the date of the conviction, the nature of the offence, the name of the court and the penalty imposed.

Further questions require information about vehicle maintenance – who is to do it, when and where is it to be done, and what facilities there are at that place. Questions are asked about the financial status of the business proprietor, his partners or the directors of the business, in particular asking whether during the past three years any of them have been made bankrupt, been involved with a company which has gone into insolvent liquidation, or been disqualified from acting as a director or taking part in the management of a company. Details about the professionally competent person supporting the application are required, where that person lives, his actual place of work and the address of the operating centre for which he is responsible.

Undertakings

When the applicant signs the form he is not only declaring that the statements of fact made on the form are true but also, in effect, he is making legally binding promises – undertakings – that statements of what he intends to do will be fulfilled. The undertakings relate to the observation of certain aspects of the law concerned with goods vehicle operation and the maintenance of vehicles included in the licence application. If at some time during the currency of the licence the TC finds that these undertakings have not been fulfilled, as evidenced by any convictions for relevant offences, he may use his powers to revoke, suspend or curtail the licence. The basis on which the applicant makes the undertakings is that he promises the following:

I, or the licensed operator, undertake to make proper arrangements so that:

- the rules on drivers' hours are observed and proper records are kept;
- motor vehicles and trailers are not overloaded;
- vehicles will operate within speed limits;

- vehicles and trailers, including hired vehicles and trailers, are kept fit and serviceable;
- drivers report promptly on defects or symptoms of defects that could prevent the safe operation of vehicles and/or trailers, and that any defects are promptly recorded in writing;
- records are kept (for 15 months) of all safety inspections, routine maintenance and repairs to vehicles and trailers and these are made available on request; and that
- in respect of each operating centre specified, that the number of authorized vehicles and trailers kept there will not exceed the maximum numbers recorded against the operating centre.

As already stated, the application form is straightforward and simple to answer and extensive explanatory notes are provided for guidance. The Traffic Area Office also sends applicants a free booklet, *Goods Vehicle Operator Licensing – Guide for Operators* (GV74–7/96), to provide further help. Further information is also available at the VOSA website: www.vosa.gov.uk. This does not, however, mean that the form should not be carefully studied or that any answer will do in an attempt to gain a licence. While it is obvious that to get an 'O' licence the undertakings must be signed, it should be remembered that the consequences of not fulfilling the stated undertakings can lead to penalties so severe as to put a small operator out of business and even to cause hardship to a large one. A warning about this in the following terms is included in the explanatory notes on the form so that applicants are left in no doubt as to the consequences of making false statements or not fulfilling statements:

'I declare that the statements made in this application are true. I understand that the licence may be revoked if the licensed operator does not comply with the undertakings made and that it is an offence to make a false declaration.'

Form GV79A

Another form is involved in making an application for an 'O' licence. This is form GV79A which is a supplementary sheet used for supplying details of vehicles, for example, registration number, maximum gross weight body type – flat or sided including skeletals, box body or van, tanker or other type such as cement mixer or livestock carrier – and whether the vehicle is articulated, a tipper or refrigerated. Certain designation letters and numbers are used to indicate body and vehicle type as follows:

- Flat or sided including skeletals
- Box body or van
- Tanker
- Other type (such as cement mixer, livestock carrier).

    T Tipper
    R Refrigerated
    A Articulated.

Additional Application Forms

Two supplementary application forms are used in connection with certain licence applications. These forms are GV79E (pale green in colour), dealing with environmental information, and GV79F (beige in colour), dealing with financial information. The forms are used only when the TC requires further information following the initial application on form GV79 on either or both of the relevant matters (ie environmental issues or finance).

Environmental Information

Form GV79E is sent to a licence applicant if the TC receives representations from local residents following publication of details of the applicant's proposals regarding his vehicle operating centre in the local newspaper. The form must be completed and returned to the TC who will then consider the application in the light of this further information, the information given by those making the environmental representations and as a result of making his own enquiries.

The form requires details of the applicant's name and address, and the address of his proposed operating centre (see p 22 for definition). It then requires information about the vehicles to be normally kept at the centre and the number and types of trailer to be kept there. Information must be given about any other parking place in the vicinity of the operating centre which is to be used for parking authorized vehicles (ie those authorized on the licence). If the applicant is not the owner of the premises he must send evidence to show that he has permission or authority to use the place for parking vehicles.

A number of further questions must be answered on the form about the operating times of authorized vehicles. In particular, what time lorries will arrive at and leave the centre, whether they will use the centre on Saturdays or Sundays, what times they will arrive and leave on these days, whether maintenance work will be carried out there and between what hours, and whether any of this work will take place on Saturdays or Sundays and if so between what hours? The TC also wants to know whether there are any covered buildings at the centre in which this work is carried out.

A plan showing the parking positions for authorized vehicles must be sent when returning the completed form. This should show entry and exit points, main buildings, surrounding roads with names and the normal parking area for the vehicles. The scale of the plan must be indicated and this is suggested as being 1:500 which is one centimetre to five metres or, roughly, one inch to 40 feet. A larger scale of one inch to 100 feet can be used if this is more convenient when the operating centre is large. If the proposed operating centre has not previously been used as such the TC must be given information about any application for or planning permission granted for the proposed use of the site as a goods vehicle operating centre.

Financial Information

Form GV79F is sometimes sent to new standard licence applicants when the TC requires additional information to enable him to consider whether the applicant meets the financial requirements for this type of licence. An application will be refused unless the TC is satisfied that the applicant has sufficient financial resources to set up and run his business both legally and safely. This fact is pointed out clearly on the form. Answers have to be given to questions about the vehicles, their average annual mileage and the estimated running cost for each individual type of vehicle.

Details must be given about the funds available to start up the business and where these are held (eg in the bank, in savings or as agreed bank overdraft or loan facilities, or in the form of share capital), and about the start-up costs for the business including the purchase price or amount of down payments on vehicles and on premises and the sum to be held in reserve as working capital. The applicant is required to give a forecast of the annual expenditure and income for his road haulage operations for a financial year. The TC expects this information to give a clear indication of the business finances for the year ahead. In certain cases the TC may ask for monthly information.

NB: Not all TCs currently use this form; some prefer to rely on supporting financial information in other forms – bank references for example.

Date for Applications

Application for an 'O' licence should be made at least nine weeks before the day on which it is desired to take effect. In some Traffic Areas the time taken to process applications is much longer than nine weeks so new operators should be aware of the fact that it is illegal to start operating vehicles until their licence has actually been granted. Where there is an urgent need to start operations before a licence is granted through the normal processes, application can be made to the TC for an interim licence (see p 29).

Offences while Applications are Pending

Applicants for licences have a duty to advise the TC if, in the period of time between the application being submitted and it being dealt with by the TC, they are convicted of a relevant offence (see p 17) which they would have had to include on the application form had the conviction been made before the application was made. Failure to notify the TC is an offence and it could jeopardize any licence subsequently granted.

Advertising of Applications

Applicants for 'O' licences who are seeking a new licence, or variation of an existing licence, are required to arrange for publication of an advertisement (following a specified format to contain the necessary information for potential environmental representors – see Figure 1.1) in a local newspaper (or newspapers) circulating in the area where the operating centre is located. If more than one operating centre is specified in the application separate advertisements must be placed for each such centre in the respective local newspapers serving those locations. The sole purpose of the advertisement is to give local residents an opportunity (given to them under the regulations) to make representations against proposals to use a particular place as a goods vehicle operating centre.

The advertisement need appear only once but it must be published during a period extending from not more than 21 days before and not more than 21 days after the licence application is made. There is no specified minimum or maximum size requirement for the advertisement but the TCs advise that it 'should not be too small and should be easy to read'. Most adverts appear in single-column format

Goods Vehicle Operator's Licence

trading as ____

of____

is applying for a licence to use ______________________________

____

____

____

as an operating centre for ________________goods vehicles and

__________________________________trailers

and to use

____

____

____

____

as an operating centre for ________________goods vehicles and

__________________________________trailers

Owners or occupiers of land (including buildings) near the operating centre(s) who believe that their enjoyment of that land would be affected should make written representations to the Traffic Commissioner at

____

____

____

____

stating their reasons within 21 days of this notice. Representors must at the same time send a copy of their representations to the applicant at the address given at the top of this notice. A Guide to making representations is available from the Traffic Commissioner's Office.

Advertisement form for use with new application on GV79

Figure 1.1 *Format which must be used for 'O' licence newspaper advertisements*

Goods Vehicle Operator's Licence

trading as ____

of____

is applying to change an existing licence as follows

* To keep an extra ____ goods vehicles and ____ trailers at the operating centre at

* To add an operating centre for ____ goods vehicles and ____ trailers at the operating centre at ____

____

*To change existing conditions or undertakings applying at the operating centre at

____

from ____

____

to ____

____

*To remove the following conditions or undertakings which reads ____________ ______________________________and which applies to the operating centre at____

Owners or occupiers of land (including buildings) near the operating centre(s) who believe that their enjoyment of that land would be affected should make written representations to the Traffic Commissioner at

____

____

stating their reasons within 21 days of this notice. Representors must at the same time send a copy of their representations to the applicant at the address given at the top of this notice. A Guide to making representations is available from the Traffic Commissioner's Office.

Advertisement form for use with major variation application on form GV81

Figure 1.1 *(continued)*

extending to a few inches of text and at an average cost estimated to be in the region of £100–150. Normally the advertisement will appear in the public or official notices section of the newspaper.

Proof that the advertisement has appeared – and is published correctly (which many are not) – must be given to the TC before he considers the application and failure to produce this proof (usually achieved by sending in the appropriate page torn from the newspaper showing the advertisement itself and the name and date of the paper) will mean that the TC, by law, must refuse to consider the application. Normally this would mean making a fresh application which, of course, delays the whole matter, adds to the costs by requiring another advertisement and could mean vehicles having to stand until the licence or variation is granted.

Interim Licences

In certain circumstances the TC may grant an interim licence pending his decision on the full licence application. The circumstances under which such a licence may be granted are not precisely specified but they may be connected with some urgent need to move goods quickly because they are perishable or for some other urgent reason. A grant of an interim licence should not be taken as a guarantee that a full-term licence will be granted by the TC. An interim licence will not be granted in any case where the main requirements for 'O' licensing appear not to be met. For example, such a licence would not be granted to an applicant for a standard 'O' licence if he has not yet passed the CPC examination nor, indeed, while examination results are being awaited, nor on the assumption that the candidate will have passed. Neither will a grant of an interim licence normally be considered before the statutory 21-day waiting period for environmental representations and objections has expired.

Interim licences are not granted for any fixed period. Normally they remain in force until the TC has made his decision on the grant of a full licence or alternatively, until they are revoked.

Duration of Licences

Operators' licences, since 1 January 1996, are valid indefinitely and will remain so unless the operator contravenes the terms under which his licence was granted or fails to pay the necessary fees by the due dates. Under normal circumstances the only reason for making a new application, and thus being subject to the risk of objection or environmental representation, is when a major variation of a licence is necessary in order to add to the number of authorized vehicles or to change an existing, or to add a new, operating centre to the licence. Despite being valid indefinitely, licences are still liable to penalty or total revocation should the operator be found to have contravened the law or breached the conditions placed on his licence. Additionally, operating centres are subject to 'review' (normally at five-yearly intervals, but not exclusively so) to determine whether they remain environmentally suitable – see pp 36–37.

Licence Fees, Refunds and Discs

Fees for 'O' licences are divided into three main categories payable on application (see item 1 below), on the grant of the licence (see item 2 below) and for specified vehicles (see item 5 below), together with other fees as follows:

1. On application for a new 'O' licence or variation of an existing licence.
   £215 payable with the application – this is non-refundable.
2. On the issue of a licence.
   £336 payable within 15 working days from the date of the grant of the licence. This fee is non-refundable.
3. On continuation of a licence after five years.
   £336 due on the fifth anniversary of the month ending before the month of issue (ie the month in which the licence documents were issued). This fee is non-refundable.
4. For interim licences or directions.
   £57 payable within 15 working days of the date of issue of the licence or direction. This fee is non-refundable. The fee for each vehicle specified is £11.30.
5. Vehicle fees:
   (i) *Paid 5-yearly in advance – £9 per vehicle per quarter year (ie £180 for 5 years).*
   (ii) *Paid annually in advance – £44 per vehicle per year. These fees are non-refundable.*
   (iii) *For interim licences/directions – £11.30 payable within 15 days of the issue of the licence or direction.*

Termination of Licences for Unpaid Fees

It is important to note that failure to pay any of the fees described above by the due date will result in a licence being automatically terminated from the date on which the fee was due. In this event, the vehicle operation will have to cease and a new licence will have to be applied for before recommencing is permitted, if indeed it is permitted at all.

Refunds

Out of the fees described above, only a proportion of vehicle fees may be refundable on the surrender, termination or revocation of a licence. However, those who pay licence fees annually will receive no such refunds. Similarly, there are no refunds of fees paid for interim licences or directions. Where refunds are payable these will relate only to full years remaining before the next payment would have been due.

Licence Discs

'O' licence discs, issued on the grant (or variation) of a licence must be displayed on the vehicle (normally in the windscreen) in a clearly visible position near to the excise duty disc and in a waterproof container. Licence discs are coloured as follows to differentiate between restricted and standard licences and between standard national and standard international licences:

- Restricted Licence - orange
- Standard Licence, National - blue
- Standard Licence, International - green
- Interim Licence - yellow
- Copy Discs - word 'COPY' in red across face of disc.

Licence discs are not interchangeable between vehicles or between operators. They are valid only when displayed on the vehicle whose registration number is shown on the disc (even if it is faded almost beyond recognition) and when that vehicle is being 'used' by the named operator to whom it was issued. Heavy fines are imposed on offenders who loan and borrow discs (this is fraudulent use and is a serious offence); their own 'O' licence may be jeopardized and the vehicle insurance could be invalidated.

Licence Surrender/Termination

The 'O' licence itself and all vehicle windscreen discs must be returned to the TC on surrender or termination of an 'O' licence.

Issue of Community Authorizations

All UK holders of standard 'O' licences covering international operations are issued (automatically) with a Community Authorization document to be kept at their main place of business together with certified copies equalling the total number of vehicles authorized on their operator's licence (which must be carried on vehicles when undertaking cross-border journeys within the EU) - more information on Community Authorizations is to be found in Chapter 32.

THE TC's CONSIDERATIONS

When an application for an 'O' licence is made, the TC will have certain points to take into consideration before deciding whether or not to grant any licence. Mainly he has to ensure that the basic legal requirements as described previously have been met and particularly that those relating to vehicle operation and maintenance will be complied with. These points are dealt with here.

Fit Persons and Good Repute

The first point, and one of the fundamental requirements for 'O' licensing (as already described on pp 17–19, but a repeat here is useful), is whether the applicant is a fit person or is of good repute and is therefore fit to hold a licence. Basically, being a fit person and being of good repute are the same but the former relates to restricted 'O' licences over which the EU has no influence while the latter is the term used by the EU in setting its requirements for the holding of a licence to carry goods for hire or reward (ie the UK system of standard 'O' licences).

The TC, when deciding this, will take into account any previous record which the applicant (or the partners or directors of the applicant's business) might have had as an operator in terms of his ability or willingness to comply with the law in respect of vehicle operations and particularly maintenance, drivers' hours and records, overloading and the like, also any previous convictions they may have for offences relating to the roadworthiness of vehicles and for other relevant offences. The regulations require that a TC take account of 'serious offences' besides just road transport offences when determining the good repute of an individual, or of company management for a corporate 'O' licence application.

Maintenance Facilities/Arrangements

The next point that the TC will consider, and one of the most important since it is at the very foundation of the 'O' licensing system, is whether the applicant has suitable facilities or has made satisfactory arrangements for the maintenance of vehicles which are to be specified on the licence in a safe and legal condition and for keeping suitable maintenance records (this is dealt with in more detail in Chapter 19). In particular the TC will be concerned to know that vehicles are being subjected to safety inspections at regular intervals of time or mileage.

The TC for the West Midland and South Wales Traffic Areas, for example, currently suggests that for operators in his area inspections should be based on a time interval only with no mileage alternative and that a period of six weeks between inspections is the maximum that he would find acceptable. He also seeks assurances that operators are using a *written* driver defect reporting system and wall charts for planning inspection and maintenance schedules. Other TCs seek similar assurances.

Drivers' Hours and Records

The TC will consider whether there are satisfactory arrangements for ensuring that the law relating to drivers' hours and records (including tachographs) will be complied with.

Overloading

Similarly, the TC will want to be sure that arrangements are made to prevent the overloading of vehicles and that vehicle weight limits in general will be observed.

Professional Competence Requirements

The TC will want to know details of the nominated professionally competent person, who may be the applicant himself or an employee who holds a certificate of professional competence covering national or both national and international transport operations (or a person who qualifies by exemption or by having passed the appropriate OCR examination) as appropriate to the type of standard 'O' licence applied for (see Chapter 2). The Transport Tribunal has ruled that the named professionally competent person in respect of a standard 'O' licence must be an active employee carrying out the functions of a transport manager in an adequate manner.

In making decisions on 'O' licence applications that involve the need to nominate a qualified transport manager, the Traffic Commissioners are concerned to ensure that the individual proposed is suitable in accordance with the Guidance Notes from the Senior Traffic Commissioner set out in September 2003 in regard to such matters as:

- the number of operator licences for which that Transport Manager will be responsible;
- the amount of time that the Transport Manager will spend in carrying out his/her duties for the operator;
- the number of operating centres and authorized vehicles for which the Transport Manager is responsible both on that licence and any other operator's licences;
- the geographical location of the Transport Manager in relation to the operator's licence and the operating centres on that licence;
- any other employment or activities in which the proposed Transport Manager is engaged which may restrict his/her ability to devote sufficient time to the duties of a Transport Manager on that operator's licence;
- the terms upon which the proposed Transport Manager is to be employed.

In order to further ensure that nominated Transport Managers carry out their duties effectively and in accordance with the law, a proposed form of Transport Manager – Contract of Employment has been devised (see illustration on p 34), which must be completed by the 'O' licence holder and the appointed Transport Manager.

Number of Qualified Persons

There is no restriction under the regulations (see Chapter 2) on the number of people in a transport department or organization who may be professionally competent and consequently hold certificates of competence. Further, although restricted 'O' licence holders have no need to specify the name of a professionally competent person in order to obtain a licence, there is no restriction on such licence holders or their employees being professionally competent if they qualify personally.

In determining how many qualified persons must be named on a standard 'O' licence the TC will take account of the management structure of applicant firms, but generally there will need to be a minimum of one qualified person per 'O' licence. The TC may require the names of more qualified persons to be specified if he considers it appropriate in view of a division of responsibilities for the operation of vehicles under the licence or if vehicles specified on the licence are located at different operating centres within the Traffic Area.

Part-Time Managers

Increasingly, TCs are questioning the role of part-time and agency-provided professionally competent transport managers. There is concern that such managers are not in a position to meet the statutory requirement for managers whereby they should have 'continuous and effective responsibility' for the fleet for which they are named. In particular, the question arises of the number of clients for whom many such agency managers are acting and the remoteness from the operating bases for which they are supposedly responsible, with some such named managers being in contact only by telephone or via a fleeting visit and thus having, as one TC put it, 'little physical contact with the drivers, the vehicles and the documentation'. This TC has stated that he intends to scrutinize closely any future applications involving agency-provided transport managers.

TRANSPORT MANAGER – CONTRACT OF EMPLOYMENT

Surname	First name(s)	Date of birth

Home address

(Postcode)

Address of place of work (if not the same as operating centre)

(Postcode)

Which operating centres will the nominated TM be responsible for?
All? Yes ☐ No ☐

If no, please list individual operating centres (by first address line only)

How many hours a week is the TM contracted to work?

Will the TM be responsible for vehicles on any other licences (in <u>any</u> Traffic Area)?
Yes ☐ No ☐

If Yes, please list details below

Licence number	Vehicles/trailers authorized	Hours worked per week

Declaration: In accordance with the requirements of the Goods Vehicles (Licensing of Operators) Act 1995, we declare that .... will carry out all necessary checks on the operation of the licence holder's business. We also confirm that we understand the requirements placed on both of us by the above Act.

We understand that these responsibilities include:

- the method of control of drivers' hours;
- the maintenance of the licence holder's vehicles, including the inspection of vehicles at the appropriate time, the action taken to remedy defects found and the recording of these events;
- the reporting and recording of vehicle defects by drivers;
- the method of compilation and the accuracy of all records kept;
- the making of arrangements to ensure that the licence holder's vehicle/s are not overloaded.

Signature (licence holder)	**date**

Signature (Transport Manager)	**date**

Representations by Local Residents

Opportunities are given to local residents individually to make representations against 'O' licence applications and variations on environmental grounds. Local residents are more precisely defined in the regulations as 'owners or occupiers of land within the vicinity' (ie of the operating centre). Those residents who wish to make representation must do so individually because group action is not permitted (although a group of individual representors may appoint a joint spokesperson to put forward their case) nor is representation by any environmental pressure group, political or other campaigning body. Similarly, parish councils, which regularly feature in such matters, have no right of objection *per se* unless they *own/occupy* land in the vicinity of an operating centre featuring in an 'O' licence application, in which case they may make representation as the owner/occupier of the land.

The grounds on which such owners or occupiers can make their representations are confined purely to environmental matters such as noise, vibration, fumes and visual intrusion but could include obstruction. They do not include road safety matters which are not an environmental issue. The grounds must be stated precisely in the written representation; to state that the representation is made for 'environmental reasons' is not sufficient. The exact wording which forms the basis of representations is specified in the legislation in the following terms, 'that place (ie the operating centre) is unsuitable on environmental grounds for (such use and) any adverse effects on environmental conditions arising from that use would be capable of prejudicially affecting the use or enjoyment of the land' (ie the land owned or occupied by the person making the representation).

One of the facts that has been difficult to establish in connection with this is a definition of the term 'within the vicinity'. It has been shown that residents living along an access road to a vehicle operating centre can be considered to be in the vicinity and adverse environmental effects of vehicles travelling along the road could be taken account of by the TC in his consideration of any environmental representations against a licence application. Each TC is left to make his own determination of whether a representor lives 'within the vicinity', but as a general rule if a representor can see, hear or smell a vehicle operating centre from his property then he will be considered to be 'in the vicinity' for the purposes of making an environmental representation.

There is no opportunity for people living near an operating centre to make representations on grounds other than environmental matters or to use the opportunity to vent long-standing grudges against the vehicle operator. The TCs will not consider any representation which falls outside the terms described above or which is considered to be vexatious, frivolous or irrelevant.

Local residents will normally become aware of their opportunity to make representations against the grant of a licence or licence variation through the local newspaper advertisement placed by the applicant (see pp 26–29). Those people wishing to make a representation must do so in writing (or have their solicitor do so on their behalf), within a period of 21 days from the date of publication of the advertisement, to the TC at the Traffic Area Office address given in the advertisement. They must also send an exact copy of their representation (ie their letter to the TC) to the licence applicant at his address which is also given in the advertisement. Their letter must clearly state the 'particulars' of the matters forming the basis of their representation so that both the TC and the licence applicant may be fully aware of the specific grounds on which the representation is made. Failure to be specific as to the facts in this letter, failure to send a copy to the licence applicant, or failure to submit the representation within the specified time-scale, will render the representation invalid.

Representors have no right of appeal should their case against use of the operating centre fail. These rights of representation are not to be confused with the rights of objection described later.

Suitability of Premises

TCs must inquire into and be satisfied that the place or places to be used as vehicle operating centres are suitable, cause no danger to the public, are environmentally acceptable and sufficiently large to accommodate all the vehicles authorized on the licence (or requested in the application). Local residents have rights (as described above) to make representations about the environmental consequences of the use of places for transport depots or vehicle operating centres and the TCs are bound to listen to these representations and make appropriate decisions about the application depending on the weight of the argument on either side – residents or operator (see p 35 and p 48). In particular the TC, when considering the suitability of premises, will take account of the following:

- whether danger to the public may be caused where vehicles first join (or last leave) a public road;
- whether danger to the public may be caused on roads (other than public roads) along which vehicles are driven between the operating centre and a public road;

- the nature or use of any other land in the vicinity of the operating centre and the effect which the granting of the licence would be likely to have on the environment of that land;
- how much granting a licence which is to materially change the use of an existing (or previously used) operating centre, would harm the environment of the land in the vicinity of the operating centre;
- for a new operating centre, any planning permission (or planning application) relating to the operating centre or the land in its vicinity;
- the number, type and size of the authorized vehicles (including trailers) which will use the operating centre;
- the parking arrangements for authorized vehicles within and near to the operating centre;
- nature and times of use of the operating centre;
- nature and times of use of equipment at the operating centre;
- how many vehicles would be entering or leaving the operating centre, and how often.

Reviews of Operating Centres

TCs are given powers to review all operating centres, normally at five-yearly intervals counting from the date when the licence was first issued – in certain circumstances more frequent reviews may be carried out. Where the TC decides to review an operating centre he must give two months' notice in writing. Once the period has passed for the TC making a decision to review an operating centre, the licence holder can rest assured that he is safe for another five years, unless he operates outside the terms of his licence, fails to pay fees or applies for a major variation of the licence.

When carrying out the review of a centre, which will most likely be as a result of written complaints by local residents (who may now write in at any time rather than just when a licence application/variation advertisement appears), the TC will be concerned to ensure that it remains environmentally suitable, meets road safety considerations and can accommodate all the vehicles authorized on the licence, or that their parking causes no adverse effect on the local environment.

When carrying out the review the TC may decide that no action is required but he has powers to act if necessary. For example, if an operating centre is found to be unsuitable he may attach conditions or vary any existing conditions for road safety or environmental reasons. However, the licence holder is given the opportunity to make representations about the effect that such conditions would have on his business before they are attached. The TC also has the power to remove an operating centre from a licence for non-environmental reasons (eg on the basis of road safety considerations), or because the operating centre is environmentally unsuitable by reason (only) of the parking of vehicles used under the licence at or near the centre.

Applications and Decisions

When an application for a new 'O' licence or a variation of an existing licence is received by the TC, details of the application (ie the name of the applicant and the number of vehicles and trailers included in the application) will be published in a

Traffic Area notice called *Applications and Decisions (As&Ds)*. As its name implies, this notice will also contain details of licences granted by the TC and details of public inquiries to be held. The notice is published fortnightly by all Traffic Areas and may be inspected at Traffic Area offices or purchased as an individual copy or on a regular basis. It is by means of this notice that statutory objectors (see below) are able to know when applications have been made, against which they may wish to object. They can do this within 21 days of publication of the relevant As&,Ds notice. As&,Ds are now available for download from the VOSA website: www.vosa.gov.uk.

Objections to the Application

Applications for 'O' licences are open to statutory objection by certain bodies listed below (and only by the listed bodies – no other individual or organization has this statutory right). Potential objectors to 'O' licences become aware of pending applications for new licences or variations to existing licences through the publication *Applications and Decisions* mentioned above.

Objections to applications can only be made by the bodies mentioned on the grounds that the applicant does not meet the essential qualifying criteria for the grant of a licence: namely that the applicant is not a fit person or is not of good repute; is not of adequate financial standing or does not meet the professional competence requirements (where appropriate); that the law in respect of those matters which the TC will be considering when he is deciding whether or not to grant a licence is not likely to be complied with, namely that the drivers' hours and records regulations will not be observed; that vehicles will be overloaded and that there are not satisfactory arrangements or facilities for maintaining the vehicles. The bodies may also object on environmental grounds (for example that the operating centre is environmentally unsuitable).

The bodies who may make statutory objection to an 'O' licence application are as follows:

- A chief officer of police
- A local authority (but not a parish council – unless it owns or occupies land)
- A planning authority
- The British Association of Removers (BAR)
- The Freight Transport Association (FTA)
- The Road Haulage Association (RHA)
- The General and Municipal Workers' Union (GMWU)
- The Rail, Maritime and Transport Union (RMTU) (formerly the National Union of Railwaymen and the National Union of Seamen)
- The Transport and General Workers' Union (TGWU)
- The Union of Shop, Distributive and Allied Workers (USDAW)
- The United Road Transport Union (URTU).

These are the only sources of objection (not to be confused with an environmental representation) to an application for an 'O' licence. If any of these bodies do make a statutory objection they are required to send a copy of their objection to the applicant at his published address at the same time as sending one to the TC and this must be within 21 days of the publication of details of the application in

Applications and Decisions. Failure to send a copy to the applicant renders the objection invalid.

Grant or Refusal of a Licence

The TC has power to grant an 'O' licence to an applicant if he considers that all the necessary requirements are met. Alternatively, he may refuse to grant a licence or he may grant a licence for fewer vehicles than the number applied for if he doubts the ability of the applicant to be able to comply with the law with more vehicles, to be able to properly maintain more vehicles or to be able to adequately finance the operation of more vehicles. He can impose environmental conditions on any licence granted and can also refuse to accept the name put forward for the professionally competent person (in the case of standard licence applications) if he believes that the person is not of good repute. The TC may be influenced in his decision by the points made by any statutory objectors or environmental representors.

Licence Grant with Conditions

The case made by those raising valid environmental representations may influence the TC either to refuse the application altogether on the grounds that the operating centre is not environmentally suitable or alternatively to grant the licence but with environmental conditions attached. Thus to prevent or minimize any adverse effects on the environment he may place conditions or restrictions on the licence granted under the following headings:

- the number, type and size of authorized vehicles (including trailers) at the operating centre for maintenance or parking;
- parking arrangements for authorized vehicles (including trailers) at or in the vicinity of the centre;
- the times when the centre may be used for maintenance or movement of any authorized vehicle; and
- how authorized vehicles enter and leave the operating centre.

As an alternative to imposing environmental conditions on the licence, the TC may seek undertakings from the operator that he will or will not follow certain practices in order to reduce environmental disturbance of local residents (eg control the number of vehicle movements into and out of the centre). The licence holder should be aware that any such undertakings he may voluntarily give to the TC become legally binding upon him and could result in penalty against his licence if he subsequently fails to observe them.

Licence holders who find they have breached environmental conditions on their licence through unforeseen circumstances must notify the TC. Failure to comply with any of these conditions during the currency of a licence may result in the TC imposing penalties on the licence such as suspension or curtailment. In serious cases the licence may be totally revoked.

Where operators find that environmental conditions placed on their licence prove too onerous to allow them to run their businesses effectively they should apply to the TC to vary the conditions rather than ignore the conditions and become liable to a licence penalty.

ADDITIONAL VEHICLES

Seeking a Margin

At the time of making an application for an 'O' licence the applicant is given the opportunity to request authorization for any additional vehicles which he may need to acquire or hire during the currency of the licence. By taking this opportunity the operator saves the problems of making a fresh application when wanting to add or hire-in vehicles on a temporary basis to meet trading peaks. It also saves facing any further environmental representations or statutory objections because once the original application is granted with additional vehicles specified, extra vehicles can be added within the number authorized by completing and submitting form GV80 to the TC when they are acquired. There will be no need for the details to be advertised in a local newspaper or published in *Applications and Decisions*.

If additional vehicles were requested and the request was granted at the time of making the original application, the operator will have a 'margin' for extra vehicles on the licence. As described above, the TC will need to be notified from the date of actually *acquiring* the additional vehicles – *not* the date of putting them into service – (by submitting form GV80) so that a windscreen disc can be issued for the vehicle and the appropriate fee charged (ie £34 per vehicle per year). From the effective date for the new TAN 21 computer system (see p 8) no goods vehicle subject to operator's licensing should be on the road without a current 'O' licence windscreen disc. Vehicles found without a current disc may be subject to impounding (see p 57).

It is useful here to clarify the terms used in connection with the numbers of vehicles for 'O' licensing purposes:

- Authorized vehicles – the maximum number of vehicles/trailers that the licence is actually granted to cover (it is illegal to operate * more than this number of vehicles at any time).
- Specified vehicles – the actual vehicles which the operator has in possession and which are specified on the licence by registration number.
- Margin – the difference between the numbers of authorized and specified vehicles on the licence, in other words the vehicles still to be acquired by the operator whether on a permanent or a temporary basis.

**NB Operate in this context means 'use' and this applies to vehicles hired without drivers. Conversely, if they are hired with drivers, then they are operated under the hire firm's 'O' licence and not within the margin of the operator's 'O' licence (see below).*

Hired Vehicles

If an operator plans to hire extra vehicles without drivers (ie where he intends to have his own or hired agency drivers to drive the vehicle/s) during the currency of his licence, whether for a short period (a day, a few days or even one or two weeks) or on a long-term contract, they must be covered by his 'O' licence and he will need to have applied for a sufficient margin of additional vehicles on his licence to cover these. If vehicles are hired within the margin the TC must be notified, and an 'O' licence disc obtained for display on the vehicle.

It is illegal to operate (ie to have employed drivers to drive) more vehicles (ie of over 3.5 tonnes gross weight) than are authorized on the 'O' licence even for a temporary period or reason (eg when an authorized vehicle is off the road for service or repairs or to cover additional delivery requirements).

Number of Extra Vehicles

When making the request for additional vehicles on the initial application for an 'O' licence, the number which may be requested is not limited in any way but it is recommended that it should be in reasonable proportion to the number of vehicles already operated (or initially required) and, most important, it should only be of a number which the applicant can maintain, and prove he can maintain (both physically and financially) on the same basis as the remainder of his fleet. If the request for additional vehicles relates to vehicles that are to be hired rather than owned it must be remembered that the person who hires a self-drive vehicle is fully responsible for the mechanical condition of the vehicle in so far as safety and legal requirements are concerned.

An applicant specifying additional vehicles on the original application should give some careful thought to the exact number of vehicles which may be needed and the reasons for needing them because the TC will ask questions about this if he calls the applicant to a public inquiry. Evidence in the form of business forecasts and trends in trade would be most useful as would figures to indicate past growth of the business; evidence also to show the financial prospects of the applicant during the currency of the licence period will help towards convincing the TC that he would be justified in granting a licence for the additional vehicles requested.

Replacement Vehicles

If for some reason an authorized vehicle ceases to be used the TC must be advised of the fact, but if at that time or later another vehicle is acquired to replace it the operator must advise the TC on form GV80 within one month of *acquiring* the replacement vehicle. This means within one month from the date of the vehicle coming into the operator's possession, not one month from the date he starts to use it. Vehicles which are not removed from the licence (even when standing in a yard or workshop smashed or cannibalized) are still counted as specified vehicles and cannot be replaced by others within the authorized number on the licence until they are removed by notifying the TC and the windscreen discs are returned to the Traffic Area Office.

LICENCE VARIATION

If the holder of a restricted 'O' licence wishes to change the licence to a standard 'O' licence it is necessary to make a 'major variation' application on form GV81 and satisfy the legal requirements for standard licences, national or international. A standard national licence holder who wishes to change to a licence covering international operations must satisfy the TC that he, or an employee, is professionally competent in international transport operations (apply on form GV81).

Where the licence holder finds, during the currency of the licence, a need to add extra vehicles to the fleet that were not specified on the original application then an application must be made to the TC by completing a major variation application on form GV81 and submitting this well in advance (minimum nine weeks). This will necessitate placing an advertisement in a local newspaper as with the original application (see pp 26–29).

Unless the variation is only of a trivial nature the TC will publish details of it in *Applications and Decisions* and it may attract objectors in the same way as a new application, and the public inquiry procedure will be the same as that already described.

The licence holder must never operate more vehicles or trailers than the total number specified on his licence. When extra vehicles are required the operator must wait until the application for an increase in the licence is granted before actually putting the vehicles on the road. This can take at least nine weeks – which is the minimum application period required – and as many as 12 to 15 weeks in some Traffic Areas.

Form GV81

Form GV81 is a five-page document called *Application to Change a Goods Vehicle Operator's Licence* and must be used by applicants who wish to:

- change the number of vehicles authorized;
- change the operating centre, add another centre or stop using a centre;
- change the type of licence (eg restricted to standard);
- change or remove a condition attached to the licence (including conditions on the use of operating centres).

Most of the changes involve the need to advertise the application in local newspapers (see p 26) and where this is necessary the fact is pointed out on the form. The applicant must send a copy of the published advertisement with the form or state the expected date of publication and the name of the newspaper in which it is to be published. The usual information regarding name and address, addresses of operating centres and the number of vehicles to be based there which are in possession now or to be acquired has to be given. Also required is similar information about vehicle maintenance arrangements to that required on the original GV79 application (see p 24).

Section 3 of the form deals with any addition or deletion to the specified operating centres and section 4 is completed if the applicant wishes to change the type of licence (eg from restricted to standard or from standard national to standard international licence). Details of the professionally competent person must be given where appropriate; so too must details of any convictions (other than those which are 'spent' – see p 19) of the applicant or the partners or co-directors.

Section 5 of the form has to be completed if the applicant wishes to change or remove any of the environmental conditions attached to the use of the operating centres. Reasons must be given as to why this change or removal is wanted and details given of any alternative proposals if the applicant has any.

The form has to be signed and the applicant is warned that the licence may be revoked if any of the statements given are false.

TRANSFER OF VEHICLES

If a vehicle is transferred from the Traffic Area in which it is licensed to a base in another Traffic Area for a period of more than three months, it must be removed from the original licence and specified on a licence in the new Traffic Area. Transfers for periods of less than three months are permitted with no need for notification to the LA provided an 'O' licence with a sufficient margin to cover the transferred vehicles is already held for that Traffic Area.

If the operator does not hold an 'O' licence in the other Traffic Area, or holds a licence in the area but it does not have a sufficient margin to accommodate the transferred vehicles, then an application for a new licence or a variation of the existing licence must be made to the TC for that Traffic Area. It is illegal to operate vehicles (ie over 3.5 tonnes gvw) from a base in a Traffic Area unless a licence is held in that area.

NOTIFICATION OF CHANGES

Licence holders should notify the TC in writing, within *one month,* of any changes in the legal entity of their business such as a change of name, address, ownership, if a new partnership has been formed, a limited company formed or the constitution of the partnership has been changed, as this makes a material difference to the information given in answer to questions on the original GV79 licence application. The TC must also be informed if the proprietor or persons concerned in the business die or if the business becomes bankrupt or goes into liquidation.

A change of business address as given in the original licence application must be notified to the TC within *three weeks.* A change of operating centre (or the use of an additional operating centre) also requires a variation application using form GV81 and the need to follow the newspaper advertisement procedure before any change actually takes place – it is illegal to change or add an operating centre without first seeking the TC's approval by the GV81 application procedure (ie to be submitted at least nine weeks prior to any change being required).

Other changes which must be notified in writing are those in maintenance facilities or arrangements and any breach of environmental conditions which the TC placed on the licence. Failure to notify the TC of such changes can have the same result as making false statements or failing to fulfil intentions stated in the original application, namely the risk of licence suspension, curtailment or revocation. The offender could also be prosecuted with the consequent penalties which can be imposed by the courts.

SUBSIDIARY COMPANIES

A holding company can include in its application for an 'O' licence vehicles belonging to any subsidiary company in which it owns more than a 50 per cent shareholding. But associate companies (ie where the shareholding arrangement is less than 50 per cent), owned by the same holding company, cannot have vehicles specified on each other's licences and separate divisions of a company are not permitted to hold separate licences unless they are separate entities in law.

The vehicles of any subsidiary company acquired during the currency of the holding company's 'O' licence can, if desired, be included in the holding company's licence either within its existing licence margin or by making application to the TC, on form GV81, to vary the licence. It is not generally likely that an application to include a subsidiary company's vehicles on the holding company's licence would be published in *Applications and Decisions* or that it would attract any objections.

Under the regulations, for the purposes of determining whether goods are carried for hire or reward in order to choose between a restricted or a standard 'O' licence, goods belonging to, or in the possession of, a subsidiary company are considered to belong to, or be in the possession of, the holding company and vice versa, so that in such cases a restricted 'O' licence would be adequate even if charges for the movement of the goods were made between the holding company and its subsidiary.

TEMPORARY DEROGATION

There are provisions in the regulations to enable a standard 'O ' licence to remain in force for up to one year initially and a further six months (maximum derogation is 18 months), if the TC feels it is appropriate, should the specified professionally competent person named on the licence die or become legally incapacitated (ie unable to carry out his duties due to reasons of mental disorder), in order to allow a replacement person to be found and specified.

The regulations enable the TC to defer revocation of, or refusal to grant, a standard 'O' licence in the event of the death or incapacity of the holder of the licence, a transport manager, or a partner whose professional competence is relied upon. Further, in the event of the death or incapacity of the licence holder the TC is empowered to authorize another person to carry on the business during the changeover period as though that person was the licence holder. Also, the TC may allow time for a transport business to be transferred to another person licensed to carry it on or for a transport manager or new partner to be appointed.

Where a person who was carrying on a business as a licence holder dies, becomes mentally incapacitated, bankrupt or goes into liquidation, or where a partnership is dissolved, the TC must be notified within two months. The person carrying on the business will be considered by the TC to be the holder of the licence if a new licence application is made within one month in the case of restricted 'O' licences and within four months in the case of standard 'O' licences.

PRODUCTION OF 'O' LICENCES

Operator's licence holders must produce their 'O' licence (form OL 1 plus forms OL 1(R) or OL 1(S) where appropriate) for examination when required to do so by the police, DoT examiners (ie certifying officers) or by the TC or a person with his authority. The holder has 14 days in which to present the licence either at one of the operating centres authorized on the licence or at his principal place of business in the Traffic Area. In the case of production to the police this can be at a police station of the holder's choice also within 14 days.

TCS' POWERS OF REVIEW

TCs have statutory powers (under the Goods Vehicles (Licensing of Operators) Act 1995, section 36) to review and, if they see fit, to vary or revoke any decision they have previously made to grant or refuse:

- an application for an operator's licence; or
- an application for the variation of an operator's licence requiring publication.

These powers to review a previous decision apply only in the following circumstances:

- If, within two months, the TC has given notice to the applicant or the licence holder that he intends to review the decision.
- If, within two months, a person who appears to the TC to have an interest in the decision has requested him to review it.
- Where neither of the above situations apply, if the TC considers that there are exceptional circumstances to justify a review.

Variation or revocation of any previous decision by the TC does not make unlawful any actions relying on that decision before the variation or revocation imposed at the review comes into force.

Normally the TC will only decide to review his previous decisions where he is satisfied that a procedural requirement was not complied with in making the decision; for example, where an environmental representation or statutory objection was overlooked, or a decision was made under the wrong section of the Act.

This facility for the TCs to review their own decisions eliminates many of the situations which, hitherto, would have required appeal to the Transport Tribunal.

LICENCE REVIEW BOARDS

Licence Review Boards (LRBs) were set up in 1990 to provide a system whereby the TCs could keep under review the 'O' licences of operators whose maintenance standards are in doubt, who have sustained relevant convictions, or who otherwise give the TCs cause for concern. The Boards are chaired by the Clerk to the TC in each of the Traffic Areas and comprise senior officials from the Vehicle Inspectorate (who deal with vehicle maintenance issues and traffic enforcement) and from the Traffic Area Office (ie managers who deal with 'O' licensing and LGV licensing matters). They meet regularly (eg fortnightly) and, where appropriate, make recommendations to the TC who may call recalcitrant operators to public inquiry.

PENALTIES AGAINST 'O' LICENCES

Warning

The TCs have warned that they are taking a harder line than ever before against 'O' licence offenders – this after being described for years as being no more than 'paper tigers'. In particular, operators are warned to pay more attention to their licence applications and what they are undertaking to do as operators. When dealing with licences the TCs are primarily concerned with road safety, fair competition and the environment.

Penalties

The maximum fine which may currently be imposed by a court for running vehicles without an 'O' licence where one is required by law is £5,000 (ie level 5). Failure to notify the TC of certain information about relevant convictions incurred by the licence holder or by his professionally competent transport manager can result in fines of up to £2,500 on summary conviction.

The TCs, as the issuing authorities for goods vehicles licences, are also given considerable legal powers to revoke, suspend or curtail an 'O' licence for a large number of reasons, of which the following are a few of the important examples:

- Contravention by the licence holder of the provision, in the case of standard 'O' licences, regarding professional competence requirements.
- Failure to notify the TC of changes in the business.
- Convictions for failure to maintain vehicles in a fit and serviceable condition.
- Contravention of speed limits, overloading or offences in connection with loading or unloading vehicles in restricted parking or waiting areas.
- Failure to ensure that drivers are correctly licensed.
- Convictions relating to the use of rebated (duty-free) fuel oil in vehicles (see Chapter 8).
- Failure to keep records relating to vehicle inspections and repairs and driver defect reports.
- Falsely stating facts on applications for 'O' licences and not fulfilling statements of intent or environmental conditions placed on the licence.
- If the licence holder becomes bankrupt or, in the case of a company, goes into liquidation.
- If a place not listed on the licence is used as a vehicle operating centre.

NB: 278 'O' licences were revoked in 2001/02 by way of disciplinary action at Public Inquiries.

Offences are committed, for which prosecution and a court appearance may follow, if a:

- windscreen licence disc is not displayed;
- change of address is not notified;
- licence is not produced for examination on request;
- duplicate windscreen disc is not returned if the original is found;
- disc is not returned when a vehicle is disposed of;
- subsidiary company featured on a holding company licence is disposed of and the TC is not advised.

Furthermore, the TCs are active in preventing speeding by lgv drivers, firstly by imposing a penalty of suspension on the LGV driving licences of offending drivers and then by penalizing the 'O' licences of firms whose drivers persistently and wilfully exceed speed limits. Evidence of such matters is mainly obtained during routine enforcement checking of tachograph charts, where recordings showing frequent instances of driving above 100 kph are clear evidence of breach of the 60 mph maximum speed limit for vehicles exceeding 7.5 tonnes maximum laden weight.

Usually, the offending licence holder will be called to public inquiry by the TC and be required to explain why the offences occurred and what action is being taken to put matters right or to ensure they will not happen again. Depending on his reaction to such explanations, the TC may initially give a warning about future conduct and the likely consequences if there is any repetition of the contraventions of the law or he will decide that an appropriate penalty should be imposed. This will be suspension, curtailment or premature termination of the licence, or revocation. If the licence holder is found to no longer comply with the basic requirements for 'O' licensing, namely good repute, financial standing or professional competence, then the TC must revoke the licence (except where a period of temporary derogation is permitted – see below). If a restricted 'O' licence holder is convicted twice in a period of five years of operating outside the terms of the licence his licence must be revoked.

Curtailment (vehicles removed from a licence) is the most commonly imposed penalty and this implies removal of one or more authorized vehicles from the licence for any period up to the expiry of the licence. Suspension involves suspension of the whole licence and this may be combined with premature termination so the TC can review the whole operation under the provisions for consideration of a new licence application. As with premature termination of an existing licence, the need to apply for a new licence places the operator at risk of objection and environmental representation. The TC can direct that a vehicle on a licence which has been suspended or limited may not be used by another operator for a maximum of six months during the period of suspension. The increasing trend by TCs to impose short, sharp two-week 'O' licence suspensions is showing to be effective against errant operators, and is generally approved by the Transport Tribunal.

When the TC revokes an 'O' licence – which is not done lightly – he may order the holder to be disqualified, for a certain period or indefinitely, from holding or obtaining an 'O' licence and the order may be limited to one or may apply to more Traffic Areas. When intending to disqualify a person from holding an 'O' licence the TC should warn him of the likelihood, explain the circumstances leading to this decision and allow the person the opportunity of making submissions against disqualification. Following an order to revoke a licence, the TC may allow the licence holder to request a 'stay' to enable the operation to continue until an appeal to the Transport Tribunal is heard.

Revocation of Licences for Smuggling

Licence holders who are convicted of smuggling and related offences – mainly resulting from the illegal importation of excessive quantities of tobacco and alcohol products – risk losing their 'O' licences in addition to facing the standard penalties such as substantial fines, imprisonment in more serious cases and confiscation of vehicles. Where 'O' licensed vehicles are involved in cross-Channel smuggling activities, Customs and Excise pass details of the operators involved to the TCs for action which could include revocation of licences.

Tough new measures to combat smuggling include £1,000 penalties to recover confiscated vehicles, and permanent confiscation in the case of repeated smuggling offences. Drivers caught smuggling may also lose their driving licences and in serious cases individuals risk a prison sentence on conviction for evading alcohol and cigarette duty.

NB: A Customs and Excise hotline is available for reporting excise duty smuggling – Freephone 0800 595000.

Revocation of Licences for Using Unlicensed Sub-Contractors

One TC may have started a trend when he revoked an operator's 'O' licence for using unlicensed sub-contractors. This ruling may be challenged in due course but in the meantime it provides a warning to other licence holders about their conduct in this respect.

It is essential for 'O' licensed hauliers to check carefully that any sub-contractor used is correctly licensed (as well as determining that he is using roadworthy vehicles and complies with the law on such matters as drivers' hours and tachographs). The use of an unlicensed operator may result in prosecution for aiding, abetting, counselling or procuring that operator to use a vehicle for hire or reward without an 'O' licence. This is a criminal offence carrying a maximum fine of £5,000 on conviction and could result in loss of the principal contractor's 'O' licence.

Loss of Licence for Unpaid Fees

It is worth repeating here that failure to pay 'O' licence fees when due will result in automatic termination of the licence.

LICENSING COURTS/TRANSPORT TRIBUNAL

Inquiries and Appeals

TCs regularly hold public inquiries (PIs) to which 'O' licence applicants are called to explain the basis of their operations. A PI enables a TC to seek more information prior to determining whether he should grant a licence or not. In the event of a representation on environmental grounds or an objection being made to a licence application the TC will hold a public inquiry at which the parties (applicant, objectors or those making representations) will have an opportunity to state their case further. If the application is refused in whole or in part or if environmental conditions are attached to a licence the applicant has rights of appeal against the TC's decision to the Transport Tribunal. Normally an existing licence will remain in force while an appeal is being heard and the TC may allow a revoked or suspended licence to continue during this time. If the TC refuses this the Tribunal can be asked to allow it to do so.

Statutory objectors also have a right of appeal to the Transport Tribunal if an application for a licence is granted and they still feel that their objection is valid. Those individuals making representations on environmental grounds have *no* similar right of appeal if their case fails.

It should be noted that the Transport Tribunal is the *only* source of appeal in regard to 'O' licensing matters.

Besides public inquiries conducted for the purposes of determining 'O' licence applications, such inquiries are also held at the TC's behest where it is necessary for him to examine the conduct of a licence holder for disciplinary purposes under the powers given him by the Goods Vehicles (Licensing of Operators) Act 1995. Section 26(1) of this Act empowers him to conduct such inquiries and impose penalties of suspension, curtailment or revocation of a licence (further details of this matter are given on pp 45–48).

Appeals are heard only in London whereas public inquiries are usually held in the town or city in which the TC's office (ie the Traffic Area Office – see Appendix I for addresses) is situated.

Public Inquiries

It is useful here to mention the way in which a public inquiry (PI) is conducted. It is presided over by the TC or his deputy and is open to members of the general public, other operators and interested persons, who may sit in and listen, and to the press, who may report all that is said.

TCs may restrict general attendance at a public inquiry to protect an operator's business, particularly in regard to personal matters, commercially sensitive information and other information obtained in confidence. Hitherto the TC could only close a PI to hear financial information. Further, TCs must disclose at a PI any information or evidence received in writing prior to the inquiry if it is intended that such information is to be taken into account in reaching a decision.

Verbal evidence is given to the TC by the applicant or by his legal representative if he has one – and this is strongly advised in most cases due to the complexities of making legal presentations and arguing points of law in a courtroom situation and possibly the need to cross-examine witnesses such as a vehicle examiner or an environmental representor, even though it is an inquiry, not a court. In fact, the 'call-up' letter to operators facing public inquiries, in which the TC states his reasons for calling the PI and states his powers to curtail, suspend or revoke 'O' licences, as well as giving details of the time and location for the inquiry, advises this. Such advice, if it is required, should be sought from an experienced transport lawyer who fully understands the legal basis of the whole licensing system as well as the intricacies of the PI system (see also pp 603–04).

Most evidence at PIs will be given in response to the TC's questions – the TC effectively playing the role of 'prosecutor' – by the licence holder/applicant, the objectors and those making representations. The evidence, unlike in criminal or civil courts, is not given under oath and statements made at a public inquiry that are defamatory or libellous of other people do not have protection by privilege. The offended person can take civil action if such statements or comments come to his notice. Similarly, if an applicant or witness lies, he will not be prosecuted for perjury, but, where this is an applicant, anything he says in support of his application may be taken by the TC to be a statement of intent to which he will be bound for the duration of any licence granted.

Evidence in some instances may be provided in writing and the TC may ask for certain supporting documents, in which case the applicant should have these to hand with extra copies for the objectors to examine. When he has heard all the evidence the TC will normally make a decision without conferring with anybody else. He may announce this at the time or defer his decision to be given later in writing. The entire proceedings of the inquiry will be recorded and transcripts can be obtained by interested parties.

TCs must now give at least 21 days' written notice of public inquiries both to operators and other parties entitled to attend, and similar notice if they intend to vary the time or place of the inquiry. However, given the consent of all parties, this requirement can be varied.

Appeals to the Transport Tribunal

The Transport Tribunal, which is under the control of the Lord Chancellor, is a completely independent body comprising at least three sitting members, one of whom is the President, and must be a lawyer. At least two of the other members must also be legally qualified and have experience in the transport industry. Currently the Tribunal comprises the President, a panel of two legal members (appointed by the Lord Chancellor) and five lay members (appointed by the Secretary of State for Transport).

An appeal may be made against a TC's decision to refuse to grant an 'O' licence; if he attaches environmental conditions to an 'O' licence; if a licence is granted authorizing fewer vehicles than the number applied for; if a licence is granted for a shorter period than that applied for; or if an existing licence is withdrawn, suspended or prematurely terminated by the TC.

A time limit of 28 days is allowed in which to make an appeal to the Transport Tribunal following a TC's decision, counting from the date of publication of the issue of *Applications and Decisions* in which the decision is published. Where the TC's decision is not published within 21 days an appeal can be made within 49 days of notification of the decision by the TC.

Where a TC makes a disciplinary decision against an 'O' licence (ie suspension, curtailment or revocation) and the licence holder wishes to appeal, he can apply for a 'stay' of the decision until the appeal is heard, in order to keep his vehicles operating. Otherwise he would have to observe the decision irrespective of the consequences (financial and operational) on his business. An initial request for a 'stay' of the decision is made direct to the TC, but failing this an application must be made immediately to the Tribunal giving details of the decision and the reason for requesting the 'stay'. Application for a 'stay' of the decision cannot be made if there is no intention to appeal.

Appeals to the Transport Tribunal must be in writing and six copies should be sent to the Tribunal stating the decision against which the appeal is made, the grounds for the appeal, and the names and addresses of every person to whom a copy of the appeal has been sent.

Copies of the appeal must be sent to the TC and to all objectors if the appeal is being made by a licence applicant, or to the applicant if the appeal is being made by an objector to the decision.

Although the Tribunal has the powers and status of the High Court, its proceedings are conducted informally and appellants may represent themselves or be represented by any person they choose (there are no wigs and gowns even for barristers present). However, in the best interests of the applicant, he should be legally represented at an appeal by a solicitor or barrister experienced in transport law to ensure that his case is fully and correctly made.

When an appeal is heard, the Tribunal examines the transcript of the public inquiry or the TC's statement of his reasons for the decision against which the appeal is lodged and then may ask further questions of the applicant or his advocate. No oath has to be taken and there is no protection by privilege. The proceedings are open to the public and the press. Tribunal appeal decisions may be announced at the hearing or later. All parties will be sent a full statement of the decision usually within three weeks of the hearing.

Generally, Tribunal decisions will fall into one of three categories. Either to uphold the TC's decision, to change the decision or to refer the matter back to the TC with a direction that he should reconsider his decision but taking account of legal guidance from the Tribunal. In exceptional circumstances the Tribunal may review its decision subject to a request to do so made within 14 days of the appeal hearing. Decisions of the Tribunal are binding from the date they are given; in other words they have immediate effect.

Further appeals against decisions of the Tribunal may be made to the Court of Appeal or the Court of Sessions in Scotland but only on points of law, not on the original decision of the TC or the subsequent ruling of the Tribunal.

No fees are payable in respect of appeals but costs may be awarded against frivolous, vexatious, improper or unreasonable appeals.

Further details of the appeals procedure can be found in a booklet, *Appeals to the Transport Tribunal,* available free from Traffic Area Offices. The address of the Tribunal is: Procession House, 55 Ludgate Hill, London EC4M 7JW (tel: 020 7029 9780; fax 020 7029 9782; e-mail: transport@dca.gsi.gov.uk).

NORTHERN IRELAND OPERATIONS

Northern Ireland has its own scheme of Road Freight Operator's Licensing administered in the Province by the Department of the Environment. The DoE in the Province is planning to introduce a full UK-style 'O' licensing system by 1 August 2007.

Goods vehicle operators in the Province do not need to obtain a short-term 'O' licence prior to entry into Great Britain. Similarly, there is no need for GB operators to obtain a short-term licence prior to entry into Northern Ireland. A goods vehicle operating on a current 'O' licence issued in Great Britain or a Road Freight 'O' licence issued in Northern Ireland is permitted to carry goods throughout the UK.

Goods vehicles from Northern Ireland engaging in own-account operations for which a Road Freight Operator's licence is not required in the Province must, while operating in Great Britain, carry a document showing details of their load and route in Great Britain.

Vehicles based and registered in England, Wales and Scotland must comply with all the normal legal requirements (eg vehicle condition, excise duty, insurance and observance of traffic rules) set out in this *Handbook* when operating in Northern Ireland, but particularly so in regard to 'O' licensing (Chapter 1), professional competence (Chapter 2), drivers' hours and record-keeping regulations (Chapters 3 and 4), tachographs (Chapter 5), driver licensing and testing (Chapters

6 and 7) and plating and testing (Chapter 16). It should be noted that in regard to road traffic and road traffic offences there are differences between the Northern Ireland requirements and those on the British mainland. A separate edition of the *Highway Code* is published for Northern Ireland.

Vehicles based and operated in Northern Ireland must comply with the law as it applies in the Province which, although substantially similar to that applicable in the rest of the United Kingdom, does vary in some respects as shown in this section.

Road Freight Operator Licensing

In order to carry goods by road for reward (but not for purely own-account operations) with vehicles over 3.5 tonnes permissible maximum weight in Northern Ireland, it is necessary to hold a road freight operator's licence (RFOL) issued under the provisions of the Transport Act (Northern Ireland) 1967 (see note above about the proposed introduction of GB-type 'O' licensing in the future).

Since 1 January 1978 regulations have been in effect to ensure that road freight operators are better qualified. These regulations, The Road Transport (Qualification of Operators) Regulations (Northern Ireland) 1977, implement EC Directive 561/74, as amended, *On admission to the occupation of road haulage operator.* Thus the requirements for obtaining a road freight operator's licence in Northern Ireland take account of this requirement and depending on the qualifications of the operator the licence may or may not be restricted to operations within the United Kingdom. The type and scope of the licence will be indicated clearly in the licence – covering either national operations or both national and international operations.

Conditions for Grant of 'O' Licence

Under the statutory requirements for the grant of a road freight operator's licence, an operator has to satisfy the issuing authority that he is:

- of good repute;
- of appropriate financial standing;
- professionally competent or that he employs a full-time manager who is professionally competent and of good repute.

The requirements of good repute and appropriate financial standing are as stated in detail in Chapter 1 of the *Handbook.* While the professional competence requirement in Northern Ireland is largely as explained in Chapter 2 there are some differences as shown below.

Professional Competence by Examination

A separate syllabus for the Royal Society of Arts examination is applicable to Northern Ireland. It differs mainly in section C, *Access to the Northern Ireland Market,* which covers vehicle and operators' licences and professional competence requirements. Copies of the syllabus *Examinations for the Certificates of Professional Competence in Road Transport (Northern Ireland)* may be obtained from: OCR, Westwood Way, Westwood Business Park, Coventry CV4 8HS. The current examination syllabus applicable to Northern Ireland national goods and national passenger operations was published in 1988 and both NI national examinations have taken the modular form, with a core examination common to both passenger and goods sectors.

Issuing Authority

For operations in Northern Ireland, the issuing authority for operators' and vehicle licences is: Department of the Environment (DoE) for Northern Ireland, Road Transport Licensing Branch, Upper Galwally, Belfast BT8 4FY.

Objections to Road Freight 'O' Licences

Objections to the grant of a road freight 'O' licence may be made to the DoE by the Road Haulage Association (RHA) and the Freight Transport Association (FTA) on the grounds that the statutory conditions for the grant of such a licence are not or will not be met.

Appeals

Appeals against refusal by the DoE to grant a road freight operator's licence or where such a licence has been suspended or revoked may be made within 28 days to the County Court.

Application for a Road Freight Operator's Licence

Application for a road freight operator's licence, which is usually valid for three years, must be made to: Department of the Environment (NI), Road Transport Licensing Branch, Upper Galwally, Belfast BT8 4FY.

Form RFL 1 is used for the application. This requires details to be supplied as follows:

1. Name and place of business.
2. Type of licence required (ie national or national and international operations).
3. Address of operating centre (ie where vehicles are to be parked when not in use). Where premises are not owned the 'original' rental agreement must be sent with a first application.
4. Number of vehicles to be operated (original receipts or hire purchase agreements for vehicles and trailers to be operated must be sent with the application).
5. Whether professional competence requirements are met.
6. If applicant is not professionally competent, the name and address of the full-time transport manager who is qualified must be given.
7. Evidence of professional competence (the original certificate must be sent).
8. Whether the applicant or his named transport manager has any convictions for offences in the past five years which are not 'spent'.
9. Details of convictions.
10. Names and addresses (and past addresses) of the applicant, the named transport manager, and all directors and/or partners.
11. Name and address of bank from which a reference can be obtained.

12. Declaration (signature, status and date).

Further information may be required in support of a road freight operator's licence regarding the applicant's financial status and a statement of estimated operating costs for vehicles specified in the application. The former requires the applicant to send the following items:

1. Bank status report.
2. Details of any credit facilities from bank or any other financial resources available (eg deposit accounts).
3. Copy of latest bank statement.
4. Rental agreement for operating premises.
5. Receipts or hire purchase agreements in respect of vehicles.
6. Original certificate of competence.
7. A statement showing a breakdown of estimated annual costs and receipts of running the proposed road transport business.

The requirement to supply statements of estimated operating costs for vehicles involves the need to complete a form designed for this purpose. The following information must be provided:

1. Details of vehicles
   (a) type and model
   (b) registration number and date of first registration
   (c) type of trailer
   (d) gross weight
   (e) unladen weight
   (f) carrying capacity
   (g) number of tyres fitted and size (including trailer)
   (h) estimated fuel consumption (miles per gallon)
   (i) estimated annual mileage
   (j) vehicle length (in feet).
2. Estimated running costs and standing costs for one year (one vehicle only)
   (a) wages (including National Insurance)
   (b) motor tax
   (c) insurance for vehicle
   (d) insurance for goods in transit
   (e) fuel
   (f) maintenance
   (g) tyres
   (h) shipping charges if applicable
   (i) hire purchase payments on vehicle and trailer (if applicable)
   (j) rent and rates for business premises
   (k) depreciation on vehicle
   (l) miscellaneous expenses (eg telephone, stationery, etc).
3. Total costs
   Estimated income from one year's operation (one vehicle only).

It is generally believed (although never officially stated) that the DoE is looking for 'O' licence applicants to have at least 25 per cent of first year operating costs in the bank at the start of the operation or assets equivalent to £7,500 for each rigid vehicle and £15,000 for each articulated vehicle either in the form of bank credits or agreed overdraft facilities.

Penalties for Improper Use

Anyone operating internationally on a road freight 'O' licence that is restricted to national operations only will be liable, on conviction, to a fine and his 'O' licence and vehicle licences may be revoked or suspended. There are also penalties laid down for using unlicensed vehicles on national operations.

Changing Licences

The holder of a road freight 'O' licence which is restricted to national transport operations may have the restriction removed to enable him to engage in both national and international operations if he can meet the additional requirements (namely professional competence in both national and international transport operations) for engaging in international operations.

Vehicle Licences

Northern Ireland motor vehicles over 3.5 tonnes gross weight may not be used to carry goods for hire or reward except under a Road Freight Vehicle Licence granted under section 17 of the Transport Act (NI) 1967 (as amended). Vehicle licences may only be granted to a person who holds a road freight operator's licence and will cover either:

1. National transport operations only (ie covering Northern Ireland and the remainder of the UK) – Blue disc.
2. National and international transport operations – Green disc.

Vehicle licences are valid for one year and are issued in the form of a disc for display in the vehicle windscreen. They are serial numbered and show the vehicle registration number, the name of the owner and the date of expiry. Application is made on form RFL 3 to the DoE Northern Ireland enclosing both the vehicle excise licence disc and the goods vehicle certificate as evidence that the vehicle is taxed and tested, if applicable, on the date of issue of the vehicle licence.

Application for Road Freight Vehicle Licences

Applications for road freight vehicle licences have to be made on form RFL 3 to the DoE. Details which must be provided by the person or firm using the vehicles are as follows:

1. Name, address and telephone number of applicant.
2. Road freight 'O' licence number.
3. Whether the RFOL is valid for international operations.
4. (a) Registration numbers of vehicles which are owned or in possession under an agreement for hire, hire purchase, credit sale or loan for which road freight licences are required

   (b) Type of vehicle
   (c) Type of body
   (d) Gross weight of vehicle.
5. Address where vehicles will be parked overnight.
6. Declaration that statements given are true.

Penalties for Illegal Use

Making a false statement to obtain the grant of a road freight operator's licence or a road freight vehicle licence is an offence punishable on conviction by a fine or imprisonment for up to six months, or both. The 'O' licence could also be suspended or revoked. Use of a motor vehicle on a road for the carriage of goods for reward without a road freight vehicle licence can result in a fine which increases for subsequent convictions.

USE OF LIGHT GOODS VEHICLES

Many existing transport operators and new entrants to the industry have sought to avoid the problems and pitfalls of 'O' licensing by using vehicles defined as 'small' vehicles – those vehicles not exceeding 3.5 tonnes maximum permissible weight. With such vehicles there is no need to obtain an 'O' licence and consequently no need to face the TC and satisfy all the conditions previously explained. The operator is also free from the legal requirements under other legislation for his drivers to operate tachographs or to keep other written records of their hours of work, and to hold LGV driving licences.

Despite this apparent freedom the operator of such vehicles does have certain obligations and responsibilities. First, if he tows a trailer with such a vehicle the combined weight of both vehicle and trailer (if over 1020 kg unladen) could exceed the 3.5 tonne weight threshold above which an 'O' licence would be needed and the provisions of the EU or British driver's hours law and the relevant record-keeping or tachograph requirements may apply (see Chapters 3, 4 and 5). Second, if he also operates, or plans to operate in the future, larger vehicles which are within the scope of 'O' licensing, his conduct as an operator of small vehicles will be taken into account by the TC when deciding whether to grant or renew his 'O' licence.

The TCs have made the point that when an operator applies to renew an 'O' licence they (the TCs) would take notice of any relevant convictions in respect of smaller vehicles belonging to the operator and could call the operator to public inquiry to show cause why the 'O' licence should not be revoked or curtailed. The operator of small vehicles still has to ensure that vehicles are not overloaded and that they are kept in a safe mechanical order under other regulations; they must be tested annually after they become three years old. Drivers of these vehicles are required to observe the drivers' hours regulations with certain exceptions. All these individual legal exemptions and requirements are discussed in later chapters.

FOREIGN VEHICLES IN THE UK

Vehicles entering Great Britain from other EU member states under valid Community Authorisations do not need an 'O' licence and no longer require cabotage authorization to operate in this country. Vehicles from certain non-EU countries where a bilateral agreement exists are exempt from the requirement to hold an 'O' licence under the Goods Vehicles (Licensing of Operators) (Temporary Use in Great Britain) Regulations 1996.

Foreign vehicles entering Great Britain under an ECMT permit do not need an 'O' licence provided the permit is being carried on the vehicle.

IMPOUNDING OF TRUCKS

Impounding of the vehicles of cowboy and illegal operators – estimated to number some 8000 vehicles – came into force on 4 January 2002. Originally envisaged in the government's White Paper on the future of transport, *A New Deal for Transport,* published in July 1998, the provision became law under the Goods Vehicles (Enforcement Powers) Regulations 2001. In the two years since impounding commenced, over 200 vehicles have been seized of which 108 were sold or scrapped. Principally, impounding applies to vehicles used for hire or reward carriage without an 'O' licence in force. Such vehicles are fitted with an immobilization device (ie a clamp) either on the spot or after removal to a more suitable site, and will be marked by an immobilization notice warning against any attempt to remove the device or move the vehicle except under proper authorization. Impounded vehicles may be returned to their owner in due course provided specified conditions are met, and loads returned to their owner where title is established. Ultimately, however, without satisfactory compliance with requirements for the release of a vehicle and its load, they will be disposed of by sale or by destruction. Obstruction of authorized examiners in the course of vehicle impounding will lead to a fine at level 3 (maximum £1,000) on conviction.

An appeals procedure via the TCs initially and the Transport Tribunal subsequently allows operators to apply for the return of an impounded vehicle, although this may be unsuccessful depending on the circumstances.

Impounding of Rental/Hired Vehicles

The new impounding legislation described above has caused concern for vehicle rental, hiring and leasing companies. However, the law is clear in that the regulations permit the seizing of vehicles hired out by such firms to unlicensed operators. The TCs too have made it clear that such firms have a duty to ensure that any hiring arrangement they enter into in relation to relevant (ie over 3.5 tonne) vehicles is with an 'O' licensed operator. The situation becomes more complex where a vehicle leased to an unlicensed operator by a vehicle leasing company is subject to a finance arrangement with a third party (such as a bank or loan company) which is therefore the statutory owner of the vehicle. In such cases, it becomes necessary for the legal owner to make application to the TC for the return of the vehicle on the basis that it, the legal owner, did not know that the vehicle had been, or was being, used without an 'O' licence in accordance with the Goods Vehicles (Licensing of Operators) Act 1995 (section 2). An application for the return of a vehicle in these circumstances has to be made within 21 days of the notification of the impounding being published in the *London Gazette* or *Edinburgh Gazette* (in which all vehicle impoundings are published).

2 Professional Competence for Transport Managers

The legal requirement for certain people employed in hire or reward road transport operations (ie in both road freight and road passenger sectors) to be professionally competent came into effect in Britain on 1 January 1978.

To date the requirement to meet professional competence standards has applied solely to those who hold supervisory or management positions where they effectively have day-to-day control of the activities of drivers and operation of vehicles. However, a new draft directive emanating from the European Commission proposes that from 1 January 2004 goods (and passenger) vehicle drivers should also undergo compulsory professional training in addition to meeting the existing requirements for driving, namely passing the relevant theory and practical driving tests. This subject is dealt with in more detail in Chapter 7.

On the road goods side of the industry, the original scheme was the final outcome of many years' work on the development of a plan to make individuals more responsible for the safe operation of vehicle fleets in their charge and better qualified to understand the legal, economic and operational requirements for safe and efficient goods vehicle operation. Previous proposals (in the Transport Act 1968) were based on a licensing system which would require transport managers to hold a 'transport manager's licence'. As a result of EU influence via EC Directive 74/561 (as amended by Directives 89/438/EC, 96/26/EC and most recently by 98/76/EC of 1 October 1998 – see below) 'On Admission to the Occupation of Road Haulage Operator in National and International Operations', the proposed British transport manager's licensing scheme, was abandoned in favour of one which provides all those people who qualify, not just those fulfilling the role of transport manager, with either a certificate of professional competence or a professional competence qualification by exemption or by examination.

Under the Goods Vehicles (Licensing of Operators) Act 1995 (section 13 and schedule 3), road haulage operators (including own-account operators who wish to carry goods for hire or reward or in connection with a business which is not their own) are required to meet the professional competence requirement in order to obtain a standard operator's licence.

Own-account transport operators who have no desire to or intention of carrying goods for hire or reward or for any purpose which is not in connection with their own trade or business (including not doing favours for customers by carrying their goods also) can apply for a restricted 'O' licence, for which there is no need to meet the professional competence requirement. The important point for road hauliers and others who carry goods for hire or reward is that in order to obtain an operator's licence they must request a standard 'O' licence covering either national or both national and international operations and specify in their application the name of a person who is professionally competent and who is responsible for the operation of the vehicles authorized on the licence – the law requires the person to have 'continuous and effective responsibility' for the management of the transport operation. This person may be the applicant, if suitably qualified, or it may be a person holding the title of transport manager (or some other person – the law is not concerned as to the person's actual job-title only that he should be the person who is responsible for the vehicle operations on a day-to-day basis) who is professionally competent and who is employed by the applicant.

Employment in this context does not mean full-time employment with the licence holder. Part-time or casual employment of a professionally competent person is permitted as a result of a regulation change in November 1991 under which a part-time qualified employee may be nominated in support of a licence application. Also, with this change in the law, a professionally competent person in one firm may be specified in support of the licence held by another, non-related, firm.

There has been a TCs' decision which indicates that a self-employed (ie freelance) transport manager may be acceptable as the professionally competent person specified on an 'O' licence. The employment requirement does not necessarily mean that the person concerned must devote all of his working time to the transport management function, he may have other duties and responsibilities in the firm (eg such as administration manager, works manager, company secretary, etc).

Concern among the TCs over the matter of part-time (ie so-called 'proxy') transport managers has led to a suggestion that seven key questions need to be considered when determining whether part-time professionally competent persons would be acceptable in support of a standard 'O' licence:

1. How many hours do they work for the licence applicant?
2. How many other employers do they have?
3. How many vehicles are they responsible for?
4. What is the distance between each of the operating centres for which they are responsible?
5. What is the nature of their duties – are they also fitters or vehicle drivers?

6. Where do they live in relation to their places of employment?
7. Are they of good repute?

Another TC has said that he would want details of exactly how 'proxy' transport managers would apportion their time between firms employing them in this role, precisely what their duties would be for each employer and how much they would be paid.

WHO MAY BECOME PROFESSIONALLY COMPETENT?

Professional competence is available and applicable only to individuals. Under the regulations, a firm or a corporate body cannot be classed as being professionally competent. Any individual, male or female, whether employed with the title of 'transport manager' or not, may become professionally competent if he or she meets the necessary qualifying conditions or otherwise passes the official examination. There is no pre-qualifying standard and no requirement that the person should have any previous experience of or is working, has been working, or plans to work in the transport industry in any capacity whatsoever. It is open to absolutely anybody to become professionally competent if they so wish.

However, only those who are actually responsible for the operation of goods vehicles on a day-to-day basis will need to be 'nominated' as the professionally competent 'transport manager' in support of an application for a standard 'O' licence. Such individuals must themselves be of good repute and of appropriate financial standing (as well as the applicant – see Chapter 1) otherwise their name may not be acceptable to the TC despite the fact that they are qualified as being professionally competent (see below).

PROOF OF PROFESSIONAL COMPETENCE

Proof that a person is professionally competent and is therefore able to satisfy the requirements of the 'O' licence system is dependent on holding one (or more) of the following:

- a 'grandfather rights' certificate of professional competence (CPC) issued by a TC (form GV203);
- a Royal Society of Arts CPC examination pass certificate;
- a membership (or exemption) certificate from one of the recognized professional institutes, which confers exemption (see pp 62–64);
- a certificate issued by another EU member state which fulfils the 'mutual recognition' requirements of EC Directive 77/796 as amended by EC Directive 89/438.

No other document provides evidence of professional competence for these purposes.

CLASSES OF COMPETENCE

Professional competence in road freight operations falls into two classes, covering national operations only or both national and international operations. A parallel scheme covers road passenger operations.

All certificates of professional competence (CPCs – form GV203) granted under the original 'grandfather rights' scheme (see below) cover both national and international operations (although it does not specifically say so on them).

In cases where people qualify for professional competence under the exemption arrangements, the level at which they qualify determines whether they are entitled to be classed as professionally competent in national transport operations only or in both national and international transport operations (see pp 62–64).

Candidates who achieve professional competence by examination will obtain appropriate documentary evidence (ie an OCR examination pass certificate) indicating that they have passed the modular examinations covering national operations only (freight or passenger) or the additional examination covering international operations (freight or passenger).

Those qualified for professional competence in national operations only will be permitted to engage in or be responsible for operations conducted on standard 'O' licences solely within the UK (ie covering national transport operations).

Holders of certificates of professional competence (form GV203) or those qualifying for professional competence in both national and international operations, by exemption or examination, will be permitted to engage in or be responsible for the operation of goods vehicles on standard 'O' licences covering national operations within the UK and international operations outside the UK.

National and International Transport Operations

Following legal decisions made in the EU the definition of what constitutes an international journey has been changed in Britain. So, for example, where loaded trailers are taken to a port for onward movement outside the UK, and the tractive unit and driver do not leave the UK, such journeys constitute national operations (previously, these were international operations), and a standard 'O' licence covering national transport operations only is needed for such operations (and the employed manager will need to be professionally competent only in national operations).

QUALIFICATIONS FOR PROFESSIONAL COMPETENCE

The qualifications needed to obtain professional competence fall into four categories as follows:

1. By experience in the industry prior to 1 January 1975 (known as 'grandfather rights').
2. By exemption.
3. By examination.
4. By holding a mutual recognition certificate issued in another EU member state (see p 72).

Grandfather Rights

NB: The issue of certificates of professional competence (form GV203 – known as CPCs) under this scheme ended on 31 December 1979, since which date the only means of qualifying for professional competence is by exemption or examination.

However, since many people continue to ask about the scheme it was felt useful to continue to include an explanation in the current edition of the Handbook.

Transport managers and other people employed in 'responsible road transport employment' prior to 1 January 1975 were able to obtain the grant of a so-called 'grandfather rights' certificate of professional competence (CPC), without examination, as of right. Once granted, the certificate continues to remain valid for as long as the professional competence requirement is in force.

For the purposes of the (now extinct) grandfather rights scheme, responsible road transport employment was defined as employment in the service of a person or a firm carrying on a road transport undertaking and was employment in a position where the individual had responsibility for the operation of goods vehicles used under an operator's licence.

Similar conditions applied to any person who was the holder of an 'O' licence prior to 1 January 1975. In other words any owner-driver or small fleet operator who held an 'O' licence in his own name or under a business name qualified, having been a licence holder. In the case of an 'O' licence held by a partnership prior to this date, all the partners qualified under the grandfather rights arrangements. Where a licence was held by a limited company, the person responsible for the day-to-day operation of the vehicles under the licence (eg the transport manager) qualified for the grandfather rights grant of a CPC.

To obtain a certificate under this arrangement, application had to be made to the TCs no later than 30 November 1979. Since that date, the opportunity to obtain a CPC other than by examination or exemption has ceased.

Issue of CPC Certificates

Certificates of professional competence (form GV203) issued under the grandfather rights scheme were obtainable from the Traffic Area Offices on application provided that the necessary form and certification of appropriate qualifying experience by an employer was supplied. Although certificates carry the name of the issuing Traffic Area, they are valid for operations in all Traffic Areas and individuals had no need to obtain separate certificates for each area in which they were responsible for the operation of vehicles.

There was no fee for the issue of a CPC, and certificates thus granted remain valid for the life of the scheme. There is no system for revocation or disqualification of CPC holders but where the holder is also the 'O' licence holder, or is the nominated professionally competent transport manager employed by an 'O' licence holder, then there is a requirement that they must be of good repute (see pp 17–19 and pp 73–74) and of adequate financial standing; otherwise the 'O' licence will be subject to penalty (see pp 45–48). Grandfather rights CPCs issued on the basis of information provided which is subsequently found to be false may be withdrawn by the TC.

Qualification by Exemption

People who did not qualify for a CPC based on previous experience in road transport operations as described above and new entrants to the industry may obtain the professional competence qualification if they satisfy certain exemption criteria.

The exemption qualifications are based on holding current and valid membership of one or more of a number of professional bodies at certain levels. Additionally, the 'Certificate in Transport' qualification offered by the Chartered Institute of Transport is recognized by the D*f*T and accepted as giving exemption.

There are two levels of exemption qualification: one covering both international and national operations and the other covering national operations only; these are as follows:

For both National and International Operations

- Fellow or Member of the Chartered Institute of Transport (CIT*) by examination and/or formal accreditation in Road Freight Transport.
- Fellow, Member, Associate Member, or Associate by examination, of the Institute of Transport Administration (IoTA) in the road transport sector, on production of the necessary exemption certificate issued by the Institute. *These exemptions are valid for 12 months only but can be reissued on application to the Institute.*
- Associate of the IoTA who has successfully completed the CIT*/IoTA Certificate in Transport.
- Fellow, Member or Associate Member, of the Institute of Road Transport Engineers (IRTE).
- Fellow, or Associate of the Institute of the Furniture Warehousing and Removing Industry (IFWRI), or from 13 May 1995, Fellow or Associate of the Movers Institute.
- Certificate in Transport, awarded by the CIT*.

For National Operations Only

- Associate Member of the CIT* by examination and/or formal accreditation in Road Freight Transport.
- Certificate in Transport, awarded by the CIT*.
- Fellow, Member or Associate Member of the IoTA on production of the necessary exemption certificate issued by the Institute.
- Associate of the IoTA who has successfully completed the CIT*/IoTA Certificate in Transport.
- Associate of the IRTE (by examination).
- General and Ordinary Certificate in Removals Management issued by the IFWRI prior to 13 May 1995, or after that date the National Certificate in Removals Management issued by the Movers Institute.
- Royal Society of Arts (RSA) Certificate in Road Freight Transport granted on or after 1 May 1984 which specifically contains a footnote granting exemption.

**NB: The CIT is now merged with the Institute of Logistics to form the Chartered Institute of Logistics and Transport (CILT).*

There are no grounds for obtaining professional competence by exemption other than those detailed above. Valid membership of the relevant body (ie subscription paid, etc) is sufficient to confirm professional competence but, if required, the institutes will issue a confirmatory certificate or statement (ie not a certificate of competence of the type issued under grandfather rights).

EXAMINATIONS

A system of examinations enables new entrants to the transport (ie road freight and road passenger) industry and those people who do not qualify under the previously mentioned grandfather rights or exemption arrangements to study for and obtain professional competence by examination. The examination scheme is organized and conducted on behalf of the D*f*T by the OCR (Oxford and Cambridge Royal Society of Arts) and examinations for both goods and passenger vehicle operations are held at main centres throughout the country four times each year.

Examination Dates

Dates for the year 2006 examinations are as follows: Friday 3 March 2006, Friday 2 June 2006, Friday 6 October 2006 and Friday 1 December 2006. Potential candidates are advised to check dates and locations for these examinations.

New Provisions since October 1999

Introduction of Council Directive 98/76/EC (which amended earlier directives) on this subject brought with it a considerable upgrading of professional competence qualification requirements via a more extensive examination syllabus and a much tougher examination regime which took effect from the October 1999 examination.

Modular Examination System

For the national examination a modular system has been adopted with a common 'core' for both road haulage and passenger subjects and then the individual modal (freight or passenger) modules. The new modular structure was introduced from the October 1999 examination as follows:

National Road Haulage CPC

- Module 1 – Core (30-minute multiple-choice examination), plus
- Module 2 – National Road Haulage (60-minute multiple-choice examination), plus
- Module 4 – National Road Haulage (90-minute case-study examination requiring short-response answers).

International Road Haulage CPC

- National Road Haulage CPC, plus
- Module 6 – International Road Haulage (60-minute case-study examination, requiring direct answers and short-response answers).

Examination Timetable

Candidates may enter for any combination of modules on any of the published examination dates. Owing to the new examination requirements, however, passenger transport modules will be timetabled to take place at the same time as the equivalent road haulage modules. Candidates cannot therefore sit a road haulage and equivalent passenger transport module on the same day.

Examinations taking place at the same time:

- Module 2 (road haulage) and module 3 (passenger transport)
- Module 4 (road haulage) and module 5 (passenger transport)
- Module 6 (road haulage) and module 7 (passenger transport).

Certification

When an examination candidate achieves the required combination of module examination successes at one sitting (ie on the same examination date) a full national or international certificate of professional competence will be issued.

Candidates who pass any individual module examination will be issued with a module certificate. When they have acquired the requisite number of individual certificates these can be exchanged for a full certificate of professional competence on application to OCR.

Examination Fees

Fees for the examinations taken from 1 September 2004 are as follows:

- Module 1 (core) – £10.80
- Module 2 (road haulage national) – £19.65
- Module 4 (road haulage national) – £31.45
- Module 6 (international road haulage) – £38.65.

These fees are payable to the OCR. Late-entry bookings are charged extra. Additional charges may be made by examining centres for the facilities they provide. Candidates who fail examinations may apply to OCR for details of their performance. They should state where and when they sat the examination and should enclose the relevant fee.

The address of OCR is as follows: OCR Examinations, Progress House, Westwood Way, Westwood Business Park, Coventry CV4 8JQ (tel: 024 7647 0033; fax: 024 7642 1944; e-mail: cib@ocr.org.uk; website: www.ocr.org.uk).

Examination Method

Examinations conducted for the purpose of providing the qualification for professional competence are based on the 'objective testing' or multiple-choice method. In the multiple-choice examinations the candidate is faced with either a number of questions, each of which is provided with a choice of possible answers (usually four), only one of which is correct, or alternatively statements to which a 'True' or 'False' answer must be indicated. The candidate must select and mark the correct answer to each question or statement. For the case study, examination candidates will be presented with a scenario (storyboard) followed by a number of questions.

They will be required to write a short-response answer applying their knowledge to the situation in the scenario and accompanying questions.

The examinations are designed to be within the grasp of candidates whose educational standard corresponds to the level normally reached in compulsory schooling supplemented either by vocational training and supplementary technical training or by secondary-level school technical training.

Examination Syllabus

Candidates for the examination are not compelled to study beforehand but a study syllabus covering the examination subjects in road haulage and passenger transport is published in the EC Directive and copies are available from OCR (see above for address).

The syllabus is designed to show:

- Assessment objectives (ie topics as specified in the Council Directive).
- Performance criteria (ie the knowledge to be tested in the examination).
- Range statement (ie the depth of knowledge required for the examination).

Study Facilities

Study may be undertaken at courses organized by group training associations, trade associations or commercial organizations. Some local technical colleges and other teaching establishments also offer part-time, usually evening, classes on the subject.

The study may involve full-time or part-time attendance at such courses or it may be by correspondence course or by training or home learning package (ie notes and cassette tapes) or with the aid of teaching manuals.

The Council Directive Syllabus

The syllabus for road haulage operations published in Council Directive 98/76/EC and effective from 1 October 1999 is detailed below. It shows the assessment objectives and (in brackets) the depth of knowledge required for the examination.

A. Civil Law

Candidates must:

1. be familiar with the main types of contract used in road transport and with the rights and obligations arising therefrom (*Contracts: legal obligations; sub-contracting; legal duties of agents, employers and employees*);
2. be capable of negotiating a legally valid transport contract, notably with regard to conditions of carriage (*Legal obligations: capacity to contract; specific performance; liability; lien; laws of agency*);
3. be able to consider a claim by his principal regarding compensation for loss of or damage to goods during transportation or for their late delivery, and to understand how such claims affect his contractual responsibility (*Performance: general and specific liabilities of principal, sub-contractors and agents for the performance of a contract. Compensation: for losses relating to damage. Settlements: interim and full payments*);

4. be familiar with the rules and obligations arising from the CMR convention on the contract for the international carriage of goods by road (*CMR liability and unwitting CMR: CMR notes to CMR convention. Successive carriers. Limits of liability. Relevance of insurance*).

B. Commercial Law

Candidates must:

1. be familiar with the conditions and formalities laid down for plying the trade, the general obligations incumbent upon transport operators (registration, keeping records) and the consequences of bankruptcy (*Trading law relating to: sole traders and partnerships; partnership agreements; rights and duties of partners; powers of partners; partners as agents; dissolution of partnerships. Company law: registered companies (private and public); AGMs; liquidation. Documentation: prospectus, memorandum of association; articles of association; certificate of incorporation*);
2. have appropriate knowledge of the various forms of commercial company and the rules governing their constitution (*Types of business organization: sole traders; partnerships; private and public limited companies; co-operatives*).

C. Social Law

Candidates must be familiar with:

1. the role and function of the various social institutions that are concerned with road transport (trade unions, works councils, shop stewards, labour inspectors) (*Role of: employment tribunals; trade unions; ACAS; arbitrators; DTI. Employees' rights: trade union membership and activities*);
2. the employers' social security obligations (*Relevant parts of current legislation relating to: health and safety; discrimination; employment protection; employment rights*);
3. the rules governing work contracts for the various categories of worker employed by road transport undertakings (form of the contracts, obligations of the parties, working conditions and working hours, paid leave, remuneration, breach of contract) (*Contracts of employment: content of written statement; time limits for the issue of contracts. Employment rights: of full and part-time employees; of self-employed; of agency staff; transfer of undertakings; remuneration and itemized pay statements; holiday entitlement; dismissal and unfair dismissal; notice to terminate employment; working time regulations*);
4. the provisions on driving periods and rest periods laid down in Regulation (EEC) 3820/85, the provisions of Regulation (EEC) 3821/85 on recording equipment in road transport, and the practical arrangements for implementing these regulations (*Community regulations: the working week; driving time; breaks; daily and weekly rest periods; emergencies. Domestic hours' law: the working week; driving time; rest periods; emergencies.*

Tachograph legislation and operation: points of law; the records; driver and employer responsibilities; enforcement and inspection; calibration and sealing; malfunctions).

D. Fiscal Law

Candidates must be familiar with the rules governing:

1. VAT on transport services (*VAT – national operations: income threshold and registration; zero-rated goods and services; VAT returns; reclaiming VAT. Turnover tax – international operations: registration for VAT; applying VAT; submitting returns; reclaiming VAT*);
2. motor vehicle tax (*Calculation of VED: basis for calculating motor vehicle taxation on general vehicles and vehicles used in special operations and conditions applied to them*);
3. taxes on certain road haulage vehicles and tolls and infrastructure user charges (*Domestic operation: toll roads and bridges and the basis on which calculation is made. International operation: rules governing tolls and taxation of vehicles on international journeys*);
4. income tax (*Corporate taxation. Double taxation. Status: rules governing the status of employees and the self-employed and the imposition of income tax regulations. Employers' responsibilities: deduction and collection of income tax and National Insurance from employees; payment of income tax and National Insurance to the Inland Revenue*).

E. Business and Financial Management of the Undertaking

Candidates must:

1. be familiar with the laws and practices regarding the use of cheques, bills of exchange, promissory notes, credit cards and other means or methods of payment (*National operation – payment methods: cash; cheques; credit cards; promissory notes; bills of exchange; debit systems and credit transfer. International operation: banking and payment systems, including the electronic transfer of funds, eg Swift, Eurocheques*);
2. be familiar with the various forms of credit (bank credit, documentary credit, guarantee deposits, mortgages, leasing, renting, factoring) and with the charges and obligations arising from them (*Different forms of credit: overdrafts; loans; documentary credit; guarantee deposits; mortgages; leases; rents; factoring*);
3. know what a balance sheet is, how it is set out and how to interpret it (*Determine: fixed assets; net current assets; current assets; long-term liabilities; current liabilities. Interpretation: calculate and interpret*);
4. be able to read and interpret a profit and loss account (*Determine: direct and indirect costs; gross (or operating or trading) profit; net profit*);
5. be able to assess the company's profitability and financial position, in particular on the basis of financial ratios (*Determine: capital employed and*

return on capital employed; return on sales and assets turnover; working capital; cash flow. Use of ratios: current ratio (working capital ratio); quick ratio (liquidity ratio or acid test ratio));

6. be able to prepare a budget (*Construct budgets from data supplied. Use of budgets: to monitor and control performance, budgetary control; variance analysis*);
7. be familiar with his company's cost elements (fixed costs, variable costs, working capital, depreciation, etc), and be able to calculate costs per vehicle, per kilometre, per journey or per tonne (*From data supplied: identify and/or calculate fixed costs, variable costs, overhead costs, depreciation. Determine: time and distance costs*);
8. be able to draw up an organization chart relating to the undertaking's personnel as a whole and to organize work plans (*Prepare an organization chart for an: organization, department, function, unit or depot. Organizing, planning and measuring work*);
9. be familiar with the principles of marketing, publicity and public relations, including transport services, sales promotion and the preparation of customer files (*Market research (primary and secondary); segmentation; product promotion, sales and publicity. Customer: relations; research files*);
10. be familiar with the different types of insurance relating to road transport (liability, accidental injury/life insurance) and with the guarantees and obligations arising therefrom (*Insurance: risk assessment; cover; claims, risk management and improvement of risk. Types of insurance: fidelity; goods in transit; employers' liability; public liability; professional negligence; motor; plant; travel; health; property; consequential loss; cash in transit. Risks: guarantees; obligations; liability and role of trustees*);
11. be familiar with the applications of electronic data transmission in road transport (*Legislation: Data Protection Act. Hardware and software: electronic vehicle status monitoring; electronic data transmission; real-time information systems; customer information systems; depot readers; GPS; route and load planning systems; vehicle and staff scheduling; data analysis; data information systems*);
12. be able to apply the rules governing the invoicing of road haulage services and know the meaning of incoterms (agreed trading terms with legal definitions, eg FOB – the International Chamber of Commerce produces a book called *Incoterms*, ref: ISBN 92-842-0087-3) (*Transactions: the purpose of quotations, orders, consignment notes. Incoterms: principal provisions of incoterms applying to any mode of road transport, including multimodal*);
13. be familiar with the different categories of transport auxiliaries, their role, their functions and, where appropriate, their status (*Agents: various agents who assist in ensuring freight is properly and correctly transferred, eg freight forwarders, shippers, clearing houses, groupage operators*).

F. Access to the Market

Candidates must be familiar with:

1. the occupational regulations governing road transport for hire or reward, industrial vehicle rental and sub-contracting, and in particular the rules governing the official organization of the occupation, admission to the occupation, authorizations for intra- and extra-Community road transport operations, inspections and sanctions (*National operation: role of Traffic Commissioners and enforcement agencies; statutory procedures concerning operator licensing; requirements for vehicle maintenance; regulations governing domestic operation. International operation: statutory procedures concerning operator licensing; regulations governing international operations*);
2. the rules for setting up a road transport undertaking (*Rules for setting up a road transport undertaking: statutory procedures and rules concerning operator licensing*);
3. the various documents required for operating road transport services and be able to introduce checking procedures for ensuring that the approved documents relating to each transport operation, and in particular those relating to the vehicle, the driver and the goods or luggage, are kept both in the vehicle and on the premises of the undertaking (*Documents and their administration: operator licences and vehicle discs; vehicle authorizations; tachograph records; waybills/consignment notes; driving entitlement; maintenance documents; insurance documents; systems for document checking and control procedures*);
4. the rules on the organization of the market in road haulage services, on freight handling and logistics (*Market organization: 1985 Transport Act, regulatory powers of the Secretary of State, the Office of Fair Trading; the Monopolies and Mergers Commission. Role of local government regarding the movement of abnormal loads. Main provisions of third country traffic and cabotage*);
5. frontier formalities, the role and scope of 'T' documents and TIR carnets, and the obligations and responsibilities arising from their use (*'T' documents, their status and rules of use. Carnets, rules governing their use. Exemptions: most practical use of various exemptions obtained by the use of various documents and the implications for operators and their customers*).

G. Technical Standards and Aspects of Operation

Candidates must:

1. be familiar with the rules concerning the weights and dimensions of vehicles in the member states of the European Union and the procedures to be followed in the case of abnormal loads that constitute an exception to these rules (*Terms used to identify the differing weight conditions. Statutory limits: weights and dimensions. Formulas used for various calculations concerned in weights and dimensions of vehicles. Main rules and most common weights and dimensions used internationally*);

2. be able to choose vehicles and their components (chassis, engine, transmission system, braking system, etc) in accordance with the needs of the undertaking (*Vehicle specifications that will improve road safety and economy, and reduce impact on the environment. Vehicle specifications to be taken into account for international operations*);
3. be familiar with the formalities relating to the type approval, registration and technical inspection of these vehicles (*Main provisions within current legislation relating to C&U and safety. Powers of enforcement agencies*);
4. understand what measures must be taken to reduce noise and to combat air pollution by motor vehicle exhaust emissions (*Main provisions of the C&U regulations; EU directives and environmental legislation. Training: awareness training*);
5. be able to draw up periodic maintenance plans for the vehicles and their equipment (*Maintenance programmes: planned preventative; methods of maintenance; operator's obligations and liabilities to maintain vehicles and equipment in a safe, roadworthy condition; responsibility for vehicles whose maintenance is contracted out*);
6. be familiar with the different types of cargo-handling and cargo-loading devices (tailboards, containers, pallets, etc) and be able to introduce procedures and issue instructions for loading and unloading goods (load distribution, stacking, stowing, blocking and chocking) (*Risk analysis and safe operations: requirements for various loads and procedures to ensure safe operations*);
7. be familiar with the various techniques of 'piggy-back' and roll-on/roll-off combined transport (*Safety requirements; vehicle specifications charging methods*);
8. be able to implement procedures for complying with the rules on the carriage of dangerous goods and waste, notably those arising from:
   - Directive 94/55/EC on the approximation of the laws of the member states with regard to the transport of dangerous goods by road;
   - Directive 96/35/EC on the appointment and vocational qualification of safety advisers for the transport of dangerous goods by road, rail and inland waterways;
   - Regulation (EEC) 259/93 on the supervision and control of shipments of waste within, into and out of the European Community;

   (*National: main provisions of regulations are identified and incorporated into a procedure to meet with current requirements. International: main provisions of regulations are identified and incorporated into a procedure to meet current requirements*);
9. be able to implement procedures for complying with the rules on the carriage of perishable foodstuffs, notably those arising from the agreement on the international carriage of perishable foodstuffs and on the special equipment to be used for such carriage (ATP) (*Procedures to ensure correct compliance with legislation and best practice*);
10. be able to implement procedures for complying with the rules on the transport of live animals (*National and International: procedures to ensure correct compliance with legislation and best practice*).

H. Road Safety

Candidates must:

1. know what qualifications are required for drivers – driving licences, medical certificates, certificates of fitness (*Vocational entitlements: different categories, types and qualifications for driving licences and entitlements. Procedures: relating to the issue, renewal, revocation and production of licences and removal of entitlements. Disciplinary matters: procedures and appeals. Driving tests: scope and conduct and sequence of theory and driving tests. International driving permits: issue and validation*);
2. be able to take the necessary steps to ensure that drivers comply with the traffic rules, prohibitions and restrictions in force in the different member states of the European Union – speed limits, priorities, waiting and parking restrictions, use of lights, road signs (*Traffic regulations: signs and signals; variation in weights, dimensions and speed of road haulage vehicles in EU member states and non-member countries. Restrictions: imposed on the movement and speeds of road haulage vehicles*);
3. be able to draw up drivers' instructions for checking their compliance with the safety requirements concerning the condition of the vehicles, their equipment and cargo, and concerning preventive measures to be taken (*Write instructions for inspection, defect reporting and the safe use of vehicles and equipment, including cargo*);
4. be able to lay down procedures to be followed in the event of an accident and to implement appropriate procedures for preventing the recurrence of accidents or serious traffic offences (*Accident procedures: introduce measures to inform appropriate authorities and personnel of accidents; take appropriate action to minimize further dangers and to relieve suffering. The use of European Accident Statements*).

The OCR states that new legislative measures will not be included in the examination for at least three months from the date of implementation.

Those wishing to study for the examination should ensure that they have an up-to-date syllabus, covering either road haulage or passenger transport operations, from the OCR (see above for address).

TRANSFER OF QUALIFICATIONS

Provisions are made to allow an interchange of professional competence qualifications and recognition of professional competence qualification certificates between the United Kingdom, Northern Ireland and other EU member states. Thus the UK Traffic Commissioners are required, under the 'mutual recognition' requirements of EU Directive 796/77/EEC, to take into account any certificate of professional competence or any alternative to such a certificate showing relevant experience in the road haulage industry issued either in Northern Ireland or in other member states when considering an application for an 'O' licence.

Similarly, the Traffic Commissioners in the UK will issue a Certificate of Qualification form GOL55 (fee £20 – payable to the Traffic Area Office) confirming the good repute, professional competence and, where relevant, the financial standing of persons or companies from the UK who wish to be admitted to the occupation of road haulage operator, or to be employed to manage the transport operations of a goods haulage undertaking in Northern Ireland or in any other EU member state.

Where a person seeking a Certificate of Qualification is not, or has not been, an 'O' licence holder in the UK, the TCs will, of course, have no knowledge of their experience, good repute or financial standing and thus will be unable to issue a certificate. In these cases application can be made to the Secretary of State (ie for the D*f*T) who is empowered to issue a certificate to such a person on payment of the requisite fee.

GOOD REPUTE

The subject of good repute has been dealt with in detail in Chapter 1 (see pp 17–19) where it is more properly located since it is an issue which arises only when an application for an 'O' licence is made by an individual, by a partnership business or by a company. However, a great deal of attention has been focused on the 'good repute' aspect of professionally competent persons. For this reason, and to catch the attention of those who may read only this chapter rather than the 'O' licensing chapter (Chapter 1), some of the salient facts are repeated here.

As stated above, the good repute of such persons is only called into question when their names are put forward in support of a standard 'O' licence application. At that time the TC will want to know if the person has previous convictions for offences relating to the operation of goods vehicles (see below).

Following on from the previous explanation of the interchange of qualifications between the UK and other EU member states, the TCs, in considering an application for an 'O' licence and the good repute of the nominated professionally competent person, are required to take into account any convictions incurred by the person in Northern Ireland or in a country or territory outside the UK.

The relevant convictions are those which correspond to the relevant convictions under the British regulations, namely offences relating to vehicle roadworthiness, speeding, overloading, safe loading, drivers' hours and record keeping, drivers' licensing, maintenance, illegal use of rebated fuel, certain traffic offences and to the International Road Haulage Permits Act 1975.

Further, the Transport Tribunal has ruled that the TC can consider other convictions when considering a professionally competent person's good repute. The Tribunal ruled that the TC does not have to regard only those convictions mentioned above (as identified in the 1968 Transport Act) in deciding good repute.

In another instance a TC has said that a nominated professionally competent person must be clear of past financial difficulties. This followed a case where a nominated person was shown to be an adjudged bankrupt and the TC refused to accept the person's name on this account.

Rights of Transport Managers

In order to protect the rights of individuals, TCs are not permitted to examine and rule upon a transport manager's good repute or professional competence unless the individual has been notified at least 28 days in advance that this is to happen (and for what reason) and has had an opportunity to make personal representation to the TC (at a public inquiry if necessary and represented by a solicitor if he so wishes) concerning any allegations made about him.

Duties of Transport Managers

Traffic Commissioners have been increasingly tightening up on the role and duties of Transport Managers nominated in support of 'O' licence applications. New Guidance Notes have been issued to the TCs and a newly devised, purpose-designed, Contract of Employment form has been brought into use in an effort to eliminate the practice whereby TMs are put forward in name only, with no concept of their statutory duties or any intention that they should perform this role as required by law (see also pp 32–35).

3 Goods Vehicle Drivers' Hours and Working Time

This chapter has been expanded to incorporate both the EU driving hours rules and the legal provisions relating to working time. With the EU's Directive on Working Time now being extended to include mobile workers (most predominantly LGV drivers), it was thought to be more appropriate to deal with the subject here along with the long-established driving hours rules, which themselves are currently the subject of consultation prior to the proposed introduction of changes (which are also outlined).

EU DRIVING AND REST TIME RULES

Any person who uses a vehicle for the carriage of goods for commercial or business purposes (ie not including those used for purely private purposes and other vehicles specifically exempted – see pp 10–13) irrespective of the vehicle weight must conform to strict rules on the amount of time he may spend driving. For most goods vehicle drivers the rules also include requirements relating to minimum breaks to be taken during the driving day and to both daily and weekly rest periods.

ENFORCEMENT AND PENALTIES

The driving hours rules are applied solely to protect public safety, especially to protect road users from the dangers of overworked and tired drivers of heavy vehicles. Enforcement is vigorously undertaken by both the police and VOSA enforcement officers, and convicted offenders are dealt with severely by the courts to emphasize the importance of these road safety measures.

Drivers who are convicted for contravening the rules risk heavy fines, while the most serious of offences can result in imprisonment. LGV driving entitlements may be revoked in consequence, losing a driver his job.

Besides prosecution for infringements and its consequent penalties, provisions brought in by the Transport Act 2000 (ie section 266 which amends section 99 of the Transport Act 1968) and effective since 1 February 2001, authorize enforcement officers of the VOSA to detain drivers found to be in breach of the drivers' hours rules.

Where it is determined during a roadside check that a UK driver has not taken sufficient break or rest periods prior to that time, he will be prohibited from continuing his journey. In the case of break period infringements the delay at the roadside will be until a full 45-minute break period has been taken. Where breaches of the daily or weekly rest period requirements are detected, the driver and his vehicle will be escorted by the police to a suitable parking area (eg service area or truck stop) where he will have to remain until he has taken a full 11-hour daily rest or 24-hour weekly rest.

The vehicle operator may send a relief driver to take over the vehicle and continue its journey, but this driver will be subject to scrutiny by the VI to ensure that he has sufficient time available within legal limits to drive the vehicle.

Employers of convicted drivers also risk prosecution and heavy fines for similar offences. Additionally they may have penalties imposed against their 'O' licences by the Traffic Commissioners (TCs) since they promised in the declaration of intent at the time of their 'O' licence application that they would make arrangements to ensure the drivers' hours law would be observed.

It is also a specific requirement of the EU rules that employers must make periodic checks to ensure the rules are observed and must take appropriate action if they discover breaches of the law to ensure there is no repetition of offences.

In mainland Europe, breaches of the rules detected in roadside checks may result in drivers incurring heavy on-the-spot fines which must be paid immediately, otherwise the vehicle may be impounded and the driver held until the fine is paid.

REPORTING OF ILLEGAL OPERATIONS

Under the Public Interest Disclosure Act 1998 employees are protected if they report corruption, wrongdoing or danger at work, or if they are unduly pressured by an unscrupulous employer to break the law.

For example, in the context of the drivers' hours law, where drivers are encouraged or pressured to breach the driving hours or tachograph rules, they would have protection under the Act if they reported such matters either to VOSA or to the TC.

The Act came into force in July 1999. This new so-called 'whistle-blowing' charter entitles workers to unlimited compensation if their employer penalizes them for exposing breaches of the law or unsafe practices. Besides this legal protection, the United Road Transport Union (URTU) provides a confidential hotline service for drivers on freephone 0800 526639.

WHICH RULES APPLY?

Goods vehicle driver employers and drivers themselves need to understand the hours rules clearly and especially which particular set of rules applies to them depending on the vehicle being driven or the nature of the transport operation on which they are engaged. Three main sets of rules apply, the EU rules, the British domestic rules and what are known as the AETR rules for international journeys outside the EU (now fully aligned with the EU rules). The specific requirements under each of these sets of rules are explained in the following pages.

EUROPEAN UNION RULES

The currently applied EU rules, contained in Regulation 3820/85/EEC, came into effect in the UK on 29 September 1986. British regulations as follows implementing these rules and modifying previous provisions came into effect on the same date:

- The Community Drivers' Hours and Recording Equipment (Exemptions and Supplementary Provisions) Regulations 1986 – which implemented the EU regulations in the UK and the derogations (ie exemptions) from them.
- The Community Drivers' Hours and Recording Equipment Regulations 1986 – which made consequential changes to the 1968 Transport Act provisions.
- The Drivers' Hours (Goods Vehicles) (Modifications) Order 1986 – which implemented changes to national and domestic driving under the 1986 Act.
- The Drivers' Hours (Harmonization with Community Rules) Regulations 1986 – which harmonize the rules for those who drive under both the EU regulations and the revised 1968 Act rules.

Vehicles Covered

EU rules always take precedence over national rules. Therefore, since all goods vehicles and other vehicles used for the carriage of goods in connection with business activities as stated above are covered by one or other of the sets of drivers' hours rules (ie EU, British or AETR), the requirements of specific EU legislation must be considered first. Mainly, this involves examination of the list of EU rules exemptions to determine whether the vehicle or the transport operation falls within scope of the rules or outside.

Where a vehicle has a maximum permissible weight not exceeding 3.5 tonnes it is exempt from the EU rules or if it is one identified on the EU exemption list or is used for a purpose which is shown in the list as exempt as mentioned previously (eg vehicles used by the public utilities and local authorities, etc) it automatically comes within the scope of the British domestic rules set out in the 1968 Transport Act as described in detail on pp 86–88.

Where vehicles are over 3.5 tonnes permissible maximum weight, including the weight of any trailer drawn, and are not exempt as shown by the list or are not used for an exempt purpose, the EU hours law applies.

Where vehicles are over 3.5 tonnes permissible maximum weight, including the weight of any trailer drawn, and are not exempt as shown by the list or are not used for an exempt purpose, the EU hours law applies. Similarly (since August 1998), the law applies to vehicles 'used for the carriage of goods', such as 'off-road' vehicles and so-called 4x4s towing trailers, and where the combined weight of the towing vehicle and the trailer exceeds 3.5 tonnes.

Exemptions to EU Rules

The EU regulations list a number of exemptions and national governments (eg the British Government) are permitted to make certain other exemptions (derogations) if they so wish, as shown below.

International Exemptions (under EU regulations)

1. Vehicles not exceeding 3.5 tonnes gross weight including the weight of any trailer drawn.
2. Passenger vehicles constructed to carry not more than nine persons including the driver.
3. Vehicles on regular passenger services on routes not exceeding 50 kilometres.
4. Vehicles with legal maximum speed not exceeding 30 kph (approx 18.6 mph).
5. Vehicles used by armed services, civil defence, fire services, forces responsible for maintaining public order (ie police).
6. Vehicles used in connection with sewerage; flood protection; water, gas and electricity services; highway maintenance and control; refuse collection and disposal*; telephone and telegraph services; carriage of postal articles**; radio and television broadcasting; detection of radio or television transmitters or receivers.

   **NB: It has been ruled (by a Divisional Court in the UK and the European Court of Justice) that privatized firms operating refuse collection services on behalf of local authorities are exempt from the EU hours rules in the same way that local authority-owned refuse vehicles drivers were exempt prior to privatization of the service. A subsequent case made it clear that this exemption applies only where refuse collection operations are carried out over short distances – not to long-haul waste operations.*

   ***NB: The DfT has ruled that this exemption from the EU driving hours rules applies equally to parcels carriers operating in competition with the Royal Mail (ie the Parcelforce service) as with the Royal Mail itself. Instead, these operations fall within the scope of the British domestic hours rules as described in this chapter. However, despite this exemption, the EU tachograph rules still apply to such operations.*
7. Vehicles used in emergencies or rescue operations.
8. Specialized vehicles used for medical purposes.
9. Vehicles transporting circus and funfair equipment.
10. Specialized breakdown vehicles.
11. Vehicles undergoing road tests for technical development, repair or maintenance purposes, and new or rebuilt vehicles which have not yet been put into service.

12. Vehicles used for non-commercial carriage of goods for personal use (ie private use).
13. Vehicles used for milk containers or milk products intended for animal feed.

National Exemptions (under British derogations)

1. Passenger vehicles constructed to carry not more than 17 persons including the driver.
2. Vehicles used by public authorities to provide public services which are not in competition with professional road hauliers.
3. Vehicles used by agricultural, horticultural, forestry or fishery* undertakings, for carrying goods within a 50 km radius of the place where the vehicle is normally based, including local administrative areas the centres of which are situated within that radius.

   **To gain this exemption the vehicle must be used to carry live fish or to carry a catch of fish which has not been subjected to any process or treatment (other than freezing) from the place of landing to a place where it is to be processed or treated.*
4. Vehicles used for carrying animal waste or carcasses not intended for human consumption.
5. Vehicles used for carrying live animals from farms to local markets and vice versa, or from markets to local slaughterhouses.
6. Vehicles specially fitted for and used:
   - as shops at local markets and for door-to-door selling;
   - for mobile banking, exchange or savings transactions;
   - for worship;
   - for the lending of books, records or cassettes;
   - for cultural events or exhibitions.
7. Vehicles (not exceeding 7.5 tonnes gvw) carrying materials or equipment for the driver's use in the course of his work within a 50 km radius of base provided the driving does not constitute the driver's main activity and does not prejudice the objectives of the regulations.
8. Vehicles operating exclusively on islands not exceeding 2300 sq km not linked to the mainland by bridge, ford or tunnel for use by motor vehicles (this includes the Isles of Wight, Arran and Bute).
9. Vehicles (not exceeding 7.5 tonnes gvw) used for the carriage of goods propelled by gas produced on the vehicle or by electricity.
10. Vehicles used for driving instruction (but not if carrying goods for hire or reward).
11. Tractors used after 1 January 1990 exclusively for agricultural and forestry work.
12. Vehicles used by the RNLI for hauling lifeboats.
13. Vehicles manufactured before 1 January 1947.

14. Steam-propelled vehicles.
15. Vintage vehicles (ie over 25 years old) not carrying more than nine persons including the driver, not being used for profit, and being driven to or from a vintage rally, museum, public display or a place where it has been or is to be repaired, maintained or tested.

Note: In exemption 2 above relating to vehicles used by public authorities, the exemption applies only if the vehicle is being used by:

(a) a health authority in England and Wales, a health board in Scotland or a National Health Service (NHS) Trust
  - to provide ambulance services in pursuance of its duty under the NHS Act 1977 or NHS (Scotland) Act 1978; or
  - to carry staff, patients, medical supplies or equipment in pursuance of its general duties under the Act;

(b) a local authority to fulfil social services functions, such as services for old persons or for physically and mentally handicapped persons;

(c) HM Coastguard or lighthouse authorities;

(d) harbour authorities within harbour limits;

(e) airports authority within airport perimeters;

(f) British Rail, London Regional Transport, a Passenger Transport Executive or local authority for maintaining railways;

(g) British Waterways Board for maintaining navigable waterways.

Regulation 3820/85/EEC

To apply the rules as required and appreciate their implications it is necessary to understand the definitions of certain words and phrases used as follows:

Driver

The regulations apply specifically to the 'driver' of the vehicle. For these purposes a 'driver' is any person who drives the vehicle, even for a short period, or who is carried on the vehicle in order to be available for driving if necessary. They do not apply to persons carried as mates (ie to help load and unload only) or to statutory attendants.

Driving

Driving is time spent behind the wheel actually driving the vehicle and relates to an accumulation of periods spent driving before a break is needed or a daily rest period is commenced. The maximum limit for driving before a break is taken is 4½ hours.

Driving Time

Driving time is the accumulation of driving between two daily rest periods, or between a daily rest period and a weekly rest period, and must not exceed the limits described in this text.

Vehicle Categories

Drivers of all types and weight categories of vehicles within the scope of the regulations are treated equally in regard to driving, break and rest periods.

A Day

For the purposes of the regulations a day is any period of 24 hours starting from the time when a driver commences work after a daily or weekly rest period (ie when the tachograph is set in motion). If a driver drives a vehicle to which the EU regulations apply on any day (no matter how short the actual time spent driving or how short the journey on the road – for example, a 10-minute drive down the road would still bring the driver within scope of the rules) then the legal requirements apply to him for the whole of that day (ie 24-hour period) and the week in which that day falls.

Fixed Week

The definition of a 'week' for the purposes of the regulations is a fixed week from 00.00 hours Monday to 24.00 hours on the following Sunday. All references in the rules to weeks and weekly limits must be considered against this fixed week. Reference to 'fortnight' means two consecutive fixed weeks as described above.

Employers' Responsibilities

Under the EU rules employers have specifically stated responsibilities as follows:

- They must organize drivers' work in such a way that the requirements of the regulations are not broken (ie on driving times, breaks and rest periods, etc).
- They must make regular checks (ie of tachograph charts – see Chapter 5) to ensure the regulations are complied with.
- Where they find any breaches of the law by drivers, they must take appropriate steps to prevent any repetition.

Additionally, of course, it goes without saying that employers should ensure their drivers do understand how the law in this respect applies to them and how to comply with its detailed provisions. Although it is no excuse in court for a driver to say he did not know the law, the court would expect the employer to have instructed the driver in its requirements and may well convict the employer for 'failing to cause' the driver to conform to the law (or for permitting offences) if it felt that insufficient attention had been given to this matter.

Driving Limits

Goods vehicle drivers are restricted in the amount of time they can spend driving before taking a break and the amount of driving they can do between any two daily rest periods (or a daily and a weekly rest period), in a week and in a fortnight. The maximum limits are as follows:

- Maximum driving before a break: 4½ hours
- Maximum daily driving normally: 9 hours

- Extended driving on 2 days in week only: 10 hours
- Maximum weekly driving: 6 daily driving shifts*
- Maximum fortnightly driving: 90 hours

**NB: The High Court ruled in 1988 that drivers can exceed the maximum of six daily driving shifts within six days as specified in the EU rules provided they do not exceed the maximum number of hours permitted in six consecutive driving periods. It should be noted that where a driver spends the maximum amount of driving time behind the wheel in one week (ie 4 × 9 hours plus 2 × 10 hours = 56 hours), during the following fixed week he may drive for a maximum of only 34 hours.*

Break Periods

Drivers are required by law to take a break or breaks if in a day the aggregate of their driving time amounts to 4½ hours or more. If the driver does not drive for periods amounting in aggregate to 4½ hours in the day there is no legal requirement for him to take a break during that day.

Break periods must not be regarded as parts of a daily rest period and during breaks the driver must not carry out any 'other work'. However, waiting time, time spent riding as passenger in a vehicle or time spent on a ferry or train are not counted as 'other work' for these purposes.

The requirement for taking a break is that immediately the 4½-hour driving limit is reached a break of 45 minutes must be taken, unless the driver commences a rest period at that time (see below). This break may be replaced by a number of other breaks of *at least* 15 minutes each distributed over the driving period or taken during and immediately after this period, so as to equal at least 45 minutes and taken in such a way that the 4½-hour limit is not exceeded. To re-emphasize the point, once the 45 minutes' break requirement has been met, the slate is effectively wiped clean and the next 4½-hour period can begin.

A break period which was otherwise due in accordance with this requirement does not have to be taken if immediately following the driving period the driver commences a daily or weekly rest period, so long as the 4½ hours' aggregated driving is not exceeded.

According to DfT examples, the driver could legally operate the following procedures:

- Drive 1 hour, 15 minutes' break, drive 3½ hours, 30 minutes' break, drive 1 hour, 15 minutes' break, drive 3½ hours, commence daily rest period.
- Drive 1 hour, 15 minutes' break, drive 1 hour, 15 minutes' break, drive 2½ hours, 15 minutes' break, drive 2 hours, 30 minutes' break, drive 2½ hours, commence daily rest period.
- Drive 3 hours, 15 minutes' break, drive 1½ hours, 30 minutes' break, drive 3 hours, 15 minutes' break, drive 1½ hours, commence daily rest period.

A European Court of Justice ruling (December 1993) is that both the previous 'wipe the slate clean' and the 'rolling 4½ hours' interpretations of the driving hours' rules are wrong. The Court ruled that where a driver has taken 45 minutes' break, either as a single break or as several breaks of at least 15 minutes each during or at the end of the 4½-hour period, the calculation should begin afresh, without taking account of the driving time and breaks previously completed by the driver. It also ruled that the calculation of the driving period begins at the moment when the driver sets his tachograph in motion and begins driving. This means that a driver cannot count any break taken in excess of the required 45 minutes after driving for 4½ hours as part of the break period legally required in respect of the next 4½-hour driving period.

NB: It is important to note that break periods (ie especially the 45-minute period as well as the alternative minimum 15-minute periods) should not be curtailed even by a minute or two. Prosecutions have been brought for offences relating to break periods which are alleged not to conform to the law even though they have been only a matter of minutes below the minimum specified in the regulations. Such matters are, of course, shown clearly on tachograph recordings which provide ample evidence for the prosecution case.

Rest Periods

Rest periods are defined as uninterrupted periods of at least one hour during which the driver 'may freely dispose of his time'. Daily rest periods, and particularly rest periods which are compensating for previously reduced rest periods, should not be confused with, or combined with, statutory break periods required to be taken during the driving day as described above.

It has been held that time spent by drivers on weekend training courses (eg Hazchem courses and such like), even where there is no direct payment of wages by the employer, breaches the requirements under EU rules for drivers to have a period of weekly rest during which they may freely dispose of their time.

Daily Rest Periods

Once each day drivers are required to observe either a normal, a reduced or a split daily rest period during which time they must be free to dispose of their time as they wish. These daily rests are to be taken once in each 24 hours commencing at the time when the driver activates the tachograph following a weekly or daily rest period. Where the daily rest is taken in two or three separate periods (see below), the calculation must commence at the end of a rest period of not less than 8 hours. Thus in each 24-hour period as defined here one or other of the following daily rest periods must be taken:

- Normal daily rest: 11 hours
  or alternatively,
- Reduced rest: 9 hours – may be taken three times in a week but the reduced time must be compensated by an equal amount of additional rest taken with other rest periods before the end of the next following fixed week
- Split rest: Where the daily rest period is not reduced (as above) the rest may be split and taken in two or three separate periods during the 24 hours, provided:
  - one continuous period is of at least 8 hours' duration;
  - other periods are of at least 1 hour's duration;
  - the total daily rest period is increased to 12 hours.

Split Daily Rest

When a daily rest period is split into two or three separate periods (shown above to be permitted under the rules) it should be noted, as a result of a European Court of Justice ruling, that the 8-hour period must be in the last portion of the rest. Previously it was thought that the separate components of a split rest period could be taken in any order.

Double-manned Vehicles

Where a vehicle is operated by a two-man crew, the daily rest period requirement is that each man must have had a minimum of 8 hours' rest in each period of 30 hours.

NB: In the case of double-manning, it should be noted that the hours' law applies to both crew members from the commencement of the journey (or their day's work if that commenced earlier).

Daily Rest on Vehicles

Daily rest periods may be taken on a vehicle provided:

- the vehicle has a bunk so the driver (but not necessarily a mate or attendant) can lie down; and
- the vehicle is stationary for the whole of the rest period.

It follows from this that a driver on a double-manned vehicle cannot be taking part of his *daily rest period* on the bunk while his co-driver continues to drive the vehicle. He could, however, be taking a *break* at this time while the vehicle is moving or he could merely spend his time lying on the bunk, with his tachograph chart recording other work.

Daily Rest on Ferries/Trains

Daily rest periods which are taken when a vehicle is carried for part of its journey on a ferry crossing or by rail may be interrupted, but once only, provided:

- part of the rest is taken on land before or after the ferry crossing/rail journey;
- the interruption must be 'as short as possible' and in any event must not be more than one hour before embarkation or after disembarkation and this time *must* include dealing with Customs formalities;
- during both parts of the rest (ie in the terminal and on board the ferry/train) the driver must have access to a bunk or couchette;
- when such interruptions to daily rest occur, the total daily rest period must be extended by two hours.

Weekly Rest Period

Once each fixed week (and after six driving shifts – see note on p 82) a daily rest period must be combined with a weekly rest period to provide a weekly rest period totalling 45 hours. A weekly rest period which begins in one fixed week and continues into the following week may be attached to either of these weeks.

While the normal weekly rest period is 45 hours as described above, this may be reduced to:

- 36 hours when the rest is taken at the place where the vehicle or the driver is based; or
- 24 hours when the rest period is taken elsewhere.

Reduced weekly rest periods must be compensated (ie made up) by an equivalent amount of rest period time taken *en bloc* and added to another rest period of at least 8 hours' duration before the end of the third week following the week in which the reduced weekly rest period is taken.

Compensated Rest Periods

When reduced daily and/or weekly rest periods are taken, the compensated time must be attached to another rest period of at least 8 hours' duration and must be granted, at the request of the driver, at the vehicle parking place or at the driver's base. Compensation in this respect *does not* mean compensation by means of payment; it means the provision of an equivalent amount of rest time taken on a later occasion but within the specified limits (ie by the end of the next week for compensated daily rest and by the end of the third following week in the case of compensated weekly rest periods and in each case added to other rest periods).

Summary of EU Rules

The following table summarizes the EU rules applicable to both national and international goods vehicle operations:

• Maximum daily driving:	9 hours 10 hours on 2 days in week
• Maximum weekly driving:	6 daily driving periods (see p 81)
• Maximum fortnightly driving:	90 hours
• Maximum driving before a break:	4½ hours
• Minimum breaks after driving:	45 minutes or other breaks of at least 15 minutes each to equal 45 minutes
• Minimum daily rest (normally):	11 hours
• Reduced daily rest:	9 hours on up to 3 days per week (must be made up by end of next following week)
• Split daily rest:	The 11-hour daily rest period may be split into two or three periods – one at least 8 hours, the others at least 1 hour each: total rest must be increased to 12 hours
• Minimum weekly rest (normally):	45 hours once each fixed week
• Reduced weekly rest:	36 hours at base – 24 hours elsewhere (any reduction must be made up *en bloc* by end of the third following week)

- Rest on ferries/trains: Daily rest may be interrupted once only if:
  - part taken on land
  - no more than 1 hour between parts
  - drivers must have access to a bunk or couchette for both parts of rest
  - total rest must increase by 2 hours

Emergencies

It is permitted for the driver to depart from the EU rules as specified above to the extent necessary to enable him to reach a suitable stopping place when emergencies arise where he needs to ensure the safety of persons, the vehicle or its load, providing road safety is not jeopardized. The nature of and reasons for departing from the rules in these circumstances must be shown on the tachograph chart.

Prohibition on Certain Payments

The EU rules prohibit any payment to wage-earning drivers in the form of bonuses or wage supplements related to distances travelled and/or the amount of goods carried unless such payments do not endanger road safety.

BRITISH DOMESTIC RULES

The current British drivers' hours rules contained in the 1968 Transport Act (as amended) came into effect on 29 September 1986. They apply to goods vehicle drivers whose activities are outside the scope of the EU requirements as described above (ie which are specified as exempt from the EU rules) and comprise only limits on daily driving and daily duty.

NB: Detailed provisions which previously applied in regard to daily and weekly duty, daily spreadover and daily and weekly rest period limits were completely abolished when these rules changes were introduced.

It is important to emphasize that drivers who are exempt from the EU rules, either because their vehicle does not exceed 3.5 tonnes permissible maximum weight or because the activity in which they are engaged falls within the scope of the EU exemptions list (see pp 78–80), must observe the British domestic rules.

NB: It should be noted that:

- *Although no records of driving or working times are required to be kept, drivers of light goods vehicles (ie not exceeding 3.5 tonnes permissible maximum weight) must still conform to the legal limits on maximum daily driving and maximum daily duty.*
- *If a trailer is attached to a vehicle not exceeding 3.5 tonnes permissible maximum weight (which itself is exempt from the EU rules on account of its weight) thereby taking the combined weight to over 3.5 tonnes then the EU rules must be followed as described in the foregoing text (unless it is exempt for other reasons – namely the nature of the operations on which it is engaged) and a tachograph must be fitted to the vehicle and used by the driver for record-keeping purposes (see Chapter 5).*

Exemptions and Concessions

The British domestic rules apply to drivers of all goods vehicles which are exempt from the EU regulations as described earlier but with the following further exceptions which are totally exempt from all hours' rules control:

- Armed forces
- Police and fire brigade services
- Driving off the public road system
- Driving for purely private purposes (ie not in connection with any trade or business).

The British domestic rules do not apply:

- to a driver who on any day does not drive a relevant vehicle; or
- to a driver who on each day of the week does not drive a vehicle within the rules for more than four hours (Note: this exemption *does not* apply to a driver whose activities fall within scope of the EU rules).

Driving and Duty Definitions

For the purposes of the British domestic rules:

- Driving means time spent behind the wheel actually driving a goods vehicle and the specified maximum limit applies to such time spent driving on public roads. Driving on off-road sites and premises such as quarries, civil engineering and building sites and on agricultural and forestry land is counted as duty time, not driving time.
- Duty time is the time a driver spends working for his employer and includes any work undertaken including the driving of private motor cars, for example, and non-driving work which is not driving time for the purposes of the regulations. The daily duty limit does not apply on any day when a driver does not drive a goods vehicle.

Summary of British Domestic Rules

- Maximum daily driving: 10 hours
- Maximum daily duty: 11 hours
- Continuous duty: no specified limit
- Daily spreadover: no specified limit
- Weekly duty: no specified limit
- Breaks during day: no requirement specified
- Daily rest: no specified requirement (but obviously minimum of 13 hours so as not to exceed the daily 11-hour duty limit)
- Weekly rest: no specified requirement

Emergencies

The daily driving and duty limits specified above may be suspended when an emergency situation arises. This is defined as an event requiring immediate action to avoid danger to life or health of one or more individuals or animals, serious interruption in the maintenance of essential public services for the supply of gas, water, electricity, drainage, or of telecommunications and postal services, or in the use of roads, railways, ports or airports, or damage to property. Details of the emergency should be entered by the driver on his record sheet when the limits are exceeded.

Light Vehicle Driving

As explained above, drivers of light goods vehicles not exceeding 3.5 tonnes permissible maximum weight – as with drivers of other goods vehicles which fall within scope of these rules – must observe the daily limits on driving (10 hours) and duty (11 hours). However, only the 10-hour daily driving limit applies when such vehicles are used:

- by doctors, dentists, nurses, midwives or vets;
- for any service of inspection, cleaning, maintenance, repair, installation or fitting;
- by a commercial traveller, and carrying only goods used for soliciting orders;
- by an employee of the AA, the RAC or the RSAC;
- for the business of cinematography or of radio or television broadcasting.

MIXED EU AND BRITISH DRIVING

It is possible that a goods vehicle driver may be engaged in transport operations which come within scope of both the EU drivers' hours rules and the British domestic hours rules on the same day or within the same week. When such a situation arises he may conform strictly to the EU rules throughout the whole of the driving/working period or he may take advantage of the more liberal British domestic rules where appropriate. If he decides on this course of action and thereby combines both British and EU rules he must beware of the following points:

- time spent driving under the EU rules cannot count as an off-duty period for the British rules;
- time spent driving or on duty under the British rules cannot count as a break or rest period under the EU rules;
- driving under the EU rules counts towards the driving and duty limits for the British rules;
- if any EU rules' driving is done in a week the driver must observe the EU daily and weekly rest period requirements for the whole of that week.

AETR RULES

Drivers on international journeys beyond the EU which take them to or through the list of countries given below are required to observe the European Agreement Concerning the Work of Crews of Vehicles Engaged in International Road Transport – 1971 (commonly known and referred to as AETR). When on such journeys the driver must observe the AETR rules for the whole of the outward and return journey including the portion travelled in Britain and through other EU member states. These rules have since 1992 been fully harmonized with the EU rules contained in EC Regulation 3820/85, as described above.

The countries beyond the EU referred to above where the AETR rules apply (ie on journeys to or through) are as follows: Belarus, Bosnia, Bulgaria, the CIS, Croatia, the Czech Republic, Estonia, Latvia, Moldova, Norway, Poland, Romania, Slovakia and Slovenia.

TAX RELIEF ON DRIVER ALLOWANCES

Sleeper Cab Allowances

The amount paid to drivers for overnight subsistence varies considerably from area to area – the national general figure is currently £26.90 (2003 figure). The Inland Revenue has agreed that LGV drivers can be paid night-out allowances on a tax-free basis amounting to £26.90 per night where the vehicle does not have a sleeper cab, reducing to £20.17 per night where a sleeper cab is available. Payment of such amounts is subject to the employer being satisfied that:

- the individual did necessarily spend the night away from his home and normal place of work and that he used a sleeper cab;
- the employee necessarily incurred extra expense in doing so; and
- the amounts paid are no more than reasonable reimbursement of the average allowable expenses of the driver (ie for payment of an evening meal, breakfast, washing facilities and the up-keep of bedding).

Owner-drivers are dealt with differently for tax purposes and may NOT claim such night-out allowances against their tax liability.

There have been recent reports about tax inspectors querying payment of night-out allowances where drivers have stayed only a few miles from base. Currently the agreement with the Inland Revenue for night-out payments is as stated above and this applies for nights-out without receipts, except where alternative figures have been agreed, in which case an annual increase of 9.3 per cent on the previously agreed figure is acceptable. These payments should only be made where the employee does actually spend the night away and incurs extra expense. If he uses the bunk in a sleeper cab the allowable amount is only that required to meet the expenses he incurs, not the full night-out allowance.

The Inland Revenue may refuse claims for tax-free payments of night-out allowances above the general limit. Employers who have paid in excess of this amount (or a locally agreed rate) without deduction of tax may find themselves liable to meet the tax due on the additional amounts paid, except where it can be proved that the expense was genuinely incurred (by production of a valid receipt); alternatively they should include it as part of the wages within PAYE.

Charts Retained for Tax Purposes

Where tachograph charts are used to justify payment of night-out and other subsistence expenses to drivers these become part of the tax record and as such must be retained for six years instead of the normal one year.

Meal Expenses

Where LGV drivers receive from their employers a contribution towards mid-day meal expenses either by the issue of meal vouchers or by means of a cash payment, these contributions may be treated as not taxable provided they:

- are of a reasonable amount only (ie not exceeding £2.00 per day); and
- are paid to long-distance drivers only whose duties oblige them to take meals away from home and their normal place of employment, but not to drivers who have fixed or local routes.

Any amount paid in excess of £2.00 per day must be subject to income tax deduction unless acceptable evidence of the expenditure is provided to the employer or where the Inland Revenue has given dispensation.

Personal Incidental Expenses

Employee drivers may be paid additional amounts by way of 'personal incidental expenses' in relation to genuine expenditure (eg newspapers, laundry and calls home) during a qualifying absence (ie when working away from home) up to a tax-free limit, without any tax consequences for the employee – or NI for the employer up to the value of £5 per day if in the UK, or £10 per day outside the UK. (Source: Road Haulage Association website: www.rha.net/public/adviceandinfo/subsistence.shtml)

Tax Relief for International Drivers

International LGV drivers who spend less than 91 days in the UK during a tax year (ie from 6 April to 5 April annually) can claim 'non-resident' status and qualify for their wages to be paid tax-free. Previously, the limit was less than 60 days but this was increased to 91 days by the March 1998 Budget.

REVISION OF EU DRIVING HOURS RULES

The European Council of Transport Ministers has been discussing proposals for amendment to the current EU driving hours regulation (3820/85/EEC) which came into force on 29 September 1986. By mid-2005, as the proposed amendments to the rules became more clear, the new provisions were being described as being likely to restrict competition, cause confusion and also likely to cost the industry millions of pounds, particularly on top of the stringent working time provisions for mobile workers (described below) which are also imminent.

According to the FTA website (www.fta.co.uk) the following key proposals were being considered in July 2004:

- The definition of a week is to be 00.00 hrs Monday to 24.00 hrs Sunday (fixed week).
- Regular daily rest requirements are to be 12 hours with reduced rest of nine hours available three times per week between any two weekly rest periods and no compensation period required.
- Daily rest may be split into two uninterrupted periods of three hours plus nine hours in a 24-hour period.

- A reduction in weekly rest to 36 hours is permitted when taken at base, or 24 hours if taken away from base – reductions to be compensated by end of the following week.
- When away from base reduced weekly rest can be taken in vehicles with suitable sleeping facilities.
- There will be a change to rest requirements that would allow weekly rest to be taken at start of week one, with next weekly rest required at latest end of week two (potentially 14 consecutive working days).
- When driving on *and* off road *all* driving is to count towards driving limits. However, driving solely off road 'on-site' is not regarded as driving time.
- The weekly driving limit is to be 56 hours and the fortnightly driving limit 90 hours.
- On any day when driving a vehicle to which EU rules apply, driving time in any other goods vehicle including light vans will count toward driving total.
- There will be exemption for breakdown vehicles within a 100 km radius of base (previously proposed 50 km).
- There will be derogation for vehicles not exceeding 7.5 tonnes used to deliver postal items as part of the universal service within 50 km radius of base.
- At roadside checks, records are to be available for the current day plus those for the 15 previous consecutive days.
- Consignors, freight forwarders, prime contractors, sub-contractors and driver employment agencies are to ensure that contractually agreed transport time schedules respect the provision of the regulations.
- There will be compulsory downloading of data from digital tachograph at a frequency determined by individual member states.
- Manual entries are to be made on record sheets and digital equipment for periods spent away from the vehicle.

At a meeting of Council Minister of Transport held on 11 June 2004 further amendments to these proposals were discussed, particularly the following issues:

- The minimum uninterrupted daily rest period is to be increased from 8 hours to 9 hours.
- During two consecutive weeks, the driver must take at least one regular weekly rest period consisting of an uninterrupted period of at least 45 hours.
- In regard to the checking of driving records by enforcement officers, the proposed 'current week and the previous 15 days period' for which drivers must be able to produce records is to be increased from 1 January 2008 to 'the current day and the previous 28 days'.

NB: It is emphasized that at the time of preparing this 2006 edition of the Handbook, no date has as yet been set for the introduction of these legislative changes so the current rules in Regulation 3820/85/EEC remain in force.

WORKING TIME

Among all the strict regulation that the road transport industry has had to contend with in recent years, there is no doubt that the imposition of the EU's Road Transport Directive (Directive 2002/15/EC – the RTD), which took effect from 4 April 2005, has had a traumatic impact on transport operators and LGV drivers alike. The predictions were that it would result in massive cost for the industry; an estimated shortfall of some 50,000 to 80,000, or even more, qualified LGV drivers and potential hardship for drivers themselves who, in the worse-case scenarios, may see their pay packets dramatically reduce with no possibility of taking up additional part-time jobs to compensate due to the legal restriction imposed by the Directive.

The first step in this particular legislative process started with the European Council Directive 93/104/EC of November 1993 commonly referred to as the working time directive (WTD). This Directive introduced a number of provisions to control working time for most employees in the interest of improving work environments and ensuring better protection for workers by limiting maximum weekly working hours, setting minimum daily and weekly rest periods and minimum in-work breaks, setting minimum annual holidays and requiring regular health checks for night workers. In particular the Directive specified a maximum 48-hour working week and limited night working in any 24-hour period. These requirements are enforced in the UK by our own domestic working time regulations (see below).

However, while this Directive originally excluded the transport sector, further proposals originally published in 1997 were intended to bring road haulage (and other previously excepted worker categories) into scope of the working time principles. From 1 August 2003 came the Horizontal Amending Directive (Directive 2000/34/EC – abbreviated to HAD) which amended the original WTD by setting rules for those employment sectors that were expressly excluded from the WTD; such as non-mobile transport workers and mobile workers not covered by the EU drivers' hours rules (for example, light vehicle drivers).

These two EU Directives are given the force of law in Great Britain by a number of regulations as follows:

- The Working Time Regulations 1998 (SI 1998 No. 1833);
- The Working Time Regulations 1999 (SI 1999 No. 3372) (amending regulations);
- The Working Time (Amendment) Regulations 2001 (SI 2001 No. 3256) (dealing mainly with annual leave);
- The Working Time (Amendment) Regulations 2002 (SI 2002 No. 3128) (dealing mainly with provisions regarding young people at work);
- The Working Time (Amendment) Regulation 2003 (SI 2003 No. 1684) (implementing the provisions of the HAD).

In the following section, the working times rules are explained in order of their relevance to the road transport industry; namely, first the RTD that applies to LGV drivers, second the HAD that applies to LGV drivers and finally the WTD itself, which is applicable to most other workers.

THE ROAD TRANSPORT DIRECTIVE (RTD)

LGV drivers who fall within scope of the existing and long-established drivers' hours law contained in EU Regulation 3820/85/EEC, namely most drivers of goods vehicles exceeding 3.5 tonnes gross weight, are covered by the RTD (Directive

2002/15/EC), which deals specifically with working time for those persons performing mobile transport activities within scope of the EU drivers' hours rules (mainly LGV drivers), but with the exception of self-employed mobile workers who will remain exempt from the Directive until 23 March 2009.

This Directive was implemented in the UK from 4 April 2005, following publication of Statutory Instrument SI 2005 No 639.

NB: It is important to note that the RTD does not replace the present EU drivers' hours rules, instead it supplements both the current Regulation 3820/85/EEC and the AETR Agreement, so drivers who fall within scope of these regulations when driving a goods vehicle must also take note of and comply with the provisions of the RTD.

Definitions

It is useful to consider the definitions for certain terms used and the specific provisions set out in the RTD as follows.

Mobile Transport Activities

Any activity carried out by workers in the course of their employment within a road transport operation.

Mobile Worker

Any worker who forms part of the travelling staff of an undertaking that operates goods or passenger transport for hire or reward or on its own account – obviously this means drivers, but also includes other crew members, trainees and apprentices.

Self-Employed Drivers

These are similarly categorized as mobile workers (ie driving in a road transport operation) except that such workers are not tied to an employer by an employment contract (see below).

Working Time

This is time spent at work and at the employer's disposal carrying out his activities or duties. This includes travelling as part of job and work-related training, but not commuting between home and work, or rest breaks where no work is done, or 'on-call' time when at home.

Periods of Availability

These periods are not working time and do not count as such, nor are they breaks or daily or weekly rest periods. They could include waiting time (for example, to load or unload or while waiting for a ferry) if the driver is free to dispose of his time during that period of waiting, but not, for example, if he has to remain with his vehicle in a queue – this would be working time. The key to this is whether the waiting period could have been foreseen in advance; if it could, it counts as non-working time, if it could not be foreseen, it is working time.

Working Week

The working week is 48 hours' work (with a maximum of 60 hours in a week) averaged over a 4-month (ie 17-week) reference period, which may be extended by agreement to 6 months. The total working time is calculated excluding breaks during the working day and waiting time (see above). It should be noted that there is *NO* opt-out from the working time limits under this Directive as there is with the other Directives (ie the WTD and the HAD).

Breaks

Breaks taken during the working period breaks must amount to a minimum of 30 minutes if working between 6 and 9 hours; or 45 minutes if working over 9 hours.

Night Work

This is defined as a 4-hour period between midnight and 4.00 am. Night working will be limited to 10 hours in any 24-hour period although an agreed derogation will allow this to be extended to 12 hours.

NB: This night work definition should not be confused with the night work definition contained in either the HAD or the WTD.

Holidays

Holidays are to comprise a minimum of 4 weeks (ie 20 days) paid at the average rate of earnings for the previous 12 weeks (average earnings in this case should include overtime payments).

It is to be an important provision of the new regulations that mobile workers must tell their employer, in writing, of any time they spend working for any other employer in any other job (even if it does not come within scope of the RTD). Any such time worked will count towards the weekly working limit of 48 hours.

Record Keeping

Irrespective of which of the three Directives (the WTD, HAD or RTD) applies in any particular work situation, employers must keep detailed records of employee working time and provide copies of such records to their employees when asked for them. There is no officially prescribed system for record keeping or special format for individual records, but whatever system is used the records must be kept for two years and be made available to VOSA enforcement officers on request.

Health Assessments

Employers will be required to provide free health assessments for night workers. Employers have a duty to determine exactly who is a night worker for this purpose and provide the health assessment accordingly. The assessment should include a questionnaire devised and monitored by a qualified health professional and, where necessary, backed up by a medical examination.

Enforcement

The RTD regulations will be enforced mainly by VOSA, most likely when they are checking compliance with the EU drivers' hours and tachograph rules, or as a result of complaints made to them.

Self-Employed Workers

As already mentioned, self-employed mobile workers are to be exempted from inclusion in the RTD until 23 March 2009. But to qualify they must be genuinely self-employed in accordance with both RTD and Inland Revenue rules on self-employment. Basically, this means:

- having as their main occupation the transport of goods by road for hire or reward under an 'O' licence;
- being entitled to work for themselves;
- not being tied to an employer under an employment contract or other hierarchical working relationship;
- having the freedom to organise their own work;
- their income must derive from the profits of their business;
- they must have the freedom, either alone or in cooperation with others, to work for more than one customer.

NB: It is emphasised that all of the above conditions must be met to establish genuine self-employed status – those LGV drivers who work solely for one employer will NOT qualify under these rules for self-employed status and will therefore remain subject to the RTD provisions as described above for those drivers who are employed in the normal way.

The Horizontal Amending Directive (HAD)

The WTD referred to above originally excluded the transport sector, but publication of an amending Directive 2000134/EC (the so-called Horizontal Amending Directive – HAD) that took effect from 1 August 2003 brought all employed road transport personnel, including office-based staff, referred to collectively as 'non-mobile workers', and other previously excepted employment categories, into line.

This Directive, which has applied to all employed road transport personnel (ie apart from those subject to the EU drivers' hours law) since 1 August 2003 as mentioned above, requires all relevant transport employees to comply with the provisions set out below. Drivers will also remain subject to the EU drivers' hours rules set out in Regulation 3820185/EEC.

Four particular provisions of the HAD apply to drivers who fall outside the scope of the Road Transport Directive – ie drivers of smaller vehicles and drivers of other vehicles that are exempt from the EU drivers' hours rules. These are:

- the 48 hour average working week;
- the requirement to have four weeks paid annual leave;
- regular health checks for night workers;
- the need for 'adequate rest'.

Under these regulations, workers are allowed to opt-out of the 48-hour average working week. However, mobile workers covered by the RTD are only entitled to the annual leave and health checks provisions (see above).

The Working Time Directive (WTD) for Non-Mobile Workers

The Working Time Directive (Directive 93/1104/EC) of November 1993, commonly referred to as the Working Time Directive or WTD, introduced a number of provisions to control working time for employees, including the concept of a maximum 48-hour working week and shift working restrictions under which night workers may not work more than a specified average number of working hours in any 24-hour period. This Directive specifically excluded the transport sector (eg road, rail, sea, air, inland waterway, etc) which are now covered by other subsequently published Directives, namely the RTD and HAD described above.

Under the WTD provisions employers are required to:

- Limit average weekly working time to a maximum of 48 hours, including overtime, calculated over successive periods of 17 weeks (ie 4 months), or for the period of employment where this is less than 17 weeks.
- The opt-out clause which would have allowed non-mobile workers to exceed the average 48-hour working week has now been withdrawn by the European Parliament.
- Limit night working to no more than 8 hours in 24 hours taken as an average over a 17-week reference period (see sections above regarding the RTD and HAD on this topic). For night workers (ie those who work for at least 3 hours at night) whose work involves special hazards or heavy physical or mental strain, the limit is a straight 8 hours in 24 hours with no averaging out.
- Provide free health assessments for night workers, and the opportunity for employees to transfer to day work if their health is affected by night working.
- Allow workers a daily rest period of at least 11 consecutive hours in each 24-hour period and an uninterrupted rest break of at least 20 minutes when their daily work exceeds 6 hours.
- Allow workers a weekly rest period of not less than 24 hours in each 7 days.
- Allow workers who have been employed continuously for 13 weeks to have at least 4 weeks' paid annual leave which may not be exchanged for payment in lieu except where it occurs on termination of the employment.
- Keep records of workers' hours of work that are adequate to show that the legal requirements are complied with, and retain these records for at least two years from the date on which they were made.

The WTD regulations set more restrictive standards for young workers and exempt certain special classes of worker such as Crown servants, the police, trainees and agricultural workers.

Unfair Dismissal

The Employment Rights Act 1996 is amended by the addition of new provisions to make it unfair to dismiss an employee for refusing to comply with a requirement contrary to the working time regulations, or to forgo his or her rights.

4 Goods Vehicle Drivers' Records

Drivers of goods vehicles and other vehicles used for the carriage of goods (eg off-road vehicles and so-called 4x4s) over 3.5 tonnes gross weight must keep records of the time they spend driving such vehicles and their working times. Most drivers are also required to record breaks taken during driving periods and their daily and weekly rest periods. Where vehicles fall within the scope of the EU drivers' hours law as described in Chapter 3 (ie unless they are specifically exempt as shown in the relevant exemptions list) the record-keeping requirement is based on the mandatory use of tachographs as described in Chapter 5.

Where a vehicle is outside the scope of the EU rules then the British domestic (ie 1968 Transport Act – as amended) driving hours rules apply (see Chapter 3) and the driver of such a vehicle is required to keep written records by means of a 'log-book' system.

To summarize the main record-keeping alternatives, these are as follows:

- Goods vehicles not exceeding 3.5 tonnes permissible maximum weight – NO RECORDS.
- Goods vehicles over 3.5 tonnes permissible maximum weight (including the weight of any trailer drawn) operating within EU rules – TACHOGRAPH RECORDS.
- Goods vehicles over 3.5 tonnes permissible maximum weight exempt from EU rules – WRITTEN 'LOG-BOOK' RECORDS (but subject to further exemption in certain cases).
- Goods vehicles over 3.5 tonnes permissible maximum weight exempt from both EU and British domestic rules (eg military vehicles) – NO RECORDS.

This chapter describes the record-keeping requirements applying to drivers falling within the third item above, namely, those operating under the British domestic rules who must keep written 'log-book' records.

EXEMPTIONS FROM RECORD KEEPING

Written records do not have to be kept in the following cases:

- By drivers of vehicles which are exempt from 'O' licensing except that the exemption does not apply to drivers of Crown vehicles which would have needed an 'O' licence if the vehicle had not been Crown property.
- By drivers of goods vehicles on any day when they drive for four hours or less and within 50 kilometres of the vehicle's base (NB this exemption is applicable only in the case of domestic operations – it does not apply to tachograph use – see pp 86–88).
- By drivers voluntarily using an EU tachograph for record-keeping purposes which has been calibrated and sealed at a DETR-approved tachograph centre.

RECORD-KEEPING SYSTEM

Prior to 29 September 1986 when the EU rules and British domestic hours rules last changed, the written record-keeping requirement related to the use of the EC diagrammatic-type International Control Book (ICB). From this date there was a transitional arrangement for phasing out the use of this record book and the introduction of a new style of simplified British record which relates specifically to the British domestic driving hours rules. Regulations brought the 'simplified' record book into use from 2 November 1987 with the old-type diagrammatic control book ceasing to have any further legal standing from that date.

Record Books

Ready-printed record books can be purchased 'off the shelf' or firms can have their own version pre-printed with their own name and logo if desired. In the latter case it is important that the specific requirements of the regulations are observed in both the format and the printing of the book.

The book must be a standard A6 format (105 mm x 148 mm) or it may be larger if preferred. It must comprise a front sheet on which is entered relevant information, a set of instructions for the use of the book, and a number of individual weekly record sheets with facilities for completing these in duplicate (ie with carbon paper or carbonless copy paper) and for the duplicate sheet to be detached for return to the employer when completed.

There is no legal requirement for the numbering of record books or for their issue against an entry in a register of record book issues as previously required.

Weekly record sheets in the book must follow the format set out in the regulations with appropriate spaces for entries to be made under the following headings:

- Driver's name.
- Period covered by sheet week commencing . . . week ending . . .
- Registration number of vehicle(s).

- Place where vehicle(s) based.
- Time of going on duty.
- Time of going off duty.
- Time spent driving.
- Time spent on duty.
- Signature of driver.
- Certification by employer (ie signature and position held).

Issue and Return of Record Books

Employers must issue their employee drivers with record books when they are required to drive vehicles to which the British domestic driving hours regulations apply and where records must be kept. Before issuing the book the employer must complete the front cover to show the firm's name, address and telephone number preferably with a rubber stamp if he has one.

When a record book is issued to the driver he should complete the front cover with his surname, first name(s), date of birth and home address and the date he first used the book. When the book is completed he should also enter the date of the last entry (ie date of last use). There is space to record the name and address of a second employer.

Books issued by an employer to an employee-driver must be returned to that employer when complete (subject to the requirement for the driver to retain it for two weeks after use) or when the employee leaves that employment. He must not take it with him to his new employer. Any unused weekly sheets and all duplicates must be included when the book is returned.

Two Employers

Where a driver has two employers who employ him to drive goods vehicles to which the British domestic hours rules apply, the first employer must issue the record book as described (completed weekly record sheets and completed record books must be returned to this employer), and the second employer must write or stamp his firm's name and address on the front cover of the record book with a statement that the holder is also a driver in his employment. When the driver does part-time driving work for another employer he must disclose to each employer, if requested, details of his working and driving times with the other employer. Similarly, when a driver changes to a new employer the former employer must give the new employer details of the driver's previous driving and working times if requested.

Record Book Entries

The driver must make entries on the weekly sheet for each day on which a record is required (instructions on the correct use of the book are printed inside the cover). Care must be taken to ensure that an exact duplicate of the entry is made simultaneously (ie two separately written repeat entries are *not* acceptable even if no carbon paper is available). When completing a daily sheet he must enter all the required details under each of the headings. If he changes vehicles during the day he must write in the registration number for each vehicle. He must then sign the sheet before returning it to his employer.

Completion of the record is straightforward, the driver having to enter the vehicle registration number, the time of coming on duty, and at the end of the day the time at which he went off-duty, and he must sign the sheet. He may enter any remarks concerning his entries, or point out corrections which should be made, in the appropriate box at the foot of the record. The employer may also use this space if required for making comments regarding the record. This space may also be used for recording the name of a second driver.

Corrections

Entries in the record book must be in ink or made by a ballpoint pen and there must be no erasures, corrections or additions. Mistakes may only be corrected by writing an explanation or showing the correct information in the remarks space. Sheets must not be mutilated or destroyed.

Return and Signing of Record Sheets

On completion of the weekly sheet, and after it has been signed by the driver, the duplicate copy must be detached from the book and handed to the employer within seven days of the date of the last entry on the sheet, and then within a further seven days the employer must have examined and signed the duplicate sheet. However, if in either case it is not reasonably practicable to do so within this time, these actions must be carried out as soon as it is possible to do so.

Retention and Production of Record Books

The driver should carry his record book with him at all times when working and must produce it for inspection at the request of an authorized examiner. The book should be shown to the employer at the end of every week or as soon as possible after the week so he can examine and countersign the entries. Following completion of the book the driver must continue to keep it with him for a further two weeks (available for inspection by the enforcement authorities) before returning it to his employer.

Completed record books must be retained by the employer, also available for inspection by the enforcement authorities, for not less than 12 months.

Record Sheets for Germany

It has been reported that the German authorities require, under their own domestic legislation, foreign drivers of goods vehicles of between 2.8 tonnes and 3.5 tonnes gross weight (ie which are exempt from record-keeping by means of the tachograph) entering the country to carry and complete AETR-type log books. These are the old-type log books long since abolished in the UK and the rest of Europe. Failure to carry such a log book could result in delays and penalties.

WEEKLY RECORD SHEETS

WEEKLY SHEETS

DRIVER'S NAME

2. PERIOD COVERED BY SHEET
WEEK COMMENCING (DATE) ....
TO WEEK ENDING (DATE)...

DAY ON WHICH DUTY COMMENCED	REGISTRATION NO. OF VEHICLE(S) 3.	PLACE WHERE VEHICLE(S) BASED 4.	TIME OF GOING ON DUTY 5.	TIME OF GOING OFF DUTY 6.	TIME SPENT DRIVING 7.	TIME SPENT ON DUTY 8.	SIGNATURE OF DRIVER 9.
MONDAY							
TUESDAY							
WEDNESDAY							
THURSDAY							
FRIDAY							
SATURDAY							
SUNDAY							

10. CERTIFICATION BY EMPLOYER

I HAVE EXAMINED THE ENTRIES IN THIS SHEET
SIGNATURE ....
POSITION HELD ....

Figure 4.1 *Simplified record sheet for British domestic transport operations*

5 Tachographs – Fitment and Use Requirements

Tachograph instruments installed in goods and passenger vehicles provide a means of recording time and the speed and distance travelled by the vehicle. This record enables drivers' working activities and driving practices to be monitored to ensure that legal requirements – especially observance of the drivers' hours rules – have been met.

'D' DAY FOR DIGITAL TACHOGRAPHS

The long-awaited digital (so-called 'smart card') tachograph regulations (Regulation 1360/2002/EC, amending Regulation 3821/85/EEC) and the technical specification in its Annex 1B were finally published in the *Official Journal* of the European Communities (the *OJ*) on 5 August 2002. This date signalled the start of an official 24-month count-down to the date when such instruments were to be legally required in new goods vehicles and certain older vehicles when their existing tachograph instruments fail and need to be replaced.

However, a series of delays for various technical and political reasons conspired to ensure that the tachograph manufacturing industry would be unable to meet the 5 August 2004 official fitment date. For this reason, in May 2004 it was announced from both Brussels and the DfT in the UK that the official fitment date would be put back one year to 5 August 2005. However, during the past year this controversial date has been subject to yet more changes: the latest being that a meeting of the EU Council of Transport Ministers held on 27 June 2005 decided to postpone the mandatory fitment of digital tachographs from 5 August 2005 until 1 January 2006. The UK government, in the meantime, has said that it will not require mandatory fitment of tachographs until such time as the situation becomes clear, but nevertheless, it has gone ahead and made new regulations effective from 5 August 2005 as required under EU law; namely, *The Passenger and Goods Vehicles (Recording Equipment) Regulations 2005* (SI 2005 No. 1904). These regulations effectively take account of EC Regulation 2135/98, which introduced the digital tachograph, and particularly added new provisions regarding enforcement by VOSA examiners and the police, powers to seize documents and copy electronic data, and new powers to enter vehicles and premises. New penalties are introduced for relevant offences such as failing to provide records or obstructing an enforcement officer, permitting the falsification of records, and supplying or installing a device designed to interfere with the function of the tachograph instrument. Maximum fines at level 5 on the standard scale (ie £5000) may be imposed on conviction for such offences.

The DVLA has announced that it will be delaying the issue of smart cards from June 2005 until September 2005. The following vehicles will be required to be fitted with digital tachographs:

- all new goods vehicles over 3.5 tonnes gross weight (apart from those specifically exempt under EU legislation); and
- existing 'in-service' vehicles over 12 tonnes gross weight in which the existing analogue tachograph fails and has to be replaced.

It is thought likely that some vehicle manufacturers will start installing these instruments in new vehicles prior to the 'due' date. Driver smart cards are to be issued by the Driver and Vehicle Licensing Agency (DVLA) in sufficient time to allow for training and familiarization prior to the official launch date.

Further details of the new provisions are included in this chapter starting on page 125.

LEGAL REQUIREMENTS FOR EXISTING ANALOGUE TACHOGRAPHS

The fitment of tachographs and their use in relevant vehicles (ie those operating within the scope of the EU driving hours rules – see Chapter 3) for record-keeping purposes, originally became a legal requirement in the United Kingdom on 31 December 1981. EU regulations (3821/85/EEC) which came into effect on 29 September 1986 amended some of the original requirements.

The legislation requires the fitment and use of tachographs in most goods vehicles and other vehicles used for the carriage of goods over 3.5 tonnes permissible maximum weight with certain EU-approved exemptions as listed in the following section of this chapter. Since the list of exemptions is limited, many categories of goods vehicle which may be thought to be exempt are not so.

It should be stressed that unless a vehicle is used for private purposes only or is specifically exempted as shown by the exemption list, the law applies to:

- any goods vehicle over 3.5 tonnes gvw, or a combination of a goods vehicle and goods-carrying trailer which together exceed 3.5 tonnes permissible maximum weight; and
- any other vehicle (such as off-road vehicles and so-called 4x4s) used for the carriage of goods which when towing a trailer has a combined permissible maximum weight exceeding 3.5 tonnes.

In particular, it should be noted that there is no exemption for short-distance operations, infrequent-use vehicles or occasional driving – once a relevant vehicle is on the highway the law applies in full.

This means that a tachograph instrument must be installed in the vehicle and that whosoever drives it must keep a tachograph record and observe the EU drivers' hours law in full both for the day on which the driving takes place and for the week in which that day falls. It should also be noted that vehicles which are exempt from the tachograph rules are not necessarily exempt from record-keeping requirements (for example, those operating under the British domestic hours rules). See Chapter 4 for details of activities where written records must be kept.

Under the EU regulations a number of specific basic requirements relating to tachograph use must be met as follows:

- The tachograph instrument must conform to the technical specification laid down in the EU regulations (3821/85/EEC Annex I – see pp 125–129).
- The instrument must be calibrated and officially sealed at an approved calibration centre to ensure that accurate (ie legally acceptable) records are made (see pp 112–115).
- The instrument must be used in accordance with the regulations, with individual responsibilities being observed by both employer and driver.
- Electronic tachographs must be capable of detecting interruptions in the power supply.
- Driving time must be recorded automatically.
- Facilities must be provided for the removal and subsequent refitting, by an approved centre, of recording equipment seals to enable speed limiters to be fitted.
- Cables connecting electronic tachographs to the transmitter (ie sender unit) must be protected by a continuous steel sheath (ie a tamper-proof armoured cable).

EXEMPTIONS

There is no requirement for the fitment and use of tachographs in vehicles in the following list or in vehicles used in connection with the particular transport operations specified in the exemption list.

Recent changes in regulations (ie The Community Drivers' Hours and Recording Equipment (Amendment) Regulations 1998 – effective from 24 August 1998) have brought within scope of the tachograph rules vehicles which are not in themselves considered to be goods vehicles (eg 4-wheel-drive off-road vehicles) and therefore were previously exempt from the rules. However, when such vehicles are drawing a trailer for the carriage of goods for commercial purposes, and the combined weight of the towing vehicle and trailer exceeds 3.5 tonnes, the fitment and use of tachographs as described in this chapter is necessary and the driver must observe the EU drivers' hours law as set out in Chapter 3.

NB: It should be noted that this is the same list of exemptions as that for the EU hours' law under Regulation 3820/85/EEC (see p 78):

National and International Exemptions

1. Vehicles not exceeding 3.5 tonnes gross weight including the weight of any trailer drawn.
2. Passenger vehicles constructed to carry not more than nine persons including driver.
3. Vehicles on regular passenger services on routes not exceeding 50 kilometres.
4. Vehicles with legal maximum speed not exceeding 30 kph (approx 18.6 mph).
5. Vehicles used by armed services, civil defence, fire services, forces responsible for maintaining public order (ie police).
6. Vehicles used in connection with sewerage; flood protection; water, gas and electricity services; highway maintenance and control; refuse collection and disposal; telephone and telegraph services; carriage of postal articles; radio and television broadcasting; detection of radio or television transmitters or receivers.

NB: Under a European Court ruling, British Gas must fit tachographs to vehicles it uses for the delivery of gas appliances, gas cylinders and meters to ensure fair competition with private transport operators. The point at issue, and contended by British Gas, was that its vehicles engaged on such deliveries were being used in connection with gas services.

Also: Vehicles used for private waste collection (ie not on behalf of local authorities) on journeys exceeding 50 km radius from the place where they are normally based must be fitted with a fully calibrated tachograph, following a European Court of Justice ruling.

7. Vehicles used in emergencies or rescue operations.
8. Specialized vehicles used for medical purposes.
9. Vehicles transporting circus and funfair equipment.
10. Specialized breakdown vehicles.
11. Vehicles undergoing road tests for technical development, repair or maintenance purposes, and new or rebuilt vehicles which have not yet been put into service.
12. Vehicles used for non-commercial carriage of goods for personal use (ie private use).
13. Vehicles used for milk containers or milk products intended for animal feed.

Further Exemptions in National Operations only

14. Passenger vehicles constructed to carry not more than 17 persons including the driver.

15. Vehicles used by public authorities to provide public services which are not in competition with professional road hauliers.
16. Vehicles used by agricultural, horticultural, forestry or fishery* undertakings, for carrying goods within a 50 km radius of the place where the vehicle is normally based, including local administrative areas the centres of which are situated within that radius.

**NB: To gain this exemption the vehicle must be used to carry live fish or to carry a catch of fish which has not been subjected to any process or treatment (other than freezing) from the place of landing to a place where it is to be processed or treated.*

17. Vehicles used for carrying animal waste or carcasses not intended for human consumption.
18. Vehicles used for carrying live animals from farms to local markets and vice versa, or from markets to local slaughterhouses.
19. Vehicles specially fitted for and used:
    - as shops at local markets and for door-to-door selling;
    - for mobile banking, exchange or savings transactions;
    - for worship;
    - for the lending of books, records or cassettes;
    - for cultural events or exhibitions.
20. Vehicles (not exceeding 7.5 tonnes gvw) carrying materials or equipment for the driver's use in the course of his work within 50 km radius of base provided the driving does not constitute the driver's main activity and does not prejudice the objectives of the regulations.
21. Vehicles operating exclusively on islands not exceeding 2300 sq km not linked to the mainland by bridge, ford or tunnel for use by motor vehicles (this includes the Isles of Wight, Arran and Bute).
22. Vehicles (not exceeding 7.5 tonnes gvw) used for the carriage of goods propelled by gas produced on the vehicle or by electricity.
23. Vehicles used for driving instruction (but not if carrying goods for hire or reward).
24. Tractors used after 1 January 1990 exclusively for agricultural and forestry work.
25. Vehicles used by the RNLI for hauling lifeboats.
26. Vehicles manufactured before 1 January 1947.
27. Steam-propelled vehicles.
28. Vintage vehicles (ie over 25 years old) not carrying more than nine persons including the driver, not being used for profit, and being driven to or from a vintage rally, museum, public display or a place where it has been or is to be repaired, maintained or tested.

NB: In the exemption above relating to vehicles used by public authorities, the exemption applies only if the vehicle is being used by:

(a) a health authority in England and Wales, a health board in Scotland or a National Health Service (NHS) Trust

- *to provide ambulance services in pursuance of its duty under the NHS Act 1977 or NHS (Scotland) Act 1978; or*
- *to carry staff, patients, medical supplies or equipment in pursuance of its general duties under the Act;*

(b) a local authority to fulfil social services functions, such as services for old persons or for physically and mentally handicapped persons;

(c) HM Coastguard or lighthouse authorities;

(d) harbour authorities within harbour limits;

(e) airports authority within airport perimeters;

(f) British Rail, London Regional Transport, a Passenger Transport Executive or local authority for maintaining railways;

(g) British Waterways Board for maintaining navigable waterways.

NB: Item 4 above includes certain works trucks and industrial tractors which have a statutory 30 kph speed limit imposed upon them but this does not include forklift trucks which may come within the scope of the rules.

Declaration of Exemption

When presenting a vehicle for the goods vehicle annual test which the operator believes is exempt from the tachograph regulations in accordance with the list above, a 'Declaration of Exemption' form has to be completed.

EMPLOYERS' RESPONSIBILITIES

The employer needs to determine whether his transport operation and his vehicles fall within scope of the EU tachograph requirements – by reference to the exemption list above – and take appropriate steps regarding the fitment and calibration of instruments as described in this chapter. He must also instruct his drivers accordingly. Additionally, the regulations place specific responsibilities on the employer of a driver who drives within the EU rules as follows:

- The employer must organize the driver's work in such a way that he is able to comply with both the drivers' hours and tachograph rules.
- The employer must supply drivers with sufficient numbers of the correct type of tachograph charts (ie one chart for the day, one spare in case the first is impounded by an enforcement officer, plus any further spares which are necessary to account for any charts which become too dirty or damaged to use), and he must ensure that completed charts are collected from drivers no later than 21 days after use.
- The employer must periodically check completed charts to ensure that the law has been complied with (ie that the driver has made a chart for the day, that he has completed it fully and properly and that he has observed the driving hours rules). If breaches of the law are found the employer must take appropriate steps to prevent their repetition.
- The employer must retain completed charts for 12 months for inspection by VOSA examiners, if required.

- The employer must give copies of the record to drivers concerned who request them.

A number of recent court cases have highlighted the extent of employer responsibilities for tachograph operation as follows:

- It has been made clear that employers who do not check tachograph records are permitting drivers' hours offences and can be prosecuted and convicted accordingly.
- Employers can be charged with failing to use the tachograph in accordance with the regulations in cases where a driver is unable to produce charts to show his driving and other work activities (eg when requested to do so in a roadside check).
- Where employers allow drivers to take their tractive units home after their day's work they must ensure that such driving is recorded on a tachograph chart and is counted as part of the driver's legally permitted driving and working time for that day – it is not part of his rest period.

DRIVERS' RESPONSIBILITIES

Drivers of vehicles operating within the EU rules must observe the tachograph requirements. In particular this means understanding what the law requires and how to comply with it. The specific responsibilities of the driver in regard to the law are as follows:

- Drivers using tachograph charts must ensure that a proper record is made by the instrument:
  - that it is a continuous record;
  - that it is a 'time right' record (ie recordings are in the correct 12-hour section of the chart – daytime or night-time hours).
- In the event of instrument failure or in circumstances where no vehicle is available when the driver is working he must make manual recordings of his activities on the chart 'legibly and without dirtying' it.
- Drivers must produce for inspection on request by an authorized inspecting officer a current chart for that day plus the charts relating to the current week and for the last day of the previous week in which he drove.
- Drivers must return completed charts to their employer no later than 21 days after use.
- Drivers must allow any authorized inspecting officer to inspect the charts and tachograph calibration plaque which is usually fixed inside the body of the instrument.

In cases where drivers take their tractive units home at the end of their working shift, this time must be recorded on the tachograph chart and counted as part of the daily maximum driving time and the driver's day's work – it is not part of his rest period.

Two-Crew Operation

Reference above to a driver also includes any other driver who is carried on the vehicle to assist with the driving. In this case a two-man tachograph must be fitted and both drivers must use it to produce records as follows:

- The person who is driving must have his chart located in the uppermost (ie number 1) position in the instrument and use the number 1 activity mode switch to enable his activities and vehicle speed and distance recordings to be made on the chart as appropriate.
- The person who is riding passenger must have his chart in the rearmost (ie number 2) position and must use the number 2 activity mode switch to record his other work activities, or break or rest periods. Only time-group recordings are made on this chart; driving, speed and distance traces are *not* produced on the second-man chart.

NB: Following a recent High Court ruling it has become clear that the second man on a double-manned vehicle operation must insert his chart in the 2nd-man position of the instrument from the commencement of the journey, not from the later time when he commences his period of driving.

TACHOGRAPH CALIBRATION, SEALING AND INSPECTION

To make legally acceptable records, tachograph installations in vehicles must be calibrated initially at an approved tachograph centre (see Appendix VI) and subsequently must be inspected every two years and fully recalibrated every six years or after repair at an approved centre.

NB: Problems have arisen over the date when two-yearly checks on tachographs are due. Operators of new vehicles have missed these dates having assumed that the two years runs from when they first took delivery of the vehicle, but in fact it runs from the date of original calibration which may have been some months earlier.

VOSA specifies and approves the premises (and the display of approved signs), equipment, staff (including their training) and procedures for the installation, repair, inspection, calibration and sealing of tachographs. Such centres must be approved to the BS 5750 Part 2 quality assurance standard before they can gain VOSA approval. No other workshops or individuals are permitted to carry out such work and any work carried out by unauthorized agents would render the installation incapable of producing legally acceptable records.

Tachograph Installation Offences

It is an offence for any unauthorized person to carry out work on tachograph installations. It is also an offence for a vehicle operator (maximum fine £5,000) to obtain and use a tachograph instrument repaired by a firm which is not BS 5750 approved.

Calibration

The calibration process requires the vehicle to be presented to an approved tachograph centre in normal road-going trim, complete with body and all fixtures, unladen, and with tyres complying with legal limits as to tread wear and inflated to manufacturer's recommended pressures. At the centre, the necessary work on the installation is carried out to within specified tolerances.

The regulations specify tolerances within which the tachograph installation must operate and valid for temperatures between 0° and 40°C as follows:

	On bench test	On installation	In use
Speed	3kph	4kph	6kph
Distance	1%	2%	4%
Time	in all cases, ± 2 minutes per day with a maximum of 10 minutes per 7 days		

In the case of both speed and distance figures shown above, the tolerance is measured relative to the real speed and to the real distance of at least one kilometre.

Calibration and Periodic Inspection Fees

Official fees charged for tachograph calibration and periodic inspections are as follows:

- Calibration £45.00 plus VAT
  (the official time for this task being 1½ hours)
- Two-yearly inspection £31.00 plus VAT.

NB: These prices are exclusive of any replacement parts used.

Sealing of Tachographs

Approved centres seal tachograph installations after calibration or two-yearly inspections with their own official seals (each of which is coded differently) and details of all seals are maintained on a register by VOSA. The seals are of the customs type whereby a piece of wire is passed through each of the connecting points between the vehicle and the tachograph itself and then a lead seal is squeezed tight on to the wire with special pliers which imprint the centre code number in the metal. Attempts to remove any of the seals or their actual removal will show and need to be accounted for.

The purpose of sealing is to ensure that there is no tampering with the equipment or any of its drive mechanism or cables which could either vary the recordings of time, speed or distance or inhibit the recording in any way. Such tampering is illegal and once seals are broken the installation no longer complies with the law and legally acceptable records cannot be made.

Besides the seals inside the body of the instrument head, the following points are sealed:

- the installation plaque;
- the two ends of the link between the recording equipment and the vehicle;

- the adaptor itself and the point of its insertion into the circuit;
- the switch mechanism for vehicles with two or more axle ratios;
- the links joining the adaptor and the switch mechanism to the rest of the equipment;
- the casings of the instrument;
- any cover giving access to the means of adapting the constant of the recording equipment to the characteristic coefficient of the vehicle.

Seal Breakage

Obviously, there are occasions when certain of the seals have to be broken of necessity to carry out repairs to the vehicle and replacement of defective parts (eg the vehicle clutch or gearbox). The only seals which may be broken in these circumstances are as follows:

- those at the two ends of the link between the tachograph equipment and the vehicle;
- those between the adaptor (ie the tachograph drive gearbox) and the point of its insertion into the circuit;
- those at the links joining the adaptor and the switch mechanism (ie where the vehicle has a two-speed rear axle) to the rest of the equipment.

With the introduction of the statutory speed-limiter fitment on many heavy vehicles, the EU now permits operators to break tachograph seals for the purpose of fitting such devices, but the seals must be replaced – at an approved tachograph centre – within seven days.

While it is permitted to break the particular seals listed above for other authorized purposes (eg in connection with vehicle maintenance), a written record must be kept of the seal breakage and the reason for it. The installation must be inspected or recalibrated and fully sealed following repair or seal breakage as soon as 'circumstances permit' and before the vehicle is used again. It is illegal to remove any of the other seals and tampering with seals by drivers is tantamount to committing fraud (ie for the purposes of making fraudulent records, which is an offence liable to lead to a prison sentence).

Under new arrangements, VOSA is planning to check that all tachograph head seals are intact when vehicles are submitted for their annual goods vehicle test. This tougher regime is intended to crack down on rogue operators who are fiddling their tachographs to avoid compliance with both statutory speed limits and the drivers' hours rules. Following trial inspections in three of its regions VOSA found that tachograph instruments in up to one in five vehicles had been illegally interfered with to produce false recordings.

Calibration Plaques

When a tachograph has been installed in a vehicle and calibrated, the approved centre must fix, either inside the tachograph head or near to it on the vehicle dashboard in a visible position, a plaque giving details of the centre, the 'turns count', and the calibration date. The plaque is sealed and must not be tampered with or the sealing tape removed. When an instrument is subjected to a two-year inspection or recalibration a new plaque must be fitted. If a vehicle is found on the road with an 'out-of-date' plaque an offence will have been committed and prosecution may follow.

The normal sequence for plaques is that one will show the initial calibration date, the next (two years later), which is fitted alongside the first plaque, will show the date of the two-year inspection and a third plaque will show the second two-year inspection. After a further two years a six-year recalibration of the installation will be due and at this time all the previous plaques will be removed, the new calibration plaque will be fitted and the procedure described above starts again. In between times, following certain repairs, a 'minor work' plaque may be fitted but this does not alter the sequence of dates for the two-year inspection and the six-year re-calibration plaques.

The two-year and six-year periods referred to above for inspections and calibrations are counted to the day/date, *not* to the end of the month in which that day/date falls.

TACHOGRAPH BREAKDOWN

If tachograph equipment becomes defective (or the seals are broken for whatever reason including authorized breakage, as described above, to carry out mechanical repairs to the vehicle, or unauthorized interference) it must be repaired at an approved centre as soon as 'circumstances permit', but in the meantime the driver must continue to record manually on the chart all necessary information regarding his working, driving, breaks and rest times which are no longer being recorded by the instrument. There is *no* requirement to attempt to record speed or distance.

Once a vehicle has returned to base with a defective tachograph, it should not leave again until the instrument is in working order and has been recalibrated (if necessary) and the seals replaced. If it cannot be repaired immediately, the vehicle can be used so long as the operator has taken positive steps (which he can satisfactorily prove later if challenged by the enforcement authorities – see paragraph below) to have the installation repaired as soon as reasonably practicable.

If a vehicle is unable to return to base within *one week* (ie seven days) counting from the day of the breakdown, arrangements must be made to have the defective instrument repaired and recalibrated as necessary at an approved centre *en route* within that time.

Defence

There is a defence in the regulations against conviction (ie not against prosecution) for an offence of using a vehicle with a defective tachograph. This has the effect of allowing subsequent use of a vehicle with a defective tachograph provided steps have been taken to have the installation restored to a legal condition as soon as circumstances permit and provided the driver continues to record his driving, working and break period times manually on a tachograph chart. In such circumstances it will be necessary to satisfactorily prove to the enforcement authorities – and to the court if they proceed with prosecution – that a definite booking for the repair had already been made at the time the vehicle was apprehended and that this appointment was for the repair to be carried out at the earliest possible opportunity.

It is also a defence to show that at the time it was examined by an enforcement officer the vehicle was on its way to an approved tachograph centre to have necessary repairs carried out. However, this defence will fail if the driver did not keep written records of his activities in the meantime.

USE OF TACHOGRAPHS

Drivers are responsible for ensuring that the tachograph instrument in their vehicle functions correctly throughout the whole of their working shift in order that a full and proper recording for a full 24 hours can be produced. They must also ensure that they have sufficient quantities of the right type of charts (see below) on which to make recordings.

Time Changes

Drivers must ensure that the time at which the instrument clock is set and consequently recordings are made on the chart agree with the official time in the country of registration of the vehicle. This is a significant point for British drivers travelling in Europe who may be tempted to change the clock in the instrument to the correct local European time rather than, for example, having it indicate and record the time in Britain. To re-emphasize the point, this means that for British drivers in British-registered vehicles the tachograph chart recording must accord with the official time in Britain regardless of the country in which that recording was made – the tachograph clock must *not* be reset to show local time when travelling abroad.

Dirty or Damaged Charts

If a chart becomes dirty or damaged in use, it must be replaced and the old chart should be securely attached to the new chart which is used to replace it.

Completion of Centre Field

Before starting work with a vehicle in which tachograph charts are to be used the driver must enter on the centre field of his chart for that day the following details:

- his surname and first name (not initials);
- the date and place where use of the chart begins;
- vehicle registration number;
- the distance recorder (odometer) reading at the start of the day.

At the end of a working day, the driver should then record the following information on the chart:

- the place and date where the chart is completed;
- the closing odometer reading;
- by subtraction, the total distance driven – in kilometres.

Making Recordings

When the centre field has been completed the chart should be inserted in the tachograph instrument ensuring that it is the right way up (it should be impossible to fit it wrongly) and that the recording will commence on the correct part of the 24-hour chart (day or night). The instrument face should be securely closed. The activity mode switch (number 1) should be turned as necessary throughout the work period to indicate the driver's relevant activities, namely driving, other work, break or rest periods. While some drivers find difficulty in getting into the habit of turning the switch to coincide with each change of activity, nevertheless this is what the law requires and it is an offence to fail to do so (ie not keeping proper records).

Other Work and Overtime Recordings

The High Court has ruled that drivers must record *all* periods of work on their tachograph charts. This makes it clear that work undertaken for the employer after the daily driving shift has been completed (eg in a yard, warehouse, workshop or office), and whether deemed part of the normal day or overtime working, must be recorded on the chart for that day – ie either by the instrument if this is convenient or otherwise manually. A contention that since tachographs were intended primarily to record driving time in the interests of road safety they should be used for that purpose only and not for recording other working activities was ruled to be contrary to the provisions of both the EU legislation and section 97 of the Transport Act 1968.

Overnight Recordings

At the end of his shift the driver can leave the chart in the instrument overnight to record the daily rest period or alternatively it can be removed and the rest period recorded manually on the chart. Generally, enforcement staffs prefer an automatic recording of daily rest made by the instrument but this is not always practicable where the vehicle may be used on night-shift work, may be driven for road testing or other purposes by workshop staff, or moved around the premises by others when the driver is at home having the rest period. Also, if the driver is scheduled to start work later on the following day, there will be an overlap recording on the chart which is illegal.

Vehicle Changes

If the driver changes to another vehicle during the working day he must take the existing chart with him and record details of the time of change, the registration number of the further vehicle(s) and distance recordings in the appropriate spaces on the chart. He then uses that chart in the next vehicle to record his continuing driving, working activities and break periods. This procedure is repeated no matter how many different vehicles (except those not driven on the public highway) are driven during the day so the one chart shows all of the driver's daily activity.

Mixed Tachographs

It is important to note that the various makes and models of tachograph currently available in the UK have different charts and they cannot all be interchanged. So the driver who switches from a vehicle with one make of instrument to a vehicle with a different make during the working day will have to make fresh entries on a second or even third chart. At the end of the day all the charts used should be clipped together to present a comprehensive (and legal) record for the whole day. However, there are some charts now available on the market suitable for dual use in different makes of tachograph. It is the employer's duty to issue drivers with the correct charts (ie with matching type approval numbers to those on the tachograph instrument in use) in sufficient numbers for the schedule which the driver has to operate.

NB: Where instruments are standardized in a fleet, one chart will suffice for the whole day and the driver must take this with him from one vehicle to the next, recording changes as previously described.

Manual Records

Drivers are responsible for ensuring that the instrument is kept running while they are in charge of the vehicle and should it fail or otherwise cease making proper records they should remove the chart and continue to record their activities manually on it as previously described. They must also make manual recordings on the chart of work done or time spent away from the vehicle (for example, periods during the day spent working in the yard, warehouse or workshop). Manual recordings must be made legibly and in making them the sheet must not be 'dirtied'.

Records for Part-time Drivers

The rules on the use of tachographs described in the foregoing text apply equally to part-time or occasional drivers such as yard and warehouse staff, office people and even the transport manager. The rules also apply fully even if the driving on the road is for a very short distance or period of time – a five-minute drive without a tachograph chart in use would be sufficient to break the law and risk prosecution. Vehicle fitters and other workshop staff who drive vehicles on the road for testing in connection with repair or maintenance are specifically exempt from the need to keep tachograph records when undertaking such activities (see exemption list on pp 78–80) but this exemption *does not* apply to them when using vehicles for other purposes (eg collecting spare parts, ferrying vehicles back and forth, taking replacement vehicles out to on-road breakdowns, taking and collecting vehicles to and from goods vehicle test stations, etc).

Retention, Return and Checking of Tachograph Charts

Drivers must retain and be able to produce, on request by authorized examiners (including the police), completed tachograph charts for each driving day of the current week and for the last day of the previous week on which they drove.

Remember, charts do not have to be made for non-driving days or rest days and therefore cannot be asked for by the police or others in respect of such days.

Charts must be returned by drivers to their employer no later than 21 days after use and, on receiving the charts, the employer must periodically check them to ensure that the drivers' hours and record-keeping regulations have been complied with. Failure by an operator to check charts for possible offences by drivers can lead to charges of 'permitting' the commission of certain tachograph and drivers' hours offences (should such offences be proven) – in one case a transport manager was held to be 'reckless' because his chart-checking system was not sufficiently thorough. Charts must then be retained, available for inspection if required, for a period of one year.

Where a driver has more than one employer in a week (eg as with agency drivers), he must return the tachograph charts to the employer who first employed him in that week. This provision clearly presents a problem for those firms which regularly employ agency drivers and which may find difficulty in securing the return of charts for driving work done with their vehicles. At the present time, there is no legal solution to this issue.

Retention of Charts for Tax Records

Where tachograph charts are used by employers to justify payment of driver night-out and other subsistence expenses for tax purposes, then the charts constitute part of the legal recording system for tax purposes and as such must be retained for six years instead of the normal one-year period as described above.

Official Inspection of Charts

An authorized inspecting officer (which means a traffic examiner of VOSA or a person authorized by a Traffic Commissioner (TC), in either case on production of their authority if requested, or a police officer) may require any person to produce for inspection any tachograph chart on which recordings have been made. Further, he may enter a vehicle to inspect a chart or a tachograph instrument (he should be able to read the recordings relating to the nine hours prior to the time of his inspection) and the calibration plaques and detain a vehicle for this purpose. At any reasonable time he may enter premises on which he believes vehicles or tachograph charts are kept and inspect the instruments in such vehicles and the completed charts. He can require (by serving a notice in writing) charts to be produced at a Traffic Area Office at any time on giving at least 10 days' notice in which to do so. Where a chart is suspected of showing a false entry he may 'seize' the chart (but not for reasons other than evidence of a false entry, or an entry intended to deceive, or an entry altered for such purposes) and retain it for a maximum period of six months, after which time, if no charges for offences have been made, the chart should have been returned. If not, the person from whom it was taken can apply to a magistrate's court to seek an order for its return.

In practice, TCs regularly ask operators to provide batches of tachograph charts covering one or more of their vehicles for a short period or possibly some months either on a routine basis or following investigations into, or leads about, hours' law or tachograph infringements – and where it is suspected that drivers are regularly exceeding speed limits. These charts are then analysed for contraventions of the law. The courts have ruled that examiners do have powers to remove tachograph charts from operators' premises when acting on the instructions of a TC – this follows the successful prosecution of a haulage firm that refused to allow charts to be removed from its premises.

Cases have been reported where the police have demanded that operators should send in to them, by post, tachograph charts required for inspection. This has been shown to be an illegal practice: there is no provision in the law which allows random collection of charts by the police or demands for charts to be submitted by post. Only in cases where an on-the-spot examination reveals possible offences can the police then request copies of the relevant chart for that day, other charts for that week and for the last day of the previous week. VOSA enforcement officers, on the other hand, may take random selections of charts away for examination.

It is an offence to fail to produce records for inspection as required or to obstruct an enforcement officer in his request to inspect records or tachograph installations in vehicles.

OFFENCES

Some tachograph-related offences have already been mentioned in connection with specific requirements of the law but there are other overriding, and very serious, offences to be considered. In particular, it is an offence to use, cause or permit the use of a vehicle which does not have a fully calibrated tachograph installed; for the driver to fail to keep records by means of a tachograph (or manually if the instrument is defective) and to make false recordings. Further, it is an offence for the driver to fail to return used charts to his employer within 21 days after use or to fail to notify his first employer of any other employer for whom he drives vehicles to which the regulations apply.

In many cases offences committed by the driver result in charges against the employer for 'causing' or 'permitting' offences. For example, where tachograph charts go missing and cannot be produced for examination by the enforcement authorities, the employer may be charged with any one of three (or even all three) relevant offences; namely, failing to cause the driver to keep a record, failing to preserve the record, failing to produce the record. It is in such cases that sound legal representation should be sought and a good defence put forward where possible.

Two other types of offence that have come to the forefront is recent times and have featured in extensively reported prosecutions concern:

- missing mileages when charts are compared;
- interference of tachograph systems by various means of wires and so-called 'magic buttons'.

Penalties on summary conviction for offences under these regulations can be a fine of up to level 4 on the standard scale (currently £2,500) and conviction for such offences can jeopardize both the employers' 'O' licence and the driver's LGV driving entitlement. Conviction for the more serious offences of making false entries on a tachograph chart and forgery can result in level 5 fines of up to £5,000 (per offence) or imprisonment for up to two years. Since April 1998 tachograph falsification has been a 'recordable' offence, meaning that a person convicted of such an offence carries a criminal record.

TCs have made it clear than heavy goods drivers convicted for using wires to interfere with their tachograph will be disqualified from holding an lgv driving licence for a period of one year.

Much heavier penalties are likely for those transport operators and LGV drivers found guilty in UK courts of forging tachograph charts and chart entries relating to international journeys. These penalties are contained in the provisions of the Forgery and Counterfeiting Act 1981 which the Court of Appeal now allows the police to use for bringing prosecutions for tachograph chart offences committed outside the UK. Fines may be in excess of the current level 5 standard scale maximum of £5,000, and custodial sentences longer than those available to the courts under other legislation.

The courts have ruled that in cases where a driver or transport operator claims not to have known that the tachograph instrument in a vehicle had been fitted with a device intended to interfere with its operation (ie a trip device to aid the production of false records) this was no excuse and an offence of 'strict liability' was committed to which there is no defence.

Tachograph Charts as Evidence

In the past it has been made clear that tachograph charts, while providing evidence of 'the facts shown' for drivers' hours purposes, would not be used by the police and enforcement authorities as evidence to bring prosecutions against drivers for speeding offences – not to be confused with the TCs' actions in imposing short-term LGV driving bans on drivers found from their charts to have regularly exceeded maximum speed limits. However, the DETR is currently considering the possibility of changing the law to permit the retrospective checking of tachograph charts for speeding and prosecution of drivers where such evidence is shown. This follows publicity surrounding a number of serious coach and lorry crashes where speeding was thought to be a contributory factor.

THE TACHOGRAPH INSTRUMENT

A tachograph is a cable or electronically driven speedometer incorporating an integral electric clock and a chart recording mechanism. It is fitted into the vehicle dashboard or in some other convenient visible position in the driving cab. The instrument indicates time, speed and distance and permanently records this information on the chart as well as the driver's working activities. Thus, the following factors can be determined from a chart:

- Varying speeds (and the highest speed) at which the vehicle was driven.
- Total distance travelled and distances between individual stops.
- Times when the vehicle was being driven and the total amount of driving time.
- Times when the vehicle was standing and whether the driver was indicating other work, break or rest period during this time.

Recordings

Recordings are made on special circular charts, each of which covers a period of 24 hours (Figure 5.1), by three styli. One stylus records distance, another records speed and the third records time-group activities as determined by the driver turning the activity mode switch on the head of the instrument (ie driving, other work, breaks and rest periods). The styli press through a wax recording layer on the chart, revealing the carbonated layer (usually black) between the top surface and the backing paper. The charts are accurately pre-marked with time, distance and speed reference radials and when the styli have marked the chart with the appropriate recordings these can be easily identified and interpreted against the printed reference marks.

Movement of the vehicle creates a broad running line on the time radial, indicating when the vehicle started running and when it stopped. After the vehicle has stopped, the time-group stylus continues to mark the chart but with an easily distinguishable thin line. The speed trace gives an accurate recording of the speeds attained at all times throughout the journey, continuing to record on the speed base line when the vehicle is stationary to provide an unbroken trace except when the

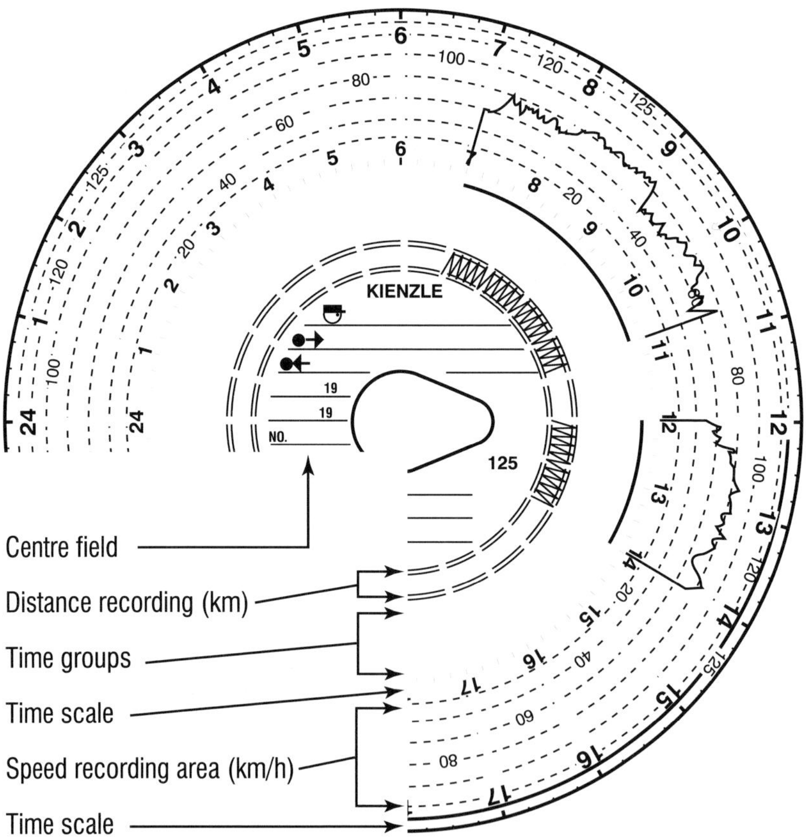

Figure 5.1 *A typical tachograph chart showing recordings of times, distance and speed*

instrument is opened. The distance recording is made by the stylus moving up and down over a short stroke, each movement representing 5 kilometres travelled; thus, every 5 kilometres the stylus reverses direction, forming a 'V' for every 10 kilometres of distance travelled. To calculate the total distance covered the 'V's are counted and multiplied by 10 and any 'tail ends' are added in, the total being expressed in kilometres.

When a second chart is located in the rear position of a two-man tachograph, only a time recording of the second man's activities (ie other work, break or rest) is shown. Traces showing driving, vehicle speed or distance cannot be recorded on this chart.

Precautions against interference with the readings are incorporated in the instrument. It is opened with a key and a security mark is made on the chart every time the instrument is opened.* When checking the chart it can be easily established at what time the instrument was opened and thus whether this was for an authorized reason or not. Interference with the recording mechanism to give false readings, particularly of speed, can be determined quite simply by an experienced chart analyst.

**NB: Changes to the EU tachograph specification to reduce the possibility of fraudulent recordings were introduced to apply to new instruments receiving Type Approval from 1991. These include a provision for the chart to be marked at every interruption of the power supply (eg when a fuse is removed).*

Faults

Tachographs are generally robust instruments, but listed below are some of the faults which may occur:

- failure of the cable drive at the vehicle gearbox;
- failure of the cable drive at the tachograph head;
- failure of the adaptor/corrector/triplex gearbox;
- cable breaking or seizure;
- electrical faults affecting lights in the instrument or the clock;
- incorrect time showing on the 24-hour clock (eg day-shift work becomes shown against night hours on the charts);
- failure of the tachograph head;
- damage to the recording styli;
- failure of the distance recorder;
- damage to charts because of incorrect insertion.

Fiddles

A key feature of tachograph recordings is that careful observation will show results of the majority of faults in recordings as well as fiddles and attempts at falsification of recordings by drivers. The main faults likely to be encountered will show as follows:

- Clock stops – recordings continue in a single vertical line until the styli penetrate the chart.
- Styli jam/seize up – recordings continue around the chart with no vertical movement.

- Cable or electronic drive failure – chart continues to rotate and speed and distance styli continue to record on base line and where last positioned respectively. Time-group recordings can still be made but no driving trace will appear.

Attempts at falsification of charts will appear as follows:

- Opening the instrument face will result in a gap in recordings.
- Winding the clock backwards or forwards will leave either a gap in the recording or an overlap. In either case the distance recording will not match up if the vehicle is moved.
- Stopping the clock will stop the rotation of the chart so all speed and distance recordings will be on one vertical line (see item 1 above about how faults in instruments show on charts).
- Restricting the speed stylus to give indications of lower than actual speed will result in flat-topped speed recordings while bending the stylus down to achieve the same effect will result in recordings below the speed base line when the vehicle is stationary.
- Written or marked-in recordings with pens or sharp pointed objects are readily identifiable by even a relatively unskilled chart analyst.

NB: This is only an outline list of a large number of possible faults and attempts at falsification likely to be encountered. Some driver fiddles are one-off attempts, crudely and clumsily executed and naively obvious; others are much more sophisticated in their execution, often as part of an ongoing violation of legal requirements. These are more difficult, but not impossible, for the transport manager or fleet operator to detect and would certainly be picked up quickly by an experienced chart analyst.

EU Instruments and Charts

Tachographs may only be used for legal record-keeping purposes if they are type-approved and comply with the detailed EU specification. Such instruments have provision for indicating to the driver, without the instrument being opened, that a chart has been inserted and that a continuous recording is being made. They also provide for the driver to select, by an activity mode switch on the instrument, the type of recording which is being made. In the UK this must be one of the following:

- driving time;
- other work time;
- break and rest periods.

Two-man instruments are also provided with a means of simultaneously recording the activities of a second crew member on a separate chart located in the rear position in the instrument.

The charts used for legal purposes must also be type-approved as indicated by the appropriate 'e' markings printed on them. It is illegal to use non-approved charts or charts which are not approved for the specific type of instrument being used. Care should be taken that charts used have accurate time registration – cheap and non-type-approved versions, which are illegal anyway, have been found in the past to be significantly inaccurate in the way they are printed, thus producing inaccurate and worthless records.

CHART ANALYSIS

Analysis of the information recorded on tachograph charts can provide valuable data for determining whether drivers have complied with the law on driving, working, break and rest period times, and have conformed to statutory speed limits. The data can also be extremely useful as a basis for finding means of increasing the efficiency of vehicle operation and for establishing productivity monitoring and payment schemes for drivers.

Many fleet operators use and rely upon the services of tachograph analysis agencies for checking their charts for conformity with the law. However, it should be noted that should such firms fail to recognize and notify the operator of deficiencies in their records, it is the operator who is at risk and his licence, not the analysis bureau. The TCs repeatedly remind operators that responsibility for driver compliance with the hours law rests entirely with them, not outside agencies. Generally, also, the checking carried out by such firms is for standard hours law infringements only, which are mainly picked up by computerized analysis and may not include identification of other irregularities or cleverly executed false entries or fraudulent recordings. Similarly, such analysis may not identify driver abuse of vehicles or frequent and excessive speeding.

Tachograph manufacturers supply accessories to enable detailed chart analysis to be carried out. A chart analyser magnifies the used chart to the extent that detailed analysis beyond the scope of a normal visual examination can be made of the vehicle's minute-by-minute and kilometre-by-kilometre progress. Journey times, average running times and speeds, delivery times, route miles, traffic delays and many other relevant factors can be readily established. With the aid of a fixed hairline cursor on the magnifier to allow precise definition of the time and speed scales and recordings on the chart, even vehicles' rates of acceleration and deceleration can be determined.

The German company, Siemens VDO (formerly Kienzle), which is the leading tachograph manufacturer, has undertaken considerable research into chart analysis and is able to offer users the service of its analysis experts both in Germany and the UK as well as in other countries to determine the activities of vehicles and particularly the progress of a vehicle immediately prior to an accident and at the point of impact. In some instances such analysis has shown that witnesses' accounts of the speed of the vehicle and its braking force just before the accident have been far from correct.

It is also claimed that by detailed analysis of the charts and by keeping drivers aware of the information obtained, driving methods can be improved, thus saving fuel and cutting down on the wear and tear on vehicle brakes, tyres, transmission and other components.

DIGITAL TACHOGRAPHS

New EC regulations (1360/2002/EC) amending the current tachograph regulation (3821/85/EC) by the introduction of a new generation of so-called digital tachographs were published on 5 August 2002 to come into effect on 5 August

2004 (now deferred to 1 January 2006). From this new date all new goods vehicles exceeding 3.5 tonnes gross weight (except those specifically exempt under the regulations) will need to be fitted with the new-type digital instruments in which the driver inserts his or her own, personalized, micro-chip 'smart' card on which his or her driving, working and rest activities will be held, along with vehicle speeds and distances driven, for up to 28 days.

There is no requirement for retrospective fitment of digital tachographs to existing vehicles except that where, after 1 January 2006, a pre-existing vehicle exceeding 12 tonnes gross weight suffers tachograph failure requiring replacement, that replacement will have to be a new-type digital instrument.

The introduction of digital tachographs will herald a new regime in the European tachograph market with at least four key players competing for a share in place of the current virtual duopoly in the UK where Kienzle and Veeder Root have been the main sellers. The principal manufacturers of EU specification digital tachographs are likely to be as follows:

- Siemens VDO – formerly Kienzle.
- Stoneridge – manufactured in Dundee, Scotland.
- ACTIA – formerly manufactured by French company Thomson.
- Delphi Grundig (a new entrant to the market).

The Regulations

The new regulations (1360/2002/EC) which are outlined here have been introduced for a number of reasons including:

- to prevent infringement and fraud in application of the driving hours rules;
- to monitor automatically driver performance and behaviour;
- to overcome problems of monitoring compliance due to the numbers of individual record sheets (charts) that have to be held in the vehicle cab;
- the need to introduce advance recording equipment with electronic storage devices and personal driver cards to provide an indisputable record of work done by the driver over the last few days and the vehicle over a period of several months; and
- to devise a system which ensures total security of the recorded data.

Definitions for Digital Tachographs and Ancillaries

The new-type equipment will feature a number of key components defined in the regulations as follows:

- *Recording equipment* is amended to mean the total equipment intended for installation in road vehicles to show, record and store automatically or semi-automatically details of the movement of such vehicles and of certain work periods of their drivers. This equipment includes cables, sensors, an electronic driver information device, one (two) card reader(s) for the insertion of one or two driver memory card(s), an integrated or separate printer, display instruments, facilities for downloading the data memory, facilities to display or print information on demand, and facilities for the input of the places where the daily work period begins and ends.

- *Data memory* means an electronic storage system built into the recording equipment, capable of storing at least 365 calendar days from the recording equipment. The memory should be protected in such a way as to prevent unauthorized access to and manipulation of the data and detect any such attempt.
- A '*driver card with memory*' means a removable information transfer and storage device allocated by the authorities of the Member States to each individual driver for the purposes of identification of the driver and storage of essential data. The format and technical specifications of the driver card must meet the requirements laid down in the technical annex to the regulations (not yet adopted).
- A '*control card*' means a removable data transfer and storage device for use in the card reader of the recording equipment, issued by the authorities of the Member States to competent authorities to get access to the data stored in the data memory or in the driver cards for reading, printing and/or downloading.
- A '*company data card*' means a removable data transfer and storage device issued by the Member State's authorities to the owner of vehicles fitted with recording equipment. The company data card allows for displaying, downloading and printing of the data stored in the recording equipment fitted in the company's vehicle(s).
- *Downloading* means the copying of a part of or a complete set of data stored in the data memory of the vehicle or in the memory of the driver card, but which does not alter or delete any stored data, allows for the origin of downloaded data to be authenticable and to be kept in a format that can be used by any authorized person, and ensures that any attempts to manipulate data are detectable.

Functions of Digital Instruments

The regulations require digital instruments to be able to record, store, display and print-out specified statutory information as follows.

Recording and Storing in the Data Memory

The instrument will be required to record and store:

- Distance travelled by the vehicle with an accuracy of 1 km.
- Speed of the vehicle:
  - momentary speed of the vehicle at a frequency of 1 s for the last 24 hours of use of the vehicle;
  - exceeding the authorized speed of the vehicle, defined as any period of more than 1 minute during which the vehicle speed exceeds 90 km/h for N3 vehicles or 105 km/h for M3 vehicles (with time, date, maximum speed of the overspeeding, average speed during the period concerned).
- Periods of driving time (times and dates), with an accuracy of 1 minute.
- Other periods of work or of availability (times and dates) with an accuracy of 1 minute.
- Breaks from work and daily rest periods (times and dates) with an accuracy of 1 minute.

- For electronic recording equipment which is equipment operated by signals transmitted electrically from the distance and speed sensor, any interruption exceeding 100 milliseconds in the power supply of the recording equipment (except lighting), in the power supply of the distance and speed sensor, and any interruption in the signal lead to the distance and speed sensor, with date, time, duration and driver card issue number.
- The driver card issue number with times and dates of insertion and removal.
- For each driver card that is inserted for the first time after it was used in another item of recording equipment:
  - current driving time since the last break or rest period;
  - driving time for the day after the last rest period of at least eight hours;
  - driving times for the day between two rest periods of at least eight hours for the preceding 27 calendar days with date, time and duration;
  - total of the driving times for the current week and the preceding week and the total of the driving times of the two completed preceding weeks;
  - rest periods of at least eight hours' duration for the day and the preceding 27 calendar days in each case with date, time and duration;
  - the VRN (vehicle registration number) of the vehicles driven.
- Date, time and duration of driving without an inserted or a functioning driver card.
- Data recorded on the places at which the daily work period began and ended.
- Automatically identifiable system faults of the recording equipment with date, time and driver card issue number.
- Faults in the driver card with date and time and driver card issue number.
- Workshop card number of the authorized fitter or workshop with date of at least the last installation inspection and/or periodic inspection of the recording equipment.
- Control card number with date of control card insertion and type of control (display, printing, downloading). In case of downloading, period downloaded should be recorded.
- Time adjustment with date, time and card issue number.
- Driving status (single/crew driving – driver/co-driver).

Storage on the Driver Card

The driver card must be capable of storing:

- the essential data for a period of at least the last 28 calendar days combined with the VRN identification of the vehicle driven and the data as required above;
- the events and faults mentioned above with the VRN identification of the vehicle driven;
- the date and time of insertion and removal of the driver card and distance travelled during the corresponding period;
- the date and time of insertion and removal of the co-driver card with issue number.

Data must be recorded and stored on the driver card in such a way as to rule out any possibility of falsification.

Recording and Storing for Two Drivers

Where vehicles are used by two drivers the driving time must be recorded and stored on the driver card of the driver who is driving the vehicle. The equipment must record and store in the data memory and on the two driver cards simultaneously, but distinctly, details of the information listed above.

Displaying or Printing for an Authorized Examiner

The equipment must be capable of displaying or printing, on request, the following information:

- Driver card issue number, expiry date of the card.
- The surname and first name of the driver who is the cardholder.
- Current driving time since the last break or rest period.
- Driving time for the day after the last rest period of at least eight hours.
- Driving times for the day between two rest periods of at least eight hours for the preceding 27 calendar days on which the driver has driven, with date, time and duration.
- Total of the driving times for the current week and the preceding week and the total times for the two completed preceding weeks.
- The other periods of work and availability.
- Rest periods of at last eight hours' duration for the day and the preceding 27 days in each case with date, time and duration.
- VRN identification of vehicles driven for at least the last 28 calendar days with the distance travelled per vehicle and day, time of first insertion and last removal of the driver card and the time of change of vehicle.
- Time adjustment with date, time and card issue number.
- Interruption of power supply to the recording equipment with date, time, duration and driver card issue number.
- Sensor interruption with date, time, duration and driver card issue number.
- The VIN and/or VRN identification of the vehicle driven.
- Driving without driver card as defined above for the last 28 calendar days.
- Details of the information stored concerning the driver.
- Recorded data on the places where the daily work period began and ended.
- The automatically identifiable system faults of the recording equipment with date, time and driver card issue number.
- The faults in the driver card with date and time and driver card issue number.
- Control card number with date of control card insertion and type of control (display, printing, downloading). In the case of downloading, period downloaded should be recorded.
- Exceeding the authorized speed as defined above, with date, time and driver card issue number for the current week and in any case including the last day of the previous week.
- Summary reports to permit compliance with the relevant regulations to be checked.

Smart Cards

Crucial to the whole system is the driver ('smart') card that is personal to the individual and carries identification information and other essential data about him. It has a capacity to store relevant data on driving and working times, breaks and rest periods covering at least 28 days, and will comprise the legal record in place of the current tachograph chart. The card itself must be tamper proof, and strict regulatory systems are to be established by national governments to prevent fraudulent issue, use and transfer of cards within their territories.

A tachograph **'smart card' (officially called a 'tachograph card')** is a plastic card similar in size to a photocard driving licence or credit card, with a microchip embedded into it. There will be a set of four such cards for use with digital tachographs as follows:

- driver cards – which the driver will insert at the commencement of his journey to record his driving and working activities;
- company cards – for use by vehicle operators to protect and download the data.
- workshop cards – available only to approved tachograph calibration centres;
- control cards – for use only by VOSA examiners and the police for carrying out enforcement activities.

Before **commencing a journey** the driver will be required to insert his driver card into the 1st (driver) or 2nd (co-driver) slot on the front of the Vehicle Unit (VU) in much the same way that a chart is put into the head of an analogue tachograph. The 'centre field' details, which the driver currently has to enter on the chart, with digital instruments, will be recorded automatically by the tachograph – ie driver name, vehicle registration number, start and finish odometer readings and name of place code.

In the same way that drivers and co-drivers currently **record their different activities** – driving, other work, breaks and rest – by changing the mode switch and by swapping the position of charts in the tachograph head, with digital tachographs the mode switch will have to be turned and digital smart cards will need to be swapped between driver and co-driver slots in the instrument when double-manning a vehicle.

Details of time spent **working away** from the vehicle that are currently written on the rear of the tachograph chart (ie record sheet) will have to be input manually into the digital tachograph. The system will also record details of any faults, interference, errors and over-speeding that occur.

All this information will be stored for at least **28 days** on the driver's personal smart card and for at least a year in the VU.

Fees for Digital Tachograph Cards (ie Smart Cards)

The following table of fees has been set by the DVLA for digital tachograph 'smart' cards.

Transaction	*Driver Card*	*Company Card*	*Workshop Card*	*Control Card*
Card application	£38	£38	£0	£0
Renewal	£19	£19	£0	£0
Replacement – lost/stolen	£19	£19	£0	£0
Exchange	£0	£38	£0	£0
Malfunction	£0	£0	£0	£0

NB: Further information on digital tachographs can be found on the official Government website: www.digitaltachograph.gov.uk

TYPICAL DIGITAL EQUIPMENT

In advance of introduction of statutory requirements for the mandatory fitment of digital tachographs, the key tachograph manufacturers have shown prototype models built to the specification included in Regulation 1360/2002/EC. Typical of such equipment is that made by Siemens VDO (formerly Kienzle) comprising a recording unit incorporating a mass memory fitted into the vehicle instrument panel. This is connected to an intelligent sensor to enable transmission of 'driving' pulses to the recording unit and to the instrument cluster for displaying road speed, time and distance travelled. Connections are provided (ie data interfaces) to allow readout of the stored data via an office PC or a laptop computer, and for it to be printed out to hard copy via a suitable printer where required. Various warning functions are incorporated and a keypad connection can be made to input additional commands and relevant data.

6 Driver Licensing and Licence Penalties

DRIVER LICENSING

Any person wishing to drive a motor (ie mechanically propelled) vehicle on a public road in the UK or within Europe must hold a licence showing a driving entitlement (either full or provisional) for the relevant category of vehicle. Specifically in the case of goods vehicle driving, drivers must hold a current licence showing a relevant lgv vocational driving entitlement (either full or provisional). This is the legal responsibility of the individual concerned and heavy penalties are imposed on any person found to be driving either without a licence, without a licence covering the correct category of vehicle, or while disqualified from driving by a court.

Furthermore, it is the responsibility of the employer of any person required to drive for business purposes to ensure that such employee-drivers, irrespective of their function, status or seniority, are correctly licensed to drive company vehicles. The fact that a driver may be disqualified or has allowed his licence to lapse without the employer knowing is no defence for the employer against prosecution on a charge of allowing an unlicensed person to drive a vehicle.

Offences

The law states that it is an offence to drive, or to cause or permit another person to drive, a vehicle on the road without a current and valid driving licence. In the case of large goods vehicles (lgvs), it is an offence to drive, or to cause another person to drive, without a valid licence (ie full or provisional) giving entitlement to drive vehicles in categories C or C+E.

Checking Licences

Haulage employers are advised by the police to check their drivers' licences on a regular basis – always initially when giving a driver a job and then at least once every three to six months – otherwise they leave themselves wide open to prosecution for a range of licensing offences and for the vehicle insurance to be invalidated. When checking licences only an original licence should be accepted, never a photocopy, check that all of a driver's Christian names shown on the licence match those in the company record, and that the date of birth shown also coincides with that recorded at the time of employment.

Employers can check the validity of a driving licence and verify the holder's driving entitlements by contacting the DVLA's Data Subject Enquiry Department at Swansea for a printout of the details currently held on file. It is necessary to obtain written authorization from the individual concerned before making application and this authorization together with a fee of £3.50 must be submitted. Regular enquirers can speed up the process by paying £35.00, for 10 enquiries, in advance. For details contact the DVLC on Tel: 01792 310075. Form P57 which lists conviction codes and what they stand for can also be obtained free from the DVLA.

Where a driver has a new-type photocard licence, it is important to examine both the photocard and the accompanying 'counterpart'. One without the other is not valid.

Invalidation of Insurance

Driving without a current and valid driving licence covering the category of vehicle being driven can invalidate insurance cover (which is itself an offence and is usually included among the charges for an unlicensed person driving) and could result in any accident or damage claim being refused by the insurance company under the terms of its policy contract – which is invariably conditional upon the law being complied with in full.

DRIVER LICENSING DIRECTIVES

Brussels has devised yet another set of proposals for a so-called 'third' driver licensing directive. This document was published in October 2003 and has been followed in the UK by a consultation document seeking views and comments on the commission's proposals. The consultation period ended in April 2004 after which the D*f*T intended formulating a response to Brussels.

The current EU framework, which sets the driving licence rules, is contained in the 'second' directive (Directive 91/439/EEC) – see below. Among other things, it sets out the knowledge, skills and behaviour required for driving a motor vehicle, the minimum requirements for driving tests and the minimum standards for physical and mental fitness to drive a motor vehicle. The new proposals are more concerned with procedural and administrative issues such as regularizing (ie harmonizing) the 80-odd different types of driving licence and entitlements across the 25 member states of the enlarged EU that are currently valid and in circulation.

However, despite the seemingly benign intent of the proposed directive, two key points that do strike home are the following two proposals. Firstly, that drivers of medium and large goods vehicles should be required to have a compulsory medical examination every five years up to the age of 65 years (currently such examinations are only required five-yearly after age 45 years) and annually after age 65 years (as now) – such a provision is estimated by the Freight Transport Association to be likely to cost the transport industry an extra £5 million annually. Secondly, that the maximum weight threshold for medium weight goods vehicles should be reduced from the present 7.5 tonnes to just 6.0 tonnes.

The Second Driver Licensing Directive

The provisions of the so-called 'second' driver licensing directive (EC Directive 439/91/EEC) were partially introduced on 1 July 1996 (ie the new theory test for car drivers and motorcycle riders) with the remainder of the provisions coming into force from 1 January 1997.

Mainly this Directive affects driving tests rather than driver licensing and as such is covered fully in Chapter 7, but certain licensing changes were included (see below), and controversial new eyesight standards for vocational drivers are part of the package – see p 152. The new measures broadly affect both existing and new licence holders as follows:

- Pre-existing holders of category C and D driving entitlements when driving such vehicles are restricted to towing trailers not exceeding 750 kg maximum weight. If they wish to haul trailers in excess of this weight they will have to take a further test to obtain a C+E or D+E entitlement.
  NB: Drivers holding licences showing category C and/or category D entitlements which permit trailers up to 5 tonnes MAM to be drawn are nevertheless restricted to towing trailers of no more than 750 kg authorized mass.
- New drivers who pass the category B car test will be restricted to driving cars and light goods vehicles up to 3.5 tonnes gross weight (and vehicles with no more than eight passenger seats) and towing a trailer up to a maximum of 750 kg gross weight (the maximum weight of the combination must not exceed 3.5 tonnes and the permissible weight of the trailer must not exceed the unladen weight of the towing vehicle).
- New drivers wishing to drive vehicles of gross weights over 3.5 tonnes and up to 7.5 tonnes will have to take a new form of category C1 driving test on a goods vehicle of 4 tonnes minimum weight and capable of a speed up to 80 kph.
  (It is stressed here that existing holders of a driving entitlement for vehicles up to 7.5 tonnes gross weight are not affected.)
- New drivers wishing to drive vehicles in category C1 (ie over 3.5 tonnes to 7.5 tonnes) with a trailer attached and for which a category C1+E entitlement is required, will have to first pass a test on a category C1 vehicle (see above) which is towing a trailer of at least 2 tonnes gross weight and making a total combination length of at least 8 metres.
- Once a C1+E entitlement is gained, and providing the person is at least 21 years of age, this allows them to drive a vehicle up to 12 tonnes gross weight (ie maximum authorized mass – see p 139–40).
- The previous 18 year age limit for driving vehicles of up to 7.5 tonnes gross weight in category C1+E is maintained.
- New drivers seeking category C1 and C1+E driving entitlements are required to meet the medical standards currently applicable for full category C and C+E entitlements.

- Driving licences are now coded to indicate whether the holder wears spectacles or contact lenses when driving.
- The new system of theory testing applies to all new drivers of cars, motorcycles, goods and passenger vehicles – see Chapter 7.

The licensing provisions described in this chapter are principally contained in:

- the Road Traffic Act 1988;
- the Road Traffic (Driver Licensing and Information Systems) Act 1989;
- the Road Traffic (New Drivers) Act 1995;

and the following regulations:

- The Motor Vehicles (Driving Licences) Regulations 1996 as amended;
- The Motor Vehicles (Driving Licences)(Large Goods and Passenger Carrying Vehicles) Regulations 1990 (which deals with entitlements to drive large goods vehicles over 7.5 tonnes gross weight and passenger-carrying vehicles which are used for hire and reward operations), plus a number of subsequent amendments;
- The Motor Vehicles (Driving Licences) Regulations 1999;
- The Motor Vehicles (Driving Licences) Amendment (No. 4) Regulations 2003.

The licence-scheme changes brought about by this legislation involved significant change in vocational (ie HGV/PSV) licensing. Particularly, the terminology changed so that heavy goods vehicles (HGV) and public service vehicles (PSV) as they were previously known are now called large goods vehicles (LGV) and passenger carrying vehicles (PCV), respectively, and goods vehicles are no longer classified by the number of axles.

Definitions

For driver licensing purposes:

- a 'large goods vehicle' is 'a motor vehicle (not being a medium-sized goods vehicle) which is constructed or adapted to carry or haul goods and the permissible maximum weight of which exceeds 7.5 tonnes';
- a medium-sized goods vehicle is defined as one having a permissible maximum weight exceeding 3.5 tonnes but not exceeding 7.5 tonnes;
- a large passenger-carrying vehicle is a vehicle constructed or adapted to carry more than 16 passengers;
- a small passenger-carrying vehicle is a vehicle which carries passengers for hire or reward and which is constructed or adapted to carry more than 8 but not more than 16 passengers.

Unified Licences

Where previously drivers had separate *licences* to drive large vehicles, their qualifications to drive them now are referred to as *entitlements* within the unified licence scheme and are shown in the single combined (ie unified) licence document along with their ordinary (ie car, light goods vehicle and any motorcycle) driving entitlements.

The previous British system of vehicle groups (shown on ordinary licences) and classes (shown on HGV licences) has been changed to the EU system of vehicle 'categories' (see table on pp 138–39). Applicants for new or renewed HGV/PSV licences are issued with a 'unified' licence showing their goods (LGV) and/or passenger carrying vehicle (PCV) driving entitlements under these new categories.

DRIVING LICENCES

Major changes to Britain's ordinary and HGV driving licence schemes were introduced in 1990/1 as the UK harmonized with EU requirements. Since 1 April 1991 the pink and green European model, 'unified' driving licence (Euro-licence), which shows all of an individual's entitlements to drive (ie for motorcycle, car, light and large goods vehicles, and passenger carrying vehicles as appropriate), has been issued from the DVLA, Swansea. Traffic Commissioners (TCs) no longer have responsibility for issuing large goods and passenger vehicle licences. The Euro-licence carries the words 'European Communities Model' and has 'Driving Licence' printed in the 11 languages of the EU (besides the language of the country of issue – eg English), including Greek and Gaelic, on the front. British-issued Euro-licences also show, where appropriate, provisional driving entitlements and any endorsements of penalty points or licence disqualification made by the courts. This part of the document is called the 'counterpart' and is coloured green.

Green Licences Still Valid

National driving licences issued in EU member states are still recognized throughout the EU and existing British 'green' ordinary driving licences continue to be valid both in Britain and in other countries which recognize British licences. Eventually, when the 'Euro-licence' is fully in use, national licences will become obsolete.

Existing UK Licence Holders

For a large proportion of British ordinary driving licence holders (ie those who have no vocational – LGV or PCV – entitlements) who currently hold 'licences-for-life', the changes mentioned above will not be noticed unless or until they apply to change the details on their licence (eg to record a new address or, in the case of a woman, applying for a new licence to show her married name), when they will be issued with the new-style Euro-licence.

All existing entitlements to drive are maintained under the new licensing scheme so that no existing licence holder is deprived of his or her rights to drive particular vehicles, either now or in the future.

Photographs on Licences

New British-issued driving licences now carry a photograph of the holder (see below) as part of a campaign to eliminate misuse and fraud involving licences. Northern Ireland-issued driving licences already carry a photograph of the holder. Plastic photocard driving licences were launched on 23 July 1998. These will eventually replace the current licence, although existing paper licences will remain valid until their expiry or revocation.

Standard information is shown on the licence, such as the individual's name, address and date of birth, the vehicles entitled to drive, and the date of issue and expiry of the licence. All this is against a pink background in one of the EU's 11 official languages (plus one other language if required, eg Welsh or Gaelic). The nationality symbol for the country of issue is shown (in the case of United Kingdom-issued licences, 'UK' will be used rather than 'GB' as shown on vehicle nationality plates) on a blue background and surrounded by the EU's 12 gold stars together with the holder's photograph.

The new-type licence will be issued to new licence applicants and to existing licence holders who apply for replacements or who wish to change the details on their licence – existing licences will not be recalled for change.

Applicants are required to complete form D750 and submit this to Swansea along with a birth certificate or valid passport (original documents must be sent – photocopies are no longer acceptable) and a recent passport-type colour photograph. Both the form and photograph must be countersigned by a professionally qualified person such as a doctor, lawyer, teacher, police officer, MP or somebody of similar standing who must have known the applicant for at least two years.

Organ Donor Option

Since 1 March 1993 UK-issued Euro driving licences have incorporated an organ donor consent section (in the green counterpart of the licence document). Previously only provisional driving licences had this facility and separate donor cards were enclosed by the DVLA when sending out full driving licences. Completion of the section is entirely voluntary and both the D*f*T and the Department of Health have jointly confirmed that a consenting donor's organs would not, in any event, be removed without the prior permission of the deceased's next of kin. The separate organ donor card remains valid for existing licence holders.

The Issuing Authority

Responsibility for the issue of vocational driving entitlements rests with the Driver and Vehicle Licensing Agency (DVLA), Swansea. However, it should be noted that the TCs still retain a disciplinary role in regard to vocational entitlements as described on pp 145–46.

All applications in connection with driver licensing (ie for both ordinary and vocational driving entitlements) should be addressed to the DVLA, Swansea followed by the appropriate postal code ie:

- SA99 1AD for first provisional licences;
- SA99 1AB for renewals, duplicate and exchange licences;
- SA99 1BJ for first full car and motorcycle licences; and
- SA99 1BR for all vocational entitlement applications (including minibuses).

Further information on driver licensing can be obtained from the Customer Enquiries Unit, DVLA, Swansea SA6 7JL (tel: 01792 772151; fax: 01792 783071 for general enquiries about the licensing system). Ring 01792 782787 to use minicom owing to hearing difficulties.

Age Minimum for Drivers

Certain minimum ages are specified by law for drivers of various categories of motor vehicle as follows:

- Invalid carriage or moped* — 16 years
- Motor cycle other than a moped* (ie over 50 cc engine capacity) — 17 years
- Small passenger vehicle or small goods vehicle (ie not exceeding 3.5 tonnes gross weight and not adapted to carry more than nine people including the driver) — 17 years
- Agricultural tractor — 17 years
- Medium-sized goods vehicle (ie exceeding 3.5 tonnes but not exceeding 7.5 tonnes gross weight) — 18 years
- Other goods vehicles (ie over 7.5 tonnes gross weight) and passenger vehicles with more than nine passenger seats — 21 years

**NB: The definition of moped is as follows:*

- *In the case of a vehicle first registered before 1 August 1977 a motorcycle with an engine cylinder capacity not exceeding 50 cc which is equipped with pedals by means of which it can be propelled.*
- *In the case of a vehicle first registered on or after 1 August 1977 a motorcycle which does not exceed the following limits:*
  (a) maximum design speed 45 km/h;
  (b) kerbside weight 250 kg;
  (c) cylinder capacity (if applicable) 50 cc.

If a goods vehicle and trailer combination exceeds 3.5 tonnes permissible maximum weight the driver must be at least 18 years of age; if such a combination exceeds 7.5 tonnes the driver must be at least 21 years of age (and will need to also hold an LGV driving entitlement).

Members of the armed forces are exempt from the 21 years age limit for driving heavy goods vehicles when such driving is in aid of the civil community (the limit is reduced to 17 years). Similarly, exemption from the 21-year minimum age limit applies to learner LGV drivers of vehicles over 7.5 tonnes gross weight if they are undergoing registered training by their employer or by a registered training establishment. In this case the minimum age is reduced to 18 years.

Disabled young people who receive a Disability Living Allowance (ie mobility allowance) may drive cars at 16 years of age, provided they can do so safely.

Road Rollers

A person under 21 but not less than 17 years old may drive a road roller if it:

- is propelled by means other than steam;
- has an unladen weight of not more than 11,650 kg*;
- is fitted with metal or hard rollers;
- is not constructed or adapted to carry a load other than water, fuel, accumulators and other equipment used for the purpose of propulsion, loose

tools, loose equipment and any object which is specially constructed for attachment to the vehicle so as to increase, temporarily, its unladen weight.

**NB: If this weight is exceeded the minimum age for driving a roller is 21 years.*

Agricultural Tractors

A person under 17 but over 16 may drive an agricultural tractor only if it is:

- of the wheeled type;
- not more than 2.45 metres wide including the width of any fitted implement;
- specially licensed for excise duty purposes as an agricultural machine;
- not drawing a trailer other than one of the two-wheeled or close-coupled four-wheeled type which is not more than 2.45 metres wide.

A 16-year-old must not drive an agricultural tractor on a road unless he has passed the appropriate test.

VEHICLE CATEGORIES/GROUPS FOR DRIVER LICENSING

For driver licensing purposes vehicles are defined according to specified groupings or categories which are shown on licences by means of capital letters as indicated in the following lists.

British Licence Groups

Pre-existing full, British-type ordinary (ie non-vocational) driving licences – many of which are still in existence – cover one or more of the following vehicle groups:

Group	Class of vehicle	Additional groups covered
A	vehicle without automatic transmission, of any class not included in any other group	B, C, E, F, K and L
B	A vehicle with automatic transmission, of any class not included in any other group	E, F, K and L
C	Motor tricycle weighing not more than 425 kg unladen, but excluding any vehicle included in group E, J, K or L, E, K and L	
D	Motor bicycle (with or without sidecar) but excluding any vehicle included in group E, K or L	C, E and motor cycles in group L
E	Moped	
F	Agricultural tractor, but excluding any vehicle included in group	K H
G	Road roller	–
H	Track-laying vehicle steered by its tracks	–
J	Invalid carriage	–
K	Mowing machine or pedestrian-controlled vehicle	–
L	Vehicle propelled by electrical power, but excluding any vehicle included in group J or K	K
M	Trolley vehicle	–
N	Vehicle exempted from duty and used for less than 6 miles per week between parts of owner's land	

EU Vehicle Categories

Under the Euro-licensing system, the old British system of vehicle groups as listed above has been replaced by EU vehicle categories listed below. All future licences will specify driving entitlements against these vehicle categories.

Category	Vehicle type
Motorcycles	
A	Motorcycles (with or without sidecar) and scooters but excluding vehicles in category K. *Additional categories covered: B1, K, P*
A1	Light motorcycles not over 125 cc and 11 kW (14.6 bhp). *Additional category covered: P*
Cars and light vans	
B	Motor vehicles not exceeding 3.5 tonnes mass and with not more than eight seats (excluding the driver's seat) including drawing a trailer not exceeding 750 kg mass. Including combinations of category B vehicles and a trailer where the combined weight does not exceed 3.5 tonnes and the weight of the trailer does not exceed the unladen weight of the towing vehicle. *Additional categories covered: F, K, P*
B1	Motor tricycles and three/four-wheeled cars and vans up to 550 kg unladen with a design speed not exceeding 50 kph and if fitted with an internal combustion engine, a cubic capacity not exceeding 50 cc. *Additional categories covered: K, P*
B+E	Motor vehicles in category B drawing a trailer over 750 kg where the combination does not come within category B.
Medium goods vehicles	
C1	Medium goods vehicles between 3.5 tonnes and 7.5 tonnes (including drawing trailer not exceeding 750 kg – maximum weight of the combination must not exceed 8.25 tonnes.
C1+E	Medium goods vehicles between 3.5 tonnes and 7.5 tonnes and drawing a trailer over 750 kg but does not exceed the unladen weight of the towing vehicle – maximum weight of the combination must not exceed 12 tonnes. *Additional category covered: B+E*
Large goods vehicles	
C	Large goods vehicles over 3.5 tonnes (but excluding vehicles in categories D, F, G and H) including those drawing a trailer not exceeding 750 kg.
C+E	Large goods vehicles in category C drawing a trailer exceeding 750 kg. *Some C+E licences, where the holder was previously qualified to drive vehicles in old HGV class 2 or 3, show a restriction limiting driving to drawbar combinations only.* *Additional category covered: B+E*
Minibuses	
D1	Passenger vehicles with between 9 and 16 seats including drawing trailer not exceeding 750 kg.
D1+E	Motor vehicles in category D1 drawing a trailer over 750 kg – the weight of the trailer must not exceed the unladen weight of the towing vehicle and the maximum weight of the combination must not exceed 12 tonnes. *Additional category covered: B+E*
Passenger vehicles	
D	Passenger vehicles with more than 8 seats including drawing a trailer not exceeding 750 kg.
D+E	Passenger vehicles in category D drawing a trailer over 750 kg. *Additional category covered: B+E*
Other vehicles	
F	Agricultural or forestry tractors but excluding any vehicle in category H.
G	Road rollers.
H	Track-laying vehicles steered by their tracks.
K	Mowing machine or pedestrian-controlled vehicle with up to three wheels and not over 410 kg.
L	Electrically propelled vehicles.
P	Mopeds.

NB: In the above table, vehicle/trailer weights, unless otherwise specified, are to be taken as the maximum authorized mass (ie mam) which is the same as the permissible maximum weight (pmw) for the vehicle/trailer – commonly referred to as the 'gross weight'.

Restricted Category for Pre-1997 Drivers

Drivers of goods vehicles with trailers in category C1+E who passed their category B test prior to 1 January 1997 are restricted to a total combination weight of 8250 kg. Driving a combination up to the 12 tonne limit requires an additional test.

Restricted Categories for Post-1997 Drivers

Since 1 January 1997, new drivers passing the car and light vehicle test (ie with vehicles not exceeding 3.5 tonnes permissible maximum weight) for the first time are not permitted to drive vehicles above this weight without securing additional driving categories on their licence. It is stressed that this restriction to 3.5 tonne driving applies *only* to those who first pass their test since this date – it will not be applied retrospectively to existing licence holders.

Drivers who pass their car test (ie category B) are not permitted to drive:

- minibuses (in category D1);
- medium-sized goods vehicles (in category C1); or
- tow large (ie over 750 kg) trailers (in categories B+E, C1+E and D1+E).

They must take a further test if they wish to drive such vehicles or vehicle combinations.

Any driver wishing to drive a vehicle towing a heavy trailer (ie one with a gross weight over 750 kg) must first pass a test in the associated rigid vehicle. Learner drivers in categories B, C1, C, D1 and D cannot drive a vehicle towing a trailer of any size.

NB: The Driving Standards Agency has warned that 'heavy trailers' in this context includes twin-axled caravans, horse box trailers and boat trailers. If such trailers are towed, a full B+E driving entitlement is required.

Towed and Pushed Vehicles

It has been ruled that a person who steers a vehicle being towed (whether it has broken down or even has vital parts missing, such as the engine) is 'driving' the vehicle for licensing purposes and therefore needs to hold current and valid driving entitlement covering that category of vehicle. Conversely, it has been held that a person pushing a vehicle from the outside (ie with both feet on the ground) is not 'driving' a vehicle, nor are they 'using' the vehicle.

Incomplete Vehicles

Drivers of incomplete goods vehicles comprising a chassis and cab only (ie before bodywork is fitted) and of articulated tractive units not yet fitted with a fifth-wheel coupling need (from 1 January 1998) must hold either a category C1 driving entitlement for such vehicles weighing between 3.5 and 7.5 tonnes, or a category C entitlement for such vehicles weighing over 7.5 tonnes. Prior to this date incomplete vehicles could be driven on a category B licence covering motor cars and light vans.

Tractive Units

Drivers of heavy (ie over 3.5 tonne) articulated tractive units with no semi-trailer attached need hold only a category C driving entitlement – contrary to popular misconception, a category C+E driving entitlement is not required.

Learner Drivers

Learner drivers must hold a provisional driving entitlement to cover them while driving under tuition. This provisional driving entitlement is shown on the green 'counterpart' of the licence.

Full category C LGV entitlement holders, can use this entitlement in place of a provisional entitlement for learning to drive vehicles in category C+E (ie drawbar combinations and articulated vehicles). But it should be noted that full entitlements in categories B and C1 *cannot* be used as a provisional entitlement for learning to drive vehicles in categories C or C+E. A proper provisional entitlement for these classes is required.

Learner drivers must be accompanied, when driving on public roads, by the holder of a full entitlement covering the category of vehicle being driven (see also below) and must not drive a vehicle drawing a trailer, except in the case of articulated vehicles or agricultural trailers. (*NB: Full licence holder must have held the licence for 3 years.*)

An 'L' plate of the approved dimensions must be displayed on the front and rear of a vehicle being driven by a learner driver (see Chapter 7). Learners driving in Wales may alternatively display a 'D' plate.

Learner drivers of category B vehicles are not allowed to drive on motorways. However, learner LGV drivers seeking a licence for category C1, C and C+E vehicles and who hold full entitlements in licence categories B and C1 may drive such vehicles on motorways while under tuition.

Compulsory Retests for Offending New Drivers

From 2 June 1997, newly qualified drivers who tot-up six or more penalty points on their licence within two years of passing the test will revert to learner status (ie with the display of 'L' plates and the need to be accompanied by a qualified driver) and have to repass both the theory test and the practical driving test before regaining a full licence.

Supervision of 'L' Drivers

Qualified drivers who supervise learner drivers in cars and in light, medium and large goods vehicles must:

- be at least 21 years old;
- have held a full driving entitlement for a continuous period of at least three years (excluding any periods of disqualification); and
- for accompanying learner LGV drivers, have held a relevant entitlement (ie for the type of vehicle on which they are supervising) for at least 3 years.

Contravention of these requirements could lead to prosecution of the supervising driver and, on conviction, a fine of up to £400, the imposition of two driving licence penalty points and possibly licence disqualification.

Insurance Scheme for New Drivers

A government scheme of insurance incentives was launched in 1995 to encourage newly qualified drivers to take additional lessons after passing their driving test. The 'Pass Plus' scheme is aimed particularly at young drivers who have not previously taken out motor insurance. It comprises six training sessions, normally taken within 12 months of passing the test, and includes driving in all weather conditions, at night, on motorways and dual carriageways and in town. Some 20 insurance companies offer preferential insurance premiums to 'Pass Plus' certificate holders.

Exemptions from Vocational Licensing

Exemptions from the need to hold an LGV driving entitlement (ie in categories C or C+E) apply when driving certain vehicles as follows (in most cases such vehicles may be driven by the holder of a category B licence):

1. Steam-propelled vehicles.
2. Road construction vehicles used or kept on the road solely for the conveyance of built-in construction machinery.
3. Engineering plant, but not mobile cranes*.
4. Works trucks.
5. Industrial tractors.
6. Agricultural motor vehicles which are not agricultural or forestry tractors.
7. Digging machines.
8. Vehicles used on public roads only when passing between land occupied by the vehicle's registered keeper and which does not exceed an aggregate of 9.7 kilometres in a calendar week.
9. Vehicles, other than agricultural vehicles, used only for the purposes of agriculture, horticulture or forestry, between areas of land occupied by the same person and which do travel more than 1.5 kilometres on public roads.
10. Vehicles used for no purpose other than the haulage of lifeboats and the conveyance of the necessary gear of the lifeboats which are being hauled.
11. Vehicles manufactured before 1 January 1960 used unladen and not drawing a laden trailer.
12. Articulated goods vehicles with an unladen weight not exceeding 3.05 tonnes.
13. Vehicles in the service of a visiting military force or headquarters as defined in the Visiting Forces and International Headquarters (Application of Law) Order 1965.
14. Any vehicle being driven by a police constable for the purpose of removing it to avoid obstruction to other road users or danger to other road users or

members of the public, for the purpose of safeguarding life or property, including the vehicle and its load, or for other similar purposes.

15. Breakdown vehicles which weigh less than 3.05 tonnes unladen, provided they are fitted with apparatus for raising a disabled vehicle partly from the ground and for drawing a vehicle when so raised, are used solely for the purpose of dealing with disabled vehicles, and carry no load other than a disabled vehicle and articles used in connection with dealing with disabled vehicles.
16. A passenger-carrying vehicle recovery vehicle other than an articulated vehicle with an unladen weight of not more than 10.2 tonnes which belongs to the holder of a PSV 'O' licence, when such a vehicle is going to or returning from a place where it is to give assistance to a damaged or disabled passenger-carrying vehicle or giving assistance to or moving a disabled passenger-carrying vehicle or moving a damaged vehicle.
17. A mobile project vehicle, which is defined as a vehicle exceeding 3.5 tonnes pmw constructed or adapted to carry not more than eight persons in addition to the driver and which carries mainly goods or burden comprising play or educational equipment for children or articles used for display or exhibition purposes.

**NB: Drivers of mobile cranes must hold a full lgv vocational entitlement covering vehicles in category C1 for driving cranes between 3.5 and 7.5 tonnes and category C for driving cranes over 7.5 tonnes mpw (applicable since 1 January 1999).*

APPLICATION FOR LICENCES AND VOCATIONAL ENTITLEMENTS

Applications for driving licences have to be made to Swansea on form D1 for ordinary licences and form D2 for vocational licences. Forms are obtainable from main post offices, direct from Swansea via the internet or from Local Vehicle Registration Offices.

Questions on the form are concerned with personal details of the applicant, the type of licence required, any previous licence held and whether the applicant is currently disqualified. LGV/PCV entitlement applicants are asked about any convictions they may have recorded against them.

Health Declaration

Applicants are asked to declare information about their health, particularly as to whether they have:

- had an epileptic event (ie seizure or fit);
- sudden attacks of disabling giddiness, fainting or blackouts;
- severe mental handicap;
- had a pacemaker, defibrillator or anti-ventricular tachycardia device fitted;
- diabetes controlled by insulin;
- angina (heart pain) while driving;
- a major or minor stroke;
- Parkinson's disease;

- any other chronic neurological condition;
- a serious problem with memory;
- serious episodes of confusion;
- any type of brain surgery, brain tumour or severe head injury involving hospital in-patient treatment;
- any severe psychiatric illness or mental disorder;
- continuing or permanent difficulty in the use of arms or legs which affects the ability to control a vehicle safely;
- been dependent on or misused alcohol, illicit drugs or chemical substances in the previous three years (excluding drink/driving offences);
- any visual disability which affects both eyes (short/long sight and colour blindness do not have to be declared).

Applicants for LGV/PCV entitlements (unless submitting a medical report – form D4 – see below) are required to state whether they have:

- sight in only one eye;
- any visual problems affecting either eye;
- angina;
- any heart condition or had a heart operation.

Where licence applicants have previously declared a medical condition they are required to state what the condition is, whether it has worsened since it was previously declared and whether any special controls have been fitted to their vehicle since the last licence was issued.

The DVLA's Considerations for Vocational Entitlements

Applicants for vocational driving entitlements must meet specified conditions as follows:

- They must be fit and proper persons.
- They must meet laid-down eyesight requirements.
- They must satisfy a medical examination and specifically must not
  - have had an epileptic attack in the previous 10 years (see below), or
  - suffer from insulin-dependent diabetes.

The decision as to whether or not an applicant will be granted an lgv driving entitlement rests entirely with the DVLA and in making this decision it will take into account any driving convictions for motoring offences, drivers' hours and record offences, and offences relating to the roadworthiness or loading of vehicles against the applicant in the 4 years prior to the application, and any offence connected with driving under the influence of drink or drugs during the 11 years prior to the application. The applicant has to declare such convictions on the licence application form (D2) but the DVLA has means of checking to ensure that applicants have declared any such convictions against them.

TCs' Powers in Respect of Vocational Entitlements

Although the issue of vocational (ie LGV/PCV) entitlements is the prerogative of the DVLA, TCs still have powers to consider the fitness of persons applying for or holding such entitlements. This disciplinary role allows a TC to call upon applicants or entitlement holders to provide information as to their conduct (and if necessary to appear before him to answer in person), to refuse the grant of an entitlement, and to suspend or disqualify a person from holding such an entitlement. The TC's decision must be communicated to the person concerned, upon whom it is binding, and to the Secretary of State for Transport (effectively the DVLA) – see also p 137.

Date for Vocational Applications

Application for an LGV driving entitlement should be made not more than three months before the date from which the entitlement is required to run. Reminders will be sent out by the DVLA to existing licence/entitlement holders two months prior to the expiry date of their existing licence/entitlement.

MEDICAL REQUIREMENTS FOR VOCATIONAL ENTITLEMENTS

Strict medical standards for vocational entitlement holders are legally established to ensure that those wishing to drive large goods or passenger-carrying vehicles are safe to do so and are not suffering from any disease or disability (especially cardiovascular disease, diabetes mellitus, epilepsy, neurosurgical disorders, excessive sleepiness, nervous or mental disorders, vision problems, or the excessive use of prescribed medicines or illicit drugs, for example) which would prevent them from driving safely.

Even tougher medical standards for vocational entitlement holders were brought in from 1 January 1997 when the EU's 'second' driver licensing Directive came into force. In particular these concern eyesight (see p 152) and a number of other serious problems which are added to the list of medical disabilities that may result in failure of the medical examination and refusal of a driving licence (see below).

UK applicants for LGV driving entitlements must satisfy such medical standards on first application and subsequently. To do so, they must undergo a medical examination and have their doctor complete the medical certificate portion of the application form D4 not more than four months before the date when the entitlement is needed to commence.

A further examination and completed medical certificate is required for each 5-yearly renewal of the entitlement after reaching age 45 years. After reaching the age of 65 years a medical examination is required for each annual renewal of the entitlement. Further medical examinations may be called for at any time if there is any doubt as to a driver's fitness to drive.

The form D4 requests applicants' consent to allow the DVLA's medical adviser to obtain reports from their own doctor and any specialist consulted if this helps to establish their medical condition.

Driving Following Medical Disqualification

Where an application for a full driving licence is made following the revocation of a licence or the refusal of its renewal on medical grounds, the applicant may be required to take a driving test (or an 'on-road assessment') to determine whether he is fit to regain his licence. A provisional licence may be granted for this purpose (ie to authorize driving on the road), but its use is restricted to driving only during the period preceding and while taking the test, and its authority ceases immediately upon conclusion of the test or assessment.

Medicals for New Category C1 Drivers

Since 1 January 1997 new drivers of vehicles over 3.5 tonnes gross weight (ie covered by driving licence category C1) require the same medical examination that previously applied only to over 7.5 tonnes vocational licence holders and must follow the same regime as described above for subsequent medical examinations (ie five-yearly after age 45 years, and annually after age 65 years).

Medical Examination Fees

Doctors charge a fee for conducting such medical examinations which the candidate must pay himself. These examinations are not available on the National Health Service in the UK – the current BMA recommended fee is £58.50, although where an employer is paying and an invoice is required, the cost may rise to around £100. The medical fee for licence/entitlement renewal can be claimed as an allowable expense for income tax purposes.

Diabetes

Normally, insulin-dependent diabetes sufferers are barred from holding an LGV entitlement, but if they held an HGV driving licence, and the TC was aware of their condition prior to 1 January 1991, an entitlement may be granted. From 5 April 2001, new rules remove the blanket limitation on driving category C1 vehicles between 3.5 and 7.5 tonnes by persons who suffered from diabetes before 1998 and held a relevant licence at the end of 1996. The new rules allow individual assessment of drivers wishing to drive vehicles of this category and extend this to cover also vehicles in category C1+E. Any person who lost his C1 licence under the 1997 rules can reapply for a licence, as can new drivers and new insulin users.

Epilepsy

A person will now be prevented from holding an LGV/PCV entitlement *only* if he has a 'liability to epileptic seizures'. Applicants must satisfy the DVLA that:

- he has not suffered an epileptic seizure during the 10 years prior to the date when the entitlement is to take effect;
- no epilepsy treatment has been administered during the 10 years prior to the starting date for the entitlement; and
- a consultant nominated by the DVLA has examined the medical history and is satisfied that there is no continuing liability to seizures.

Car, light vehicle and certain other drivers (ie in licence categories A, B, B+E, F, G, H, K, L and P), but not LGV/PCV drivers, who suffer from epilepsy can obtain a licence to drive such vehicles provided they:

- have been free from an epileptic attack during the period of one year from the date the licence is granted; or,

- if not free from such an attack, had an asleep-attack more than three years before the date on which the licence is granted and have had attacks only while asleep between the date of that attack and the date when the licence is granted; and
- provided the DVLA is satisfied that driving by them will not cause danger.

Coronary Health Problems

Drivers who have suspected coronary health problems are permitted (since April 1992) to retain their LGV driving entitlements while medical enquiries are made. (Previously the rule was to revoke the licence pending enquiries into the holder's health.) Such drivers no longer have to submit to coronary angiography (ie angiogram testing). The DVLA says that ECG exercise tests will be undertaken no earlier than three months after a coronary event and providing the driver displays no signs of angina or other significant symptoms, he is allowed to keep his driving entitlement while investigations are made, but subject to the approval of his own doctor.

Drivers who have suffered, or are suffering from, the following heart-related conditions must notify the DVLA:

- heart attack (myocardial infarction, coronary thrombosis);
- coronary angioplasty;
- heart valve disease/surgery;
- coronary artery by-pass surgery;
- angina (heart pain);
- heart operation (other than a heart transplant).

The DVLA's Drivers Medical Branch has the following advice for heart sufferers:

- Following a heart attack or heart operation, driving should not be recommenced for at least one month following the attack or operation. Driving may be resumed after this time if recovery has been uncomplicated and the patient's own doctor has given his or her approval.
- A driver suffering from angina may continue to drive (whether or not he is receiving treatment) unless attacks occur while driving, in which case he must notify the DVLA immediately (see below) and *stop driving*.
- A driver who suffers sudden attacks of disabling giddiness, fainting, falling, loss of awareness or confusion must notify the DVLA immediately (see below) and *stop driving*.

Any driver who has doubts about his ability to continue to drive safely is advised to discuss the matter with his own doctor, who has access to medical advice from the DVLA.

Alcohol Problems

Persons with repeated convictions for drink-driving offences may be required to satisfy the DVLA (with certification from their own doctor) that they do not have an 'alcohol problem' before their licence is restored to them (see also pp 165–66).

Other Medical Conditions

Other disabilities which may cause failure of the driver's medical examination include:

- sudden attacks of vertigo ('dizziness');
- heart disease which causes disabling weakness or pain;
- a history of coronary thrombosis;
- the use of hypertensive drugs for blood pressure treatment;
- serious arrhythmia;
- severe mental disorder;
- severe behavioural problems;
- alcohol dependency;
- inability to refrain from drinking and driving;
- drug abuse and dependency;
- psychotrophic medicines taken in quantities likely to impair fitness to drive safely.

A licence will be refused to a driver who is liable to sudden attacks of disabling giddiness or fainting unless these can be controlled.

Those who have had a cardiac pacemaker fitted are advised to discontinue LGV driving, although driving vehicles below the 7.5 tonnes LGV threshold is permitted if a person who has disabling attacks which are controlled by a pacemaker has made arrangements for regular review from a cardiologist and will not be likely to endanger the public.

Notification of New or Worsening Medical Conditions

Once a licence has been granted (ie whether ordinary or covering vocational entitlements), the holder is required to notify the Drivers Medical Group, DVLA at Swansea SA99 1TU of the onset, *or worsening*, of any medical condition likely to cause him to be a danger when driving – *failure to do so is an offence*. Examples of what must be reported are:

- giddiness;
- fainting;
- blackouts;
- epilepsy;
- diabetes;
- strokes;
- multiple sclerosis;
- Parkinson's disease;
- heart disease;
- angina;
- 'coronaries';
- high blood pressure;
- arthritis;
- disorders of vision;
- mental illness;
- alcoholism;
- drug-taking;
- loss, or loss of use, of any limb.

In many cases the person's own doctor will either advise reporting the condition to the DVLA himself, or the doctor (or hospital) may advise the DVLA direct. In either case the driving licence will have to be surrendered until the condition clears.

There is no requirement to notify the DVLA of temporary illnesses or disabilities such as sprained or broken limbs where a full recovery is expected within three months.

Enquiries about medical conditions can be raised with the Drivers Medical Group at the DVLA. Telephone the Customer Enquiries Unit on 08702 411879.

Medical Appeals and Information

The final decision on any medical matter concerning driving licences rests with the Drivers Medical Group of the DVLA. However, there is the opportunity of appeal, within six months, in England and Wales to a magistrate's court, and within 21 days in Scotland to a sheriff's court. In other cases the refused driver may be given the opportunity to present further medical evidence which the medical adviser will consider.

Further information on medical conditions relating to driving are to be found in a booklet, *Medical Aspects of Fitness to Drive*, price £6, available from the Medical Commission on Accident Prevention, 35–43 Lincoln's Inn Fields, London WC2A 3PN (Tel 020 7242 3176). Useful information for diabetic drivers may be obtained from the British Diabetic Association, 10 Queen Anne Street, London W1M 0BD (tel 020 7636 6112; fax 020 7462 2732).

Cover Against Loss of Licence/Entitlement on Medical Grounds

In view of the risk of losing their lgv entitlement in later life due to the onset or worsening of a medical condition and therefore jeopardizing their employment prospects, it is possible for drivers (or their employers) to insure against loss of driving licence/vocational entitlement. Special insurance schemes to cover such an eventuality are offered by, among others, the FTA, RHA and the drivers' union, the TGWU; the last promotes a scheme provided by Unity Trust. Various other independent schemes are offered by district offices of the TGWU.

Drugs and Driving

Official sources say that drugs are a major cause of one in five fatal road accidents. Another source says that driving after smoking cannabis could be a greater danger than drink-driving, and that as many as three million people could be driving under the influence of this drug. Yet another report has highlighted the fact that drivers who use tranquillizers are involved in 1600 road accidents every year – 110 of them fatal.

Illegal Drugs

Among the drugs which may be detected and which can affect driving are:

- cannabis – produces slow reaction times;
- cocaine – may increase reaction times, but severely affects accuracy and judgement; has potential to cause hallucination;
- amphetamines – may increase reaction times in the short term, but severely affects accuracy and judgement;

- ecstasy – may increase reaction times, but severely affects accuracy and judgement;
- heroin – produces reduced reaction times and causes drowsiness and sleep.

Prescribed Drugs

Prescribed tranquillizers, sedatives and antidepressants, as well as diabetes and epilepsy drugs, may have an adverse effect on a driver's judgement and reactions, and therefore increase the risk of an accident. These include a number of anti-anxiolytic benzodiazepines (prescribed to reduce stress and anxiety), including:

- Valium;
- Librium;
- Ativan.

The sedatory effect of these drugs is substantially compounded by the addition of alcohol, even when taken in relatively small quantities, resulting in a potentially significant loss of co-ordination. Similarly, sleeping tablets (eg diazepam, temazepam and nitrazepam) including the new drug zopiclone may also have a continuing sedatory effect on a driver the following morning. Furthermore, a whole range of other proprietary medicines such as painkillers, antihistamines, cold and flu remedies, eye drops, cough medicines and common painkillers taken in sufficient quantities may have similar effects.

If a driver feels drowsy, dizzy, confused, or suffers other side effects that could affect reaction times or judgement, *he should not drive.*

Drug Testing

Trial drug testing of drivers in random roadside spot checks was carried out by the police in spring 1998. Principally the scheme was designed to test the 'Drugwipe' testing kit, which is wiped across the driver's forehead to pick up any traces of drugs in his sweat. This is a foretaste of what may become standard practice.

Drug testing of LGV drivers by their employers is becoming an increasing practice in the UK, especially among tanker fleet operators, following the pattern in the USA which has had mandatory testing since 1992. In fact, Unilabs UK, one of the leading drug-testing laboratories, has reported a substantial increase in requests for testing by haulage companies in recent times. While there is no suggestion at this stage that the practice should become mandatory in this country, most of Britain's major oil companies now carry out random testing for both alcohol and drug problems – Shell has produced a staff booklet identifying 11 banned substances (including amphetamines) and warning of the consequences of drink or drug abuse.

In mid-2000, emergency tests were introduced for road users because of a sharp rise in drug-related road deaths. Drivers suspected of having taken drugs are asked, among other physical tests, to walk in a straight line and count their steps. Suspects are then taken to a police station for further tests. Fines of up to £5,000 may be imposed on convicted offenders, together with a mandatory driving ban and risk of a prison sentence.

Useful guidance on this matter is contained in Unilabs' brochure *Drug Abuse – The Facts* which can be obtained from the company at: Bewlay House, 32 Jamestown Road, London NW1 7BY (tel 020 7267 2672; fax 020 7267 2551).

Join the Fight Against Drugs

If you have any information about drugs or drug smugglers, H.M. Customs request that you ring the 24-hour hotline **0800 59 5000**. You don't have to tell them who you are, but for important information you may be eligible for a cash reward.

Eyesight Requirement

The statutory eyesight requirement mentioned above for ordinary (ie car and light goods vehicle) licence holders is for the driver to be able to read, in good daylight (with glasses or contact lenses if worn), a standard motor vehicle number plate from 20.5 metres (ie 67 feet). A 120-degree-wide field of view is also required. It is an offence to drive with impaired eyesight and the police can require a driver to take an eyesight test on the roadside. If glasses or contact lenses are needed to reach these vision standards they must be worn at all times while driving. It is an offence to drive with impaired eyesight. There are proposals for drivers to undergo regular eye tests.

NB: A proposal to reduce this distance (20.5 metres) to just 20 metres when reading the new (ie since September 2001) narrower fonts used on number plates is under consideration, as is a reduction from 12.3 metres to 12 metres where such number plates are fitted to category K vehicles (ie mowing machines and pedestrian controlled vehicles).

Eyesight Standards for Vocational Licence Holders

Tougher eyesight standards for LGV and PCV drivers were introduced from 1 January 1997. Specifically, drivers of vehicles in categories C, C1, C+E, C1+E, D, D1, D+E and D1+E (effectively trucks over 3.5 tonnes and passenger vehicles with more than nine seats) must have eyesight which is at least:

1. 6/9 on the Snellen scale in the better eye*, and
2. 6/12 on the Snellen scale in the other eye*, and
3. 3/60 in each eye without glasses or contact lenses.

**NB: these standards may be met with glasses or contact lenses if worn.*

To achieve these standards means being able to read the top line of an optician's chart (ie Snellen chart) with each eye from a distance of *at least* three metres without the aid of glasses or contact lenses – if it can only be read from, say, 2.5 m or less the test is failed. Wearers of spectacles or contact lenses must have vision of at least 6/9 in the better eye and at least 6/12 in the weaker eye, which means being able to read the sixth line of an optician's chart at 6 metres. Besides these requirements, all drivers must meet existing eyesight standards which include having a field of vision of at least 120° (horizontal) and 20° (vertical) in each eye with no double vision.

Drivers who held a licence before 1 January 1997 and who do not meet these higher standards are advised to check their licensing position with the Drivers Medical Group at the DVLA (see above for address and telephone number).

LICENCE FEES AND VALIDITY

Current fees for driving licences are as follows:

Licence type	
First provisional	
Car	£38.00
Bus or lorry	FREE
First full	
Car	£9.00
Bus or Lorry	FREE
Renewal	
From age 70	FREE
For medical reasons	FREE
Bus or lorry	FREE
After disqualification	£50.00
HRO* disqualification	£75.00
After revocation	£38.00
Duplicate	
Replace lost or stolen licence	£19.00
Paper licence for photocard licence	£19.00
Full Northern Ireland licence	FREE
Full EC/EEA or other foreign licence (inc. Channel Islands or IoM)	£38.00
Exchange	
Add or remove entitlement	FREE
Remove expired endorsements	£19.00
Replacement	
Change of name and/or address	FREE

* High risk offender disqualified for drink/driving convictions

Licence Validity

An LGV/PCV driving entitlement is normally valid for 5 years or until the holder reaches the age of 45 years, whichever is the longer. After the age of 45 years, 5-year entitlements are granted subject to medical fitness, but may be for lesser periods where the holder suffers from a relevant or prospective relevant disability. From the age of 65 years, vocational entitlements are granted on an annual basis only.

Tax Deductions

The cost of renewing LGV driving entitlements and for undergoing medical examinations in connection with licence renewals is income tax deductible against Schedule E earnings – but not the costs of first obtaining such an entitlement.

Lost or Mislaid Licences

Drivers who lose or mislay their driving licences should apply for a duplicate licence in the normal way using the standard application form, form D2, at a cost of £19.00.

UK drivers moving to live abroad who have mislaid their driving licence may obtain a temporary 'Certificate of Entitlement' (commonly referred to as a cover note), valid for one month, either from the DVLA at Swansea (free of charge) or from local Vehicle Registration Offices (VROs), subject to proof of their identity. This document is valid for proving entitlement to drive to enable such persons to apply for and obtain an equivalent driving entitlement in their new country of residence. Applicants for these certificates will be able to complete an application for a duplicate licence at the VRO and this will be sent on to the DVLA in Swansea for processing.

NOTIONAL GROSS WEIGHTS

For the purposes of determining driving licence requirements in cases where a goods trailer does not have a manufacturer's or 'official' gross weight marked on it, a system of 'multipliers' is used to calculate the notional maximum gross weight. The unladen weight of the trailer is multiplied by the number given in the table below and the resulting figure is taken to be the gross weight, but only for the purposes of deciding what driving licence is required and not for any other purpose.

Trailers

Class of Trailer		*Number*
1.	Engineering plant	1
2.	Trailers which consist of drying or mixing plant designed for the production of asphalt or of bituminous or tarmacadam	1
3.	Agricultural trailers	1
4.	Works trailers	1
5.	Living vans	1.5
6.	Any trailers not mentioned above.	3

PRODUCTION OF DRIVING LICENCES

Both the police and enforcement officers of VOSA can request a driver – and a person accompanying a provisional entitlement holder – to produce his licence showing ordinary and vocational entitlements to drive. If he is unable to do so at the time, they may be produced without penalty:

- if the request was by a police officer, at a police station of his choice within 7 days; or
- if the request was by an enforcement officer, at the Traffic Area Office within 10 days.

In either case, if the licence cannot be produced within the 7 or 10 days it can be produced as soon as reasonably practicable thereafter. A TC can also require the holder of an LGV/PCV driving entitlement to produce his licence at a Traffic Area Office for examination within 10 days.

Failure to produce a licence on request is an offence.

A police officer can ask a driver to state his date of birth – British ordinary driving licences carry a coded number which indicates the holder's surname and date of birth. The name and address of the vehicle owner can also be requested.

NB: In the case of the new-type photocard driving licence, it is necessary to produce also the paper 'counterpart' that accompanies it. One without the other is not valid.

When required by a VOSA examiner to produce his licence, an LGV entitlement holder may be required to give his date of birth and to sign the examiner's record sheet to verify the fact of the licence examination. This should not be refused.

Licence holders apprehended for endorsable fixed penalty (ie yellow ticket) offences are required to produce their driving licence to the police officer at that time or later (ie within seven days) to a police station and surrender the licence, for which they will be given a receipt. Failure to produce a licence in these circumstances means that the fixed penalty procedure will not be followed and a summons for the offence will be issued requiring a court appearance. Drivers summoned to appear in court for driving and road traffic offences must produce their driving licence to the court on, at least, the day before the hearing.

INTERNATIONAL DRIVING PERMITS

Certain foreign countries do not accept British ordinary driving licences – (see the AA website for a list of such countries: www.theaa.com/getaway/idp) – in which case an international driving permit (IDP) will be required by British licence holders wishing to drive in those countries. These permits are obtainable from the RAC, RSAC, AA or Green Flag. The fee is £5.50 and a passport-type photograph is required for attachment to the permit. Applicants must be UK residents, over 18 years of age and hold full driving entitlements for the category of vehicle which the IDP is required to cover.

The AA warns drivers to beware of bogus IDPs on sale mainly on the Internet, but also in magazines and newspapers. Such documents are counterfeit and are not recognized by countries that are party to the United Nations Convention on Road Traffic.

EXCHANGE OF DRIVING LICENCES

British driving licence holders can exchange their licence if necessary in order to:

- add new categories to a full licence;
- remove out-of-date endorsements or suspension details;
- add or take off provisional motorcycle entitlement;
- exchange an old-style pink or green licence for a new-style one.

Exchanging a Foreign Licence for a GB Licence:

Northern Ireland Licences

Full NI driving licences or a test pass can be exchanged for a GB licence. Alternatively, a driver can continue to use a NI licence in Britain until it expires.

EU/EEA Licences

A valid full licence issued in any EU or EEA country (ie all EC countries plus Liechtenstein, Iceland and Norway) need not be changed immediately for a licence. So long as such a licence remains valid, the holder can drive in Great Britain until he reaches 70 years of age or for three years after becoming resident in Great Britain, whichever is the longer period.

Foreign drivers of large vehicles can drive in Great Britain until aged 45 or for five years after becoming resident, whichever is the longer period. Drivers aged over 45 years but under 65 can drive until their 66th birthday or for 5 years after becoming a GB resident, whichever is the shorter. Drivers aged 65 years or older may drive for 12 months after becoming a resident in Great Britain.

In order to continue driving after this time, a British driving licence must be obtained by making application on form D2. Application can be made for a British licence at any time, even after expiry of the foreign national licence.

Non-EC (Designated) Countries and Gibraltar

A full valid car licence issued in any of the following countries can be exchanged for a British licence: Australia, Barbados, British Virgin Islands, Gibraltar, Hong Kong, Japan, Kenya, Malta, New Zealand, Cyprus, Singapore, Switzerland and Zimbabwe.

Holders of lorry and bus licences from any of the above countries should contact the DVLA's Customer Enquiry Unit (at Swansea SA6 7JL) for further information.

Full Jersey or Isle of Man car, lorry or bus licences, or a full Guernsey car licence, can be exchanged for a British licence if they were valid within the past 10 years.

Certain driving licences for vehicles up to 3.5 tonnes issued in South Africa and Canada may now be exchanged for an equivalent GB licence.

A foreign car, lorry or bus driving licence can be used in Great Britain for one year only provided it remains valid, but exchange of a driving licence issued in any of the above-listed countries can be made up to five years after taking up residence in Great Britain.

Visitors from other countries not mentioned above cannot exchange their national licence for a British equivalent but they can drive in Great Britain on such licences (or on an International Driving Permit) for up to one year.

All foreign licence holders who, for various reasons cannot obtain an exchange licence or who wish to drive vehicles which their national licence does not cover, must apply for provisional entitlement in the normal way using form D1.

Full GB driving licences can only be issued to foreign nationals who become normally resident in Great Britain.

Visitors Driving in the UK

Visitors to the UK may drive vehicles in the UK provided they hold a domestic driving licence issued in their own country (ie outside the UK and the EU) or a Convention Driving Permit (issued under the 1949 Geneva Convention on Road Traffic by a country outside the UK). Holders of such permits are entitled to drive vehicles of a class which their own national or international licence covers for a period of 12 months from the date of their entry into the UK.

Driving While Tired

The tragic consequences of the Selby Rail disaster in 2001, caused predominantly by a tired Land Rover driver falling asleep at the wheel and crashing on to a main railway line in front of a speeding express train, have highlighted the great dangers of driving when tired. LGV drivers who work through the night are at particular risk in the early morning hours when the body is at its lowest ebb and nodding off, even only fleetingly, is a considerable danger. This is all it takes – a momentary loss of concentration as the eyes close may be sufficient to lead to tragedy.

Drivers must be encouraged to STOP when they feel tired, whether during the night-time hours or during daytime, and rest for a short while, no matter what the pressures of the job or the particular journey in which they are engaged. Employers have responsibilities under health and safety legislation to assess such potential risks and take proper action to eliminate them. A VOSA prosecutor, dealing with a case of a haulier permitting a driver to take insufficient rest, said, 'Haulage bosses should become amateur policemen and keep a close watch on their driver to make sure they have sufficient rest.'

Think – Don't Drive Tired

Since tiredness has been established as the principal factor in around 10 per cent of all accidents the D*f*T has launched a new campaign to combat tiredness among both car and LGV drivers, *Think – Don't Drive Tired.* The main points for drivers to observe are that they should:

- make sure they are fit to drive, particularly before undertaking any long journeys (over an hour) – avoid such journeys in the morning without a good night's sleep or in the evening after a full day's work;
- avoid undertaking long journeys between midnight and 6 am, when natural alertness is at a minimum;
- plan their journey to take sufficient breaks. A minimum break of at least 15 minutes after every two hours driving is advised;
- if they feel at all sleepy, stop in a safe place and either take a nap for not more than 15 minutes, or drink two cups of strong coffee.

The research leading to these conclusions is summarized in two Transport Research Laboratory (TRL) reports (published in September 1995): *Driver Sleepiness as a Factor in Car and HGV Accidents* and *Falling Asleep at the Wheel.* These are available from the TRL at Crowthorne, Berkshire RG11 6AU (tel 01344 773131).

THE 'WELL DRIVEN' CAMPAIGN

The 'Well Driven' campaign is an indication that the vehicles on which the sign appears (with a freephone number 0800 22 55 33) belong to operators who have signed up to observe the voluntary 'Good Lorry Code' which promotes high standards of driving, vehicle maintenance and consideration for other road users. The freephone number is an invitation for members of the public (and indeed other drivers and transport operators) to have their say about the standard of driving (bad or good) or the condition of the vehicle. A help desk, manned round the clock, takes calls which, provided sufficient information is given, are passed on to the operator concerned.

The Good Lorry Code is promoted by the Freight Transport Association (FTA), the Road Haulage Association (RHA) and the Confederation of British Industry (CBI). Further information on the campaign may be obtained from the FTA, tel: 01892 552355.

DRIVING LICENCE PENALTY POINTS AND DISQUALIFICATION

Driving licence holders may be penalized following conviction by a court for offences committed on the road with a motor vehicle. These penalties range from the issue of fixed penalty notices for non-endorsable offences, which do not require a court appearance unless the charge is to be contested and incur no driving licence penalty points, although the relevant fixed penalty has to be paid; to those for endorsable offences when penalty points are added on the licence counterpart and the fixed penalty is incurred or a heavy fine imposed on conviction if a court appearance is made.

In other cases, licence disqualification for a period (extending to a number of years in serious cases – especially for drink-driving-related offences), and in very serious instances imprisonment of the offender may follow conviction in a magistrates' court or indictment for the offence in a higher court. Holders of vocational driving entitlements may be separately penalized for relevant offences which could result in such entitlements being suspended or revoked and in serious circumstances the holder being disqualified from holding a vocational entitlement – see below.

The Penalty Points System on Conviction and Disqualification

Driving licence endorsement of penalty points following conviction for motoring offences is prescribed by the Road Traffic Offenders Act 1988 with further provisions relating to driving offences being contained in the Road Traffic Act 1991 – see pp 159–60.

The penalty points system grades road traffic offences according to their seriousness by a number or range of penalty points, between 2 and 10, imposed on the driving licence of the convicted offender. Once a maximum of 12 penalty points has been accumulated within a three-year period counting from the date of the first offence to the current offence (not from the date of conviction), disqualification of the licence for at least six months will follow automatically.

Most offences rate a fixed number of penalty points to ensure consistency and to simplify the administration; but a discretionary range applies to a few offences where the gravity may vary considerably from one case to another. For example, failing to stop after an accident which only involved minor vehicle damage is obviously less serious than a case where an accident results in injury.

Unless the court decides otherwise, when a driver is convicted of more than one offence at the same hearing, only the points relative to the most serious of the offences will normally be endorsed on the licence. Once sufficient points (ie 12) have been endorsed on the driving licence and a period of disqualification has been imposed (six months for the first totting-up of points), the driver will have his 'slate' wiped clean and those points will not be counted again. Twelve more points would have to be accumulated before a further disqualification would follow, but to discourage repeated offences the courts will impose progressively longer disqualification periods in further instances (minimum 12 months for subsequent disqualifications within three years and 24 months for a third disqualification within three years).

Licence Endorsement Codes and Penalty Points

Following conviction for an offence, the driver's licence (ie the green counterpart) will be endorsed by the convicting court with both a code (to which employers and prospective employers should refer so they can assess the offences which drivers have committed) and the number of penalty points imposed as follows:

Code		*Penalty points*
	Accident offences	
AC 10	Failing to stop after an accident	5–10
AC 20	Failing to report an accident within 24 hours	5–10
AC 30	Undefined accident offence	4–9
	Disqualified driver	
BA 10	Driving while disqualified	6
BA 20	Driving while disqualified on age grounds	2
BA 30	Attempting to drive while disqualified	6
	Careless driving	
CD 10	Driving without due care and attention	3–9
CD 20	Driving without reasonable consideration for other road users	3–9
CD 30	Driving without due care and attention or without reasonable consideration for other road users	3–9
CD 40	Causing death through careless driving when unfit through drink	3–11
CD 50	Causing death by careless driving when unfit through drugs	3–11
CD 60	Causing death by careless driving with alcohol level above the limit	3–11
CD 70	Causing death by careless driving then failing to supply a specimen	3–11
	Construction and use offences	
CU 10	Using a vehicle with defective brakes	3
CU 20	Causing or likely to cause danger by reason of unsuitable vehicle or using a vehicle with parts or accessories (excluding brakes, steering or tyres) in a dangerous condition	3
CU 30	Using a vehicle with defective tyre(s)	3
CU 40	Using a vehicle with defective steering	3
CU 50	Causing or likely to cause danger by reason of load or passengers	3
CU 60	Undefined failure to comply with C&U Regulations	3
	Reckless/dangerous driving	
DD 40	Dangerous driving	3–11
DD 60	Manslaughter or culpable homicide while driving a vehicle	3–11
DD 80	Causing death by dangerous driving	3–11
	Drink or drugs	
DR 10	Driving or attempting to drive with alcohol level above limit	3–11
DR 20	Driving or attempting to drive while unfit through drink	3–11
DR 30	Driving or attempting to drive then failing to supply a specimen for analysis	3–11
DR 40	In charge of a vehicle while alcohol level above limit	10
DR 50	In charge of a vehicle while unfit through drink	10
DR 60	Failure to provide a specimen for analysis in circumstances other than driving or attempting to drive	10
DR 70	Failing to provide a specimen for breath test	4
DR 80	Driving or attempting to drive when unfit through drugs	3–11
DR 90	In charge of a vehicle when unfit through drugs	10

Code		Penalty points
	Insurance offences	
IN 10	Using a vehicle uninsured against third-party risks	6–8
	Licence offences	
LC 10	Driving without a licence	2
LC 20	Driving otherwise than in accordance with a licence	3–6
LC 30	Driving after making a false declaration about fitness when applying for a licence	3–6
LC 40	Driving a vehicle having failed to notify a disability	3–6
LC 50	Driving after a licence has been revoked or refused on medical grounds	2
	Miscellaneous offences	
MS 10	Leaving a vehicle in a dangerous position	3
MS 20	Unlawful pillion riding	3
MS 30	Contravention of the Road Traffic Regulations provisions on street playgrounds	2
MS 40	Driving with uncorrected defective eyesight or refusing to submit to a test	3
MS 50	Motor racing on the highway	3–11
MS 60	Offences not covered by other codes as appropriate	
MS 70	Driving with uncorrected defective eyesight	3
MS 80	Refusing to submit to an eyesight test	3
MS 90	Failure to give information as to identity of driver, etc	3
	Motorway offences	
MW 10	Contravention of Special Roads Regulations (excl speed limits)	3
	Pedestrian crossings	
PC 10	Undefined contravention of Pedestrian Crossing Regulations	3
PC 20	Contravention of Pedestrian Crossing Regulations with moving vehicle	3
PC 30	Contravention of Pedestrian Crossing Regulations with stationary vehicle	3
	Provisional licence offences	
PL 10	Driving without 'L' plates	3–6
PL 20	Not accompanied by a qualified person	3–6
PL 30	Carrying a person not qualified	3–6
PL 40	Drawing an unauthorized trailer	3–6
PL 50	Undefined failure to comply with conditions of a provisional licence	3–6
	Speed limits	
SP 10	Exceeding goods vehicle speed limits	3–6
SP 20	Exceeding speed limit for type of vehicle (excluding goods or passenger vehicles)	3–6
SP 30	Exceeding statutory speed limit on a public road	3–6
SP 40	Exceeding passenger vehicle speed limit	3–6
SP 50	Exceeding speed limit on a motorway	3–6
SP 60	Undefined speed limit offence	3–6
	NB: Disqualification is obligatory where the relevant speed is in excess of 30 mph over the statutory limit.	
	Traffic directions and signs	
TS 10	Failing to comply with traffic light signals	3
TS 20	Failing to comply with double white lines	3
TS 30	Failing to comply with a 'Stop' sign	3
TS 40	Failing to comply with direction of a constable or traffic warden	3
TS 50	Failing to comply with a traffic sign (excluding 'Stop' signs, traffic lights or double white lines)	3
TS 60	Failing to comply with a school crossing patrol sign	3
TS 70	Undefined failure to comply with a traffic direction or sign	3
	Special Code	
TT 99	To signify a disqualifaction under 'totting up' procedure. If the total of penalty points reaches 12 or more within 3 years, the driver is liable to be disqualified	
	Theft or unauthorized taking	
UT 10	Taking and driving away a vehicle without consent or an attempt thereat	8
UT 20	Stealing or attempting to steal a vehicle	8
UT 30	Going equipped for stealing or taking a vehicle	8
UT 40	Taking or attempting to take a vehicle without consent; driving or attempting to drive a vehicle knowing it to have been taken without consent; allowing oneself to be carried in or on a vehicle knowing it to have been taken without consent	8
UT 50	Aggravated taking of a vehicle	3–11

Where the offence is one of aiding or abetting, causing or permitting or inciting, the codes are modified as follows:

Aiding, Abetting, Counselling or Procuring

Offences as coded, but with zero changed to 2, eg UT 10 becomes UT 12.

Causing or Permitting

Offences as coded, but with zero changed to 4, eg LC 20 becomes LC 24.

Inciting

Offences as coded, but with zero changed to 6, eg DD 30 becomes DD 36. The length of time for periods of disqualification is shown by use of the letters D = days, M = months and Y = years. Consecutive periods of disqualification are signified by an asterisk (*) against the time period. The symbol + means that 3 to 11 points are added to a licence if for exceptional reasons disqualification is not imposed.

Disqualification

The endorsing of penalty points will also arise on conviction for offences where disqualification is discretionary and where the court has decided that immediate disqualification is not appropriate (for example if acceptable 'exceptional' reasons are put forward – see also below). In this case the offender's driving licence will be endorsed with four points. The courts are still free to disqualify immediately if the circumstances justify this.

Offences carrying obligatory disqualification are shown in the following list:

- Causing death by dangerous driving and manslaughter.
- Dangerous driving within three years of a similar conviction.
- Driving or attempting to drive while unfit through drink or drugs.
- Driving or attempting to drive with more than the permitted breath-alcohol level.
- Failure to provide a breath, blood or urine specimen.
- Racing on the highway.

Driving while disqualified is a serious offence which can result in a fine at level 5 on the standard scale (see below), currently £5,000 maximum, or six months' imprisonment, or both.

Disqualification for Anti-Social Crime

Since January 2004 the courts have had authority under Section 146 of the Powers of Criminal Courts (Sentencing) Act 2000 to disqualify any person from driving for any period of time and for any relevant offence – this means, not just for drink driving or dangerous driving. Basically, the provision is intended to target 'anti-social' crime involving the use of vehicles where disqualification would not otherwise be available as a penalty.

Special Reasons for Non-Disqualification

The courts have discretion in exceptional mitigating circumstances (ie when there are 'special reasons') not to impose a disqualification. The mitigating circumstance must not be one which attempts to make the offence appear less serious and no account will be taken of hardship other than exceptional hardship. Pleading that you have a wife and children to support or that you will lose your job is not generally considered to be exceptional hardship for the purposes of determining whether or not disqualification should be imposed.

If account has previously been taken of circumstances in mitigation of a disqualification, the same circumstances cannot be considered again within three years.

Where a court decides, in exceptional circumstances as described above, not to disqualify a convicted driver, 3 to 11 penalty points will be added to the driver's licence in lieu of the disqualification.

Driving Offences

Dangerous Driving

The previous charge of reckless driving was replaced from 1 July 1992 by a charge of 'dangerous' driving. A person is to be regarded as driving dangerously if the way he drives 'falls far short of what would be expected of a competent and careful driver, and it would be obvious to a competent and careful driver that driving in that way would be dangerous'. Driving would be regarded as dangerous 'if it was obvious to a competent and careful driver that driving the vehicle in its current state would be dangerous' – this obviously applies to the vehicle's mechanical condition or the way it is loaded. Also 'dangerous' refers to danger either of injury to any person or of serious damage to property. The principal offences to which this relates are dangerous driving and causing death by dangerous driving.

Interfering with Vehicles, etc

It is an offence for any person to cause danger to road users by way of intentionally and without lawful authority placing objects on a road, interfering with motor vehicles, or directly or indirectly interfering with traffic equipment (eg road signs, etc).

Penalties

Penalties for these offences are heavy. For example, causing death by careless driving while under the influence of drink or drugs carries a maximum penalty of up to five years in prison and/or a fine. For causing a danger to road users the maximum penalty is up to seven years' imprisonment and/or a fine.

In addition to disqualification and the endorsement of penalty points on driving licences, courts may impose fines and, for certain offences, imprisonment. The maximum fine for most offences is determined by reference to a scale set out in the Criminal Justice Act 1991 as follows:

Level 1	£200
Level 2	£500

Level 3	£1000
Level 4	£2500
Level 5	£5000

Offences such as dangerous driving, failing to stop after an accident or failure to report an accident and drink-driving offences, carry the current maximum fine of £5000, as do certain vehicle construction and use offences (eg overloading, insecure loads, using a vehicle in a dangerous condition, etc) and using a vehicle without insurance.

Under new provisions included in the Criminal Justice Act 2003, the maximum penalty for the offence of causing death by dangerous driving has been increased from a 10-year prison sentence to 14 years. This measure has been taken, according to the Home Secretary, to ensure that drivers who kill can be 'properly punished'.

New Driver Penalties

The Road Traffic (New Drivers) Act 1995 concerns new drivers who first passed their driving test on or after 1 June 1997. Where such drivers acquire six or more penalty points on their licence within two years of passing that test (the so-called 'probationary period') the DVLA will automatically revoke the licence on notification by a court or fixed penalty offence. Such drivers have to surrender their full licence and obtain a provisional licence to start driving again as a learner. They will have to pass both the theory and practical tests again in order to regain their full driving licence.

Penalty points counting towards the total of six include any incurred before passing the test, if this was not more than three years before the latest penalty point offence. Points imposed after the probationary period will also count if the offence was committed during that period.

Passing the retest will not remove the penalty points from the licence; these will remain and if the total reaches 12, the driver will be liable to disqualification by a court.

Short-Period Disqualification (SPD)

When drivers are disqualified for less than 56 days, the court will stamp the counterpart of their driving licence and return it to them. The stamp will show how long disqualification is to last. The licence does not have to be renewed when the Short-Period Disqualification ends – it becomes valid again the day following expiry of the disqualification.

Removal of Penalty Points and Disqualifications

Penalty points endorsed on driving licences can be removed (by application to the DVLA Swansea on form D1 and on payment of a fee of £13.50).

The waiting period before which no such application would be accepted is 4 years from the date of the offence, except in the case of reckless/dangerous driving convictions when the 4 years is taken from the date of conviction. Endorsements for alcohol-related offences must remain on a licence for 11 years.

Licences returned after disqualification will show no penalty points but previous disqualifications (within 4 years) will remain and if a previous alcohol/drugs driving offence disqualification has been incurred, this will remain on the licence for 11 years.

Application may be made by disqualified drivers for reinstatement of their licence after varying periods of time depending on the duration of the disqualifying period as follows:

- Less than 2 years – no prior application time.
- Less than 4 years – after 2 years have elapsed.
- Between 4 years and 10 years – after half the time has elapsed.
- In other cases – after 5 years have elapsed.

The courts are empowered to require a disqualified driver to retake the driving test before restoring a driving licence, and following the introduction of provisions contained in the Road Traffic Act 1991, it is now mandatory for them to impose 'extended' retests following disqualification for the most serious of driving offences, namely, dangerous driving, causing death by dangerous driving and manslaughter by the driver of a motor vehicle (in Scotland, the charge is culpable homicide).

The fee charged for replacement licences following disqualification is £35.00, but where the disqualification was for drink-driving-type offences the fee for a replacement is £50.00.

Retests for Offending Drivers

Where drivers are convicted of the offences of manslaughter, causing death by dangerous driving or dangerous driving, and mandatory disqualification is imposed, an 'extended' retest (involving at least one hour's driving) must be taken before the driving licence is restored. This also applies to drivers disqualified under the penalty points totting-up procedure. Courts may also order drivers disqualified for lesser offences to take an appropriate (ie ordinary) driving test.

Drink-Driving and Breath Tests

It is an offence to drive or attempt to drive a motor vehicle when the level of alcohol in the breath is more than 35 micrograms per 100 millilitres of breath. This is determined by means of an initial breath test, conducted on the spot when the driver is stopped, and later substantiated by a test on a breathtesting machine (eg Lion Intoximeter) at a police station. The breath/alcohol limit mentioned above equates to the blood/alcohol limit of 80 milligrams of alcohol in 100 millilitres of blood or the urine/alcohol limit of 107 milligrams of alcohol in 100 millilitres of urine.

Failure to Produce a Breath Sample and Low Breath-Test Readings

If those suspected of an alcohol-related offence cannot, due to health reasons, produce a breath sample, or if a breath test shows a reading of not more than 50 micrograms of alcohol per 100 millilitres of breath, they are given the opportunity of an alternative test, either blood or urine, for laboratory analysis. This test can only be carried out at a police station or a hospital and the decision as to which alternative is chosen rests with the police (unless a doctor present determines that for medical reasons a blood test cannot or should not be taken).

Similarly, if a breath test of a driver shows the proportion of alcohol to be no more than 50 micrograms in 100 millilitres of breath, the driver can request an alternative test (ie blood or urine) as described above for those who cannot provide a breath sample for analysis.

A House of Lords ruling says that a driver recording a breath sample that does not exceed 50 micrograms in 100 millilitres of breath must be given a police statement to this effect.

Prosecution for Drink-Driving Offences

Prosecution will follow a failure to pass the test which will result in a fine or imprisonment and automatic disqualification from driving. Failure to submit to a breath test and to a blood or urine test are serious offences, and drivers will find themselves liable to heavy penalties on conviction and potentially long-term disqualification or driving licence endorsement (endorsements for such offences remain on a driving licence for 11 years).

The police *do not* have powers to carry out breath tests at random but they *do have* powers to enter premises to require a breath test from a person suspected of driving while impaired through drink or drugs, or who has been driving, or been in charge of a vehicle which has been involved in an accident in which another person has been injured.

In a highly publicized case in 1991, charges against an LGV driver breathalysed by police having been woken from sleep on the sleeper berth of his vehicle cab (and with his tachograph set to record rest) were dismissed by the court. Although presumably over the statutory limit (otherwise no charge would have been brought), the driver was clearly making no attempt to drive his vehicle and the vehicle was parked in a proper (ie non-public) place. Since the court dismissed the case there was no opportunity to obtain a ruling from a higher court as to the precise legal position of drivers in such circumstances. It is considered that drinking and then sleeping in a vehicle cab does present risk of prosecution for drivers.

Drink-Driving Disqualification

Conviction for a first drink-driving offence will result in a minimum one year period of disqualification and for a second or subsequent offence of driving or attempting to drive under the influence of drink or drugs longer periods of disqualification will be imposed by the court. If the previous such conviction took place within 10 years of the current offence the disqualification must be for at least 3 years.

Drivers convicted twice for drink-driving offences may have their driving licence revoked altogether. Offenders who are disqualified twice within a 10-year period for any drink-driving offences and those found to have an exceptionally high level of alcohol in the body (ie more than 2½ times over the limit) or those who twice refuse to provide a specimen will be classified as high-risk offenders (HROs) by the Driver and Vehicle Licensing Agency. They will be required to show that they no longer have an 'alcohol problem' by means of a medical examination (including blood analysis of liver enzymes) by a DVLA-approved doctor before their licence will be restored to them. A higher than normal fee (ie £75) will be charged for the renewal of such licences.

Drink-Driving Courses

The Road Traffic Act 1991 contained provisions whereby certain (but not all) drink-driving offenders may have the period of their disqualification reduced if they agree to undertake an approved rehabilitation course and satisfactorily complete it. The provision applies only where the court orders the individual (who must be over 17 years of age) to be disqualified for at least 12 months following conviction under the Road Traffic Act 1988 for:

- causing death by careless driving when under the influence of drink or drugs;
- driving or being in charge of a motor vehicle when under the influence of drink or drugs;
- driving or being in charge of a motor vehicle with excess alcohol in the body; or
- failing to provide a specimen (of breath, blood or urine) as required.

The court has a duty to ensure that a place on an approved course is available for the offender. It must explain to the offender 'in ordinary language' the effect of the order (to reduce the disqualification period), the amount of the fees and that these must be paid in advance, and it must seek the offender's agreement that the order should be made – in other words, it is a completely voluntary scheme and there is no question of force.

Given these provisos, the court can order the period of disqualification to be reduced by not less than 3 months and not more than one-quarter of the unreduced period (for example, with a 12-month disqualification, the reduction will be 3 months, leaving a 9-month disqualification period to be served).

The latest date for completion of one of these rehabilitation courses is two months prior to the last day of the reduced disqualification period (so, in the nine month example given above, the course would have to be completed by the last day of the seventh month). On completion of the course the offender is given a certificate which must be returned to the clerk of the supervising court (named in the order) in order to secure the reduction. Failure to complete the course or to produce the certificate on time will result in the loss of this facility.

A number of approved courses have now been established around the country (with fees of between £50 and £200) and these will require attendance for between 16 and 30 hours, made up of a number of separate sessions, during which the offender will learn about the effects of alcohol on the body and on driving performance and behaviour, about drink-driving offences, and about the alternatives to drinking and driving. A range of relevant advice will also be given.

New Drug Driving Code of Practice

A new Code of Practice has been established (effective from 9 December 2004) to tackle the dangers caused by drivers under the influence of drugs – this includes prescription drugs and medicines as well as illegal drugs. The Code requires suspected drivers to submit to a roadside impairment test by the police. This test will be carried out in accordance with the Code and will include the so-called pupilliary test, the walk and turn test and the one-leg stand. Refusal to participate in the test will render a driver liable to the same penalty as refusal to undertake a breath test.

Penalties against Vocational Entitlements

Where a licence holder is disqualified from driving following conviction for offences committed with cars or other light vehicles, or as a result of penalty point totting-up, any vocational entitlement which that person holds is automatically lost until the licence is reinstated. Additionally, the holder of an LGV/PCV vocational entitlement may have this revoked or suspended by the DVLA without reference to the TC – see below – and be disqualified from holding such entitlement, for a fixed or an indefinite period, at any time on the grounds of misconduct or physical disability. Furthermore, a person can be refused a new LGV/PCV driving entitlement following licence revocation, again either indefinitely or for some other period of time which the Secretary of State (ie via the DVLA) specifies. A new vocational test may be ordered before the entitlement is restored – see further below.

Disqualification from holding an LGV vocational entitlement as described above does not prevent a licence holder from continuing to drive vehicles within the category B and C1 entitlements that he holds.

The TCs continue to play a disciplinary role under the new licensing scheme in regard to driver conduct, but only at the request of the DVLA. They have powers under the new provisions to call drivers to public inquiry (PI) to give information and answer questions as to their conduct. Their duty is to report back to the DVLA if they consider that an LGV/PCV entitlement should be revoked or the holder disqualified from holding an entitlement – the DVLA must follow the TCs' recommendation in these matters.

Failure to attend a PI when requested to do so (unless a reasonable excuse is given) means that the DVLA will automatically refuse a new vocational entitlement or suspend or revoke an existing entitlement.

LGV drivers who have been off the road for a period of time after being disqualified are having to prove themselves capable of driving small goods vehicles legally and safely for a period of time before their LGV driving entitlement may be restored by the TCs. One TC has said that banned LGV drivers must prove themselves on vehicles up to 7.5 tonnes before being allowed to drive vehicles up to 38 tonnes again.

Rules on disciplining LGV entitlement holders require TCs to follow a set of recommended guidelines in imposing penalties against such entitlements. Under these rules, and where there are no aggravating circumstances, a driver being disqualified for 12 months or less should be sent a warning letter, with no further disqualification of the LGV entitlement. Where a driving disqualification is for more than one year, offenders should be called to appear before the TC and should incur an additional suspension of their LGV entitlement, amounting to between one month and three months. The intention here is to allow them to regain their driving skills and road sense in a car before driving a heavy vehicle again. Where two or more driving disqualifications of more than eight weeks have been incurred within the past five years, and the combined total of disqualification exceeds 12 months, the driver should be called to public inquiry and a further period of LGV driving disqualification imposed amounting to between 3 and 6 months.

In the case of new LGV entitlements, for applicants who already have nine or more penalty points on their ordinary licence, the guidelines recommend that the

TC should issue a warning as to future conduct or suggest that the applicant tries again when the penalty points total on his licence has been reduced.

Removal of LGV Driving Licence Disqualification

Drivers disqualified from holding an LGV/PCV entitlement, as described above, may apply to have the disqualification removed after 2 years if it was for less than 4 years, or after half the period if the disqualification was for more than 4 years but less than 10 years. In any other case including disqualification for an indefinite period an application for its removal cannot be made until 5 years have elapsed. If an application for the removal of a disqualification fails another application cannot be made for three months.

The DVLA will not necessarily readily restore LGV/PCV driving entitlements on application following disqualification of a driving licence. Applicants may be called by an TC to public inquiry when he will inquire into the events which led to the disqualification and at which he may also decide that the applicant must wait a further period before applying again, must spend a period driving small (ie up to 7.5 tonnes) vehicles or must take a new LGV/PCV driving test in order to regain the vocational entitlement.

Appeals

If the DVLA refuses to grant an application for an LGV/PCV driving entitlement or revokes, suspends or limits an existing entitlement, the applicant or entitlement holder may appeal against the decision under the Road Traffic Act 1988. The first step is for them to notify the DVLA, and any TC involved in consideration of the applicant's conduct, of their intention to appeal. The appeal can then be made to a magistrates' court acting for the petty sessions in England or Wales, or in Scotland to the local sheriff.

7 Driver Testing, Training and Development

Greater emphasis than ever before is being placed on the issues of driver training and development in the light of the significant projected shortfall in qualified drivers to meet current needs. Figures vary, but most estimates suggest that some 80,000 new drivers will be needed within the next few years. In an industry already suffering shortages, the forthcoming implementation of the Road Transport Directive (see pp 92–96) which will reduce driver working times from the current average of 60/66 hours to just 48 hours, and the proposed Driver Training Directive that will take existing drivers out of service for training purposes on at least one day each year, the prospects for recovery to normal staffing levels currently look somewhat bleak. Added to this will be the prolonged period of training required by newcomers before they can get behind the wheel of a heavy vehicle even for training and the test, never mind starting work for an employer.

This chapter deals with the legal requirements for driving tests (including the Advanced Commercial Vehicle Driving Test and the Master Driving Certificate), the New Young Driver training scheme, training for the use of lorry loaders and for dangerous goods carrying. Additionally, it covers in detail the requirements for compulsory training of LGV drivers under the proposed EC Directive (still currently in draft form and subject to further consultation).

DRIVING TESTS

The main purpose of driver testing is to ensure that all drivers taking a vehicle on the road:

- are safe and competent to do so;
- know the rules of the road and the significance of traffic signs and signals; and
- appreciate the dangers arising from moving vehicles.

The purpose of the additional tests for LGV/PCV entitlements is to ensure that such drivers are competent to drive these large vehicles on the roads in safety – especially when carrying heavy loads or passengers. Vocational testing also provides a measure of professionalism among goods and passenger vehicle drivers.

Approximately 1.2 million ordinary driving tests and 62,000 lorry and bus tests were carried out in 2000/01.

New driver testing procedures, following the requirements of the EU's 'second' driver licensing Directive, were introduced for car and light vehicle drivers from 1 July 1996 and from 1 January 1997 for large goods and passenger vehicle drivers – including car and LGV theory tests. The main objective of the LGV/PCV test remains, which is to ensure that goods and passenger vehicle drivers are competent to drive large vehicles on the roads in safety. It also provides a measure of professionalism among commercial vehicle drivers. The test is more comprehensive and more complex than the ordinary (ie car and light goods) driving test and, consequently, demands greater skill and knowledge from the driver who wishes to pass.

Proof of Identity

Candidates for both ordinary and vocational (ie LGV and PCV) driving tests must produce satisfactory photographic evidence of identity when arriving for a test. Acceptable identity documents for this purpose include existing driving licences (ordinary, LGV or PCV or an overseas driving licence), a passport or an employer-issued identity card bearing the holder's name, signature and photograph. If a test candidate cannot produce satisfactory means of identification the test will not be conducted and the fee forfeited.

This measure is necessary to combat a rising incidence of persons with false identities taking multiple LGV driving tests on behalf of others, who cannot themselves pass the test, but who are prepared to pay large sums to acquire a test pass certificate.

Ordinary Driving Test

Before persons can be granted a licence to drive a motor vehicle on the road they must pass both a written theory test and a practical driving test on the class of vehicle for which they require the licence. An entitlement to drive will not be gained until both parts of the test are passed – the theory test having to be passed before the practical driving test can be taken.

The Theory Test

Driving test candidates must take and pass a 'touch screen' theory test – which replaces the old, verbal, questioning to test the candidate's knowledge of the *Highway Code* – carried out on behalf of the Driving Standards Agency (DSA – an Executive Agency of the Department for Transport, (D*f*T) at a network of practical test centres across Britain.

The test comprises 35 multiple-choice questions of which the candidate must get at least 30 correct to achieve a pass certificate. The questions concern such matters as; driver attitude, traffic signs and regulations, the effects of alcohol and drugs, driver fatigue and safety, and environmental aspects of vehicles. All these topics are covered in the *Highway Code* and the DSA's *Driving Goods Vehicles: The Official DSA Syllabus*. The DSA has also published *The Official Theory Test for Drivers of Large Vehicles* which explains the test and lists the 900-odd bank of questions from which test papers are set, together with the correct answers. All these publications are available from The Stationery Office or main booksellers.

The theory test is computerized with the questions appearing on screen, which the candidate has only to touch to indicate the correct answer. This system makes the test more accessible to the general public, and is particularly helpful for people with special needs. Candidates are advised of the result of their theory test immediately.

Practical Driving Tests

The practical driving test (lasting approximately 38–40 minutes) is carried out by examiners from the Driving Standards Agency. Candidates must present a theory test pass certificate – practical driving tests will not be carried out without the candidate having first passed the theory test.

Test candidates have to meet the following requirements:

- They must show that they are fully conversant with the contents of the *Highway Code*.
- They must prove that they are able to read in good daylight (with the aid of spectacles, if worn) a motor vehicle's registration number in accordance with the vision requirements (see p 152).
- They must show that they are competent to drive without danger to and with due consideration of other users of the road, including being able to:
  - start the engine of the vehicle;
  - move away straight ahead or at an angle;
  - overtake, meet or cross the path of other vehicles and take an appropriate course;
  - turn right-hand and left-hand corners correctly;
  - stop the vehicle in an emergency and in a normal situation, and in the latter case bring it to rest at an appropriate part of the road;
  - drive the vehicle backwards and while doing so enter a limited opening either to the left or to the right;
  - cause the vehicle to face the opposite direction by the use of forward and reverse gears;
  - carry out a reverse parking manoeuvre which involves stopping the vehicle next to and parallel with a parked vehicle, then reversing to position and park the vehicle in front of or behind the other vehicle, level with and reasonably close to the kerb;
  - indicate their intended actions at appropriate times by giving appropriate signals in a clear and unmistakable manner (in the case of a left-hand drive vehicle or a disabled driver for whom it is impracticable or undesirable to give hand signals there is no requirement to provide any signals other than mechanical ones);

- act correctly and promptly on all signals given by traffic signs and traffic controllers and take appropriate action on signs given by other road users.

The driving test may be terminated in the event of a serious driving error which jeopardizes public safety.

Cars used for the ordinary (ie category B) driving test must be fitted with a seatbelt and a head restraint for the front seat passenger and an additional rear view mirror for use by the examiner.

Candidates who fail the driving test are given an oral explanation of the reasons for their failure.

Extended Driving Tests

Since 6 January 1992 the Courts have been vested with powers to impose an extended driving test on any person who has been convicted of dangerous driving offences, or convicted of other offences involving obligatory disqualification.

Loss of a driving licence for dangerous driving or related offences involving obligatory disqualification means that the licence holder returns to 'learner' status and has to conform to the normal rules applicable to provisional licence holders. A theory test must be taken and passed for the category of vehicle on which the extended practical driving test is to be taken. This extended driving test is longer and therefore more demanding than the normal, lasting for about 70 minutes and covering a wide variety of roads, usually including dual carriageways. Candidates for the extended retest are advised to prepare by taking suitable instruction from an Approved Driving Instructor (ADI). The higher fee (as indicated below) reflects the length of the test.

The purpose of re-testing is not intended to be punitive; rather it is a road safety measure, designed to re-check the driver's competence.

Driving Test Fees

The fee for all theory tests is £20.50.

The fee for an ordinary driving test (conducted on weekdays up to 4.30 pm) is £42, or £51 if conducted on a weekday evening or on a Saturday.

Driving tests for lorries and buses cost £80 on weekdays (up to 4.30 pm) and £98 if conducted in the evening or on a Saturday.

Higher test fees of £84 on weekdays (up to 4.30 pm) and £102 on Saturdays are payable where an extended retest is ordered following obligatory disqualification of a driving licence (see above).

Test fees are forfeited unless at least 10 clear days' notice of cancellation is given.

Driving Instruction

Only approved instructors (ADIs – approved by the Driving Standards Agency) are permitted to give driving instruction for payment on vehicles legally defined as 'motor cars' which basically means the following classes of vehicle:

- Private cars.
- Goods vehicles not exceeding 3050 kg unladen weight.

Tuition given for payment on heavy goods vehicles does not come within the scope of this legislation despite a certain amount of transport industry opinion that it should do so. However, to ensure that the level of driving instruction available to learner LGV drivers is of a consistently high standard a new scheme for the voluntary registration of LGV driving instructors has been established. Entry to the register is currently via a three-part examination of both driving and instructional abilities.

LGV Driver Testing

In order to drive a large vehicle (ie goods or passenger) it is necessary to pass both the large vehicle theory test and a practical driving test on either a large goods or passenger vehicle of the appropriate category for which a licence is required.

Theory Testing

Since 1 January 1997 large vehicle (ie LGV/PCV) driving test candidates must take and pass a 'touch screen' theory test – which has replaced the old, verbal, questioning to test the candidate's knowledge of technical and safety matters and the *Highway Code*. This test has been expanded with the inclusion of a wider range of topics, such as fuel economy, environmentally sensitive driving and safety issues.

Theory tests are carried out on behalf of the Driving Standards Agency (DSA) at their dedicated test centres nationwide. The test comprises 35 multiple-choice questions – a pass is achieved with 30 correct answers. Full details of the theory test are contained in *The Professional LGV Drivers' Handbook* (by David Lowe, and available from Kogan Page), which explains the test and lists sample questions together with the correct answers.

The LGV Driving Test

In order to undertake a large goods or passenger vehicle (LGV/PCV) driving test, candidates must, since 1 January 1997, produce a large vehicle theory test pass certificate – without this the examiner will not conduct the practical driving test.

The 90-minute large goods vehicle practical driving test is conducted by DSA examiners and booking has to be made through the Driving Standards Agency (see Appendix V for addresses). Certain test centres offer Saturday morning LGV and PCV driver testing.

The staged system of testing means that:

- Applicants for LGV tests must already hold a full category B (car and light vehicle) driving entitlement before taking a test to obtain a category C entitlement.
- Category B entitlement holders have to pass a test on a rigid goods vehicle in category C before being able to take a test to qualify for driving articulated vehicles and drawbar combinations in category C+E.
- Category C1 entitlement holders wishing to drive vehicles in category C1+E must take a further test for this type of vehicle combination.

In each case the driver must hold a provisional entitlement for the category of vehicle on which he or she wants to be tested.

Application and Fees

Large vehicle theory tests cost £20.50. This test must be passed before the practical driving test is taken.

Applications for the large vehicle practical driving test have to be made on form DLG26 obtainable from local DSA offices (see Appendix V). The current test fee of £80 (£98 for Saturday and weekday evening tests) must be sent with the application. Test applications can also be made by telephone with payment by credit or debit card.

Applicants are warned to apply for a test in good time. This is important if an lgv driving entitlement is required – subject to passing the test – from a particular date; and applicants must also ensure that their driving is of a sufficiently high standard to be able to pass the test.

Test Cancellation

Should a candidate need to cancel a test appointment this should be done at least 10 clear working days in advance, otherwise the fee will be forfeited.

Identification

Test candidates must be able to produce satisfactory photographic identification on arrival at the test centre, otherwise the examiner may refuse to conduct the test and the fee will be forfeited (see p 170).

Trainer Bookings

The Trainer Booking scheme introduced in 1995 allows driver training schools to make block bookings for LGV tests. Test fees have to be paid in full (ie £76 for a weekday test and £94 for a Saturday test) at the time of making the booking. Such bookings can be made up to 10 weeks in advance of the required test date with the individual driver's name and vehicle details being given no later than two days before the test. Cancellation of a block-booked test has to be made at least 10 days before the test date to avoid losing the fee.

Vehicles for the LGV Driving Test

The candidate has to provide the vehicle (or arrange for the loan of a suitable vehicle) on which he wishes to be tested and it must comply with the following requirements:

- it must be unladen and of the category (ie a 'minimum test vehicle') for which an LGV driving entitlement is required – see below;
- it must display ordinary 'L' plates front and rear*;
- it must be in a thoroughly roadworthy condition;
- seating accommodation in the cab must be provided for the examiner;
- it must have sufficient fuel for a test lasting up to two hours.

**NB: The old-type orange-background hgv 'L' plate is no longer valid. Learner drivers in Wales may display a 'D' plate instead of the usual 'L' plate.*

Minimum Test Vehicles

Higher minimum test vehicle standards have been required for vehicles first registered since 1 October 2003. For any vehicle first registered before 1 October 2003 the pre-existing minimum test vehicle standards will apply until 1 July 2007, when the higher standards will apply.

Minimum Test Vehicles (registered before 1 October 2003)

Category C1 (subject to notes 2 and 3)

- A medium sized lorry with a maximum authorized mass (MAM) of at least 4 tonnes; and
- capable of 80 km/h (50 mph).

Category C1 + E (subject to notes 1 and 3)

- A drawbar combination comprising a Category C1 vehicle towing a trailer of at least 2 tonnes MAM;
  - with a combined length of at least 8 metres; and
  - capable of 80 km/h (50 mph); or
- A medium sized articulated lorry with a MAM of at least 6 tonnes;
  - at least 8 metres in length; and
  - capable of 80 km/h (50 mph).

Category C (subject to notes 2 and 3)

- A rigid goods vehicle with a MAM of at least 10 tonnes;
  - at least 7 metres in length;
  - capable of 80 km/h (50 mph); and
  - with a maximum length of 12 metres (39 ft 4 in).

Category C + E (subject to notes 1 and 3)

- An articulated lorry with a MAM of at least 18 tonnes;
  - at least 12 metres in length;
  - capable of 80 km/h (50 mph); and
  - with a maximum length of 16.5 metres (54 ft);
- A drawbar outfit comprising a combination of a Category C vehicle with a trailer of at least 4 metres platform length;
  - of at least 4 tonnes MAM;
  - with a combined weight of at least 18 tonnes;
  - capable of 80 km/h (50 mph);
  - with a combined length of at least 12 metres; and
  - with a maximum length of 18.75 metres (61 ft 5 in).

Minimum Test Vehicles (first registered from 1 October 2003)

Category C1 (subject to notes 2, 3 and 4)

- A medium sized lorry with a MAM of at least 4 tonnes;
  - at least 5 metres in length;
  - capable of 80 km/h (50 mph); and
  - with a closed box* cargo compartment at least as wide and as high as the cab.

Category C1 + E (subject to notes 1, 3 and 4)

- A drawbar combination comprising a Category C1 vehicle;
  - towing a trailer of at least 2 tonnes MAM;
  - with a combined length of at least 8 metres;
  - capable of 80 km/h (50 mph);
  - with a closed box* trailer slightly less wide than the towing vehicle, but rear-view must be by external mirrors only; or
- A medium sized articulated lorry with a MAM of at least 6 tonnes;
  - with a combined length of at least 8 metres;
  - capable of 80 km/h (50 mph);
  - with a closed box trailer slightly less wide than the towing vehicle, but rear-view must be by external mirrors only.

Category C (subject to notes 2, 3 and 4)

- A rigid goods vehicle with a MAM of at least 12 tonnes;
  - at least 8 metres in length;
  - at least 2.4 metres in width;
  - capable of 80 km/h (50 mph);
  - with at least 8 forward gears;
  - a closed box* cargo compartment at least as wide and as high as the cab; and
  - a maximum length of 12 metres (39 ft 4 in).

Category C + E (subject to notes 1, 3 and 4)

- A drawbar outfit comprising a combination of a Category C vehicle and trailer with a MAM of 20 tonnes;
  - with a combined length of at least 14 metres;
  - with a trailer at least 7.5 metres in length from coupling eye to extreme rear;
  - at least 2.4 metres in width;
  - capable of 80 km/h (50 mph);
  - with at least 8 forward gears;
  - with a closed box* cargo compartment at least as wide and as high as the cab; and
  - a maximum length of 18.75 metres (61 ft 5 in); or
- An articulated lorry with a MAM of at least 20 tonnes;
  - with a minimum length of 14 metres;
  - at least 2.4 metres in width;
  - capable of 80 km/h (50 mph);
  - with at least 8 forward gears;

- with a closed box* cargo compartment at least as wide and as high as the cab; and
- a maximum length of 16.5 metres (54 ft).

Notes:

1. All vehicle combinations must operate the appropriate service brakes and utilize a heavy duty coupling arrangement suitable for the weight.
2. A tractive unit with a fifth wheel coupling is not a suitable vehicle for Category C or C1 test.
3. All vehicles presented for test must be fitted with a fully working tachograph.
4. All vehicles presented for test must be fitted with an anti-lock braking system (ABS).

**A closed box body cargo compartment or closed box trailer can be curtain sided.*

The LGV Driving Test Syllabus

The recommended syllabus to be studied by candidates preparing for the LGV driving test is as follows. See also DSA publication, *Driving Goods Vehicles: The Official DSA Syllabus Manual* (available from The Stationery Office) and *The Professional LGV Drivers' Handbook* (by David Lowe, available from Kogan Page).

The syllabus comprises seven main sections containing advice and detailing specific requirements for the test as well as an introductory section outlining the prior 'thorough' knowledge that a candidate should acquire before attempting the test.

The syllabus contains the following main headings:

Part One: Getting Started

- Selecting an instructor.
- Applying for your licence.
- The theory test.
- Medical requirements.
- Professional standards.
- Responsibility and attitude.

Part Two: Understanding Large Goods Vehicles

- Understanding LGVs.
- Forces affecting your vehicle.
- Vehicle characteristics.
- Vehicle limits.
- Braking systems.
- Gear systems.
- Load restraint.

Part Three: Limits and Regulations

- Environmental impact.
- Drivers' hours and records.
- Other regulations.

Part Four: Driving Skills

- Professional driving.
- Driving at night.
- Motorway driving.
- All-weather driving.
- Accidents.
- First aid.
- Breakdowns.

Part Five: Preparing for the Driving Test

- About the driving test.
- The theory test.
- How to apply for your test.
- Before attending your test.
- Legal requirements.
- At the test centre.
- The official syllabus.

Part Six: The LGV Driving Test

- The reversing exercise.
- The braking exercise.
- The vehicle controls.
- Other controls.
- The gear-changing exercise.
- Moving off.
- Using the mirrors.
- Giving signals.
- Acting on signs and signals.
- Awareness and anticipation.
- Making progress.
- Controlling your speed.
- Separation distance.
- Hazards.
- Selecting a safe place to stop.
- Uncoupling and recoupling.
- If you pass.
- If you don't pass.

Part Seven: Additional Information

- Disqualified drivers.
- DSA services.
- DSA offices.
- LGV test centres.
- Traffic Commissioners and Traffic Area Offices.
- Other useful addresses.
- Categories of LGV licences.

- Minimum test vehicles.
- Hazard labels.
- Cone positions.
- Glossary.

New LGV Tests from September 2003

A number of significant changes to the vocational (LGV) driving test were made from September 2003 as follows.

For a start, drivers seeking vocational licence entitlement are expected to be experienced and technically expert. Test candidates are asked five 'show me' and 'tell me' type questions, with a driving fault being recorded for each incorrect answer up to a maximum of four driving faults. If all five answers are given incorrectly, a serious fault will be recorded.

Examples of safety check questions are as follows.

Identify where:

- you would check the engine oil level and tell me how you would check that the engine has sufficient oil.
- you would check the engine coolant level and tell me how you would check that the engine has the correct level.
- the windscreen washer reservoir is and tell me how you would check the windscreen washer level.

Tell me how you would check:

- your tyres to ensure that they are correctly inflated, have sufficient tread depth and that their general condition is safe to use on the road.
- that the brake lights are working.
- the condition of the reflectors on this vehicle.
- the condition of the windscreen and windows on this vehicle.
- the condition of the windscreen wipers on this vehicle.
- the condition of the suspension on this vehicle.
- that the condition of the body is safe on this vehicle.
- that the power assisted steering is working.

Tell me:

- how you would operate the loading mechanism on this vehicle (vehicle specific ie tail lift).
- the main safety factors involved in loading this vehicle.

Show me how you would check:

- that the headlamps, sidelights and tail lights are working.
- that the direction indicators are working.
- the operation (specify) of the audible warning devices on this vehicle.
- the wheel nuts are secure on this vehicle.
- the condition of the mudguards on this vehicle.
- for the correct air pressure on this vehicle.
- for air leaks on this vehicle.

- that your cargo doors are secure.
- that your cab locking mechanism is secure.

Show me how you would:

- replace the tachometer disc on this vehicle.

Show me:

- what instrument checks you would make before and after starting the engine on this vehicle.
- where the emergency exits are and how you would check that they are operating correctly.
- where the first aid equipment is on this vehicle.
- where the fire extinguishers are on this vehicle.

The Safe Parking Test

Candidates will be asked to carry out the current reversing exercise, but at the end of the reversing bay there will be a simulated loading platform to up to which the candidate will be required to reverse (stopping with the extreme rear of the vehicle within a stopping area marked by black and yellow hatched lines) and park safely for loading and unloading.

Stopping short of the marked stopping area will be assessed as a serious fault. Reversing through the marked stopping area and dislodging the simulated loading/unloading platform will be assessed as a serious fault. Stopping within the marked stopping area but unacceptably short of the platform will be assessed as a driving fault. A total loss of vehicle control, which causes actual danger to the driver, examiner or another road user, will be assessed as a dangerous fault.

It is recognized that some professional drivers reverse, under control, up to a loading/unloading platform until they gently touch the platform. This will be acceptable on test as long as the platform is not dislodged.

The Uncoupling and Recoupling Exercise

After uncoupling the candidate will be required to park the towing vehicle alongside the trailer and then realign the towing vehicle with the trailer before recoupling. The competencies of control, accuracy and effective observation will be assessed during the exercise. A fault involving actual danger to the driver or another road user will be assessed as a dangerous fault. A fault that would be potentially dangerous to the driver or another road user will be assessed as a serious fault. Faults that reflect that the driver does not have the required competencies to carry out the exercise following industry best practice and recognized procedures will be assessed as driving or serious faults, depending upon the severity of the fault.

New EU LGV Driving Test Standards

Under European Commission Directive 2000/56/EC, which amended an earlier Directive on driving licences (91/439/EEC), the conduct of the practical driving test itself was revised and applied to tests conducted from September 2003.

The following text summarizes Annex II to the Directive in so far as it applies to LGV driving tests only.

Driving Skills and Behaviour for LGV Drivers

Driving test candidates must demonstrate the following key skills:

- Preparation and technical check of the vehicle for the purposes of ensuring road safety. Applicants must demonstrate that they are capable of preparing to drive safely by satisfying the following requirements:
  - adjusting the seat as necessary to obtain a correct seated position;
  - adjusting rear-view mirrors, seat belts and head restraints if available;
  - random checks on the condition of the tyres, steering, brakes, lights, reflectors, direction indicators and audible warning device;
  - checking the power-assisted braking and steering systems; checking the condition of the wheels, wheel-nuts, mudguards, windscreen, windows and wipers, fluids (eg engine oil, coolant, washer fluid); checking and using the instrument panel including the recording equipment as defined in Regulation (EEC) No 3821185 [ie the tachograph];
  - checking the air pressure, air tanks and the suspension;
  - checking the safety factors relating to vehicle loading: body, sheets, cargo doors, loading mechanism (if available), cabin locking (if available), way of loading, securing load;
  - checking the coupling mechanism and the brake and electrical connections (categories C+E, C1+E only);
  - reading a road map, route planning, including the use of electronic navigation systems (optional).
- Special manoeuvres to be tested with a bearing on road safety:
  - coupling and uncoupling, or uncoupling and recoupling a trailer from its motor vehicle; the manoeuvre must involve the towing vehicle being parked alongside the trailer (ie not in one line) (categories C+E, C1+E, only);
  - reversing along a curve;
  - parking safely for loading/unloading at a loading ramp/platform or similar installation.
- Behaviour in traffic. Applicants must perform all the following actions in normal traffic situations, in complete safety and taking all necessary precautions:
  - driving away: ie after parking, after a stop in traffic, when exiting a driveway;
  - driving on straight roads; passing oncoming vehicles, including in confined spaces;
  - driving round bends;
  - crossroads: approaching and crossing of intersections and junctions;
  - changing direction: left and right turns; changing lanes;
  - approach/exit of motorways or similar (if available): joining from the acceleration lane; leaving on the deceleration lane;
  - overtaking/passing: overtaking other traffic (if possible); driving alongside obstacles, eg parked cars; being overtaken by other traffic (if appropriate);

- special road features (if available): roundabouts; railway level crossings; tram/bus stops; pedestrian crossings; driving up/downhill on long slopes;
- taking the necessary precautions when alighting from the vehicle.

Marking of the Skills and Behaviour Tests

In assessing driver competence, the examiner will be looking for the ease with which the applicant handles the vehicle controls and his capacity to drive in complete safety in traffic. Importantly, the examiner must be made to feel safe throughout the test. The examiner will also pay special attention as to whether the applicant shows defensive and social driving behaviour, in particular ensuring the safety of all road users, especially those who are the weakest and most exposed, by showing due respect for others. Overall, drivers are expected to make progress in an economically and environmentally friendly manner.

During the test, the examiner will be checking to ensure that the test candidate makes proper use of safety belts, rear-view mirrors, head restraints; driving seat; lights and other equipment; clutch, gearbox, accelerator, braking systems (including third braking system, if available) and the steering. He will also be checking the driver's control of the vehicle under different circumstances and at different speeds; his steadiness on the road (this means no fast acceleration, smooth driving and no hard braking); the weight and dimensions and characteristics of the vehicle; the weight and type of load.

Additionally, drivers must show that they are capable of detecting any major technical faults in their vehicles, especially those posing a safety hazard, and know how to have them remedied in an appropriate fashion.

LGV Test Passes and Failures

A driver who passes the LGV/PCV driving test is issued with a certificate to that effect, valid for a period of two years, and the holder can apply for an LGV driving entitlement of the appropriate category to be added to his unified driving licence.

A driver who fails the LGV/PCV driving test is given a written statement of failure and an oral explanation of the reasons for his failure. He may apply for an immediate retest to be taken after an interval of at least 3 days.

Advanced Commercial Vehicle Driving Test

For many years the advanced driving test organized by the Institute of Advanced Motorists has been looked upon as a severe test of driving skills requiring a high degree of knowledge of the 'rules of the road' for private car drivers. Commercial vehicle drivers who wish to show that they too have attained an exceptional level of proficiency can take the advanced driving test designed specially for commercial vehicles and on passing they may display the coveted IAM badge.

The advanced commercial vehicle driving test is organized by the Institute of Advanced Motorists (IAM House, 510 Chiswick High Road, London W4 5RG; (tel: 020 8996 9600; fax: 020 8996 9601; website: www.iam.org.uk) and is open to any heavy goods vehicle driver, subject to certain conditions as follows:

- Loads, if carried on test vehicles, must be properly secured.
- A safe seat at the front of the vehicle must be available for the examiner.
- The driver must not, by taking the test, contravene the drivers' hours and record keeping rules.

On passing the test, the applicant becomes eligible for admission to membership of the Institute of Advanced Motorists.

Fees

The IAM has introduced a combined fee structure under its *Skill For Life* programme which includes the following:

- a copy of the IAM's Advanced Driving/Motorcycling Manual;
- associate membership of one of the many local IAM Groups;
- a place on an IAM course to prepare for the Advanced Driving Test;
- the fee for the IAM Advanced Test;
- a First-Year Membership of the IAM after passing the Advanced Test.

Exemption from the Test

Certain specially qualified drivers may apply to become Members of the Institute without taking the advanced driving test.

- Royal Navy, Army and Royal Air Force lgv instructors and qualified testing officers who have passed an LGV instructor's course and whose application is supported by the recommendation of the applicant's commanding officer.
- Holders of the Road Transport Industry Training Board Instructor's Certificate.
- Fire Service LGV instructors who have completed an LGV instructor's course, and whose application is supported by the senior instructor.
- LGV driving examiners.

The Advanced Test

To pass the test the driver should show 'skill with responsibility' and any driver of reasonable experience and skill should be able to pass without difficulty. The Institute examiners are all ex-police drivers holding a Class 1 Police Driving Certificate, and they test candidates on routes located all over Britain.

The test lasts about two hours, during which the test route of some 35–40 miles is covered. The route incorporates road conditions of all kinds including congested urban areas, main roads, narrow country lanes and residential streets. Candidates are not expected to display elaborate driving techniques. Examiners prefer to see the vehicle handled in a steady workmanlike manner without exaggeratedly slow speeds or excessive signalling. Speed limits must be observed (driving in excess of any limit results in test failure) and the driving manner must take into consideration road, traffic and weather conditions. However, the examiners expect candidates to drive briskly within the limits and to cruise at the legal limit (ie either on the road or the vehicle, whichever is lower) whenever circumstances permit.

Drivers will be asked to reverse around a corner and to make a hill start. There will be spot checks on the driver's power of observation (ie the examiner will ask questions about road signs or markings recently passed or about other significant landmarks).

Examiners ask a number of questions of candidates but these are not trick questions. The previous requirement for the driver to give a running commentary during a portion of the route no longer exists – although the test regulations do state that candidates are free to give a commentary if they wish to make extra clear their ability to 'read the road'.

Test Requirements

The examiner will consider the following aspects of driving:

- *Acceleration*: must be smooth and progressive, not excessive or insufficient, and must be used at the right time and place.
- *Braking*: must be smooth and progressive, not fierce. Brakes should be used in conjunction with the driving mirror and signals. Road, traffic and weather conditions must be taken into account.
- *Clutch control*: engine and road speeds should be properly co-ordinated when changing gear. The clutch should not be 'ridden' or slipped and the vehicle should not be coasted with the clutch disengaged.
- *Gear changing*: should be smooth and carried out without jerking.
- *Use of gears*: the gears should be correctly selected and used and the right gear engaged before reaching a hazard.
- *Steering*: the wheel should be correctly held with the hands and the 'crossed-arm' technique should not be used except when manoeuvring in confined spaces.
- *Driving position*: the driver should be alert and should not slump at the wheel. Resting an arm on the door while driving should be avoided.
- *Observation*: the driver should 'read' the road ahead and show a good sense of anticipation and the ability to judge speed and distance.
- *Concentration*: the driver should keep his attention on the road and should not be easily distracted.
- *Maintaining progress*: taking account of the road, traffic and weather conditions, the driver must keep up a reasonable pace and maintain good progress.
- *Obstruction*: the candidate must be careful not to obstruct other vehicles by driving too slowly, taking up the wrong position on the road or failing to anticipate and react correctly to the traffic situation ahead.
- *Positioning*: the driver must keep to the correct part of the road especially when approaching and negotiating hazards.
- *Lane discipline*: the driver must keep to the appropriate lane and be careful not to straddle white lines.
- *Observations of road surfaces*: the driver must keep an eye on the road surface especially in bad weather and should watch out for slippery conditions.
- *Traffic signals*: signals, signs and road markings must be observed, obeyed and approached correctly and the driver should show courtesy at pedestrian crossings.
- *Speed limits and other legal requirements* these should be observed. The examiner cannot condone breaches of the law.

- *Overtaking*: must be carried out safely and decisively maintaining the right distance from other vehicles and using the mirror, signals and gears correctly.
- *Hazard procedure and cornering*: road and traffic hazards must be coped with properly, and bends and corners taken in the right manner.
- *Mirror*: the mirror must be used frequently especially in conjunction with signals and before changing speed or course.
- *Signals*: direction indicator, and hand signals when needed, must be given at the right place and in good time. The horn and headlamp flasher should be used as per the *Highway Code*.
- *Restraint*: the driver should show reasonable restraint, but not indecision, at the wheel.
- *Consideration*: sufficient consideration and courtesy should be shown to other road users.
- *Vehicle sympathy*: the driver should treat the vehicle with care, without overstressing it by needless revving of the engine and by fierce braking.
- *Manoeuvring*: manoeuvres (reversing) should be performed smoothly and competently.

DRIVER TRAINING

Young LGV Driver Scheme

The Young LGV Driver Scheme has been devised by Skills for Logistics in conjunction with the D*f*T, the Road Haulage Association (RHA), the Freight Transport Association (FTA) and the transport trades unions with the initial objective of providing a route for young people aged between 16 and 21 years who would like to follow a career in the road transport and distribution industry as a large goods vehicle driver. The scheme makes sure that they have the required skills and knowledge to competently perform all the duties and responsibilities involved in transporting goods by road. Additionally, the scheme aims to effectively promote the safe operation of vehicles at all timesfor the benefit of the employer, employees, customers and the general public.

The scheme provides structured training and development for young people between 16 and 21 years and enables them to obtain an LGV category C licence at 18 years of age and progress to an LGV category C+E licence by 20 years of age.

In addition to thorough driver training, the scheme also requires the completion of a Level '2' Road Haulage and Distribution National/Scottish Vocational Qualification (N/SVQ), in either Transporting Goods by Road or Driving Goods Vehicles.

Membership of the scheme is open to young people between their 16th and 21st birthdays who are employed by a haulage company, and to employers who operate at least one category C LGV and are able to provide approved training and assessment for a trainee. A Skills for Logistics Approved Training Organisation must carry out all the training and assessments required within the scheme.

Progress through the Scheme follows the stages listed below:

- Selection of a trainee by an employer/prospective trainee finding suitable employment.
- Contacting an Approved Training Provider.

- Registering the company as an approved employer.
- Registering the trainee.
- Developing a training plan with the Approved Training Provider.
- Induction and safety training (Units 1–4 of Transporting Goods by Road or units 6–7 of Driving Goods Vehicles).
- Category C driver training.
- Completion of the N/SVQ.
- Further driving assessments.
- Award of large goods vehicle Young Driver Certificate.

Further information may be obtained from Skills for Logistics at 14 Warren Yard, Warren Farm Office Village, Stratford Road, Milton Keynes MK12 5NW (tel: 01908 313360; fax: 01908 313006; website: www.skillsforlogistics.org).

Driver Training for Lorry Loaders

No mandatory requirements exist at the present time for goods vehicle drivers to hold certificates of competence to operate lorry-mounted cranes. However, under the Health and Safety at Work etc Act 1974 employers have a statutory duty to provide adequate instruction and safety training for all employees.

A voluntary certification scheme is run by the Construction Industry Training Board (CITB) to improve safety on construction sites. This scheme is strongly supported by the construction industry which may refuse entry to its own sites to non-certified lorry drivers. Mainly, the Board's scheme is concerned with ensuring a sound understanding of safety procedures for the use of a wide range of equipment including lorry-mounted cranes and skip loaders. Under the scheme, the Board provided certification of existing lorry-loader and skip-loader operators who could show by means of employer confirmation that they were experienced in the use of such equipment.

Existing operatives prior to June 1993 who could produce the employer declaration could obtain the Board's safety certificate for lorry-mounted crane operation under a grandfather rights arrangement. A similar arrangement applied to skip-loader operatives until the end of 1994. Newcomers seeking first-time certification and drivers renewing grandfather rights certificates have to undergo (re-) training and site-based assessment to show that they can operate such equipment with complete safety. CITB Safety certificates are renewable at five-yearly intervals.

Another scheme is run by the Contractors' Mechanical Plant Engineers (CMPE), although this requires no specific training, certification being based solely on employers' references. Training is also provided by most member firms of the Association of Lorry Loader Manufacturers and Importers (ALLMI). Additionally, the Association itself publishes a Code of Practice for safe application and use of loaders, and is currently preparing a training programme. Details are available from the Association at 14 Manor Close, Droitwich, Worcestershire WR9 8HG (tel: 01905 770892 or 01905 451040). Other organizations providing suitable lorry-loader training include the Freight Transport Association, Centrex (previously the RTITB) and a number of commercial training firms.

Driver Training for Dangerous Goods Carrying

From 10 May 2004 the UK's multiplicity of domestic dangerous goods legislation was replaced by a single regulatory document containing, among many other provisions, those relating to the training requirements for drivers of dangerous goods vehicles. These new regulations, The Carriage of Dangerous Goods and Use of Transportable Pressure Equipment Regulations 2004 (regulation 24), replace the previous regulations, namely; The Carriage of Dangerous Goods by Road (Driver Training) Regulations 1996, which are completely revoked. They require that dangerous goods drivers are trained in accordance with section 8.2.1 of the European Agreement Concerning the International Carriage of Dangerous Goods by Road (commonly referred to as the ADR) and that they hold Vocational Training Certificates issued by the Driver and Vehicle Licensing Agency (DVLA), Swansea, and gained by attending an approved course and passing a written examination set by the City and Guilds of London Institute (C&G).

Relevant Vehicles

The regulations apply broadly to the carriage of specified dangerous goods in:

- road tankers with a capacity exceeding 1000 litres or,
- tank containers with a capacity greater than 3000 litres (with certain exceptions), or
- vehicles exceeding 3.5 tonnes permissible maximum weight carrying specified dangerous goods in packages.

Driver Responsibilities

Vocational training certificates are valid for a period of five years and are renewable, subject to the holder attending an approved refresher course and taking a further examination within the 12 months prior to expiry date of an existing certificate. Drivers must carry the certificate with them when driving relevant vehicles and must produce it on request by police or a goods vehicle examiner. It is an offence to drive a dangerous goods vehicle without being the holder of a certificate, or to fail to produce such a certificate on request.

Employer Responsibilities

It is the responsibility of the employer to ensure that dangerous goods drivers receive training so they understand the dangers arising from the products they are carrying and what to do in an emergency situation, and that they hold relevant certificates covering the vehicle being driven and the products carried. The employer has a duty to provide necessary training leading to the certificate and the official C&G examination. The employer must retain records of all instruction and training given to drivers.

Approved Training and the City and Guilds Examination

The Department for Transport has approved suitable establishments where dangerous goods driver training can be obtained and the C&G examination taken (see Appendix X for a list of training facilities).

The syllabus for the examination involves both theoretical sessions and practical exercises. The examination itself comprises a core element designed to assess candidates' practical and legal knowledge plus a specialist element for either road tanker and tank container drivers or packaged goods drivers (or both if required). Additionally, candidates have to pass individual 'dangerous substance' examination papers, which cover each of nine classes of dangerous goods, to test their specialist knowledge of the products they carry in their work.

Candidates who pass the examination (by achieving a pass mark of at least 75 per cent in each element – ie core, tanker/package and substance) will receive their certificate direct from the DVLA, Swansea. Those who fail can apply to resit the examination without further training within a period of 16 weeks from receipt of the notification of failure.

Further information on the training and examinations may be obtained from Area Offices of the Heath and Safety Executive (see local telephone directory).

Master Driver Certificate

The Centrex training organization has introduced a new 'Master Driver' qualification for heavy goods drivers. Drivers are put through tests and have their driving ability and overall competence assessed to gain certification.

Content of the Master Driver Test

Master Driver is assessed during the working day by an RTITB accredited assessor. It is renewable every year and drivers are reassessed every three years. Specific marking criteria ensure a uniform, national standard.

Daily Vehicle Check (duration: 25 minutes)

- Against a comprehensive Master Driver checklist.
- Competence of the driver is observed by a qualified assessor.

Vehicle Load and Associated Documentation Check (duration: 10 minutes)

- Ensures the driver knows how to check a load is distributed and secured safely.
- Ensures that associated documents are completed correctly.
- Encourages driver responsibility for load.

Driving Assessment (duration: 120 minutes)

- Single in-cab assessment.
- Identification of faults.
- Defensive driving.
- Economic driving.
- Assessment over all types of road including motorway.

Route Planning and Map Reading (duration: 20 minutes)

Requires driver to plan 150-mile journey taking account of:

- Driving a laden vehicle.
- Environmental issues.
- Fuel economy.
- Journey time.

Associated Knowledge (duration: 40 minutes)

- First aid.
- Customer care.
- Drivers' hours.
- Highway Code.
- Fire control.
- Vehicle safety.
- Vehicle weight distribution.
- Fuel economy.

Full details of the scheme are available from RTITB Ltd at: Master Driver, Ercal House, 8 Pearson Road, Central Park, Telford TF2 9TX (tel: 01952 520200; fax: 01952 520201; website: www.rtitb.masterdriver.co.uk).

DRIVER DEVELOPMENT

Draft European Directive on Compulsory Training for LGV Drivers

The EU is intent upon raising the standards of professionalism for LGV (and PCV) drivers by means of extended study/training periods and tough theory and practical testing, leading to a certificate of professional competence for such drivers.

A draft Directive on the qualification and training requirements for most LGV drivers sent out for consultation in 2001 was followed by publication of a revised version in April 2002 amending aspects of the original proposals. Despite the amendments, however, the objectives remain the same, namely to improve road safety and the safety of the driver, and additionally to 'arouse young people's interest in the profession, contributing to the recruitment of new drivers at a time of shortage'.

The EC's view is that, with no more than 5–10 per cent of professional lorry drivers in Europe having undertaken any training other than what was needed to pass the relevant driving test, better trained drivers would lead to fewer road accidents and a better quality of service being offered to customers, and harmonized training standards (which the Directive aims to achieve) would help promote growth and competition in the European transport market. Road accidents involving lorries in Great Britain in 2000 resulted in 3033 people being killed or seriously injured and 12,161 people being slightly injured. (Source: *Road Accidents in Great Britain 2000: The casualty report*, DTLR.)

The Road Haulage and Distribution Training Council (RHDTC) estimates that no less than 100,000 additional LGV drivers will be needed by 2010 to meet driver shortfall. In response to this and other demands, the government has announced that it is planning to spend £5 million over the next three years in support of driver training.

The Directive's aim is for a system of certification of professional competence for drivers who satisfy the training requirements for 'initial' or 'accelerated initial' training and 'periodic' training. However, it is not the intention that the need to obtain an initial qualification should apply to existing drivers who have held a category C1, C1+E, C or C+E licence (or a licence recognized as being equivalent) that was issued no later than three years after the final date for this Directive coming into force. They will, nevertheless, have to undergo the periodic training amounting to 35 hours every five years (which may be spread over the period by taking one 7-hour training session each year).

Exemptions will also apply to drivers of the following vehicles:

- vehicles with a maximum authorized speed not exceeding 45 kilometres an hour;
- vehicles used by, or under the control of, the armed forces, civil defence, the fire service and forces responsible for maintaining public order;
- vehicles undergoing road tests for technical development, repair or maintenance purposes, or of new or rebuilt vehicles which have not yet been put into service;
- vehicles used in states of emergency or assigned to rescue missions;
- vehicles used in the course of driving lessons for any person wishing to obtain a driving licence or a certificate of professional competence (CPC);
- vehicles used for non-commercial carriage of passengers or goods for personal use;
- vehicles carrying material or equipment to be used by the driver in the course of his work, provided that driving the vehicle is not the driver's principal activity.

Initial Training

Initial training may be accomplished either by:

1. A course attendance, amounting to 280 hours of tuition (including 20 hours of driving in an approved test vehicle of a relevant category), followed by:
   - a 4-hour, two-part theory test of multiple choice and/or direct answer questions and case studies; and
   - a two-part practical test of driving ability (minimum 90 minutes) on an approved test vehicle, and a test of loading a vehicle and various health and safety issues (minimum 30 minutes);

or alternatively by;

2. A test alone comprising:
   - a 4-hour, two-part theory test of multiple choice and/or direct answer questions and case studies; and
   - a two-part practical test of driving ability (minimum 90 minutes) and of loading a vehicle and various health and safety issues (minimum 30 minutes).

NB: Where a driver already holds a CPC qualification on either goods or passenger vehicles, and wishes to obtain the necessary qualification to drive alternative vehicle types (ie passenger or goods), the length of study for the initial qualification will be reduced to 70 hours, including 5 hours of individual driving.

The theory test referred to above will consist of a written or an oral test which must include at least one question on each of the objectives set out in the syllabus below.

The practical driving test referred to above is aimed at assessing training in rational driving based on safety regulations. The test must take place, whenever possible, on roads outside built-up areas, on fast roads and on motorways (or similar), and on all kinds of urban highways presenting the different types of difficulties that a driver is liable to encounter. It would be desirable for this test to take place in different traffic density conditions. The driving time on the road must be used optimally in order to assess the candidate in all traffic areas likely to be encountered.

The practical test may be supplemented by a third test taking place on special terrain or on a top-of-the-range simulator so as to assess training in rational driving based on safety regulations, in particular with regard to vehicle handling in different road conditions and the way conditions change with different atmospheric conditions and the time of day or night.

The duration of this optional test is not fixed, but the time (not exceeding 30 minutes) may be deducted from the 90 minutes of the driving test referred to above.

Accelerated Initial Training

A system of accelerated initial training will apply to drivers over 18 years of age who drive or wish to drive certain lower weight vehicles – such as those in licence categories C1 and C1+E.This qualification must include the teaching of all subjects in the syllabus over a study period of 140 hours, including at least 10 hours driving in a relevant approved test vehicle, followed by tests as described above.

Four hours of the 10 hours of individual driving may be on special terrain or on a top-of-the-range simulator so as to assess training in rational driving based on safety regulations, in particular with regard to vehicle handling in different road conditions and the way conditions change with different atmospheric conditions and the time of day or night.

NB: Where a driver already holds a CPC qualification on either goods or passenger vehicles, and wishes to obtain the necessary qualification to drive the alternative vehicle types (ie passenger or goods), the length of study for the initial qualification will be reduced to 35 hours, including two and a half hours of individual driving.

Periodic Training

Periodic training courses, organized by an approved training centre, must be of 35 hours duration. These must be undertaken by all licensed LGV drivers, at least every five years, and in periods of at least seven hours. This effectively means that drivers must undertake either one day of training each year or a week's training once every fifth year. Training courses for this purpose will follow the same subject areas covered in the published syllabus (see below).

MINIMUM TRAINING AND QUALIFICATION REQUIREMENTS

List of Subjects

The knowledge to be taken into account when establishing the driver's qualification and training must include at least the subjects in this syllabus.

Trainee drivers must reach the level of knowledge and practical competence necessary to drive the relevant categories of vehicle with safety.

The minimum level of a driver's knowledge must not be less than the level reached during compulsory education, supplemented by professional training.

Syllabus

1. Advanced training in rational driving based on safety regulations

1.1. Objective: to learn the characteristics of the vehicle transmission system in order to use it to the best possible effect:
- curves relating to engine torque and power;
- specific consumption of an engine;
- area of optimum use of revolution counter;
- gearbox-ratio cover diagrams.

1.2. Objective: to learn the technical characteristics and operation of the safety controls in order to control the vehicle, minimize wear and tear and prevent disfunctioning:
- specific features of hydraulic, vacuum and servobrake circuits;
- limits to the use of brakes and retarder;
- combined use of brakes and retarder;
- making better use of speed and gear ratios;
- making use of vehicle inertia;
- using ways of slowing down and braking on downhill stretches;
- action in the event of failure.

1.3. Objective: to learn how to optimize fuel consumption:
- optimization of fuel consumption by applying know-how as regards the above mentioned points in 1.1 and 1.2.

1.4. Objective: to acquire the ability to load a vehicle with due regard for safety rules and proper vehicle use:
- forces affecting vehicles in motion;
- use of gearbox ratios according to vehicle load and road profile;
- calculation of payload of vehicle or assembly;
- calculation of total volume and load distribution;
- consequences of overloading the axle;
- vehicle stability and centre of gravity;
- types of packaging and pallets;
- main categories of goods needing securing;
- damping and securing techniques;
- use of securing straps;
- checking of securing devices;

- use of handling equipment;
- placing and removal of tarpaulins.

2. **Application of regulations**

2.1. Objective: to understand the social environment of road transport and the rules governing it:
- maximum working periods specific to the transport industry;
- principles, application and consequences of Regulations (EEC) Nos. 3820/85 and 3821/85;
- penalties for failure to use, improper use of and tampering with the tachograph;
- knowledge of the social environment of road transport;
- rights and duties of drivers as regards initial qualification and periodic training.

2.2. Objective: to learn the regulations governing the carriage of goods:
- transport operating licences;
- obligations under standard contracts for the carriage of goods;
- drafting of documents which form the transport contract;
- international transport permits;
- obligations under the Convention on the Contract for the International Carriage of Goods by Road (ie CMR);
- drafting of the international consignment note;
- crossing borders;
- freight forwarders;
- special documents accompanying goods.

3. **Health, road safety and environmental safety, service, logistics**

3.1. Objective: to make drivers aware of the risks of the road and of accidents at work:
- types of accidents at work in the transport sector;
- road accident statistics;
- involvement of lorries;
- human, material and financial consequences.

3.2. Objective: to learn how to prevent trafficking in illegal immigrants:
- trafficking in illegal immigrants and illegal immigration;
- general information;
- implications for drivers;
- preventive measures; check list;
- legislation on transport operator liability.

3.3. Objective: to learn how to prevent physical risks:
- ergonomic principles;
- movements and postures that pose a risk;
- physical fitness;
- handling exercises;
- personal protection.

3.4. Objective: to understand the importance of physical and mental ability:
- principles of healthy, balanced eating;
- effects of alcohol, drugs or any other substance likely to affect behaviour;

- symptoms, causes and effects of fatigue and stress;
- fundamental role of the basic work/rest cycle.

3.5. Objective: to learn how to assess emergency situations:
- behaviour in an emergency situation;
- assessment of the situation;
- avoiding complications of an accident;
- summoning assistance;
- assisting casualties and giving first aid;
- reaction in the event of fire;
- evacuation of occupants of a lorry;
- ensuring the safety of all passengers;
- reaction in the event of aggression;
- basic principles for the drafting of an accident report.

3.6. Objective: to learn how to adopt behaviour to help enhance the image of a service undertaking:
- behaviour of the driver and company image;
- importance for the company of the standard of service provided by the driver;
- the roles of the driver and people with whom the driver will be dealing;
- vehicle maintenance;
- work organization;
- commercial and financial effects of a dispute.

3.7. Objective: to understand the economic environment of road haulage and the organization of the market:
- road transport in relation to other modes of transport (competition, shippers);
- different road transport activities (transport for hire or reward, own account, auxiliary transport activities);
- organization of the main types of transport company and auxiliary transport activities;
- different transport specializations (road tanker, controlled temperature, etc.);
- changes in the industry (diversification of services provided, rail–road, subcontracting, etc).

8 Vehicle Registration, Excise Duty and Trade Licences

VEHICLE REGISTRATION

New vehicles must be registered with Local Offices of the DVLA or at Swansea and a registration number obtained for each. A one-off first registration charge of £38 is made to cover the DVLA's administrative costs. See Appendix XI for a list of DVLA Local Offices and new vehicle registration marks, also described on pp 370–71.

The registration number must be displayed on plates mounted on the front and rear of the vehicle. These must conform to legal requirements as specified in the Road Vehicles (Display of Registration Marks) Regulations 2001 as amended (see p 370 for full details).

Annual registration prefix changes were replaced by six-monthly changes from 1 March 1999 (prefix 'T') until 1 September 2001, when an entirely new format for registration plates was introduced. This comprises two letters representing the place of registration (ie 1 of 19 areas followed by a letter for the office of registration). Two numbers denote the year of registration. The last three letters are randomly selected.

Form V55, which is used for motor vehicle registrations, consists of a single sheet used for first-licensing and registration purposes and a two-sheet section with carbon paper inserts, so that details of the vehicle and the dealer's name and town (but nothing more) will copy through on to those sheets. These are sent to an agency acting for the motor industry and the DfT for the assembly of official vehicle registration statistics.

A feature of these additional sheets is a voluntary statistical section in which applicants are invited to voluntarily supply information of value for government and motor industry statistical purposes. It includes questions about the purchaser's occupation and previous vehicle and about the main use to which he intends to put the new vehicle. The information given will help the industry to obtain a better knowledge of the market and therefore to give the best possible service to its customers. Also, if purchasers choose to give their name and address, this will enable the manufacturer to get in touch with them direct over any safety matters that may arise. Any information given in this section will be treated as confidential.

Most of the main vehicle manufacturers and importers will have entered the details of each vehicle on the form before it reaches the dealer. This has two advantages, both for motor dealers and for purchasers. First, they no longer have to fill in the details themselves, and second, the association of one form with one vehicle right from the start is intended to make it more difficult to register a stolen vehicle as new. If the manufacturer or importer has not completed the form in advance, copies are available on demand at licensing offices.

Documents

A completed form V55, a certificate of insurance, a copy of the supplier's invoice, and an appropriate Type Approval Certificate (TAC) for goods vehicles subject to the type approval regulations, are required when first registering a goods vehicle. When a excise licence is required for vehicles exempt from annual testing a completed form V112G (Declaration of Exemption) must also be produced.

Vehicle Registration Documents

A Registration Document/Certificate shows the registered keeper of a vehicle. The registered keeper is the person who keeps the vehicle on a public road, but is not necessarily the legal owner. The Certificate gives the keeper's name and address, the registration mark and other details of the vehicle, including engine, VIN and Type Approval numbers, the category into which it falls, its body type, maximum mass (ie weight) etc. A new Registration Certificate is issued each time the DVLA updates the record with any change to the existing details. The vehicle keeper should make sure that the details on this certificate are accurate and must tell the DVLA immediately if any changes are necessary. It may be difficult to sell the vehicle if any of the information is inaccurate. However, it is important to note that the Registration Document/Certificate is *not* a document of title (ownership).

Since January 2004 anyone applying for a V5 Registration Document will have been issued with the new-style Vehicle Registration Certificate (V5C). This new Registration Certificate has been developed to comply with European Directive 1999/37/EC, which required member states to introduce a common format for Registration Certificates no later than June 2004, with the aim of helping to identify vehicles in international traffic and to allow for the re-registration of vehicles throughout the EU.

Between June 2004 and June 2005 anyone who has either:

- purchased a new or used vehicle;
- applied for a duplicate registration document; or
- notified the DVLA of a change of details

will have been or will be issued with a new-style Registration Certificate.

Also between these dates new-style Registration Certificates will be automatically sent to all vehicle keepers shortly after they tax their vehicle or notify DVLA that a vehicle is being kept off the road (ie by the Statutory Off Road Notification – SORN procedure).

The intention is that by June 2005 all traceable vehicles will have been issued with the new-style Registration Certificate and from 1 July 2005 all existing Registration Documents will cease to be valid.

On receipt of the new Registration Certificate, vehicle keepers should check that the details are correct, put the document in a safe place (together with the guidance notes) and destroy the existing old-style Registration Document if it is still in possession.

NB: A statutory fee of £19 is made by the DVLA for replacing a lost, stolen, destroyed or damaged vehicle registration document.

Police officers and certain officers of VOSA may request production of the registration document for inspection at any reasonable time.

Issue of Licence Disc

On payment of the relevant amount of vehicle excise duty (VED) for a vehicle, the DVLA issues a circular licence disc indicating the period of validity, the vehicle registration number, and the fee paid. This licence must be displayed in the vehicle windscreen where it can be clearly seen from the nearside.

From October 2003 a new-style windscreen disc has been issued with enhanced anti-fraud security features, particularly a barcode, which will enable licensing information to be electronically transferred to the DVLA.

Number Plates

As part of the DVLA's measures to combat vehicle crime, it has been a requirement since the beginning of 2003 that vehicle number (registration) plates may be produced only by registered manufacturers who must keep records of the plates they sell, to whom they are sold and whether that person is authorized to require the plates. Purchasers of plates need to produce evidence as to who they are (a passport or photocard driving licence) and that they have a right to the plates for the vehicle in question (generally by producing the vehicle registration document (Form V5) or a letter of authorization from the vehicle owner).

Registration of 'O' Licence Vehicles

The Traffic Commissioners have ruled that vehicles specified on British 'O' licences must be registered in the UK.

The DVLA has also warned that British VED must be paid on vehicles unless the operator has a genuine business abroad.

VEHICLE EXCISE LICENCES

All mechanically propelled vehicles, whether used for private or business purposes, which are used or parked on public roads in Great Britain, must be covered by, and display, a vehicle excise licence indicating that the appropriate amount of vehicle excise duty (VED) has been paid – except when being driven, by previous appointment, to a place to have their annual test.

It is an offence to use or keep an unlicensed vehicle on the road for any period, however short. Conviction for such an offence will result in heavy fines and back duty may be claimed (see also p 208). Furthermore, new vehicle licensing rules from 1 January 2004 mean that the registered keeper of a vehicle (the person named on DVLA's record) remains responsible for taxing a vehicle (or making a SORN declaration) until that liability is formally transferred to a new keeper. To avoid liability for taxing it, the keeper needs to inform DVLA when the vehicle is off the road, or has been sold, transferred, scrapped or exported. Once DVLA has been notified about a sale/transfer or that the vehicle is off-road they will issue an acknowledgement letter, which should be kept as proof that the vehicle record has been changed. Failure to re-license a vehicle or make a SORN declaration will incur an automatic penalty of £80 (reduced to £40 if paid within 28 days). Determined offenders will be faced with prosecution and could be fined a minimum of £1,000. Operators' licence holders also jeopardize their 'O' licences when committing excise duty offences.

The present system of excise licences for most goods vehicles is based on gross weights (ie the 'revenue weight') and axle configurations, while goods vehicles not exceeding 3500 kg gross weight are in the same class as private vehicles which is called private/light goods (PLG) (see p 533 for full details).

Exemptions

Exemption from vehicle excise duty (as specified in the Vehicle Excise and Registration Act 1994) applies to the following:

1. Vehicles used for police, fire brigade, ambulance or health services (including veterinary ambulances).
2. Mines rescue vehicles.
3. Vehicles used for the haulage of lifeboats and lifeboat gear.
4. Vehicles for disabled people.
5. Vehicles used solely for forestry, agriculture or horticultural purposes travelling on public roads to pass between land occupied by the same person and the distance travelled on public roads is not more than 1.5 kilometres per journey.
6. Vehicles travelling to or from a place where they are to have an annual roadworthiness test by prior appointment.
7. Vehicles for export.
8. Vehicles in the service of a visiting force or headquarters.

The special concessionary rate which formerly applied has been abolished and the following vehicles that were included in this class are now free of vehicle excise duty and are thus added to the exempt class along with the above-listed vehicles:

- An agricultural or off-road tractor which is a tractor used on public roads only for purposes relating to agriculture, horticulture, forestry or cutting verges, hedges or trees bordering public roads. An off-road tractor is one which is not an agricultural tractor but which is designed and constructed primarily for use otherwise than on roads and is incapable of exceeding 25 mph on the level.
- A light agricultural vehicle is a vehicle with a revenue weight not exceeding 1000 kg which is designed and constructed to seat only the driver and primarily for use otherwise than on roads; and is used only for agriculture, horticulture or forestry.
- An agricultural engine.
- Mowing machines.
- Steam powered vehicles.
- Electrically propelled vehicles.
- Vehicles when used, or going to or from a place for use, or kept for use, for clearing snow on public roads by means of a snow plough or similar device.
- Vehicles constructed or adapted and used only for the conveyance of machinery for spreading material on roads to deal with frost, ice or snow.

Special Vehicles

Special vehicles over 3.5 tonnes – current and annual rate of duty £165. This class includes mobile cranes, digging machines, works trucks, road rollers and showmen's goods and haulage vehicles. Vehicles classed as 'mobile pumping vehicles' have been added to this category.

Rates of Duty

The rate of duty payable varies depending on the way in which the vehicle is constructed, the way that it is used and, in the case of light vehicles, their unladen weight, and for heavy goods vehicles, their 'revenue' weight and the number of axles. The current rates of duty shown have been effective since 1 December 2001.

Revenue Weight

Revenue weight means either the 'confirmed maximum weight' as determined under the plating and testing regulations (ie by the issue of a plate and plating certificate), or the 'design weight' for vehicles not subject to plating and testing and currently known as restricted hgvs (this is the maximum laden weight at which non-plated vehicles can legally operate on roads in the UK).

Down-Plating

For various operational reasons heavy goods vehicles are specified to a higher standard and gross weight than payload weights demand. However, to avoid the higher rates of vehicle excise duty that would normally be payable, such vehicles may be down-plated without making any technical change to the specification – see p 393 for details of how this can be done.

Reduced Pollution Vehicles

Reduced rates of vehicle excise duty (by up to £1,000 for the heaviest vehicles) apply to certain buses, haulage vehicles and heavy goods vehicles which have been built or adapted to ensure reduced pollution exhaust emissions under the provisions of the Vehicle Excise Duty (Reduced Pollution) Regulations 1998. These regulations:

- set out the reduced pollution requirements to be satisfied in order for a vehicle to qualify for reduced rates of excise duty;
- provide for the issue of reduced pollution certificates where the specified requirements have been met;
- set out the procedure for applying for a vehicle examination in order to obtain a reduced pollution certificate.

Graduated Duty

Graduated rates of duty benefit smaller-engine, less polluting cars and light goods vehicles with an engine size not exceeding 1549 cc. VED on new cars is based on carbon dioxide (CO_2) emissions as follows:

Band	*CO_2 emission figure (g/km)*	*Alternative fuel car (1)*		*Petrol car (2)*		*Diesel car (3)*	
		12mth £	*6mth £*	*12mth £*	*6mth £*	*12mth £*	*6mth £*
A	Up to 100	75.00	41.25	65.00	35.75	55.00	30.25
B	101 to 120	85.00	46.75	75.00	41.25	65.00	35.75
C	121–150	115.00	63.25	105.00	57.75	95.00	52.25
D	151–165	135.00	74.25	125.00	68.75	115.00	63.25
E	166–185	160.00	88.00	150.00	82.50	140.00	77.00
F	Over 185	170.00	93.50	165.00	90.75	160.00	88.00

Articulated Combinations

In the case of articulated tractive units which are used with a variety of trailers, the determining factor for taxation purposes is the number of axles on the trailer likely to be used with it. This is so that the maximum amount of duty is paid to prevent a vehicle operating on the road at a rate less than that which is applicable. In general terms, the system operates so that the fewer the axles the greater the duty payable and vice versa. This is because the duty relates to the road damage caused and greater damage arises with fewer axles. In consequence of this it is illegal to operate a vehicle on a road for which the incorrect (ie insufficient) excise duty has been paid.

Trailers

Where drawbar trailers are drawn, if the gross weight of the trailer exceeds 4 tonnes and the gross weight of the towing vehicle exceeds 12 tonnes, additional duty is payable in accordance with published duty tables but not otherwise. So, if the towing vehicle or the trailer falls below these weights, no trailer duty is payable.

Goods Carrying Vehicles

The goods vehicle rate of duty is payable if the vehicle is built or has been converted to carry goods, and is actually used to carry goods in connection with a trade or business.

In general this rate of duty applies to all types of lorries, vans, trucks, estate cars, dual-purpose vehicles and also passenger vehicles if they are converted for carrying goods. If, however, dual-purpose vehicles (see below) and goods vehicles (over 3.5 tonnes) are never used for carrying goods in connection with a trade or business they may be licensed at the private heavy goods rate of duty which is currently £165 per year.

Following a High Court ruling in March 1999, heavy vehicles carrying wide loads in excess of 4.3 metres must be taxed at the 'Special Types' rate of duty, currently £2,585 per year (£2,085 for reduced pollution vehicles), irrespective of their gross weight. This duty is based on the use of a vehicle rather than its construction and when carrying a wide load under the Motor Vehicles (Authorization of Special Types) General Order 2003, the relevant Special Types duty rate must be applied.

In yet another interpretation case, the Crown Court has ruled that concrete-pumping vehicles must be taxed at the goods vehicle rate of duty, and not at the reduced rate applicable to mobile cranes.

Driver Training and Non-Goods-Carrying Vehicles

Heavy goods vehicles which are used exclusively for driver training purposes may be licensed at the private heavy goods rate of duty even if they carry ballast (eg concrete blocks) to simulate driving under loaded conditions – *but not other loads.* Similarly, other goods vehicles used for private purposes (ie carriage of goods but not in connection with a trade or business – for example, privately used horse-boxes) may be licensed at the private heavy goods vehicle rate of duty – currently £165 for vehicles over 3.5 tonnes.

Private-Rate Taxation of Heavy Vehicles on International Work

The use of heavy vehicles for goods carrying in the UK or within other EU member states while taxed only at the private heavy goods rate of duty is illegal both in the UK and in Europe (see also section on penalties, p 208). Furthermore, running a vehicle on the Continent with an expired UK tax disc, on the assumption that the tax can be renewed on the vehicle's return to the UK, is also illegal. Under the provisions of bilateral agreements and under International Conventions on the Taxation of Road Vehicles, vehicles on which the correct rate of vehicle excise duty has been paid in their country of registration are exempt from payment of further duty on being temporarily imported into the territories of other parties to the agreements. Where the correct duty is not paid in the 'home' country, there is a liability for payment of additional duty in each other country through which the vehicle travels.

Dual-Purpose Vehicles

For the purposes of the regulations a dual-purpose vehicle (as referred to above) is defined as a vehicle, built or converted to carry both passengers and goods of any description, which has an unladen weight of not more than 2040 kg and has either four-wheel drive or:

- has a permanently fitted roof;
- is permanently fitted with one row of transverse seats (fitted across the vehicle) behind the driver's seat (the seats must be cushioned or sprung and have upholstered back-rests);
- has a window on either side to the rear of the driver's seat and one at the rear.

The majority of so-called estate cars, shooting-brakes, station wagons, hatchbacks, certain Land Rovers and Range Rovers are dual-purpose vehicles under this definition. It should be noted, however, that vehicles used for *dual operations* are not dual-purpose vehicles in terms of the legal requirements.

VED Tables

These rates of duty apply to all licences taken out from 1 March 2001.

- *Private/light goods vehicles* (ie goods vehicles not over 3,500 kg revenue weight)

	12-month rate £	*6-month rate £*
Private vehicles: light vans, cars, taxis, etc, not over 1549 cc	110.00	60.50
Private vehicles: light vans, cars, taxis, etc, over 1549 cc	165.00	90.75

- *Euro 4 Light Goods Vehicles (vehicles not over 3,500 kg) (registered on or after 1 March 2003)*

	12-month rate £	*6-month rate £*
Euro 4 Light Goods Vehicles	110.00	60.50

- *Buses*

			Reduced pollution vehicle	
Seating capacity (excluding driver)	*12-month rate £*	*6-month rate £*	*12-month rate £*	*6-month rate £*
9–16	165.00	90.75	165.00	90.75
17–35	220.00	121.00	165.00	90.75
36–60	330.00	181.50	165.00	90.75
61 and over	500.00	275.00	165.00	90.75

- *General haulage vehicles*

		Reduced pollution vehicle	
12-month rate £	*6-month rate £*	*12-month rate £*	*6-month rate £*
350.00	192.50	165.00	90.75

- *Trade Licences*

	12-month rate £	*6-month rate £*
Trade Licences available for all vehicles	165.00	90.75
Trade Licences available only for:		
Bicycles and tricycles not over 450 kg	60.00	33.00

- *Recovery vehicles*

Over	*Not over*	*12-month rate £*	*6-month rate £*
3500 kg	25,000 kg	165.00	90.75
25,000 kg	–	410.00	225.50

- *Private HGV* (exceeding 3500 kg)

	12-month rate £	*6-month rate £*
	165.00	90.75

Annual VED Rates from 1 December 2001

A new, simplified structure for heavy vehicle excise duty rates was introduced from 1 December 2001 with significant reductions in the duty rates payable – the highest being £1,850 per year for the heaviest trucks.

Rigid vehicles

	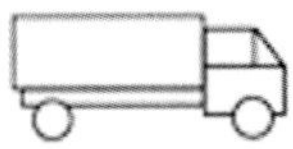	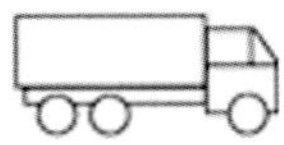	
Vehicle weight	*2-axled rigid*	*3-axled rigid*	*4 or more-axled rigid*
Not over 7,500 kg	A	A	A
Not over 15,000 kg	B	B	B
Not over 21,000 kg	D	B	B
Not over 23,000 kg	D	C	B
Not over 25,000 kg	D	D	C
Not over 27,000 kg	D	D	D
Not over 44,000 kg	D	D	E

Trailer Duty

NB: Where vehicles draw trailers weighing 4,000 kg or below they do not come within a trailer taxation class and no additional trailer duty is payable.

Trailer duty Where the drawing vehicle has a weight of over 12,000 kg and draws laden trailers over 4,000 kg, additional trailer duty is payable				Reduced pollution trailer duty	
Over	***Not over***	***12 months rate £***	***6 months rate £***	***12 months rate £***	***6 months rate £***
4,000	12,000	165.00	90.75	165.00	90.75
12,000	–	230.00	126.50	230.00	126.50

Example: A two-axled vehicle with a weight of 16,260 kg that draws trailers with a weight of 12,130 kg would pay £650 plus £230.00 annual rate.

Two-axled tractive unit articulated vehicles

	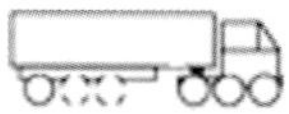	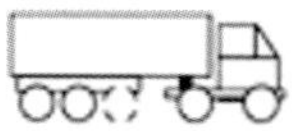	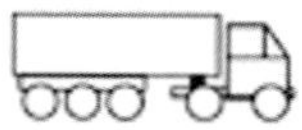
Vehicle weight	*1 or more-axled semi-trailer*	*2 or more-axled semi-trailer*	*3 or more-axled semi-trailer*
Not over 25,000 kg	A	A	A
Not over 28,000 kg	C	A	A
Not over 31,000 kg	D	D	A
Not over 34,000 kg	E	E	C
Not over 38,000 kg	F	F	E
Not over 44,000 kg	G	G	G

Three-axled tractive unit articulated vehicles

		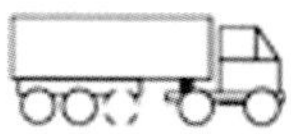	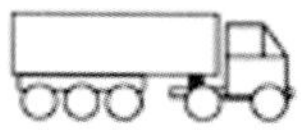
Vehicle weight	*1 or more-axled semi-trailer*	*2 or more-axled semi-trailer*	*3 or more-axled semi-trailer*
Not over 28,000 kg	A	A	A
Not over 31,000 kg	C	A	A
Not over 33,000 kg	E	C	A
Not over 34,000 kg	E	D	A
Not over 36,000 kg	E	D	C
Not over 38,000 kg	F	E	D
Not over 44,000 kg	G	G	E

Key to VED bands				
Lorry VED band	*Standard*		*Reduced pollution*	
	12 months rate £	*6 months rate £*	*12 months rate £*	*6 months rate £*
A	165.00	90.75	*160.00	*88.00
B	200.00	110.00	*160.00	*88.00
C	450.00	247.50	210.00	115.50
D	650.00	357.50	280.00	154.00
E	1,200.00	660.00	700.00	385.00
F	1,500.00	825.00	1,000.00	550.00
G	1,850.00	1,017.50	1,350.00	742.50

*Where two or more bands show the same rate of duty the remainder renewal may display one band only.
Source: Reproduced from DVLA website with acknowledgement.

- *Special Types vehicles*

			Reduced pollution vehicle	
	12-month rate £	*6-month rate £*	*12-month rate £*	*6-month rate £*
Used to carry abnormal indivisible loads under the Special Types General Order	2585.00	1421.50	2085.00	1146.50

Payment of Duty

Excise duty is payable once annually, or every six months. Application for initial registration or renewal of duty is made on the appropriate form as follows:

- First registration
  - – form VE 55/1 for new vehicles
  - – form VE 55/5 for reimported vehicles
- Renewal of duty (12 or 6 months)
  - – form V10 for all vehicles up to 3500 kg
  - – form V11 renewal reminder which enables vehicles to be relicensed at a post office
  - – form V85 for heavy goods vehicles which may only be relicensed at VROs.

Display of Licence Discs

A circular licence disc is issued by the Driver and Vehicle Licensing Agency (DVLA) – or a post office when renewing licences here – when the duty is paid and this must be displayed in the vehicle windscreen where it can be clearly seen from the near side. It is an offence to fail to display a current and valid vehicle excise licence disc – those not displaying discs will be reported to the DVLA and the operator may face a fine of up to £25,000 plus back duty and costs.

Reduced, Extended Periods and Refunds

A vehicle may be licensed for only six months at a time if this is preferred but this method of licensing means that a higher amount of duty is paid annually. All licences are valid only from the first day of the month in which they come into force. Alternatively, new vehicles may be licensed part-way through a month (but only at local VROs) with the duty period commencing on the 10th, 17th or 24th day of the month and continuing through to the end of the 12-month period from the first day of the following month.

In cases where a vehicle is to be used for a shorter period than either 6 or 12 months or is taken out of service during the currency of the licence, a licence has to be taken out for one or other of these periods and then surrendered to the VRO (see local telephone directory for addresses of local VROs) when it is no longer required; the VRO will refund the duty for each complete month remaining on the licence. To gain a full month refund the licence must be surrendered at least by the last day of the previous month. Application for a refund should be made on form V14.

Statutory Off-Road Declaration Scheme

As part of its initiative to recover the estimated £175 million lost each year to road tax evasion, a new scheme requires registered vehicle keepers to notify the Driver and Vehicle Licensing Agency (DVLA) whenever a vehicle is off the road untaxed. By law, vehicle keepers must provide a Statutory Off-Road Notice (SORN) declaring to the DVLA that the vehicle is not being used and is not being kept on a public road and is therefore not liable to vehicle excise duty.

The Government introduced new vehicle tax rules from 1 January 2004 under which keepers of vehicles who have not paid their tax (VED) within the month will be identified from the DVLA computer database and will receive an *automatic* £80 penalty. This is reduced to £40 if the tax is paid within 28 days; but determined offenders could face prosecution and a minimum fine of £1,000. The changes mean that for the first time the registered keeper of the vehicle is now legally responsible for taxing it. If they do not tax their vehicle, or fail to inform the DVLA that the vehicle is off-road or has had a change of keeper, then they will be liable for the £80 penalty – even if the vehicle belongs to someone else.

This scheme will enable the Government to crack down further on 'vehicle cheats' who do not tax their vehicles and many of whom also drive without insurance or a valid test certificate. Currently there are estimated to be some 1.75 million untaxed vehicles on UK roads. As a test certificate (for light vehicles over 3 years old and LGVs over one year old) and insurance certificate are needed to purchase a vehicle tax disc, this measure, according to the Government, will not only cut the number of untaxed vehicles, but will also cut the number of unsafe and illegal vehicles on our roads.

Making a false 'off-road' declaration will result in prosecution and a maximum fine of £5,000 or two years' imprisonment on conviction.

Renewal of Licences

When the licence expires after 6 months or 12 months it has to be renewed by completing renewal form V10 or form V85. Renewal reminders are sent out from the DVLA on form V11 and this may be used as an alternative to form V10 to renew the licence at certain post offices (provided the vehicle is not over 3500 kg gross weight where form V10 is used, there is no change in ownership or address which has not been recorded and no change to the vehicle or its use), or at DVLA Local Offices. Form V85 is used for goods vehicles over 3500 kg gross weight. The vehicle registration document, the appropriate fee, a valid certificate of insurance and, if the vehicle is subject to annual testing, a current test certificate, must accompany the application.

Where to Apply

Applications for renewal of duty may be made as follows:

• If you have the official licence renewal reminder, form V11 (providing there is no change in the details on the form)	In person to main post offices or Local Offices, or by post to Local Offices.
• If you have form V11 but there are changes in the details	In person or by post to Local Offices.
• If the vehicle is a heavy goods vehicle for which no form V11 has been received	In person or by post to Local Offices.

Replacement Licences and Discs

Duplicates or replacements for lost or defaced vehicle licence windscreen discs are available on application, using form V20 at VROs. They cost £7 each (except in certain exceptional cases – eg if stolen with the vehicle, if for a VED-exempt vehicle, if lost in the post). Replacements for lost, stolen or destroyed vehicle registration documents/certificates (or the new keeper supplement) can be obtained by applying to the DVLA using Form V62 (Vehicle Registration Certificate V5C Application Form), which is available from DVLA Local Offices, certain Post Office branches or can be downloaded from the DVLA website at: www.dvla.gov.uk/vehicles. An existing registered vehicle keeper on the DVLA computer may apply by telephoning 0870 240 0010 using a debit/credit card for payment, provided there are no changes in the registration details. A statutory fee of £19 is charged.

Alteration of Vehicles

If a vehicle is altered during its life by adding or removing equipment, by changing the type of body or even the colour of the vehicle, by changing the plated weight or increasing the unladen weight by fitting a heavier body or heavier components, the DVLA must be advised of the changes at once. If the changes mean that the vehicle goes into a higher weight range, then the additional licence duty becomes payable from the date of the changes.

If a van is converted to carry passengers or has side windows fitted to the body behind the driver's seat the owner becomes liable to pay car tax to H.M. Customs and Excise. The amount of tax payable is based on the current wholesale value of the vehicle. Any person making such a conversion must report the fact to the Customs and Excise authorities immediately.

Sale of Vehicles

When a vehicle is sold, the seller must notify the DVLA by completing the bottom tear-off portion of the registration document (form V5), which is perforated for this purpose, with the name and address of the new owner of the vehicle. It is an offence to fail to notify a change of vehicle ownership (maximum fine £1,000) and it can lead to the original owner being prosecuted for offences committed with the vehicle by the new owner and leaving the previous owner to pay any fixed penalty fines incurred. The new owner has to fill in his name and address in the changes section of the registration document and send the document to the DVLA for registration in his name. If a vehicle is sold for scrap or is broken up, the registration document has to be sent to the DVLA with a note advising them of this.

Under a new vehicle keepership scheme, the police and the DVLA will combine reporting of the offences of failing to register a vehicle (ie by the new owner) and vehicle excise duty evasion. This scheme is intended to clamp down on those who fail to notify disposal or acquisition of a vehicle.

Un-Repairable Vehicles

Where a motor vehicle is un-repairable and the insurer notifies the DVLA of this fact, or the vehicle registration document has been surrendered to the DVLA, it must undergo an examination (ie a Vehicle Identity Check) to confirm it is the originally registered vehicle before a new registration document can be issued. A charge of £35 is made for the VIC.

Production of Test Certificates

A valid goods vehicle test certificate has to be produced at the time of re-licensing any goods vehicle over one year old (ie vehicles over 3500 kg gross weight and articulated vehicles which are subject to the goods vehicle annual test), and a light vehicle or private car type (ie MoT) test certificate when re-licensing any vehicle not over 3500 kg gross weight which is over three years old and is subject to the light vehicle annual test scheme.

Vehicles Exempt from Plating and Testing

When applying to license or relicense goods vehicles which are exempt from plating and testing, a declaration has to be made on form V112G (available from VROs) to cover the non-production of a valid test certificate.

Where a person does not intend to use a vehicle during the period of an excise licence except for a purpose which is specifically exempt from testing requirements (see list on p 395), a 'Form of Declaration' must be made as specified in *The Goods Vehicles (Evidence of Test Certificates) Regulations 2004* (SI 2004 No. 2577).

Penalties and Payment of Back Duty

Offenders prosecuted for evasion of vehicle excise duty face fines of up to £1,000 for cars and £25,000 for heavy goods vehicles. This also applies to payment at an incorrect (ie too low) rate of duty or payment by means of a cheque which defaults while making a false 'off-road' declaration will lead to a fine of up to £5,000 or two years' imprisonment on conviction, as previously mentioned. Additionally, an offender may be ordered to pay an amount of back duty. Previously the amount of back duty payable could be reduced if the offender could prove non-use of the vehicle during any particular month, but this provision has been rescinded. Failure by a vehicle keeper to notify the DVLA of a change of name or address could result in a £1,000 fine on conviction.

Data Protection

In accordance with the provisions of the Data Protection Act the DVLA at Swansea is registered as a Data User and as such must make available, on request, details held on file concerning individual persons. Such information is normally shown on driving licences and vehicle registration documents but an enquiry to establish details held on file can be made (on payment of a fee) to the: Vehicle Enquiry Unit (Data Protection Queries), DVLA, Swansea SA99 1AN.

TRADE LICENCES

Trade licences (trade plates) are available for use by motor traders and vehicle testers to save them the inconvenience of having to licence individually every vehicle which passes through their hands. Probationary licences are available to those setting up in business where a trade licence can be legally used.

Fees and Validity

A trade licence can be obtained from the local VRO in whose area the applicant has his business; it is valid either for one year or for six months (licences are issued on 1 January and 1 July) and the licence fee is £165 for one year and £90.75 for six months. Replacements for lost or defaced trade plates (which remain the property of the local VRO) cost £13.50 per set of two plates or £18 for a set of three plates. A replacement holder for the licence costs £2 and a replacement (ie duplicate) licence is £7. Besides the 12-month and 6-month licences, it is now possible to obtain such a licence for periods between 7 months and 11 months at pro-rata rates to the six monthly rate (ie 55 per cent of the annual rate for six months, plus one sixth of this amount for every month in excess of six months).

Display of Trade Licences

The VRO will, when they have approved the application, issue a pair of special number plates with the registration number in red letters on a white background – an additional plate is available in certain circumstances. One of the plates has a licence holder attached to it and the licence affixed (it does not have to be displayed on the windscreen); this plate must always be carried at the front of the vehicle on which the plates are being used. The other plate is carried at the rear.

Issue of Licences

There are considerable restrictions on the issue and use of trade licences; as already mentioned they are issued only to:

- motor traders, defined as 'manufacturers or repairers of, or dealers in mechanically propelled vehicles' (this also includes dealers who are in business consisting mainly of collecting and delivering mechanically propelled vehicles), plus those who modify vehicles (eg by fitting accessories) and those who provide valet services for vehicles who may use the licence for all vehicles which are from time to time temporarily in their possession in the course of their business as motor traders;
- vehicle testers for all vehicles which are from time to time submitted to them for testing in the course of their business as vehicle testers.

Vehicles such as service vans or general runabout vehicles owned by motor traders cannot be used under trade licences; the full rate of duty has to be paid for such vehicles.

Use of Trade Licences

The following are the purposes for which vehicles operated under a trade licence may be used by a motor trader or vehicle tester:

- For test or trial in the course of construction or repair of the vehicle or its accessories or equipment and after completing construction or repair.
- Travelling to or from a weighbridge to check the unladen weight or travelling to a place for registration or inspection by the council.
- For demonstration to a prospective customer and for travelling to or from a place of demonstration.
- For test or trial of the vehicle for the benefit of a person interested in promoting publicity for the vehicle.
- For delivering the vehicle to a purchaser.
- For demonstrating the accessories or equipment to a prospective purchaser.
- For delivering a vehicle to, or collecting it from, other premises belonging to the trade licence holder or another trader's premises.
- For going to or coming from a workshop in which a body or equipment or accessories are to be, or have been fitted, or where the vehicle is to be or has been valeted, painted or repaired.
- For delivering the vehicle from the premises of a manufacturer or repairer to a place where it is to be transported by train, ship or aircraft or for returning it from a place to which it has been transported by these means.
- Travelling to or returning from any garage, auction room or other place where vehicles are stored or offered for sale and where the vehicle has been stored or offered for sale.
- Travelling to a place to be tested (and returned), dismantled or broken up.

It should be noted that the use of a vehicle on trade plates does not exempt the driver or operator from the need to ensure that it is in sound mechanical condition when on the road, even if being driven for the purposes of road testing or fault finding prior to repair or after repair. The police will prosecute if they find trade licensed vehicles on the road in an unsafe or otherwise illegal condition.

Carriage of Goods on a Trade Licence

Goods may only be carried on a vehicle operating under a trade licence:

- When a load is necessary to demonstrate or test the vehicle, its accessories or its equipment – the load must be returned to the place of loading after the demonstration or test unless it comprised water, fertilizer or refuse.
- When a load consists of parts or equipment designed to be fitted to the vehicle being taken to the place where they are to be fitted.
- When a load is built in or permanently attached to the vehicle.
- When a trailer is being carried for delivery or being taken to a place for work to be done on it.
- If the goods are another fully licensed vehicle being carried for the purpose of travel from or to the place of collection or delivery (ie the driver's own transport to get him out or back home).

Carriage of Passengers on a Trade Licence

The only passengers who are permitted to travel on a trade licensed vehicle are:

- The driver of the vehicle, who must be the licence holder or his employee – other persons may drive the vehicle with the permission of the licence holder but they must be accompanied by the licence holder or his employee (this latter proviso does not apply if the vehicle is only constructed to carry one person).

- Persons required to be on the vehicle by law; a statutory attendant, for example.
- Any person carried for the purpose of carrying out his statutory duties of inspecting the vehicle or trailer.
- Any person in a disabled vehicle being towed including persons from the disabled vehicle being carried provided this is not for hire or reward.
- A prospective purchaser or his servant or agent.
- A person interested in promoting publicity for the vehicle.

NB: It is illegal for transport fleet operators to road test their own vehicles on trade plates. This has been established on the grounds that the vehicles are not 'temporarily' in their possession and are therefore outside the permitted terms of trade licence use.

RECOVERY VEHICLES

A separate class of VED at an annual rate of duty of £165 or £410* currently applies to these vehicles. Any vehicle used for recovery work which does not conform to the definition given below must be licensed at the normal goods vehicle rate according to its class and gross weight.

**The annual rates of duty for recovery vehicles based on design weights are as follows:*

- *3.5 to 25 tonnes* *£165.00 (£90.75 for 6 months)*
- *over 25 tonnes* *£410.00 (£225.50 for 6 months)*

Definition of Recovery Vehicle

For the purpose of this taxation class, a recovery vehicle is one which is 'either constructed or permanently adapted primarily for the purpose of lifting, towing and transporting a disabled vehicle, or for any one or more of those purposes'. A vehicle will no longer be a recovery vehicle under the regulations (ie the Vehicles Excise and Registration Act 1994, Schedule 1 part V) if at any time it is used for a purpose other than:

- the recovery of a disabled vehicle (a maximum of two disabled vehicles may be recovered at any one time);
- the removal of a disabled vehicle from the place where it became disabled to premises at which it is to be repaired or scrapped;
- the removal of a disabled vehicle from premises to which it was taken for repair to other premises at which it is to be repaired or scrapped;
- carrying fuel and other liquids required for its propulsion and tools and other articles required for the operation of, or in connection with, apparatus designed to lift, tow or transport a disabled vehicle;
- travelling to a place where it will be available to recover or remove a disabled vehicle and to go from a place where it has recovered a disabled vehicle or to any place to which it has removed a disabled vehicle;

- repairing a vehicle at the place at which it became disabled or to which it had been taken for safety;
- towing or carrying one trailer which had previously been towed or carried by the vehicle immediately prior to it becoming disabled;
- removing a vehicle from the road to a nominated place on the instruction of a police constable or a local authority under their statutory powers.

The Act allows the carriage of people and/or goods on a recovery vehicle under the first two items listed above, provided they were either a driver or passenger in the vehicle immediately prior to it becoming disabled, or goods being carried on the vehicle immediately prior to it becoming disabled. The driver and/or passenger of a disabled vehicle (and his personal effects) may also be carried from the place where the disabled vehicle is to be repaired or scrapped to his original destination.

The use of a recovery vehicle, subject to VED at the recovery vehicle rate, for purposes other than those described above is illegal and may lead to prosecution and demands for payment of duty (and back duty) at the full goods vehicle rate.

Operation of Recovery Vehicles

Recovery vehicles licensed under the recovery vehicle taxation class are not exempt from goods vehicle plating and testing unless they satisfy the definition of a 'recovery vehicle' (see above). Such vehicles, however, are exempt from 'O' licensing, the EU drivers' hours rules and the tachograph requirements but those persons who drive them must comply with the British domestic driving hours rules (see pp 86–88 for details).

REBATED HEAVY OIL

Commercial vehicles powered by diesel (heavy oil) engines must use diesel fuel on which the full rate of duty has been paid. A lower rate of duty (ie 5.22p/lit against 47.1p/lit) is payable on fuel used for purposes other than driving road vehicles, such as driving auxiliary equipment, for contractors' plant which does not use public roads, bench-testing of engines and space heating. Fuel on which the lower rate of duty has been paid is known as rebated heavy oil but is more commonly called gas oil or red diesel.

Rebated heavy oil must be marked, when delivered from bonded oil warehouses, with a red dye so that its use can easily be detected, and the supplier must deliver to the recipient a delivery note bearing a statement that the oil is 'not to be used as road fuel'. If both rebated and unrebated oils are stored in the same place a notice bearing the same wording must be placed at the outlet of the rebated oil supply.

Supplies of so-called 'green' diesel have become available from the Irish Republic. This fuel is to be regarded in the same way as 'red' diesel. In other words, it is illegal for normal goods vehicle use on UK roads and in Europe.

Road fuel testing units staffed by officers of Customs and Excise operate throughout the UK to test fuel in vehicles and in storage tanks. Under The Hydrocarbon Oil Regulations 1973 Customs and Excise officers are empowered to examine any vehicle and any oil carried in it or on it and may also enter and inspect any premises and inspect, test or sample any oil on the premises whether the oil is in a vehicle or not. Vehicle owners and drivers must give the officers facilities for inspecting oils in vehicles or on premises.

The following vehicles may use rebated heavy oil as fuel. All other vehicles must use unrebated (full-duty paid) oil at all times:

- vehicles not used on public roads and not licensed for road use;
- road rollers;
- road construction machinery (vehicles used or kept on a road solely for carrying built-in road construction machinery);
- vehicles exempted from excise licence duty which use public roads for not more than six miles in a week;
- agricultural machines;
- trench digging and excavating machines;
- mobile cranes;
- mowing machines;
- works trucks.

Heavy penalties, including fines of up to £500, repayment of duty and even arrest and impounding of vehicles, are imposed on offenders convicted of using illegal diesel fuel.

It should be noted particularly by international hauliers that use of 'red diesel' (ie untaxed diesel fuel) carried in reserve or 'belly' tanks or fed from diesel tanks on refrigerated semi-trailers and used once outside the UK is an illegal practice (NB: Red diesel can legally be used for powering refrigeration units on vehicles). It is also illegal to carry red diesel in unconnected or disconnected additional vehicle tanks. Although Customs checks abroad on vehicle fuel tanks are limited, any operator found running on or illegally carrying such fuel within the EU is likely to face heavy penalties.

Customs may question operators with unusually large diesel tanks mounted on reefer semi-trailers to determine whether such fuel is being illegally used to power the vehicle as well as the fridge unit.

The maximum legal capacity for an engine-connected diesel tank on a vehicle is 1500 litres. Customs and Excise policy is that vehicle manufacturer-fitted fuel tanks are acceptable, but any additional tank fitted to the vehicle, or the exchange of a standard manufacturer tank for a large capacity tank, infringes the rules and will incur a penalty. Above 1500 litres it is considered that the vehicle would have to operate under ADR regulations (see Chapter 23). It should also be noted that transferring fuel purchased abroad for use in another vehicle is illegal and may result in charges for evasion of duty.

OIL STORAGE

Many haulage firms need to store oil in large quantities and will therefore need to be aware of the new Control of Pollution (Oil Storage)(England) Regulations 2001 which came into effect in 2003, but with certain key provisions coming into force from 1 September 2005. These regulations apply to all types of oil (with the exception of waste mineral oil) stored above ground in containers of more than 200 litres at industrial or commercial premises. The regulations set minimum standards for oil storage (ie tanks to have a 110 per cent bund capacity) with the objective of preventing pollution and protecting the environment, oil being a major pollutant of watercourses and toxic to plants and animals. Fines for non-compliance are up to £5,000 but if pollution is caused they could be up to £20,000 or three months' imprisonment. Further information can be obtained from the following websites: www.defra.gov.uk and www.environment-agency.gov.uk/business/ppg.

9 Insurance (Vehicles, Premises and Business) and Security

Owners and operators of motor vehicles using the public highway must insure against third-party* injury and passenger claims. Further, an essential part of any investment in property (buildings, vehicles, plant, etc) is to obtain protection by insurance against loss or damage by theft, fire or other eventuality. It is also wise to be protected against claims made by third parties for compensation following injury to themselves or damage to their property as a result of some occurrence involving you, your employees, your property, or taking place on your premises.

**The insurance company is the first party, the insured person(s) is the second party, and anybody else involved (particularly if they make a claim for compensation) is termed the third party.*

MOTOR VEHICLE INSURANCE

Third-Party Cover

The Road Traffic Act 1988 (s143) requires that all motor vehicles, except invalid carriages and vehicles owned by local authorities or the police, used on a road must be covered against third-party risks. This can be achieved by means of a conventional insurance policy or, alternatively, by a deposit of £500,000 in cash or securities to the Accountant-General of the Supreme Court but this is only applicable if authorization is granted by the Secretary of State for Transport. Normally, such authorization is only granted to public bodies and authorities and to major organizations with access to the substantial funds which may be needed to meet major accident claims.

Where the cover is obtained by conventional insurance means the Road Traffic Act 1988 stipulates that such cover is valid only if taken out with insurers who are members of the Motor Insurers' Bureau (MIB) which is a body established to meet claims for compensation (in respect of death or personal injuries only) by third parties involved in accidents with motor vehicles which subsequently prove to be uninsured against third-party risks.

Sections 145 (3a) and (3c) of the 1988 Act state that the insurance policy:

> . . . must insure such person, persons or classes of persons as may be specified in the policy in respect of any liability which may be incurred by him or them, in respect of the death or bodily injury to any person or damage to property caused by, or arising out of, the use of the vehicle on a road in Great Britain . . . (and) . . . must also insure him or them in respect of any liability which may be incurred by him or them . . . relating to payment for emergency treatment.

NB: Since April 2000, the term 'on a road' used above has been extended to include 'or other public place'.

The emergency treatment referred to is that provided at the scene of an accident by a doctor or hospital authority and that provided by and charged for by a hospital for in-patient or out-patient care. The maximum amounts payable for such treatment, effective from 1 April 2005, are £593 per day with a maximum charge of £35,500 for persons treated as in-patients and £483 for persons treated as out-patients.

Passenger Liability

Passenger liability insurance cover for motor vehicles is compulsory. This requirement applies to all vehicles which are required by the Road Traffic Act to have third-party insurance and the cover must extend to authorized passengers (other than employees of the insured who are covered separately by the compulsory employers' liability insurance), other non-fare-paying passengers and also to what may be termed 'unauthorized passengers' such as hitch-hikers and other people who are given lifts.

Unauthorized Passengers

The display in a vehicle of a sign which says 'No passengers' or 'No liability' does not indemnify a vehicle operator or driver from claims by so-called 'unauthorized' passengers who may claim for injury or damage received when travelling in or otherwise in connection with the vehicle resulting from the driver's or vehicle operator's negligence. The law ensures that such liabilities are covered within the vehicle policy of insurance.

Property Cover

In accordance with an EU Directive (EC 5/84), from 1 January 1989 all UK motor insurance policies have been required to cover liability for damage to property (up to a maximum liability of £250,000 arising from one accident or a series of accidents from one cause). Damage to property in this context includes that caused by the weight of the vehicle (eg to road surfaces, paving slabs, etc) and by vibration which may damage services (eg gas and water mains, gullies and sewers, telephone cables, etc) below the road surface, and third parties whose property is damaged in a vehicle accident have the right to request details of the vehicle insurance.

Certificate of Insurance

A policy of insurance does not provide the cover required by the Act until the insured person or organization has in their possession a Certificate of Insurance. Possession means, in this context, exactly what it says: 'promised' or 'in the post' is not sufficient to satisfy the law. The policy itself is not *proof* of insurance cover; it only sets out the terms and conditions for the cover and the exclusion and invalidation clauses.

The Certificate (or a temporary cover note proving cover until the Certificate is issued) which is *proof* (or evidence) of cover must show the dates between which the cover is valid, give particulars of any conditions subject to which the policy is issued (eg the permitted purposes for which the vehicle may be used and those which are not permitted) and must relate to the vehicles covered, either individually by registration number or by specification, and to the persons who are authorized to drive them.

Production of Insurance Certificate

It is necessary to produce a current Certificate of Insurance when making application for an excise licence (road tax) for a vehicle. Alternatively, a temporary cover note may be produced and this will be accepted, but the insurance policy document itself is not acceptable.

New regulations effective from 21 February 1997 and applying only where a fleet comprises 250 vehicles or more, remove the normal requirement for the production of individual Certificates of Insurance for fleet vehicles when applying for vehicle excise licences.

Further regulations, effective from 17 September 1999, exempt an applicant for a vehicle licence from producing evidence of insurance in cases where that applicant's insurer maintains a database from which the DVLA can verify that a policy of insurance is in force when the licence has effect.

The owner (ie registered keeper) of a motor vehicle must produce a Certificate of Insurance relating to the vehicle if required to do so by a police officer. If he is not able to produce the Certificate on the spot, or if an employed driver is required to produce a Certificate of Insurance for the vehicle he is driving, it may be produced for inspection, no later than seven days from the date of the request by the police officer, at any police station which the owner or driver chooses. The person to whom the request is made does not have to produce the Certificate personally, but may have somebody else take it to the nominated police station for him. A valid temporary cover note would suffice instead of the Certificate if this has not yet been issued.

Duty to Give Information

If requested to do so, the owner of a vehicle must give the police any information they request to help determine whether on any particular occasion a vehicle was driven without third-party insurance cover in force. The owner must also give information about the identity of a driver who may at any time have been driving a vehicle which is registered in his name, or information which may lead to identification of a driver if he is asked to do so by the police.

When the vehicle or vehicles concerned in such a request are the subject of a hiring agreement, the term 'owner' for the purposes of these insurance provisions includes each and every party to the hiring agreement.

The Motor Insurance Database

Under European law (the Fourth Motor Insurance Directive), vehicle fleet operators are required to register with the Motor Insurance Database (MID) which will hold details of the insurance arrangements for every vehicle on our roads. Failure to register will result in prosecution for the fleet operator, whose duty it is to provide details of the current fleet and any subsequent changes including the use of any temporary vehicles. Information about privately owned vehicles is provided to the MID directly by the insurer.

Registration of details may be made via the Internet to, and further details obtained from, the Motor Insurers' Information Centre at: www.miic.org.uk.

Checking Uninsured Vehicles

With over one million uninsured people reportedly driving on UK roads, the police are anxious to identify and seize uninsured vehicles, for which they have been given powers. However, under another new D*f*T initiative it is suggested that vehicles must be insured at all times, whether on a road or not. If this proposal is implemented, it would become an offence to be the registered keeper of any vehicle for which there was no certificate of insurance in force, initially carrying a fixed penalty of £100. But if the vehicle remained uninsured, then prosecution would follow, resulting in a fine at level 3 (ie up to £1,000), with liability resting upon the registered keeper of the vehicle, although it would not be an endorsable offence. Exemption would be provided for vehicles genuinely 'off the road' for repairs or 'laid-up', but in that case a SORN declaration would need to have been made to the DVLA beforehand. With this system, an offender would not have to be caught driving an uninsured vehicle, simply that their vehicle does not show up as a match when the DVLA's records are compared with the Motor Insurers' Database – as is now the practice when re-licensing vehicles via the Internet; ie no documentary proof of insurance is required, it is simply achieved by comparing the computer databases.

Invalidation of Cover

Insurance cover may be invalidated and claims for compensation refused if policy conditions are not strictly adhered to. In particular these circumstances may arise if the vehicle is operated illegally (for example, in excess of its permissible weight, without a valid test certificate, in an unsound mechanical condition, outside the terms of an 'O' licence, with an incorrectly or unlicensed driver or one who is disqualified, or if replacement components fitted to the vehicle are not to manufacturer's specification).

By way of example, a case was reported where liability was rejected by an insurance company when a fast sports saloon motor car was found to be fitted with tyres not rated for the top speed of which the car was capable (well in excess of 100 mph), although the claim arose out of an accident at less than 30 mph.

It is important to stress the need to examine carefully all the clauses contained in a motor insurance policy and to take steps to avoid any action which may invalidate the policy. The employment of unlicensed or incorrectly licensed drivers or the use of unroadworthy vehicles (ie vehicles which do not comply with legal requirements or are found to be on the road in a dangerous condition) are two examples of the most likely ways of invalidating a motor insurance policy. Similarly, the policy should cover *all persons* who may be required (or may need in an emergency) to drive vehicles, not just employees.

Use of Unfit Drivers

It is important not to use drivers who are, or who are believed to be, medically unfit to drive. Insurance companies have a duty to notify the Secretary of State for Transport of the names and addresses of people refused insurance cover on medical grounds so their driving licences can be withdrawn.

Payment to Travel

Previously, motor insurance cover could be invalidated if passengers paid towards the cost of car-running expenses. As a result of provisions in the Transport Act 1980, the receipt of travel expenses contributions from passengers in private cars does not invalidate insurance policies provided no profit is made.

Cancellation of Insurance

When an insurance policy is cancelled, the Certificate of Insurance – there may be one or more depending on the number of vehicles covered by the policy – relating to that policy must be surrendered to the insurer within seven days of the cancellation date.

Cover in EU Countries

It is a requirement that every motor insurance policy issued in an EU country, including Britain, must include cover against those liabilities which are compulsorily insurable under the laws of every other EU member state (*Article 7 (2) of the EEC Directive on Insurance of Civil Liabilities arising from the use of Motor Vehicles* (*No 72/166/CEE*)) and some non-EU states (eg Switzerland).

Motor policies issued in the UK contain provisions for such cover but this only provides very limited legal minimum cover and, while an international motor insurance 'green card' is no longer essential to enable EU member state boundaries to be crossed by vehicles (private or commercial), it is wise to obtain a green card when travelling or sending goods vehicles abroad, in order to obtain the much wider cover provided by the policy. Possession of a green card provides adequate evidence of insurance when abroad and it also eliminates problems of language and different procedures in foreign countries.

International Accident Report Form

A special accident report form has been devised by the European Insurance Committee (CPA) for use as an agreed statement and accident report to be completed by drivers at the time of an accident when travelling in a foreign country (ie country other than that in which the vehicle is insured). The use of such a form (available from insurers), particularly when dealing with persons from other countries who cannot speak your language, can help to resolve matters later.

FLEET INSURANCE

Most large fleet operators obtain insurance cover on a 'blanket' basis. Under this arrangement vehicles are not specified on the Certificate of Insurance by registration number but there is a statement on the Certificate to the effect that cover is provided for any vehicle owned, hired or temporarily in the possession of the insured person or company. With blanket insurance it is normal to advise the insurance company by means of a quarterly return of the registration numbers of all vehicles added to or deleted from the fleet strength during that period.

The basic insurance premium is calculated on the total fleet at the beginning of the policy year and adjustments are made by the insurance company issuing debit or credit notes as necessary following receipt of the quarterly returns. This system saves the insurance companies having continually to issue and cancel cover notes and Certificates for vehicles in fleets where there may be many changes during a year because of staggered replacement programmes. The insured company also benefits by not having to get in touch with its insurers every time a vehicle is obtained or disposed of and, further, by being able to obtain an excise licence for a new vehicle without having to wait to receive a cover note from the insurance company.

ADDITIONAL VEHICLE INSURANCE COVER

Extended Cover

The minimum cover against third-party risks mentioned above is not sufficient protection for the owner of a vehicle in the event of it being involved in an accident, damaged in any way (eg by vandals or by another vehicle when the driver was not present) or stolen. To obtain extra protection against such contingencies it is necessary to extend the insurance cover beyond the third-party legal minimum. This can be done in varying stages depending on what the vehicle owner considers necessary for his purpose. The basic policy can be extended to cover loss of the vehicle or damage to it as a result of fire or theft. The insurance can be further extended to give comprehensive cover which provides protection against third-party claims, fire and theft risks and accidental damage to the vehicle itself.

Loading and Unloading Risks

Goods vehicle insurance policies should include clauses which give protection against claims arising from the loading or unloading of vehicles or the activities of employees engaged on such work.

Loss of Use

Most motor vehicle policies do not include cover for the loss of use of a vehicle or for the hire of a replacement vehicle following an accident. If the accident proves to be the fault of the third party, a claim has to be made against the third party for the loss of use or the hiring charges incurred but such claims are often difficult to substantiate (particularly the value of loss of use of the vehicle) and may result in only meagre awards. An extension to the policy covering such eventualities is the most satisfactory means of protection against this type of loss.

Mechanical Failure

Mechanical failure is another item which is not normally included in motor insurance policies and generally it is not possible to obtain this type of cover for motor vehicles (although it is for some items of heavy engineering plant). Damage to engines caused by frost is covered in the majority of commercial vehicle insurance policies although there are certain qualifications. It is necessary, for example, if a claim is to be met, for the vehicle to have been sheltered in a properly constructed garage between specified hours of the night. It is a condition of all policies that all reasonable steps should be taken to safeguard the vehicle from such loss or damage, and this clause particularly is one which the insurance company can use to escape a claim if it feels the policy conditions were not complied with.

Windscreen Breakage

Insurance companies normally provide cover for windscreen breakage within the standard motor insurance policy. Claims made for broken windscreens are generally limited to a fixed amount but are paid to the policy holder without detriment to any existing no-claims bonus and irrespective of whether or not an 'excess' clause is in force on the policy. Similar cover applies on most goods vehicle policies providing for the cost of replacement of the broken windscreen and for repairs to paintwork damaged by the broken glass.

Towing

Insurance cover for towing a vehicle which has broken down is normally provided under a goods vehicle policy, but the cover does not extend to damage caused to the vehicle while it is being towed, or to loss or damage of any goods being carried by the broken-down vehicle.

Damage by Weight

A goods vehicle insurance policy should provide cover against claims for damage caused to roads, bridges, manhole covers and such like by the weight of the vehicle passing over them. Some policies have a limit on the maximum liability acceptable for damage to property and this amount should be checked to ensure that it is adequate to meet likely claims in this respect.

Defence Costs

A motor insurance policy can be extended to cover legal costs incurred in defending a driver faced with manslaughter or causing death by reckless driving charges.

GOODS IN TRANSIT INSURANCE

Motor vehicle insurance does not provide cover for claims made for damage or loss to goods carried on or in the vehicle. Goods in Transit (GIT) insurance cover is needed to provide protection for this eventuality.

Most GIT insurance policies provide cover in accordance with the limits included in the Road Haulage Association Conditions of Carriage which is normally a maximum liability of £1,300 per tonne for goods carried within the UK. If goods are of relatively low value (bulk traffics, such as coal, gravel and other excavated materials, for example) a lower limit of liability and consequently a lower premium can be considered. However, in many cases the £1,300 per tonne limit can be totally inadequate. Many loads these days are valued at tens of thousands of pounds with an equivalent value per tonne way in excess of the RHA level and it is necessary to ensure that the insurance is adequate to cover the value of such loads. Owner-drivers, in particular, are advised to examine their goods-in-transit policies with great care – many such policies become automatically invalidated if high value loads are left unattended (some loads must be accompanied *at all times*).

When goods of this level of value are carried regularly the insurance company will provide suitable annual cover but in some instances goods vehicle operators may find that a lower level of cover is suitable for most of their activities since they only occasionally carry high-value loads. It is important when this happens that the haulier makes himself aware of the load value and that the insurance company is advised of such loads and the appropriate cover obtained. Failure to do so could leave a haulier facing expensive loss or damage claims from his own pocket.

Some GIT policies specifically exclude certain high-risk loads such as cigarettes, tobacco, spirits, livestock, computers, etc, so the operator faced with a request to carry such a load should check that his policy covers the value and consult his insurers before accepting an order to move the goods.

The GIT policy can be on an 'All-Risks' basis but it is usual for the policy to meet the particular requirements of the operator to give him protection against the liabilities he assumes when he accepts goods for carriage. Such liabilities may be accepted under conditions of carriage (see later in this chapter) or under a contract or agreement or, in the absence of any specific contract or conditions, at common law. If the operator is carrying his own goods in addition to other people's he should make sure that these are also covered under the policy.

Liability for Goods

Most transport managers are aware that it is essential that they should effect a GIT insurance policy in respect of the goods carried, but it is most important to consider very carefully the liabilities which are assumed for the goods handled on behalf of customers and that there is full understanding of the GIT insurance contract which has been arranged.

If transport contractors for commercial reasons assume total responsibility for very high-value loads and do not in any way limit their liability by contract or by the application of conditions of carriage (see below), they will soon realize that the claims which are being handled by their GIT insurers become so expensive that the premium subsequently demanded will be far too high to bear.

It is for this reason that the majority of transport contractors find it sensible to limit their liability in accordance with Conditions of Carriage such as those published by the RHA (its most recent Conditions are dated 1998 – these are the copyright of the RHA and may not be used by non-members) where the liability is based on a value of £1,300 per tonne on the actual weight of the goods carried or on the computed weight if the volume of the goods exceeds 80 cubic feet per tonne. This limit can be varied on the insurance policy to suit individual demands but otherwise retaining for the operator the legal liability limitations.

New conditions have also been drawn up by the FTA in conjunction with the Institute of Purchasing and Supply, and these specify a liability limit of £2,000 per tonne. These are not copyright and may be used by hauliers although legal difficulties may arise where RHA conditions and the new FTA/IPS conditions are applied to the same movement contract (ie one set by the customer and one by the haulier).

High-Value Loads

There may be a temptation for haulage contractors to accept high-value loads because the freight rate being offered by the consignor is generous in comparison to normal haulage rates. Such loads, however, are notoriously attractive to thieves so it is important before accepting them to examine the GIT insurance policy to make sure that such high-value goods are not specifically excluded and that any special requirements which the insurance company may have imposed regarding overnight parking and general vehicle security and protection are complied with.

Night Risk and Immobilizer Clauses

Insurance policies frequently contain clauses requiring vehicles carrying high-value goods to be securely parked in locked or guarded premises overnight (known as the 'Night Risk' clause) or to be fitted with approved vehicle protection devices such as steering column locks, engine immobilizers and alarm systems (known as the 'Immobilizer' clause). It is a condition of the insurance that such devices must be maintained in good working order and must be put into effect when the vehicle is left unattended. Failure to comply with such conditions can render the cover invalid.

Sub-Contracting

Before valuable loads are sub-contracted to other hauliers, operators should take considerable pains to satisfy themselves as to the genuineness of any driver calling at their premises for a load (telephoning the driver's employer is one suggested method of checking). Experience shows that drivers with criminal intent will state that they are employed by a certain firm and that they require a return load; documents are frequently handed to the driver in such cases; he picks up the load and disappears. A few days later, when investigations are made, it is only then discovered that the driver obtained the load by false pretences. GIT insurers may not accept responsibility for such losses, or indeed any losses involving sub-contracted loads unless they have had prior notification of the loads and the circumstances.

GIT on Hired Vehicles

The increasing use of vehicles on contract hire raises an important issue regarding liabilities for goods carried. Normally under the terms of the hire contract it is made quite clear by the hire company supplying the vehicle that it assumes no responsibility for loss or damage to the goods which are carried on the vehicle.

Vehicles Hired with Drivers

If, however, under the terms of the contract the hire company offers to provide a driver, the driver acts under its instructions. Should he act in a way which would be considered contrary to normal reasonable action (for example, leaving a fully laden vehicle overnight in the open when he had been specifically instructed to empty the vehicle or to place it in a locked garage), the hire company may find itself held liable at law for a 'fundamental breach of contract' and be faced with having to pay the full amount of any loss incurred. A method of overcoming this difficulty is for the hire company to arrange with the owner of the goods for a GIT insurance policy to be effected in their joint names and for the owners of the goods to pay the premium in the contract hire agreement.

CONDITIONS OF CARRIAGE

An operator carrying goods for hire or reward is advised to set out conditions of carriage under which he contracts to carry goods. In these conditions the carrier can define his liabilities by stipulating limits on the value of goods for which he will normally accept responsibility, with goods of higher value being carried only on special terms, and stating circumstances and provisions under which no compensation is payable. For example, an operator could make it a condition that he accepts no liability under the following circumstances:

- Act of God.
- Act of war or civil war.
- Seizure under legal process.
- Act or omission of the trader, his employees or agents.
- Inherent liability to wastage in bulk, or weight; latent defect, inherent defect, vice or natural deterioration of the merchandise.
- Insufficient or improper packing.
- Insufficient labelling or addressing.
- Riots, civil commotions, strikes, lock-outs, stoppage or restraint of labour from whatever cause.

- Consignee not taking or accepting delivery within a reasonable time.
- Loss of a particular market whether held daily or at intervals.
- Indirect or consequential damages.
- Fraud on the part of the trader. In this context trader means either consignor or consignee.
- If non-delivery of a consignment, whether in part or whole, is not notified in writing within a specified number of days of despatch and a claim made in writing within a further specified number of days of despatch.
- If pilferage or damage is not notified in writing within a specified number of days of delivery, and a claim made in writing within a further specified number of days of delivery.

A note to the effect that goods are carried only under the Conditions of Carriage should be made on all relevant business documents, particularly consignment and delivery notes, invoices and quotations. Conditions of Carriage should always be drawn to the customer's attention and copies made available for customers to examine before they give orders for movements to commence.

RHA Conditions of Carriage

The Conditions of Carriage used by members of the Road Haulage Association are an excellent example of the sort of conditions which could be used by a haulage contractor to define his responsibilities. The Association's current Conditions are dated 1998 – these limit liability to £1,300 per tonne. Further sets of conditions for livestock carrying and sub-contracting are also prepared for members.

NB: The RHA Conditions of Carriage are the copyright of the Association and their use by a non-member would be a breach of copyright and therefore illegal – the RHA has taken legal action in cases of unauthorized use of its Conditions.

COVER FOR INTERNATIONAL HAULAGE JOURNEYS

The GIT cover described above is not sufficient or even legally acceptable where vehicles are engaged on international haulage work. In most cases, such operations are governed by the provisions of the *Convention on the Contract for the International Carriage of Goods by Road,* commonly known and referred to as the CMR Convention. This Convention automatically applies where an international haulage journey takes place between different countries at least one of which is party to the Convention (with the exception of UK–Eire and UK mainland–Channel Islands journeys which are ruled not to be international journeys for this purpose).

Road hauliers who carry goods on any part of an international journey, whether they they choose to know it or not, fall within the legal confines of the CMR Convention under which compensation levels for loss or damage to goods are much higher than the standard Conditions of Carriage GIT cover applicable in national transport operations. CMR levels of cover vary according to a set standard which is published daily in the financial press. For this reason it is important to obtain adequate cover when involved in international transport.

Additionally, where hauliers undertake cabotage operations they should discuss the levels of cover required with their insurers – indeed they should see if extended cover is available to cover certain liabilities such as losses from unattended vehicles. Difficulties may arise where local conditions of carriage are imposed, and claims and legal wrangling arise under law other than English law; under French law, for example, minimum liability is set at a value equivalent to approximately £12,000 per tonne and as this is a domestic requirement it may not be covered by a UK haulier's CMR policy.

Unfair Contract Terms

The Unfair Contract Terms Act 1977 affects such contracts as conditions of carriage. The effect of this legislation is to increase the liability of transport operators, particularly in respect of instances where liability for negligence is disclaimed by contract or by notice. Further, it prevents a business from excluding its liabilities for breach of contract when dealing with the general public. Consequently, any term in a contract purporting to exclude liability for personal injury by negligence is void. Any term excluding liability for damage to property by negligence is also void unless the term used is reasonable as between the parties to the contract.

SECURITY

Insurance claims relating to vehicle thefts have increased in recent years and this is a serious problem of concern to insurance companies as well as to vehicle operators and the police. It is important that when away from base drivers should be encouraged to park their vehicles, especially if they are loaded with valuable goods, in guarded security parks. While the number of suitable security parks is limited and they are not conveniently located, it is nevertheless in everybody's interest that vehicles should not be left parked overnight on the roadside or on pieces of wasteland. Some insurers, Norwich Union among them, are now refusing to pay out on loss or theft claims if keys are left in or on a vehicle.

Security Warning

The RHA issues a security warning to members as follows:

- Make every effort to ensure you are employing honest staff. Take up references over at least the previous five years and be suspicious of unexplained gaps. When checking references by telephone be sure to look in the telephone directory yourself for the number. A number supplied by a dishonest applicant could connect you to his accomplices. A staff enrolment form is available on application to RHA Area Offices and this form or a similar one should be completed.
- Until you have seen his driving licence and have in your possession his P45 tax form and photograph, do not allow a newly engaged driver to take out a vehicle.
- Fit a vehicle immobilizer and/or alarm in as inaccessible a position as possible. Choose one which provides protection without the driver having to perform any operation which he normally would not have to do to stop his vehicle. Inspect the device frequently.

- Drivers of vehicles carrying valuable loads should not get out of their cab if stopped. Even if a police officer requests them to do so, they should offer to go to the nearest police station. Bolts on the inside of the cab doors give added protection against hijackers.
- A trouble-free cash bonus, from which a driver can be fined if he does not observe your security drill, is helpful.
- Vehicles should not be left unattended for long periods, especially at night. At no time should keys be left in an unattended vehicle. Remember, a stationary vehicle with its windscreen wipers or indicators operating gives a clear signal to any watching criminal that it is his for the taking.
- Discourage drivers from using the same cafés at the same time each day, particularly where their vehicles are not parked within sight.
- Starter or ignition switches, security lock keys: remove numbers and keep the keys for each vehicle on a ring which is welded so that they cannot be separated.
- If a vehicle's keys are lost, change switches and locks. It is much cheaper than losing a load.
- Invite drivers to report to the police any suspicious circumstances, such as transfer of goods from one vehicle to another without apparent reason, which they might see on their travels, or the registration number of any vehicle which is persistently following them.
- When disposing of a vehicle, remove the name of your firm so that a thief cannot use it to secure a load by false pretences.

National Stolen Lorry Load Desk

The Metropolitan Police has taken over the National Stolen Lorry Load Desk from Essex Police, which had run the scheme for a number of years. The Desk is funded jointly by the Home Office and the road haulage industry. It is now under the control of the Assistant Commissioner, Specialist Operations at New Scotland Yard and is headed by DCI David Ryan of the Serious and Organized Crime Branch. The Desk can be contacted via the following website: www.truckpol.com

Also of interest in this context is the Eurowatch service that is concerned with crime against commercial vehicles throughout Europe. This can be contacted via www.eurowatchcentral.com

OTHER INSURANCE – BUSINESS AND PREMISES

While vehicle operators may be particularly concerned about obtaining appropriate insurance cover for their vehicles and the loads carried, they also need adequate insurance protection for other business contingencies in exactly the same way as any other employer or firm. Some of these insurances are listed and briefly described below.

Employers' Liability

It is a legal requirement under the Employers' Liability (Compulsory Insurance) Regulations 1988 for employers to cover their liabilities for any bodily injury incurred or disease contracted by their employees during or arising from their employment in the employer's business activities. In the case of a haulage contractor the business activities for which cover is required would be both as a haulage contractor and as an owner or occupier of property (ie the business premises). A Certificate of Insurance must be displayed in workplaces where employees have access to it and retained for 40 years (in any 'eye readable' form).

The minimum cover required by law is £5 million, but usually policies are issued with cover extending to £10 million. Premiums are based on the total payroll of the firm divided into categories (eg clerical staff, drivers, maintenance staff). Extensions to the basic policy can provide cover for employees engaged on private work for directors or senior management, against liability incurred in work/employment-related sporting, social, first aid or welfare activities and against liabilities arising through sub-contracting work. Legal costs are recoverable in addition to any compensation awards.

Public Liability

Public liability policies provide cover against legal liability to third parties for bodily injury or illness or loss of or damage to their property arising out of the insured's business activities because of the insured's negligence. The policy can be extended to cover contractual liability which is essential for hauliers operating under printed Conditions of Carriage.

The policy conditions should be such that they cover contingencies not covered by the motor vehicle policies for damage caused by the vehicle or driver. It should also provide cover for liability arising from goods or vehicles being sold, supplied, altered, serviced or repaired.

Indemnity is usually set at a fixed figure per occurrence. A minimum of £500,000, or preferably £1 million, is recommended as court awards have reached this figure for injuries to just one person.

Money Cover

Insurance cover can be obtained for the loss of cash, bank notes, currency notes, cheques, postal orders, postage or revenue stamps, national insurance stamps (now available only for limited use by self-employed people), 'holiday with pay' stamps, luncheon vouchers, trading tokens, credit vouchers, travellers' cheques, VAT vouchers and other negotiable instruments.

Extensions to the policy should include assault benefits to employees and other persons lawfully carrying these items and provide cover in respect of damage to and loss from slot machines, franking machines and safes. Premiums are usually based on estimated carryings of money to and from banks in a year.

Credit Insurance

Cover is available for financial loss because of customers defaulting on payment or their insolvency.

Indirect or Consequential Loss

When operating under RHA Conditions of Carriage which exclude liability for indirect or consequential loss and delay, the owner of the goods is not covered under a Goods in Transit policy. Cover is available under indirect or consequential loss insurance at a premium amounting to approximately 20 per cent of the GIT insurance premium.

Theft

Theft cover provides protection in the event of loss or damage resulting from entry to or exit from property by violent or forcible means. Usually, no cover is provided in such policies against larceny if there is no damage or visible sign of entry.

Petrol Installations and Oil Storage Tanks

Cover for accidental damage to fuel pumps, surface tanks and piping (but not for the loss of the contents, which is covered under the normal theft or fire insurance) is provided by such insurance. It also covers collapse, rupture or weld failure of these items.

Storm Damage

These policies compensate for losses resulting from the entry of water into petrol and diesel tanks as a result of heavy rain storms.

Fire and Special Perils

Cover can be obtained under this form of insurance for damage caused by fire, aircraft falling on the premises, explosion, riot and civil commotion, lightning, impact (including by own vehicles which is important because you cannot claim for this under your own vehicle insurance), and burst pipes.

Particular attention has been drawn to the need for adequate cover in respect of riot and civil commotion following well-publicized disturbances in some of Britain's inner-city areas. Unless a fire insurance policy is extended to provide such cover, fire damage by rioters is not covered. It is possible to claim on the local authority under the Riot Damages Act 1886 but this would hardly provide adequate compensation for full reinstatement of premises.

Glass

This cover provides compensation for broken windows and for the cost of temporary boarding up.

Pressure Vessels and Boilers

This type of policy includes provision for the regular and statutory inspection of compressors and air receivers, together with compensation in the event of their explosion or collapse.

Lifting Equipment

The law requires such equipment to be regularly inspected and certificated. Insurance cover can be obtained which provides the inspection and protection against claims arising from the use of hoists, lifts, cranes, forklift trucks, pulley blocks, chains, and so on.

Personal Accident

A firm can cover its principals, directors, staff, drivers and maintenance staff against personal accident while driving or while away from the premises (not on the premises). This will provide set levels of regular income for various contingencies or lump-sum damages for loss of limbs. Where the firm is large enough there are special schemes available which have significant tax advantages to the employer.

Medical Expenses

When drivers are required to travel abroad they should be covered for medical expenses incurred in foreign countries, for compensation for taking relatives out to visit them if they are detained in hospital and for bringing the patient back to the UK for further treatment if necessary. Compensation for the loss of drivers' personal effects and baggage can usually be included in this type of policy.

For 25 years, the Transmed scheme, purpose-designed for the international road transport industry, has been rescuing drivers from abroad. The firm can be contacted on: 020 8399 6003.

Legal Expenses

Cover may be obtained to protect against the legal costs incurred in contesting unfair dismissal claims and other breach of employment contract issues such as pension rights and matters arising from legislation on equal pay, sex discrimination and race relations. Such policies also usually provide cover for legal expenses incurred in other disputes over liability or responsibility for the action of individuals or firms and for legal expenses following criminal prosecution under any statute. Other schemes are available to cover actual compensation awards.

Loss of Profit

This type of insurance would apply if a transport operator lost his vehicles and warehouse as a result of, for example, a fire. While the fire insurance would cover the cost of damage incurred, a loss of profit policy would compensate the operator for his business losses to competitors during the period of disruption until he got his business back on its feet again.

Usually the policy would provide this cover for at least 12 months and possibly even longer in view of the delays experienced in obtaining new vehicles and in getting premises replanned and rebuilt.

Book Debts

This insurance provides cover for debts caused by loss of records as a result of fire or other physical causes which leave no trace of amounts owed.

Fidelity Guarantee

Fidelity guarantee insurance provides cover against fraud or dishonesty by employees in connection with their employment during or within 18 months of the period of the fraud occurring or termination of the employment, whichever comes first.

Hired-In Plant

This type of policy covers all the liabilities imposed by hiring agreements including damage to the plant (vehicles, forklift trucks, cranes, etc), losses in hiring revenue incurred by the owner and claims made by third parties. 'The Contractor's Plant Association Conditions of Hire' also impose such liabilities while hired plant is on the highway. A motor insurance certificate is also required in these instances.

Motor Contingency

Operators will find that their existing motor policies and other insurances exclude such things as liabilities incurred when hiring-in vehicles and drivers. A motor contingency policy would protect the insured by covering the liabilities of the person or firm owning the vehicle or employing the driver if its insurance was not current or was invalidated for some reason (eg premiums not paid). Similarly, a motor contingency policy would cover employees' use of their own cars on company business. The cover would be for any liability of the employer resulting from the use of the car on his business (eg if the employee had not paid his premium or, for example, if the vehicle was not taxed or not in roadworthy condition thus causing his cover to be invalid), but it would not provide any cover for the liability of the employee himself.

All Risks

This is a policy which could be used to cover any eventuality which is not specifically covered in any other policy held by a business. For example, it would provide compensation for loss or damage to any valuable paintings or antiques on company premises or in directors' offices, the firm's sporting trophies, awards of merit, and so on.

Computers

Policies are available to provide cover for material damage to computer systems resulting from accidental causes, electrical or mechanical breakdown. Such policies also cover consequential loss. Further policies cover software against loss or damage – even coffee spillage – and accidental or malicious erasure of data, and provide for the costs of re-establishing information.

OBTAINING THE BEST COVER

When negotiating with an insurance company for cover the premiums and policy conditions should be carefully considered and it is usually advisable to compare the terms offered with those available from other insurers. The services of an insurance broker can be helpful in finding the most satisfactory terms and competitive premium rates. Brokers retained for this purpose will give advice on the terms and conditions, handle claims and generally ensure that you are getting the best possible cover at economic rates. No payment is made to insurance brokers; they obtain their payment by way of commissions or discounts on premiums paid to the insurance company.

INSURANCE CLAIMS

When making claims on insurance policies there are a number of points which deserve particular attention if a broker has not been engaged to handle these problems. The first and most important point in regard to claims following motor vehicle accidents is that insurance companies must be given immediate notice of an accident to any vehicle for which they supply cover, followed by a properly and fully completed accident report and claim form. A time limit is specified for this, normally seven days.

Completion of Claim Forms

If the claim is being made as a result of a vehicle accident the driver, if possible, should complete a claim form giving as much detail of the accident as he can: time, place, conditions of the weather and road, his position on the road and his speed, his direction of travel and the location of identifying objects, and the names and addresses of other parties involved and of any witnesses. A description of events leading up to the occurrence and a sketch of the position of the vehicles involved, both before and after the collision, should also be made on the report form, together with an indication of the damage to vehicles and property. The driver should not make any statement at the scene of the accident admitting or indicating liability (eg by apologizing for his mistake).

Processing Claims

Once the insurance company has received the claim form they will get on with the business of deciding where the responsibility for the accident lies and how it should be apportioned. They will arrange for their motor vehicle assessor to examine the damaged vehicle and give permission for the repairs to be carried out if the estimate which the repairer has submitted is acceptable and, of course, if the vehicle is repairable. If the vehicle is beyond economical repair the assessor will authorize a 'write-off'.

Recovery of Uninsured Losses

In the event of the third party being at fault in an accident and the insured having an excess on the policy (by which the insured person volunteers to pay part of the cost of the repairs to their own vehicle, usually the first £25 to £500, for which there is usually a reduction in premium), he will need to make a claim against the third party for recovery of the excess (usually termed the 'uninsured loss'). The insurance company (or brokers) will deal with this matter if the damage is more than the excess and they are meeting the difference, but if the damage is slight and it is not intended to make a claim on the insurance company, a claim must be made direct to the third party for the 'uninsured loss'.

10 Road Traffic Law

Road traffic regulations are very complex and are to be found in a number of Acts and statutory instruments relating to all aspects of road use by pedestrians, cyclists and motorcyclists, motorists and, of course, large goods and passenger vehicle drivers and operators. In recent years new traffic offences have been introduced along with tougher measures to deal with drink-drivers (with five years' imprisonment for drink-drivers who cause a death) and with vandals who place road users' lives at risk by placing dangerous objects on a road or interference with traffic signs or signals (with up to seven years' imprisonment for convicted offenders).

At the time of preparing the 2005 edition of this *Handbook* Parliament is considering a new Traffic Management Bill. While the Bill is predominantly concerned with traffic management matters and roads issues, one of the key provisions that will relate more specifically to transport operators is the establishment of a uniformed on-road traffic officer service to manage the traffic consequences of such incidents as accidents, vehicle breakdowns, obstructions and debris on the carriageway, etc. These traffic officers will operate under the auspices of the Highways Agency and will have the power to stop and direct traffic in the interests of keeping traffic flowing.

Definition of Roads

For the purposes of most aspects of road traffic law a 'road' is defined in the Road Traffic Act 1988 (section 192) for England and Wales as: 'any highway and any other road to which the public has access, and includes bridges over which a road passes'. In Scotland, the Roads (Scotland) Act 1984 applies and defines a public road as 'any road and any other way to which the public has access, and includes bridges over which a road passes'.

NB: It has been held that so-called 'private' dock roads at ports from which ferry services operate, are in fact 'open to the public' and therefore public roads on which drivers are subject to the normal road traffic rules applicable to all other public roads.

The term 'highway' has its meaning in common law as a way over which all members of the public have the right to pass and repass. This might be on foot, on horseback, accompanied by a beast of burden or with cattle, or with a vehicle. For the purposes of the Highways Act 1980 a 'highway' includes the whole or part of the highway other than a ferry or waterway, and where it passes over a bridge or through a tunnel, the bridge or tunnel is part of the highway. This Act makes it an offence for a person, without lawful authority or excuse, in any way wilfully to obstruct free passage along the highway of whatever type.

The Highway Code

Many of the particular legal requirements relating to the use of vehicles on the road are identified in the *Highway Code* (a new edition of which was published in 1999), along with much useful advice on driving and road usage, although it should be remembered that the *Code* is intended for guidance and is not, in itself, a book of definitive traffic law. However, many of the rules in the *Code* are legal requirements and failure to comply with these will render a driver liable for prosecution, and if convicted he may facing a fine, penalty points being imposed on his driving licence, and even disqualification from driving. In any case, even where a specific offence has not been committed, such as in the event of an accident, failure to follow advice given in the *Code* on safe driving may still result in attachment of blame.

The new edition of the *Highway Code** is intended to be essential reading for everybody, not just learner drivers. It has been extended to 100 pages of vital information (with more diagrams and a quick-reference index) and includes a great deal of advice on safe driving and responsible road use, covering such topics as the use of seat belts, child safety in cars, fitness to drive and the reporting of health conditions to the DVLA, vision requirements, and alcohol and drugs – it reiterates the message that you must not drink and drive or drive while under the influence of drugs – and it also advises road users about the use of mobile phones and in-cab technology (ie drivers must exercise proper control of their vehicles at all times). Other useful advice is given on control of the vehicle, including driving in adverse weather, what to do in the event of a skid, the use of an ABS braking system when making an emergency stop, what to do if the vehicle brakes are affected by water, and *not* coasting out of gear or with the clutch disengaged under any circumstances.

**NB: Available from The Stationery Office, most booksellers and many newsagents, price £1.49.*

SPEED LIMITS

Excessive speed is said to be a contributory factor in one in three road casualties. How fast we drive is crucially important to how safe our roads are, according to a Government statement in March 1999, when launching a 'far reaching' review of road safety strategy. In the meantime the Government's message is to 'Kill your Speed', and a new system of speed enforcement is currently being introduced – see p 239.

Three levels of speed limit for vehicles are imposed on road users:

- Limits applying to vehicles using particular roads.
- Limits applying to particular classes of vehicle (including limits imposed by the mandatory fitment of speed limiter devices).
- Temporary speed limits on vehicles introduced for special reasons such as in potentially hazardous situations and in times of fuel shortages.

Speed Limits on Roads

On roads where street lights are positioned at intervals of not more than 200 yards (defined as a 'restricted' road), an overall speed limit of 30 mph applies to all classes of vehicle unless alternatively lower speeds are indicated by signs or unless the vehicle itself is subject to a lower limit by reason of its construction or its use. In some instances, speeds in excess of 30 mph are permitted on such roads and this is indicated by appropriate signs showing the higher maximum limits.

The present maximum speed limits on roads outside built-up areas are 60 mph on single-carriageway roads and 70 mph on dual-carriageway roads and motorways, except where specified temporary or permanent lower limits are in force. In certain high-accident risk areas (eg housing estates) 20 mph speed limit zones are being introduced in conjunction with road humps. The first was started in Norwich in January 1991 and others have followed. Other so-called 'traffic calming' measures being increasingly used to improve road safety include road-narrowing chicanes.

Advisory speed limits on motorways should be observed. These are shown by illuminated signs which indicate hazardous situations and roadworks ahead and by temporary speed limits signs at roadworks. The amber flashing warning lights positioned on the nearside of motorways (two lights, one above the other) indicate an advisory slowing down until the danger, and the next non-flashing light, is passed. Mandatory speed limits may also be seen at roadworks' sites on motorways (indicated by white signs with black letters and a red border). Failure to comply with these mandatory motorway speed warning signs can result in prosecution.

Speed Limits on Vehicles

Vehicles are restricted to certain maximum speeds according to their construction, weight or use but when travelling on roads which themselves are subject to speed restrictions it is the lowest permitted speed (ie of the vehicle or of the section of road) which must be observed.

Private Cars

Motor cars and dual-purpose vehicles must observe the 70 mph limit on motorways and dual-carriageway roads and 60 mph on single-carriageways and the appropriate lower limits on all other occasions.

Car-Derived Vans

Light vans derived from private-type motor cars (ie car-derived vans including car-derived open-back, pick-up trucks, up to 2000 kg gross weight) may travel at the same speeds as private cars on these roads (ie 70 mph on motorways and dual-carriageways and 60 mph on other roads unless lower limits are in force).

Private Cars and Car-Derived Vans Towing Trailers

The speed limit for motor cars (including dual-purpose vehicles), car-derived vans and motor caravans towing trailers and caravans is 50 mph on single-carriageway roads and 60 mph on dual-carriageways and motorways. There is no longer any legal requirement for such vehicles towing trailers and caravans to display a '50' mph plate at the rear.

Light Goods Vehicles

Light goods vehicles (apart from car-derived vans mentioned above) for the purposes of speed limits are vehicles up to and including 7.5 tonnes maximum laden weight (mlw). Speed limits for rigid vehicles in this category are 50 mph on single-carriageway roads, 60 mph on dual-carriageway roads and 70 mph on motorways.

The maximum speed for rigid vehicles up to 7.5 tonnes maximum laden weight drawing trailers and articulated vehicles up to 7.5 tonnes maximum laden weight on single-carriageway roads is 50 mph and on dual-carriageways and motorways the limit is 60 mph.

Large Goods Vehicles

There is no distinction in terms of maximum speed between rigid and articulated large goods vehicles and those with drawbar trailers. Large goods vehicles are those over 7.5 tonnes maximum laden weight (mlw) which, under other regulations, are required to display rear reflective markers so making them readily identifiable for speed limit enforcement purposes. Speed limits for all vehicles in this category are 40 mph on single-carriageway roads, 50 mph on dual-carriageway roads and 60 mph on motorways.

Many new goods vehicles are now required under UK legislation to be fitted with speed limiters set to restrict their top speed to 60 mph while under EU legislation certain other heavy vehicles are restricted to 90 kph (approx 56 mph)(see Chapter 13 for full details).

NB: In an effort to further reduce road accidents, the EC is planning to extend the 90 kph speed-limiter scheme downwards to include goods vehicles in the weight range 3.5 tonnes to 12 tonnes gross weight.

Passenger Vehicles

The maximum speed limit for buses and coaches over 3.05 tonnes unladen weight and with more than eight passenger seats is dependent upon the overall length. For those not exceeding 12 metres length the limits are 50 mph on single-carriageway roads, 60 mph on dual-carriageway roads and 70 mph on motorways (but see below). For those over 12 metres in length the limits are 50 mph on single-carriageway roads and 60 mph on dual-carriageways and motorways.

Speed limiter legislation has restricted coaches over 7.5 tonnes gross weight to a maximum speed of 65 mph since 1 January 1996.

NB: In all cases mentioned above, it must be stressed that these limits only apply where no lower limit is in force.

Special Types Vehicles

Vehicles operating outside the Construction and Use (C&U) regulations for the purposes of carrying abnormal indivisible loads come within scope of the Special Types General Order (STGO) as described in Chapter 22 and must conform to specified speed limits depending on their category. These speed limits are stated on p 486 but are repeated here with other speed limits for ease of reference:

Vehicle category	Motorways	Dual-carriageways	Single carriageways
Category 1	60 mph	50 mph	40 mph
Category 2 and 3	40 mph	35 mph	30 mph

Works Trucks and Industrial Tractors

The maximum speed limit for works trucks and industrial tractors is 18 mph but the latter are not permitted on motorways.

Agricultural Vehicles

Agricultural vehicles are limited to 40 mph. They are not permitted on motorways.

Motor Tractors/Locomotives

Where such vehicles (including their trailers) are fitted with springs and wings their maximum permitted speeds are 40 mph on motorways and 30 mph on other roads. When they do not have springs and wings the maximum speed limit is 20 mph on all roads.

Track-Laying Vehicles

These vehicles are limited to maximum speeds of 20 mph or 5 mph depending on their construction (ie whether they have springs and wheels fitted with pneumatic or resilient tyres – see below).

Vehicles with Non-Pneumatic Tyres

Vehicles with resilient, non-pneumatic (ie solid) tyres are restricted to a maximum speed of 20 mph on all roads. Those with non-resilient tyres (eg traction engines) are restricted to 5 mph – see also above, track-laying vehicle.

Emergency Vehicles

Fire, police and ambulance service vehicles are exempt from all speed limits if, by observing the speed limit, they would be hampered in carrying out their duties. However, drivers of such vehicles have a duty to take particular care when exceeding statutory limits and could face proceedings if an accident results while exceeding the limits.

Table of Vehicle Speed Limits

	Motorway mph	Dual-carriageway mph	Other roads mph
Private cars			
– solo	70	70	60
– towing caravan or trailer	60	60	50
Buses and coaches			
– not over 12 metres length	70 (65)*	60	50
– over 12 metres length	60	60	50
**NB: Coaches over 7.5 tonnes subject to speed limiter legislation are restricted to 65 mph.*			
Goods vehicles			
Car-derived vans			
– solo	70	70	60
– towing caravan/trailer	60	60	50
Not Exceeding 7.5 tonnes mlw			
– solo	70	60	50
– articulated	60	60**	50
– drawbar	60	60**	50
*** In Northern Ireland the speed limit for vehicles in these two categories is 50 mph only.*			
Over 7.5 tonnes mlw			
– solo	60	50	40
– articulated	60	50	40
– drawbar	60	50	40

Note: mlw means maximum laden weight (ie maximum gross weight for a vehicle as specified in construction and use regulations).
NB: See above for speed limits for Special Types and other vehicles.

Speed-Enforcement Cameras

The Road Traffic Act 1991 legally authorized the use of photographs as evidence by courts in cases of alleged speeding (and traffic-light jumping). Cameras have been installed at key sites since 1 July 1992 to catch speeding drivers and traffic-light offenders. The so-called Gatso cameras (named after their inventor, Dutch ex-racing driver Maurice Gatsonides) produce film showing the vehicle and its number plate (which can be enhanced by scientific means for purposes of clarity – and can even decipher the registration number of vehicles fitted with plates which have been photo-reflective sprayed), date, time and the vehicle speed.

However, digital speed-enforcement cameras are being progressively introduced to overcome the physical constraints of the present system which requires the manual loading, unloading, removal and storage of film from the roadside units – a costly and time-consuming operation for already hard-pressed police forces.

Legislation allows for a prescribed device which 'captures and records images of motor vehicles at two positions on the road and calculates the average speed of the vehicle between those positions'. The new device, announced in August 1999, and called a Speed Violation Detection Deterrent (SVDD), was first installed on the M2 between junctions 4 and 5. Two cameras are set one mile apart and linked by digital signal to a computer centre where the time a vehicle takes to pass the two points is measured and the average speed instantly calculated. Where the statutory speed limit has been exceeded the vehicle and registered keeper are identified from DVLA records and a relevant penalty is imposed.

New digital SPECS speed-enforcement cameras were launched in Nottingham on 28 July 2000. The system measures average speeds between pairs of cameras mounted on distinctive SPECS columns a distance apart which create Speed Controlled Zones.

NB: The blue poles that have sprouted with amazing rapidity on Britain's roadsides are not speed cameras – they are part of a £10 million development of the Trafficmaster network of traffic monitoring sensors (see p 264).

Eight pilot schemes have been launched nationwide, under which revenues raised from the speeding fines imposed will go directly into the funding and maintenance of speed cameras.

LIGHTING-UP TIME

All mechanically propelled vehicles must display front and rear position lights and headlamps (where required by regulations) between sunset and sunrise and during daytime hours when visibility is seriously reduced (see Chapter 15 for lighting details).

Night Parking

Goods vehicles not exceeding 1525 kg unladen do not require lights at night when standing on restricted roads (ie on which a 30 mph speed limit – or lower limit – is in force), if they are parked either in a recognized parking place (ie outlined by lamps or traffic signs) or on the nearside, close to and parallel to the kerb, facing the direction of travel and with no part of the vehicle within 10 metres of a junction (ie on the same side as the vehicle or on the other side of the road). On any road where these conditions are not met lights must be shown (ie front and rear position lights).

All goods vehicles exceeding 1525 kg unladen weight must display lights at all times when parked on roads between sunset and sunrise. Trailers and vehicles with projecting loads must not be left standing on roads at night without lights.

Vehicles should be parked on the nearside of the road when left standing overnight except when parked in a one-way street or in a recognized parking place and they must not cause obstruction.

Increasing attention is being given by the police and local authorities to drivers sleeping overnight in heavy vehicles with sleeper cabs while parked in lay-bys. Drivers should be warned against this practice which is usually considered illegal on the grounds that the vehicle is causing an obstruction. A similar situation applies when drawbar trailers and semi-trailers are left in lay-bys.

Parking in lay-bys, separated from the main carriageway only by a broken white line, without sidelights and other obligatory lights as appropriate being lit after lighting-up time, is an offence and the driver will be prosecuted if caught. This does not apply where the lay-by is segregated from the highway.

STOPPING, LOADING AND UNLOADING

Leaving Engine Running

Whenever a driver leaves his vehicle on a road, the engine must be stopped (except in the case of fire, police or ambulance service vehicles or when the engine is used to drive auxiliary equipment or to power batteries to drive such equipment). Hitherto, the requirement for stopping a vehicle engine when stationary was to prevent noise, but an amendment to the C&U regulations, effective from 2 February 1998, makes preventing exhaust emissions an additional reason for the requirement.

Obstruction

A vehicle must not be left in a position where it is likely to cause obstruction or danger to other road users, eg near an entrance to premises, near a school, a zebra crossing or a road junction (also see p 246).

Trailers (including articulated semi-trailers) must not be left on a road when detached from the towing vehicle.

Loading and Unloading Restrictions

Vehicles must not stop or park on clearways to load or unload. In some areas loading and unloading restrictions are indicated by yellow lines painted on the kerb at right angles to it as follows:

- A single yellow line at intervals indicates a ban on loading and unloading between the times shown on a nearby plate (eg Mon–Sat 8.30 am – 6.30 pm).
- Double yellow lines at intervals indicate a complete ban on loading and unloading at any time.

The precise terms of the restriction are indicated on signs mounted on nearby lamp posts, walls, etc. Delivery drivers should check these carefully to avoid any infringement of the law.

NB: Previous kerb markings consisting of three yellow lines at intervals which indicated a ban at all times have been abolished.

Waiting and Parking Restrictions

Single, double or broken yellow lines painted on the road parallel to the kerb apply to waiting and parking at various times, but they do not indicate a ban on loading or unloading and the same applies to 'no waiting' prohibitions indicated by 'no waiting' signs (see *Highway Code* for full details of waiting and parking restrictions).

Parking Meter Zones

Loading and unloading in parking meter zones during the working day (the times are indicated on signs) is not allowed unless a gap between meter areas or a vacant meter space can be found. A vehicle using a meter space for loading or unloading can stop for up to 20 minutes without having to pay the meter fee (this does not apply when parking for any purpose other than loading or unloading the vehicle).

MOTORWAY DRIVING

Motorway driving requires special care and observance of the motorway regulations. In particular, vehicles not capable of exceeding 25 mph on the level are prohibited from using motorways, including vehicles operating under the Special Types General Order (see also Chapter 22) if they cannot exceed this speed.

Vehicles must not stop on motorways except through mechanical defect or lack of fuel, water or oil, due to an accident, the illness of a person in the vehicle or for other emergency situations (including giving assistance to other persons in an emergency), or to permit a person from the vehicle to recover or remove objects from the carriageway. It is illegal to drive on the hard shoulder or the central reservation, to reverse or to make a 'U-turn' on a motorway. Vehicles which must use the hard shoulder for emergency reasons as described above must remain there only for so long as is necessary to deal with the situation.

Use of Lanes

Goods vehicles with maximum laden weights in excess of 7.5 tonnes and vehicles drawing trailers (and certain other heavy motor cars not included in the categories mentioned) must not use the outer or offside lane of three- and four-lane motorways.

On some steep slopes of two-lane motorway sections large goods vehicles (ie over 7.5 tonnes) are banned from using the outside lane; these bans are clearly signposted on the approaches to the appropriate section indicating the extent of the banned section and the vehicles prohibited from using the outer lane.

From 1 January 2005 the range of vehicles prohibited from using the outer (offside) lane of three-lane motorways is extended to include goods vehicles with a maximum laden weight exceeding 3.5 tonnes which are fitted with speed limiters as required by the Road Vehicles (Construction and Use) Regulations, also with effect from 1 January 2005 (see also p 341).

Temporary Speed Limits

Where carriageway repairs take place on motorways or where contraflow traffic systems are used an *advisory* 50 mph speed limit is usually imposed. This is considered by the police to be a maximum speed, and they may prosecute drivers found speeding in these sections for a 'driving without due consideration' type of offence. However, it is becoming more common for a *mandatory* temporary speed limit to be imposed in such cases and where drivers are detected speeding in these sections they will be prosecuted for this offence.

Speed Limits for Recovery Vehicles on Motorways

For the purposes of motorway speed limits, recovery vehicles may travel at up to 60 mph. This follows a High Court ruling (on an appeal by the Director of Public Prosecutions) that such vehicles are constructed to carry a load and therefore may travel at the same maximum speed as other heavy goods vehicles. Previously such vehicles were required to observe the 40 mph maximum limit applicable to vehicles classified as motor tractors, light and heavy locomotives.

Other Vehicles on Motorways

Light and heavy locomotives which do not comply with C&U regulations, dump trucks, engineering plant and vehicles for export which do not comply with the C&U regulations, may be driven on motorways provided they are capable of attaining a speed of 25 mph on the flat when unladen and not drawing a trailer.

Learner Drivers on Motorways

Learner drivers are not allowed to drive on motorways, but holders of provisional lgv driving entitlements may drive heavy goods vehicles on motorways provided they hold a full ordinary driving entitlement (ie category B) and are accompanied by a qualified driver.

Lights, Markings and Signs on Motorways

Hazard Warning

Motorways are equipped with amber hazard warning lights located on the nearside verge and placed at one mile intervals. When these lights flash, vehicles must slow down until the danger which the lights are indicating, and a non-flashing light, has been passed.

Rural Motorways

Rural motorways have amber lights, placed at not more than two mile intervals and usually located in the central reservation, which flash and indicate either a maximum speed limit or, by means of red flashing lights, that one or more lanes ahead are closed. The speed limit indicated applies to *all* lanes of the motorway and should not be exceeded.

Urban Motorways

Urban motorways have overhead warning lights placed at 1000-yard intervals. Amber lights flash in the event of danger ahead and indicate a maximum speed limit or an arrow indicating that drivers should change to another lane. If red lights flash above any or all of the lanes, vehicles in those lanes must stop at the signal. It is as much of an offence to fail to stop at these red lights as it is to ignore automatic traffic signals.

Information Signs

A new system of overhead information signs is to be installed on the M1 motorway between Junctions 4 (Elstree) and 6a (Abbotts Langley). The 18 overhead driver information signs operated from the National Traffic Control Centre in Birmingham and by the police from the Highway's Agency regional control room at South Mimms will notify incidents and conditions on the road ahead and warn of delays. The system should be in operation by the summer of 2006.

Motorway Road Markings

Experimental road markings designed to reduce the risk of nose-to-tail collisions are being tried out on some sections of motorway (eg the M1 near Leicester). The chevron-shaped markings are painted on the road surface at 10-metre intervals for five kilometres (ie three miles). Drivers are advised to keep at least two chevrons (ie two seconds) between themselves and the vehicle in front.

Local Radio Station Frequency Signs

Under new arrangements with the D*f*T, local radio stations can have their broadcasting frequencies indicated on motorway signs. It is a condition of such signposting that the station in question provides traffic news relevant to the location of the signs and of benefit to long-distance travellers 24 hours a day, seven days a week with at least four broadcasts per hour at peak times and two per hour during off-peak times, with programme interruptions for important announcements. The signs do not carry the station name or logo; only the broadcasting frequency enabling drivers quickly to tune into the appropriate wavelength.

Emergency Telephones on Motorways

Emergency telephones are located at one-mile intervals on the hard shoulders of each side of motorways – there is no need (and it is both dangerous and illegal) to cross the carriageways to reach an emergency telephone. Arrows on the back of posts on the hard shoulder indicate the direction to the nearest telephone. The use of the telephone is free, and it connects directly to the police who should be given full details of the emergency, who is calling and the vehicle involved. A woman travelling alone is advised to tell the police of this fact. After making the call you should return immediately to your vehicle to await help – a police patrol may arrive within minutes.

Driving in Fog

To help drivers avoid the grave hazards of driving in fog, the *Highway Code* clearly advises, in Rule 210, that drivers should:

- use headlights as required by the *Code* (ie switch them on when visibility is reduced to 100 metres – 328 ft);
- keep a safe distance behind the vehicle in front, remembering that the other vehicle's rear lights can give a false sense of security;
- be able to pull up within the distance they can see clearly, especially on motorways and dual carriageways, where vehicles are travelling much faster;
- use the windscreen wipers and demisters;
- beware of other drivers not using headlights;
- not accelerate to get away from a vehicle which is being driven too close behind;
- check mirrors before slowing down and use the brakes so the brake lights warn drivers behind that their vehicle is slowing down;
- stop in the correct position at a road junction with limited visibility and listen for traffic until it is safe to cross, then do so positively without hesitating in a position directly in the path of approaching vehicles.

The *Code* also states that drivers MUST NOT use front or rear fog lights unless visibility is seriously reduced as they dazzle other road users and can obscure the brake lights. These lights must be switched off when visibility improves.

M25 Automatic Fog-Warning System

An automatic fog-warning system operates on the M25 motorway. Detectors installed alongside the motorway identify when visibility falls below 300 metres and automatically switch on the existing matrix signals to display the message 'FOG'. The signals are located at strategic points where unexpected pockets of fog may occur as identified by the Meteorological Office (30 such zones are identified and 54 danger spots within those zones). When the fog signs are on drivers should slow down and proceed at a speed where they can safely stop within their range of vision.

HAZARD WARNING FLASHERS

Four-way direction-indicator flasher systems fitted to vehicles may be legally used to indicate that a vehicle is temporarily obstructing the road or any part of the carriageway either while loading or unloading, when broken down or for emergency reasons (previously their use was only permitted in emergencies). Further details of vehicle lighting requirements are to be found in Chapter 15.

TEMPORARY OBSTRUCTION SIGNS

New regulations (ie The Traffic Signs (Temporary Obstructions) Regulations 1997) from 1 March 1998 authorize drivers to place on the road, behind a broken down vehicle:

- a minimum of either four traffic cones, traffic delineators (a flattened cone) or traffic pyramids;
- a red warning triangle; and
- a flashing amber warning lamp with any of these warning devices.

They can also place, on the vehicle itself, a 'Road Vehicle Sign' described as a highly visible flexible yellow sheet depicting a red warning triangle.

The person in charge of, or accompanying, an emergency or breakdown vehicle that is causing an obstruction is authorized to place a 'keep right' sign to indicate a route past the vehicle.

LIGHTS DURING DAYTIME

If visibility during the daytime is poor, because of adverse weather conditions, drivers of all moving vehicles must switch on both front position lights and headlamps or front position lights and matched fog and spot lights. This applies in the case of heavy rain, mist, spray, fog or snow or similar conditions. When vehicles are equipped with rear fog lights (see p 365) these should be used when the other vehicle lights are switched on in poor daytime visibility conditions. Further details of vehicle lighting requirements are to be found in Chapter 15.

PARKING

Drivers who park their vehicles in a position which causes danger or obstruction to other road users can be prosecuted and their driving licence endorsed with penalty points on conviction (usually three) or they can be disqualified from driving.

The *Highway Code* lists the many places where drivers should not or must not park, in paragraphs 213 to 226. Drivers should study these lists to avoid causing danger or committing an offence.

There are not many parking places for the goods vehicle driver who has collections or deliveries to make, particularly in town, and for this reason drivers should be instructed to take reasonable care when parking in congested areas to avoid causing obvious obstruction or danger. For example, drivers should not double park, block entrances and exits of business or private premises, park near dangerous junctions or near pedestrian crossings, as well as avoiding the areas mentioned in the *Highway Code*.

PARKING ON VERGES

The Road Traffic Act 1988 (sections 19 and 20) makes it an offence to park a heavy commercial vehicle (ie a vehicle over 7.5 tonnes maximum laden weight including the weight of any trailer) on the verge of a road, on any land between two carriageways or on a footway whether the vehicle is totally parked on those areas or only partially so.

There are exemptions to this: when a vehicle is parked on such areas with the permission of a police officer in uniform, or in the event of an emergency, such as for the purposes of saving life or extinguishing fire, or for loading and unloading, provided that the loading or unloading could not have been properly performed if the vehicle had not been so parked and that the vehicle was not left unattended while it was parked.

It is an offence (under the Road Traffic Act 1988 section 34) for any person to drive a motor vehicle on to common land, moorland or other land which does not form part of a road, or on any footpath or bridleway beyond a distance of 15 yards, except where legal permission exists to do so but then only for the purposes of parking or to meet an emergency such as saving life or extinguishing fire.

LORRY ROAD USER CHARGING

The Lorry Road User Charge (LRUC)

Details of the highly controversial LRUC have not been included in previous editions of this *Handbook* for the simple reason that sufficient positive information on the government's proposals was not available. Indeed, over the past year this topic has been the subject of much abuse and criticism from many quarters, culminating in a very sudden and surprising decision of the government, announced by the Secretary of State for Transport, Alistair Darling, on 5 July 2005, that its plans for a separate Lorry Road User Charge scheme were abolished. Mr Darling said that the proposed scheme would now form part of a wider national road-pricing scheme – by implication, applying to all vehicles on all roads – with an initial suggestion that a pilot scheme would be developed first, covering one or more local areas. The road haulage industry saw this step as a let-down; firstly, by allowing hundreds of foreign haulage vehicles to continue entering the country at no extra cost for their UK operations and, secondly, by missing the opportunity to charge fuel duty for commercial vehicles at lower rates than for cars and more in line with prices payable by their Continental road haulage competitors.

LKW-Maut and GO-Box Systems

Germany, Austria and Switzerland have their own lorry road toll (Maut) systems. Germany's LKW-Maut scheme, which applies solely to heavy trucks over 12 tonnes gross weight using the country's motorways, was finally inaugurated on 1 January 2005 after two previous false starts due to technological hiccups. Austria's Maut system, colloquially known as the GO-Box system, was introduced in 2004 to replace both that country's former Eco-points and 'vignette' systems. Switzerland's system is a mileage-based truck tax that goes by the acronym LSVA. Further details of these systems can be found in Chapter 32.

LORRY ROUTES AND BANS

Local authorities identify preferred routes for heavy vehicles passing through their areas and display on them appropriate signs that:

- Mark the most suitable route between dock areas and the nearest convenient connection with the primary route/motorway network.
- Mark a suitable alternative route at any place on the primary route network where drivers of goods vehicles might be advised to avoid a particular part of that route, but where it is not appropriate to direct all traffic on to the alternative route, or to the primary route itself.
- Mark routes from the primary/motorway route system to local inland centres which generate a high level of goods vehicle traffic (industrial estates, for example).

Certain areas, and especially London, impose bans on the movement of goods vehicles and the parking of goods vehicles. These bans and restrictions are always marked with appropriate signs and operators are advised to ensure that their drivers observe them.

London Bans

Vehicles over 18 tonnes pmw are prohibited from travelling along many routes through Greater London at certain times unless the operator holds an exemption permit (issued under the London Boroughs Transport Scheme) which must be carried on the vehicle in a conspicuous position. Vehicles are also required to be fitted with air-brake silencers (hush kits).

The routes on which the ban applies are well signposted and the times at which it applies are also given on the signs. It is an offence for a driver of a goods vehicle over 18 tonnes pmw to travel on the banned routes at the relevant times unless a valid exemption permit has been issued and is carried on the vehicle. Application for exemption permits and vehicle plates should be made to the ALG Transport and Environment Committee, Traffic Enforcement Unit, New Zealand House, 80 Haymarket, London SW1Y 4TE (tel: 020 7747 4770; fax: 020 7930 2719).

The London ban applies (to vehicles without permits) at the following times:

- Sunday — at all times
- Monday to Friday — midnight to 7 am and 9 pm to midnight
- Saturday — midnight to 7 am and 1 pm to midnight

Westminster Council, London, intends to ban from its designated low emission zones (LEZs) heavy vehicles that cannot meet either: 1) Euro 2 emission levels with retro-fitted exhaust after-treatment; or 2) Euro 3 emission levels. Operators with such vehicles wishing to enter the LEZs will be required to pay an £80 annual fee.

London 'Red Route' Scheme

The 'Red Route' scheme to prevent traffic congestion on certain primary routes in London has been extended. Originally about 6.5 miles of roadway between Archway Road and the Angel, Islington, was designated as a red route but other sections have been, and are being, progressively introduced until the full 315-mile red route network is complete. Each section is identified by single or dual red road markings and accompanying red route signs. Vehicles may only stop to collect or deliver at specified times (or not at all), only for limited periods (ie 20 minutes only between the hours of 10 am and 4 pm), and only in marked (ie white-painted) loading bays. Generally the red route bans will apply between 7 am and 7 pm.

Failure to comply with the restrictions can lead to severe penalties. A red parking ticket (penalty currently £100) will be issued by police or traffic wardens, but if the driver argues or refuses to move, an impoundment order may be issued and the vehicle taken away. This will cost the driver or operator £150 to get the vehicle back, plus the £100 penalty and a £25 daily storage charge.

London Congestion Charging

Congestion charges of £8 (but only £7 for fleet vehicles) per day apply in a central area of London on weekdays only (Monday to Friday) between the hours of 7 am and 6.30 pm excluding public holidays.

The charging zone covers any route that crosses the River Thames by Lambeth, Westminster, Waterloo, Blackfriars, Southwark or London Bridges. Not included in the charging area are Euston Road, Tower Bridge, Elephant and Castle, Vauxhall Bridge, Victoria and Park Lane – gridlock is still likely on these key arterial routes. Cameras are located at 85 per cent of the boundary intersections and at random points within the zone. These cameras record vehicle registration numbers and cross-check via computer with records of charges paid.

Vehicles running on gas or electricity are exempt after paying a £10 registration fee. All other commercial vehicles are included, unless they are emergency service vehicles or are vehicles on municipal duty, not competing with private firms.

Drivers who regularly enter the charging zone may pre-register and pay a fee, otherwise payment of the daily fee may be made by credit card – a call centre will process credit-card numbers – or by buying carnets from newsagents and convenience stores and from ticket terminals in car parks, hospitals and other local amenities. Any driver who has not paid prior to entry into the charging zone may pay later that day – up to 10 pm.

Failure to pay the daily charge will result in evaders incurring an £80 penalty, issued to the vehicle owner (not necessarily the driver), but this is halved to £40 if paid within two weeks. There will be an appeals procedure.

BUS LANES

Traffic lanes on urban roads reserved solely for use by buses are a common feature in many towns and cities. Uniform traffic signs and road markings indicate bus lanes. A single wide solid white line is used to mark the edge of the reserved lanes. Upright signs incorporating international symbols combined with arrows show to other traffic the number of lanes available for their use. When the signs and restrictions are in operation on a road, all other vehicles, except for pedal cycles (and taxis if signed to this effect), are prohibited from using the bus lane.

LEVEL CROSSINGS

Most railway level crossings are now fitted with automatic half-barrier crossing gates and appropriate warning signs are given in advance. When a train is approaching such crossings, red lights flash and a bell rings to warn drivers and pedestrians. Once these warnings start the barrier comes down immediately, and drivers should not zig-zag around the barriers. When the train has passed, the barriers will rise unless another train is following, in which case the warnings will continue.

Drivers of vehicles which are large or slow (ie with their loads that are more than 2.9 metres (9 ft 6 in) wide or more than 16.8 metres (55 ft) long or weighing more than 38 tonnes gross or incapable of a speed of more than 5 mph) wishing to cross one of these crossings must, before attempting to cross, obtain permission to do so from the signalman by using the special telephone which is provided at the crossing. Failure to do this is an offence. In the event of a vehicle becoming stuck on the crossing the driver should advise the signalman immediately by using the telephone.

WEIGHT-RESTRICTED ROADS AND BRIDGES

Where signs indicate that a particular section of road or a bridge is restricted to vehicles not exceeding a specified weight limit or axle weight limit, unless otherwise expressly stated, the weight limit shown relates to the actual weight of the vehicle or to an individual axle of the vehicle, not the relevant plated weights.

Signs protect weak bridges by restricting vehicles according to their maximum authorized gross weight (ie their plated weight) – indicated as 'mgw'. These have replaced existing signs as described above where the weight shown relates only to the actual weight of the vehicle and load at the time. The new signs apply even if the vehicle, with a plated weight greater than the limit shown, is unladen at the time and therefore well below the maximum weight limit for the bridge (unless the sign permits 'empty vehicles'). Where doubt exists about any particular sign it is advisable to consult the local authority responsible for its erection and to determine the precise wording of the Traffic Management Order under which authority the sign would have been erected.

OWNER LIABILITY

Under the Road Traffic Offenders Act 1988 responsibility for payment of fixed penalty fines or excess parking charges rests with the registered vehicle owner (ie the keeper of the vehicle, not necessarily the legal owner), if the driver who committed the offence cannot be identified or found. The registered owner of the vehicle is sent details of the alleged offence and is obliged to pay the fine or submit a 'Statutory Statement' of ownership in which he states whether he was the vehicle owner at the time of the alleged offence (in which case he should name the driver), had ceased to be the owner at that time or had not yet become the owner at that time. Where the person was not the owner he must give the name of the previous owner or the new owner to whom he transferred the vehicle, if he knows it.

When a vehicle is hired out for less than six months, and such an incident arises, the hiring company can declare that the vehicle was on hire and send a copy of the hiring agreement together with a signed statement of liability from the hirer accepting responsibility for the fine or excess parking charge. Such a clause is normally included in the hiring agreements which the hirer signs. Failure to pay a fixed penalty or excess charge, or to give information as required by the police in such matters, can result in a fine of up to £1,000 on conviction (or even £5,000 in certain circumstances).

FIXED PENALTIES

In order to reduce the pressure on the courts, a system of fixed penalties exists by which both traffic wardens and the police can issue fixed penalty notices requiring the vehicle driver or owner to pay the fixed penalty or to elect to have the case dealt with in court in the normal way. The fixed penalty system operates on two levels: non-endorsable offences (mainly dealt with by traffic wardens), and driving licence endorsable offences which only the police can deal with since traffic wardens have no general authority to request the production of driving licences (see p 253). Additionally, two London area parking offences are included in the fixed penalty system.

Non-Endorsable Offences

For non-endorsable offences a white ticket/notice (penalty £30) is issued either to the driver if present or is fixed to the vehicle windscreen. Since no driving licence penalty points are involved for such offences there is no requirement to examine the licence. Traffic wardens have authority to issue fixed penalty tickets for the following non-endorsable offences:

- Leaving a vehicle parked at night without lights or reflectors.
- Waiting, loading, unloading or parking in prohibited areas.
- Unauthorized parking in controlled parking zone areas.
- Contravention of the Vehicle Excise and Registration Act 1994 by not displaying a current licence disc.
- Making 'U' turns in unauthorized places.
- Lighting offences with moving vehicles.

- Driving the wrong way in a one-way street.
- Overstaying on parking meters, returning to parking places before the expiry of the statutory period, or feeding meters to obtain longer parking facilities than those permitted in a meter zone.
- Parking on pavements or verges by commercial vehicles exceeding 3050 kg unladen weight (see p 246).

Endorsable Offences

The extended fixed penalty system covers driving licence endorsable offences which can be dealt with only by the police (ie not traffic wardens) – this includes some 250 driving and vehicle use offences. For endorsable offences a yellow ticket/notice with a different level of penalty applies as described below.

For driving licence endorsable offences the police issue a yellow ticket/notice for which a penalty of £60 is payable. These tickets are only issued after the police officer has seen the offender's driving licence and has established that the addition of penalty points appropriate to the current offence, when added to any points already on the licence, will not result in automatic disqualification under the 12-point totting-up procedure. If this is the case, the ticket will be issued and the driving licence will be confiscated (an official receipt, covering the holder for non-possession or production of his licence, will be given – valid for two months) being returned to the holder with the appropriate penalty points added when the penalty has been paid.

If the offender does not have his driving licence with him at the time the penalty notice will not be issued on the spot but will be issued at the police station if the driving licence is produced there within seven days – subject again to the number of penalty points already on the licence.

Where the addition of further points in respect of the current offence would take the total of penalty points on the licence to 12 or more, thus leading to automatic disqualification, the ticket will not be issued and the offence will be dealt with by the offender being summoned to appear in court in the normal manner.

London Area Parking Offences

The fixed penalty system includes two specific London area parking offences, namely parking on a 'Red Route' (see p 248), for which the penalty is currently £60, and parking in other prohibited places, for which the penalty is £40.

Payment or Election to Court

Fixed penalty notices must be paid in accordance with the instructions on the notice and within the specified time limit of 28 days. Alternatively, the offender can elect to have the charge dealt with by a court so he has the opportunity of defending himself against the charge or, even if he accepts that he is guilty of the offence, of putting forward mitigating circumstances which he feels may lessen any penalty which may be imposed.

The address of the fixed penalty office to which the penalty payment should be sent is given in the notice together with instructions for making application for a court hearing if this course of action is chosen.

Failure to Pay

With both the white and yellow ticket systems, failure to pay the statutory penalty within the requisite period of 28 days will result in the offender being automatically considered guilty and the penalties being increased by 50 per cent (ie to £45 and £90 respectively). These increased amounts become fines and continued non-payment will lead to the arrest of the offender and appearance before a court in the district where the offence was committed. This could be many miles from where the offender lives and may necessitate him being transported there under arrest and possibly held overnight.

Summary of Offences

Among the many offences covered by the fixed penalty scheme are the following:

- Parking at night without lights or reflectors.
- Waiting, parking, loading or unloading.
- Breach of controlled parking zone regulations.
- Failing to display a current excise licence disc.
- Making 'U' turns in unauthorized places.
- Lighting offences with a moving vehicle.
- Driving the wrong way in a one-way street and making banned right turns.
- Contravening traffic regulation orders.
- Breach of experimental traffic orders.
- Breach of experimental traffic scheme in Greater London.
- Contravening motorway traffic regulations.
- Using a vehicle in contravention of a temporary prohibition or restriction of traffic on a road.
- Driving in contravention of an order prohibiting or restricting driving on certain classes of roads.
- Breach of pedestrian crossing regulations.
- Contravention of a street playground order.
- Breach of parking orders on roads, and of parking place designation orders and other offences committed in relation to them, except failing to pay an excess charge.
- Contravening minimum speed limits.
- Speeding.
- Driving or keeping a vehicle not showing a registration mark.
- Driving or keeping a vehicle with a registration mark or hackney carriage sign obscured.
- Failing to comply with traffic directions or signs.
- Leaving a vehicle in a dangerous position.
- Failing to wear a seat belt.
- Breach of a restriction on carrying children in the front or rear of vehicles.
- Driving a vehicle elsewhere than on the road.
- Parking a vehicle on a footpath or verge.
- Breach of construction and use regulations.
- Contravening lighting restrictions on vehicles.
- Driving without a licence.

- Breach of provisional driving licence conditions.
- Failing to stop when required to do so by a uniformed police officer.
- Obstructing the highway with a vehicle.
- Overtaking on a pedestrian crossing.
- Failure to display a registration mark (ie number plate) on a vehicle.
- Driving on a footpath.

Table of Fixed Penalties

The following is the current list of fixed penalties for both road traffic and other offences.

- Fixed penalty under section 143 of the Road Traffic Act 1988 £200
- Fixed penalty offence under section 172 of the Road Traffic Act 1988 £120
- Any other fixed penalty offence involving obligatory disqualification £ 60
- Fixed penalty parking offence committed in Greater London on a red route £ 60
- Any other fixed penalty parking offence committed in Greater London £ 40
- Fixed penalty offence under section 47 of the Road Traffic Act 1988 £ 60
- Fixed penalty offence under section 33 of the Vehicle Excise and Registration Act 1994 £ 60
- Any other fixed penalty offence £ 30

TRAFFIC WARDENS

In addition to the powers of traffic wardens to issue fixed penalty tickets as described in the previous section, they also have powers to act as parking attendants at street parking places, to carry out special traffic control duties, to inquire into the identity of drivers of vehicles, to act in connection with the custody of vehicles at car pounds and to act as school crossing patrols.

They may demand to know the names and addresses of those believed to have committed parking, obstruction, traffic sign and excise licence offences and to see the driving licence of any person who is reasonably suspected of such offences. If the licence cannot be produced at that time the warden may issue a form HO/RT 1 requiring its production at a police station within seven days. Wardens have no powers to request the production of insurance certificates or vehicle test certificates.

Traffic wardens' powers are now extended (since December 2002) to include stopping vehicles for testing and other purposes and escorting abnormal loads – duties previously the prerogative of the police only.

PEDESTRIAN CROSSINGS

There are two types of pedestrian crossing. Zebra crossings are bounded on either side by areas indicated by zig-zag road markings in which overtaking, parking and waiting are prohibited. The marked areas extend to about 60 ft on either side of the crossing. A 'give way' line 3 ft from the crossing is the point at which vehicles must stop to allow pedestrians to cross. Pelican crossings are controlled by traffic lights which vehicle drivers must observe and pedestrians should cross only when the green light signal indicates that they should do so (at many pelican crossings an audible bleeper is provided to assist blind people to cross with safety). With this type of crossing, if there is a central refuge for pedestrians, each side of the refuge is still considered to be part of a single crossing. Only if the two parts of the crossing are offset does it become two separate crossings.

BUILDERS' SKIPS

Provisions are contained in the Highways Act 1980 to control the placing of builders' skips on the road. Before such a skip is placed on the road, permission must be obtained from the local authority. This will be given subject to conditions relating to the size of the skip, its siting, the manner in which it is made visible to oncoming traffic, the care and disposal of its contents, the manner in which it is lit or guarded and its removal when the period of permission ends.

Owners must ensure that skips carry proper reflective markers (see p 374), are properly lit at night, are clearly marked with their name and telephone number or address, and that they are moved as soon as is practical after they have been filled.

The police and highway authorities have powers to reposition or remove a skip from the road and recover the cost of doing so from the owner, and a penalty may be imposed.

For the purposes of the Act the definition of a builder's skip is 'a container designed to be carried on a road vehicle and to be placed on a highway or other land for the storage of builders' materials, or for the removal and disposal of builders' rubble, waste, household and other rubbish or earth'.

ABANDONED MOTOR VEHICLES

It is an offence under the Road Traffic Regulation Act 1984 to abandon a motor vehicle or any part of, or part removed from, a motor vehicle in the open air or on any other open land forming part of a highway. Such offences, on conviction, can lead to fines of up to £200 for a first offence and up to £400 and a term of up to three months' imprisonment or both for a subsequent offence.

Vehicles which are illegally or obstructively parked can be removed and a statutory charge of £105 imposed. An additional charge for storage of a removed vehicle is £12 per day and £50 is charged for its disposal. Removed and impounded vehicles are not released until all relevant charges have been paid.

RETENTION AND DISPOSAL OF SEIZED VEHICLES

Where a police officer has seized and removed a vehicle as authorized under section 165A of the Road Traffic Act 1988, certain legal procedures must be followed; namely, the constable must give a seizure notice to the driver (unless it is impracticable to do so) and must also give a seizure notice to the registered keeper and the owner (where that appears to be someone different) requiring him to claim the vehicle within a specified period (but not less than seven working days). The notice must indicate the charges payable to reclaim the vehicle – amounting to £105 for the seizure and £12 for each 24 hours or part thereof that the vehicle is in custody, counting from noon on the first day after the seizure – and the person claiming the vehicle must produce a valid driving licence and certificate of insurance at a specified police station, the vehicle being retained until these conditions are met. Where the police cannot issue a seizure notice or the conditions of the notice are not met, the vehicle may be disposed of after 14 days. If disposal is by means of a sale, the net proceeds must be paid to the owner if he makes a claim within one year of the sale.

WHEEL CLAMPS

Vehicles which are illegally parked or which cause obstruction in a wide area of Central London will be immobilized by the Metropolitan Police or by contractors on their behalf. A wheel clamp, known as the 'Denver Boot', will be fixed to one wheel of the vehicle thereby preventing it being driven away. A notice will be stuck to the vehicle giving the driver notice of the offence committed and instructions for securing release from the clamp.

It is an offence to try to remove a wheel clamp or to attempt to drive off with one fitted. Vehicle drivers finding a clamp fixed to their vehicle must go to the Metropolitan Police pound in Hyde Park underground car park and request removal of the clamp. Both a removal charge (currently £38) and a fixed penalty (currently £30) have to be paid before the clamp is removed. If the vehicle has been removed by the police a removal charge of £105 is payable, in addition to any fixed penalty, to secure its release.

OVERLOADED VEHICLES

It is an offence to drive an overloaded vehicle on a road. Under the Road Traffic Act 1988 (sections 70 and 71) an authorized examiner or police officer may prohibit the use of an overloaded vehicle on the road until the weight is reduced to within legal limits and may direct, in writing, the person in charge of an overloaded vehicle to remove the vehicle to a specified place. See also pp 316–17.

A driver may be instructed to drive for a distance of up to five miles to a weighbridge for the weight of his vehicle and load to be checked. If he is directed to drive more than five miles to the weighbridge and his vehicle is found to be within the maximum permitted weights then a claim may be made against the appropriate highway authority for the costs incurred.

The maximum fine for an overloading offence is £5,000 but any one instance of an overloaded vehicle could result in conviction for more than one offence, each of which carries this maximum penalty. Subsequent convictions for such offences could lead to higher fines. Convictions for overloading offences also jeopardize the operator's licence (see also Chapter 12 on vehicle weights and overloading).

ROAD TRAFFIC ACCIDENT PROCEDURE

Road accidents are said to have caused more than 41,000 deaths and 1.7 million reported injuries throughout the EU in 2000 at a cost of £124 billion. The relevant figures for the UK alone in 2001 were 3443 people killed (1 per cent up on 2000) and 37,094 people seriously inured (3 per cent down on 2000).

Any driver involved in a road accident in which personal injury is caused to any person other than himself, or damage is caused to any vehicle other than his own vehicle, or damage is caused to any animal* other than animals carried on his own vehicle, or to any roadside property (see below for definition) MUST STOP. Failure to stop after an accident is an offence and fines of up to £5,000 can be imposed on conviction.

The driver of a vehicle involved in an accident must give to anybody having reasonable grounds for requiring them his own name and address, the name and address of the vehicle owner and the registration number of the vehicle.

If the accident results in injury or damage to any person other than the driver himself or to any other vehicle or to any reportable animal,* or to roadside property, then the details of the accident must be reported to the police *as soon as reasonably practicable afterwards, but in any case no later than 24 hours after the event.* This obviously does not apply if police at the scene of the accident take all the necessary details. Failure to report an accident is an offence which also carries a maximum fine of £5,000.

**NB: For these purposes an animal means any horse, ass, mule, cattle, sheep, pig, goat, dog (in Northern Ireland only, a 'hinnie' is added to this list).*

Reportable incidents include those that occur 'in any public place' as well as on the road. Any public place would include, for example, supermarket, rail station and multi-storey car parks.

Under the Road Traffic Act 1988 the need to stop following accidents extends to cover any damage caused to any property 'constructed on, fixed to, growing on, or otherwise forming part of the land in which the road is situated or land adjacent there to'. This means that if a vehicle runs off the road and no other vehicles or persons are involved the driver still has to report damage to fences, hedges, gate-posts, street bollards, lamp-posts, and so on.

Third parties injured in accidents who find when making claims for damages that the vehicle was uninsured at the time can make a claim for their personal injuries to the Motor Insurers' Bureau (MIB) (see also Chapter 9). More recently the MIB compensation scheme has been extended to cover claims for damage to property by uninsured vehicles. Such claims for property damage will only be accepted if the vehicle driver is traced – not otherwise – and provided no claim for the damage can be made elsewhere. There is a limit of £250,000 on claims and they are subject to a £175 excess clause.

ROAD HUMPS – TRAFFIC CALMING

Regulations permit the construction of road humps on sections of the highway where a 30 mph speed limit is in force. The road hump will be treated as part of the highway provided it complies with the regulations. New regulations were introduced in 1989 to enable local authorities to reduce pedestrian casualties by the introduction of more road humps (see also p 236 regarding 20 mph speed limit zones).

Increasing use is being made of a variety of so-called 'traffic-calming' measures to reduce vehicle speeds and cut accident risks. Besides road humps (often called 'sleeping policemen'), other measures include rumble strips which draw the driver's attention by extra road noise and vibration, mini-roundabouts and artificial chicanes (ie pinch points). However, these steps are seen in some quarters as being a countermeasure, driving fast traffic on to other vulnerable routes and impeding emergency vehicles.

SALE OF UNROADWORTHY VEHICLES

It is an offence to sell, supply, offer to sell or expose for sale a vehicle in such a condition that it does not comply with the construction and use regulations, and is therefore legally unroadworthy. Under the Road Traffic Act 1988 offenders are liable to a fine of up to £5,000.

This means that to display a vehicle for sale which needs attention to bring it up to the required standard is an offence, even though the intention would have been to remedy any defects before a purchaser paid for or took the vehicle away. However, it would not be an offence if the buyer was made aware of the defects and he intended to remedy them or have them remedied before using the vehicle on the road.

It is also an offence to fit any part to a vehicle which, by its fitting, makes the vehicle unsafe and causes it to contravene the regulations. For example, fitting a tyre which is below the limits regulating tread depth would be to commit such an offence.

Under the Road Traffic Act 1991 the law relating to the sale of unroadworthy vehicles has been tightened: in particular, the seller of such a vehicle now has a statutory duty to take steps to ensure that the buyer is aware that the vehicle is in an unroadworthy condition. Previously the seller could rely on the defence that he believed that the vehicle was not to be used on the road until made roadworthy.

SEAT BELTS

Legislation specifying the compulsory wearing of seat belts in motor vehicles came into force on:

- *31 January 1983*
  for drivers and front-seat passengers of motor cars, light vehicles not exceeding 1525 kg unladen weight registered on or after 1 April 1967 and vehicles not exceeding 3500 kg gross weight registered since 1 April 1980.
- *1 July 1991*
  for adults (ie persons aged 14 years or over) travelling in the rear of motor cars to wear seat belts where they are fitted – *irrespective of the age of the car.*
- *2 February 1993*
  for the driver and any person sitting in a seat which is equipped with a seat belt (even on a voluntary basis) to wear the belt provided – this includes drivers of goods vehicles over 3.5 tonnes gross weight which are fitted with seat belts on a voluntary basis since there is no legal requirement for such vehicles to be so fitted.
  Also, from this date:
  - It is illegal to carry any unrestrained child in the front seat of a motor vehicle.
  - Children under three years of age travelling in the front seat of a vehicle must be restrained by a suitable child restraint.

- Children under 12 years of age and under 150 centimetres (4 ft 11 in) in height travelling in vehicles must use a restraint if there is a suitable one anywhere in the vehicle.

NB: Seat belts have been fitted to all new cars and taxis since 1987.

Legislation requiring compulsory fitment of seat belts in new goods vehicles over 3.5 tonnes gross weight came into effect from 1 October 2001. Belts must be fitted to all forward-facing front seats and exposed rear seats.

Responsibility for Seat Belt Wearing

It is the individual responsibility of the vehicle driver and any adult passengers (ie persons 14 years of age and over) to wear the seat belt provided for the seat in which they are sitting (ie front or rear). The driver is *not* liable where adult passengers fail or refuse to comply with the law in this regard. However, in the case of children under 14 years of age it is the driver's responsibility to ensure that the law is complied with, irrespective of whether the child's parents, guardians or other responsible person in whose charge they are, are in the vehicle.

Bench-Type Seats

Where a light goods vehicle is fitted with a bench-type or double front passenger seat, it is illegal to occupy the centre part of the seat (ie next to the driver), where a belt may not be provided, if the outer part of the seat with the belt provided is unoccupied.

Failure to Wear Seat Belts

Failure by a person to wear a seat belt as required by law could result in a fixed penalty (currently £20) or a fine of up to £500 if convicted by a court. In a case relating to illegally carrying an unrestrained child a fine of up to £200 could be imposed.

A court has ruled that if, as a result of an accident, injuries were sustained which may have been prevented or lessened had the injured person been wearing a seat belt, then the damages awarded to that person in any claim should be reduced by an appropriate amount. Subsequently, other cases involving motor accident claims have followed the same lines.

Exemptions

Exemption from seat belt wearing applies:

- When holding a valid medical certificate giving exemption (see further details below).
- When driving a vehicle constructed or adapted for the delivery or collection of goods or mail to consumers or addresses, while engaged in making local rounds of deliveries or collections, but only where such journeys do not exceed 50 metres in length.
- When driving a vehicle at the time of carrying out a manoeuvre which includes reversing.

- When accompanying a learner driver as a qualified driver and supervising the provisional entitlement holder while that person is performing a manoeuvre which includes reversing.
- In the case of a driving test examiner (but *not* an instructor) who is conducting a test of competence to drive and who finds that wearing a seat belt would endanger himself or any other person.
- In the case of a person who is driving or riding in a vehicle being used for fire brigade or police purposes, or for carrying a person in lawful custody, including a person being so carried.
- In the case of a driver of a licensed taxi who is seeking hire, answering a call for hire, or carrying a passenger for hire; or of a driver of a private hire vehicle which is being used to carry a passenger for hire.
- When *riding* in a vehicle being used under a trade licence for the purposes of investigating or remedying a mechanical fault in the vehicle.

NB: this particular exemption refers specifically to 'riding' in a vehicle and does not include 'driving' a vehicle for the same or similar purposes – therefore, it must be concluded that the driver of a vehicle using it for the purpose described would not be exempt from wearing a seat belt whereas a passenger riding in the vehicle for the same purpose would be exempt.

- In the case of a disabled person, wearing a disabled person's seat belt.
- In the case of a person *riding* (see note above) in a vehicle while it is taking part in a procession organized by or on behalf of the Crown. This exemption also applies to a person riding in a vehicle which is taking part in a procession held to mark or commemorate an event which is commonly or customarily held in the police area in which it is being held, or for which a notice has been given under the Public Order Act 1986.

The regulations also do not apply to a person who is:

- driving a vehicle if the driver's seat is not provided with an adult seat belt;
- riding in the front of a vehicle in which no adult belt is available to him;
- riding in the rear of a vehicle in which no adult belt is available to him.

It should be noted that these exemptions relate to the non-wearing of seat belts where they are not provided, but this circumstance may involve other infringements of the law relating to the non-fitment of seat belts.

STOWAWAYS

Concern about the number of illegal immigrants (legally termed 'clandestine entrants') entering this country stowed away in heavy goods vehicles has resulted in severe penalties being imposed on drivers and operators caught with such stowaways hidden in their vehicles. The Immigration and Asylum Act 1999 sets penalties of £2,000 for each immigrant found in a lorry trailer with confiscation of the vehicle as the sanction should the fine not be paid. However, under existing legislation (ie the Immigration Act 1971, section 25) it is already an offence to aid illegal immigrants to enter Great Britain. On conviction, an offending driver could face a heavy fine or up to seven years' imprisonment.

Despite a Court of Appeal ruling in February 2002 that these illegal immigrant fines are contrary to human rights and therefore illegal, the Home Office has continued to impose them on drivers and operators caught with illegal stowaways in their vehicles.

In a further contrary step, the government proposes, in amendments to the Nationality, Immigration and Asylum Bill before Parliament in June 2002, to increase the civil penalty from £2,000 to £4,000.

Proof of 'due diligence' may be accepted in defence of any charges under the legislation, but it is important for an accused driver to clearly show that he followed the Code of Practice, which has yet to be formally established. In the meantime, the Road Haulage Association provides the following advice to its members to pass on to their drivers:

- Never leave ignition keys in the vehicle. Lock cab doors and secure the vehicle's load space whenever the vehicle is unattended.
- Avoid routine stops for papers, cigarettes, etc, particularly within 100 km of Channel ports.
- Always ensure windows are closed when away from your vehicle.
- If sleeping in the vehicle, lock all doors and try to block access to the rear doors by parking up against a wall or other secure barrier.
- Be on the lookout for bogus officials or staff.
- Never leave keys hidden for a relief driver.
- On arrival at your destination, do not leave your vehicle in someone else's care.
- Look out for and report any security defects on your vehicle, such as faulty locks or straps.
- Use pre-planned secure overnight parking wherever possible.
- Try to keep your vehicle in sight if you leave it unattended.
- Make sure your vehicle is correctly loaded.
- If you make the same journey frequently, consider whether the route or schedule can be varied.
- Report any irregularity of loading, sealing or documentation.
- Never accept unsolicited offers of assistance.
- Avoid talking about your route over the radio.
- When returning to a vehicle, check for suspicious vehicles/people nearby and if concerned note descriptions, registration numbers, etc.
- After every stop look for signs of tampering with doors, straps, curtains, etc.
- Managers should ensure that operational procedures are constantly reviewed.
- Report to base when you arrive at an unoccupied site, or when you see suspicious activity.
- If you are uncomfortable opening your vehicle for potentially bogus authorities, ask for their details so you can check up on them and offer to open your vehicle at the nearest police station.

Hauliers are advised to turn in any stowaways found in their vehicles, despite the onerous civil penalties involved. Facilitating the entry of illegal immigrants is a serious criminal offence that carries a maximum prison sentence of 10 years.

Home Office Civil Penalty Code of Practice

This Code of Practice is issued in accordance with section 33 of the Immigration and Asylum Act 1999. This sets out the measures to be taken and the procedures to be followed by persons operating a system for preventing the carriage of clandestine entrants to the United Kingdom in respect of road vehicles.

Under section 34(3) of the 1999 Act, where it is alleged a person is liable to a penalty under section 32 of that Act for bringing a clandestine entrant to the United Kingdom, it is a defence to show that:

- he did not know and had no reasonable grounds for suspecting that a clandestine entrant was, or might be, concealed in the transporter;
- there was an effective system in operation in relation to the transporter to prevent the carriage of clandestine entrants;
- on the occasion concerned, the person or persons responsible for operating that system did so properly.

Regard will be had to this Code of Practice in determining whether such a system is effective (section 34(4) of the 1999 Act).

Part 1: Road Haulage and Other Commercial Vehicles

'Commercial vehicle' means any vehicle excluding buses, coaches, cars, taxis, mobile homes and caravans.

'Vehicle' refers to the entire vehicle, including any attached trailer(s) and any container carried. It also refers to a detached trailer, in the case of which any reference to 'driver' is to be read as 'operator'.

1.1 Measures to be taken to secure vehicles against unauthorized entry

1.1.1 Before final loading takes place, all existing cuts or tears in the outer shell or fabric of the vehicle that exceed 25 centimetres in length must be repaired and sealed so as to prevent unauthorized entry.

1.1.2 If present at the time of final loading, the owner, hirer or driver of the vehicle must check it to ensure that no persons have gained entry and are concealed within. It must then be locked, sealed, or otherwise made secure to prevent unauthorized entry. If not present at the time of final loading, the owner, hirer or driver must, where possible, ensure that such checks are conducted at that point by reputable persons. The owner, hirer or driver must then obtain written confirmation from those persons that these checks were properly conducted and that the vehicle did not contain concealed persons at the time of final loading and securing.

1.1.3 When the final loading has been completed, the load space must be secured immediately by lock, seal or other security device that will prevent unauthorized entry.

1.1.4 Tilt cords and straps, where used, must be undamaged, pass through all fastening points, made taut and be secured by lock, seal or other security device.

1.1.5 There must be no means of entry to the load space, other than via access points which have been secured by lock, tilt cord/strap and seal, or other security device.

1.1.6 Locks, tilt cords, straps and other devices used to secure the load space must be effective and of robust quality.

1.1.7 Seals, other than Customs' seals, must be distinguished by a number from a series that is unique to the owner, hirer or driver. This must be recorded in the documentation accompanying the vehicle.

1.1.8 When a sealed container (except a container sealed by Customs) is loaded on to a vehicle, the owner, hirer or driver must, where possible, check to ensure that it does not contain unauthorized persons. It must then be resealed and made secure in accordance with the above requirements. These actions and the number of the new seal used must be recorded in documentation accompanying the vehicle.

1.1.9 The same checking, securing and recording procedure detailed in paragraph 1.1.8 above must be followed when the load space in the vehicle has been opened by the owner, hirer, driver, or any other person before the final checks detailed in section 1.2 below are carried out.

1.1.10 Where a new driver becomes responsible for the vehicle en route to the United Kingdom, he should ensure that it does not contain unauthorized persons and that the requirements detailed above have all been met.

1.1.11 Paragraphs 1.1.1–1.1.10 above will not apply in relation to any vehicle that it is not possible to secure by means of lock, seal or other security device. However, in such circumstances, it will be for the owner, hirer or driver concerned to establish alternative arrangements to prevent unauthorized entry, and to be able to demonstrate that such arrangements have been made and complied with.

1.2 Measures to be taken immediately prior to the vehicle boarding the ship, aircraft or train to the United Kingdom, or before arrival at the UK immigration control at Coquelles.

1.2.1 Where used, check tilt cords and straps for evidence of tampering, damage or repair.

1.2.2 Where used, check that seals, locks or other security devices have not been removed, damaged or replaced. In order to ensure that there has been no substitution, numbers on seals must be checked to confirm that they correspond with those recorded on the documentation accompanying the vehicle.

1.2.3 Check the outer shell/fabric of the vehicle for signs of damage or unauthorized entry, paying particular attention to the roof, which may be checked from either inside or outside the vehicle.

1.2.4 Check any external storage compartments, toolboxes, wind deflectors and underneath the vehicle.

1.2.5 Check inside the vehicle. Effective detection devices may be used for this purpose at the discretion of the owner, hirer or driver, but this will not obviate the requirement that the other checks be carried out (detailed above). Where it is not possible to secure a vehicle lock, seal or other security device, a thorough manual check of the load and load space must be conducted.

1.3 General Principles

1.3.1 Vehicles should be checked regularly en route to the United Kingdom to ensure that they have not been entered, particularly after stops when left unattended.

1.3.2 A document detailing the system operated to prevent unauthorized entry must be carried with the vehicle, so that it may be produced immediately on demand to an immigration officer in the event of a possible liability to a penalty.

1.3.3 A report detailing the checks that were carried out must be carried with the vehicle. If it is possible to arrange it, the report should be endorsed by a third party, who has either witnessed or carried out the checks himself by arrangement with the owner, hirer or driver, as the report will then be of greater evidential value.

1.3.4 While owners, hirers or drivers may contract with other persons to carry out the required checks on their behalf, they will nevertheless remain liable to any penalty incurred in the event of failure to have an effective system in place or to operate it properly on the occasion in question.

1.3.5 Where the checks conducted suggest that the security of the vehicle may have been breached, or the owner, hirer or driver otherwise has grounds to suspect that unauthorized persons have gained entry to the vehicle, it must not be taken on to the ship, aircraft or train embarking for the United Kindom, or to the UK immigration control at Coquelles. Any such circumstances must be reported to the police in the country concerned at the earliest opportunity, or at the latest, to the passport control authorities at the port of embarkation. In the event of difficulties arising, owners, hirers or drivers should contact the UK Immigration Service at the proposed port of arrival for advice.

USE OF RADIOS AND TELEPHONES IN VEHICLES

The *Highway Code* contains advice against using a hand-held microphone or telephone handset while the vehicle is moving except in an emergency. Drivers should not stop on the hard shoulder of a motorway to answer or make a call no matter how urgent. The *Code* recommends that a driver should only speak into a fixed, neckslung, or clipped-on microphone when it would not distract his attention from the road.

TRAFFIC AND WEATHER REPORTS

Drivers concerned about prevailing or likely weather conditions prior to making a journey, or even while en route, can check forecasts from local and national newspapers and get an update of the current situation by listening to radio and television bulletins. In particular, the following are useful for traffic and weather forecasts:

Radio

BBC Radios 1, 2, 4 and 5 and local BBC and commercial radio stations broadcast regular weather and traffic bulletins.

The D*f*T signposts show tuning frequencies for local radio stations on some motorways. These wavelengths broadcast 24 hours a day, 7 days a week, giving as many as four local traffic and weather reports an hour at peak times, and two during off-peak times, with programme interruptions for important traffic announcements.

National radio covers major traffic and weather problems and the BBC's Radio Data System (RDS) enables drivers listening to BBC radio to be interrupted with relevant travel news. Receivers using additional data, Enhanced Other Networks (EON), can switch drivers from a BBC national radio station, cassette or CD, to a BBC local radio station for a travel bulletin.

Television

BBC 2 Ceefax (pp 501–07) and ITV Teletext (pp 191–93) provide up-to-date travel information.

Telephone

AA Roadwatch and Weatherwatch (tel: 0990 500 600) provide national motorway information, while the RAC Travel News (tel: 0891 500242) provides information on traffic in London and on UK motorways. Telephone calls are currently charged at premium rates.

On-Board Traffic Information

Sophisticated modern technology can now provide drivers with on-board (ie in-vehicle) visual information on traffic routes and conditions. Three systems are currently in use:

Autoguide

This system, operating within the M25 London orbital motorway, gives drivers recommended routes to their destination via a dashboard-mounted display screen activated by roadside mounted beacons which also update the central computer – it is self-compensating, so does not shift traffic jams from one location to another.

AA Roadwatch Pager

The AA Roadwatch Pager displays dynamic traffic news on the screen of its pocket-sized pager unit, taking account of the current location of the unit. For more information call 0800 300 300 (toll free).

Trafficmaster

Despite what many road users think, the blue poles sprouting up on Britain's roadsides are not speed cameras. They are part of the Trafficmaster live traffic-monitoring network of infra-red sensors and roadside beacons on over 7500 miles of motorway and trunk roads. The system provides information on traffic conditions and real-time information on journey times on over 95 per cent of the country's main roads and can predict when traffic is at its worst – and therefore enable drivers to determine when is the best time to travel.

The system works by using specially developed infra-red sensors, installed approximately four miles apart, to measure the time taken for a vehicle to travel between each sensor site. Computers at each site continuously transmit this information back to Trafficmaster's National Control Centre in Milton Keynes, where it is processed and transmitted directly to vehicles fitted with Trafficmaster receivers.

With traffic congestion increasing – an average of 570,000 vehicles are affected every day on motorways alone and within the next 10 years this is likely to rise to 623,000 according to Trafficmaster – the annual cost to the economy by the year 2010 is likely to be £7.7 billion.

Trafficmaster claims that its technology is playing a key role in helping to cut the cost and time wasted in congestion by providing accurate, high-quality, dynamic information directly to drivers, resulting in significant reductions in both journey times and stress levels.

Trafficmaster is available on annual subscription. For more information call 01582 484 414.

SMUGGLING

The Customs authorities are becoming increasingly concerned about the volumes of smuggled goods, particularly tobacco products, being brought into the UK. It is estimated that tobacco smuggling alone amounts to some £3.5 billion annually.

Besides new measures to detect smuggling such as the use of sophisticated X-ray equipment, tougher new penalties have been introduced which could result in convicted offenders losing their vehicle (all goods vehicles found to be carrying contraband goods are seized: 300 HGVs were impounded in the first year of operation) and their driving licence, and facing up to seven years in prison.

The maximum legal limits for personal importation of goods into the UK are as follows:

From within the EU:

3200 cigarettes	400 cigarillos
200 cigars	3 kg of smoking tobacco
110 litres of beer	10 litres of spirits
90 litres of wine	20 litres of fortified wine (eg port or sherry)

From outside the EU:

200 cigarettes, or	600cc/ml of perfume
100 cigarillos, or	250cc/ml of toilet water
250 grams of tobacco	£145 worth of all other goods including gifts and souvenirs

NB: For these purposes, the EU is the enlarged EU of 25 countries, but in the case of Cyprus goods from any area of Cyprus which is not under the effective control of the Government of the Republic of Cyprus are treated as non-EU imports. Similarly, Spain does not include the Canary Islands and neither the Channel Islands nor Gibraltar are counted as EU territories.

Any amounts of such goods brought in exceeding these limits may be confiscated and penalties imposed.

The High Court (in August 2002) has ruled that UK Customs and Excise policing of cross-channel shopping expeditions by means of random searches and routine confiscations causes 'injustice and hardship'. The court established that there should be a presumption in favour of the free movement of individuals, and their goods, across internal EU borders. Instead of random checks, the court said that customs officers should act only when there are reasonable grounds for suspicion that excise goods have been acquired for commercial purposes. It also held that travellers should no longer bear the burden of proving that their excise goods fall within acceptable limits 'and the mere fact that a traveller is transporting a significant cargo does not constitute a "persuasive presumption" that he has acquired them for commercial purposes'.

'Transporting large quantities of excise goods into this country may suggest a possible commercial intent. But that should be for Customs to prove, beyond reasonable doubt, in court. Why should citizens be forced to justify their appetite for wine, consumption levels of beer or fondness for cigarettes to outside authority to prevent their purchases being impounded? Citizens have a right to be presumed dissolute before any court may find them guilty of criminality.'

In a more recent ruling, the European Court of Justice has determined that truck drivers caught with smuggled goods on their vehicles are to be held responsible for the duty payable even if they are proven to be the innocent victims of deceit.

Customs has a special telephone number for reporting suspected cases of smuggling: Customs Confidential 0800 59 5000.

ROAD SAFETY BILL 2005

Despite Great Britain having one of the best road safety records in the world, the government introduced its Road Safety Bill into the House of Lords on 24 May 2005 with the objective of achieving further casualty reduction targets and further improving safety on our roads.

The Bill targets six main areas:

- Drink-driving – with mandatory re-testing for repeat offenders, an improved Drink-Drive Rehabilitation Scheme, changes to the High Risk Offenders Scheme and the introduction of an Alcohol Ignition Interlock Programme of the type used successfully in North America and Australia (ie that prevents a vehicle being driven by somebody with alcohol on their breath).
- Speeding – introducing graduated fixed penalties for offenders, retraining courses for speeding offenders, a ban on speed camera jammers and detectors, and increased penalties for failing to identify an offending driver.
- Penalties – introducing increased fines for careless driving (ie from level 4 at £2500 to level 5 at £5000), making driving while using a mobile phone an endorsable offence, making a second offence of driving a dangerous vehicle into a disqualification offence and generally increasing various penalties.
- Re-training – introducing improved Drink-Drive Rehabilitation courses (as mentioned above) and other re-training courses.
- Driver training and testing – introducing better driving test arrangements, modernizing driving instruction, publishing performance data on driving

instructors and extending the rules on paid driving instruction to LGV instruction.

- Foreign drivers – introducing new systems for ensuring that foreign driving offenders do not escape UK penalties.

11 Employment Law and Transport Training

This chapter examines existing, recently implemented and proposed employment law provisions that have a particular relevance to transport managers and operators. It also covers training related to road transport, though dangerous goods training for drivers and some other aspects of driver training are dealt with in Chapter 7.

EMPLOYMENT LAW

INTRODUCTION

Besides being constrained by strictly enforced goods vehicle legislation, transport employers must also contend with a mass of exacting law on employee rights and discrimination. Failure to comply with these provisions may result in an employer being called before an industrial tribunal to answer complaints by an aggrieved employee, and being required to pay compensation if he loses the action, or even having to reinstate an adjudged 'unfairly dismissed' person.

EMPLOYMENT RELATIONS BILL

A new Employment Relations Bill was introduced into parliament in December 2003 and became law in the autumn of 2004. The Bill is mainly concerned with collective labour law and trade union rights. It will implement the findings of the review of the Employment Relations Act 1999 (see page 270). Briefly, the Bill contains:

- measures to improve the operation of the statutory recognition procedure;
- measures to simplify the law on industrial action ballots and ballot notices;
- measures to widen the ability of unions to expel or exclude racist activists and others whose political behaviour is incompatible with trade union membership;
- a power for the Secretary of State to make funds available to independent trade unions and federations of trade unions to modernize their operations;
- measures to tackle the intimidation of workers during recognition and de-recognition ballots;
- measures to implement the European Court of Human Rights judgment which ensures that trade union members have clear rights to use their union's services, and cannot be bribed by employers to relinquish essential union rights;
- measures to improve the operation of some individual employment rights, such as clarification of the role of the companion in grievance and disciplinary hearings; and a technical change to flexible working legislation concerning protections from unfair dismissal;
- a power to make regulations to introduce information and consultation in the workplace (in Great Britain and Northern Ireland), by implementing the EC Directive on Information and Consultation (Council Directive 2002/14/EC);
- measures to improve the enforcement regime of the national minimum wage;
- measures to give the Certification Officer greater powers to strike out weak or vexatious claims;
- measures to improve trade union regulation, and a power to allow the Secretary of State to include non-postal methods of balloting in statutory union elections and ballots.

THE EMPLOYMENT ACT 2002

This Act mainly covers paternity and adoption leave and pay, maternity leave and pay, flexible working, employment tribunal reform and resolving disputes between employers and employees. It amends current legislation in many areas, particularly:

- The Employment Rights Act 1996: making provision for statutory rights to paternity and adoption leave and amending the law relating to statutory maternity leave.
- The Social Security Contribution Benefits Act 1992, by introducing statutory paternity and adoption pay and amending the law relating to maternity pay and maternity allowance.
- The Employment Tribunals Act 1996, in relation to costs and expenses, conciliation, powers to delegate the prescription of forms, determination without a hearing, practice directions and pre-hearing reviews.
- The Employment Rights Act 1996: making provision on the use of statutory procedures in employment disputes, and by introducing new provisions relating to procedural unfairness in unfair dismissal cases. Amendments are

also made to the Act's provisions on the particulars of employment that employers must give to employees, and to provisions relating to dismissal procedures agreements.

- The Equal Pay Act 1970, making provision for questionnaires in relation to equal pay issues.
- The Trade Union and Labour Relations (Consolidation) Act 1992, making provision for time off for trade union learning representatives in organizations where trades unions are recognized.
- The Social Security Administration Act 1992, making provision for work-focused interviews for partners of benefit claimants and the use of information for, or relating to, employment and training.

Certain sections of the Act such as those relating to pay and administration of statutory paternity and adoption pay, fixed term work, and statutory dispute resolution procedures do not amend existing legislation.

THE EMPLOYMENT RELATIONS ACT 1999

The Employment Relations Act, which received Royal Assent on 27 July 1999, forms part of a package of reforms to employment and trade union law that had been outlined in the Government's White Paper, *Fairness at Work*, published in May 1998.

This Act contains wide-ranging provisions of concern for all employers, not least those in road haulage and associated businesses. The following text outlines the main provisions, some of which are covered in greater detail elsewhere in this chapter.

The Act comprises:

- new statutory procedures for the recognition and de-recognition of trade unions for collective bargaining, to apply when unions and employers are unable to reach agreement voluntarily, and a requirement for employers to inform and consult unions recognized under the statutory procedure on their training policies and plans;
- changes to the law on trade union membership, to prevent discrimination by omission and the blacklisting of people on grounds of trade union membership or activities;
- changes to the law on industrial action, in particular to certain aspects of the ballot requirements, and to enable dismissed strikers to complain of unfair dismissal in certain circumstances;
- new rights and changes in family-related employment rights, aimed at making it easier for workers to balance the demands of work and the family;
- a new right for workers to be accompanied in certain disciplinary and grievance hearings;
- other changes in individual employment rights;
- other miscellaneous provisions.

Most of the Act consists of amendments to the Trade Union and Labour Relations (Consolidation) Act 1992 and the Employment Rights Act 1996.

Trade Union Recognition and De-Recognition

Effect is given to the Government's proposals for a new system of trade union recognition, as set out in chapter 4 of the *Fairness at Work* White Paper. The statutory procedure for which the Act provides is intended for use only if attempts to reach a voluntary agreement on recognition fail, and to act as a means of resolving the dispute without recourse to industrial action.

Trade Unions and Individual Employment Rights Related to Trade Unions

The Act contains provisions on three other matters to do with trade unions and related individual employment rights:

1. To protect workers against discrimination by omission on the grounds of trade union membership or non-membership or trade union activities.
2. To protect workers against blacklisting on the grounds of trade union membership or activities.
3. To remove the requirement for trade unions in certain circumstances to identify to employers those workers who will be balloted or called out on industrial action.

Although the Act does not affect the right of employers and employees to agree individual contracts, even where a union is recognized for collective bargaining in respect of the bargaining unit where the employees work, it does provide for employees to be protected against being forced into accepting individual contracts in such circumstances.

The Act contains provisions aimed at clarifying and simplifying the present law on industrial action ballots and notice in a number of respects.

Family-Related Employment Rights

In chapter 5 of *Fairness at Work*, the Government sets out proposals aimed at helping employees to combine work and family life satisfactorily. This was done by simplifying and extending existing maternity rights, and introducing new rights to the three months' parental leave for men and women when they have a baby or adopt a child, and to time off for urgent family reasons, in line with the provisions of the Parental Leave Directive. The Act implements these proposals by conferring basic rights to maternity and parental leave and to time off for dependents. See pp 277–79 for further details of this provision.

Right to be Accompanied in Disciplinary and Grievance Hearings

The Act provides that employees should have the right to be accompanied by a fellow employee or trade union representative of their choice in disciplinary and grievance procedures.

Other Rights of Individuals

The Act:

- provides in certain circumstances for employees to be protected from dismissal for taking part in lawfully organized official industrial action;

- prohibits the use of waivers for unfair dismissal but not redundancy payments in fixed-term contracts;
- contains regulation-making powers for the Secretary of State to ensure that part-time workers receive no less favourable treatment than full-time workers (see pp 286–88);
- exempts residential members of religious and other similar communities from the National Minimum Wage Act 1998, giving effect to a recommendation from the Low Pay Commission;
- contains order-making powers for the Secretary of State to extend the coverage of some or all existing employment rights by regulation to specified categories of individuals.

Other Miscellaneous Provisions

The Act makes a number of provisions on institutional matters, as follows:

- The new statutory recognition and de-recognition procedures create a new role for the Central Arbitration Committee. The Act amends the arrangements for the appointment of CAC members and for the proceedings of the CAC in respect of cases under the new procedures.
- It abolishes the Commissioner for the Rights of Trade Union Members (CRTUM) and the Commissioner for Protection Against Unlawful Industrial Action (CPAUIA). (The Commissioner's main functions are to provide assistance to individuals taking or contemplating certain legal proceedings against unions or union officials.) The Act gives new powers to the Certification Officer to hear complaints involving most aspects of the law where CRTUM is currently empowered to provide assistance.
- It provides that the general duty of ACAS should be 'to promote the improvement of industrial relations', repealing the current emphasis given by the 1992 Act to ACAS's functions in relation to the settlement of trade disputes.
- It provides for the annual reports of ACAS, the CAC and the Certification Officer to be produced on a financial rather than calendar year basis.

The Act also:

- provides statutory authority for the Secretary of State to make funding available to assist and develop partnerships at work;
- amends and extends the order-making power in the Employment Agencies Act 1973 under which the Secretary of State regulates the industry; amends the prohibition on charging fees to people seeking work; amends and extends the range of premises that may be entered by inspectors; lengthens the time limit for prosecutions under the 1973 Act; and provides for information obtained under the compulsory powers in the 1973 Act to be disclosed for the purposes of any criminal proceedings and not just those relating to offences under that Act;
- removes limits restricting certain employment rights, broadly, to those who ordinarily work in the United Kingdom;
- increases the limit on the compensatory award for unfair dismissal from £12,000 to £50,000; consolidates the existing system of special and additional awards into a single award, to apply in cases where an employer

fails to comply with a re-employment order; and provides that compensation will not be subject to the monetary limit where an individual has been dismissed unfairly under the provisions of the Public Interest Act 1998, or for reasons relating to health and safety matters;

- gives the Secretary of State power to make regulations requiring employees to be given the same protections in specified circumstances failing outside the scope of the EC Acquired Rights Directive as they have under the UK's implementing legislation in circumstances that do fall within the scope of that Directive;
- provides for Inland Revenue officers to be able to pass information obtained in respect of tax and National Insurance contributions to their National Minimum Wage colleagues and others for use in enforcing the NMW;
- makes a technical amendment to the School Standards and Framework Act 1998 to reflect the reduction in July 1999 of the unfair dismissal qualifying period from two years to one year;
- amends the law under which ministers can exclude Crown servants from certain employment rights on grounds of national security, giving staff of the security and intelligence services in particular better access to employment tribunals.

EMPLOYEE RIGHTS

Current law is established by the following Acts:

- Employment Protection (Consolidation) Act 1978, as amended.
- Trade Union and Labour Relations (Consolidation) Act 1992.
- Trade Union Reform and Employment Rights Act 1993.
- Employment Relations Act 1999.

These Acts consolidate employees' rights in regard to:

- trade union membership;
- trade disputes and picketing;
- contracts and written statements of terms and conditions of employment;
- itemized pay statements;
- guaranteed payments;
- suspension on medical grounds;
- time off for trade union duties and activities, and for public duties;
- maternity rights;
- statutory sick pay (SSP);
- rights on termination of employment;
- dismissal and unfair dismissal;
- redundancy – time off for job hunting or to arrange training.

Trade Union Membership

A 'recognized' trade union is one recognized by an employer for collective bargaining purposes. It is also one recognized by the parties to a union agreement, or one that ACAS recommends as being recognized.

Employees have the right not to be victimized or discriminated against (short of dismissal) for the purpose of:

- preventing or deterring them from becoming members of a trade union;
- preventing or deterring them from taking part in trade union activities;
- compelling them to become or remain a member of a trade union.

Employees can complain to an industrial tribunal (within three months) if such action is taken against them. The tribunal may award compensation which is just and equitable in relation to any loss suffered by an employee.

Trade Disputes and Picketing

Dispute Procedures

New regulations intended to help resolve workplace disputes without recourse to employment tribunal proceedings were introduced in October 2004. The *Employment Act 2002 (Dispute Resolution) Regulations 2004* impose statutory minimum procedures that must be followed by both employers and employees before employment tribunal proceedings begin; failure to do so will result in increased penalties at the tribunal.

A trade dispute is a dispute between an employer and his employees, or between employees themselves, relating to:

- the terms and conditions of employment or the physical conditions in which employees are required to work;
- the engagement or non-engagement of one or more workers, or the termination or suspension of their employment;
- the allocation of work or the duties of employment between employees or groups of employees;
- matters of discipline among employees;
- membership or non-membership of a trade union;
- the provision of facilities for trade union officials;
- the machinery for negotiation or consultation and other procedures relating to the above matters, including trade union recognition by employers or employers' associations.

Trade disputes must be agreed in a secret ballot requiring workers to make a straight 'Yes' or 'No' vote to the action. Without a properly conducted ballot the trade union and its officers have no legal immunity if they take or incite industrial action.

Secondary industrial action where a person is induced to take industrial action against an employer who is not party to a trade dispute may result in legal action for damages.

Picketing staged in furtherance of a trade dispute is legal if its purpose is peacefully to communicate information or peacefully to persuade a person to refrain from working (or, conversely, to work) and provided it is carried out at or near the place of work.

Contracts and Written Statements of Employment

A contract of employment exists as soon as a job has been offered and accepted by the prospective employee. It does not have to be in writing. Acceptance of the terms described at the interview or in a letter of confirmation is assumed if the employee reports for work at the due time.

Within two months of starting employment an employee must be given a written statement detailing:

- the parties to the contract;
- the job title and the place of work;
- the date when employment began (and if for a fixed term, when it will end), and the date when continuous employment began (eg after any probationary period);
- rates of pay and when payment is to be made (eg weekly or monthly);
- hours of work (normal and otherwise);
- holiday entitlements and pay;
- sick pay provisions;
- pension rights;
- notice of termination required;
- disciplinary rules, including the name of the person to whom appeals on such matters can be referred and how appeals must be made.

If the terms of employment set out in the written statement are changed, the employer must give the employee details in writing within one month.

Itemized Pay Statements

Employees are entitled (on or before pay day) to an itemized pay statement in writing showing:

- the gross amount of pay;
- any deductions made and for what reason (unless regular fixed deductions are made and a standing statement is given plus a cumulative statement of deductions);
- the net amount payable;
- where payment is by different means (eg part cheque/part cash), the method of paying.

Guaranteed Payments

If an employer cannot find work on any day for an employee who has been employed continuously for at least one month, the employee must receive a guaranteed payment for that day, but not if the workless day occurs as a result of a trade dispute. Also payment need not be made if:

- the employee was offered alternative work which was unreasonably refused;
- the employee does not comply with reasonable requirements to ensure that his services are available.

Guaranteed pay is limited to a maximum of five workless days in any three-month period for a five-day week worker. Payment of guaranteed pay does not affect the right of employees to receive any other pay under their contract of employment.

Employees can complain to an industrial tribunal (within three months) if an employer fails to pay any guaranteed pay due. If the tribunal upholds the complaint, the employer will have to pay the amount due.

Suspension on Medical Grounds

If employees are suspended from work on medical grounds as a result of legal requirements or recommendations based on Health and Safety at Work Act Codes of Practice, they must be paid for up to 26 weeks.

Employees are not entitled to such payment for any time when they are medically suspended if:

- they had not been continuously employed for at least one month prior to the suspension;
- they are incapable of work owing to disease or bodily or mental disability;
- the employer has offered suitable alternative work which is refused;
- they do not comply with reasonable requirements to make their services available;
- they are employed under a fixed-term contract of no more than three months' duration.

An employee can complain to an industrial tribunal if the employer fails to make such payment. If the tribunal upholds the complaint, the employer will have to pay the amount due.

Time Off for Union Activities

In certain circumstances employees must be given time off work either with or without pay to undertake:

- *Trade union duties*
  Paid time off (which is reasonable in all the circumstances) must be given to an official of a recognized trade union to:
  - carry out the duties concerned with industrial relations between employer and employees;
  - undergo training in industrial relations: (a) relevant to the duties mentioned; (b) approved by the union or the Trades Union Congress (TUC).
- *Trade union activities*
  Unpaid time off (ie which is reasonable in all the circumstances) to take part in the activities of a recognized union (*excluding* for industrial action) must be given to employee members of the union (ie *not* union officials). In both the above cases employees can complain to an industrial tribunal if the employer fails to permit such time off. If the tribunal upholds the complaint, it will award 'just and equitable compensation' to the employee.
- *Public duties*
  A reasonable amount of unpaid time off must be given to employees to fulfil certain public duties, such as being a:

- Justice of the Peace;
- member of a local authority;
- member of a statutory tribunal;
- member of a National Health Service Trust, an area, district or regional health authority or a family practitioner committee;
- member of a police authority;
- member of a board of visitors for prisons, remand centres or young-offender institutions;
- governor of a grant-maintained school, higher education corporation or educational establishment maintained by a local education authority; or
- member of the National Rivers Authority.

The amount of time off to be given is that which is reasonable in all the circumstances, having regard to how much time is required to fulfil the duties; the amount of time off already taken in this respect; and the effect on the employer's business of the employee's absence. A complaint can be made to an industrial tribunal (within three months) if the employer fails to allow such time off. Compensation may be ordered. Although not part of this legislation, employees listed for jury service must be allowed appropriate time off for such purposes – without pay.

- *Looking for work or arranging for training*
  A reasonable amount of paid time off during working hours must be given to an employee who is to become redundant to look for new employment or to arrange training for future employment. A complaint can be made to an industrial tribunal (within three months) if the employer refuses such time off and compensation of up to two-fifths of a week's pay may be awarded.

Maternity Rights for the Expectant Mother

A qualifying employee who is absent from work owing wholly or partly to pregnancy or confinement is entitled to:

- 26 weeks' 'ordinary maternity leave' before and/or after childbirth;
- paid time off for ante-natal examinations;
- no risk of dismissal during the period of pregnancy and maternity leave except in exceptional circumstances wholly unconnected with the pregnancy;
- preservation of her contractual rights during the maternity leave;
- pay at the sick-pay rate during the period of maternity.

The employee is entitled to return to work after her confinement (up to a maximum of 29 weeks (ie 'extended maternity leave') from the week of the birth), provided:

- she continues working until the eleventh week before the expected week of confinement;
- she has been continuously employed prior to this for not less than one year;
- she informs her employer in writing 21 days in advance of her absence (or as soon as is reasonably practicable) that:
  - she will be (or is) absent wholly or partly due to pregnancy and confinement (and the week in which it is due), and
  - that she intends to return to work;

- she produces a medical certificate (if the employer requests it) stating the expected week of her confinement.

Maternity Pay

There are two main maternity benefits:

1. Statutory Maternity Pay (SMP), which is paid by employers to women who qualify. It has minimum earnings and continuous employment conditions.
2. Maternity Allowance (MA) is paid by the social security/Jobcentre Plus office to qualifying women. It is based on the woman's recent employment and earnings record.

If an employee does not qualify for SMP, her employer should explain to her why she does not qualify by giving her Form SMP1 (*Why I cannot pay you SMP* – available from Jobcentre Plus/social security offices or the Department for Work and Pensions website) which the employee must send to her social security office with her completed MA form to claim MA. An MA form can be obtained from Jobcentre Plus/social security offices. For more information employees should see the Department for Work and Pensions website: www.dwp.gov.uk.

Employee's Right to Return to Work

An employee's right to return to work after pregnancy or confinement is subject to specified conditions. It must be:

- to the original employer or his successor;
- up to the end of 29 weeks from the week of the confinement;
- in the job in which she was previously employed;
- under the original contract of employment;
- on terms and conditions no less favourable than those applicable had she not been absent.

If, because of changed circumstances, the employer cannot offer her the original job, she must be offered any suitable alternative vacancy that exists under a new contract of employment which is not substantially less favourable than the original contract.

The employee wishing to return to work must notify the employer in writing at least 28 days beforehand. The employer can postpone the return date for not more than 4 weeks providing the employee is notified to this effect.

The employee, after giving notice of return, can postpone the return date for not more than 4 weeks – even if this goes beyond the 29 weeks. She must produce a medical certificate showing why she cannot return before the notified day of return or before the end of the 29 weeks.

Parental Leave

Parental leave is a right for parents to take time off work to look after a child or make arrangements for the child's welfare. Parents can use the leave to spend more time with children and strike a better balance between their work and family commitments.

The leave is available to employees who have, or expect to have, parental responsibility for a child provided they have one year's continuous service with their current employer. Special rules apply for parents of children born, or placed for adoption, before 14 December 1999. They will be eligible for parental leave from their current employer if they completed one year's continuous service with another employer between 15 December 1998 and 9 January 2002.

Employees get 13 weeks' leave for each child (18 weeks for disabled children, ie for whom an award of disability living allowance has been made), which may be taken in short or long blocks depending on a workplace agreement. If twins are born each parent gets 13 weeks' leave for each child and 18 weeks in the case of each disabled child.

Statutory Sick Pay (SSP)

The Social Security Contributions and Benefits Act 1992 and the Statutory Sick Pay Act 1994 legislate for the statutory sick pay (SSP) scheme under which employers are required to make sickness payments to employees. Employees must satisfy rules on periods of incapacity, entitlement periods and qualifying days. Employers may reclaim the full payments made to employees by deduction from payments of National Insurance Contributions (NIC) to the Inland Revenue where these exceed 13 per cent of the employer's gross NIC payments in a month. State sickness benefit still applies for certain special cases – such as those who have to be regularly off work for long-term medical treatment and those who do not qualify for or who have exhausted their SSP entitlements.

Employees qualify for SSP after a specified period of incapacity for work (PIW) – four consecutive days – due to physical or mental illness or disablement. The first *three* days of each PIW are waiting days – sickness payment is made only in respect of qualifying days, which count from the fourth day onwards. The maximum entitlement to SSP in any period of incapacity for work is 28 weeks, after which a new PIW will start.

Notification of sickness can be by use of the DSS self-certification form, by submitting a doctor's medical certificate, or the employer can institute his own scheme of self-certification.

The amount of SSP that must be paid is based on the employee's average gross weekly pay and is specified by the Government.

Rights on Termination of Employment

Employment may be terminated by specified minimum periods of notice being given by the employer to the employee or vice versa.

Termination by the Employer

A person who has been employed continuously for 4 weeks or more must be given the following periods of notice of termination of employment:

Continuous employment	*Notice*
• Less than 2 years	1 week
• More than 2 years but less than 12 years	1 week for each continuous year
• More than 12 years	12 weeks

Termination by the Employee

Employees terminating their employment after four weeks or more of employment must give at least one week's notice, irrespective of the length of service.

If a contract of employment stipulates shorter notice, the legal minima (above) still apply but either party may waive the right to notice or the employee may accept payment in lieu of notice.

Dismissal and Unfair Dismissal

Where an employer has to dismiss an employee certain conditions must be observed, depending on the circumstances of the case. Dismissal includes:

- termination of employment with or without notice;
- the term of a contract expiring without renewal;
- termination by the employee with or without notice due to the employer's conduct;
- failure by an employer to let an employee return to work after pregnancy or confinement.

Statement of Reasons for Dismissal

Employees are entitled to be provided on request, and within 14 days, with a written statement giving reasons for dismissal if:

- they are given notice of termination (unless the employment was for less than 26 weeks);
- they are dismissed without notice;
- a fixed-term contract expires without being renewed.

Where the employer fails to give such a written statement (or the reason given in writing is not adequate or is known to be incorrect) the employee can complain to an employment tribunal (within three months of dismissal), which will, if the case is well founded:

- declare the reason for dismissal;
- make an award that requires the employer to pay an amount equal to two weeks' pay.

Fair and Unfair Dismissal

Employees have a right not to be unfairly dismissed. For an employee to be fairly dismissed the employer must show:

- the reason for dismissal;
- that the reason was within one of the following areas:
  - the capability or qualification of the employee for the work involved including fitness to perform the work;
  - the conduct of the employee;
  - redundancy;
  - contravention of legal restrictions by the employee in connection with his work.

If the reason for dismissal was within the first two reasons above, the employer must show that he acted reasonably in treating it as a sufficient reason for dismissing the employee.

Dismissal for Trade Union Membership

It is unfair to dismiss an employee because that employee:

- was, or proposed to become, a trade union member;
- had taken part, or proposed to take part, in trade union activities;
- refused to become a trade union member.

Dismissal on Redundancy

It is unfair to dismiss an employee on the grounds of redundancy if the employee:

- was selected for dismissal for an inadmissible reason;
- was selected in contravention of customary arrangements or agreed procedures.

Dismissal on Pregnancy

It is unfair to dismiss an employee who is pregnant or for any reason connected with pregnancy except if the employee:

- is incapable of doing her job because of the pregnancy;
- cannot do her job owing to her pregnancy without contravention of legal restrictions.

Dismissal on Replacement

If a new employee is engaged to replace a pregnant employee it is *not* unfair to dismiss that employee on the return to work of the former employee if the replacement employee was told that dismissal would then occur.

Dismissal following Industrial Action

It is unfair to dismiss an employee where at the date of dismissal:

- the employer was conducting a lock-out, or the employee was taking part in industrial action.

Qualifying Period and Age Limit for Dismissal

Claims for unfair dismissal can be made to an employment tribunal within three months of the date of dismissal. Unfair dismissal cannot be claimed if the employee:

- was employed for less than one year; or
- was the normal retiring age for an employee or if:
  - a man had attained the age of 65 years;
  - a woman had attained the age of 60 years.

Remedy for Unfair Dismissal

Employees can (within three months) make a complaint of unfair dismissal against an employer to an employment tribunal, which, if the case is well founded:

- will order reinstatement or re-engagement (subject to the employee's wishes); or
- award compensation, up to a maximum of £50,000.

Redundancy

Employees may be made redundant when a particular job ceases to exist, when the demand for a particular product or service falls or ceases, or when a business closes down altogether. They are not redundant if they are replaced by new or other employees.

For employees to claim redundancy payment they must have been continuously employed for at least two years since reaching the age of 18 years, and must not have unreasonably refused any alternative offer of suitable employment.

The redundancy payment depends on the employee's age, length of service with the employer and present weekly rate of pay. The scale of pay rates is as follows:

- half a week's pay for each year continuously employed between the ages of 18 and 21 years;
- one week's pay for each year continuously employed between the ages of 22 and 40 years;
- one-and-a-half week's pay for each year continuously employed between the ages of 41 and 64 years.

The following employees are among those not entitled to redundancy payments:

- self-employed people;
- those employed for less than two years continuously with their employers since reaching the age of 18;
- people (ie both men and women) aged over 65 years;
- the husband or wife of an employer;
- people outside Britain at the time of being made redundant unless they normally work in Great Britain.

Consultation – Disclosure

When an employer wishes to make one or more employees redundant he must *consult* their trade union about the proposed dismissal. Consultation should begin 30 days before dismissal when 10 or more employees are to be made redundant in a period of 30 days, and 90 days before dismissal when 100 or more employees are to be made redundant in a period of 90 days.

An employer is required to disclose (in writing) to trade union representatives:

- the reason why the employees have become redundant;
- the number and description of employees whom it is proposed to make redundant;
- the total number of comparable employees at the establishment;
- the proposed method of selecting the employees to be dismissed;
- the proposed methods of dismissal, including the period over which they will take effect.

The employer is obliged to consider any representations made by the trade union representative during consultation, to reply to those representations and, if he rejects them, to state his reasons.

Failure to comply with these provisions enables the trade union to complain to an industrial tribunal, which will make a protective award to the employee(s) if it finds the complaint well founded.

An employer is also required to notify the Secretary of State for Education and Employment of proposed redundancies:

- 30 days in advance when proposing to dismiss 10 employees or more;
- 90 days in advance when dismissing 100 employees or more.

Flexible Working Rights

Since 6 April 2003, parents of children aged under six years or of disabled children aged under 18 years have had the right to apply to their employer to work on a flexible time basis (eg by varying working times or shift patterns, or working from home). Employers must consider such requests seriously and favourably, provided they are properly made – in writing giving details of the employee's responsibility for a relevant child and of the flexible working pattern being sought.

Grievance Procedure

Amendments to the Employment Relations Act 1999 (see pp 270–73) effective from 4 September 2001 give employees the statutory right to be accompanied at disciplinary and grievance hearings.

The new law provides that when an employee is required or invited by his employer to attend certain disciplinary or grievance hearings he may make a reasonable request to be accompanied by a fellow worker or trade union official. This right is additional to any contractual rights and applies to all workers, not just employees.

For these purposes it is necessary to determine precisely what is a disciplinary hearing. An employer may choose to deal with disciplinary matter initially by means of an informal interview or counselling session. Provided this informal interview or counselling session does not result in a formal warning or other form of disciplinary action it may not be necessary for the individual to be accompanied, since such matters are best resolved informally by the individual and the company representative (eg manager) concerned. If it is obvious that formal disciplinary action will be necessary, the informal interview should be terminated and a formal hearing convened at which the worker should be afforded the statutory right to be accompanied.

ACAS has launched a revised draft *Code of Practice on Disciplinary and Grievance Procedures* due to come into effect from October 2004 which is available on the ACAS website at: www.acas.org.uk

THE INFORMATION AND CONSULTATION DIRECTIVE

Firms employing more than 150 workers will be subject to a European Directive on informing and consulting with the workforce, anticipated to become effective by about 2004/5. Progressively the minimum numbers of employees where this legislation applies will be reduced during the period from 2004/5 to about 2008, but will not fall below 50 employees, thus leaving many small haulage firms outside the scope of the directive. Many large firms covered by this legislation already have information and consultation procedures in being, regularly talking to their workforce about all relevant key issues.

CRIMINAL DISCLOSURE

Employers can check on the past criminal records of employees and potential employees via the Criminal Records Bureau (CRB). The bureau offers a disclosure service to employers, voluntary organizations, professional bodies and certain licensing authorities. Disclosure cannot be sought directly by an employer: the individual whose record is to be searched must make a request to the CRB, so the process relies upon prospective employees cooperating with employers.

There are two levels of disclosure:

Basic Disclosure

Any employer or organization may request a prospective employee or volunteer to obtain a Basic Disclosure from the CRB. This will provide details of any criminal convictions held at national level on the Police National Computer that are not considered to be 'spent' under the Rehabilitation of Offenders Act 1974 (ROA). The results of a Basic Disclosure will be sent directly to the individual and will not be revealed automatically to the employer; it will be up to the individual whether he/she discloses the record.

Standard Disclosure

This will not be available to all employers. It will only apply to positions and/or professions covered by the Rehabilitation of Offenders Act 1974 (Exceptions) Order 1975. This disclosure will contain details of all convictions on record, together with any cautions, reprimands or warnings given to that individual by the police.

To obtain a Standard Disclosure the employer must first apply to the CRB to become a Registered Body: that is, an employer, voluntary organization or individual who is entitled to ask exempted questions under ROA. The individual concerned must apply for his or her own disclosure, but a Registered Body must countersign the application. The results of the disclosure in this case will be sent both to the individual and the Registered Body.

Code of Practice

A Code of Practice for registered persons and other recipients of disclosure information is designed to ensure that information revealed by disclosure is fairly and properly used by recipients. Where a registered person fails to comply with the Code of Practice, or countersigns an application for a body or individual that has failed to comply with the Code of Practice, the CRB may refuse to issue a disclosure.

DISCRIMINATION IN EMPLOYMENT

Besides the prohibition on discrimination against disabled workers described elsewhere, there are a number of other relevant pieces of anti-discrimination legislation (see below) on the statute book, about which the transport employer should be aware. In particular, employers should beware of discriminating against or between employees in areas such as: recruitment and selection; pay, terms and conditions of employment; promotion, transfer and training opportunities; and selection for redundancy or short-term working.

- Under the Asylum and Immigration Act 1998 it is an offence for an employer to offer employment to a person who is not legally entitled to work in the UK. Employers must verify the status of an immigrant or suspected immigrant before confirming any offer of employment – ie by demanding to see a P45, P60 or previous payslip showing a National Insurance number, or a British passport or UK birth certificate; though none of these items is conclusive proof of the individual's right to work in the UK, having requested and seen one of them is a defence against prosecution for employing an illegal worker. The penalty for contravention of this provision is a fine of up to £5,000 for each offence.
- The Equal Pay Act 1970 and Equal Pay (Amendment) Regulations 1983 require that both men and women must receive equal pay for work described as like work, work that is broadly similar, work that is rated as equal under a job evaluation scheme and work that is of equal value.
- The Fair Employment (Northern Ireland) Act 1989 is intended to ensure that employers in Northern Ireland do not discriminate against employees on religious grounds, and most specifically that they do not because they are either of Catholic or Protestant persuasion.
- The Race Relations Act 1976 (as amended by the Race Relations Act 1976 (Amendment) Regulations 2003) prohibits discrimination by employers on grounds of race, colour, creed or ethnic origin.
- Under the Sex Discrimination Acts of 1975 and 1986 employers must not discriminate against men or women, or married persons. Employers should be particularly careful when dealing with allegations of sexual harassment among their workforce. Any complaint should be treated as being of a serious nature and should be dealt with immediately. The employer should also be aware that where an employee is alleged to have sexually harassed another employee, under the Act, the employer may himself be considered to have been party to the harassment whether or not it was done with his knowledge or approval. The Protection from Harassment Act 1997 is also relevant in this context.
- Under the Trade Union and Labour Relations (Consolidation) Act 1992 employers must not discriminate against employees on grounds of their trade union membership or non-membership of a trade union.
- The Police Act 1996 introduced the concept of the Criminal Records Agency, through which employers will be able to establish the criminal record or otherwise of potential employees. This facility came into operation in 2000.

The Road Haulage Association warns its members of the substantial penalties that may be imposed for failing to comply with discrimination law and suggests the following code of best practice:

- Recruit from all sections of the community.
- Identify the best available person for the job.

- Consider flexible working arrangements.
- Ensure that all employees have equal access to opportunities for training development.
- Let your employees know that you are an equal opportunities employer.
- Include an equal opportunities statement in job advertisements.
- Provide a clear and simple complaints procedure and review the effects of good practice.

A European Directive (2000/43/EC) demands that by 2006 much tougher laws on discrimination in the workplace must be enacted by the UK Government and all other EU Member States. These will particularly include anti-discriminatory measures on grounds of age and of sexual orientation.

EMPLOYMENT OF PART-TIME WORKERS

Many part-time workers are employed in transport operations. Consequently, the new part-time workers regulations may have serious implications for road hauliers and other transport operators. For this reason the new regulations are outlined below.

The Part-time Workers (Prevention of Less Favourable Treatment) Regulations 2000, which came into force from 1 July 2000, implemented European Council Directive 97/81/EC (normally referred to as the Part-time Work Directive), which is extended to the United Kingdom by Directive 98/23/EC.

These Regulations give part-time workers the right in principle not to be treated less favourably than full-time workers of the same employer who work under the same type of employment contract. Additionally, the right is given to workers who become part-time or, having been full-time, return part-time after absence, to not be treated less favourably than they were before going part-time.

Rates of Pay

Under these regulations, part-timers must not be paid a lower basic rate of pay than comparable full-timers, except where this is justified on objective grounds, for example, within a performance-related pay scheme where employees are shown to have different levels of performance measured by a fair and consistent appraisal system.

To comply with the law it is clear that part-timers should receive the same hourly rate as comparable full-timers.

Overtime

As case law currently stands, part-timers do not have an automatic right to overtime payments once they work beyond their normal hours. However, once they have worked up to the full-time hours of comparable full-timers, they do have a legal right to overtime payments where these apply.

To comply with the law, part-timers should receive the same hourly rate of overtime pay as comparable full-timers, at least once they have worked more than the normal full-time hours.

Contractual Sick and Maternity Pay

The Regulations apply directly to contractual sick and maternity pay, meaning that there is an obligation for employers not to treat a part-timer less favourably than a comparable full-timer. For example, the benefits that a full-timer receives must also apply to part-timers on a pro-rata basis. The only exception is if the different treatment is justified on objective grounds.

To comply with the law, part-timers should not be treated less favourably than comparable full-timers in terms of:

- calculating the rate of sick pay or maternity pay;
- the length of service required to qualify for payment;
- the length of time the payment is received.

Occupational Pensions

Pension scheme rules may need to be revised, to ensure that they comply with the new legislation.

To comply with the law, employers must not discriminate between full-timers and part-timers with regards to access to pension schemes, unless different treatment is justified on objective grounds.

Access to Training

The regulations place an obligation on employers not to exclude part-timers from training, which will need to be structured wherever possible to be at the most convenient times for the majority of staff, including part-timers.

To comply with the law:

- employers should not exclude part-time staff from training simply because they work part-time;
- training should be scheduled so far as possible so that staff, including part-timers, can attend.

Leave/Holidays/Breaks: Annual Leave, Maternity and Parental Leave, Career Breaks

Part-timers, like their full-time colleagues, are entitled to a minimum of statutory annual leave, maternity leave, parental leave, and time off for dependents.

In some cases, companies extend these statutory entitlements with enhanced contractual conditions. When this occurs, part-timers should have the same entitlements as their full-time colleagues, on a pro-rata basis where appropriate.

In the case of career breaks, part-time staff should be treated no less favourably than comparable full-time staff.

To comply with the law:

- the contractual holiday entitlement of part-time staff should be pro rata to that of comparable full-timers;
- contractual maternity leave and parental leave should be available to part-timers in the same way as for comparable full-timers;

- careers break schemes should be available to part-timers in the same way as for comparable full-timers, unless their exclusion is objectively justified on grounds other than their part-time status.

Redundancy

In a redundancy situation, part-timers should be treated no less favourably than their full-time equivalents. Different treatment of part-timers and full-timers will only be lawful if it can be justified on objective grounds.

To comply with the law, the criteria used to select jobs for redundancy should be objectively justified, and part-timers must not be treated less favourably than comparable full-timers.

Objective Justification

The right of part-timers not to be treated less favourably than a comparable full-timer applies only if the treatment is not justified on objective grounds.

Less favourable treatment will only be justified on objective grounds if it can be shown that it is:

- to achieve a legitimate objective, for example, a genuine business objective;
- necessary to achieve that objective;
- an appropriate way to achieve the objective.

EMPLOYMENT OF DISABLED PERSONS

The Disabled Persons (Employment) Acts of 1944 and 1958 and the Disability Discrimination Act 1995 (commonly abbreviated to the DDA) make provisions to enable persons handicapped by disablement to:

- secure employment; or
- work on their own account; and
- be protected from discrimination on account of their disability.

A disabled person for these purposes is either:

- an individual who was registered as disabled under the Disabled Persons (Employment) Act 1944 on both 12 January 1995 and 2 December 1996; or
- a person who has or has had a disability as defined in the DDA as 'a physical or mental impairment which has a substantial and long-term adverse effect on their ability to carry out normal day-to-day activities'.

NB: For persons to be protected under the Act, they must satisfy the four main conditions of disability under the Act.

Physical and Mental Impairment

Physical impairment includes impairments that affect hearing or sight, while mental impairment is defined as an impairment resulting from or consisting of a mental illness, but only if the illness is a *clinically well-recognized* illness (ie one that includes manic depression, schizophrenia and severe and extended depressive psychoses). A clinically well-recognized illness is one that is recognized by a body of respected medical practitioners such as the World Health Organization.

Impairments for these purposes do not include such conditions as nicotine and substance addictions, alcoholism, arsonism, thieving, exhibitionism and voyeurism, and physical or sexual abuse of other persons.

Discrimination against the Disabled

Discrimination against a disabled person by an employer may occur in two ways:

1. If for a reason that relates to a disabled person's disability the employer treats that person less favourably than he treats or would treat others who do not have a disability and he cannot show that the treatment is justified.
2. If the employer fails to make a 'reasonable adjustment' in relation to the disabled person and cannot show that such failure is justified.

Complex definitions relating to the circumstances in which less favourable treatment of a disabled person may be justified, and to the employer's duty to make 'reasonable adjustments' for such persons, are included in the 1995 Act, as are examples of reasonable adjustments, such as:

- making adjustments to premises;
- allocating some of the disabled person's duties to another person;
- transferring him to fill an existing vacancy;
- altering his working hours;
- assigning him to a different place of work;
- allowing him to be absent during working hours for rehabilitation, assessment or treatment;
- giving, or arranging for him to be given, training;
- acquiring or modifying equipment;
- modifying instructions or reference manuals;
- modifying procedures for testing assessment;
- providing a reader or interpreter;
- providing supervision.

Remedies for Discrimination against Disabled Persons

A disabled person may make a complaint of discrimination on account of their disability to an Employment Tribunal within three months of an act of such discrimination occurring (or longer if the circumstances are 'just and equitable'). In cases where other employees discriminate against a disabled person, the employer is held liable unless he can show that he has taken reasonably practicable steps to prevent discrimination by other employees.

In the event of a successful claim for discrimination by a disabled person, the discriminator will be required to take reasonable action to cease such discrimination and compensation may be awarded (with no ceiling on the amount).

The previous exemption from this legislation for firms employing no more than 15 persons was removed under amendments that took effect from October 2004.

ILLEGAL WORKING

Illegal working is an issue that is very relevant to UK haulage employers who may be faced with the temptation to employ LGV drivers from the new EU member states. Changes to certain provisions of the Asylum and Immigration Act 1996, effective from 1 May 2004, mean:

- the risk of an employer committing a criminal offence by employing a person who has no right to work in the UK;
- added responsibilities for employers to make more stringent document checks on potential employees in order to create a statutory defence against conviction for employing an illegal worker;
- employers being obliged to ensure fair recruitment practices to avoid discrimination against individuals on racial grounds.

A copy of the Home Office guidance on this subject can be downloaded from its website at: www.ind.homeoffice.gov.uk.

PUBLIC DISCLOSURE

It is now possible for an employee to officially 'shop' an employer who, in various ways, contravenes the law. Under the so-called 'whistleblowers' Public Interest Disclosure Act 1998, effective from 1 January 1999, employees are legally protected from detrimental treatment when they disclose information in the reasonable belief that the employer is contravening the law, and particularly so where such disclosure may be seen to be in the public interest. For example, in road transport this might be the disclosure of information about such matters as:

- the use of ill-maintained or unsafe vehicles;
- the use of unlicensed, untaxed or uninsured vehicles;
- encouragement by an employer for drivers to speed or overload vehicles;
- the employer setting unreasonable work/delivery schedules that contravene the law;
- the use of illegal 'red' diesel fuel in road vehicles;
- drivers being encouraged or coerced into exceeding the drivers' hours limitations;
- the falsification or fraudulent use of tachographs;
- contravention of other aspects of transport or employment legislation.

Disclosure may be by any worker, including agency workers, and need not be of a case where positive proof is available. It is sufficient for the person disclosing the information to have a reasonable belief that:

- a criminal offence has been, is being or is likely to be committed;
- a person has failed, is failing or is likely to fail to comply with a legal obligation;
- a miscarriage of justice has occurred, is occurring or is likely to occur;
- the health and safety of any individual has been, is being or is likely to be endangered;
- the environment has been, is being or is likely to be damaged;

- information showing the above has been, is being or is likely to be deliberately concealed.

A 'protected disclosure' is one that is made:

- in good faith to an employer;
- to another person, where the malpractice relates to the other person's conduct;
- to another person with legal responsibility for the subject matter of disclosure;
- where the worker uses an employer's authorized procedure;
- where the worker seeks legal advice;
- where the worker, in good faith, makes a disclosure to regulatory authorities;
- to anybody:
  - if the disclosure is made in good faith;
  - if the disclosure is made without the motive of personal gain;
  - if the disclosure contains true information;
  - if disclosure to the employer might result in the complainant suffering detrimental treatment;
  - if the complainant believes that disclosure to his employer would result in evidence being concealed or destroyed; and
  - if the complainant has reported the matter to his employer previously without any response.

Employers are not permitted to prevent their employees or workers from making or to persuade their employees/workers to make public interest disclosures, nor may they make contracts with employees/workers that prohibit disclosure. Any such arrangement would be legally invalid. Instead, they must establish procedures to deal with complaints of malpractice, including proper investigation and action to remedy any such complaints.

Dismissal of any employee or worker for making a public disclosure would be considered to have been unfair and lead to a compensatory award by an industrial tribunal for an unlimited amount.

THE NATIONAL MINIMUM WAGE

The national minimum wage provisions introduced by the National Minimum Wage Regulations 1999 under the National Minimum Wage Act 1998, which came into force in the UK on 1 April 1999, have been further amended by the National Minimum Wage Regulations 1999 (Amendment) Regulations 2004. These require that workers (other than self-employed persons) aged 22 years or over must be paid at least £5.05 per hour, and workers aged from 18 to 21 years at least £4.25 per hour. The minimum wage applies to most workers in the UK including agency workers (eg agency drivers), part-time and casual workers and those paid on a commission basis.

NB: These rates were increased on 1 October 2005.

New workers aged 22 years or over who receive accredited training must be paid at least £4.25 per hour for the first six months. In this case, the employer will have to come to an agreement with the worker, committing him to providing training on at least 26 days during that six-month period.

Assessing Minimum Pay

For the purposes of assessing minimum pay, payments to employees comprising bonuses, incentives and performance-related awards count as part of the pay package, but other allowances not consolidated into an employee's pay are not counted. Similarly, overtime payments and shift payments do not count. Benefits in kind, such as the provision of overnight subsistence, meals, uniforms and workwear allowances, are also excluded. Gross pay, with all deductions and reductions subtracted, should be divided by the number of hours worked to determine whether the resulting hourly pay rate at least matches, if not exceeds, the national minima stated above.

Types of Work

The work hours for which an employer has to pay are calculated according to the type of work employees are engaged upon. Mainly this is as follows:

- time work, where an employee is paid for working a set number of hours or a set period of time;
- salaried work, where an employee has a contract to work a set number of basic hours annually in return for an annual salary paid in equal instalments;
- unmeasured work, where employees are paid to do specific tasks (eg driving), but are not set specific hours for the work.

NB: In this case the employer must agree with the employee (in writing) a daily average of hours to be spent carrying out the assigned tasks. The employer must be able to show that the number of hours agreed is realistic.

Enforcement and Penalties

Enforcement of the minimum wage provisions is by the Inland Revenue and by the employees themselves, who have a right to complain if they are not being paid the national minimum wage.

Employers obviously need to keep accurate records of the hours worked and hourly rates paid to employees in case such information is called into question later – for a minimum of three years. An employee (or any other qualifying worker) may make a written request for access to his own records, and this must be allowed within 14 days unless extended by agreement. Should a dispute arise, the burden of proof is on the employer to show that the national minimum wage has been paid, not on the employee to prove that it has not.

Refusal to pay the national minimum wage is a criminal offence carrying a maximum fine of up to £5,000 on conviction. Dismissal of an employee who becomes eligible for the national minimum wage or for a higher rate of pay will constitute unfair dismissal with no qualifying period to be served by workers to secure protection against this form of unfair dismissal.

SELF-EMPLOYMENT

The road haulage industry is renowned for the number of self-employed workers, especially owner-driver lorry operators. However, not all such so-called self-employed persons are genuinely self-employed in terms that meet legal requirements of the Inland Revenue (IR) in regard to payment of tax and National Insurance contributions.

To satisfy the IR a self-employed person needs to meet a series of so-called 'tests' under which the self-employed person:

- decides, broadly, how and when specified work is to be carried out, the actual hours he works and when breaks and holidays are taken, and is not subject to the disciplinary provisions of an employer;
- provides his own tools and equipment and is free to send another person (or sub-contractor) in his place to carry out work where necessary;
- has no entitlement to payment for public or annual holidays or sickness; is not included in an employer's pension scheme, and has no rights to claim redundancy payments, unfair dismissal or any entitlement to unemployment benefit if his services are no longer required;
- takes financial risk with the aim of making a profit, is responsible for paying his own income tax and national insurance contributions, and charges for the services provided by submitting an invoice;
- is free to work for other employers as required if he so wishes (a self-employed person who works for only one employer is likely to be considered to be an employee of that employer).

NB: See definition of self-employment under the Road Transport Directive on page 95.

A number of instances have arisen in transport where the self-employed status of owner-drivers has been questioned by the Inland Revenue (mainly because they work for only one firm). In such cases, where the IR has ruled that owner-driver agreements are merely employment contracts, the employer becomes liable for back tax and NI contributions for its sub-contractors. This is especially so where the owner-driver works under the 'O' licence of the employing company – and if he is genuinely self-employed, then, in any event, this practice is illegal under 'O' licensing legislation.

Registration of Small Businesses

Those who set up, or have recently set up, in business, whether as a small employer or on a self-employed basis, are required to register their existence and the nature of their activities with the 'relevant enforcing authority'. In the case of road haulage businesses this would be the local HSE office (address in a local telephone directory). Further information can be obtained from the Health and Safety Executive's website: www.hse.gov.uk/startup/index.htm or by telephoning the HSE's infoline: 08701 545500.

THE HUMAN RIGHTS ACT 1998

The Human Rights Act 1998, most of the provisions for which came into effect from 2 October 2000, is expected to have far-reaching implications in many areas of law, and not least in regard to individual liability for road traffic offences.

In this particular case, as has already been established in Scotland where the law was applied earlier than in England and Wales, it appears that the mandatory right of the police to require the owner of a vehicle to name the person who was driving on any particular occasion may be contrary to Article 6 of the Act, and consequently, may prejudice the vehicle owner's right to silence. This point and many others will undoubtedly become clearer as the provisions of the Act take effect and cases are brought before the courts. It has been suggested that the Act will provide a bonanza for human rights' lawyers.

The Act, which gives effect to the rights and freedoms guaranteed under the European Convention on Human Rights, comprises 22 articles and four schedules dealing with the following:

- interpretation;
- compatibility between primary legislation and rights established under the Convention;
- the rights of the courts to intervene in proceedings under the Act;
- the illegality of public authorities acting in a manner that is incompatible with a Convention right (ie where the police require information that may incriminate a person, as described above);
- proceedings, judicial remedies and remedial action.

Other sections deal with safeguarding existing human rights, freedom of expression and freedom of thought, conscience and religion. Various derogations and reservations are specified together with provisions for these to be periodically reviewed. Matters relating to the appointment of judges to the European Court of Human Rights are also included.

The schedules deal with basic human rights, such as:

- the right to life;
- the right not to be subjected to torture, inhuman or degrading treatment or punishment;
- the right not to be held in slavery or servitude;
- the right to liberty and security;
- the right to a fair trial;
- the right to respect for private and family life;
- the right to freedom of thought, conscience and religion;
- the right to freedom of expression;
- the right to freedom of assembly and association;
- the right to marry;
- the prohibition of discrimination;
- restrictions on the political activity of aliens;
- prohibition of the abuse of rights;
- limitation on the use of restrictions on rights;
- protection of property;
- the right to education;

- the right to free elections;
- abolition of the death penalty;
- the imposition of the death penalty in time of war.

PENSION RIGHTS

New legislation came into force in April 2001 giving rights to all employees (including those in road haulage) to be included in the Government's new Stakeholder Pension scheme. The law applies to firms with five or more employees that do not already have their own pension scheme. Employers have six months in which to provide employees with access to a suitable scheme from one of a list of pension providers offering stakeholder pensions, otherwise they face fines of up to £50,000. Individual employees can contribute up to £3,600 per year by way of a tax-free deduction from their pay.

VETTING OF E-MAILS

The Telecommunications (Lawful Business Practice) (Interception of Communications) Regulations – the so-called 'snoopers' charter – which came into force on 24 October 2000 permits employers to read e-mails sent and received by employees on company computers and monitor their use of their internet access to determine whether there has been unlawful and/or improper use of the firm's communication system. It is implicit that the employer is required to make 'reasonable efforts' to inform staff that communications through a firm's computers may be monitored.

Existing law gives firms powers to dismiss staff for misconduct, which would include abuse of any communications equipment (eg excessive use of the telephone for private calls as well as misuse of e-mail for sending smutty messages or the internet for accessing pornographic sites).

THE EU VIBRATION DIRECTIVE

Millions of workers in the mining, building, manufacturing and transport industries and in agriculture are exposed to mechanical vibrations which can lead to health problems such as loss of feeling in the fingers, pains in the shoulders and limbs, back problems, slipped disks and premature degeneration of the spinal column. Quantifying the precise damage is very difficult.

Conscious of these problems, the Council of the European Union published a Directive (Directive 2002/44/EC, of 25 June 2002) on the minimum health and safety requirements regarding the exposure of workers to the risks arising from physical agents (vibration). The so-called 'Physical Agents Directive' will, principally, require road hauliers to:

- Carry out risk assessments (and measurements) to determine whether their drivers are exposed to vibration from mechanical equipment (which will include road vehicles) on a sufficient scale to come within scope of the specified limits of the Directive.
- Change equipment (or face prosecution if they do not and their drivers become ill or infirm from vibration related causes) if it is found to cause levels of vibration above the set limit.

- Conform to specified exposure limit values for:
  - hand-arm vibrations, and
  - whole body vibrations.

A five year transitional period will be allowed from 6 July 2005 (where work equipment was in use before 6 July 2007).

With hand–arm vibration syndrome now reported to be a serious and widespread occupational disease affecting hundreds of thousands of people in Britain, new regulations to control its effects in the workplace have been introduced, coming into force from 6 July 2005. *The Control of Vibration at Work Regulations 2005* (SI 2005 No. 1093) require employers to take action to prevent their employees from developing diseases caused by exposure to vibration at work from equipment, vehicles and machines. While the risks are obviously greater in vehicle repair and bodybuilding workshops, for example, where power tools are used, nevertheless it is a fact that goods vehicle drivers can also be adversely affected by vibration and it is the employer's duty to assess such risks and take appropriate action to avoid them. Employers could face compensation claims from employees who are exposed to hand–arm vibration and/or whole body vibration causing, for example, Raynaud's disease or 'White Finger Syndrome'.

TRANSPORT TRAINING

NVQs in Transport

Occupational training standards for the road haulage industry are established by the Road Haulage Industry and Distribution Training Council (RHDTC), an employer-led, independent body. These standards underpin a number of National Vocational Qualifications (NVQs) which have been developed for use within the industry as follows:

- NVQ level 1
  Assisting in road haulage and distribution operations
  - covers basic operational skills which may involve helping other more skilled staff.
- NVQ level 2
  Transporting goods by road
  - aimed at the driver of a small or large goods vehicle.

  Distribution and warehousing
  - aimed at someone engaged in the storage and despatch of goods.

  Organizing road transport operations
  - aimed at the traffic planner or clerk who is concerned with traffic operations.
- NVQ level 3
  Supervisory management
  - aimed at supervisors within the road haulage and distribution industry performing road haulage and distribution operations.
- NVQ level 4
  Management
  - aimed at managers in the road haulage and distribution industry.

Continuing Professional Development

Constant changes in business, industry and technology mean that executives need to keep their skills and knowledge up-to-date in order to maximize professional opportunities and their market value. Continuing Professional Development (CPD) schemes, such as those operated by the Chartered Institute of Logistics and Transport in the UK (CILT) for its members, enables individuals to improve their knowledge and competence systematically throughout their working life – a concept supported by the Government.

Professional development is the key to career enhancement. Although it is possible to train as and when suitable courses become available, this approach can leave the individual with key needs unfulfilled. A CPD scheme offers personal development solutions structured to fit around a certain career and lifestyle. Planning career development in this systematic way brings the following developments:

- It makes the best use of training and learning opportunities.
- It encourages the individual to look for alternative ways of meeting training/learning needs.
- It enables the individual to become aware of less obvious knowledge or skills gaps.
- By periodic evaluation of development individuals can judge whether they are meeting their personal objectives.
- Records are kept to track progress.

Learning Activities

Formal learning usually comprises:

- company training;
- study courses including open learning;
- teaching and making presentations;
- attendance at conferences, seminars and meetings related to the individual's work or professional interests.

Informal learning includes:

- reading;
- networking;
- mentoring and coaching;
- project work.

CPD places emphasis on helping individuals to plan and progress their professional future. Normally, schemes focus on outputs without prescribing a minimum number of hours or particular courses to attend; professional persons are expected to take responsibility for their own learning. The CILT, for example, helps members to identify opportunities and provides accreditation of their learning log.

Further details of the CILT and its CPD scheme can be obtained from:

Professional Standards Section
The Chartered Institute of Logistics and Transport in the UK
PO Box 5787
Corby
Northants NN17 4XQ
E-mail: enquiry@iolt.org.uk
Website: www.iolt.org.uk

12 Goods Vehicle Weights and Dimensions

European and British law relating to the weights and dimensions of goods vehicles and trailers is extremely complex. Much of it is difficult for operators to comprehend and apply, and, even worse, for their drivers who may, when loading, have to make on-the-spot decisions which could later prove to be wrong, thereby breaching the law and bringing possible prosecution for themselves and their employers. We have seen many weight and dimensional changes in recent years which have been documented in previous editions of this *Handbook*.

Interest is still being shown in 2004 for increased overall lengths for vehicle and trailer combinations up to 25.25 metres. Currently vehicles at this length can operate legally in Sweden and Finland.

The maximum permitted dimensions and weight limits for goods vehicles and trailers in Great Britain are set out in *The Road Vehicles (Construction & Use) Regulations 1986* (C&U regulations), as amended, and *The Road Vehicles (Authorised Weight) Regulations 1998* which introduced from 1 January 1999 the following specific weight increases:

- two-axle rigids (including buses) increase from 17 to 18 tonnes;
- three-axle articulated buses increase from 27 to 28 tonnes;
- four-axle combinations increase from 35 to 36 tonnes;
- four-axle articulated vehicles fitted with road-friendly suspension on the drive axle increase from 35 to 38 tonnes;
- five-axle combinations increase from 38 to 40 tonnes;

- six-axle (3+3) combinations with road-friendly suspension on the motor vehicle and trailer increase to 41 tonnes;
- five-axle (3+2) articulated vehicles at 44 tonnes permitted for the carriage of 40-foot ISO containers on international combined transport operations;
- drive axles at 11.5 tonnes for all the above except 41-tonne combinations.

These changes are detailed in this chapter, along with existing C&U regulation limitations on vehicle weights which remain in force. Information on weights and dimensions for vehicles carrying abnormal indivisible loads under other legislation is to be found in Chapter 22.

Since 1 February 2001, 44-tonne lorries have been permitted for general haulage operations within the United Kingdom. A 10.5 tonne maximum drive axle weight applies to these vehicles, and the maximum rate of vehicle excise duty for a 44-tonne, 3+3 combination is reduced from 1 December 2001 to £1,200 per annum (£700 if a Reduced Pollution Certificate is in force for the vehicle).

European limits on vehicle weights and dimensions are to be found in EC Directive 96/53/EC of 25 July 1996 which repeals previous directive 85/3/EEC on this subject and its various amendments.

DEFINITIONS

The following definitions apply when considering the lengths, widths and weights of goods vehicles and trailers:

Overall length – the distance between the extreme forward and rearward projecting points of the vehicle/trailer inclusive of all parts, but excluding load-securing sheets and flexible coverings, receptacles for customs seals and tailboards (provided they are not supporting the load in which case they are included in the overall length). In the case of drawbar trailers overall length *excludes* the length of the coupling.

Overall width – the distance between the extreme projecting points on each side of the vehicle/trailer inclusive of all parts but excluding driving mirrors, distortion of the tyres caused by weight, receptacles for customs seals, load-securing sheets and flexible coverings.

Weight – the maximum gross (or design) weight (gvw) for a vehicle/trailer is that at which it has been designed to operate. The permissible maximum weight (pmw) for a vehicle/trailer is the limit set by law (and is shown on the 'Ministry' plate attached to the vehicle) which must not be exceeded on the road in Great Britain.

LENGTH

Rigid Vehicles

The maximum overall length permitted for rigid vehicles is 12 metres.

Articulated Vehicles

For certain articulated vehicles the maximum permitted length is 16.5 metres (see also below), provided the combination can turn within minimum and maximum swept inner and outer concentric circles of 5.3 metres radius and 12.5 metres radius respectively (see Figure 12.1); otherwise the maximum permitted length is 15.5 metres. The swept circle requirements do not apply to:

- low loader or step-frame low-loader combinations;
- car transporters;
- articulated vehicles constructed to carry indivisible loads of exceptional length;
- articulated vehicles with semi-trailers built or converted to increase their length prior to 1 April 1990;
- articulated vehicles not exceeding 15.5 metres overall length;
- articulated vehicles when an axle of the trailer is raised to aid traction.

Articulated vehicles first used since 1 June 1998 which are fitted with a lift axle must meet the turning circle requirement (described above) both with and without all the wheels in contact with the ground.

For the purposes of enforcement of the turning circle requirements, VOSA has notified vehicle manufacturers that it will take a notional measurement from the kingpin to the centre-line of the semi-trailer bogie. Where such a dimension does not exceed 7.8 metres the combination will be assumed to comply. Where this dimension exceeds 7.8 metres VOSA reserves the right to demand a turning circle demonstration on a steering pad.

The maximum overall length for an articulated vehicle incorporating a low-loader semi-trailer (but not a step-frame semi-trailer) is 18 metres. Such vehicles do not have to meet the turning circle requirements described above.

Where an articulated vehicle is designed to carry indivisible loads of exceptional length there is no length restriction (an indivisible load means 'a load which cannot without undue expense or risk of damage be divided into two or more loads for the purpose of conveyance on a road').

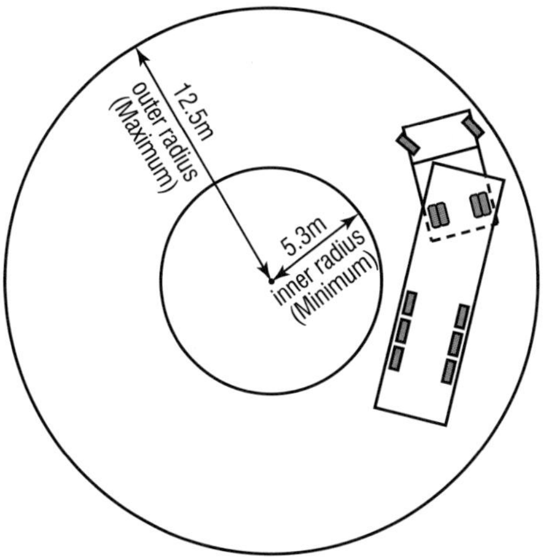

Figure 12.1 *The maximum and minimum outer and inner swept circles within which a 16.5 metre long articulated vehicle must be able to turn*

Vehicle and Trailer Drawbar Combinations (Road Trains)

When a rigid motor vehicle is drawing a trailer the maximum overall length for the combination is 18.75 metres, subject to certain other minimum dimensional requirements being met (see below).

The 18.75 metre maximum length for road train combinations incorporates:

- a maximum loadspace of 15.65 metres to be shared between the two bodies;
- a minimum coupling dimension of 0.75 metres (to provide a 16.4 metre 'envelope' of load and coupling space); and
- a minimum cab length of 2.35 metres (see Figure 12.2).

NB: These dimensions do not apply to any combination which does not exceed 18 metres overall length – this will apply until 2006 by which time any road train exceeding the above dimensions will have to be withdrawn.

When a trailer is designed for carrying indivisible loads of exceptional length the length of the drawing vehicle must not exceed 9.2 metres and the whole combination must not exceed 25.9 metres.

When two or more trailers are drawn the overall length of the combination must not exceed 25.9 metres unless an attendant is carried and two days' notice is given to the police. When two trailers are drawn within the 25.9 metre limit mentioned here (ie only legally permissible with a vehicle classed as a motor tractor or locomotive), only one of the trailers may exceed an overall length of 7 metres. When three trailers are drawn (ie only legally possible with a vehicle classed as a locomotive) none of the trailers may exceed a length of 7 metres.

The limits do not apply when a broken-down vehicle (which is then legally classed as a trailer) is being towed.

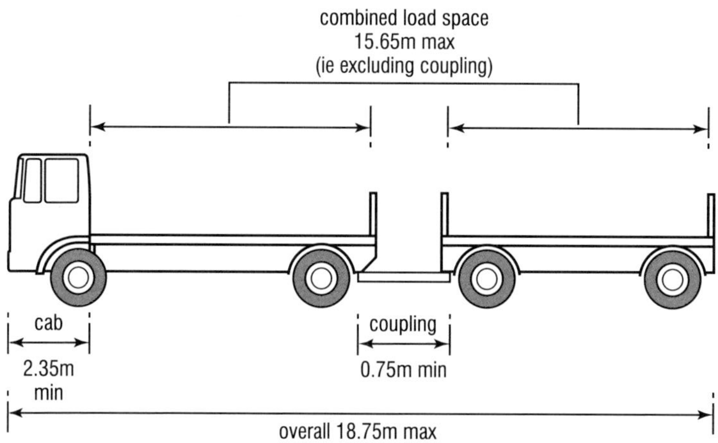

Figure 12.2 *The maximum dimensions for drawbar vehicle combinations*

Trailers

The maximum length for any drawbar trailer which has four or more wheels and is drawn by a vehicle which has a maximum gross weight exceeding 3500 kg, is 12 metres. The same 12-metre maximum length limit also applies to agricultural trailers.

NB: Although the maximum individual lengths for both rigid drawing vehicles and trailers are 12 metres as stated, two such maximum length units obviously cannot be combined within the overall 18.75 metre limit for drawbar combination described above.

The maximum permitted length for all other drawbar trailers is 7 metres.

Composite trailers (see p 322) having at least four wheels and drawn by a goods vehicle over 3500 kg permissible maximum weight or by an agricultural vehicle, may be up to 14.04 metres long.

Articulated Semi-Trailers

The maximum permitted length for certain articulated vehicles is 16.5 metres. This applies where such vehicles include a semi-trailer with a distance from the centre-line of the kingpin to the rear of the trailer which does not exceed 12 metres and where the distance from the kingpin (or foremost kingpin if more than one) to any point on the front of the trailer does not exceed 2.04 metres (4.19 metres for car transporters) – see Figure 12.3.

Articulated semi-trailers built since 1 May 1983 are limited to a maximum length of 12.2 metres. There is no specified length limit for semi-trailers built prior to this date. In measuring the 12.2 metre dimension no account need be taken of the thickness of front or rear walls or any parts in front of the front wall or behind the rear wall or closing device (ie door, shutter, etc). The thickness of any internal partitions must be included in the length measurement. This means effectively that the dimension relates only to load space between the front and rear walls (Figure 12.4).

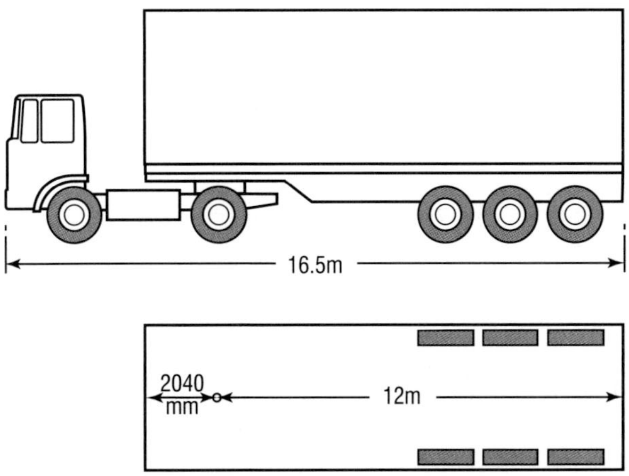

Figure 12.3 *The maximum dimensions for articulated vehicles and semi-trailers*

The 12.2 metre length limit for semi-trailers as described does not apply to a trailer which is normally used on international journeys, part of which are outside the UK. Similarly, articulated vehicles operating within the 15.5 metre limit on international journeys do not have to meet the turning circle requirements described on p 301 for 16.5 metre long vehicles.

The 45 ft Container Problem

The carriage of standard 45 ft pallet-wide ISO-type shipping containers cannot be carried out legally on maximum-length articulated vehicles on account of the limitation on swing radius as set out in Community Directive 96/53/EC. This was particularly worrying for European intermodal road hauliers who needed to use the 45 ft pallet-wide container, the only size longer than 40 ft that would enable them to compete effectively against the traditional road hauliers using 13.60-metre-long semi-trailers. However, this problem has been overcome by use of the Geest (ie Geest North Sea Line b.v.)-designed 'Euro-Casting' fitted to the front end of a 45 ft container, which successfully meets the crucial swing radius of 2.04 metres – measured from the centre of the kingpin to any point at the front end of the container or skeletal semi-trailer – as well as the 12.0-metre length requirement measured from the centre of the kingpin to the rearmost point of the semi-trailer (see Figure 12.3).

Measurement of Length

In measuring vehicle or trailer length account must be taken of any load-carrying receptacle (eg demountable body or container) used with the vehicle. Excluded from the overall length measurement are such things as rubber or resilient buffers and receptacles for customs seals. In the case of drawbar combinations the length of the drawbar is excluded from overall length calculations. With dropside-bodied vehicles the length of the tailboard in the lowered (ie horizontal) position is excluded unless it is supporting part of the load in which case it must be included in the overall length measurement and for the purposes of establishing overhang limits (see below).

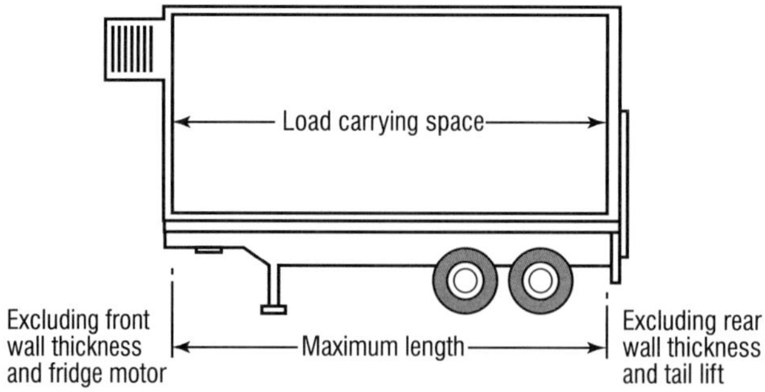

Figure 12.4 *Measurement of the maximum length dimension for pre-1983 and other semi-trailers*

Overhang

Overhang is the distance by which the body and other parts of a vehicle extend beyond the rear axle. The maximum overhang permitted for rigid goods vehicles (ie motor cars and heavy motor cars) is 60 per cent of the distance between the centre of the front axle and the point from which the overhang is to be measured. The point from which overhang is measured is, in the case of two-axled vehicles, the centre line through the rear axle, and in the case of vehicles with three or more axles two of which are rear axles, 110 mm to the rear of the centre line between the two rear axles (Figure 12.5).

This regulation does not apply to vehicles used solely in connection with street cleansing; the collection or disposal of refuse; the collection or disposal of the contents of gullies or cesspools; works trucks; or tipping vehicles, provided the total overhang does not exceed 1.15 metres (3 ft 9 in approx). There is no specified overhang limit on trailers.

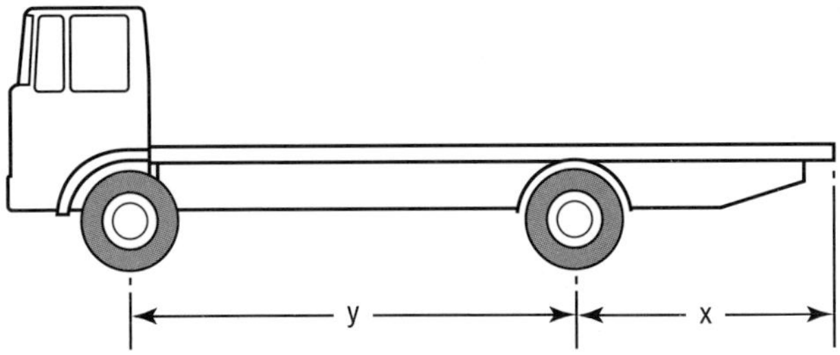

Two-axled vehicles: Overhang 'x' must not exceed 60 per cent of length 'y'

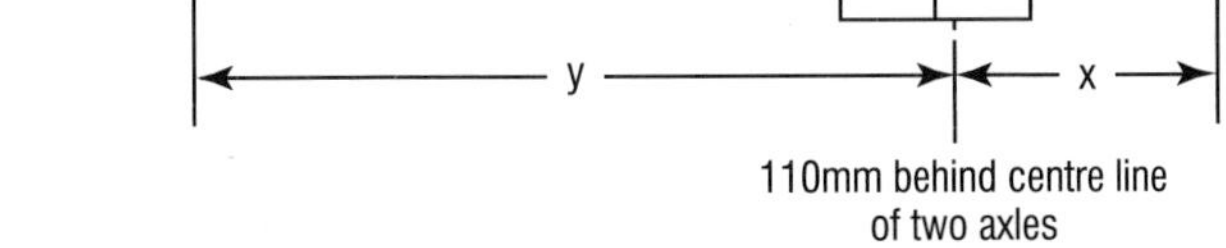

Vehicles with three axles or more: Overhang 'x' measured from 110mm behind the centre-line of the two rear axles must not exceed 60 per cent of the length of 'y' which is the distance beween the centre-line of the front wheel to the centre-line of the two rear axles plus 110mm

Figure 12.5 *How to measure overhang on vehicles with two axles, three axles and more (this measurement applies equally to two- and three-axle tractive units)*

WIDTH

Motor Vehicles

The overall width of motor tractors, motor cars and heavy motor cars (most goods vehicles are included in these classifications) must not be more than 2.55 metres (see note below about refrigerated vehicles) and the maximum width of locomotives must not be more than 2.75 metres.

Trailers

The maximum permissible width for trailers is 2.55 metres provided the drawing vehicle has a maximum permissible weight exceeding 3500 kg. If the towing vehicle is below this weight, the width of the trailer must not exceed 2.3 metres.

Refrigerated Vehicles

The maximum permitted width for refrigerated (ie reefer) vehicles, semi-trailers and drawbar trailers is 2.60 metres provided that the thickness of the side walls (inclusive of insulation) is at least 45 mm. For the purposes of this regulation 'refrigerated vehicle' means a vehicle (or trailer) specially designed to carry goods at low temperature.

HEIGHT

There are currently no legal maximum height limits for goods vehicles or for loads in Britain (see below) but these are, obviously, governed by the height of bridges on the routes on which the vehicles are operated. For general information, the minimum height of bridges on motorways is normally 16 ft to 16 ft 6 in and the maximum heights for buses is 4.57 metres (see also item regarding high loads on p 488).

NB: The Blackwall Tunnel (on the A102(M)), the River Thames crossing to the east of London, is limited to a maximum height of 4 metres northbound and 4.72 metres southbound.

The operator must bear in mind, however, that if he loads vehicles to a height which could cause danger he would be liable to prosecution under the construction and use regulations. Also, if a vehicle were to be loaded to a height whereby the load hit a bridge on the route being used, the operator could be accused under these regulations of using a vehicle on a road for a purpose for which it was so unsuitable as to cause, or to be likely to cause, danger.

EU regulations specify a height limit of four metres but this does not apply in the UK.

Height Limit on Vehicles over 35 Tonnes

The maximum height limit of 4.2 metres on both articulated vehicles and drawbar combinations operating at (ie laden to) weights in excess of 35 tonnes (previously 32.5 tonnes) was abolished from 1 June 1995.

Height Marking and Route Descriptions

Since 1 October 1997 all vehicles with an overall travelling height of more than three metres must have a notice prominently displayed in the driver's cab indicating the actual travelling height of the vehicle, its load or equipment in feet and inches (or in both feet and inches and metres).

The height marking, where this is shown in feet and inches only, must be in letters and figures at least 40 mm tall, But where the height is shown in both imperial and metric measure the figures shown must not differ by more than 50 mm.

As an alternative to the height marking described above, in circumstances where the driver is operating on a particular journey during which he is unlikely to be confronted with any bridge or overhead structure which does not exceed the maximum travelling height by at least one metre, a document may be carried on the vehicle (within easy reach of the driver) which describes the route, or a choice of routes, which the driver must follow in order to avoid any risk of the vehicle, its load or equipment colliding with any bridge or overhead structure.

Where vehicles are fitted with 'high-level equipment' (defined as power-operated equipment) with a maximum height of more than three metres they must have a warning device which gives visible warning to the driver if the height of the vehicle exceeds a predetermined level. This applies to vehicles first used and trailers first made from 1 April 1998.

Bridge Bashing

Bridge bashing remains a major problem despite new legislation on height marking as described above. In 1990, 800 such accidents were recorded, of which 11 resulted in train derailments, and in a recent (1995) instance an owner-driver was left facing a compensation claim for more than £100,000. Some 20 or so key railway bridges have been identified as providing risk of potential disaster, should they be struck by a large vehicle. Infra-red warning systems are to be installed at these key sites, which will trigger alarms when over-height vehicles approach.

The *AA Trucker's Atlas of Britain* (scale 3 miles to 1 inch), recommended by both Railtrack and the RHA, shows 1500 road- and rail-bridge heights. It contains useful advice on how not to be a 'bridge basher' and suggests that by not knowing their vehicle height, drivers risk the consequences of bridge bashing from causing rush-hour chaos to losing their job, or even causing loss of life. It warns that drivers should not hit and run but should report incidents – failure to do so being an offence. Most rail bridges have a sign indicating the Railtrack contact number (ie 020 7928 4616) to be telephoned following contact with a bridge structure and advising that the police should also be notified immediately by dialling 999.

WEIGHT

Maximum permitted weights (the total weight of the vehicle and load, including the weight of fuel, and the driver and passenger if carried) for goods vehicles and trailers depend on their wheelbase, the number of axles, the outer axle spread (the distance between the centre of the wheels on the front and rearmost axles) or the relevant axle spacing in the case of articulated vehicles (see pp 308–309).

All rigid goods vehicles over 3500 kg gross weight, articulated vehicles and trailers over 1020 kg unladen weight should be fitted with a 'Ministry' plate (see p 393) on which is shown, for that vehicle, the maximum permissible axle weights and gross weight (or in the case of articulated vehicles, the combined weight of the tractive unit and trailer) for that vehicle in Great Britain.

The maximum permitted laden weights for different types of vehicle are shown in the following tables:

Rigid Vehicles and Trailers (not fitted with road-friendly suspensions)

	Description	*kg*
1.	Trailer with two closely spaced axles and with a distance between the foremost axle of the trailer and the rearmost axle of the drawing vehicle of at least 4.2 metres	18,000
2.	Trailer with three closely spaced axles and with a distance between the foremost axle of the trailer and the rearmost axle of the drawing vehicle of at least 4.2 metres	24,000
3.	Two-axled *vehicle* with a distance between the foremost and rearmost axles of at least 3.0 metres	17,000
4.	Two-axled *trailer* with a distance between the foremost and rearmost axles of at least 3.0 metres	18,000

Rigid Vehicles (not falling within table above)

	No of axles	*Distance between outer axles*	*kg*
1.	2	less than 2.65 metres	14,230
2.	2	at least 2.65 metres	16,260
3.	3 or more	less than 3.0 metres	16,260
4.	3 or more	at least 3.0 but less than 3.2 metres	18,290
5.	3 or more	at least 3.2 but less than 3.9 metres	20,330
6.	3 or more	at least 3.9 but less than 4.9 metres	22,360
7.	3	at least 4.9 metres	25,000
8.	4 or more	at least 4.9 but less than 5.6 metres	25,000
9.	4 or more	at least 5.6 but less than 5.9 metres	26,420
10.	4 or more	at least 5.9 but less than 6.3 metres	28,450
11.	4 or more	at least 6.3 metres	30,000

Rigid Vehicles (not including articulated tractive units) with Drive Axles Fitted with Twin Tyres and Road-Friendly Suspensions

	No of axles	*Distance between outer axles*	*kg*
1.	2	less than 2.65 metres	14,230
2.	2	at least 2.65 metres	16,260
3.	3 or more	less than 3.0 metres	16,260
4.	3 or more	at least 3.0 but less than 3.2 metres	18,290
5.	3 or more	at least 3.2 but less than 3.9 metres	20,330
6.	3 or more	at least 3.9 but less than 4.9 metres	22,360
7.	3 or more	at least 4.9 but less than 5.2 metres	25,000
8.	3	at least 5.2 metres	26,000
9.	4 or more	at least 5.2 but less than 6.4 metres	the distance in metres between the foremost and rearmost axles multiplied by 5000 and rounded up to the next 10 kg
10.	4 or more	at least 6.4 metres	32,000

Articulated Tractive Units

	No of axles	Minimum distance between outer axles	Intermediate axle weight (kg)	Total laden weight kg
1.	2	at least 2.0 metres	–	14,230
2.	2	at least 2.4 metres	–	16,260
3.	2	at least 2.7 metres	–	17,000
4.	3 or more	at least 3.0 metres	8,390	20,330
5.	3 or more	at least 3.8 metres	8,640	22,360
6.	3 or more	at least 4.0 metres	10,500	22,500
7.	3 or more	at least 4.3 metres	9,150	24,390
8.	3 or more	at least 4.9 metres	10,500	24,390

Articulated Vehicle Weights

Current maximum weights for different classes of articulated vehicles, except those used in combined transport operations (see p 299), based on the number of axles (subject to plated weights and axle spacings) are as shown in the table below:

	Type of articulated vehicle	Maximum weight (kg)
1.	Motor vehicle first used on or after 1 April 1973 and semi-trailer having a total of five or more axles	38,000
2.	Motor vehicle with two axles first used on or after 1 April 1973 and semi-trailer with two axles being used on international transport	35,000
3.	Motor vehicle with two axles first used on or after 1 April 1973 with driving axles having twin tyres and road-friendly suspension, and two-axle semi-trailer	35,000
4.	Motor vehicle and semi-trailer with a total of four axles (not covered in items 2 and 3 above)	32,520
5.	Motor vehicle with two axles first used on or after 1 April 1973 with driving axles having twin tyres and road-friendly suspension, and single-axle semi-trailer	26,000
6.	Motor vehicle with two axles and single-axle semi-trailer (not covered in item 5 above)	25,000

Maximum Permitted Laden Weights for Articulated Vehicles

The maximum permitted laden weight for *complete articulated vehicles* (ie tractive unit and semi-trailer) depending on the number of axles of the tractive unit and the relevant axle spacing (ie rearmost axle of tractive unit to rearmost axle of semi-trailer – see Figure 12.6) is as follows:

Relevant axle spacing (m)		
Where motor vehicle has 2 axles	Where motor vehicle has at least 3 axles	Maximum weight (kg)
at least 2.0	at least 2.0	20,330
at least 2.2	at least 2.2	22,360
at least 2.6	at least 2.6	23,370
at least 2.9	at least 2.9	24,390
at least 3.2	at least 3.2	25,410
at least 3.5	at least 3.5	26,420
at least 3.8	at least 3.8	27,440
at least 4.1	at least 4.1	28,450
at least 4.4	at least 4.4	29,470
at least 4.7	at least 4.7	30,490
at least 5.0	at least 5.0	31,500
at least 5.3	at least 5.3	32,520
at least 5.5	at least 5.4	33,000
at least 5.8	at least 5.6	34,000
at least 6.2	at least 5.8	35,000
at least 6.5	at least 6.0	36,000
at least 6.7	at least 6.2	37,000
at least 6.9	at least 6.3	38,000

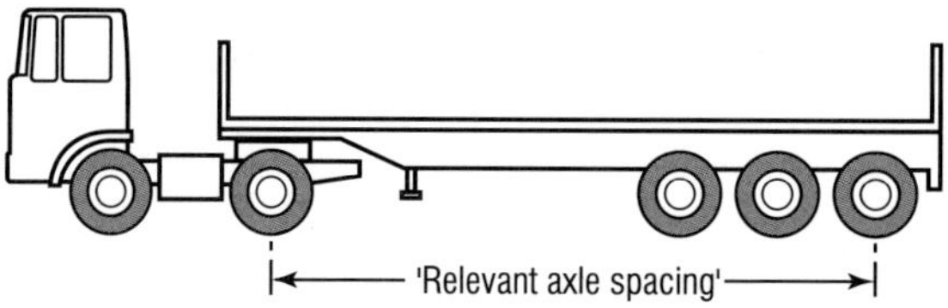

Figure 12.6 *Measurement of axle spacing for articulated vehicles*

Axle Spacing for Articulated Vehicles

For the purposes of determining articulated vehicle weights a dimension referred to as the 'relevant axle spacing' is used. This is defined as 'the distance between the rearmost axle of the drawing vehicle (ie the tractive unit) and the rearmost axle of the semi-trailer' (Figure 12.6).

Lorry and Trailer Combination Weights

Lorry and trailer (ie drawbar or road train) combinations must not exceed 24,390 kg gross weight unless the trailer is fitted with power-assisted brakes which remain operative even when the drawing vehicle's engine is not running, and a brake pressure warning device in the driver's cab in which case the permitted maximum train weight for the combination is 32,520 kg or more depending on specification – see list below.

- Where a drawbar combination was first used on or after 1 April 1973 and complies with the braking requirement mentioned above it may operate at up to 35,000 kg gross train weight.
- Where a four-axle drawbar combination first used on or after 1 April 1973 complies with the braking requirements mentioned above and its driving axles have twin tyres and road-friendly suspension it may operate at up to 35,000 kg.
- Where a drawbar combination with five or more axles first used on or after 1 April 1973 complies with the braking requirements mentioned above and all its driving axles have twin tyres and road-friendly suspension (or do not exceed 8.5 tonnes) it may operate at up to 38,000 kg.
- Where a drawbar combination with at least six axles complies with the braking requirements mentioned above and all its driving axles have twin tyres and road-friendly suspension (or do not exceed 8.5 tonnes) it may operate at up to 44,000 kg in combined transport operations only – see p 312.

Trailer Weights

The maximum laden weight permitted for unbraked trailers is not more than half the unladen weight of the towing vehicle.

Trailers with overrun brakes are limited to a maximum laden weight of 3500 kg unless they were first manufactured before 27 February 1977 in which case their maximum laden weight is 3560 kg (these limits do not apply to agricultural trailers).

The maximum permissible weight for trailers used in drawbar combinations is 18 tonnes provided the trailer has an axle spacing of at least three metres and is plated for 18 tonnes.

Axle and Wheel Weights

Maximum permitted axle and wheel weights (ie the total weight which may be imposed by an axle or a wheel on the road surface when the vehicle is fully loaded) for vehicles is as follows:

Axle/wheel type	*Maximum permitted weight transmitted to road (kg)*
1. Two wheels in line transversely each of which is fitted with a wide tyre or with two pneumatic tyres having the centres of their areas of contact with the road, not less than 300mm apart, measured at right angles to the longitudinal axis of the vehicle	
(a) if the wheels are on the sole driving axle of a motor vehicle (ie not a bus)	10,500
(b) if the vehicle is a bus which has 2 axles and of which the weight transmitted to the road surface by its wheels is calculated in accordance with C & U regulations	10,500
(c) in any other case	10,170
2. Two wheels in line transversely otherwise than as mentioned in item 1	9,200
3. More than two wheels in line transversely	
(a) in the case of a vehicle manufactured before 1 May 1983 if the wheels are on one axle of a group of two closely spaced axles or on one of three adjacent axles as mentioned in the C & U regulations	10,170
(b) in the case of a vehicle manufactured on or after 1 May 1983	10,170
(c) in any other case	11,180
4. One wheel not transversely in line with any other wheel	
(a) if the wheel is fitted as described in item 1	5,090
(b) in any other case	4,600

Wheel and Axle Weights for Motor Vehicles and Trailers not included above:

5. More than two wheels transmitting weight on to a strip of the road surface on which the vehicle rests contained between two parallel lines at right angles to the longitudinal axis of the vehicle	
(a) less than 1.02m apart	11,180
(b) 1.02m or more apart but less than 1.22m apart	16,260
(c) 1.22m or more apart but less than 2.13m apart	18,300
6. Two wheels in line transversely	9,200
7. One wheel, where no other wheel is in the same line transversely	4,600

NB: An increase in maximum drive axle weights from 10.5 tonnes to 11.5 tonnes has been permitted for certain articulated vehicles since 1 January 1999 – see p 313.

Maximum Permitted Weights for Vehicles with Two Closely Spaced Axles

	Description of vehicle	*Maximum permitted weight for two closely spaced axles (kg)*
1.	Motor vehicle or trailer with a distance between the two closely spaced axles in either case being less than 1.3 metres	16,000
2.	Motor vehicle with a distance between the two closely spaced axles of at least 1.3 metres, or a trailer with a distance between the two closely spaced axles of at least 1.3 but less than 1.5 metres (and not being a vehicle described in item 3 or 4 below)	18,000
3.	Motor vehicle with a distance between the two closely spaced axles of at least 1.3 metres and with every driving axle fitted with twin tyres and either road-friendly suspension or neither of them with an axle weight above 9500 kg	19,000
4.	Trailer with two closely spaced axles driven from the motor vehicle, and fitted with twin tyres and either road-friendly suspension or neither of them with an axle weight over 9500 kg	19,000
5.	Trailer with a distance between the two closely spaced axles of at least 1.5 and less than 1.8 metres	19,320
6.	Trailer with a distance between the two closely spaced axles of at least 1.8 metres	20,000

Maximum Permitted Weights for Vehicles with Three Closely Spaced Axles

	Description of vehicle	*Maximum permitted weight for three closely spaced axles (kg)*
1.	Vehicle where smallest distance between any two of the three closely spaced axles is less than 1.3 metres	21,000
2.	Vehicle where smallest distance between any two of the three closely spaced axles is at least 1.3 metres and at least one of the axles does not have air suspension	22,500
3.	Vehicle where smallest distance between any two of the three closely spaced axles is at least 1.3 metres and all three axles have air suspension	24,000

Vehicles for Combined Transport

Swap-body and container-carrying goods vehicles running to and from rail terminals on intermodal operations may operate at up to 44 tonnes gross weight provided they are:

- articulated vehicles and drawbar combinations equipped with:
  - at least six axles; and
  - road-friendly suspensions (or having no axle exceeding 8.5 tonnes); and
- articulated vehicles comprising specially built bimodal semi-trailers (ie capable of running on road or rail).

For these purposes, containers and swap-bodies are defined as being receptacles at least 6.1 metres long designed for repeated carriage of goods and for transfer between road and rail vehicles.

To comply with the law, the driver must carry with him documentary evidence to show that the swap-body or container load is on its way to a rail terminal (the document must show the name of the rail terminal, the date of the contract and the parties to it), or is on its way back from a rail terminal (in which case the document must show the terminal and the date and time that the unit load was collected). There is no restriction on the distance that may be travelled to or from a rail terminal for the purposes of complying with this legislation.

Braking Efficiencies

It is a requirement that to achieve the maximum weights set out in these tables vehicles and vehicle combinations must meet specified braking efficiencies, otherwise the maximum permitted weight in each case is as follows:

- Rigid 4-wheel vehicle 14,230 kg
- Rigid 6-wheel vehicle 20,330 kg
- Rigid vehicle with more than 6 wheels 24,390 kg
- Articulated vehicle with trailer having less than 4 wheels 20,330 kg
- Articulated vehicle with trailer having 4 wheels or more 24,390 kg

AUTHORIZED WEIGHT REGULATIONS

UK lorry weights have been aligned with EC Directive 96/53/EC since 1 January 1999 under The Road Vehicles (Authorized Weight) Regulations 1998 (SI 1998 No 3111) which specify maximum authorized weights for vehicles (ie weights not to be exceeded in any circumstances). These regulations include the widely publicized new weight limits for driving axles (ie 11.5 tonnes), for rigid vehicles (ie 18 tonnes) and for articulated vehicles and vehicle and trailer combinations (ie 40 and 41 tonnes, respectively) subject to specified conditions.

NB: These regulations were amended in 2000 (SI 3224/2000) to permit 44-tonne operation, and again in 2001 (SI 1125/2001) to correct errors in the former regulations.

The pre-existing vehicle and axle weight limits set out in the Road Vehicles (Construction and Use) Regulations 1986 (as amended) as described above remain in force largely unchanged, apart from three particular provisions listed below which were implemented when these regulations were amended in consequence of the making of the new authorized weight regulations.

- An increase in the maximum total weight of all trailers drawn at any one time by a locomotive.
- An increase in the maximum weight permitted for uncompensated steering axle from 7120 kg to 8500 kg.
- Amendments to the provisions relating to vehicles used in combined transport operations.

Maximum Authorized Weights for Rigid Vehicles and Tractor Units

Description of vehicle	*Number of axles*	*Maximum authorized weight (kg)*
Rigid motor vehicle	2	18,000
Tractor unit	2	18,000
Trailer (not a semi-trailer or centre-axle trailer)	2	18,000
Trailer (not a semi-trailer or centre-axle trailer)	3 or more	24,000
Rigid motor vehicle	3	25,000
Rigid motor vehicle(1)	3	26,000
Tractor unit	3 or more	25,000
Tractor unit(1)	3 or more	26,000
Articulated bus	any number	28,000
Rigid motor vehicle	4 or more	30,000
Rigid motor vehicle(1)	4 or more	32,000

(1) Applies to vehicles with either:
(a) the driving axle, if it is not a steering axle fitted with twin tyres and roadfriendly suspension; or
(b) each driving axle fitted with twin tyres and not exceeding 9500 kg.

Maximum Weight determined by Axle Spacing

The authorized weight (in kilograms) for the vehicles shown below shall equal the product of the distance measured in metres between the foremost and rearmost axles of the vehicle multiplied by the factor specified in the third column and rounded up to the nearest 10 kg, if that number is less than the 'maximum authorized weight' (ie in accordance with the table immediately above).

Description of vehicle	*Number of axles*	*Factor to determine maximum authorized weight*
Rigid motor vehicle	2	6000
Tractor unit	2	6000
Trailer which is not a semi-trailer or centre-axle trailer	2	6000
Rigid motor vehicle	3	5500
Tractor unit	3 or more	6000
Trailer which is not a semi-trailer or centre-axle trailer	3 or more	5000
Rigid motor vehicle	4 or more	5000
Articulated bus	any number	5000

Maximum Authorized Weights for Vehicle Combinations

Description of combination	*Number of axles*	*Maximum authorized weight (kg)*
Articulated vehicle	3	26,000
Rigid motor vehicle towing a trailer	3	22,000
Rigid motor vehicle towing a trailer(1)	3	26,000
Articulated vehicle	4	36,000
Articulated vehicle(2)	4	38,000
Rigid motor vehicle towing a trailer	4	30,000
Rigid motor vehicle towing a trailer(1)	4	36,000
Articulated vehicle	5 or more	40,000
Rigid motor vehicle towing a trailer	5 or more	34,000
Rigid motor vehicle towing a trailer(1)	5 or more	40,000
Articulated vehicle(3)	6 or more	41,000
Rigid motor vehicle towing a trailer(1 & 3)	6 or more	41,000
Rigid motor vehicle towing a trainer (1, 3 & 4)	6 or more	44,000

(1) Applies to vehicles that have a distance between the rear axle of the motor vehicle and the front axle of the trailer of not less than 3 m.
(2) Applies to combinations where:
(a) the combination is a 2-axle tractor unit and 2-axle semi-trailer;
(b) the weight of the tractor unit comprised in the combination does not exceed 18,000 kg;
(c) the sum of the axle weights of the semi-trailer does not exceed 20,000 kg; and
(d) the driving axle is fitted with twin tyres and road-friendly suspension.
(3) Applies to vehicles where:
(a) the weight of each driving axle does not exceed 10,500 kg; and
(b) either
(i) each driving axle is fitted with twin tyres and road friendly suspension; or
(ii) each driving axle that is not a steering axle is fitted with twin tyres and does not exceed 8500 kg;
(c) each axle of the trailer is fitted with road-friendly suspension; and
(d) each vehicle comprised in the combination has at least 3 axles.
(4) Applies to vehicles powered by low pollution engines (ie gas or diesel complying with emission requirements).

Weight by Reference to Axle Spacing

The maximum authorized weight in kilograms for an articulated vehicles 'shall be the product of the distance measured in metres between the king-pin and the centre of the rearmost axle of the semi-trailer multiplied by the factor specified and rounded up to the nearest 10 kg, if that weight is less than the authorized weight' (ie in accordance with table above).

Description of vehicle combination	*Number of axles*	*Factor to determine maximum authorized weight*
Articulated vehicle	3 or more	5500

Maximum Authorized Axle Weights

Description of axle	*Maximum authorized weight (kg)*
Single driving axle	11,500
Single non-driving axle	10,000
Driving tandem axle	18,000
Driving tandem axle(1)	19,000
Non-driving tandem axle	20,000
Tri-axle	24,000

(1) Applies to axles where:
(a) the driving axle is fitted with twin tyres and road-friendly suspension; or
(b) each driving axle is fitted with twin tyres and no axle has an axle weight exceeding 9500 kg.

Where maximum weights are less than those in the fourth table above.

Description of axle	*Specified dimension*	*Length metres*	*Maximum authorized weight (kg)*
Driving tandem axle	Distance between the 2 axles	Less than 1	11,500
Driving tandem axle	Distance between the 2 axles	Not less than 1 but less than 1.3	16,000
Non-driving tandem axle	Distance between the 2 axles	Less than 1	11,000
Non-driving tandem axle	Distance between the 2 axles	Not less than 1 but less than 1.3	16,000
Non-driving tandem axle	Distance between the 2 axles	Not less than 1.3 but less than 1.8	18,000
Tri-axle	Distance between any 1 axle and the nearer of the other 2 axles	1.3 or less	21,000

Overall Weight Limits

The total weight of the load on a vehicle, together with the weight of the vehicle itself, must not exceed the maximum permitted weight for each individual axle or for the vehicle.

Load Movements – C&U Regulations versus Special Types Order

The majority of general haulage loads moved in this country fall within scope of the weights, lengths and width limitations set out in *The Road Vehicles (Construction and Use) Regulations 1986* (C&U) as described in this chapter. However, an increasing number of long, heavy and/or wide loads are being transported which, due to C&U restrictions on load lengths, widths and weights, are required to be moved under the abnormal indivisible loads (AIL) provisions of the *Road Traffic Act 1988* Section 44(1) detailed in *The Motor Vehicle (Authorization of Special Types) General Order 2003* (STGO) as described in Chapter 22 of this *Handbook*. In particular, it should be noted that maximum vehicle and load lengths permitted under Regulation 7 of C&U must not be confused with those authorized under the STGO.

WEIGHT OFFENCES

It is an offence on the part of both the driver and the vehicle operator (ie the driver's employer) to operate a goods vehicle on a road laden to a weight above that at which it has been plated by VOSA (ie above the maximum permitted gross and individual axle weights) and both are liable to prosecution. Such offences are 'absolute' in that once the actual overweight has been established the fact that it was a deliberate action to gain extra revenue or purely accidental, unintentional, outside the driver's or vehicle operator's control or loading was in a place where no weighing facilities existed, is of no consequence in defending against the charge.

Currently, the policy adopted by VOSA is that if a vehicle exceeds its weight limits by more than 5 per cent (up to a maximum of one tonne) it will be prohibited from proceeding on its journey until the weight is reduced to within legal limits – see below. Where an overload exceeds 10 per cent of the vehicle's legal limits or a maximum of one tonne, a prosecution will follow with heavier penalties being imposed on conviction. Trading standards officers may not take such a lenient view in their dealing with overloaded vehicles.

Defence

There is no defence of 'due diligence' against charges of overloading, something for which the Road Haulage Association has been campaigning, but it is a defence under the Road Traffic Act 1988 (section 42) to prove that at the time the contravention was detected the vehicle was proceeding to the nearest available weighbridge or was returning from such weighbridge to the nearest point at which it was reasonably practicable to remove the excess load. There have also been instances reported where successful defences have been made against conviction where the operator was able to show that he had no way of knowing or controlling the weight placed on a vehicle.

It is a further defence, where the weight exceeds maximum limits by not more than 5 per cent, to prove that the weight was within legal limits at the time of loading the vehicle and that no person had since added anything to the load.

Penalties

Overloading offences are looked upon very seriously by the enforcement authorities and by the courts and very heavy penalties are imposed on offenders. In addition to punitive fines (currently a maximum of £5,000 per offence – an overloaded vehicle could result in a number of individual offences related to gross and axle weights), the operator risks losing his 'O' licence and the driver could put his lgv driving entitlement in jeopardy.

Prohibition of Overweight Vehicles

Any vehicle on a road found by a vehicle examiner to be overloaded to the extent that it could endanger public safety will be ordered off the road immediately. The necessary powers to enable this step to be taken are included in section 70 (2) and (3) of the Road Traffic Act 1988 and they empower an authorized officer to prohibit the driving of a goods vehicle on a road if after having it weighed it appears to him that the vehicle exceeds the relevant weight limits imposed by the Construction and Use Regulations, and as a result would be an immediate risk to public safety if it were used on a road. The officer may be one of VOSA's examiners or a specially authorized weights and measures inspector or a police constable.

A prohibition notice, form TE160, will be issued to the driver of a vehicle found to be overweight and it is the driver's responsibility to remove the excess weight to his own satisfaction and clear the TE160 before proceeding on his journey. The penalty for ignoring a prohibition notice is a fine of up to £5,000.

Official Weighing of Vehicles

Under the Road Traffic Act 1988 an authorized officer can request the person in charge of a vehicle to drive it to a weighbridge to be weighed. If the journey is more than five miles and on arrival the vehicle is found to be within the legal limit then the vehicle owner can claim for the loss involved in making the journey. It is an offence to refuse to go to a weighbridge if requested (maximum fine currently £5,000). Once a vehicle has been weighed the official in charge should give the driver a 'certificate of weight' (whether it is overloaded or not) and the vehicle will be exempt from further requests for weighing while carrying the same load on that journey.

The two following Codes of Practice are currently in use by VOSA for vehicle weighing. However, a new combined Code covering all types of weighing, including the use of weigh pads, was due to be published by the end of 2000.

Code of Practice for Conventional Weighing

The procedure will be as follows:

- Normally only one authorized officer will check-weigh a vehicle.
- The driver and any passengers should remain in the vehicle during weighing as they are part of the weight transmitted to the road.
- When requested the driver should move the vehicle smoothly on to the weighbridge plate.
- The engine should be switched off but left in gear during weighing. When a vehicle's individual axles are being check-weighed both the hand and foot brakes should be released.
- If the vehicle is too long to go on the weighing plate or is over the weight capacity of the machine it may require two or more weighings to get a total weight for the vehicle. (See note below.)
- The driver may be required to have the axle weights checked, and in positioning the vehicle he should carefully follow instructions given by the authorized officer.
- A certificate of weight will be issued for each vehicle weighed and will be handed to the driver. In some cases, where circumstances permit, he may be invited to see the weight recorded on the indicator. The certificate of weight exempts a vehicle from being weighed again on the same journey *with the same load* if stopped at another weight check. The driver should give the certificate of weight to his employer as soon as he reaches his base.
- The driver should also note that previous overloading offences may be detected by checks on weighbridge records.

NB: Operation of Weighing Equipment
Weighbridges will be operated only by or under the supervision of a duly authorized officer fully conversant with the operation of the weighing equipment and with approved methods of weighing. Weighing by the method known as 'double-weighing' will only be carried out by a Trading Standards Officer whose professional qualifications and experience will enable him to give expert evidence as to the accuracy of this weighing procedure; or by a duly authorized officer at sites which have been examined and approved as suitable for double-weighing by the local Trading Standards Officer, and who will if necessary be available to give expert evidence in court. NOTE: double-weighing occurs when the results of two or more individual weighings of a vehicle's axles are summated to produce the total gross weight of the vehicle.

Dynamic Weighing

The use of dynamic axle weighing machines is permitted under regulations (The Weighing of Motor Vehicles (Use of Dynamic Axle Weighing Machines) Regulations 1978), which specify that an enforcement officer can require a vehicle to be driven across the weighing platform of a machine for this purpose. The permitted weights for the vehicle are measured to within plus or minus 150 kg for each axle and within plus or minus 150 kg multiplied by the total number of axles to determine the tolerance on the total vehicle weights.

The Department of Environment, Transport and the Regions has a 'Code of Practice for Weighing Goods Vehicles on Dynamic Axle Weighers' (see below), which must be followed by enforcement officers to secure a successful prosecution for vehicle overloading. This Code supplements the regulations and provides enforcement officers as well as vehicle operators and drivers with information on the correct setting up and operational use of weighing machines and on periodical (ie maximum six-monthly) accuracy tests which must be carried out. Among other things, the Code specifies standards for the approach and exit from the concrete apron of the equipment (ie plus or minus 3 mm for a distance of 8 metres on each side of the weighbeam, and the finished surface must not deviate from the level by more than 3 mm under a 3-metre long straight edge).

Portable Weighers

There have been reports of Trading Standards Officers in some areas using portable axle weigh-pads as a means of checking vehicle/axle weights and have brought overloading prosecutions against operators based on such weighings. There is considerable doubt about the accuracy of such machines and the Road Haulage Association has expressed its concern about their use.

Code of Practice on Dynamic Weighing

The following procedure should be adopted for weighing vehicles:

- The vehicle to be weighed is to be stopped a minimum distance of 6 metres from the weighbeam on a level approach; avoid stopping on uneven ground, eg with one or more wheels on the kerb.
- Possible errors caused by surges in the liquid load of a single compartment unbaffled tanker should be taken into account.
- The enforcement officer operating the console is to recheck and reprint the Low and High indication, set the equipment in the dynamic mode and, where appropriate, set the direction selector switch to suit the vehicle approach. He must also press the totalizer button to ensure that any residual information in the totalizer has been cleared.
- Where vehicles under three tonnes unladen weight are to be checked, and the equipment has a Low Weight mode, this button should be depressed before weighing commences.
- An enforcement officer should then instruct the driver of the vehicle to drive across the weighbeam at a steady speed not exceeding 2.5 miles per hour; during this run the driver must neither accelerate nor use his brake. This can normally be achieved by engaging lowest forward gear and driving at tick-over speed across the weighbridge.
- The vehicle is to be observed at all times during the weighing procedure to ensure that a consistent speed is maintained. In the case of a foreign driver who cannot speak English it is desirable for him to be guided by an enforcement officer walking alongside the vehicle.
- If at any time during the weighing a driver accelerates above the permitted speed, thus causing a red printout, or no printout, to be registered, or if he uses his brakes to cause a sharp deceleration, the weighing should be disregarded and the vehicle weighed again until a satisfactory weighing and an all-black printout is achieved.
- At the conclusion of a satisfactory weighing, with an all-black printout showing axle weights, the operator should press the appropriate button to

give a summation of the axle weights. The registration number of the vehicle should then be written on the printout adjacent to the readout.

- Each weighing is to be followed by a zero and high indication check which will be shown on the print-roll.
- Following weighing of the vehicle, normal Road Traffic Act procedures are to be followed in the light of the Weighing of Motor Vehicles (Use of Dynamic Axle Weighing Machines) Regulations 1978. The prescribed Certificate of Weight should be issued to the driver. The recorded weights should be assessed in the light of the presumed accuracy limits, laid down in the regulations, of ± 150 kg per axle, with a consequent accuracy limit on gross vehicle weight of ± 150 kg, multiplied by the number of axles; compensating axles as usual should continue to be assessed as a combined weight against the combined plated weights.
- The print-roll should be left intact for the complete day's weighings, then removed, dated and retained by the enforcement staff.

13 *Construction and Use of Vehicles*

In constructing goods vehicles and trailers, manufacturers and bodybuilders must observe requirements regarding the specification and standards of construction of components and the equipment used in the manufacture. While some of these items are covered by the Type Approval scheme (see Chapter 14), the majority are included in The Road Vehicles (Construction and Use) Regulations 1986, Statutory Instrument (SI) 1078/1986 (available from The Stationery Office) and subsequent amendments to these regulations which, collectively, are commonly referred to as the C&U regulations. Certain EU legislation also applies, as indicated in the text.

Once a goods vehicle or trailer has been built and put into service, it is the operator as the vehicle user (see pp 13–15) who must then ensure that it complies fully with the law regarding its construction and use when on the road. It is worth pointing out that where a vehicle on the road is found to contravene the constructional aspects of the regulations, it is the operator who will be prosecuted and if convicted he will be liable to meet the penalty and may find that it jeopardizes his 'O' licence. It is no defence or excuse to say that the fault with the vehicle rests with the manufacturer, bodybuilder or even the supplying dealer.

Only the main items from the regulations and amendments which concern the goods vehicle operator and transport manager are included in this chapter. Those aspects of the regulations dealing with the limitations on vehicle weights and dimensions are covered in Chapter 12. The provisions of the C&U regulations dealing with safe operation and safety of vehicles and loads are explained in Chapter 20. Many other points relating to a wide variety of other types of vehicle (motor bicycles and invalid carriages, for example) are not included because they are not generally thought to be relevant to the reader of this *Handbook*.

DEFINITIONS OF VEHICLES

For the purpose of these regulations the following definitions, as given in the Road Traffic Act 1988, apply:

- A **goods vehicle** is a vehicle or a trailer adapted or constructed to carry a load.
- A **motor car** is a vehicle which, if adapted for the carriage of goods, has an unladen weight not exceeding 3050 kg but otherwise has an unladen weight not exceeding 2540 kg.
- A **heavy motor car** is a vehicle constructed to carry goods or passengers with an unladen weight exceeding 2540 kg.
- A **motor tractor** is a vehicle which is not constructed to carry a load and has an unladen weight not exceeding 7370 kg.
- A **light locomotive** is a vehicle which is not constructed to carry a load and which has an unladen weight of more than 7370 kg but not exceeding 11,690 kg.
- A **heavy locomotive** is a vehicle which is not constructed to carry a load and which has an unladen weight exceeding 11,690 kg.
- An **articulated vehicle** as defined in the C&U regulations is a motor car or heavy motor car with a trailer so attached that when the trailer is uniformly loaded at least 20 per cent of the weight of the load is imposed on the drawing vehicle.
- A **composite trailer** is a combination of a converter dolly and a semi-trailer, and is treated as one trailer only when considering the number of trailers which may be drawn.
- **Engineering plant** means movable plant or equipment in the form of a motor vehicle or trailer which is specially designed and constructed for the purposes of engineering operations and which cannot, for this reason, comply with the C&U regulations. Also it is constructed to carry only materials which it had excavated from the ground and which it is specially designed to treat while being carried. It also means mobile cranes, which do not conform in all respects with the C&U regulations.
- An **agricultural motor vehicle** means a motor vehicle that is constructed or adapted for use off roads for the purposes of agriculture, horticulture or forestry and which is primarily used for one or more of those purposes.
- A **pedestrian controlled vehicle** means a motor vehicle which is controlled by a pedestrian and which is not constructed or adapted to carry a driver or passenger.
- A **works truck** means a motor vehicle (other than a straddle carrier) designed for use in private premises and used on a road only in delivering goods from or to such premises, or from a vehicle on a road in the immediate neighbourhood, or in passing from one part of the premises to another or to other private premises in the immediate neighbourhood, or in connection with road works or in the immediate vicinity of the site of such works.
- A **works trailer** means a trailer used for the same purposes as a works truck.

INTERNATIONAL VEHICLE CLASSIFICATIONS

Goods vehicles are classified according to the following international categories:

- Category M: Motor vehicles having at least four wheels, or having three wheels when the maximum weight exceeds 1 metric ton, and used for the carriage of passengers:
  - *Category M1*: Vehicles used for the carriage of passengers and comprising no more than eight seats in addition to the driver's seat;
  - *Category M2*: Vehicles used for the carriage of passengers, comprising more than eight seats in addition to the driver's seat, and having a maximum weight not exceeding 5 metric tons;
  - *Category M3*: Vehicles used for the carriage of passengers, comprising more than eight seats in addition to the driver's seat, and having a maximum weight exceeding 5 metric tons.
- Category N: Motor vehicles having at least four wheels, or having three wheels when the maximum weight exceeds 1 metric ton, and used for the carriage of goods:
  - *Category N1*: Vehicles used for the carriage of goods and having a maximum weight not exceeding 3.75 metric tons;
  - *Category N2*: Vehicles used for the carriage of goods and having a maximum weight exceeding 3.75 but not exceeding 12 metric tons;
  - *Category N3*: Vehicles used for the carriage of goods and having a maximum weight exceeding 12 metric tons.
- Category O: Trailers (including semi-trailers):
  - *Category O1*: Trailers with a maximum weight not exceeding 0.775 metric ton;
  - *Category O2*: Trailers with a maximum weight exceeding 0.775 metric ton but not exceeding 3.75 metric tons;
  - *Category O3*: Trailers with a maximum weight exceeding 3.75 but not exceeding 10 metric tons;
  - *Category O4*: Trailers with a maximum weight exceeding 10 metric tons.

CONSTRUCTIONAL AND MAINTENANCE REQUIREMENTS

Brakes

All goods vehicles must meet specified braking efficiencies. The regulations state minimum efficiencies for the service brake, for the secondary brake, and for the parking brake or handbrake. On pre-1968 vehicles the secondary brake can be the handbrake and on post-1968 vehicles it can be a split or dual system operated by the footbrake. If it is the latter it must be capable of meeting the secondary requirement if part of the dual system fails. The parking brake must achieve the required efficiency by direct mechanical action or by the energy of a spring without the assistance of stored energy.

Every vehicle must have a parking brake system to prevent at least two wheels from turning when it is not being driven. All vehicles first used after 1 January 1968 must have an independent parking brake.

Anti-lock (ie anti-skid) braking systems are required on certain new articulated vehicles and draw-bar trailer combinations under EU legislation (for which Category 1 – wheel-by-wheel systems are necessary) and under British C&U regulations. The vehicles affected are articulated tractive units and rigid goods vehicles over 16 tonnes equipped to draw trailers and first used from 1 April 1992, and trailers over 10 tonnes built on or after 1 October 1991.

From 1 March 2002* anti-lock braking systems (ABS) will be required on all newly registered goods vehicles over 3.5 tonnes gross weight. In due course the requirement will be for electronic braking systems (EBS) on heavy vehicles.

**NB: This date has been further deferred.*

Specified Braking Efficiencies

Vehicles used before 1 January 1968:

Two-axle rigid vehicles	Service brake	45 per cent
	Secondary brake	20 per cent
Multi-axled rigid vehicles, trailer combinations and articulated vehicles	Service brake	40 per cent
	Secondary brake	15 per cent

Vehicles First Used on or after 1 January 1968:

All vehicles	Service brake	50 per cent
	Secondary brake	25 per cent
	Parking brake must be capable of holding the vehicle on a gradient of at least 1 in 6.25 without the assistance of stored energy (1 in 8.33 with a trailer attached).	

Maintenance of Brakes

The braking system on a vehicle, including all of its components and means of operation, must be maintained in good and efficient working order and must be properly adjusted at all times.

Braking Standards

The regulations reflect the requirements for braking standards laid down in EC Directives 320/1973, 524/1975 and 489/1979 (as amended). These call for the fitment of load sensing valves or anti-lock braking on drive axles – most modern tractive units already comply with these requirements – and the overall emphasis is on stability, eliminating jackknifing and trailer swing. EU rules permit the use of two-line air braking systems instead of the traditional British three-line systems. Most older tractive units in the UK have three-line braking systems fitted with yellow, blue and red couplings.

Vehicles built to the European standard have only two lines (red and yellow or two black). Coupling three-line tractive units to three-line trailers, two-line tractive units to two-line trailers and two-line tractive units to three-line trailers presents no difficulties. Problems arise when coupling three-line tractive units to two-line trailers. Such combinations must not be used unless they are specially designed or modified by fitting a fourth coupling or internal valves and connecting pipework.

On tractive units first registered before 1 April 1983 a notice should be displayed stating that the system is suitable for coupling. If no such notice is displayed the driver should check the system to see if it is suitable for coupling.

Anti-Lock Brakes

The following goods vehicles must be fitted with anti-lock brakes:

- Motor vehicles first registered on or after 1 April 1992 with design gross vehicle weight greater than 16,000 kg and authorized to tow trailers with total design axle weight exceeding 10,000 kg.
- Trailers with total design axle weight exceeding 10,000 kg manufactured on or after 1 October 1991.
- New Category N vehicles from 1998 were required to have anti-lock braking fitted as part of their type approval.
- All Category N2 and N3 vehicles and Category O3 and O4 trailers registered from April 2002 were required to have anti-lock braking systems fitted.

Anti-lock systems must conform to Annex 10 of the EU braking directive (Council Directive 71/320/EEC).

A journey in a vehicle with a faulty anti-lock braking system may be completed if the fault arose *en route*, or the vehicle may be driven to a place where the fault is to be repaired. However, this does not apply if the defective braking system does not meet specified braking efficiency requirements.

Endurance Braking Systems

From 1 January 1995 UK braking requirements were harmonized with those of the EU thereby requiring new vehicles from this date carrying dangerous goods to be fitted with endurance braking systems. Older vehicles are likely to have to be retro-fitted in due course*. EU vehicles over 16 tonnes gross weight on cross-border ADR work have required such braking systems since 1 July 1993.

**NB: A 2002 planned date for implementation has been further deferred.*

The purpose of endurance braking systems is to relieve excess loadings on the normal vehicle service brakes, particularly, for example, on long downhill gradients, where they tend to fade due to overheating. A variety of proprietary retarders are available to meet this requirement.

Brakes and Couplings on Trailers

Trailers constructed before 1 January 1968 must have an efficient braking system on half the number of wheels. Trailers constructed after this date must be fitted with brakes operating on all wheels which are capable of being applied by the driver of the drawing vehicle and having maximum efficiencies matching the braking requirement for the drawing vehicle, emergency brakes operating on at least two wheels and a parking brake capable of holding the trailer on a gradient of at least 1 in 6.25.

New lightweight trailers (ie weighing not more than 750 kg) made after 1 January 1997 must be fitted with a secondary safety coupling. Such trailers must also be marked with their date of manufacture.

Overrun Brakes

Overrun brakes may be fitted to trailers not exceeding 3500 kg gross weight (or 3560 kg if made before 27 February 1977). Overrun brake couplings must be damped and matched with the brake linkage. Normally, to ensure that these standards are met, the coupling design needs to be type approved. Trailer braking efficiency must be at least 45 per cent and the parking brake must be capable of holding the laden trailer on a gradient of 1 in 6.25 (ie 16 per cent). Modern braked trailers must also be fitted with an emergency device which automatically applies the brakes if the trailer becomes uncoupled from the towing vehicle. This does not apply to single axle trailers up to 1500 kg gross weight provided they are fitted with a safety chain or cable to stop the coupling head touching the road if the trailer becomes detached.

Light Trailer Brakes

Light trailers must be fitted with brakes if:

- their maximum gross weight exceeds 750 kg and their unladen weight exceeds 102 kg;
- their maximum gross weight exceeds 750 kg and they were built on or after 1 October 1982; or
- their laden weight on the road exceeds half the towing vehicle's kerb side weight (this does not apply to agricultural trailers or to trailers whose unladen weight does not exceed 102 kg and which were built before 1 October 1982).

Unbraked trailers must have their maximum gross weight marked in kilograms in a conspicuous position on the nearside.

Parked Trailers

When trailers are detached from the towing vehicle they must be prevented from rolling by means of a brake, chain or chock applied to at least one of their wheels.

Braking Coils

Braking coils to ISO 7638 standard must be used when both drawing vehicle and trailer are fitted with connections. A prohibition notice will be issued where these coils are found unconnected in roadside checks.

Exhaust Emissions

Since 1 April 1991 newly registered diesel-engined goods vehicles exceeding 3.5 tonnes gross weight have been required to comply with EC Directive 88/77, which sets gaseous emission (ie exhaust emission) limits (see Chapter 24 for details of emission requirements for fleet cars and light vans). These limits have already been applied in Type Approval regulations to vehicles first used from 1 April 1991 (later for cars/vans with engines over 1400 cc and those with diesel engines).

The purpose of the EU's legislative programme (under EC Directive 91/542) is to reduce the amount of nitrous oxide (NOx), carbon monoxide (CO) and unburned hydrocarbons (HC) blown into the atmosphere from vehicle exhausts. The first stage of this programme (ie the so-called Euro-1 standard) applied to all new Type Approved goods vehicles from 1 July 1992 (and to earlier Type Approved vehicles from 1 October 1993). Tougher Euro-2 standards have applied to new vehicles over 3.5 tonnes pmw since 1 October 1996; and more stringent Euro-3 emission controls have applied since 1 October 2001. Euro-4 limits will be in force from 1 October 2006 and Euro-5 from 1 October 2008.

The following table indicates the relevant emissions standards with the date (or proposed date) of implementation, the test cycle applied and the emissions and smoke limits.

Emissions Standards

Standard (date)	Test cycle (& application)	Emission limits g/kwh						Smoke
		CO	THC	NMHC	CH4	NOX	PM	M^{-1}
Euro-2 (1995/96)	R49 (diesel engines)	4.0	1.1	–	–	7.0	0.15	–
Euro-3 (2000/01)	ESC (diesel engines)	2.1	0.66	–	–	5.0	0.10	0.8
	ETC (gas engines)	5.45	–	0.78	1.6	5.0	(0.16)	–
Euro-4 (2005/06)	ESC (diesel engines)	1.5	0.46	–	–	3.5	0.02	0.5
	ETC (diesel & gas)	4.0	–	0.55	1.1	3.5	0.03	–
Euro-5 (2008)	ESC (diesel engines)	1.5	0.46	–	–	2.0	0.02	0.5
	ETC (diesel & gas)	4.0	–	0.55	1.1	2.0	0.03	–
EEVs (10/1999)	ESC (diesel engines)	1.5	0.25	–	–	2.0	0.02	0.15
	ETC (diesel & gas)	3.0	–	0.40	0.65	2.0	0.02	–

ESC, European steady-state cycle; ETC, European transient cycle; EEVs, enhanced environmental vehicles; THC, total hydrocarbons; NMHC, non-methane hydrocarbons. *Acknowledgement: TRUCK* magazine

The vehicle user is required by law to keep the engine of his vehicle and any emission control equipment (ie catalytic converter) in good working order and in tune.

Local Authority Emissions Testing

Seven local authorities, namely Birmingham, Bristol, Canterbury, Glasgow, Middlesbrough, Swansea and the City of Westminster, are currently operating a pilot scheme of roadside emissions testing. Drivers of vehicles that fail the test will be issued with a fixed penalty notice requiring payment of £60, rising to £90 if unpaid within 28 days. The test levels are the same as those that vehicles would be expected to meet at annual test. Commercial vehicles that fail an emission test invariably have poorly maintained engines. The scheme covers all types of vehicles, including commercial vehicles, and if the tests prove successful they will be extended nationally.

Smoke

Vehicles must not emit smoke, visible vapour, grit, sparks, ashes, cinders or oily substances that might cause damage to property or injury or danger to any person.

Excess Fuel Devices

Excess fuel devices must not be used on diesel vehicles while the vehicle is in motion. Such devices are incorporated in the vehicle fuel pump to enable extra fuel to be fed to the engine to aid cold-starting. Their use when the engine is warm slightly increases the power of the engine, but in doing so black smoke is emitted from the exhaust. For this reason their use is forbidden while the vehicle is in motion.

Smoke Opacity Limits

Diesel-engined vehicles first used after 1 April 1973 (but not manufactured before 1 October 1972) must comply with smoke opacity limits specified in BS AU 141a/1971. Engines fitted to such vehicles must be of a type for which a type test certificate in accordance with the British Standard Specification for *The Performance of Diesel Engines for Road Vehicles* has been issued by the Secretary of State. The certificate will indicate that engines of that type do not exceed the emission of smoke limits set out in the BS Specification.

Exemptions

Land tractors, industrial tractors, works trucks and engineering plant propelled by diesel engines with not more than two cylinders are exempt from this requirement; so too are vehicles fitted with the Perkins 6.354 engine manufactured before 1 April 1973.

Offences

It is an offence to use a vehicle to which this type test applies if the fuel injection equipment, the engine speed governor or other parts of the engine have been altered or adjusted in such a way that the smoke emission of the vehicle is increased. An offence is committed if a vehicle emits black smoke or other substances even without alteration or adjustment of the parts (eg as a result of lack of maintenance). VOSA has promised tougher enforcement of goods-vehicle smoke emissions, particularly at the time of submitting vehicles for annual test.

Control of Fumes

Petrol-engined vehicles first used after 1 January 1972 must be fitted with a means of preventing crank-case gases escaping into the atmosphere except through the exhaust system.

Fuel Tanks

Vehicle fuel tanks must be constructed so as to prevent any leakage or spillage of fuel. In particular, there is concern about leakage from heavy vehicle fuel tanks on to the road where it causes exceptional danger to cyclists and motorcyclists. Failure to maintain diesel fuel tanks in good condition (and especially filler caps) is now an offence in its own right and gives the police the opportunity to prosecute without having to actually observe spillage from the tank. Fines of up to £2,000 could be imposed on conviction for such offences.

Vehicles first used since 1 July 1973 and manufactured since 1 February 1973 that are propelled by petrol engines must have metal fuel tanks fitted in a position to avoid damage and prevent leakage. This provision does not apply where the vehicle complies with relevant EU regulations (EC 221/70) and is marked accordingly.

NB: See page 213 for information on fitting long-range fuel tanks.

Ground Clearance for Trailers

Minimum ground clearances are specified for goods-carrying trailers manufactured since 1 April 1984. Such trailers must have a minimum ground clearance of 160 mm if they have an axle interspace of more than 6 metres and not more than 11.5 metres. If the interspace is more than 11.5 metres, the minimum clearance is 190 mm (Figures 13.1 and 13.2).

Measurement of the axle interspace is taken from the point of support on the tractive unit in the case of semi-trailers or the centre line of the front axle in other cases to the centre line of the rear axle or the centre point between rear axles if there is more than one (Figure 13.2).

In determining the minimum ground clearance no account should be taken of any part of the suspension, steering or braking system attached to any axle, any wheel and any air skirt. Measurement of the ground clearance is taken in the area formed by the width of the trailer and the middle 70 per cent of the axle interspace (Figure 13.3).

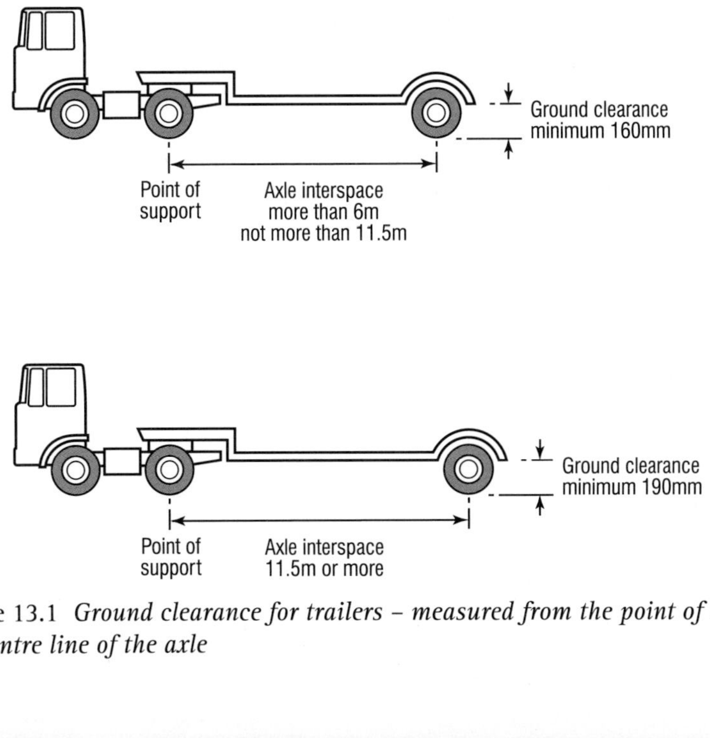

Figure 13.1 *Ground clearance for trailers – measured from the point of support to the centre line of the axle*

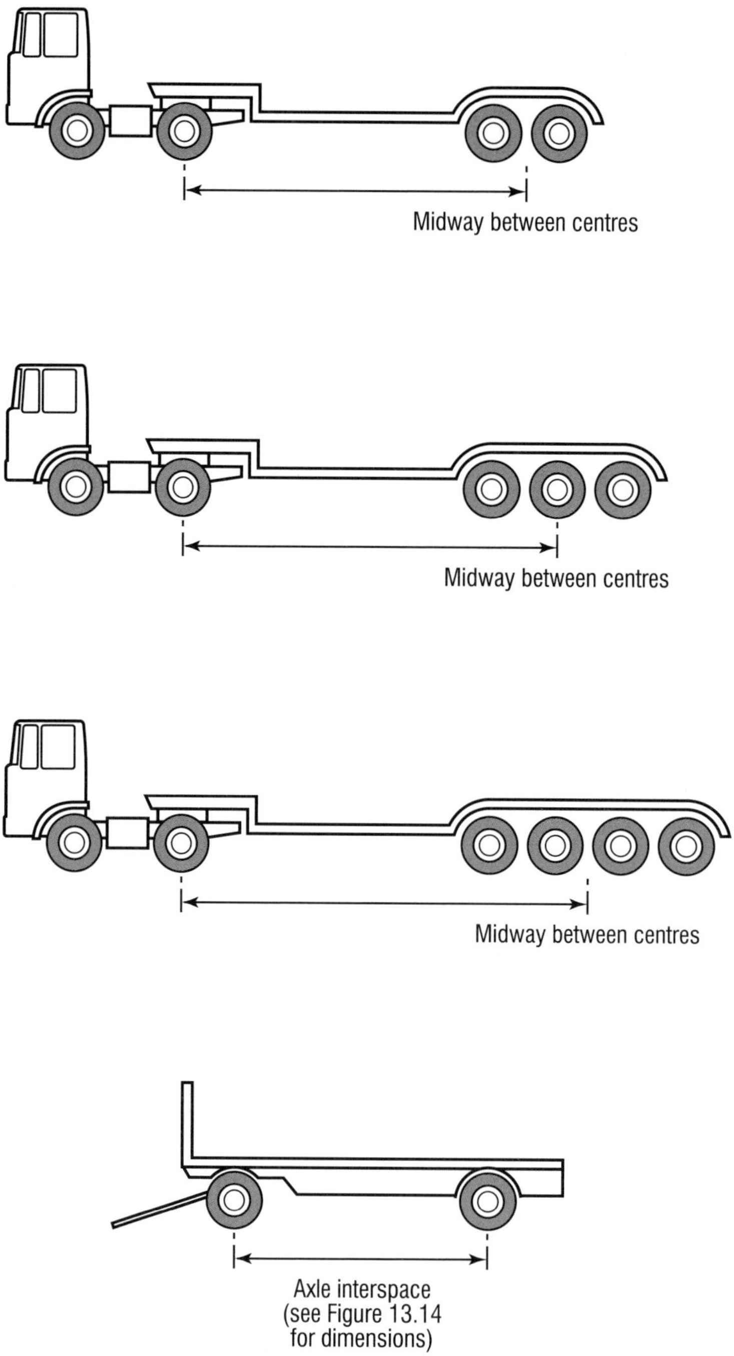

Figure 13.2 *The point for measuring axle interspace on multi-axle semi-trailers and on other trailers*

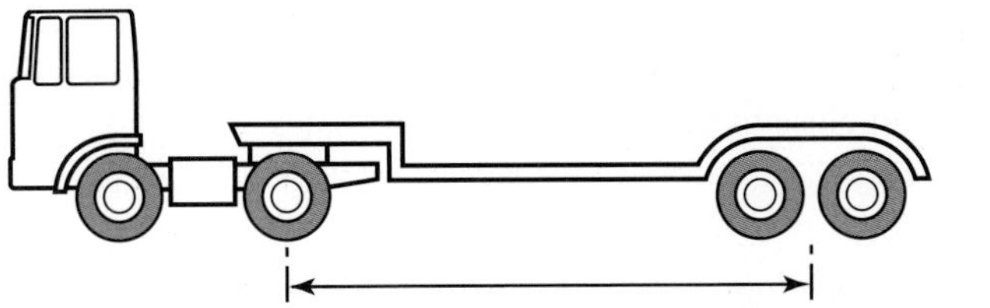

Figure 13.3 *The area in which minimum ground clearance is measured*

Horn

All vehicles with a maximum speed exceeding 20 mph, except works trucks and passenger-controlled vehicles, must be equipped with an audible warning instrument. The sound emitted by a horn must be continuous and uniform and not strident. Gongs, bells, sirens and two-tone horns are only permitted on emergency vehicles, though a concession allows similar instruments, except two-tone horns, to be used on vehicles from which goods are sold to announce the presence of the vehicle to the public. Any vehicle first used since 1 August 1973 must not be fitted with multi-toned or musical horns.

Restriction on Sounding Horns

Audible warning instruments must not be sounded at any time while the vehicle is stationary or in a built-up area (ie where 30 mph speed restriction is in force) between 11.30 pm and 7.00 am (see also item about reversing alarms on p 354). The use of the horn on a stationary vehicle in an emergency situation (ie 'at times of danger due to another, moving vehicle on or near the road') is allowed.

Horns Used as Anti-Theft Devices

Audible warning instruments that are gongs, bells or sirens may be used to prevent theft or attempted theft of a vehicle provided a device is fitted that will stop the warning sounding continuously for more than 5 minutes. Hazard lights and interior lights may be set to operate continuously for (a maximum of) 5 minutes if the vehicle is tampered with.

Mirrors

Goods vehicles and dual-purpose vehicles must be fitted with at least two mirrors. One of these must be fitted externally on the offside and the other must, in the case of vehicles first used since 1 June 1978, be fitted in the driver's cab or driving compartment. When an interior mirror does not provide an adequate view to the rear, a mirror must be fitted externally on the near side. The mirrors must show traffic to the rear or on both sides rearwards. Mirrors fitted to vehicles over 3,500 kg must conform to Class II and those fitted to other vehicles with Class II or III as described in EC Directive 127/71.

From 2007 new vehicles exceeding 7.5 tonnes gross weight will need to be fitted with an additional convex mirror, mounted above the windscreen to give a view of the ground close to the front and on both sides.

Fitment of Mirrors

External mirrors with a bottom edge less than 2 metres from the ground (when the vehicle is loaded) must not project more than 20 cm beyond the overall width of the vehicle or vehicle and trailer. Mirrors fitted on the offside must be adjustable from the driving seat unless they are of the spring-back type. Internal mirrors fitted to vehicles first registered on or after 1 April 1969 must be framed with some material (usually plastic beading) which will reduce the risk of cuts to any passenger who may be thrown against the mirror.

Type Approval for Mirrors

Since 1 June 1978 vehicles must be fitted with rear-view mirrors bearing the EU Type Approval 'E' mark. The relevant date for implementation of this requirement for Ford Transit vehicles was 10 July 1978, rather than 1 June 1978 which was the date for all other vehicles.

Wide-Angle Mirrors

Goods vehicles over 12 tonnes maximum permissible weight first used since 1 October 1988 must be fitted with additional mirrors that provide close proximity and wide-angle vision for the driver in accordance with EC Directives 205/85 and 562/86.

New Requirements for Mirrors

A draft EU Directive in circulation during 2001/02 proposes changes to the current legal requirements for mirrors on goods vehicles and the introduction of new measures to provide improved vision for drivers using new technologies. These changes include requirements for:

- additional mirrors to be fitted to the front of heavy vehicles to increase all-round safety;
- the fitment of aspherical mirrors (ie two-part mirrors) on light commercial vehicles; changes to the curvature of mirrors on all new vehicles; and
- the replacement of some mirrors currently fitted or required with so-called 'indirect vision systems' (defined as devices to enable the driver to observe the traffic area adjacent to the vehicle which cannot be observed by direct vision). These systems can include either conventional mirrors or camera-monitor type systems capable of submitting information about the indirect field of vision to the driver.

Safety Glass

Goods vehicles must be fitted with safety glass (ie toughened or laminated glass which, when fractured, does not fly into fragments likely to cause severe cuts) for windscreens and windows in front of and on either side of the driver's seat. The windscreen and all windows of dual-purpose vehicles must be fitted with safety glass. Glass bearing an approval mark under EC Directive 92/22 is regarded as meeting the requirement stated above for safety glass.

The glass must be maintained so as not to obscure the vision of the driver while the vehicle is being driven on the road. This means that a driver could be prosecuted for having a severely misted up, iced up or otherwise dirty windscreen.

Seat Belts

Seat belts for the driver and one front-seat passenger must be fitted to goods vehicles not exceeding 1525 kg unladen registered since 1 April 1967 and goods vehicles not exceeding 3500 kg gross weight first used since 1 April 1980. From 1 October 1988 goods vehicles over 3500 kg must be fitted with seat belt anchorage points for each forward-facing seat to which lap-strap type seat belts can be fixed (see Chapter 10 for details of seat-belt fitment requirements in fleet cars).

Since 1 March 2001 seat belts have been required by law to be fitted in new goods vehicles over 3.5 tonnes gross weight (although many new vehicles were fitted with seat belts by manufacturers on a voluntary basis long before this date).

Vehicles to which this regulation applies, first used since 1 April 1973, must be fitted with belts that can be secured and released with one hand only and also with a device to enable the belts to be stowed in a position where they do not touch the floor. Vehicles to which this requirement applies will fail the annual test if seat belts are not fitted, are permanently obstructed or are not in good condition.

The legal requirement for drivers and passengers to wear seat belts came into effect in January 1983 with certain exemptions (see p 257–59).

Since 1 February 1997 it has been illegal to use a minibus or coach for carrying a group of three or more children (aged between 3 and 16 years) on an organized trip unless as many forward-facing passenger seats as there are children are fitted with seat belts.

Sideguards

Most heavy vehicles and trailers must be fitted with sideguards to comply with legal requirements except certain vehicles and trailers that are exempt from the fitting requirement as listed at the end of this section.

Sideguards must be fitted to the following vehicles and trailers:

- Goods vehicles exceeding 3.5 tonnes maximum gross weight manufactured since 1 October 1983 and first used since 1 April 1984.
- Trailers exceeding 1020 kg unladen weight manufactured since 1 May 1983 and which, in the case of semi-trailers, have a distance between the foremost axle and the centre line of the kingpin (or rearmost kingpin if there is more than one) exceeding 4.5 metres (Figure 13.4).
- Semi-trailers made before 1 May 1983 with a gross weight exceeding 26,000 kg and used in an articulated combination with a gross train weight exceeding 32,520 kg.

Sideguards are not required on vehicles and trailers, other than semi-trailers, where the distance between any two consecutive (ie front and rear) axles is less than three metres (Figure 13.5).

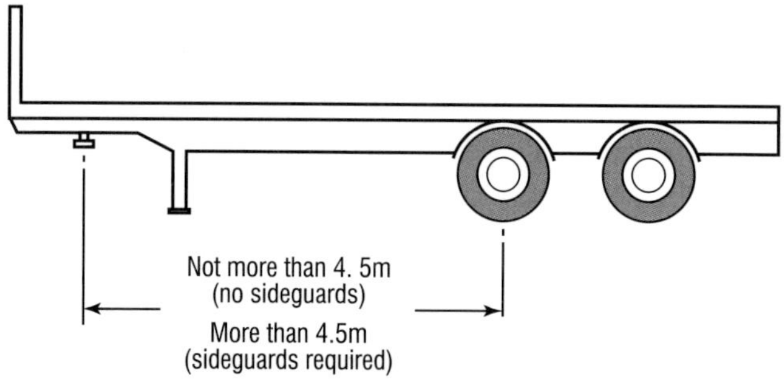

Figure 13.4 *Measurements of relevant distance between foremost axle and centre of kingpin for semi-trailers to determine if sideguards must be fitted*

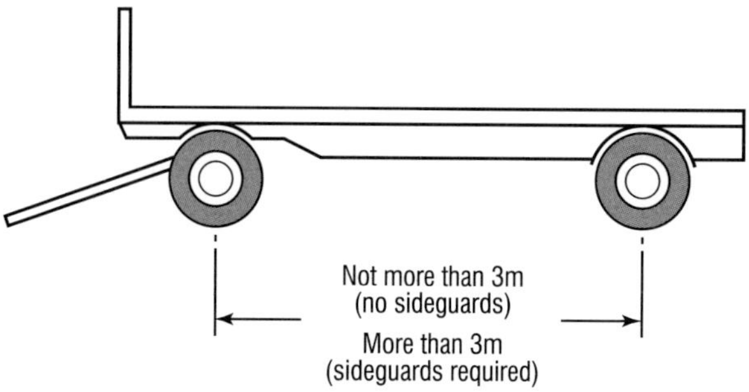

Figure 13.5 *Measurements of two consecutive axles on drawbar trailers – the same dimension applies to rigid vehicles – to determine if sideguards must be fitted*

Strength of Sideguards

Sideguards must be constructed so they are capable of withstanding a force of 200 kg (two kilonewtons) over their length, apart from the rear 250 mm, without deflecting more than 150 mm. Over the last 250 mm the deflection must not be more than 30 mm under such force (Figure 13.6). These force resistance requirements *do not* apply where sideguards were fitted to existing semi-trailers (ie those built before 1 May 1985, and which were used at weights above 32,520 kg).

Fitment of Sideguards

The fitting position for sideguards depends on the type of vehicle or trailer as follows:

- Rigid vehicles: at front – not more than 300 mm behind the edge of the nearest tyre and the foremost edge of the sideguard; at rear – not more than 300 mm behind the rearmost edge of the sideguard and the edge of the nearest tyre (Figure 13.7).
- Trailers: at front – not more than 500 mm behind the edge of the nearest tyre and the foremost edge of the sideguard; at rear – not more than 300 mm behind the rearmost edge of the sideguard and the edge of the nearest tyre (Figure 13.8).

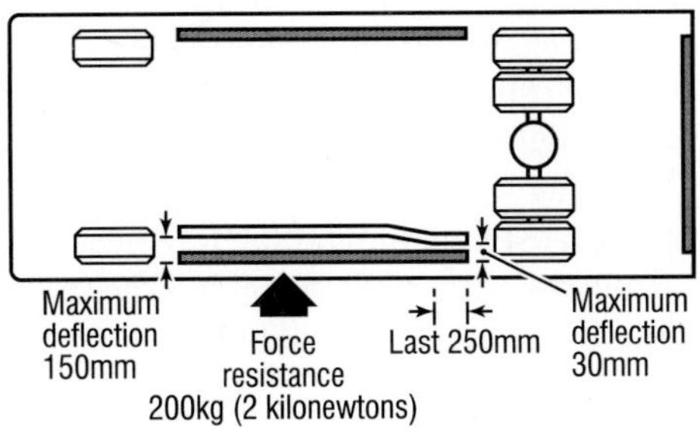

Figure 13.6 *How force resistance applies to sideguard on new vehicles and trailers (It does not apply to sideguard fitted to existing semi-trailers)*

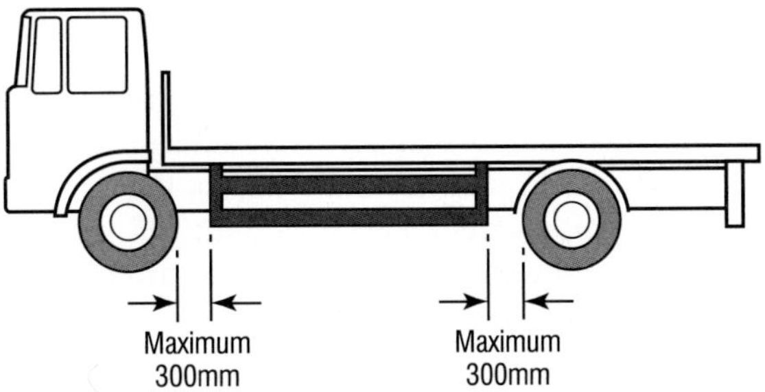

Figure 13.7 *Fitting position for sideguard on rigid vehicles*

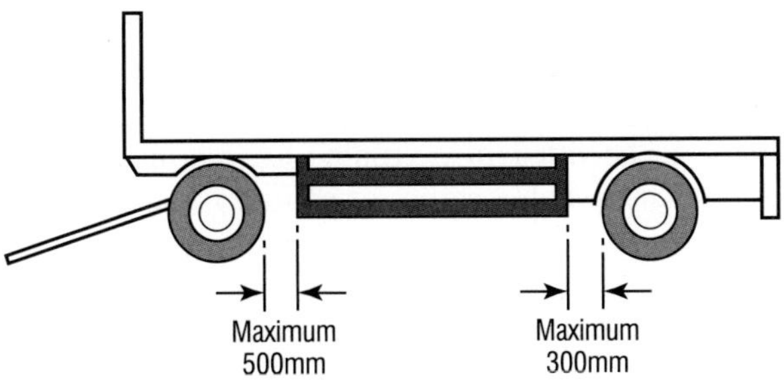

Figure 13.8 *Fitting position for sideguard on trailers*

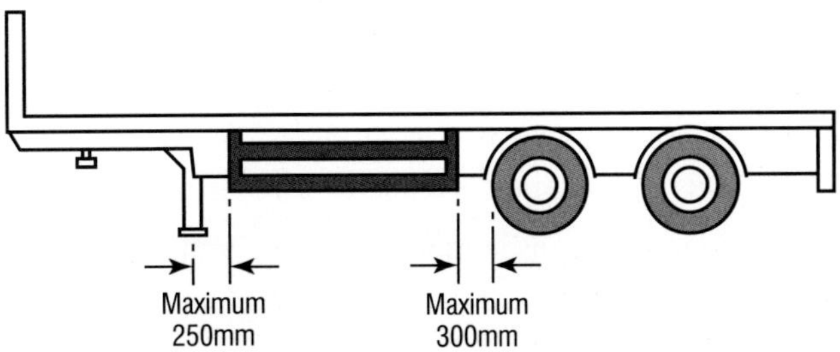

Figure 13.9 *Fitting position for sideguard on semi-trailers with landing legs*

- Semi-trailers with landing legs: at front – not more than 250 mm behind the centre line of the landing legs and the foremost edge of the sideguard; at rear – not more than 300 mm behind the rearmost edge of the sideguard and the edge of the nearest tyre (Figure 13.9).
- Semi-trailer without landing legs: at front – not more than 3 metres behind the centre line of the rearmost kingpin and the foremost edge of the sideguard; at rear – not more than 300 mm behind the rearmost edge of the sideguard and the edge of the nearest tyre (Figure 13.10).

In all cases sideguards must be fitted so they are not inset more than 30 mm from the external face of the tyre, excluding any distortion due to the weight of the vehicle (Figure 13.11).

The upper edge of sideguards must be positioned as follows:

- In the case of vehicles or trailers with a body or structure that is wider than the tyres, no more than 350 mm from the lower edge of the body or structure (Figure 13.12).
- In the case of vehicles or trailers with a body or structure that is narrower than the tyres or that does not extend outwards immediately above the wheels, a vertical plane taken from the outer face of the tyre must be measured upwards for 1.85 metres above the ground. If this plane is dissected by the vehicle structure within 1.85 metres from the ground, the sideguard must extend up to within 350 mm of the structure where it is cut by the vertical plane (Figure 13.13); if the vertical plane is not dissected by the vehicle structure, the upper edge of the sideguard must extend to be level with the top of the vehicle structure to a minimum height of 1.5 metres from the ground (Figure 13.14).

The lower edge of sideguards must not be more than 550 mm from the ground. This dimension is to be measured on level ground and, in the case of a semi-trailer, when its load platform is horizontal.

When sideguards are to be fitted to extendible trailers and to vehicles and trailers designed to carry demountable bodies or containers, the following fitting provisions apply:

- Sideguards must be fitted to extendible trailers in compliance with the original fitting specifications in regard to spacings from the nearest wheel, kingpin or landing leg when the trailer is at its shortest length. When the trailer is extended beyond its minimum length the spacings between the front edge of the sideguard and the semi-trailer landing legs or kingpin (if it has no landing legs) and the rear edge of the sideguard and the foremost edge of the tyre nearest to it are no longer applicable.
- Sideguards must be fitted to vehicles and trailers that are designed and constructed (not merely adapted) to carry demountable bodies or containers so that when the body or container is removed the sideguards remain in place.

This means that if the vehicle runs without a body or container it must still comply with the sideguard requirements.

Vehicles that are fitted with sideguards complying with EC Directive 89/279 do not have to comply with the fitting requirements specified in the UK regulations as detailed above.

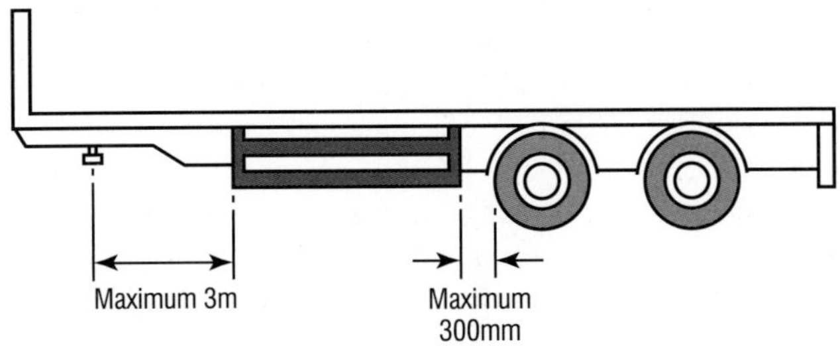

Figure 13.10 *Fitting position for sideguard on semi-trailers without landing legs*

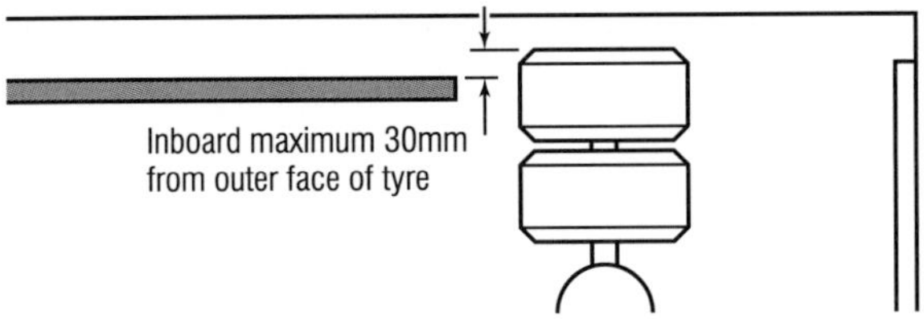

Figure 13.11 *Inboard mounting position for sideguards – all vehicles and trailers*

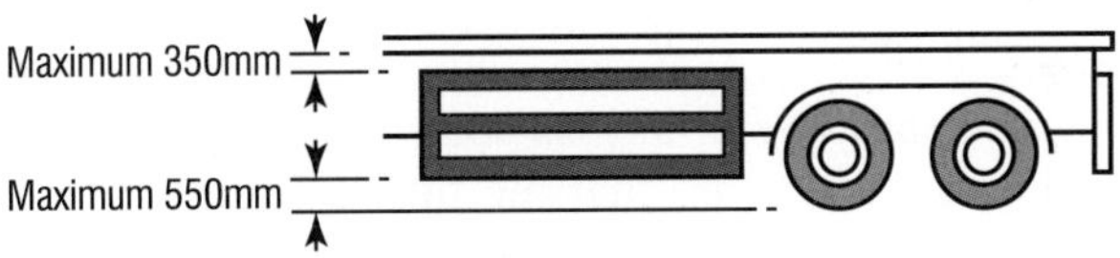

Figure 13.12 *Fitting position for sideguards where body or structure is wider than the tyres*

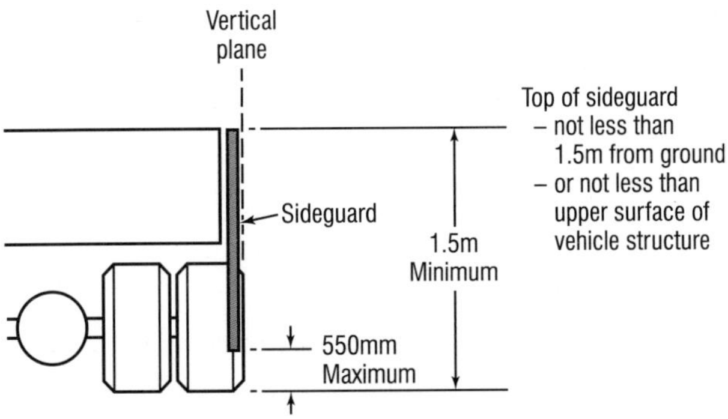

Figure 13.13 *Fitting position for sideguards where body or structure is narrower than the tyres up to a height of 1.85 metres*

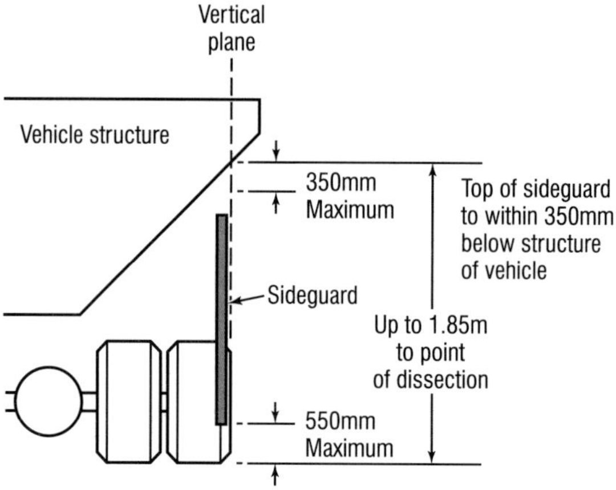

Figure 13.14 *Fitting position for sideguards where body or structure is narrower than the tyres*

Construction of Sideguards

All parts of the sideguard that face outwards must be 'smooth, essentially rigid and either flat or horizontally corrugated'. Each face of the guard must be a minimum of 100 mm wide (including the inward face at the forward edge) and the vertical gaps between the bars must not be more than 300 mm wide.

Maintenance of Sideguards

Sideguards must be maintained free of any obvious defect that would impair their effectiveness. It is important to ensure that the fitting dimensions are observed, particularly when the sideguards are damaged (eg by forklift truck impact).

Exemption to Fitting Dimensions

The specific requirements relating to the fitting positions for sideguards as previously described only apply so far as is practicable in the case of the following vehicles and trailers:

- Those designed solely for the carriage of a fluid substance in closed tanks permanently fitted to the vehicle and provided with valves and hose or pipe connections for loading and unloading.
- Those vehicles that require additional stability during loading and unloading or while working and which are fitted with extendible stabilizers on either side (eg lorry-mounted cranes, tower wagons, inspection platforms).

Exemptions

Sideguards do not have to be fitted to vehicles and trailers in the following list:

- Vehicles incapable of a speed of more than 15 mph on the level under their own power.

- Agricultural trailers.
- Engineering plant.
- Fire engines.
- Land tractors.
- Side and end tipping vehicles and trailers.
- Vehicles with no bodywork fitted and being driven or towed for the purposes of a quality or safety check by the manufacturer, distributor or dealer in such vehicles, or being driven by prior arrangement to have bodywork fitted.
- Vehicles being driven or towed to a place by prior arrangement to have sideguards fitted.
- Vehicles designed solely for use in connection with street cleansing, the collection or disposal of refuse or the collection or disposal of the contents of gullies or cesspools.
- Trailers specially designed and constructed to carry round timber, beams or girders of exceptional length.
- Articulated tractive units.
- Naval, military or air-force vehicles.
- Trailers specially designed and constructed (not merely adapted) to carry other vehicles loaded from the front or rear (eg car transporters).
- Temporarily imported foreign semi-trailers.
- Low-loader trailers where:
  - the upper surface of the load platform is not more than 750 mm from the ground; and
  - no part of the edge of the load platform is more than 60 mm inboard from the external face of the tyre (discounting the distortion caused by the weight of the vehicle).

Silencer

An adequate means of silencing exhaust noise and of preventing exhaust gases escaping into the atmosphere without first passing through a silencer must be fitted to all vehicles. Silencers must be maintained in good and efficient working order and must not be altered so as to increase the noise made by the escape of exhaust gases.

Speedometer

Speedometers (and/or tachographs, as appropriate – see Chapter 5) must be fitted to all vehicles registered since 1 October 1937 except those that cannot or are not permitted to travel at more than 25 mph, agricultural vehicles that are not driven at more than 20 mph and works trucks first used before 1 April 1984. In the case of vehicles first used since 1 April 1984 the speedometer must indicate speed in both miles per hour and kilometres per hour. The instrument must be maintained in good working order at all material times and kept free from any obstruction that might prevent it being easily read.

Defence

It is a defence to be able to show that a defect to a speedometer or a tachograph occurred during the journey when the offence was detected or that at that time steps had been taken to get the defect repaired with all reasonable expedition (ie as soon as reasonably practicable).

Speed Limiters

In a move to reduce the number and severity of road accidents, legislation requires certain heavy vehicles to be fitted with speed limiters. The Government says that this will also have beneficial effects for operators in terms of fuel economy – possibly saving as much as 150 million litres annually when all relevant vehicles are fitted – and benefit the environment by an annual reduction of an estimated 0.5 million tonnes of carbon monoxide (CO) pumped, by way of exhaust emission, into the atmosphere. At the same time, the EU requires certain vehicles also to be fitted with speed limiters set to a maximum of 90 kph (approx 56 mph), and this has led to no end of confusion since the two sets of requirements do not align.

The UK regulations, which came into effect on 1 August 1992, limit relevant vehicles to a maximum speed of 60 mph (ie 96.5 kph). They apply to all new goods vehicles exceeding 7.5 tonnes permissible maximum weight that are capable of a speed in excess of 60 mph on the flat and which were first registered on or after 1 August 1992.

Vehicles not capable of travelling at or above 60 mph (eg refuse collection and certain highway maintenance vehicles) are exempt from the regulations, as are the following vehicles:

- those being taken to a place to have a speed-limiter device fitted or calibrated;
- those owned and being used by the army, navy or air force;
- those being used for military purposes while driven by a person under military orders;
- those being used for fire brigade, ambulance or police purposes;
- those exempt from excise duty because they do not travel more than six miles per week on public roads).

Speed limiter equipment must comply with BS AU 217 (or an acceptable equivalent), be calibrated to a set speed not exceeding 60 mph and be sealed by an 'authorized sealer'. Existing speed limiters, fitted on a voluntary basis prior to 1 August 1992, were permitted and did not have to be sealed by an 'authorized sealer' as stated above.

Speed limiters must be maintained in good working order, though it is a defence to show that where a vehicle is driven with a defective limiter, the defect occurred during that journey or that at the time it is being driven to a place for the limiter to be repaired.

EU Requirements

The speed limiter requirements of EC Directive 92/6/EEC apply from 1 January 1994 and require new goods vehicles first registered from this date and which exceed 12 tonnes gross weight to be fitted with speed limiters. Since 1 September 1996 speed limiters should be set at a speed of not more than 85 kph (52.8 mph), allowing a stabilized speed of not more than 90 kph.

Speed Limiter Plates

Vehicles that are required to be fitted with speed limiter equipment must carry a plate (fitted in a conspicuous and readily accessible position in the vehicle cab) on which is shown the words:

- 'SPEED LIMITER FITTED' (in large letters at the top);
- the Standard with which the installation complies (eg BS AU 217 Part 1A 1987);
- the speed setting in mph/kph;
- the vehicle registration number;
- the name/trade mark of the firm which carried out the calibration; and
- the place where the speed limiter was fitted and the date of fitment.

Plates that refer to a maximum speed of 56 mph must be changed (as of 1 September 1997) to show 85 kph. Normally replacement plates will be supplied (free of charge) by VOSA when relevant vehicles are presented for their annual test after this date.

Warning – Penalties in France

It is reported that the French authorities are imposing stringent penalties where vehicles are found to have defective or incorrectly set speed limiters (ie above the 85 kph legal limit). These include substantial fines and the need for the vehicle to undergo a French-style annual goods vehicle test – this is resulting in long delays for those operators affected.

Speed Limiters for Light Trucks

An amendment to the C&U regulations, effective from 1 January 2005, requires light goods vehicles over 3.5 tonnes gross weight to be fitted with speed limiters restricting their speed to 100 kph (62 mph). The law previously applied only to vehicles over 7.5 tonnes and up to 12 tonnes gross weight, which are restricted to 60 mph (96.5 kph) and to over-12-tonne vehicles which are restricted to 90 kph (56 mph). Implementation of this provision will take effect from 1 January 2006 for vehicles used in both national and international transport operations and 1 January 2007 for vehicles used solely for national transport operations.

Spray Suppression

Regulations require certain vehicles and trailers to be equipped with anti-spray devices. The following vehicles and trailers must be fitted with approved equipment from the dates shown:

- Motor vehicles over 12 tonnes gross weight made on or after 1 October 1985 and first used on or after 1 April 1986. Fitment required from date when vehicle first used on road from 1 April 1986.
- Trailers over 3.5 tonnes gross weight made on or after 1 May 1985. Fitment required from date trailer first used on road from new.
- Trailers over 16 tonnes gross weight with two or more axles.

Exemptions

Anti-spray requirements do not apply to those vehicles and trailers that are fitted with such devices in accordance with EC Directive 91/226 (and marked accordingly) and to those that are exempt under the C&U regulations from the need for wings. Further exemptions are as follows:

- Four-wheel and multi-wheel drive vehicles.
- Vehicles with a minimum of 400 mm (approximately 16 in) ground clearance in the middle 80 per cent of the width and the overall length of the vehicle.
- Works trucks.
- Works trailers.
- Broken-down vehicles.
- Vehicles that cannot exceed 30 mph on the level under their own power owing to their construction.
- Vehicles specified in regulations as exempt from sideguards:
  - agricultural trailers and implements;
  - engineering plant;
  - fire engines;
  - side and end tippers;
  - military vehicles used for military, naval or air-force purposes;
  - vehicles with no bodywork fitted being driven on road test or being driven by prior appointment to a place where bodywork is to be fitted or for delivery;
  - vehicles used for street cleansing, or the collection or disposal of the contents of gullies or cesspools;
  - trailers designed and constructed to carry round timber, beams or girders of exceptional length;
  - temporarily imported foreign semi-trailers.
- Concrete mixers.
- Vehicles being driven to a place by prior arrangement to have anti-spray equipment fitted.
- Land locomotives, land tractors and land implement conveyors.
- Trailers forming part of an articulated vehicle or part of a combination of vehicles having, in either case, a total laden weight exceeding 46,000 kg.

British Standard

The British Standard on spray suppression was originally contained in two documents, BS AU 200 (parts 1 and 2) 1984, which applied to vehicles fitted before 1 May 1987 and has now been replaced by BS AU 200 (parts 1a and 2a) 1986, which apply to fitment since this date. The regulations require anti-spray devices to conform to the Standard set out in these documents. However, there is no requirement for existing equipment fitted to the requirements of the original BS AU Standard to be changed to equipment that meets the new BS AU Standard.

In order to comply with the law, relevant vehicles and trailers must be fitted with anti-spray systems that fall into one of two main categories:

- a straight valance across the top of the wheel and a flap hanging vertically behind the wheel, all made from approved spray suppressant material; or

- a semicircular valance following the curvature of the wheel with either:
  - air/water separator material round the edge; or
  - a flap of spray suppressant material hanging from the rear edge.

Spray Suppressant Material

Two types of material are referred to in the Standard. These are generally identifiable as follows:

- Spray suppressant material – designed to absorb or dissipate the energy of water thrown up from the tyre in order to reduce the degree to which water shatters into fine droplets on hitting a surface.
- Air/water separator – 'a device forming part of the valance and/or wheel flap which permits air to flow through while reducing the emission of spray'.

Maintenance of Anti-Spray Equipment and Devices

The regulations stipulate that all devices fitted to comply with the legal requirement (and every part of such a device) must be maintained, when the vehicle is on the road, so that they are free from 'any obvious defect which would be likely to affect adversely the effectiveness of the device'. It is also important that fitting dimensions are maintained, especially if the flaps are damaged.

Fitment – Valances and Flaps

Where the choice is for spray suppression to be achieved by the use of valances and flaps (particularly on rear vehicle wheels and trailer wheels) the specific requirements for fitment are as follows:

NB: Capital letters in brackets in the following text refer to items on the adjacent diagrams illustrating fitment details.

- Valances of spray suppressant material must extend across the top of the tyre from a line vertical with the front edge of the tyre (A) to a line beyond the rear wheel which will allow the rear flap to be suspended no more than 300 mm from the rear edge of the tyre (B). The valance must be at least 100 mm deep (C).

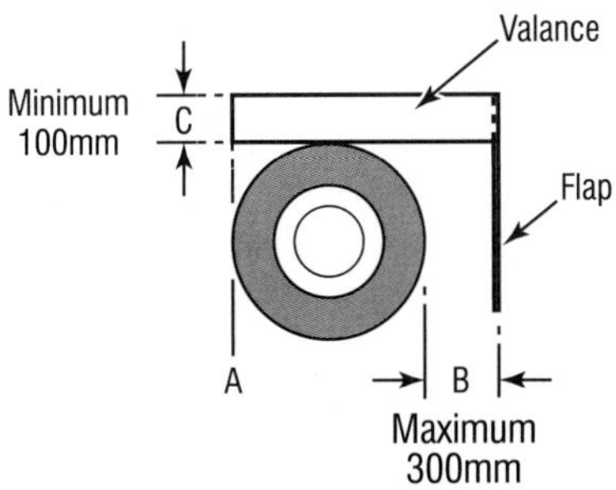

- The valance must extend downwards to be level with the top of the tyre (D) or it may overlap the top of the tyre (E).

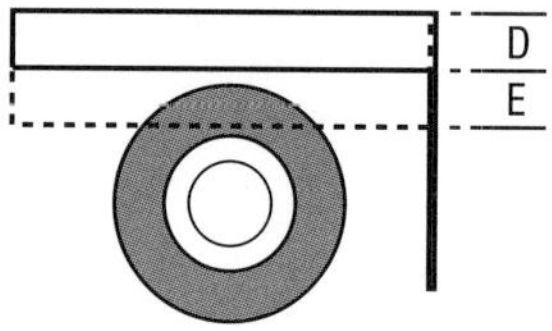

- In the case of multiple-axle bogies the relevant dimensions are shown below with the additional requirement that where the gap between the rear edge of the front tyre and the front edge of the rear tyre is greater than 250 mm (F) a flap must be fitted between the two. *Note: No middle flap is required if the distance does not exceed 250 mm.*

NB: The top of the valance may be in two separate sections (see shaded part) so long as it otherwise conforms to the dimensions.

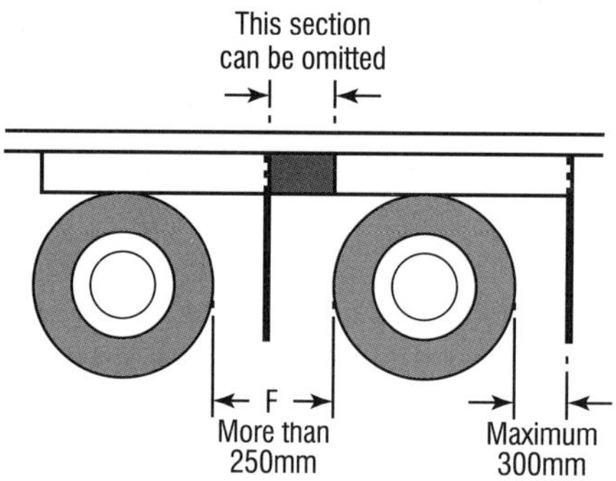

- Valances must extend the full width of the tyre and beyond to a maximum of 75 mm (G) in the case of the rear wheels (non-steerable) and 100 mm (H) in the case of steered wheels.

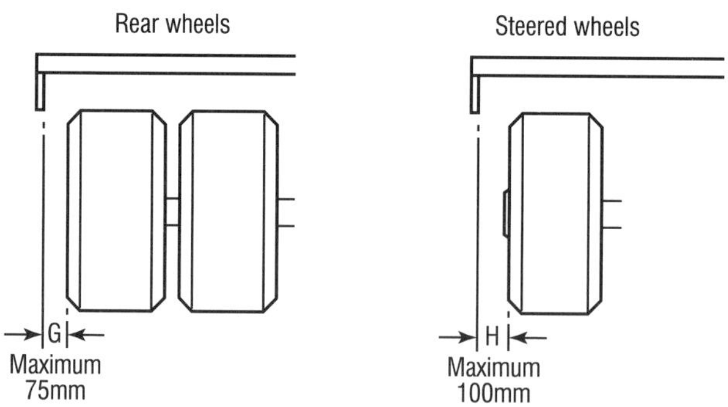

- If the valance extends below the level of the tyre on fixed wheels, the gap between the tyre face and the valance can be extended to 100 mm (J). There must be no gaps between the valance and the vehicle body.

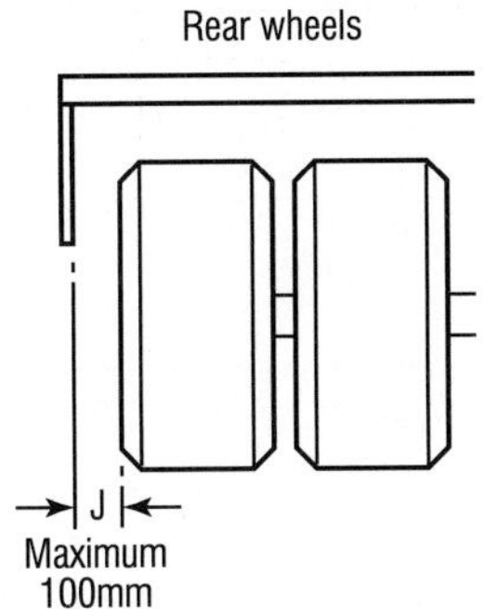

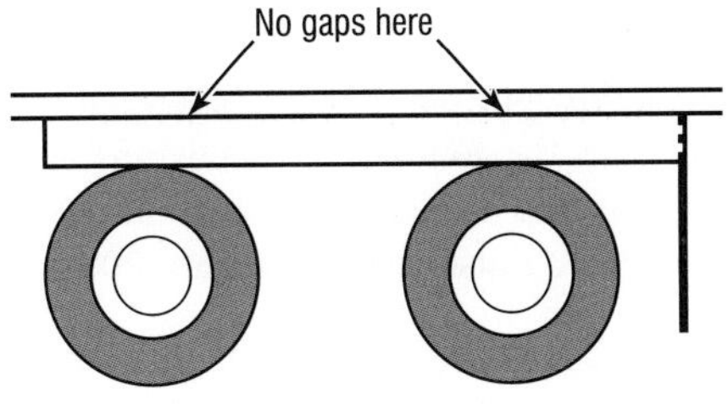

- Flaps used in conjunction with valances as described above must conform to the following dimensions:
  - They must extend to the full width of the tyre/tyres (K).
  - They must reach down to within 200 mm of the ground (L) when the vehicle is unladen (300 mm on rearmost axles of trailers used on roll-on/roll-off ferries or on any axle where the radial distance of the lower edge of the valancing does not exceed the radius of the tyres fitted).

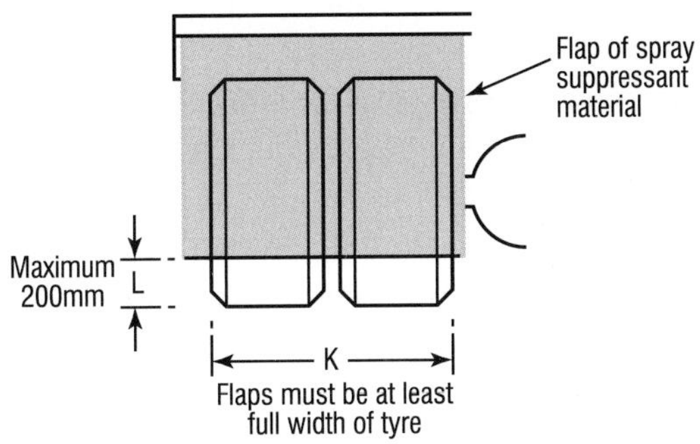

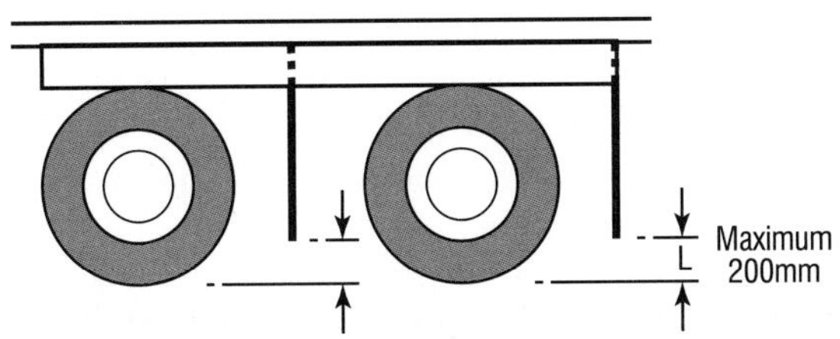

  - When flaps are used in conjunction with mudguard valances the top of the flap must extend upwards at least to a point 100 mm above the centre line of the wheel irrespective of the position of the lower edge of the mudguard.

Flap used in conjunction with mudguard valance

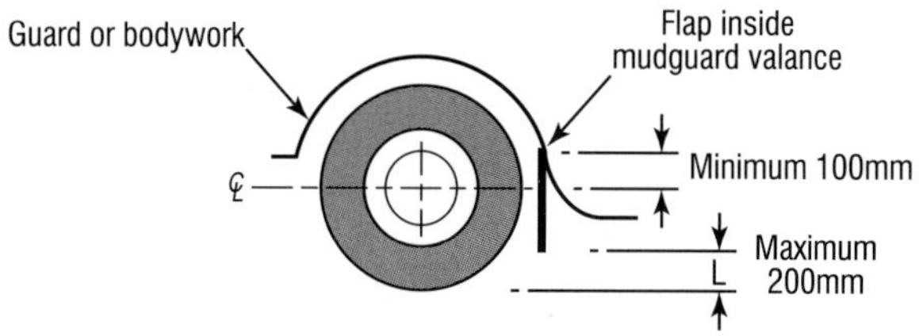

- Where the flap extends inside the guard it must be at least the width of the tyre tread pattern.

- If the flap used is of a type with an air/water separator device (ie bristles) fitted to the bottom edge, the following dimensions apply:
  - Rear edge of tyre to flap – maximum distance 200 mm (M).
  - The edge of the device must come to within 200 mm of the ground (N).

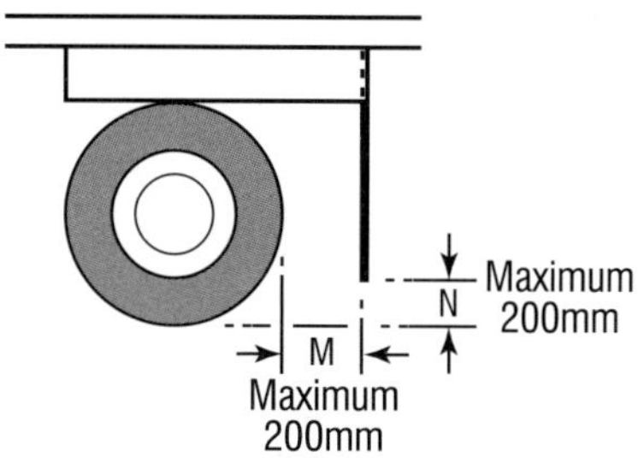

Deflection of Flaps

Wheels flaps must not be capable of being deflected rearwards more than 100 mm when subjected to a force of 3N (ie 4 lbs) applied near the bottom of the flap.

Fitment – Mudguards and Air/Water Separator Devices

Where the choice for compliance with the regulations is by means of conventional mudguarding there are specific dimensions to be observed:

- If the mudguard is covering a steerable wheel (see later note about steerable axles on drawbar trailers), the radius of the edge of the valance must be not more than 1.5 times the radius of the tyre measured at three points (P) and in the case of non-steerable wheels, 1.25 times the same radius.
  - Vertically above the centre of the tyre.
  - A point at the front of the tyre 20 degrees above the horizontal centre line of the tyre (non-steerable wheels) or a point 30 degrees above the horizontal centre line of the tyre (steerable wheels).
  - A point at the rear of the tyre 100 mm above the horizontal centre line of the tyre.

Three points of measurement of radial for mudguards (P)

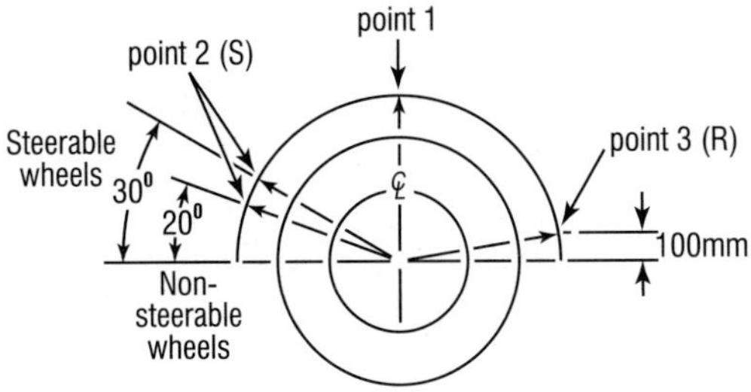

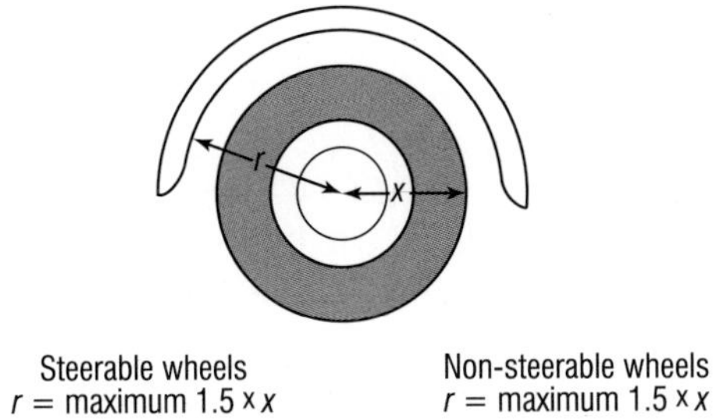

- Mudguard valances must be at least 45 mm deep behind a point vertically above the wheel centre. They may reduce in depth forward of this point (Q).

Depth of mudguard valances

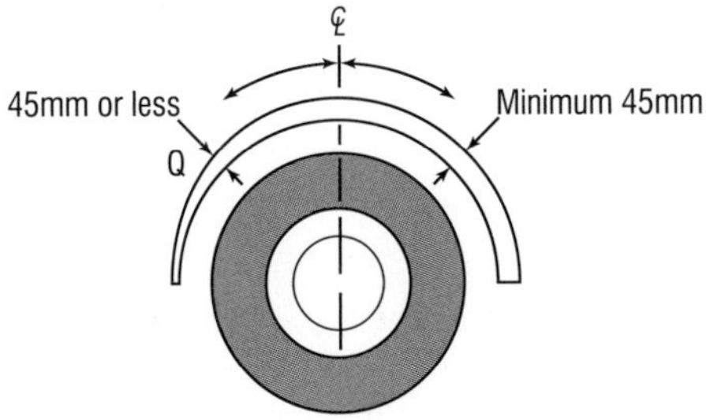

- In the case of drawbar trailers, the 1.5 times radius dimension applies as above for the front steerable axle unless the mudguards are fitted to the turntable and thus turn with the wheels, in which case the maximum radius for the valance is 1.25 times the tyre radius.
- If the valancing on fixed-wheel mudguards is provided by means of air/water separator material (ie bristles), the edge must follow the periphery of the tyre. On steerable wheels the edge must be not more than 1.05 times the tyre radius.

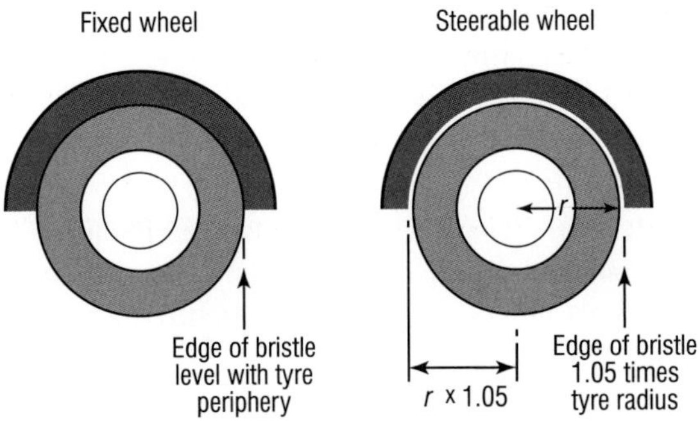

The valances on mudguards must extend downwards at the front and rear to at least the following dimensions:
- at rear – to within 100 mm above the centre line of the axle (point 3) (R).
- at front – to within a line 20 degrees above the centre line of the axle. In the case of steerable wheels this dimension is raised to 30 degrees (point 2) (S).

- In the case of multi-mudguarding over tandem axles or bogies, the intersection of the guards between the wheels must conform to one of the two dimensions:
  - The gap between the guards at the valance edges must not exceed 60 mm (T); or
  - The edges must come down to within 150 mm of the horizontal centre line across the wheels (V).
- If the gap between the tyre edges is greater than 250 mm, a flap must be provided between the wheels. If the gap is more than 300 mm, the wheels should be treated as though separate for mudguarding purposes (W).

Notes

- All dimensions in the regulations are to be taken when the vehicle is unladen, when steerable wheels are straight ahead and when the load platforms of articulated semi-trailers are level.
- All suppression material or devices and air/water separator material or devices must be permanently and legibly marked with the following mark: BS AU 200/2, plus 'the name, trademark or other means of identification of the responsible manufacturer'.

Tyres

It is an offence to use, or cause or permit to be used on a road, a vehicle or a trailer with a pneumatic tyre which is unsuitable for the use to which the vehicle is being put. It is also an offence to have different types of tyres fitted to opposite wheels of the vehicle or trailer; for example, radial-ply tyres must not be fitted to a wheel on the same axle as wheels already fitted with cross-ply tyres and vice versa. Tyres must be inflated to the vehicle or tyre manufacturers' recommended pressures so as to be fit for the use to which the vehicle is being put (for example, motorway work or cross-country work). No tyre must have a break in its fabric or a cut deep enough to reach the body cords, more than 25 mm or 10 per cent of its section width in length, whichever is the greater; also there must be no lump, bulge or tear caused by separation or partial fracture of its structure; neither must there be any portion of the ply or cord structure exposed.

Approval Marks on Tyres

It is an offence to sell motor car tyres unless they carry an 'E' mark to show compliance with EU load and speed requirements. It is also an offence to sell retreaded car or lorry tyres unless they are manufactured and marked in accordance with British Standard BS AU 144b 1977.

Since 1 October 1990 it has been a requirement for tyres on heavy goods vehicles to show load and speed markings in accordance with UN ECE Regulation 30 or 54. It is a legal requirement that these limits of both loading and speed performance are strictly observed. Failure to do so can result in prosecution and could invalidate insurance claims in the event of an accident to a vehicle loaded above the weight limit of the tyres or travelling at a speed in excess of the tyre limit.

Tread Depth

All tyres on goods vehicles of over 3500 kg gross weight must have a tread depth of at least 1 mm across three-quarters of the breadth of the tread and around the entire circumference of the tyre. This 1 mm tread depth must be in a continuous band around the entire circumference of the tyre. Further, on the remaining one-quarter of the width of the tyre where there is no requirement for the tread to be 1 mm deep, the base of the original grooves must be clearly visible.

The minimum tread depth for cars, light vans (not exceeding 3500 kg gross weight) and light trailers was increased to 1.6 millimetres from 1 January 1992, and this applies across the central three-quarters of the width of the tyre and in a continuous band around the entire circumference. The 1 mm limit stated above remains in force for heavy goods vehicles.

Recut Tyres

Recut tyres may be fitted to goods vehicles of over 2540 kg unladen weight which have wheels of at least 405 mm rim diameter and to trailers weighing more than 1020 kg unladen weight and electric vehicles. They must not be used on private cars, dual-purpose vehicles, goods vehicles or trailers of less than the weight or wheel size specified.

Run-Flat and Temporary Use Spare Tyres

The regulations permit the legal use of 'run-flat' tyres in a partially inflated or flat condition and what are described as temporary use spare tyres. This is of more consequence to motor car users because the only tyres of the former (ie run-flat) type currently available are designed for a limited range of private cars (eg Dunlop Denovo tyres). Where a temporary use spare tyre is being used the vehicle speed must not exceed 50 mph otherwise the legal provision which permits their use ceases to apply. The temporary use spare tyre or the wheel to which it is fitted must be of a different colour to the other wheels on the vehicle and a label must be attached to the wheel giving clear information about the precautions to be observed when using the wheel.

Lightweight Trailer Tyres

Tyres fitted to lightweight trailers since 1 April 1987 must be designed and maintained to support the maximum axle weight at its maximum permitted speed (ie 60 mph).

Front Underrun Protection

From August 2003 it is a requirement under EC Directive 2000/40/EC that all goods vehicles exceeding 3.5 tonnes gross weight should be fitted with an approved form of front underrun protection (ie bumper) system (FUPS). This bumper must extend to the full width of the vehicle, must be fitted no higher than 400 mm from the ground when the vehicle is unladen and must meet the same strength requirements specified for rear underrun bumpers (see below). Vehicles over 16 tonnes gross weight used for off-road purposes (ie principally tippers – in European category N2 and N3) and vehicles whose 'use is incompatible with the provisions of front underrun protection' are exempt from the legal requirements.

Rear Underrun Bumpers

Rear underrun bumpers (referred to in legislation as rear underrun protection) must be fitted to most rigid goods vehicles over 3.5 tonnes gross weight manufactured since 1 October 1983 and first used since 1 April 1984. Trailers, including semi-trailers, over 1020 kg unladen weight manufactured since 1 May 1983 must also be fitted with bumpers. Certain vehicles and trailers are exempt (see below) from the fitting requirements and there were no retrospective fitting requirements for existing vehicles.

Strength of Bumpers

Rear underrun bumpers must be constructed so they are capable of withstanding a force equivalent to half the gross weight of the vehicle or trailer or a maximum of 10 tonnes, whichever is the *lesser*, without deflecting more than 400 mm measured from the rearmost point of the vehicle or trailer – not from the original vertical position of the bumper.

Fitment of Bumpers

Bumpers must be fitted as near as possible to the rear of the vehicle and the lower edge must be not more than 550 mm from the ground (see Figure 13.15). Normally, only one bumper would be fitted, but where a tail-lift is fitted or the bodywork or other parts of the vehicle make this impracticable, two or more bumpers may be fitted. When a single full-width bumper is fitted it must extend on each side of the centre to within at least 100 mm from the outermost width of the rear axle, but must not in any case extend beyond the width of the rear axle measured across the outermost face of the tyres. When two or more bumpers are fitted, for the reasons mentioned above, the space between each part of the bumper must not exceed 500 mm and the outermost edge of the bumpers must extend to within at least 350 mm from the outermost width of the rear axle (see Figure 13.16). Bumpers must not protrude beyond the width of the vehicle or trailer and the outside ends of the bumper must not be bent backwards.

Maintenance of Bumpers

Rear bumpers must be maintained free from any obvious defect that would adversely affect their performance in giving resistance to impact from the rear. It is

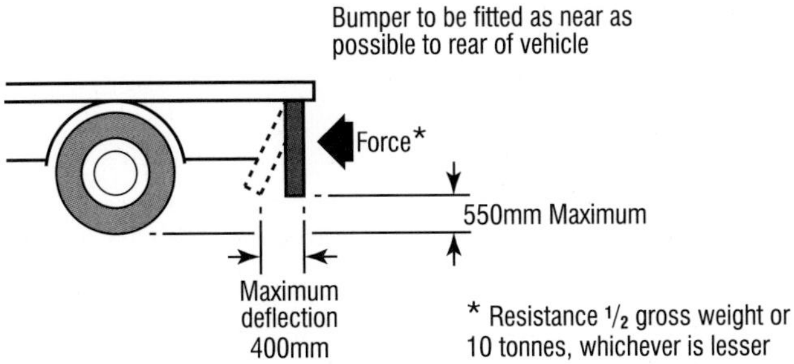

Figure 13.15 *Illustration of the rear underrun bumper force resistance requirements and ground clearance dimension*

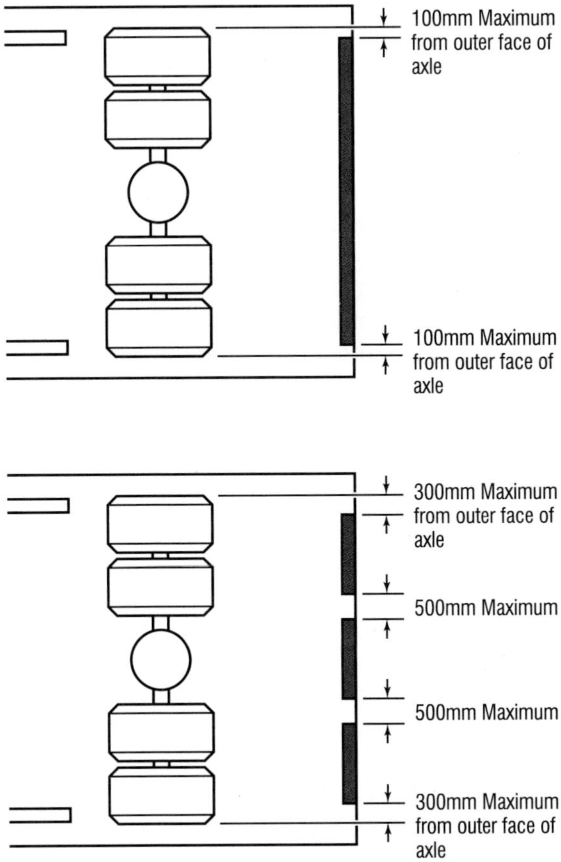

Figure 13.16 *Illustrations of the fitting dimensions for single and multiple rear underrun bumpers*

also important to ensure that the dimensional requirements are met, particularly if the bumper is damaged (for example by forklift truck impact or reversing on to loading bays).

Exemptions

Rear underrun bumpers do not have to be fitted to vehicles and trailers in the following list:

- Vehicles incapable of a speed exceeding 15 mph on the level under their own power.
- Tractive units of articulated vehicles.
- Agricultural trailers, trailed appliances and agricultural motor vehicles.
- Engineering plant.
- Fire engines.
- Road spreading vehicles (ie for salt and grit).
- Rear tipping vehicles.
- Military, naval or air-force vehicles.
- Vehicles being taken to have bodywork fitted, or being taken for quality or safety checks by the manufacturer, distributor or dealer in such vehicles.
- Vehicles being driven or towed to a place to have a rear underrun bumper fitted by prior arrangement.
- Vehicles designed to carry other vehicles which are loaded from the rear (eg car transporters).
- Trailers designed and constructed (not just adapted) to carry round timber, beams or girders of exceptional length.
- Vehicles fitted with tail-lifts where the tail-lift forms part of the floor of the vehicle and extends to a length of at least one metre.
- Temporarily imported foreign vehicles and semi-trailers.
- Vehicles specially designed (not just adapted) for carrying and mixing liquid concrete.
- Vehicles designed and used solely for the delivery of coal by means of a conveyor fixed to the vehicle so as to make the fitment of a rear underrun bumper impracticable.

View to the Front

Drivers must have a full view of the road and traffic ahead at all times when driving. Obstructing the windscreen with mascots, stickers, stone guards and other such things could result in prosecution and/or failure of the vehicle when it is presented for annual test.

Vehicles with closed-circuit television (CCTV) rear-view monitors mounted in the cab in such a position that the driver's view to the front is obscured, or partly obscured, will be failed their annual test by VOSA. This particularly applies to refuse collectors and ambulances which traditionally use such equipment. (See also page 355.)

Windscreen Wipers and Washers

Windscreen Wipers

All vehicles must be fitted with one or more efficient automatic windscreen wipers capable of clearing the windscreen to provide the driver with an adequate view to the front and sides of the vehicle. They must be maintained in good and efficient working order and must be adjusted properly. This provision does not apply if the driver has an adequate view of the road without looking through the windscreen.

Windscreen Washers

Vehicles required to be fitted with windscreen wipers must be fitted with a windscreen washer that is capable, in conjunction with the wipers, of clearing the area of the windscreen swept by the wipers of mud or dirt. Washers are not required on land tractors, track-laying vehicles and vehicles that cannot travel at more than 20 mph.

A vehicle should not be driven on the road with defective windscreen wipers or washers. This is an offence which could result in a fine of up to £1,000 (as with many other C&U regulation offences).

Wings

Goods vehicles and trailers must be fitted with wings to catch, as far as practicable, mud and water thrown up by the wheels unless adequate protection is provided by the bodywork.

Articulated vehicles and trailers used for carrying round timber are exempt from these requirements in respect of all except the front wheels of the tractive unit. Vehicles and trailers in an unfinished condition that are proceeding to a bodybuilder for work to be completed and works trucks are also exempt from the need to have wings.

USE OF VEHICLES

In addition to the foregoing constructional requirements which are mainly the responsibility of the vehicle manufacturer or the person building the bodywork, there are requirements regarding the use of vehicles which are the responsibility of the operator. However, as mentioned earlier, the vehicle user carries full legal responsibility for the mechanical condition of a vehicle on the road and its compliance with the constructional requirements even if the fault which led to an offence could be laid at the door of the chassis manufacturer, the bodybuilder, an ancillary equipment supplier or the dealer.

These requirements cover such items as vehicle weights (which were dealt with earlier), towing, fumes, the condition and maintenance of vehicles and their components, noise, smoke, and general safety in the use of vehicles. They also include the regulations regarding the number of trailers which a vehicle may draw.

Gas-Powered Vehicles

Regulations specify technical standards for fuel tanks or containers, the filling system and valves and general requirements for gas propulsion systems in motor vehicles. The regulations permit the use of LPG only in gas-propelled vehicles, though this may be combined with petrol fuel systems, and the use of methane or hydrogen is prohibited. The D*f*T has published a free guide *Gas Installations in Motor Vehicles and Trailers*, which is available from: Department of Transport, B3, Victoria Road, South Ruislip, Middlesex HA4 0NZ.

A new article (17A) and new schedules (5A and 5B) to the Motor Vehicles (Authorisation of Special Types) General Order 1979 – inserted by virtue of Amendment No 2 of 1 December 1998 to the Order – authorize the use on roads of vehicles propelled by compressed natural gas (CNG), notwithstanding that they do not comply with the requirements of the C&U regulations. They must, however, comply with the provisions of the new schedules as to constructional requirements and the requirements for testing gas containers used in these systems.

Noise

It is an offence to use, or cause or permit to be used on a road a motor vehicle or trailer which causes an excessive noise because of a defect, lack of repair or faulty adjustment of components or load. Also, no motor vehicle must be used on a road in such a manner as to cause any excessive noise which could have been reasonably avoided by the driver. Noise for these purposes is the combined noise emitted by the exhaust plus that from the tyres, engine, bodywork and equipment, and the load. Noise levels for goods vehicles are measured by special meters either by the police or by VI examiners at goods vehicle testing stations and occasionally on roadside tests.

Noise Limits

It is no longer a legal requirement for goods vehicles to maintain the noise standards set when the vehicle was manufactured. The regulations covering noise emissions are adequately covered under other provisions (see above).

Reversing Alarms

It is legally permissible to fit reversing alarms to certain goods and passenger vehicles if desired – *the regulations do not make it mandatory to do so.* Such alarms may be voluntarily fitted and used on the following vehicles:

- commercial vehicles over two tonnes gross weight;
- passenger vehicles with nine or more seats;
- engineering plant;
- works trucks.

Time Restriction on Use of Reversing Alarms

The alarms are subject to the same night-time restrictions that apply to the sounding of horns in built-up areas (ie not after 11.30 pm and before 7.00 am – 23.30 to 07.00) and the sound emitted must not be capable of being confused with the pelican crossing 'safe to cross' signal.

Restriction on Fitment of Reversing Alarms

Such alarms *must not* be fitted to light goods vehicles below two tonnes gross weight or to motor cars.

Advice on Use of Alarms

A number of cases have arisen following reversing accidents resulting in death or injury where the Health and Safety Executive has prosecuted the vehicle operators concerned for not voluntarily fitting reversing alarms. In other words, the HSE line is that the offenders had not taken sufficient steps to ensure safety when their vehicles were reversing by fitting equipment which the law permits, but not mandatorily requires, them to do.

Televisions in Vehicles

It is illegal for a vehicle to be fitted with television-receiving apparatus where the driver can see the screen either directly or by reflection, except where such equipment displays nothing other than information:

- about the state of the vehicle or its equipment;
- about the location of the vehicle and the road on which it is located;
- to assist the driver to see the road adjacent to the vehicle (eg to the rear when reversing); or
- to assist the driver to reach his destination.

NB: See also page 352.

Towing

Goods vehicles may draw (ie tow) only one trailer. An exception to this is when a rigid goods vehicle tows a broken-down vehicle on a towing ambulance or dolly in which case although this is counted as towing two trailers it is allowed. In a case where an articulated vehicle has broken down, this may be towed by a rigid goods vehicle so long as the articulated vehicle is not loaded. In these circumstances the outfit is treated as one trailer only, but if it is loaded an articulated outfit being towed is considered to be two trailers and it would be illegal for a normal goods vehicle (ie a heavy motor car) to tow it. Only a locomotive can tow a broken-down articulated vehicle which is laden (see below).

Motor tractors may draw one laden or two unladen trailers and locomotives may draw three trailers (see p 322 for definitions).

Composite Trailers

The C&U regulations make it permissible for rigid goods vehicles (apart from locomotives and motor tractors) to draw two trailers instead of only one, when one of the trailers is a towing implement (ie a dolly) and the other is an articulated-type semi-trailer secured to and resting on, or suspended from, the dolly. This combination of dolly and semi-trailer is known as a composite trailer (Figure 13.17).

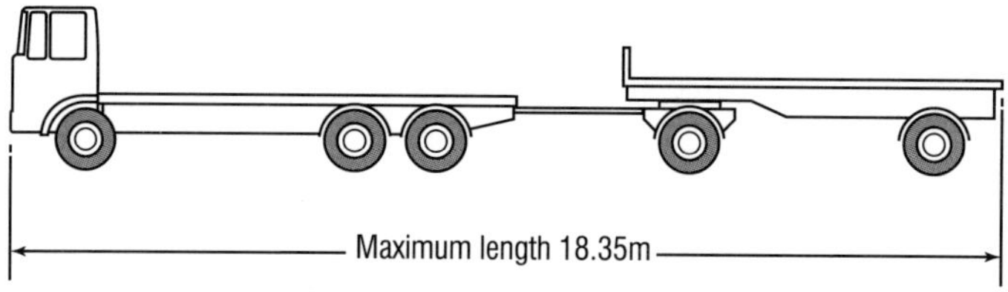

Figure 13.17 *A conventional six-wheeled rigid vehicle drawing a dolly mounted semi-trailer for which the maximum overall length is 18.35 metres*

To comply with the regulations, the dolly needs to have two or more wheels and be specifically designed to support a superimposed semi-trailer. Dollies must display a manufacturer's plate and they are subject to the annual heavy goods vehicle test.

Towing Distance

The distance between the nearest points of two vehicles joined by a tow rope or chain must not exceed 4.5 metres. When the distance between the two vehicles exceeds 1.5 metres the rope, chain or bar must be made clearly visible from both sides of the vehicles. There is no specified maximum distance limit if a solid tow-bar is used for towing.

14 Type Approval

Type Approval is a scheme which requires vehicle manufacturers to submit new vehicles (ie new designs, new models and changes of specifications for existing approved models) for approval before they are put on the market. The Department for Transport examines the vehicle submitted to ensure that it meets all legal requirements and also meets minimum standards of construction and performance. When a vehicle has been approved the manufacturer is then required by law to build all vehicles of a similar type to exactly those standards and certify this fact to the customer by the issue of a Certificate of Conformity.

An EU directive lays down the basic procedures for the Type Approval scheme for vehicles and components. Subsidiary directives have also been issued setting out agreed standards on some aspects of vehicle safety or pollution. They cover the same ground as existing national regulations. The directives do not yet cover all vehicle features which need to be regulated so until the programme is complete both EU directives and national regulations apply to relevant items. The UK established a non-compulsory scheme to enable exporting vehicle manufacturers to gain the necessary Type Approval in order to sell their products in EU countries.

VEHICLES COVERED

Type Approval requirements apply to all passenger cars first licensed for use in the UK from 1 April 1978. The Motor Vehicles (Type Approval for Goods Vehicles) (Great Britain) Regulations 1982 (as amended) make the application of the Type Approval scheme compulsory for goods vehicles, motor caravans, motor ambulances and bi-purpose vehicles constructed for the carriage of both goods and passengers (but not more than eight passengers) which are not included in the car Type Approval scheme and which have been manufactured since 1 October 1982 and first used since 1 April 1983. Applications for Type Approval could be made from 1 October 1981 onwards.

While there is currently no Type Approval scheme for trailers or semi-trailers the EU is proposing that they should be included in Type Approval requirements and is currently endeavouring to harmonize national legislation to provide for the introduction of an EU-wide package covering marking, brakes, lighting, underrun protection and sideguards, spray suppression and bulkhead load strength by 2002.

Exemptions from Type Approval

Type Approval does not apply to the following vehicles:

- Vehicles manufactured before 1 October 1982, whenever they are first registered.
- Vehicles manufactured on or after 1 October 1982, providing they are first licensed before 1 April 1983.
- Temporarily imported vehicles.
- Vehicles proceeding for export from the UK.
- Vehicles in the service of visiting forces or headquarters.
- Certain vehicles which are, or were formerly, in use in the public service or the Crown.
- Prototypes which are not intended for general use on the roads.
- Motor tractors, light locomotives and heavy locomotives.
- Engineering plant, pedestrian-controlled vehicles, straddle carriers, works trucks and track-laying vehicles.
- Vehicles specially designed and constructed for use in private premises for moving excavated materials, vehicles fitted with movable platforms and vehicles designed and constructed for the carriage of abnormal indivisible loads.
- Tower wagons.
- Fire engines.
- Road rollers.
- Steam-propelled vehicles.
- Vehicles constructed for the purpose of preventing or reducing the effect of snow or ice on roads.
- Two-wheeled motorcycles with or without sidecars.
- Electrically propelled vehicles.
- Breakdown vehicles.
- Any vehicle not exceeding 3500 kg maximum gross weight which is constructed or assembled by a person not ordinarily engaged in the manufacture of goods vehicles of that description.
- Vehicles not exceeding 3500 kg maximum gross weight providing that:
  - the vehicle has been purchased outside the UK for the personal use of the individual importing it or his dependants;
  - the vehicle has been so used by that individual or his dependants on roads outside the UK before it is imported;
  - the vehicle is intended solely for such personal use in the UK; and
  - the individual importing the vehicle intends, at the time when the vehicle is imported, to remain in the UK for not less than 12 months from that date.

RESPONSIBILITY FOR COMPLIANCE

Responsibility for complying with the complex construction standards rests with the manufacturer although users are still responsible for maintaining vehicles in roadworthy condition. The construction standards applied by the scheme are limited to those which can be approved during the primary stage of manufacture. The standards are identical to those already required under the Construction and Use Regulations but under this scheme vehicles have to be approved before they can be used on the road.

Effects on Plating and Testing

The scheme requires plated weights for heavy goods vehicles to be set during the Type Approval process instead of waiting until the first annual plating and testing examination. This means that heavy vehicle operators need a Type Approval Certificate in order to get a Ministry plate for display in the vehicle cab (see Chapter 16). However, annual testing is retained so as to check the condition of vehicles and to ensure that plated weights are accurate.

RESPONSIBILITY FOR TYPE APPROVAL

All aspects of Type Approval are the responsibility of VOSA.

THE STANDARDS CHECKED

To obtain goods vehicle national Type Approval it is first necessary to obtain individual systems approvals for the following items:

- Power-to-weight ratio (not applicable to petrol-engined vehicles or dual-purpose vehicles).
- Gaseous exhaust emissions (petrol-engined vehicles only).
- Particle emission (ie exhaust smoke) (diesel-engined vehicles only).
- External noise level.
- Radio-interference suppression (petrol-engined vehicles only).
- Brakes.

ARRANGEMENTS FOR FIRST LICENSING OF VEHICLES

For vehicles over 1525 kg unladen weight or which form part of an articulated vehicle, application for first licensing on form V55 must be accompanied by two copies of the Type Approval Certificate, which should have been supplied with the vehicle. On one of these the applicant must complete a declaration saying whether or not the vehicle is exempt from the plating and testing regulations (see Chapter 16 for full details) and whether it has been altered in any way that has to be notified to VOSA under the Type Approval regulations and, if so, whether any action arising from the notification has been satisfactorily completed.

Issue of Plates

When application is made for first licensing a vehicle which is subject to plating and testing, the local Vehicle Registration Office (VRO) will send a copy of the Type Approval Certificate with the applicant's declaration to the Goods Vehicle Centre (GVC) at Swansea. The second copy will be stamped and returned to the applicant to serve as a temporary Ministry plate. When the GVC receives the copy of the certificate, and if the details compare satisfactorily with those on the copy sent direct by the vehicle manufacturers, it will issue a Ministry plate and laminated plating certificate. These will be sent direct to the person or company in whose name the vehicle is registered. Thus, operators buying new vehicles receive their first plate and plating certificate for the vehicle from the GVC at the time of licensing rather than from the goods vehicle testing station when the vehicle is presented for its first annual test as under previous arrangements. When application is made for licensing a vehicle which is exempt from plating and testing, a copy of the certificate and the declaration will be sent to the GVC so that it is aware that it is exempt.

Refusal to Licence

Since 1 April 1983 no vehicle subject to the Type Approval regulations will be first licensed unless the DVLA registration form V55 has a valid Type Approval number on it or it is an exempt vehicle.

ALTERATION TO VEHICLES

If a vehicle, which has been issued with an Approval Certificate and supplied to a dealer or direct to an operator, is modified by them, prior to first licensing, they must notify the Vehicles and Components Approvals Division (VCA) at Bristol and send the certificate for the vehicle together with full technical details, drawings of the alterations and details of the weights on the certificate which need, or may need, changing.

The VCA will judge whether the alterations affect the vehicle's compliance with the regulations; if they do not affect compliance the certificate will be returned so the vehicle can be licensed. If they do contravene compliance, the certificate will be cancelled and fresh approval will need to be obtained before the vehicle can be licensed. This is a complex and costly procedure that most operators will want to avoid; they can do so by registering and licensing the vehicle before any alterations are carried out.

15 Vehicle Lighting and Marking

The legal requirements for vehicles to be fitted with and to display lights at night and other times, and for the fitment of reflectors and other markings on vehicles, are contained in the Road Vehicles Lighting Regulations 1989 (as amended). These regulations specify in considerable detail all the requirements for the position of lamps and reflectors and the angles from which they must be visible. This chapter can only include a summary of the relevant requirements and dimensions as they apply to goods vehicles which, for most normal purposes, is satisfactory because vehicles and trailers are generally ready-fitted with lamps and reflectors conforming to legal requirements when supplied from new. However, for a variety of reasons, operators may find it necessary to replace and relocate lamps from time to time when carrying out repairs and conversions, and at this time they are advised to check the regulations carefully to ensure strict compliance with the law. Not only can prosecution follow for incorrectly positioned or non-functioning lights and reflectors, but vehicles could fail their annual test on this account which adds to operating costs and wastes time. There are also the safety considerations with the lives of both vehicle drivers and other road users at risk if vehicles are not showing correct or adequate lights.

The regulations require that lights and reflectors which are fitted to vehicles must be maintained so as to enable them to be driven on a road between sunset and sunrise (times are published in most daily and local newspapers), or in seriously reduced visibility between sunrise and sunset, or to be parked on a road between sunset and sunrise without contravening the regulations. All lights must be kept clean and in good working order. It is an offence to cause undue dazzle or discomfort to other road users by the use of lights or through their faulty adjustment, or to have defective or obscured lighting on a vehicle at any time. There is no longer a defence to a charge of having defective lights on a vehicle (see below).

OBLIGATORY LIGHTS

Between sunset and sunrise vehicles used on a public road must display the following obligatory lights, other lights and reflectors:

- Two front position lamps (ie sidelamps) showing white lights to the front.
- Two rear position lamps (ie rear lamps) showing red lights to the rear.
- Two headlamps showing white lights to the front (alternatively, the light may be yellow).
- Illumination for the rear number (ie registration) plate when the other vehicle lights are on.
- Certain goods vehicles and trailers additionally require side marker lamps plus side-facing reflectors and rear reflective markings – see below.
- One or two red rear fog lamps on post-1 April 1980 vehicles.
- Two red reflex retro-reflectors at the rear.
- End-outline marker lamps.
- Any other lights or lighting devices with which the vehicle is fitted (eg stop lamps, hazard warning signals, running lamps, dim-dip devices and headlamp levelling devices).

It is illegal except in certain specified cases for a goods vehicle to show a white light to the rear (showing such a light when reversing and indirect illumination of the rear registration plate are permitted, for example) or a red light to the front.

Headlamps

Motor vehicles must be fitted with two headlamps capable of showing a white or yellow light to the front – both lamps must emit the same colour light. Headlamps must be either permanently dipped or fitted with dipping equipment. Vehicles first used since 1 April 1987 must have dim-dip lighting devices unless their lighting equipment complies with EU requirements (see below).

Headlamps must be mounted so that they are not lower than 500 mm from the ground and not higher than 1200 mm. They must be placed on either side of the vehicle with their illuminated areas not more than 400 mm from the side of the vehicle. They must be equipped with bulbs or sealed-beam units of not less than 30 watts in the case of vehicles first used before 1 April 1986. For vehicles used since this date no minimum wattage requirement is specified.

Headlamp Exemptions

Certain vehicles are exempt from the headlamp requirements. These include vehicles with less than four wheels, pedestrian-controlled vehicles, agricultural implements, land tractors, works trucks, vehicles not capable of travelling at a speed of more than 6 mph and military vehicles.

Headlamps on Electric Vehicles

Electrically propelled goods vehicles with four or more wheels registered before October 1969 and electric vehicles with two or three wheels first used before 1 January 1972 and capable of a speed of more than 15 mph are required to comply with the headlamp requirements. Those electrically propelled vehicles which are incapable of speeds of more than 15 mph are exempt from the headlamp requirements.

Use of Headlamps

Headlamps must be adjusted so that they do not cause undue dazzle or discomfort to other road users. When vehicles which require headlamps are being driven on unlit roads between sunset and sunrise and in seriously reduced daytime visibility the headlamps must be illuminated. They must be switched off when the vehicle is stationary except at traffic stops.

Incorrect aim of vehicle headlights is reported to be one of the most common reasons for annual goods vehicle test failure. Correction can be effected by simple adjustment, or by changing bulbs – inconsistent filament positioning on some low-quality bulbs being a possible cause of incorrect headlight aim.

NB: Unlit roads are roads on which there are no street lamps or on which the street lamps are more than 200 yds apart.

Headlamps in Daylight

It is a legal requirement for vehicles to use side-position lights (ie sidelights) and dipped headlights when travelling in seriously reduced daytime visibility conditions such as in fog, smoke, heavy rain, spray or snow. If matching fog or fog and spot lights are fitted in pairs these may be used instead of headlights, but sidelights must still be used and the other vehicle lights must be on (eg side marker lights).

There is no specific definition of 'seriously reduced visibility' in the regulations. It is left to drivers to judge whether it is advisable and sensible to use their lights to enable their vehicle to be seen by others.

Dim-Dip Lighting

Since 1 April 1987 newly registered vehicles must be fitted with dim-dip lighting devices which operate automatically when the obligatory lights of the vehicle are switched on and ensure that either 10 per cent (with halogen) or 15 per cent (with grading filament lamps) of the normal dipped beam intensity shows when the vehicle ignition key is switched on or the engine is running. The European Court of Justice has ruled that it is unfair for the British Government to legislate for the fitment of dim-dip lighting devices for vehicles which already comply with the EU Lighting Directive (EC 756/1976 as amended by EC 663/1991) which has no dim-dip requirement.

Front Position Lamps (sidelamps)

Two front position lamps (ie sidelamps) emitting a white light through a diffused lens must be fitted to all motor vehicles (with three or more wheels) and trailers (except those not more than 1600 mm wide), those no longer than 2300 mm (excluding the drawbar) built before 1 October 1985 and those used for carrying and launching boats. If such lamps are incorporated within a headlamp showing a yellow light then the side position lamps may be yellow. No minimum wattage is specified for these lights. The lights must be equal in height from the ground and mounted not more than 1500 mm from the ground (in exceptional circumstances this height can be increased to 2100 mm) in the case of vehicles first used on or after 1 April 1986 and 2300 mm in other cases, and not more than 400 mm from the outer edge of the vehicle for vehicles first used since 1 April 1986 and 510 mm in other cases. No minimum height above the ground is specified.

Rear Position Lamps (rear lamps)

Two red rear position lamps (ie rear lamps) must be fitted to all motor vehicles and trailers. There is no specified wattage for these lights. They must be mounted not less than 350 mm and not more than 1500 mm (2100 mm in exceptional circumstances) from the ground. They must be at least 500 mm apart (no specified distance on pre-1 April 1986 registered vehicle) and not more than 400 mm (800 mm on pre-1 April 1986 registered vehicle) from the outside edge of the vehicle.

Stop Lamps

All goods vehicles (except those not capable of more than 25 mph) must be fitted with red stop lamps which are maintained in a clean condition and in good and efficient working order. Vehicles registered before 1 January 1971 need only one such lamp which must be fitted at the centre or to the offside of the vehicle, although a second matching lamp may be fitted on the near side. Vehicles registered since that date need two such lamps (specified wattage 15 to 36 watts except with pre-1 January 1971 registered vehicles) mounted not less than 350 mm from the ground and not more than 1500 mm (in exceptional circumstances this may be increased to 2100 mm) and they must be at least 400 mm apart. Such lamps must be visible horizontally from 45 degrees on either side and normally from 15 degrees above and below vertically (from only 5 degrees below where fitted less than 750 mm from the ground and only 10 degrees below when fitted not more than 1500 mm from the ground).

Reversing Lamps

White reversing lamps (not more than two) may be fitted to vehicles provided they are only used while the vehicle is reversing and operate automatically only when reverse gear is selected. Alternatively, they may be operated manually by a switch (which serves no other purpose) in the driver's cab provided that a warning device indicates to the driver that the lights are illuminated. The lights must be adjusted so as not to cause dazzle to other road users. Such lamps when bearing an 'e' approval mark do not have to meet minimum wattage requirements but those without approval marks must not exceed 24 watts.

Number Plate Lamp

Rear number plates (ie registration plates) on vehicles must be indirectly illuminated when the other obligatory lamps on the vehicles are lit. The light must be white and must be shielded so that it only illuminates the number plate and does not show to the rear.

Rear Fog Lamps

Rear fog lamps (at least one, but two may be fitted) must be fitted to new vehicles and trailers manufactured on or after 1 October 1979 and first used since 1 April 1980. There is no legal requirement to fit such lamps on pre- 1 October 1979 registered vehicles but if they are fitted voluntarily they must comply with the regulations in regard to mounting position, method of wiring and use.

Rear Fog Lamps on Articulated and Towing Vehicles

In the case of articulated combinations, the relevant date in this connection is the date of the older of the tractive unit or semi-trailer. Thus, if a post-April 1980 registered tractive unit is coupled to a pre-October 1979-built trailer there appears to be no legal requirement for the vehicle to carry rear fog lamps. This means there is no retrospective fitting requirement for such lamps on older trailers. A broken-down vehicle being towed does not need rear fog lamps. However, it is important to remember the dangers which arise if such a combination is used when the tractive unit or towing vehicle itself has rear fog lamps which would, in some instances, be visible to following motorists who, in bad visibility, might not be aware of some 40 feet of trailer or another vehicle on tow behind the lights.

Mounting of Rear Fog Lamps

Rear fog lamps must be mounted either singly in the centre or on the offside of the vehicle or in a matched pair not less than 250 mm and not more than 1000 mm from the ground (in the case of agricultural vehicles this height limit is increased to 1900 mm or in cases where, because of the shape of the vehicle, 1000 mm is not practical it may be increased to 2100 mm). The lamps must be at least 100 mm from existing stop lamps.

Restriction on Wiring of Rear Fog Lamps

The lights must be wired so that they only operate when the other statutory lights on the vehicle are switched on; they must not be wired into the brake/stop light circuit and the driver must be provided with an indicator to show him when the lights are in use.

Use of Rear Fog Lamps

The lights should only be used in conditions affecting the visibility of the driver (ie in fog, smoke, heavy rain or spray, snow, dense cloud, etc), when the vehicle is in motion or during an enforced stoppage (a motorway hold-up, for example). They must not cause dazzle. The *Highway Code* recommends these lights should not be used unless visibility is below 100 metres.

SIDE MARKER LAMPS

Side marker lamps must be fitted on vehicles and trailers as follows:

- Vehicles first used on or after 1 April 1991 and trailers made from 1 October 1990 and being over six metres long:

- one lamp on each side within four metres of the front of the vehicle;
- one lamp on each side within one metre of the rear of the vehicle;
- additional lamps on each side at three metre intervals (or if impracticable four metres) between front and rear side marker lamps.

- Vehicles (including a combination of vehicles) over 18.3 metres long (including the length of the load):
  - one lamp on each side within 9.15 metres of the front of the vehicle;
  - one lamp on each side within 3.05 metres of the rear of the vehicle;
  - additional lamps on each side at 3.05 metre intervals between front and rear side marker lamps.
- Vehicles in combination between 12.2 metres and 18.3 metres long (but not articulated vehicles) carrying a supported load:
  - one lamp on each side within 1530 mm of the rear of the rearmost vehicle in the combination;
  - one lamp on each side within 1530 mm of the centre of the load, if the load extends further than 9.15 metres to the rear of the drawing vehicle.
- Trailers more than 9.15 metres long (6 metres for post-1 October 1990 trailers):
  - one lamp on each side within 1530 mm of the centre of the trailer length.

Side marker lamps fitted to pre-1 October 1990-built trailers may show white side marker lights to the front and red lights to the rear; in all other cases such lights must be amber. They must be positioned not more than 2300 mm from the ground.

END-OUTLINE MARKER LAMPS

Vehicles (except those less than 2100 mm wide and those first used before 1 April 1991) and trailers (except those less than 2100 mm wide and those built before 1 October 1990) must be fitted with two end-outline marker lamps visible from the front and two visible from the rear. They must be positioned no more than 400 mm in from the outer edges of the vehicle/trailer and mounted at the front at least level with the top of the windscreen. They must show white lights to the front and red lights to the rear.

LIGHTING SWITCHES

On vehicles first used since 1 April 1991 a single lighting switch only must be used to illuminate all front and rear position lamps, side and end-outline marker lamps and rear number plate lamp although one or more front or rear position lamps may be capable of being switched on independently.

VISIBILITY OF LIGHTS AND REFLECTORS

A part, at least, of each front and rear position light, front and rear-mounted direction indicator lamp and rear retro-reflector required to be fitted to a vehicle/trailer must be capable of being seen from directly in front or behind the lamp or reflector when the vehicle doors, tailgate, boot lid, engine cover or other movable part of the vehicle is in a fixed open position.

LIGHTS ON PROJECTING LOADS

Details of the requirements for the display of lights on projecting loads are given fully in Chapter 22 which covers this subject.

DIRECTION INDICATORS

All goods vehicles must be fitted with amber-coloured direction indicators (on pre-September 1965 registered vehicles, indicators can be white facing to the front and red facing to the rear) at the front and rear which must be fixed to the vehicle not more than 1500 mm (2300 mm in exceptional cases) and not less than 350 mm above the ground, at least 500 mm apart and not more than 400 mm from the outer edges of the vehicle. Side repeater indicators are required on vehicles first used since 1 April 1986 and these must be fitted within 2600 mm of the front of the vehicle. Normally vehicles should have one indicator on each side at the front and rear but may have two on each side at the rear. They must not have more than one on each side at the front.

Indicators bearing approval marks do not have to meet minimum wattage requirements but those without such marks must be between 15 and 36 watts. They must flash at a rate of between 60 and 120 times a minute and a visible or audible warning must indicate to the driver when they are operating. The indicators must be maintained in a clean condition and in good and efficient working order.

Semaphore Arm Indicators

Vehicles first registered before 1 September 1965 are allowed to have either semaphore arm or flashing indicators. The semaphore arm type must be amber in colour but the flashing indicators may show a white light to the front and a red light to the rear.

Hazard Warning

Direction indicators operating on both sides of the vehicle simultaneously as a hazard warning to other road users are required by law on all vehicles first used since 1 April 1986. They must be actuated by a switch solely controlling that device and a warning light must indicate to the driver that the device is being operated. The hazard indicators may be used when the vehicle is stationary on a road, or any part of the road (ie not just the carriageway), because of a breakdown of it or another vehicle, an accident or other emergency situation, or when the vehicle is causing a temporary obstruction on a road when loading or unloading.

Emergency Warning Triangles

As an additional warning of a hazard drivers *may* (ie it is not compulsory to do so) place a red warning triangle on the road to the rear of a vehicle causing a temporary obstruction (eg through breakdown). The triangle must be made and marked to British Standard Specification BS AU 47: 1965. It must be placed upright on the road, 45 metres to the rear of the obstruction and on the same side.

Other safety devices may be used to warn of vehicles broken down on the roadside. These include traffic cones, warning lamps and traffic pyramids, as well as conventional warning triangles mentioned above.

OPTIONAL LAMPS

Optional main-beam headlamps (which includes spot lamps) may be fitted to a vehicle but they must be capable of being dipped, and if fitted as a matched pair they must also be capable of being switched off together – not individually. They must emit either a white or yellow light, must be adjusted so that they do not cause dazzle to other road users and must not be lit when the vehicle is parked. If optional front fog lamps are fitted and used singly, the headlamps must also be illuminated.

These optional lamps should be positioned not more than 1200 mm from the ground and not more than 400 mm from the sides of the vehicle. They should be aligned so that the upper edge of the beam is, as near as practicable, 3 per cent below the horizontal when the vehicle is at its kerbside weight and has a weight of 75 kg on the driver's seat.

Vehicles first registered since 1 April 1991 may be fitted with only one pair of extra dipped-beam headlamps and then only on vehicles intended to be driven on the right-hand side of the road. They must be wired so that only one pair of dipped-beam headlamps can be used at any one time. Pre-April 1991 registered vehicles may have any number of additional dipped-beam headlamps.

Any number of extra main-beam headlamps (including spot/driving lamps) may be fitted and there is no restriction on their fitment or use except that they must not cause dazzle to other road users.

Blue Lights

There is an increasing trend towards the use of blue lights on the front of, or in the windscreens of, both cars and commercial vehicles, giving rise to questions about their legality. As the law stands at present it appears that it is not illegal to fit or use such lights as long as they do not flash in the same manner as those on emergency service vehicles. The police, however, are opposed to their use and are anxious to see them prohibited by law, because of the risk of confusion with emergency service vehicles and because of their potential use by criminals purporting to be police. A number of prosecutions and fixed penalty notices have been issued to drivers displaying such lights, and others can expect to be challenged, although it will take a test case to fully determine the legal position.

Drivers variously believe that the ultraviolet light emitted by such lamps reduces in-cab white-light glare from oncoming vehicle headlights and that they help to 'confuse' roadside enforcement cameras.

WARNING BEACONS

Amber warning beacons must be fitted to vehicles with four or more wheels and having a maximum speed no greater than 25 mph when using unrestricted dual-carriageway roads except where such use is merely 'for crossing the carriageway in the quickest manner practicable in the circumstances'.

Amber warning beacons may also be fitted to vehicles used at the scene of an emergency, when it is necessary or desirable to warn of the presence of a vehicle on the road (eg Special Types vehicles carrying abnormal loads) and to breakdown vehicles used at the scene of accidents and breakdowns and when towing broken-down vehicles.

Green warning beacons may be used on vehicles by medical practitioners registered with the General Medical Council when travelling to or dealing with an emergency.

Blue warning beacons and other special warning lamps may only be used on emergency vehicles (ie ambulance, fire brigade or police service vehicles; Forestry Commission fire fighting vehicles; military bomb disposal vehicles; RAF Mountain Rescue vehicles; Blood Transfusion Service vehicles; British Coal mines rescue vehicles; HM Coastguard and Coast Life Saving Corps vehicles; RNLI vehicles; and those used primarily for transporting human tissue for transplanting).

In all cases such beacons should be fitted with their centres no less than 1200 mm from the ground and visible from any point at a reasonable distance from the vehicle. The light itself must show not less than 60 and not more than 240 times per minute.

SWIVELLING SPOTLIGHTS (WORK LAMPS)

White swivelling spotlights may be used only at the scene of an accident, breakdown or roadworks to illuminate the working area or work in the vicinity of the vehicle provided it does not cause undue dazzle or discomfort to the driver of any vehicle.

REAR RETRO-REFLECTORS

Motor vehicles must be fitted with two red reflex retro-reflectors facing squarely to the rear. Reflectors must be fitted not more than 900 mm and not less than 350 mm from the ground. For normal goods vehicles and trailers they must be within 400 mm of the outer edge of the vehicle or trailer and not less than 600 mm apart. Reflectors must be capable of being seen from an angle of 30 degrees on either side.

Triangular Rear Reflectors

Triangular rear reflectors must not be used other than on trailers or broken-down vehicles being towed.

SIDE RETRO-REFLECTORS

Vehicles more than six metres long first used since 1 April 1986 (more than eight metres long if first used before 1 April 1986) and trailers more than five metres long must be fitted with two (or more as necessary) amber side retro-reflectors on each side. One reflector on each side must be fitted not more than one metre from the extreme rear end of the vehicle and another no more than four metres from the front of the vehicle with further reflectors at minimum three metre intervals (or can be four metre intervals) along its length. They must be mounted not more than 1500 mm and not less than 350 mm from the ground. On pre-April 1986 vehicles one reflector must be positioned in the middle third of the vehicle length and the other within one metre of the rear. Where such reflectors are mounted within one metre of the rear of the vehicle/trailer they may be coloured red instead of amber.

FRONT RETRO-REFLECTORS

Trailers built since 1 October 1990 must be fitted with two obligatory front retro-reflectors, white in colour and mounted facing forwards, at least 350 mm but not more than 900 mm from the ground and no more than 150 mm in from the outer edges of the trailer and at least 600 mm apart.

VEHICLE MARKINGS

Number (registration) Plates

All vehicles first registered since 1 January 1973 must be fitted with number plates made of reflecting material complying with BS AU 145a. This requirement does not apply to goods vehicles over 7.5 tonnes gross weight which are required to display rear reflective markers (see pp 367–74) or works trucks, agricultural machines and trailers or pedestrian-controlled vehicles. If a vehicle over 3050 kg unladen is exempt from the requirement to fit rear reflective markers then it must be fitted with reflective number plates. Regulations specify the precise style of letters and numerals to be used on vehicle registration plates, in particular the spacing between each individual character. It is an offence to alter, rearrange or misrepresent the characters or to alter the spacings to form words or names. It is also an offence to use a number plate with a patterned or textured background. Three possible repercussions may result from the illegal presentation of vehicle registration numbers:

- A fine of up to £1,000 may be imposed on conviction.
- The registration number (mark) may be withdrawn.
- The vehicle may fail its MoT test.

From this it will be seen that the use of so-called 'fun' number plates in which spacings between letters and numbers are rearranged to produce key words is prohibited and will result in police action (usually an order to get the plates changed within five days, but possibly more serious consequences as listed above). Italicized, computer-type and shadow lettering on vehicle number plates will also invite police action. It is useful to note that the firms making and selling these plates commit no offence – only the person who uses them on the road.

A new scheme introduced by the DVLA requires number plate manufacturers to be registered as part of the campaign to curb vehicle crime. Eventually number plates will carry an electronic 'chip' which will be read by roadside microwave beacons and cameras.

Vehicle Registration Numbers

From 1 September 2001 a new system of vehicle registration numbers was introduced in the UK requiring that number plates conform to a new British Standard (BS AU 145d) which replaces the previous standard referred to above. Since that date all new number (ie registration) plates must comply with the following format:

AB51 DVL

The letters 'AB' identify the local area in which the vehicle is registered (see Appendix XI). The number '51' identifies the date of first registration (ie 51 representing September 2001) and will change every six months in September and March each year (eg 02 being March 2002 and 52 being September 2002, etc). The last three letters (in the above example DVL) are chosen at random.

Also from 1 September 2001, all new number plates must display the mandatory fonts (in both style and size – see below) specified in The Road Vehicles (Display of Registration Marks) Regulations 2001. No lettering other than in the specified standard font may be used (ie computer, script-type and other 'fun' lettering will be banned). The name and trade mark of the maker and the postcode of the supplying outlet may be used.

Mandatory Character Font

From 1 September 2001 all new number plates must display the new mandatory font as shown here and characters must be of the regulation size:

123456789
ABCDEFGH
JKLMNOPQ
RSTUVWXYZ

Height	79mm
Width	50mm
Stroke	14mm
Side margin	11mm
Top & bottom margin	11mm
Space between characters	11mm
Space between groups	33mm

(Applicable to all vehicles except motorcycles and vehicles manufactured before 1973)

Existing number plates will not have to be changed to conform to the new style provided the font used is 'substantially the same' as the new style. However, number plates must be replaced if they have been customized with either:

- stylized letters and figures such as italics;
- or number plate fixing bolts that alter the appearance of the letters or numbers.

Failure to replace such number plates may result in the vehicle keeper risking prosecution.

Nationality Symbols

It is a legal requirement (under the Vienna Convention on Road Traffic) for vehicles travelling in a country other than that in which they are registered to display at the rear a nationality plate showing the official symbol for the country of registration (eg GB for Great Britain, F for France and D for Germany). These plates, usually in the form of a self-adhesive sticker, should be of an approved pattern, oval in shape (at least 6.9 in by 4.5 in) and contain the relevant national symbol in black letters on a white background.

However, under new provisions contained in EC Regulation 2411/98/EC, vehicles travelling within EU member states are exempted from the above requirement provided they are fitted with so-called 'Euro-plates'. These are the new-style number plates (as described above) which incorporate, in a panel on the left-hand side, the EU symbol comprising a circle of 12 gold stars on a blue background surrounding the nationality letters of the country in which the vehicle is registered (eg GB). Vehicles displaying this symbol on the number plate will no longer need to use the traditional black GB on a white oval background sticker mentioned above

Switzerland (not an EU member state) has been asked by the EU to recognize these plates as being legal.

Weight Markings

Goods vehicles must not display any weight markings other than their plated weights, any weights required to be displayed under the C&U regulations, or weights required under other regulations (eg the Motor Vehicles (Authorisation of Special Types) General Order). This means that maximum weights shown on 'Ministry' plates and maximum laden weights (so long as these do not exceed 'Ministry' plated weights) may be shown on either one or both sides of a vehicle.

Motor tractors and locomotives must have their unladen weight shown in a conspicuous place on the outside of the vehicle where it can easily be seen.

Dimensions Marking

Goods vehicles over 3.5 tonnes and trailers made since 31 May 1998 must be fitted with a plate showing their length and width, unless this information is already shown on either the manufacturer's plate or a 'Ministry' plate.

'Ministry' Plates

Goods vehicles over 3500 kg maximum gross weight and goods carrying trailers over 1020 kg unladen weight must display a 'Ministry' plate and/or a manufacturer's plate showing the maximum permissible gross vehicle weight and individual axle weights (see Chapter 16).

Special Types Plates

Vehicles carrying abnormal loads must display a manufacturer's plate showing the maximum weights at which the vehicle can operate and the relevant speeds for travel at those weights. The weights shown must be those approved by the vehicle or trailer manufacturer and the vehicle must not exceed specified Special Types speed limits when travelling loaded to the weight shown on the plate (see details of speeds for different classes of Special Types vehicles, p 238).

Food Vehicles

Vehicles which are used in connection with a food business or from which food is sold must display in a clearly visible place the name and address of the person carrying on the business and the address at which the vehicle is kept or garaged. If the vehicle bears a fleet number and is kept or garaged on that person's premises the garage address is not required but the local authority must be notified.

Height Marking

The travelling height of vehicles and trailers carrying engineering equipment, containers and skips, where the height of the vehicle and load exceeds three metres, must be marked in the cab where the driver can see it. Further provisions extend this by requiring the fitment of warning devices where vehicles carry 'high-level equipment' (see pp 306–07 for further details).

Hazard Marking

Vehicles which carry hazardous, radioactive or explosive loads must display appropriate hazard warning symbols on the vehicle whether a bulk tanker, a tank container or a normal delivery vehicle used for carrying hazardous consignments, and on the individual packages too in the latter case. Further details are given in Chapter 23.

Outline (contour) Marking of Large Vehicles

Increasing use is being made of reflective outline marker tape on the sides (yellow tape) and rear (red tape) of large goods vehicles (ie contour marking) to make them more visible at night, thus enhancing road safety. Although this is not yet a statutory requirement, it is anticipated that it will become so in due course under EC legislation. A European specification for such markings already exists, namely ECE104.

Rear Reflective Markings

All vehicles with a maximum permissible weight exceeding 7500 kg and trailers with a maximum permissible weight exceeding 3500 kg (see also p 307) must be fitted with rear reflective markers which make them more conspicuous at night and in poor visibility (Figures 15.1 and 15.2). The markers may also be displayed on loads such as builders' skips (see below).

For new vehicles first used from 1 April 1996 and trailers manufactured on or after 1 October 1995, which under the regulations require the fitment of rear reflective markings as described below, fitment of a new type of rear marking (in accordance with ECE Regulation 70) is necessary.

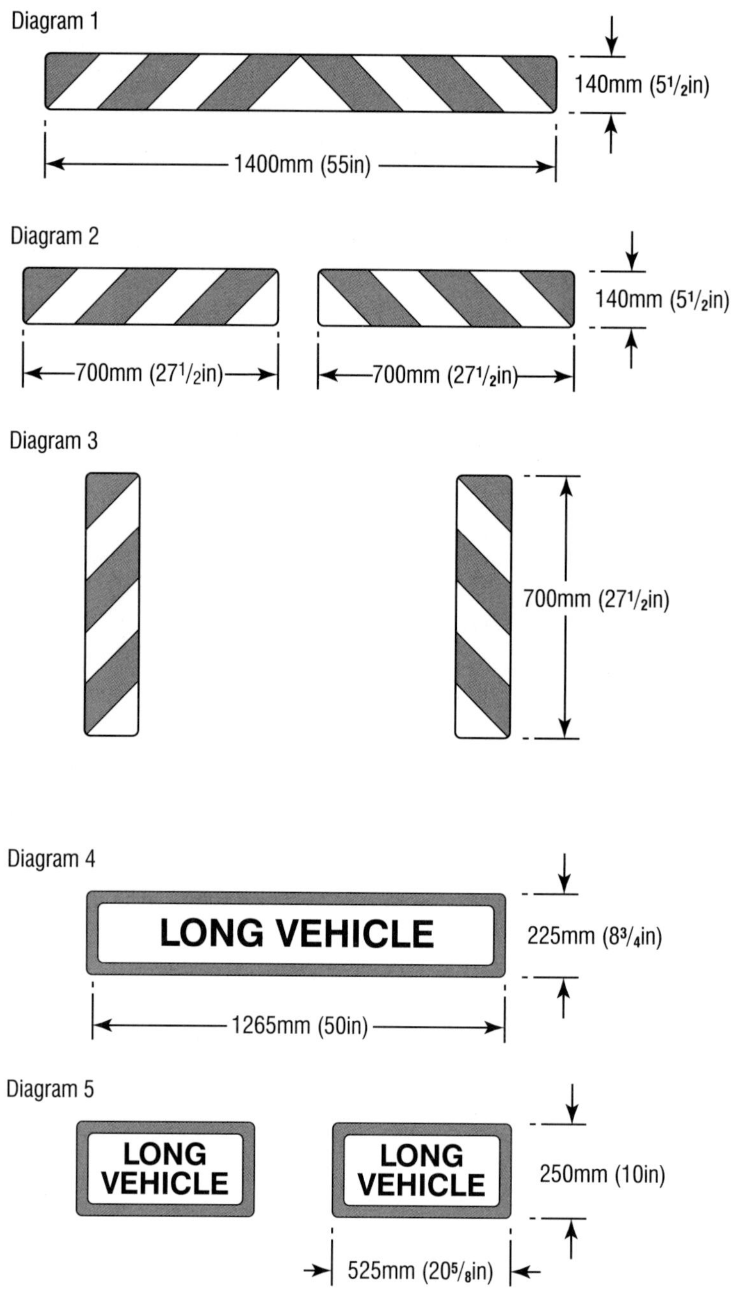

Figure 15.1a *Rear reflective markers required on certain goods vehicles. The plates comprise red fluorescent material background. The lettering is in black on yellow reflex reflecting material*

Types of Markers

For pre-1 April 1996-used vehicles and pre-1 October 1995-built trailers, there are two types of markings, each in two sizes:

- Alternating red fluorescent and yellow reflective diagonal strips – diagrams 1, 2, and 3 in Figure 15.1a.
- A central yellow reflective panel overprinted with the words LONG VEHICLE and having a red fluorescent surround – diagrams 4 and 5 in Figure 15.1a.

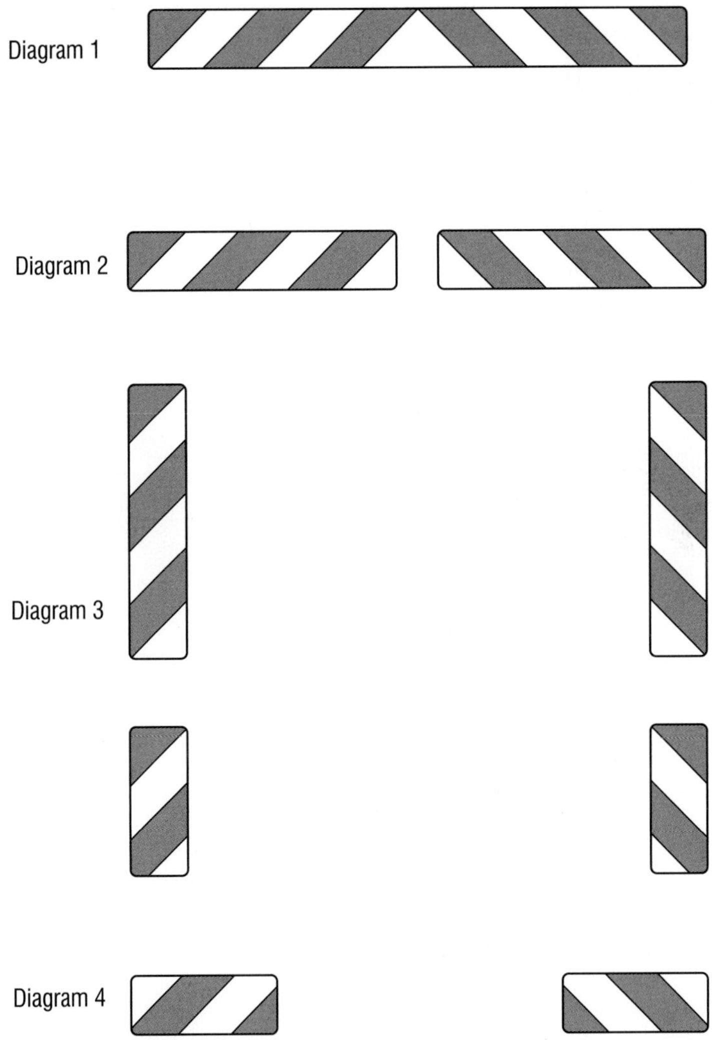

Figure 15.1b *Rear markings for post-1 April 1996-used vehicles and post-1 October 1995-built trailers*

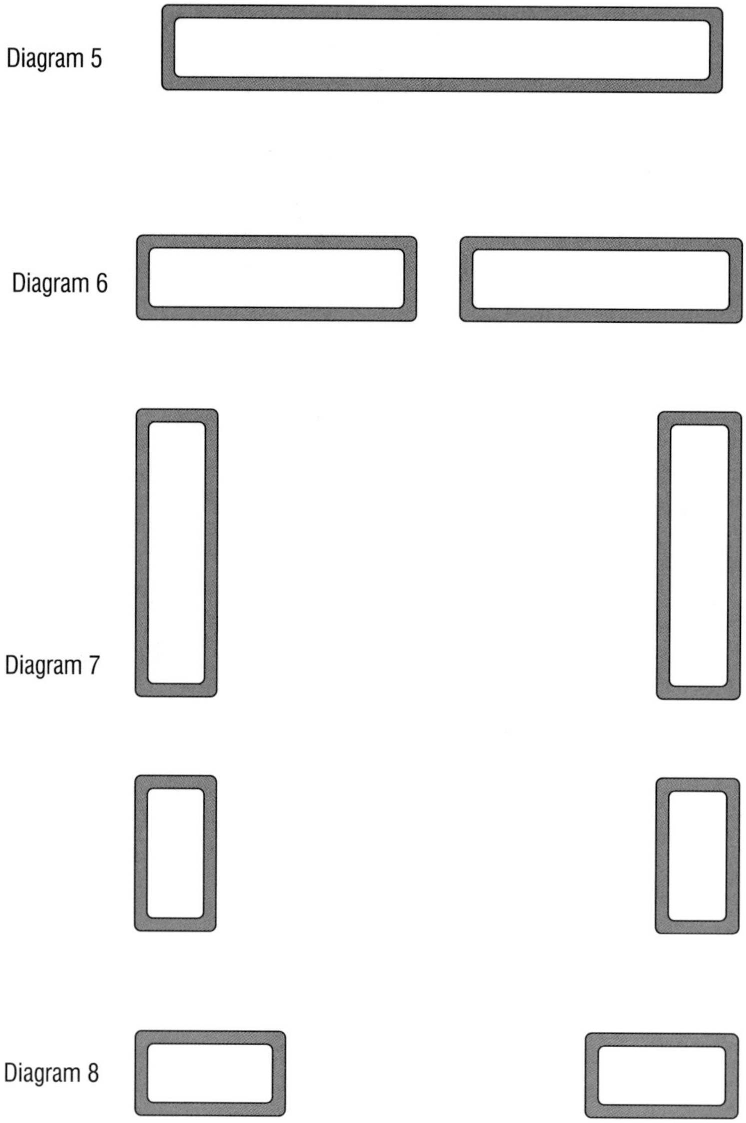

Figure 15.1c *New-type rear markers for use on long vehicles first used from 1 April 1996 and trailers built since 1 October 1995*

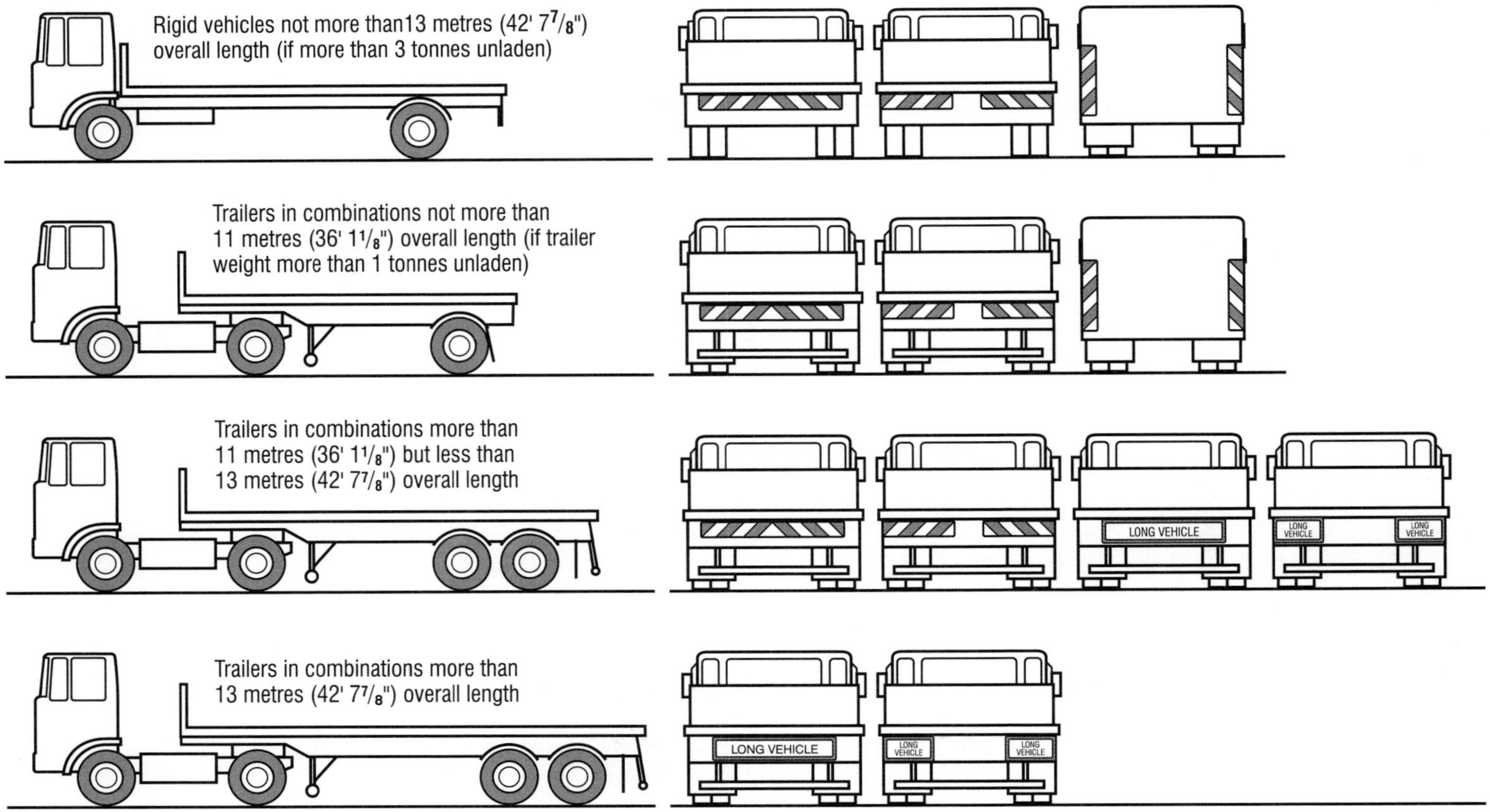

Figure 15.2 *Vehicles which must carry reflective rear markers and the alternative fitting arrangements*

For post-1 April 1996-used vehicles and post-1 October 1995-built trailers, the markings comprise two types:

- Alternating red fluorescent and yellow retro-reflective diagonal strips – diagrams 1, 2, 3 and 4 in Figure 15.1b.
- A central yellow retro-reflective panel surrounded by a red fluorescent border – diagrams 5, 6, 7 and 8 in Figure 15.1c.

Specification for Markers

Markers of the type illustrated in Figure 15.1a must comply with the regulations regarding size and colour and they must be in the form of durable plates stamped with the mark BS AU 152/1970. Those of the type illustrated in Figures 15.1b and 15.1c must also comply with the regulations as above and be stamped with a designated approval mark. They must not be simulated by being painted on the vehicle and the plates must not be defaced, cut or modified to aid fitting to the vehicle.

Which Markers to be Fitted

Pre-1 April 1996-used Vehicles and Pre-1 October 1995-built Trailers

Vehicles not exceeding 13 metres in length and trailers in combinations not exceeding 11 metres must be fitted with the markers shown in diagrams 1, 2 or 3 in Figure 15.1a. Alternatively they may be fitted with the markers shown in diagrams 1, 2, 3, or 4 in Figure 15.1b.

Vehicles more than 13 metres long and trailers in combinations more than 13 metres long must be fitted with the markers shown in diagram 4 or 5 in Figure 15.1a (or alternatively those shown in diagrams 5, 6, 7 or 8 in Figure 15.1c).

Trailers in combinations of more than 11 metres but not more than 13 metres may be fitted with the markers shown in diagram 1, 2, 3, 4 or 5 in Figure 15.1a or those in diagrams 1 to 8 in Figures 15.1b and 15.1c.

Post-1 April 1996-used Vehicles and Post-1 October 1995-built Trailers

Vehicles not exceeding 13 metres in length and trailers in combinations not exceeding 11 metres must be fitted with a marker or set of markers of the types shown in diagram 1, 2, 3 or 4 in Figure 15.1b.

Vehicles more than 13 metres long and trailers in combinations more than 13 metres long must be fitted with the markers shown in diagrams 5, 6, 7 or 8 in Figure 15.1c.

Trailers in combinations of more than 11 metres but not more than 13 metres must be fitted with a marker or set of markers of the types shown in diagrams 1 to 8 in Figures 15.1b and 15.1c.

Fitting Position

The height from the ground to the lower edge of the marker when fitted must not exceed 1700 mm but must be at least 400 mm. It must be fitted parallel to the ground and be facing square to the rear.

Alternative Fitting Position

When a vehicle, which by law requires rear reflective markers to be displayed, is carrying a load which obscures partly or wholly the markers so that they are not clearly visible from the rear, the reflective markers may be fitted to the rear of the load.

Exemptions

Certain vehicles as indicated in the following list are exempt from the requirement to fit these markers. Previously it was illegal to fit the markers to these vehicles, but a change in the regulations permits the fitting of such markers to exempt vehicles on a voluntary basis, provided the vehicles exceed the specified weight limit. It is illegal to display these markers on vehicles which do not require them by law except as mentioned below.

Exempt Vehicles

- vehicles with a maximum gross weight not exceeding 7500 kg;
- passenger vehicles other than articulated buses;
- land tractors, land locomotives, land implements, land implement conveyors, agricultural tractors or industrial tractors;
- works trucks or works trailers;
- vehicles in an unfinished condition proceeding to a works for completion or to a place where they are to be stored or displayed for sale;
- motor vehicles constructed or adapted for the purpose of forming part of articulated vehicles;
- broken-down vehicles while being drawn in consequence of the breakdown;
- engineering plant;
- trailers, not being part of an articulated bus, drawn by public service vehicles;
- vehicles designed for fire fighting or fire salvage purposes;
- vehicles designed and used for the purpose of servicing or controlling aircraft;
- vehicles designed and used for the transportation of two or more motor vehicles carried thereon, or of vehicle bodies or two or more boats;
- vehicles proceeding to a place for export;
- vehicles brought temporarily into Great Britain by persons residing abroad;
- vehicles in the service of a visiting force or of a headquarters;
- motor vehicles first used before 1 January 1940;
- vehicles owned or in the service of the Army, Navy or Air Force;
- vehicles designed for heating or dispensing tar or similar material for road construction or maintenance;
- trailers being drying or mixing plant designed for the production of asphalt, bitumen or tarmacadam;
- trailers made before 1 August 1982 with an unladen weight not exceeding 1020 kg;
- trailers with a gross weight not exceeding 3500 kg.

Builders' Skips

Rear reflective markings of the type described above (as shown in Diagram 3, Figure 15.1a) must be fitted to the ends of builders' skips which are placed on the highway. They must be fitted as a matched pair as near to the outer edge as possible, mounted vertically and no more than 1.5 metres from the ground to the top edge. They must be kept clean, in good order and be visible from a reasonable distance. Such skips are required to be illuminated when standing on roads at night.

16 Goods Vehicle Plating, Annual Testing and Vehicle Inspections

Most goods vehicles are required to be tested annually to ensure they are safe to operate on the road and meet the legal requirements relating to mechanical condition. In particular, this annual inspection is intended to determine whether vehicles and trailers meet the standards specified in the Road Vehicles (Construction and Use) Regulations 1986 (as amended). Additionally, it is necessary for certain goods vehicles and trailers to be 'Ministry plated' to show the maximum permissible gross weight and maximum axle weights at which they may be operated on roads in Great Britain. The requirement for annual plating and testing of goods vehicles is contained in the Road Traffic Act 1988 and is detailed in The Goods Vehicles (Plating and Testing) Regulations 1988 as amended.

The Vehicle and Operator Services Agency (VOSA) is responsible for the operation of goods vehicle testing stations – of which there are currently 76 full-time and 15 part-time stations plus 174 HGV 'designated premises' (DPs) in the UK. It is also the main enforcement authority involved in roadside checks on goods vehicles.

ANNUAL TESTING

Articulated tractive units, rigid goods vehicles over 3500 kg gross weight, goods carrying semi-trailers and drawbar trailers over 1020 kg unladen weight and converter dollies must be tested annually at either a goods vehicle testing station (see Appendix IV for list), or at one of the 45 (currently) designated operator workshops (ie designated premises – DPs) – for which a test fee surcharge applies, plus any additional charge made by the DP.

Certain specialized vehicles are exempt from the test, as shown on pp 396–398, and the regulations do not apply to vehicles used under a trade licence. Goods vehicle test stations also carry out the Class V test, which is the MoT-type test, on large passenger vehicles which cannot get into normal MoT test garages (see also Chapter 17).

The Class VII test provides for MoT testing of goods vehicles of more than 3000 kg but not more than 3500 kg design gross weight which were formerly subject to goods vehicle plating and testing when they had an unladen weight exceeding 1525 kg.

Types of Test

There are various types of goods vehicle test as follows:

- *First test:* the first annual test of the vehicle or trailer (carried out no later than the end of the anniversary month in which it was first registered) at which the information shown on the vehicle's official (ie 'Ministry') plate (issued from Swansea) is verified.
- *Parts 2 and 3 retests*: re-examination of vehicles which have failed their first test or a retest.
- *Periodical test*: the annual test which applies to all relevant vehicles after the first test.
- *Part 4 test*: a test provided for in the regulations which may be required if a notifiable alteration has been made to the vehicle, or if the operator wants the plating certificate amended to show different weights as a result of changes to the vehicle or, for example, if different tyre equipment has been fitted.
- *Retest following appeal*: retests carried out following an appeal to the Test Station Manager or subsequently the VOSA Regional Manager.

Test Dates

Vehicles may be submitted for test at any one of the full-time or part-time goods vehicle test stations selected by the vehicle operator. Vehicles are due for test each year no later than the end of the anniversary month in which they were first registered (eg a vehicle registered on 1 January 1991 would be due for its first test no later than 31 January 1992 and for subsequent tests by 31 January in each following year).

Trailer Test Dates

The first test for trailers is due by the end of the first anniversary month from when they were sold. When application is made for a first trailer test a serial number is given, and this number must be stamped on or permanently attached to the trailer prior to its test. It is a condition of acceptance for test that this number is on the trailer.

Year of Manufacture/Registration

There is an anomaly with due test dates when a vehicle or trailer is manufactured in one year and is not registered (or sold in the case of a trailer) before 1 July of the following year. In this case it must be tested by the end of December in the year in which it was first registered or sold.

Phased Programmes and Missed Test Dates

VOSA allows vehicle operators the facility of having vehicles tested before their due date to accommodate phased programmes of test preparation, rather than having a large number of vehicles due for preparation and test in any particular month of the year.

Test Applications

Initial application for a first test of vehicles and trailers has to be made direct to the Goods Vehicle Centre, Welcombe House, 91/92 The Strand, Swansea (not to be confused with the DVLA at Swansea). Application for subsequent tests can be made either to Swansea or direct to any test station. The following forms, which may be obtained from good vehicle test stations or Traffic Area Offices, are used for making the application:

VTG1 First test of a vehicle
VTG2 First test of a trailer
VTG40 For subsequent tests of both vehicles and trailers
VTG15 To be completed and submitted every time a dangerous goods vehicle is presented for test.

Time for Application

Applications for test should be made during the two months prior to the month in which the test is required to take place (ie the last day on which the vehicle may legally operate without a test certificate). Applications should ideally be made at least one month before the date on which the test is preferred by the operator (see note below about the Saturday testing facility).

Saturday and Out-of-Hours Testing

Saturday and out-of-hours testing is available at most test stations at an additional fee. Operators wishing to take advantage of this facility must mark their applications very clearly 'SATURDAY TEST' and show the appropriate date.

Test Fees

The appropriate test fee, as follows, should be sent to the Goods Vehicle Centre with the first test application form (VTG1 or VTG2), while fees for subsequent tests should be sent direct to the test station. (Operators can open a direct debit account with VOSA to enable any fee to be collected for any test station.) The test fees listed here were effective from 30 September 2005.

LGV annual test fees – effective from 30 September 2005

	Normal	Out of hours	Designated premises	Out of hours at DPs
Motor vehicles				
with 2 axles	£46.00	£78.00	£57.00	£89.00
with 3 axles	£66.00	£98.00	£77.00	£109.00
with 4 or more axles	£86.00	£118.00	£97.00	£129.00
Trailers				
with 1 axle	£23.00	£43.00	£29.00	£49.00
with 2 axles	£34.00	£54.00	£40.00	£60.00
with 3 or more axles	£44.00	£64.00	£50.00	£70.00

LGV retest fees
Retests within 14 days

	Normal	Out of hours	Designated premises	Out of hours at DPs
Motor vehicles				
with 2 axles	£30.00	£46.00	£53.00	£69.00
with 3 axles	£44.00	£60.00	£67.00	£90.00
with 4 or more axles	£57.00	£73.00	£80.00	£103.00
Trailers				
with 1 axle	£15.00	£26.00	£29.00	£40.00
with 2 axles	£22.00	£33.00	£36.00	£47.00
with 3 or more axles	£29.00	£40.00	£43.00	£54.00

Part paid retests
These are retests that apply only when certain specified items are marked as a failure on the Inspection Document. For vehicles (with any number of axles) the fee is £10.00 and for trailers (with any number of axles) the fee is £5.00

Retests after 14 days
Retests carried out after 14 days are chargeable at the full test rate shown above.

Trailer Testing

Many operators have more semi-trailers and trailers than tractive units or drawing vehicles and in order to have these additional trailers tested it may be necessary for them to be submitted for test with a vehicle which has already been tested and has a current valid test certificate. In these cases only the trailer will be examined and the fee payable will be the trailer fee only.

Test Appointments

Following application to the Goods Vehicle Centre for a first or subsequent test (see note above about Saturday testing facility), the test station selected by the operator will confirm the test booking in due course with an appointment letter, and all further communications regarding the test must be made with the test station, not with Swansea. If, owing to excessive workload or staff shortage, the chosen test station cannot accommodate the test an appointment will be made at the nearest alternative test station and a letter will be sent from that station. The appointment letter and the vehicle registration document must be produced on arrival at the test station.

It is essential that vehicles arrive at the test station at the appointed time. Late or non-arrival of a vehicle can mean cancellation of the test and the fee will be forfeited unless an acceptable reason citing 'exceptional circumstances' is put forward. Exceptional circumstances include accident, fire, epidemic, severe weather, failure in the supply of essential services or other unexpected happenings. Breakdown, mechanical defect or non-availability of the vehicle because of shortages in spare parts supply or for operational reasons for example are not looked upon as exceptional circumstances.

Cancellations

If it is necessary to cancel a test booking after making application, provided seven days' notice is given to the test station, either a new test date will be arranged or the fee will be refunded after deduction of a £1.50 charge. In exceptional circumstances, such as an accident to the vehicle on the way to the test station, if notification is given to the station within three days of the accident the fee will be carried forward or refunded less £1.50. If other exceptional circumstances arise within seven days of the due date for the test, providing satisfactory evidence is given to the test station, the fee will be similarly carried forward or refunded less £2.50.

Refusal to Test

Test station officials have the right to refuse to test a vehicle or trailer for the following reasons, in which circumstances form VTG12 will be issued:

- Arrival after the appointed time.
- Appointment card or vehicle registration document not produced.
- If it is found that the vehicle brought to the test station does not conform to the details given on the application form.
- If the vehicle was booked for the test with a trailer but the trailer is not taken to the test station.
- If the chassis number cannot be found by the examiner or if the serial number given for the trailer by VOSA is not stamped on it.
- If the vehicle is in a dirty or dangerous condition.
- If the vehicle does not have sufficient fuel or oil to enable the test to be carried out.
- If the test appointment card specified that the vehicle should be loaded for the test and it is taken to the test station without a load. Under normal circumstances the decision whether the vehicle is to be tested in a laden or unladen condition is left to the owner to suit his convenience but in some circumstances the test station may request that the vehicle is fully or partially loaded to enable the brakes to be accurately tested on the roller brake tester.
- In the case of a trailer if the vehicle submitted with it is not suitable to draw it.
- If the vehicle breaks down during the test.
- If the vehicle is submitted for its annual test (ie not for the first test) or a retest and the previous test and plating certificates are not produced.

Test Procedure

Goods vehicle test stations vary in size and in the number of examination staff. Testing normally takes approximately 45 minutes during which the driver must be available to assist and move the vehicle as required. Examination of vehicles is carried out by VOSA examiners based at the station in accordance with the *Heavy Goods Vehicle Inspection Manual* published by The Stationery Office. All items which have to be inspected are listed in the *Manual* together with, where necessary, details of how the inspection of each item should be carried out and the reasons for failing the item. Under the 'reasons for rejection' column in the *Manual*, where the item inspected is one that is subject to wear, the maximum tolerance will be indicated.

The examination is conducted in four stages:

- Items 1 to 39 in the *Inspection Manual* are inspected on the hard standing outside the test building.
- Items 40 to 59 are inspected at the second stage of the examination over the test pit inside the building.
- Items 60 to 70 are inspected at the next point, where a beam setter is used for checking headlamp alignment.
- Items 71 to 74, which comprise the brake tests, are the last to be carried out at the end of the test line. A roller brake tester is used for these purposes; the machine indicates on dials on a console the braking force of all the wheels on an axle and each individual wheel (or a pair of twin wheels) when the vehicle's footbrake, handbrake and emergency brake are applied by the driver when directed to do so.

The Smoke Test

Free acceleration smoke emission testing of diesel engines by means of a calibrated smoke meter (to measure the density of smoke in a vehicle exhaust) was introduced as part of the goods vehicle annual test from 1 September 1992. Where it is necessary to carry out a purely visual test, the driver is asked to depress the accelerator pedal firmly from the engine idle position (preferably with the engine already warm) to the maximum fuel delivery position and, immediately the governor operates, release the pedal until the engine slows to a steady idling speed. Smoke emission from the first attempt at this procedure is ignored but the procedure has to be repeated for a maximum of 10 times until the smoke emission is considered to be of equal density for two successive accelerations.

Inspection Card

During the test the examiner has an inspection card (Form PG14/2) on which all items in the *Inspection Manual* are listed and in the case of failure of any item the card is marked accordingly.

A computerized test card is to be introduced which will allow the examiner to produce:

- a test 'pass' certificate;
- an advice note (for action required);
- a failure notice;
- a PG9 prohibition notice; or
- a PG14 refusal of test certificate.

Items to be Inspected

The list of items shown in the *Heavy Goods Vehicle Inspection Manual* (1997 edition) to be inspected is as follows:

1. –
2. –
3. Seat belts *
4. –
5. Exhaust emissions *^
6. Road wheels and hubs ^
7. Size and type of tyres
8. Condition of tyres ^
9. Sideguards, rear underrun devices and bumper bars
10. Spare wheel and carrier
11. Vehicle to trailer coupling ^
12. Trailer parking and emergency brakes and air line connections ^
13. Trailer landing legs ^
14. Spray suppression, wings and wheel arches ^
15. Cab security *^
16. Cab doors *
17. Cab floor and steps *
18. Driver's seat *
19. Security of body
20. Condition of body
21. –

22 Mirrors *

23. Glass and view of the road *

24 –

25. Windscreen washers and wipers*
26. Speedometer/tachograph *^
27. Horn *
28. Driving controls *^
29. –
30. Steering control *^
31. –
32. –
33. Speed limiter *^
34. Pressure/vacuum warning and build-up *
35. –
36. Hand lever operating mechanical brakes *^
37. Service brake pedal *^
38. Service brake operation ^
39. Hand-operated brake control valves ^
40. –
41. Condition of chassis
42. Electrical wiring and equipment
43. Engine and transmission mountings *
44. Oil leaks *^

45. Fuel tanks and systems *^
46. Exhaust systems *
47. –
48. Suspension ^
49. –
50. –
51. –
52. –
53. Axles, stub axles and wheel bearings ^
54. Steering ^
55. –
56. –
57. Transmission *^
58. –
59. Brake system and components ^
60. –
61. –
62. Rear markings and reflectors ^
63. Lamps ^
64. –
65. –
66. Direction indicators and hazard warning lamps
67. Aim of headlamps *
68. –
69. –
70. –
71. Service brake performance ^
72. Secondary brake performance *^
73. Parking brake performance ^
74. Other dangerous defects

* These items do not apply when trailers are being inspected.
^ These items have been amended in the First Amendment to this Inspection Manual published in 2002 (copies available from The Stationery Office).
– The blank spaces are left to enable VOSA to add any further items at a later time.

Additional Test Items

From 1 April 2005, the goods vehicle annual test has been extended to include, where relevant, an inspection of shipping container location and securing devices (commonly referred to as twistlocks) and stabilizer arms, where these items are fitted.

Tachograph Testing

Tachograph testing is part of the annual test under item 26 above and comprises an inspection of the:

- tachograph scale marking in kilometres per hour (additional mph marking is allowed);

- presence of the tachograph serial number and appropriate 'e' marking;
- tachograph installation/calibration plaque (its presence, condition and calibration date);
- 'K factor' plaque to ensure its presence, condition and that the 'K factor' is clearly visible;
- condition and presence of seals (checking particularly for evidence of interference).

Note Regarding Tachograph Exempt Vehicles Using a Tachograph as a Speedometer

A vehicle must have a tachograph fitted unless the person presenting the vehicle for test claims that it is exempt from the tachograph regulations. If a tachograph is fitted on an exempt vehicle and used in place of a speedometer, the tachograph must be calibrated and sealed in accordance with the Community Recording Equipment Regulations. However, providing the seals remain intact there will be no requirement for two-yearly inspections or further calibration. If the above vehicle legally requires a speed limiter and has the speed signal operating the limiter transmitted from the tachograph head, then all the seals including the gearbox sender unit must remain intact to avoid the requirement for re-calibration. If the vehicle does not legally require a speed limiter (or the speed limiter does not receive the speed signal from the tachograph head) then the tachograph head seals (excluding the gearbox sender unit seal) must remain intact to avoid the requirement for re-calibration.

Speed Limiter Checks

An electronic device is used during testing to check the function and setting of the speed limiter under test item 33 above and failure of the item will jeopardize the whole vehicle test.

Dedicated ABS/EBS Cables

Where tractive units and trailers are fitted with ISO 7638 sockets to power ABS and and/or EBS braking systems, these sockets must be connected by a dedicated ABS cable (where one or the other does not have an ISO 7638 socket, the ABS is powered through the stop light circuit). Vehicles presented for test after 1 April 2002 without such a cable will receive notice VTG 12, 'refusal to test a vehicle'. Also, from April 1 2002, vehicles stopped at roadside checks that are found to have an ISO 7638 socket on both tractive unit and trailer, but do not have an ABS cable connecting them, will receive a prohibition notice (PG9 – see p 402).

Where a vehicle and trailer are fitted with an electric braking system (EBS) as well as ABS, a seven-core cable connection must be used instead of the normal ABS five-core cable. Vehicles found during roadside checks to have only a five-core instead of the requisite seven-core cable will receive a Defect Note (PGDN 35). On presentation at the test station such vehicles will receive an advisory notice to replace the cable.

Test Pass

When a goods vehicle is found to be in satisfactory order a test certificate is issued by the test station. For goods vehicles the certificate is form VTG5 and for goods-carrying trailers it is VTG5A which is issued together with a trailer test disc (VTG5B). The trailer test disc is included with the certificate and this must be removed and fixed on to the trailer in a protective holder in a position where it is conspicuous, readily accessible and clearly visible from the nearside.

The heavy goods vehicle test certificate must be produced when applying for an excise licence for a vehicle which comes within the scope of these regulations and otherwise at the request of a competent authority.

Replacement Documents

Replacement test certificates and trailer test date discs and replacement plates and plating certificates (see later in chapter) may be obtained from local test-stations or the Goods Vehicle Centre at Swansea at a cost of £11.00 each. Application in both cases should be made on form VTG59 obtainable from goods vehicle test stations or Traffic Area Offices. Automatic replacement of lost or defaced documents is not guaranteed. The Secretary of State has powers to order a retest before issuing such replacements, in which case full test fees become payable.

Test Failure and Retests

When the vehicle is sent for test, it is recommended that a mechanic with a tool-kit and minor spares items (light bulbs, for example) accompanies it so that any minor defects can be rectified on the premises and the test can be completed. The examiner may allow certain minor defects to be repaired during the test if they do not take up too much time (usually no more than 20 minutes) and delay the test schedule. In some instances the vehicle may be allowed to be taken out of the test line for minor repairs to be carried out but this again is at the discretion of the examiner.

If it is necessary to take the vehicle away to get the defects rectified and it is submitted again later that day or during the next working day, no additional charge is made. These free retests are restricted to those cases where the vehicle failed because of certain prescribed defects in items as follows:

- Legal plate position
- Legal plate details
- Bumper bars
- Spare wheel carrier
- Cab doors
- Mirrors
- View to front
- Speedometer
- Audible warning
- Oil leaks
- Fuel tanks pipes and system
- Obligatory sidelamps
- Obligatory rear lamps

- Reflectors
- Direction indicators
- Headlamps – vertical aim
- Obligatory headlamps
- Obligatory stop lamps.

If the vehicle which fails the test is submitted to the same test station again within 14 days, a reduced retest fee is charged and only the items on which the test was failed are re-examined. The retest fee is £23.50 for vehicles and £12.50 for trailers. Arrangements for retests have to be made with the manager of the test station concerned.

Appeals against Test Failure

If a vehicle or trailer undergoing test or retest failed for a reason which the operator believes is not justified, he has a right of appeal to the test station manager or VOSA Regional Manager, by completing form VT17 and paying the relevant fee (ie the same as the test fee for that vehicle). The person who submitted the appeal will receive notification as to where and when he must submit the vehicle for re-examination.

Reduced Pollution Examination of Vehicles

Operators wishing to gain the reduced rate of VED for 'reduced pollution' vehicles (see page 200) may submit their vehicles to a goods vehicle testing station for examination by VOSA.

The reduced pollution requirement may be satisfied by:

- a new vehicle meeting the required standard;
- the fitting of a new engine to a vehicle; or
- the fitting of a type-approved device for which there is a Certificate of Conformity issued by the vehicle manufacturer.

The reduced pollution requirements are satisfied if the rate and content of a vehicle's particulate emissions do not exceed the number of grams per kilowatt-hour specified in the table below:

Item	*Instrument setting standard to which vehicle was first used*	*Rate and content of particulate emissions after adaptation (grams per kilowatt-hour)*
1.	Directive 88/77/EEC	0.16
2.	Directive 91/542/EEC (limits A)	0.16
3.	Directive 91/542/EEC (limits B)	0.08
4.	European Commission Proposal (Com [97] 627) for a European and Council Directive amending Council Directive 88/77/EEC	0.04

NB: A vehicle first used before 1 April 1991 shall be taken to be a vehicle falling within item 1 above.

Examinations of goods vehicles for the purposes of determining whether they meet the reduced pollution requirements are carried out by an authorized examiner, following which a 'reduced pollution certificate' is issued, providing conclusive evidence that the statutory requirements have been met. A reduced pollution certificate must be produced on application for a vehicle licence at a reduced rate of duty.

Fees for examinations conducted for this purpose are:

- £15 if carried out as part of the annual goods vehicle test;
- £25 in any other circumstances.

Procedures are established for:

- the re-examination of a certified vehicle;
- issue of a rectification certificate where a vehicle is found no longer to satisfy the requirements;
- the revocation, surrender and cancellation of certificates;
- making appeals against the refusal or revocation of a reduced pollution certificate.

Hauliers whose reduced pollution certificated vehicles are found to be emitting black smoke will lose the certificate and the benefit of reduced duty.

PLATING OF GOODS VEHICLES AND TRAILERS

Manufacturer's Plating

All new goods vehicles and new trailers over 1020 kg unladen weight must be fitted with a plate by the manufacturer which shows specified information as follows:

- The manufacturer's name
- The date of manufacture
- Vehicle type
- Engine type and power rating
- Chassis or serial number
- Number of axles
- Maximum weight allowed on each axle
- Maximum gross weight for the vehicle (including the weight imposed on the tractive unit by a semi-trailer in the case of articulated vehicles)
- Maximum train weight.

The plate containing this information is normally fitted inside the driver's cab on the nearside.

For trailers, the information shown on the plate is:

- The manufacturer's name
- Date of manufacture
- Chassis or serial number
- Number of axles
- Maximum weight allowed on each axle
- Maximum weight imposed on the drawing vehicle in the case of semitrailers
- Maximum gross weight for the trailer.

The plate for trailers is usually riveted to the chassis frame on the nearside.

Design Weights

The weights stated are those at which the manufacturer has designed the vehicle to operate. Where these weights exceed those permitted by law (ie in the C&U regulations) for the type of vehicle in question then, until such time as the vehicle is plated by VOSA, the lower statutory weight limits apply. Conversely, if the manufacturer's design weight is lower than that permitted by law for the type of vehicle then it is the lower limit which applies.

Vehicle Dimensions

Goods vehicles over 3.5 tonnes gross weight and trailers manufactured since 31 May 1998 must be fitted with a plate showing their length and width, unless this information is already shown on either the manufacturer's plate or a 'Ministry' plate.

'Ministry' Plating

When Type Approved goods vehicles over 3500 kg gross weight are first registered an official plate (ie commonly referred to as a 'Ministry' plate) is issued showing the maximum permissible gross vehicle weight and individual axle weights at which the vehicle or trailer is allowed to operate within Great Britain.

A plating certificate (form VTG7 – for all vehicles and trailers) and a plate (form VTG6 blue – for pre-1 April 1983 vehicles; VTG6T red – for post-1 April 1983 vehicles; VTG6A green – for all trailers and, on special request (apply on form VTG101), for goods vehicles used in international operations), giving similar details to those on the manufacturer's plate, are issued by the Goods Vehicle Centre at Swansea after receipt of the necessary documents (including the Type Approval Certificate) when a new vehicle is first registered. The plating certificate must be retained by the vehicle operator but the plate (made of paper and protected in a laminated casing, not metal as is the manufacturer's plate) which is also issued must be fixed to the vehicle in an easily accessible position. Generally it should be fitted inside the cab of vehicles on the nearside (but not affixed to the door) and in a suitable position on the nearside of trailers (it is usually fitted to the chassis frame). In all cases the plates should be protected against the weather, kept clean and legible and secure against accidental loss.

NB: The red, blue and green plates referred to above are being phased out and replaced by computer-generated plates.

Tractive units and semi-trailers in articulated outfit combinations are plated separately and the individual plates must be fixed separately to the tractive unit and the trailer.

Down-Plating of Vehicles

Down-plating to obtain the benefit of lower rates of Vehicle Excise Duty (VED), will be carried out on request using form VTG10.

Where no alteration has been made to the vehicle, inspection by VOSA will not be necessary. However, it should be noted that in no circumstances will a vehicle that has an inefficient braking performance, is not maintained to required safety standards or which otherwise does not comply with legal requirements be plated at lower weights to compensate for these defects. In such cases examiners at the Goods Vehicle Testing Station will refuse to issue the vehicle with a test (pass) certificate.

The procedure for down-plating is to complete form for Notifiable Alterations (VTG10), obtainable from goods vehicle test stations and the Goods Vehicle Centre at Swansea and then:

- hand it in at the time of presenting the vehicle for its annual goods vehicle test;
- contact the local goods vehicle test station at any time for an appointment to submit the vehicle for down-plating; or
- send the form and the fee (see below) to the Goods Vehicle Centre at Swansea.

On payment of the statutory fee of £15.00 a new vehicle plate (to be fitted in the vehicle cab) and plating certificate will be issued.

International Plates

International Proof of Compliance plates are available and may be voluntarily fitted to vehicles by international hauliers. These plates show compliance with the weights and dimension requirements of EC Directive 96/53/EC. They have been introduced to speed up clearance times through Customs for vehicles on international journeys, showing that the vehicles to which they are fitted do meet the regulations on weights, widths and lengths. Operators wishing to fit these plates should apply to the Goods Vehicle Centre at Swansea using application forms obtainable from Swansea direct or from heavy goods vehicle testing stations. Eventually the new-style plate will replace the present style of 'Ministry' plates.

Vehicles and trailers manufactured since 31 May 1998 should be fitted with these 'dimensions' plates, unless the information is already shown on the manufacturer's plate or on a 'Ministry' plate issued for the vehicle or trailer.

Notifiable Alterations

Operators who make any alteration to the structure of their vehicles must notify VOSA before the vehicle is used on the road. Details of the alterations which require notification on form VTG10 are as follows:

(a) *Alterations to the structure or fixed equipment of a vehicle which vary its carrying capacity*

These include alterations to any of the following items:

(i) *Chassis frame or structure*

Any alteration which increases or decreases the front or rear overhang by more than one foot. Any structural alteration (other than normal adjustment of an extensible structure) which reduces or extends the wheelbase (or in the case of a semi-trailer the equivalent distance). Any other extension, deletion or alteration including cutting, welding, riveting, etc, which materially weakens the chassis frame or structure or changes its torsional stiffness.

(ii) *Steering suspension, wheels and axles (including stub axles and wheel hubs)*
The fitting of steering gear, axles, hubs or road springs of a different design or load-bearing capacity. The fitting of additional wheels and axles, or the removal of such items. Any addition, deletion or alteration which reduces the inherent strength of the above components.

(iii) *The fitting of an alternative body of different design, construction or type*
Any alteration which reduces materially the strength of the body structure or the means by which it is attached to the chassis. Any alteration which causes the body to extend beyond the rear of the chassis frame.

(b) *Alterations to braking system*
These comprise alterations which adversely affect either the braking system or the braking performance of the vehicle. They include the addition or deletion of components such as reservoirs, servo motors, brake actuators, exhausters and compressors. They would also include the addition of any equipment which it is necessary to connect to any part of the braking system, and the fitting of different brake drums or shoes or liners of a smaller contact area.

(c) *Other alterations to the structure or fixed equipment*
Any other alteration made in the load-bearing structure or fixed equipment of the vehicle, eg the coupling gear, which could make the vehicle unsafe to travel on roads at any weight shown on the plate and plating certificate. In the case of a motor vehicle this could include such alterations as changing the type of engine or repositioning the engine or its mountings (eg petrol to diesel, normal control to forward control, etc).

NB: Any trailer submitted for examination under the notifiable alterations procedure that is over eight years old must have its braking system brought up to the latest standard applicable at the time of its rebuilding.

Exemptions from Plating and Testing

Vehicles which are subject to the Goods Vehicle (Plating and Testing) Regulations 1988 (as amended) are exempt from the need to hold current plating and testing certificates while being taken to a test station, when used on a road during the test, returning to base from the test station after a test and being taken (unladen) to a working or repair centre for work to be carried out on them in connection with the test.

Temporary Exemption

Temporary exemption from the need to hold a test certificate for not more than three months can be granted by the test station manager if for some special reason it is not possible for a vehicle or trailer to be submitted for test by the last day of the month in which it was due to be tested.

Temporary exemption may be granted in the event of severe weather, fire, epidemic, a failure in the supply of essential services, an industrial dispute, or other unexpected happenings to either a vehicle or the test station. A normal mechanical breakdown of the vehicle will not be considered a circumstance for temporary exemption. A certificate (form VTG33) confirming the exemption will be issued and this can be produced either to the police or examiners of VOSA if they ask for the test certificate, or when applying for an excise licence for the vehicle.

Exempt Vehicles

Vehicles to which the regulations do not apply are as follows:

- Dual-purpose vehicles not constructed or adapted to form part of an articulated vehicle.
- Mobile cranes as defined in Schedule 1 of the Vehicle Excise and Registration Act 1994.
- Breakdown vehicles.
- Engineering plant and movable plant and equipment specially designed and constructed for the special purposes of engineering operations.
- Trailers being drying or mixing plant designed for the production of asphalt or of bituminous or tarmacadam.
- Tower wagons as defined in Schedule 2 of the Vehicle Excise and Registration Act 1994.
- Road construction vehicles as defined in Schedule 2 of the Vehicle Excise and Registration Act 1994 and road rollers.
- Vehicles designed for fire fighting or fire salvage purposes.
- Works trucks, straddle carriers used solely as works trucks, and works trailers.
- Electrically propelled motor vehicles.
- Motor vehicles used solely for clearing frost, ice or snow from roads by means of a snow plough or similar contrivance, whether forming part of the vehicle or not.
- Vehicles constructed or adapted for, and used solely for, spreading material on roads to deal with frost, ice or snow.
- Motor vehicles used for no other purpose than the haulage of lifeboats and the conveyance of the necessary gear of the lifeboats which are being hauled.
- Living vans not exceeding 3500 kg design weight.
- Vehicles constructed or adapted for, and used primarily for the purpose of, carrying equipment permanently fixed to the vehicle which equipment is used for medical, dental, veterinary, health, educational, display or clerical purposes, such use not directly involving the sale, hire or loading of goods from the vehicle.
- Trailers which have no other brakes than a parking brake and brakes which automatically come into operation on the overrun of the trailer.
- Vehicles exempted from duty because they do not travel on public roads for more than six miles in any week and trailers drawn by such vehicles (Vehicle Excise and Registration Act 1994, Schedule 3).
- Agricultural motor vehicles.
- Agricultural trailers and trailed appliances drawn on roads only by a land tractor.
- Passenger-carrying vehicles and hackney carriages.
- Vehicles used solely for the purpose of funerals.
- Goods vehicles proceeding to a port for export and vehicles in the service of a visiting force.

- Vehicles equipped with new or improved equipment or types of equipment and used solely by a manufacturer of vehicles or their equipment or by an importer of vehicles, for or in connection with the test or trial of any such equipment.
- Motor vehicles temporarily in Great Britain.
- Motor vehicles for the time being licensed in Northern Ireland.
- Vehicles having a base or centre in any of the following islands, namely Arran, Bute, Great Cumbrae, Islay, Mull or North Uist, from which the use of the vehicle on a journey is normally commenced.
- Trailers temporarily in Great Britain, a period of 12 months not having elapsed since the vehicle in question was last brought into Great Britain.
- Track-laying vehicles.
- Steam-propelled vehicles.
- Motor vehicles manufactured before 1 January 1960 used unladen and not drawing a laden trailer, and trailers manufactured before 1 January1960 and used unladen.
- Three-wheeled vehicles used for street cleansing, the collection or disposal of refuse, and the collection or disposal of the contents of gullies.
- Vehicles designed and used for the purpose of servicing or controlling aircraft, while so used on an aerodrome within the meaning of the Civil Aviation Act 1982 or on roads to such extent as is essential for the purpose of proceeding directly from one part of such an aerodrome to another part thereof or, subject as aforesaid, outside such an aerodrome unladen and not drawing a laden trailer.
- Vehicles designed for use, and used on, an aerodrome mentioned in the last preceding paragraph, solely for the purpose of road cleansing, the collection or disposal of refuse or the collection or disposal of the contents of gullies or cesspools.
- Vehicles provided for police purposes and maintained in workshops approved by the Minister as suitable for such maintenance, being vehicles provided in England and Wales by a police authority or the receiver for the Metropolitan Police or, in Scotland, by a police authority or a joint police committee.
- Heavy motor cars or motor cars constructed or adapted for the purpose of forming part of an articulated vehicle which are used for drawing only a trailer of a type described above or a trailer being used for, or in connection with, any purpose for which it is authorized to be used on roads under the Special Types General Order.
- Play buses.

NB: The following definitions apply in the above exemptions:

- 'Breakdown vehicle' means a motor vehicle on which is permanently mounted apparatus designed for raising one disabled vehicle and for drawing that vehicle when so raised, and which is not equipped to carry any load other than articles required for the operation of, or in connection with, that apparatus, or for repairing disabled vehicles.
- 'Engineering plant' means movable plant or equipment being a motor vehicle or trailer (not constructed primarily to carry a load) specially designed and constructed for the purposes of engineering operations.

- 'Works truck' means a motor vehicle designed for use in private premises and used on a road only in delivering goods from or to such premises, to or from a vehicle on a road in the immediate neighbourhood, or in passing from one part of any such premises to another or to other private premises in the immediate neighbourhood, or in connection with road works while at or in the immediate neighbourhood of the site of such works.

Taxing Exempt Vehicles

When applying for a vehicle excise licence for a vehicle exempt from plating and testing, it is necessary to complete declaration form V112G (available from VROs) in order to obtain the licence without a valid test certificate.

Tachograph Testing

Inspection of tachographs is now included in the annual test, but only in respect of the following items:

- that a tachograph is fitted (ie where the vehicle requires to have it fitted by law);
- that it can easily be seen from the driving seat;
- the condition of the instrument;
- that the instrument can be illuminated;
- that all seals are present and intact.

Operators submitting vehicles for test, which are exempt from the requirement for tachograph fitment, must declare the exemption on an appropriate form which lists the exempt categories (see pp 107–110 for list of tachograph exemptions).

Production of Documents

The police and examiners of VOSA can request production of both test and plating certificates for goods vehicles when such vehicles have been involved in an accident or if they believe an offence has been committed. If these documents cannot be produced at the time they may be produced within seven days at a police station convenient to the person to whom the request was made, or as soon as reasonably practicable thereafter.

ENFORCEMENT CHECKS ON VEHICLES

Euro-Wide Enforcement Link-Up

Enforcement agencies from most EU member countries are teaming up to harmonize spot checks on vehicles and legal documentation. Under the EU's *Karolus* programme, national civil servants are participating in exchange programmes throughout the Community for the purposes of enforcing a range of single market legislation, including road transport matters such as drivers' hours and tachograph use, mechanical condition of vehicles and overloading. In time, harmonized enforcement documentation will be issued by way of standardized prohibition notices and enforced rest notices.

Roadside Checks

In addition to carrying out annual vehicle tests at the goods vehicle testing stations, examiners of VOSA operate roadside checks on commercial vehicles. These roadside checks are carried out at intervals on main roads by examiners who usually take over a lay-by which will accommodate several large vehicles. VOSA has stepped up its normal system of roadside checks in recent times and occasionally carries out checks during the night and during weekends. A police officer, in uniform, standing on the road directs vehicles (*NB: Only a police officer in uniform can stop a moving vehicle*) which are required for examination into the lay-by, where they are inspected mainly for visible wear and defects of the brakes, steering gear, silencers, tyres, lights and reflectors and for the emission of black smoke when the engine is revved up. The examiners only have a limited amount of equipment on these checks so the extent of their examination is likewise limited but nevertheless the inspection is undertaken by skilled and observant people and very little escapes their attention.

NB: VOSA examiners use a guide called Categorisation of Defects when considering the prohibition of vehicles found to be defective during roadside checks or while undergoing their annual test. Copies of this guide (price £21.32) can be obtained from goods vehicle testing stations – see Appendix IV.

Vehicle Inspections on Premises

Besides roadside checks, VOSA examiners and police officers in uniform are, at any reasonable time, free to enter any premises on which goods vehicles are kept and examine the vehicles. The owner's consent is not needed in this case but to examine any vehicle (ie other than goods vehicles) the owner's consent has to be obtained to carry out the examination or he must be given at least 48 hours' notice of such a proposal to carry out an examination; *if the notice is sent by recorded delivery post the period is increased to 72 hours*. In the latter case the consent of the owner of the premises, if different from the vehicle owner, must also consent. If on these inspections defects are found on the vehicles examined, the same procedure applies regarding the prohibition of their use as explained for roadside checks.

A police officer in uniform or a VOSA examiner, on production of suitable identification, can instruct a driver in charge of a stationary goods vehicle on a road (by the issue of form GV3) to take the vehicle to a suitable place to be examined but this must not be for a distance of more than five miles.

Prohibition notices can be issued by examiners at a goods vehicle test station if defects of a serious enough nature are found. Also, the police may be notified if prosecution is warranted, although it is not usual for vehicles in such a precarious state to be submitted for test.

Powers of Police, VOSA Examiners and Certifying Officers

The Road Traffic Act 1988 gives authorized examiners (specifically, examiners of VOSA, London taxi examiners, authorized police officers and persons appointed for the purpose by a chief officer of police and certifying officers appointed under the Road Traffic Act 1991) powers to test and inspect vehicles (and examine vehicle records), on production of their authority, as follows:

- They can test any motor vehicle or trailer on a road to check that legal requirements regarding brakes, silencer, steering, tyres, lights and reflectors, smoke and fumes are complied with, and may drive the vehicle for this purpose.
- They can test a vehicle for the same purposes on premises if the owner of the premises consents or has been given at least 48 hours' notice, except where the vehicle has been involved in a notifiable accident, when there is no requirement to give notice. If the notice given is in writing it must be sent by recorded post and the time limit is extended to 72 hours.
- They may at any time enter and inspect any goods vehicle and goods vehicle records, and may at a reasonable time enter premises on which they believe a goods vehicle or goods vehicle records are kept.
- They can request a driver of a stationary goods vehicle to take the vehicle to a place for inspection up to five miles away.
- They can at any reasonable time enter premises where used vehicles are sold, supplied or offered for sale or supply or exposed or kept for sale or supply, to ensure that such vehicles can be used on a road without contravening the appropriate regulations. They may drive a vehicle on the road for this purpose.
- They may enter at any reasonable time premises where vehicles or vehicle parts are sold, supplied, offered for sale or supply, exposed or kept for sale or supply.
- They may require the person in charge of any vehicle to take it to a weighbridge to be weighed. If the vehicle is more than five miles from the place where the request is made and the vehicle is found not to be overloaded, the operator can claim against the highway authority for any loss sustained.
- When a goods vehicle has been weighed and found to exceed its weight limit and its use on a road would be a risk to public safety, they can prohibit its road use by the issue of form TE160 until the weight is reduced to within the legal limit.
- If they find that a goods vehicle is unfit or likely to become unfit for service they can prohibit the driving of the vehicle on the road either immediately or from a later date and time by the issue of a prohibition notice (form PG9 – see below).
  NB: Police powers in this respect are restricted under the provisions of the Road Traffic Act 1991 (effective from 1 July 1992) to the issue only of immediate prohibitions in circumstances where they consider that the driving of a defective vehicle would involve a 'danger of injury to any person' as opposed to the VOSA examiner's right to prohibit a vehicle which is 'unfit for service'.
- Where a prohibition order has been placed on a vehicle for various reasons, they are empowered to remove the prohibition (by the issue of form PG10) when they consider the vehicle is fit for use. *Since 1 July 1992 clearance of prohibitions has required the vehicle to be subjected to a full roadworthiness test by VOSA at a goods vehicle testing station – and payment of the full test fee (currently £68.00 and £39.00 respectively for the largest vehicles and trailers).*
- They can ask the driver of a goods vehicle registered in an EU member state, fitted with a tachograph, to produce the tachograph record of the vehicle when it is used in this country and ask to examine the official calibration plaque in the instrument. They can at any reasonable time enter premises

where they believe such a vehicle is to be found or that tachograph records are kept, and may inspect the vehicle and records (ie tachograph charts).

Powers of Police Constables

- A police constable *in uniform* can stop a moving vehicle on a road ('constable' includes any rank of uniformed police officer). They can test any motor vehicle or trailer on a road to check that legal requirements regarding brakes, silencer, steering, tyres, lights and reflectors, smoke and fumes are complied with. They may drive the vehicle for this purpose.
- They can test a vehicle for the same purposes on premises if the owner of the premises consents or has been given at least 48 hours' notice (or 72 hours if given by recorded post), except that consent is not necessary where the vehicle has been involved in a notifiable accident.

NB: This authority applies to all constables even if not specially authorized under the Road Traffic Act 1988 as mentioned above.

New Enforcement Powers for VOSA Examiners

Specially selected and trained enforcement officers of VOSA have powers to stop moving vehicles. These officers, known as Enforcement Support Officers (colloquially referred to as 'stopping officers') and accredited by Chief Constables under provisions contained in the Police Reform Act 2002, are uniformed and operate with liveried and fully marked vehicles (silver Ford Galaxy cars with amber beacons and 'Follow Me' lighting).

Powers of Trading Standards Officers

Trading standards officers (ie employed by local authorities) can request the driver of a vehicle which is carrying goods that need an official conveyance note (ie ballast which includes sand, gravel, shingle, ashes, clinker, chippings, including coated material, hardcore and aggregates) to take that vehicle to a weighbridge to be weighed. Goods may have to be unloaded if necessary.

INSPECTION NOTICES AND PROHIBITIONS

Vehicle Inspection Notice – Form PGDN35

Following an inspection of a vehicle which is found to have no serious road safety defects, form PGDN35 (commonly known as an inspection notice) is issued to indicate to the user either that one or more minor defects were found which are in his interest, and the interests of other road users, to have rectified at an early date – it is not actually a prohibition – or that the vehicle has no apparent defects – subject to a disclaimer – to save further inspection later on that journey or that day.

Direction Notice

Form GV3 is authorization for VOSA examiners to direct a vehicle to proceed to a specified place to be inspected (normally not more than five miles away).

Prohibition Notices

The driver of any vehicle found by VOSA examiners to have serious defects is given a form PG9 on which these are listed. The form PG9 is the examiner's authority to stop the use of the vehicle on the road for carrying goods. Depending on how serious the defects are, the prohibition will either take effect immediately, in which case the vehicle, if loaded, has to remain where it is until either it is repaired or has been unloaded and then taken away for repair, for which the examiner will give authority by issuing form PG9B, or it may be delayed for 12 to 24 hours or more depending on the seriousness of the fault (see p 403). In this case the vehicle may continue to operate until the limit of the period of exemption of the prohibition by which time, if it is not repaired and cleared, it must be taken off the road. If defects recorded as requiring immediate attention are repaired quickly on the roadside (either by the driver or by a mechanic or repair garage staff who come out to the vehicle) the examiner may issue a variation to the PG9 notice with form PG9A which then allows the vehicle to be removed and used until the new time specified on the variation notice (the forms are described further individually below).

C&U Offences

If defects are found at a roadside check which make the vehicle unsafe to be on the road (usually brakes, steering or suspension defects) or if the defects are such that an offence under the Construction and Use Regulations is committed (particularly in respect of lights, reflectors, smoke emission or the horn), VOSA examiners will report these items to the police in attendance for consideration for prosecution.

Prohibition Forms

A number of official forms as described below are used by VOSA examiners in the process of inspecting vehicles, recording defects and prohibiting the use of those which are defective.

Form PG9

When an inspection by a VOSA examiner reveals defects of a serious nature form PG9 will be issued, specifying the defects and stating the precise time at which the prohibition preventing further use of the vehicle comes into force (which could be the time when the notice is written out – ie with immediate effect – or later). A copy of the PG9 is given to the driver and this must be carried on the vehicle until the prohibition is removed. Further copies of the notice are sent to the vehicle operator and, if the vehicle is specified on an 'O' licence, to the relevant Traffic Commissioner (TC).

If the PG9 has immediate effect, this means that the vehicle cannot be driven or towed away (see below).

Form PG9A

This form (Variation in the Terms of a Prohibition . . .) is issued if the VOSA examiner wishes to vary the terms of a PG9 notice by either suspending the PG9 until a future time (eg midnight on the day of issue), altering the time (which is effectively the same thing as suspending the notice as mentioned above) or altering the list of defects shown on the PG9 notice.

Form PG9B

A VOSA examiner may, after issuing a PG9, exempt the vehicle (Exemption from a Prohibition . . .) from the terms of the prohibition and permit its movement provided that the vehicle:

- is unladen;
- proceeds at a speed not in excess of a specified figure;
- does not tow a trailer;
- is towed on a rigid tow-bar;
- is towed on a suspended tow;
- is not used after lighting-up time (if it has lighting defects);
- proceeds only between two specified points.

Other conditions may be added as appropriate.

Form PG9C

When a vehicle which is subject to a PG9 notice is presented to a VOSA examiner for clearance of the defect and the examiner is not satisfied that it is fit for service, he may issue form PG9C (Refusal to Remove a Prohibition . . .), which means that the original PG9 notice remains in force until the defects are satisfactorily rectified.

NB: Forms PG9A/B/C are usually computer-generated these days.

Form PG10

If the defects specified in a PG9 notice have been repaired to the satisfaction of the VOSA examiner to whom the vehicle is presented for clearance and the examiner is satisfied that the whole vehicle is in a satisfactory condition for use on the road, he will issue a PG10 notice (Removal of Prohibition . . .) which removes the prohibition. The TC must be notified of the clearance if the vehicle is specified on an 'O' licence.

Form TE160

This notice (*which does not have a PG prefix*) relates to prohibition on the use of overweight vehicles. It effectively requires the driver of the vehicle to take the vehicle to a weighbridge and, if found to be overloaded, reduce the gross weight to legal limits before proceeding on his journey. *NB: It is a defence to a charge of overloading that the vehicle was on its way to the nearest practicable weighbridge or that at the time of loading the weight was within legal limits and was subsequently not more than 5 per cent heavier despite not having any additions to the load en route.*

Effects of Prohibition

A vehicle must not, under any circumstances, be used to carry goods while it is the subject of a PG9 prohibition notice, but despite the prohibition notice a vehicle may be driven unladen to a goods vehicle test station or to a place agreed with a goods vehicle examiner (both by previous appointment only) in order to have the vehicle inspected. The vehicle may also be driven on the road for test purposes, provided it is unladen, within three miles of where it has been repaired.

Forms PG9 and PG9A have a panel of boxes identified by letters of the alphabet from A to M. These boxes are used by enforcement staff to codify certain aspects of the vehicle check as follows:

A Whether the vehicle was laden or unladen.

B Whether examination took place on the roadside, at a test station, during the annual test, following an accident, during a fleet check or on any other occasion.

C If defects appeared to be from a significant failure in maintenance arrangements a letter 'S' is entered; 'X' is used if the examiner can attach no undue significance to the defect (ie it could have happened on the journey); and the box is left blank if the examiner is not able to form an opinion as to the cause of the defect.

D To indicate whether a prosecution will follow (the letter 'P' in this box indicates prosecution by the police and 'V' by VOSA).

E If vehicle is issuing smoke ('S' equals smoking in service; 'S/A' equals smoking under free acceleration when vehicle is stationary).

L Indicates number of pages of prohibition notice served at the time (ie page 1 of 1, page 1 of 2). This indicates to the examiner who is asked to clear the prohibition that there was another sheet(s) to the prohibition notice issued at the time.

F, G, H, K and M Not currently used.

Appeals against the Issue of Prohibition Notices

There is no appeal against the imposition of a PG9 prohibition notice, but there is a right of appeal against refusal to remove a prohibition after repair.

Clearance of Prohibition

An operator having a vehicle placed under an immediate or a delayed prohibition notice has to get the defects repaired and then submit the vehicle to his local goods vehicle testing station either for a full roadworthiness examination in the case of serious defects (with payment of the relevant fees – currently up to £68.00 for vehicles and up to £39.00 for trailers; in each case depending on the number of axles; see full list of fees on p 384) or, in the case of minor defects, a 'partial clearance' or 'mini' test (for which a reduced fee of £45.00 for vehicles and £26.00 for trailers is payable). A clearance certificate, PG10, is issued if the examiner is satisfied with the repair. If the examiner is not satisfied with the repair he will issue a form PG9C 'Refusal to Remove a Prohibition', or form PG9A if some of the defects are cleared and others are not. If because of the better inspection facilities at the test station he finds further defects another form PG9 may be issued.

VDRS for Trucks

A vehicle defect rectification scheme for heavy commercial vehicles similar to the one operated by the police for defective cars and light vehicles (see p 416) was introduced nationwide (late in 1996) following the success of pilot schemes in Durham and Essex. Operators of vehicles found with minor defects are given 14 days in which to rectify them, and produce evidence of the repair (ie by means of an invoice).

LIGHT VEHICLE TESTING

Light goods vehicles up to 3500 kg gross weight and dual-purpose vehicles (see p 201 for definition) under 2040 kg unladen weight are required to be tested on the third anniversary of the date of their original registration and annually thereafter (commonly known and referred to as the MoT test). The tests are carried out at private garages approved by VOSA and displaying the blue and white triple-triangle sign. For full details of the testing scheme for these vehicles, see Chapter 17 which deals with light vehicle testing.

NORTHERN IRELAND CERTIFICATION OF VEHICLES

The United Kingdom system of goods vehicle plating and annual testing does not apply in Northern Ireland. The Province has its own Goods Vehicle Certification scheme (commonly known and referred to in the Province as the PSV test) which requires heavy goods vehicles to be submitted to the Department of the Environment (NI) for an annual mechanical examination. Under the Goods Vehicles (Certification) Regulations (Northern Ireland) 1982, owners of goods vehicles (other than those specifically exempted – see p 408) must obtain a test certificate for each vehicle no later than one year from the date of first registration and annually thereafter (see p 408).

Application for a certificate must be made to the following address at least one month before the date on which the certificate is to take effect: Department of the Environment, Vehicle and Driving Test Centre, Balmoral Road, Belfast BT12 6QL.

Retest fees apply where application is made within 21 days from the date of service of the notice and the vehicle is presented for re-examination within 28 days.

Applications by Non-NI Based Bodies

Where an application is made by a corporate body with its principal or registered office outside Northern Ireland or by a person residing outside Northern Ireland the following conditions must be observed:

- During the currency of the certificate a place of business must be retained in NI.
- They must be prepared to accept, at such a place of business, any summons or other document relating to any matter or offence arising in NI in connection with the vehicle for which the certificate is applied for.
- They must undertake to appear at any court as required by such a summons or by any other document.
- They must admit and submit to the jurisdiction of the court relative to the subject matter of such summons or other document.

Failure to comply with any of the above-mentioned requirements will involve immediate revocation of the certificate.

Examination of Vehicle

When notified by the Department the applicant must present the vehicle for examination, in a reasonably clean condition, together with the registration book and previous certificate, if any, at the time and at the centre specified in the notice.

Issue of Certificate

If, after examining the vehicle, the Department is satisfied that it complies in all respects with the regulations in respect of the construction, use, lighting and rear marking of vehicles, a certificate will be issued.

Refusal of Certificate

If the vehicle does not meet the requirements of the regulations a certificate will be refused and the applicant will be notified of the reasons why.

Re-examination of Vehicles

When a certificate has been refused and the defects specified in the notice have been put right, an application may be made for a further examination of the vehicle. A reduced retest fee will be payable if the retest is conducted within 21 days of the original test.

Refund of Fees

Prepaid test fees may be refunded in the following circumstances:

- If an appointment for an examination of a vehicle is cancelled by the Department.
- If the applicant cancels the appointment by giving the Department (at the centre where the appointment is made) three clear working days' notice.
- If the vehicle is presented to meet the appointment but the examination does not take place for reasons not attributable to the applicant or the vehicle.
- If the applicant satisfies the Department that the vehicle could not be presented for examination on the day of the appointment because of exceptional circumstances which occurred no more than seven days before the day of the appointment, and providing notice is given to the centre where the examination was to take place within three days of the occurrence.

Duplicate Certificates

Duplicate certificates may be issued in replacement of those which have been accidentally lost, defaced or destroyed. A fee is payable for replacement certificates. If subsequently the original certificate is found, it must be returned to the nearest examining centre or to any police station.

Display of Certificates

The certificate issued on satisfying the examiners must be attached to the vehicle in a secure, weather-proof holder and must be displayed on the nearside windscreen or on the nearside of the vehicle not less than 610 mm and not more than 1830 mm above the road surface so that the particulars of the certificate are clearly visible (ie at eye level) to a person standing at the nearside of the vehicle.

Conditions of Certificate

It is a condition of the certificate that the vehicle owner:

- Must not permit the vehicle to be used for any illegal purpose.
- Must not deface or mutilate the certificate or permit anybody else to do so.
- Must, at all reasonable times, for the purpose of inspection, examination or testing of the vehicle to which the certificate relates:
  - produce the vehicle at such a time and place as may be specified by any Inspector of Vehicles;
  - afford to any Inspector of Vehicles full facilities for such inspection, examination or testing including access to his premises for that purpose;
  - must ensure that the vehicle and all its fittings are maintained and kept in good order and repair and must take all practical steps to ensure that all parts of the mechanism, including the brakes, are free from defects and are in efficient working order;
  - must immediately notify the nearest examination centre of any alteration in design or construction of the vehicle since a certificate was issued.

Transfer of Certificates

If a vehicle owner sells or changes the ownership of a vehicle, he must return the certificate for the vehicle to the nearest examination centre and notify the Department of the name and address of the transferee. The Department may then transfer the certificate on request by the new owner.

If a vehicle owner dies or becomes infirm of mind or body, on application of any person the Department may transfer the certificate to such a person.

Change of Address

If a certificate holder changes his address during the currency of a certificate, he must notify details of such changes to the nearest examination centre.

Markings on Vehicles

When certificates have been issued for vehicles, those vehicles must be marked with:

- the name and address of the owner in legible writing and in a conspicuous position on the nearside of the vehicle; and
- where the unladen weight of the vehicle exceeds 1020 kg, the unladen weight should be painted, or otherwise clearly marked, in a conspicuous position on the offside of the vehicle. In the case of articulated vehicles the unladen

weight of the tractive unit and the trailer must be marked on the respective unit and trailer.

Offences

It is an offence to operate when a certificate has expired, to alter, deface, mutilate or fail to display a certificate. Failure to observe such rules will result in the certificate being declared invalid. It is also an offence to assign or to transfer a certificate to another person with the same resultant penalty. Fines or six months' imprisonment may be imposed on summary conviction for such offences or up to two years' imprisonment upon any further conviction or indictment.

Renewal of Certificates

At least one month before the expiry date of a certificate the holder should apply for a new one using an application form obtainable from any examination centre or Local Vehicle Licensing Office of the Department.

Exemptions from Certification

The following vehicles are exempt from the requirements of NI certification:

- Vehicles constructed or adapted for the sole purpose of spreading material on roads or used to deal with frost, ice or snow.
- A land tractor, land locomotive or land implement.
- An agricultural trailer drawn on a road only by a land tractor.
- A vehicle exempted from duty under section 7(i) of the Vehicles (Excise) Act (Northern Ireland) 1972 and any trailer drawn by such a vehicle.
- A motor vehicle for the time being licensed under the Vehicles (Excise) Act 1971, paragraph (a).
- A trailer brought into NI from a base outside NI if a period of 12 months has elapsed since it was last brought into NI.
- A pedestrian-controlled vehicle.
- A track-laying vehicle.
- A steam-propelled vehicle.
- A vehicle used within a period of 12 months prior to the date of it being registered for the first time in NI or the UK; or, where a vehicle has been used on roads in NI or elsewhere before being registered, the exemption applies for the period of 12 months from the date of manufacture rather than from the date of registration. For this purpose any use before the vehicle is sold or supplied retail is disregarded.

17 Light Vehicle (MoT) Testing

Private cars, motor caravans (irrespective of weight), dual-purpose vehicles (see p 201 for definition) under 2040 kg unladen weight and light goods vehicles not exceeding 3500 kg gross weight are subject to annual testing (commonly referred to as the MoT test) at VOSA-approved commercial garages, starting on the third anniversary of the date of their first registration and each year thereafter.

Cars and light goods vehicles may be tested up to one month before the due date of their first test and the test certificate will extend for 13 months to expire on the anniversary of the first registration date.

Testing is carried out at garages displaying the blue and white MoT triple-triangle symbol, and application is made direct with the garage for a suitable appointment. Some centres claim to provide MoT testing 'while you wait'. There is no application form to be completed.

A new computer system is being installed in MoT testing stations which will facilitate checks on a vehicle's MoT status and reduce theft and forgery of MoT test certificates. Records of all MoT tests will be held on a secure central database, which will comprise the definitive proof of a valid test pass. MoT certificates will be produced as A4-format receipts which will have no value to thieves and there will be no blank certificates to steal.

Fees for the light vehicle test – payable to the garage at the time of the test – are as follows:

Class IV & IVa (Class IVa includes seat belt installation check)

Vehicle type	Age first test certificate required (years)	Fee
Cars (up to 8 passenger seats)	3	£44.15
Motor caravans	3	£44.15
Dual purpose vehicles	3	£44.15
PSV's (up to 8 seats)	3	£44.15
Ambulances and taxis	1	£44.15
Private passenger vehicles and ambulances (9–12 passenger seats)	1	£46.15
Class IVa	- -	£51.55

Class V (with more than 13 passenger seats)

Vehicle type	Age first test certificate required (years)	Fee
13–16 passenger seats	1	£47.95
More than 16 passenger seats	1	£65.00

Class Va (includes seatbelt installation check)

Vehicle type	Age first test certificate required (years)	Fee
13–16 passenger seats	–	£64.85
More than 16 passenger seats	–	£100.40

Class VII

Vehicle type	Age first test certificate required (years)	Fee
Goods vehicles (over 3,000 kg up to 3,500 kg DGW)	3	£47.20

Duplicate test certificates cost half of the above amounts depending on the class of vehicle.

VEHICLE CLASSES

Vehicles subject to the MoT test are classified as follows:

Class I Light motor bicycles not exceeding 200 cc cylinder capacity with or without sidecars.

Class II All motor bicycles (including Class I) with or without sidecars.

Class III Light motor vehicles with three or more wheels (excluding Classes I and II) not exceeding 450 kg unladen weight.

Class IV Heavy motor cars and motor cars (excluding Classes III and V); ie any vehicle with an unladen weight of more than 450 kg which is:

(a) a passenger vehicle (ie private car, taxi, vehicle licensed as private, with 12 passenger seats or less, or small public service vehicle with less than 8 passenger seats);

(b) a dual-purpose vehicle not exceeding 2040 kg unladen weight (see p 201 for definition);
(c) a goods vehicle not exceeding 3000 kg gross weight;
(d) a motor caravan irrespective of weight;

Class V Large passenger carrying vehicles; ie motor vehicles which are constructed or adapted to carry more than 12 seated passengers in addition to the driver, and which are not licensed as public service vehicles.

Class VI Public service vehicles other than those in Class V above.

Class VII Goods vehicles of 3001 kg to 3500 kg gross weight (these vehicles may be tested at MoT garages – where the facility to test such vehicles is available – or at goods vehicle testing stations).

Exemptions

Public service vehicles with seats for eight or more passengers excluding the driver, track-laying vehicles, vehicles constructed or adapted to form part of an articulated vehicle, works trucks and all trailers are excluded from the above classes and the following vehicles are also exempted from the test:

- Heavy locomotives.
- Light locomotives.
- Goods vehicles over 3.5 tonnes gross weight.
- Articulated vehicles other than articulated buses.
- Vehicles exempt from duty under the Vehicle Excise and Registration Act 1994.
- Works trucks.
- Pedestrian-controlled vehicles.
- Vehicles kept and used by invalids.
- Vehicles temporarily in Great Britain.
- Vehicles proceeding to a port for export.
- Vehicles provided for use by the police force.
- Imported vehicles owned or in the service of HM Navy, Army or Air Force.
- Vehicles which have Northern Ireland test certificates.
- Electrically propelled goods vehicles not exceeding 1525 kg unladen weight.
- Certain hackney carriages.

Many of these vehicles are subject to the goods vehicle testing and plating scheme (see Chapter 16 for details).

There is also an exemption which applies to vehicles which come within the MoT test scheme while they are being driven to a place by previous arrangement for a test or bringing it away, if it fails, to a place to have work done on it. This means that such vehicles can be driven on the road without a valid test certificate being in force, but only in the circumstances mentioned and no other.

THE TEST

Testing is carried out in accordance with *The MoT Tester's Manual* – copies available from The Stationery Office. (*Note: This publication should not be confused with* The Heavy Goods Vehicle Inspection Manual *which relates solely to the goods vehicle annual test and which is also available from The Stationery Office.)*

When presenting a vehicle for test the following conditions must be observed:

- The vehicle registration document must be produced.
- The vehicle must be sufficiently clean so as not to make the test unreasonably difficult.
- The vehicle must have sufficient petrol and oil to enable the test to be completed.
- If the vehicle is presented for the test in a loaded condition (ie in the case of light goods vehicles) the load must be secured or else removed.

Refusal and Discontinuance of Test

Failure to observe any of the above conditions can lead to a refusal to test the vehicle, and the test fee will be refunded. Furthermore, if the tester finds a defect of a serious nature which, in his opinion, makes it essential to discontinue the inspection on the grounds of risk to his own safety, risk to the test equipment or to the vehicle itself, he may do so and issue form VT30 showing the defects which caused the test to be discontinued.

Items Tested

The following items are tested and must meet the conditions specified in *The MoT Tester's Manual*:

Section I	Function of obligatory front and rear lamps
	Function of obligatory headlamps
	Function of stop lamps
	Obligatory rear reflectors
	Function of direction indicators
	Aim of headlamps
Section II	Steering wheel and column
	Steering mechanism
	Power steering
	Front wheel bearings
	Suspension
	All suspension types
	Suspension assemblies (springs, torsion bars, etc)
	Shock absorbers
Section III	Parking brake and operating lever
	Parking brake mechanism (under vehicle)
	Service brake operating pedal
	Service brake mechanism under vehicle
	Brake performance test
Section IV	Tyres
	Roadwheels
Section V	Seat belts

Section VI Function of windscreen washers
Function of windscreen wipers
Exhaust system
Function of audible warning device
Condition of the vehicle structure.

Additional Test Items

Since 1 November 1991 the examination has covered additional items such as exhaust emissions (see below) – particularly in light goods vehicle tests; ABS (ie anti-lock braking system) warning lights; rear seat belts (ie where fitted); rear wheel bearings; more extensive corrosion checks.

Since 1 January 1993 additional items have been included in the MoT test in accordance with EU requirements. These include checks on the following vehicle components and systems:

- mirrors – a motor car must have a mirror on the driver's side and either an exterior mirror on the passenger side, or an interior mirror; they must be in good condition, capable of adjustment and be secure;
- windscreens – damage (ie a crack) to a windscreen, exceeding 10 mm in zone A (an area 290 mm wide centred on a vertical line passing through the centre of the steering wheel and within the area swept by the windscreen wipers), will result in test failure as will obscuring clear vision with stickers or other obstructions;
- fuel tanks and feed pipes – must be in good condition and secure (including the tank filler cap);
- bodywork and body security – must be sound;
- seat fixings – must be secure and backrests lockable;
- security of doors and other openings – all doors must be capable of being opened from the inside and outside and they must latch securely;
- number (ie registration) plates – must be secure and in good condition and the letters and numbers must be correctly spaced (incorrect spacing of letters and numbers to make up names or other words will result in test failure);
- vehicle identification number – must be present and capable of being clearly read;
- number plate lamps (at rear) – must be fitted and working correctly;
- rear fog lamps – vehicles first used after 1 April 1980 must have one and it must work and an interior warning light must be provided and must work;
- hazard warning devices – must work with the ignition on or off and there must be an interior warning light which works.

Exhaust Emission Checks

A check is made on emissions from diesel-engined vehicles subject to the test. These tests were reintroduced in 1994 after being temporarily suspended.

Cars, light goods vehicles and particularly goods vehicles within the weight range 3001 kg to 3500 kg gross weight which fall within the *Class VII* MoT test are subjected to additional exhaust emission checks during their annual roadworthiness test, as well as checks for excessively smoking exhausts. This involves measurement of the levels of carbon monoxide (CO) and hydrocarbons (HC) to ensure that engines are properly tuned. The current maximum level of carbon monoxide acceptable for post-1 August 1983 registered vehicles is 4.5 per cent (6 per cent for vehicles first registered prior to this date). The limit for hydrocarbons for post-1 August 1975 registered vehicles is 1200 parts per million (ppm).

Higher standards (ie lower permissible emissions), originally intended for introduction in 1996, were brought forward to come into effect from 1 September 1995, as follows:

- light goods vehicles in Class II (ie between 1250 and 1700 kg gross weight):
  carbon monoxide (CO) limit - 5.17 kg/km
  hydrocarbon/nitrous oxide - 1.4 g/km
  particulates - 0.19 g/km
- light goods vehicles in Class III (ie 1700 kg to 3500 kg gross weight):
  carbon monoxide (CO) limit - 6.9 kg/km
  hydrocarbon/nitrous oxide - 1.7 g/km
  particulates - 0.25 g/km

Emission standards tests as described above do not apply to motorcycles, three-wheelers or diesel-engined vehicles.

Regulations permit tests for exhaust emissions to be carried out on the roadside as well as at approved MoT test garages.

Seat Belt Checks

Additional seat belt checks have been included in the MoT scheme since 1 August 1998. From this date the condition of seat belts in all vehicles (whether required to be fitted mandatorily or fitted on a voluntary basis) will be examined to determine whether the belts are likely to fulfil their intended function in the event of an accident. Similar checks on seat belts will be carried out during roadside and other enforcement spot checks.

Certain vehicles are to be subject to a once-only check of the quality of their seat belt installations – during subsequent MoT checks only the condition of the belts, as normal, will be checked. Vehicles affected by these additional checks are those falling into Class IV, Class V and Class VI (see pp 410–11). The checks will be carried out only at Class V MoT testing stations, VOSA testing stations and VOSA PSV designated premises by appropriately trained vehicle examiners.

A one-off additional fee is to be charged for these installation-quality inspections.

Test Failure

If the vehicle fails to reach the required standard of mechanical condition a 'notification of refusal of a test certificate' (form VT30) is issued. This indicates the grounds on which the vehicle failed the test (ie it names the faulty components or component area and the actual fault).

Retests

If a vehicle fails the test, no further test fee is payable if it is left at the garage for the necessary repairs to be carried out before the end of the 10th working day from the day of the initial test. However, if the vehicle is taken away following the test failure, a three-tier system of retests and fees applies as follows:

- Tier 1 Full Test. Where a vehicle is presented for retest more than 10 working days after the day of the initial test or if the original MoT tester is not available for a tier 2 test, a full test must be carried out and the full test fee is payable.
- Tier 2 Partial Retest. Where a retest is necessary that requires the use of a major piece of testing equipment (including hoist or pit, brake performance tester, exhaust gas analyser (EGA), diesel smoke meter (DSM) and headlamp aim equipment), or where a repair has been carried out by welding, or there has been a repair to the steering or braking system (excluding stop lamps), provided that the retest is carried out by the same tester at the same MoT garage before the end of the 10th working day from the day of the initial test a partial retest will be carried out and only two-thirds of the full test fee will be payable.
- Tier 3 Partial Retest. Where a retest does not require the use of a major piece of testing equipment, provided it is carried out at the same MoT garage before the end of the 10th working day from the day of the initial test, only one-third of the full test fee will be payable.

Where any vehicle does not fall into tier 2 or 3 above, a full re-examination in accordance with tier 1 must be carried out. Only one partial re-examination is permissible per full examination.

Issue of Test Certificate

On completion of the test, if the vehicle is found to be in satisfactory condition and in compliance with the law, a test certificate (form VT20) is issued. Additionally, a copy of form VT29 is issued to indicate to the vehicle owner the general state of the vehicle and to point out components which may need attention in the future to keep it in good, safe working order. Comments on form VT29 are made against the following headings:

I. Lighting equipment
II. Steering and suspension
III. Braking system
IV. Tyres and wheels
V. Seat belts
VI. General items.

Appeals

A vehicle owner can appeal if he is not satisfied with the result of the test. He must complete form VT17 and send it to the local Traffic Area Office within 14 days of taking the test.

PRODUCTION OF TEST CERTIFICATES

It is necessary to produce a current, valid test certificate when taxing a vehicle to which the regulations apply, otherwise an excise licence will not be issued. A police officer can ask to see a vehicle test certificate. If the driver does not have it with him he can be asked to produce the certificate at a police station nominated by him within seven days.

VEHICLE DEFECT RECTIFICATION SCHEME

Operators of light vehicles (up to 3500 kg gross weight) found on the road with non-endorsable minor vehicle defects (ie lights, wipers, speedometer, silencer, etc) in certain areas may be offered the VDRS procedure by the police whereby no prosecution will result if:

1. Immediate arrangements are made for the repair of the defect.
2. The repaired vehicle and VDRS notice are presented to an MoT garage for examination and certification that the defects have been rectified.
3. The certificate is sent to a local Central Ticket Office.

Failure to follow the procedure on receipt of a VDRS notice will result in prosecution.

18 Vehicle Maintenance

The requirements of the Road Traffic Act 1988 for the annual testing of goods vehicles and trailers, the requirements of the current Construction and Use Regulations, and the parts of the Transport Act 1968 setting out the conditions relating to vehicle maintenance under which an 'O' licence will be granted create a situation whereby operators must ensure that their vehicles and trailers are always safe, are in a fit and roadworthy condition and that their maintenance, vehicle inspection and maintenance record systems meet the requirements laid down in the legislation.

In order for operators to meet these requirements they must maintain their vehicles and trailers (no matter how old they are) to a sufficiently high standard to enable them to pass the stringent annual goods vehicle test which they should be able to do on the test day *and on every other day when they are on the road.* To be able to do this, it is not sufficient merely to take the vehicle off the road for a short period once a year just before test day and work frantically to get it up to scratch, and for the rest of the year allow it to run on the road in a condition which is something below the required standard.

The risk of the vehicle encountering a roadside check, or being on the operator's premises when examiners of the Vehicle and Operator Services Agency (VOSA) or the police decide to make an inspection, as they are empowered to do, is too great to take when the penalties for failure to maintain vehicles are so high. Besides the risk of heavy fines, there is the possibility that the operator may lose his 'O' licence since a satisfactory state of maintenance is one of the factors taken into account by the Traffic Commissioner in considering applications and renewals for such licences.

MAINTENANCE ADVICE

A VOSA team along with operator associations have produced a Code of Practice on vehicle maintenance for transport operators – *Guide to Maintaining Roadworthiness: Commercial Goods and Passenger Carrying Vehicles* – available from The Stationery Office, Traffic Area Offices and Heavy Goods Vehicle Testing Stations. Additionally, there is the D*f*T's free guide *A Guide to Goods Vehicle Operators' Licensing* (reference GV74), which contains some specific advice on vehicle maintenance arrangements for licence applicants. The following notes are included as an appendix to the *Guide*.

There are two separate vehicle checks and inspections which should be carried out:

- daily running checks;
- vehicle safety inspections and routine maintenance at set intervals on items which affect vehicle safety, followed by repair of any faults.

Daily running checks are normally carried out by drivers before a vehicle starts its daily journey. They are checks on such basic items as engine oil, brakes, tyre pressures, warning instruments, lights, windscreen wipers and washers and trailer coupling.

Vehicle safety inspections and routine maintenance should be carried out at set intervals of either time and/or mileage whichever occurs first. How often these inspections are done should be decided by the nature of the operator's business. A vehicle used on long-distance work will need inspecting at different intervals from one employed in heavy traffic on local work with frequent stops and starts. The items inspected should include wheels, tyres, brakes, steering, suspension, lighting, and so on. More detailed information can be found in VOSA's publication *The Heavy Goods Vehicle Inspection Manual* (available from The Stationery Office). Vehicle checks and inspections are extra to a routine maintenance schedule. It is vital to the vehicle's safety that both types of checks and inspections are done.

Staff doing inspection checks must be able to recognize faults they find, such as excessive wear of components. They should also be aware of the acceptable standard of performance and wear of parts. Trade associations offer regular inspections for their members' vehicles.

Records

Records must be kept of all safety inspections to show the history of each vehicle. These records must be kept for at least 15 months. If vehicles from several operating centres are inspected and repaired at a central depot, the records may be kept at that depot, although VOSA examiners are entitled to request inspection of records at the operating centres where vehicles are based.

If an outside garage does the inspections and repairs, you must still keep maintenance records. You are responsible for the condition of any vehicle or trailer on your licence.

Facilities

These will depend on the number, size and types of vehicles to be inspected. It must be possible to inspect the underside of a vehicle with sufficient light and space to examine individual parts closely. Ramps, hoists or pits will usually be necessary, but may not be needed if the vehicles have enough ground clearance for a proper underside inspection to be made on hard-standing. Creeper boards, jacks, axle stands and small tools should be available.

As well as providing facilities for checking the underside of vehicles, operators should whenever possible use equipment for measuring braking efficiency and setting headlights. If many vehicles have to be inspected, it may be worthwhile providing a roller brake tester.

Drivers' Reports

Drivers must report vehicle faults to whoever is responsible for having them put right. The maintenance system should allow for these reports, which must be recorded in writing either by the driver himself or by the person responsible for maintaining the vehicle. Owner-drivers must note faults as they arise and keep these notes as part of their maintenance record.

Hired Vehicles and Trailers

In the case of hired, rented or borrowed vehicles or those belonging to other operators used in inter-working arrangements, it is the user who is responsible for their mechanical condition on the road. If disciplinary action is taken as a result of a mechanical fault, it is against the user's licence, not the company from whom the vehicle is hired, or the owner.

THE CHOICE: TO REPAIR OR CONTRACT OUT

The first decision an operator has to make when planning his vehicle maintenance is to determine whether it is to be carried out in his own workshops, by his own staff, or whether it is to be contracted out to a repair garage. The size of the fleet and what existing facilities and premises he already has will usually determine the method to be used. It is unlikely to be an economic proposition for a very small operator with a fleet of less than five or six vehicles to establish his own workshop unless he actually does the work himself. On the other hand it is unlikely to be an economic proposition for a large operator to contract out the work. There are, however, many examples in industry where the opposites apply, and very successfully too; so it remains very much a matter for operators to assess their own requirements, balance out the costs of the alternatives and make arrangements accordingly.

Choice of Repairer

When choosing a repairer to do the work preference should be given to a garage which is a main distributor for the make of vehicle operated or an agent for that make of vehicle. The reason for this is that such a firm, as part of its arrangement with the manufacturer it represents, will have had to send some of its mechanics to the factory for training in the repair and servicing of that particular make of vehicle. This ensures that skilled staff are working on the vehicle. Moreover, the distributors are also usually required by the manufacturer to hold considerable stocks of spare parts, with a predominance of fast-moving items. By using such a garage the risk of a vehicle being kept off the road waiting for spare parts is therefore considerably reduced, and since vehicle downtime is a heavy cost burden these days this is an important consideration. If a garage of this type is not available locally and a second choice has to be made, this should be a firm which is experienced in heavy vehicle repair work and which has suitably trained staff and the necessary equipment. A garage which normally only handles motor car repair work should not be used.

Repair Arrangements

Any arrangements made with a repairer should be in writing. The 'O' licence application form requests that copies of any maintenance contract should be sent with the application but in any event the Traffic Commissioner (TC) will want to know what arrangements have been made with a repairer. Verbal arrangements or the practice of sending vehicles in for repair as necessary on an ad hoc basis are not acceptable to the TCs, some of whom have said that a verbal agreement for these purposes is no agreement at all.

Maintenance Agreements

The agreement should include provisions for the repairer to be responsible for supplying the operator with suitable records of the inspections and repair work carried out. The operator should ensure that he can escape reasonably quickly and easily from any agreement in the event of the standard of work deteriorating and thereby placing his 'O' licence at risk.

Care should be taken when discussing or making an agreement that the repairer is aware of the consequences of any negligent action on his part on the livelihood of the operator, remembering that it is the vehicle 'user' who remains responsible at all times for the safe and satisfactory mechanical condition of the vehicle, and it is he or she who is responsible for keeping and producing records of vehicle inspections and other maintenance work.

There is no standard or recommended form of maintenance agreement in universal use; it is left to the parties concerned to agree on terms.

NB: A suitable format for agreements is suggested by the DfT *in its guide to 'O' licensing. A copy of this is reproduced here for reference.*

The agreement should clearly set out in detail the work to be done, the intervals at which it is to be done, the responsibilities of the operator and the garage, the form which records should take and the action to be taken if defects are discovered or repairs over a certain value are found to be necessary.

MAINTENANCE CONTRACTS

In recent years there has been a proliferation of contract maintenance schemes, many of them offered by truck manufacturers who are looking to retain future business. However, the advice given in this chapter remains relevant even with such sophisticated schemes. It is worth pointing out here that no matter how good the contract scheme or how reputable and reliable the repairer, it remains the responsibility of the operator to ensure that vehicles are in a fit, serviceable and safe condition when on the road and that they comply fully with the law. (This is covered in more detail later in the chapter.)

Some of the prominent commercial vehicle repair specialists offer contract maintenance schemes. There are usually a number of options in such schemes which, for instance, give the operator a choice of having his vehicles inspected only at set intervals; inspected and serviced according to the manufacturer's recommendations; or inspected, serviced and all repair work carried out (excluding damage caused by the vehicle having been involved in an accident) including the supply of materials except for such things as tyres and batteries.

The charges (both labour and materials) for the various options are usually incorporated in the contracts and these remain fixed for the period of the contract with a clause which enables the garage to make additional charges if the vehicle exceeds, by a large margin, the mileage estimated by the operator at the time of negotiating the contract. The additional charges are usually based on the excess mileage at an agreed figure per mile.

In offering a fixed charge agreement to include all normal repair work, the garage is dependent to quite an extent on the good faith of the operator in using his vehicles in a manner which is not likely to involve the garage in excessive costs above what may be normally estimated in advance. Again, a clause may be included in the contract stating that additional charges will be raised for the repair of damage or defects because of misuse by the owner.

If a full maintenance contract under one of these schemes is negotiated, the garage usually accepts the responsibility for ensuring that the vehicle is always in a fully roadworthy condition, able to pass through VI roadside checks and pass its annual test without difficulty. In the event of a vehicle getting a PG9 prohibition for defects, the operator has some grounds for a claim against the garage but, unfortunately, he has no defence to present to the TC if he is called to explain why the vehicle was not maintained to the required standard because of the condition of 'O' licensing which makes the vehicle user totally responsible for the condition of his vehicles on the road (see below).

The garage offering a full fixed-price contract maintenance scheme will prefer, wherever possible, to negotiate a contract for a vehicle from the day it is new because it is much easier to plan and cost the amount of work likely to be necessary for an estimated annual mileage and the spare parts required.

The effective working of a contract agreement needs the co-operation of both parties if it is to be successful. Operators may feel that in return for paying the garage a fixed price for full maintenance they are left with little responsibility. But they must make it their business to ensure that whatever other considerations may be pressing, a vehicle is sent to the garage when it is required if the contract is one in which the garage calls the vehicle in at specified intervals for the work to be done.

A model agreement between the operator and a garage or agent for safety inspections and/or repair of vehicles and trailers subject to operators' licensing.

This Agreement is made the .... day of .... 19 .... between

(a) .... whose address [registered office] is

.... ("the operator") of the one part, and

(b) .... whose address [registered office] is

.... ("the contractor") of the other part.

1. The Contractor agrees that he [it] will, in relation to every vehicle mentioned in the Schedule below, on every occasion when that vehicle is submitted by the operator as mentioned in Article 2 below on or after the date of this Agreement –

(a) inspect all the items specified in the maintenance record in the form for the time being approved by the Department of Transport which relate to the vehicle:

(b) if the operator so consents, carry out such renewals and repairs as may be necessary to ensure that the vehicle and every part of it specified in that maintenance record is in good working order and complies with every statutory requirement applying to it; and

(c) complete that maintenance record to show –

(i) which items were in good working order and complied with the relevant statutory requirements when the vehicle was submitted and which remain in that condition;

(ii) which (if any) items were not in good working order or failed to comply with those requirements when the vehicle was submitted but have been replaced or repaired so that those requirements are satisfied; and

(iii) which (if any) items were not in good working order or failed to comply with those requirements when the vehicle was submitted and which has not been so replaced or repaired.

2. The operator agrees that he [it] will –

(a) submit to the contractor each vehicle mentioned in the Schedule below in order that the contractor may, as regards that vehicle, comply with the provisions of Article 1 above –

(i) within .... weeks of the date of this Agreement, and, thereafter;

(ii) within .... weeks of the last submission or, in the the case of a motor vehicle, when the mileage shown on the odometer has increased by .... miles, whichever is the sooner;

(b) pay to the contractor such reasonable charges as the contractor may make pursuant to his [its] obligations under Article 1 above; and

(c) retain, and make available for inspection by an officer mentioned in section 82(1) of the Transport Act 1968, every maintenance record mentioned in Article 1 above for a period of at least 15 months commencing with the date of its issue.

3. This Agreement shall be determinable by either party giving to the other .... months' written notice of his [its] intention to determine it.

Figure 18.1 *A model agreement between the operator and a garage or agent for safety inspections and/or repair of vehicles and trailers subject to operators' licensing. With acknowledgement to the Department for Transport (reproduced from GV74)*

Schedule

(Motor Vehicles and trailers which [are authorised vehicles] [it is intended shall become authorised vehicles] under an operator's licence [held] [applied for] by the operator under Part V of the Transport Act 1968)

1. Motor Vehicles (give registration numbers and descriptions).

2. Trailers (give brief descriptions).

As witness etc

(Signature(s), or seal operator)

(Signature(s), or seal, or contractor)

Figure 18.1 *(continued)*

Although armed with what appears to be a fair agreement with a reputable repairer, operators should not become complacent. It is strongly recommended that wherever possible they should arrange for a physical check of the work that the repairer claims to have carried out. While not suggesting that a repairer is likely to claim to have done work that he knows has not been done, it must be remembered what is at stake and an operator should therefore doubly check the quality of the work.

Contract Maintenance Schemes

A number of package contract maintenance schemes are available to operators. Whichever they choose, if indeed they choose any, will depend on a number of considerations: the make of the vehicles they own, the trade association to which they belong, the availability of a suitable repairer, etc.

Freight Transport Association Maintenance Services

Under the FTA quality control maintenance inspection service member operators contract to have their vehicles inspected one or more times a year by Association inspectors (only skilled and highly experienced people are appointed) to see that they meet the requirements of the law. The inspections can be used by operators either as a second check on their own inspection and maintenance system, or as the sole means of inspection of their vehicles to comply with the law in this respect; the actual maintenance work being carried out by their own staff. Alternatively, the scheme can be used as a means of checking the maintenance work carried out by a repair garage or agent.

The FTA vehicle check system is not purely a maintenance scheme as such. It is an inspection and maintenance advisory service available to Association members as a do-it-yourself scheme using the documents, checklists and advice provided. It can include contract inspections of vehicles by a qualified inspector.

After carrying out a vehicle check, the inspector completes a checklist indicating whether items are satisfactory, within reasonable tolerances, or whether attention is needed. When the inspector returns to carry out the next check on the vehicle he will expect to see that his previous recommendations have been followed. If they have not, the operator is told that there is little point in paying for a service which is not being used to its fullest advantage.

Following the experience of many operators who have received PG9s on brand new vehicles, the FTA offers a service whereby it will conduct a full inspection of a brand new vehicle before it goes into service. One of the obvious advantages of this independent check is that it provides the operator with evidence to support claims to the vehicle supplier and manufacturer for new vehicles delivered in faulty condition.

Responsibility for Maintenance

The major point to remember when making any arrangements with a garage is that it is the operator (ie the vehicle 'user' – see pp 13–15 for definition), not the repairer, who remains responsible for the mechanical condition of the vehicle, even where defects are due to negligence by the repairer. In *A Guide to Operators' Licensing* available from local Traffic Area offices the following warning is given:

'Operators who contract out their inspections and maintenance work are still the legal "users" of their vehicles and as such will be held fully responsible by the TC for the arrangements they make and the condition of their vehicles. If either are unsatisfactory it is the operator's licence which will be placed in jeopardy.'

The same responsibility applies in the case of vehicles hired without drivers and trailers even if the hire company, as part of the agreement, carries out the inspection and repair work.

Responsibility for Records

Operators are also responsible for ensuring that proper records of maintenance work are kept. Even if the repairer makes the records and holds them on file it is up to the operator to ensure, first, that they are properly kept with all the necessary information recorded and, second, that they are retained on file, available for inspection, for a minimum period of 15 months. In completing the 'O' licence application 'declaration of intent' (see p 429) the vehicle operator promises to 'keep records'. This can be interpreted as meaning that the operator keeps the record rather than the repair garage keeping it on his behalf (see also Chapter 19).

Negligence by Repairers

Unfortunately, however satisfactory the arrangements made with a garage may be in other ways, there is very little that operators can do contractually to protect themselves completely from negligence on the part of the repairer or the repairer's employees. It is unlikely that the repairer would agree to be party to a contract in which he has fully to indemnify the operator against failure of his employees to carry out work to a required standard, even though he may agree that morally he should be held responsible (moral responsibility, incidentally, has no standing in law). Moreover, most small operators are not likely to be in a position to have sufficiently persuasive powers to get the repairer to agree to such terms.

Operators have no protection against poor workmanship other than relying on the reputation of the garage. However, if they are members of one of the trade associations (the Freight Transport Association for own-account operators and the Road Haulage Association for haulage contractors) they could try to enlist their help in pressing a claim.

IN-HOUSE REPAIRS

The advantages of operators having their own workshop and being able to do their own safety inspections and repairs are many, provided that sufficient vehicles are operated to justify the overheads involved. Principal among them is that by employing the staff they have direct control over the work carried out, the standard of the work, and the record keeping which is so important.

If operators provide their own maintenance facilities they must be of a suitable standard, although once again no specific details are given in the regulations. The main requirement is for a covered area with hard-standing and facilities including adequate lighting for conveniently inspecting the underside of vehicles. Ideally this means that either a pit in the ground or an hydraulic lift should be provided. The former is the most commonly used and it is very much the cheaper of the two alternatives. Suitable lighting, either fixed in the pit shining upwards to the underside of the vehicles or by means of portable inspection lamps, is necessary.

The remainder of the tools and machines with which a workshop should be equipped are left entirely to the operator's choice, but such items as a beam-setter and a portable Tapley brake-efficiency recorder are useful to check that vehicles comply with the test requirements. Servicing equipment such as jacks, high-pressure greasing equipment, high-pressure washing or steam cleaning equipment make maintenance work much less of a chore. Hand tools are, of course, essential and in general the better the equipment available (including the availability of the special tools often needed to carry out work on today's sophisticated vehicles) the more likelihood there is of the work meeting the required standard.

VEHICLE INSPECTIONS

Vehicle inspections are an essential part of the maintenance programme, and legislation covers this aspect. Section 74 of the Road Traffic Act 1988 requires operators of goods vehicles to have them regularly inspected by a 'suitably qualified person' to ensure that they comply with Construction and Use Regulations (Chapter 13) and, of course, it is in the operator's best interests to have vehicles regularly inspected to ensure that they are kept in a fully safe and roadworthy mechanical condition.

Frequency of Inspection

Legislation does not specify the intervals at which vehicles should be inspected, what form the inspection should take or what is meant by a 'suitably qualified person'. In the case of the first-mentioned, it is very much a question of the type of operation on which the vehicle is used. A tipping vehicle, for example, which spends much of its time on rough sites, with perhaps a fairly high mileage on the road as well, certainly should be inspected at least weekly. The same applies to a vehicle which, although it remains on normal roads, does perhaps 800 to 1200 miles a week. On the other hand, a monthly inspection may be quite sufficient for a local delivery vehicle doing low weekly mileages on good roads and spending a great deal of its time standing while deliveries are being made.

Operators' own experience of their type of operation should indicate the intervals between which wear and tear takes place and defects become apparent. It has even been suggested that vehicles standing out of use in depots should be checked at least monthly. Although there is some suggestion that inspection intervals can be based on either a time or miles/kilometres alternative it is becoming clear that some TCs at least will not accept anything other than a time-based frequency from 'O' licence applicants. The VOSA Code of Practice on vehicle maintenance referred to on page 418 recommends that the maximum time interval between safety inspections should be six weeks with no mileage alternative. It also recommends operators to have flow charts covering at least 12-monthly periods to indicate when vehicle safety inspections are due.

Items for Inspection

A full list of the items to be inspected at regular intervals is not laid down in regulations but it is obviously necessary that the inspection at the very least covers all the items set out in *The Heavy Goods Vehicle Inspection Manual* (see pp 387–88), with particular emphasis on those items (eg brakes, steering, wheels, tyres, suspension systems and lights) that have special relevance to the safe operation of the vehicle. Tyres on vehicles used on site work or on local delivery work should receive particularly careful examination for damage.

The Inspector

There are no specified qualifications for a vehicle inspector. Clearly, the most obvious one is wide experience in the repair of heavy commercial vehicles. A person with such experience would know where and how to look for wear and for defects and would recognize the symptoms of hidden faults, such as uneven tyre wear indicating that the steering is out of alignment or that kingpins or wheel bearings are worn.

It is possible, however, to train a person specifically as a vehicle inspector, and this is being done in the industry. The emphasis in training in such cases must be on following a predetermined list, such as *The Heavy Goods Vehicle Inspection Manual*, examining every individual item carefully and methodically, testing the wear in components and measuring the tolerances of moving parts accurately.

It is desirable, although not a legal requirement, that the person carrying out the vehicle inspection should not be expected to carry out repairs, however small, at the time of making the inspection. To have to do this would cause a lack of concentration and could lead to other items being missed if repairs took up too much of the inspector's time. In the case of an owner-driver carrying out his own inspections and repairs this can be a difficult situation and in such instances it is useful to have the work verified and an audit-type check carried out by an outside agency, such as the FTA, to ensure that the vehicle is kept up to a high standard.

Authority to Stop Use of Vehicles

The inspector, besides being suitably qualified and experienced to spot defects, should have the authority to prevent a vehicle being taken on the road in the event of a serious or potentially dangerous defect being found.

VEHICLE SERVICING

Regular servicing as opposed to specific inspections and repairs is an important part of vehicle maintenance and as such it should be carried out with unfailing regularity at predetermined intervals of time or mileage. The importance of servicing cannot be too highly emphasized for two reasons. First, because the vehicle must meet the requirements of the law and second, because operators will benefit by always having their vehicles ready for work and able to carry out a job without breakdowns and delays. It also increases the life of the working parts and consequently of the whole vehicle as well as reducing down-time costs and disrupted delivery schedules.

The intervals at which servicing should be carried out are left to the owner's discretion depending on the work on which the vehicle is employed, in much the same way as the intervals for inspection are decided. To give owners some guidance, however, vehicle manufacturers usually provide a service schedule which owners can use in their own workshop or which their agent will use when vehicles are sent in for servicing. A useful guide to service intervals is 5000/6000 mile services carried out at least monthly with more extensive services at 15,000/20,000 mile/three-monthly intervals and 30,000/40,000 mile/six-monthly intervals. Progressively, service intervals are being extended with the use of longer-life components and particularly improved filters and lubricating oils which can go for very much longer periods these days without detriment to their lubricating properties

CLEANING OF VEHICLES

Particular reference is made in the instructions to operators submitting vehicles for VOSA heavy goods vehicle tests, that those which are not sufficiently clean will be refused the test. Again, it is difficult to specify a standard of cleanliness for the underside of a vehicle, but the main point is that all the components listed for the examination must be easily visible so that inspectors can see without difficulty if wear or damage exists. To achieve and keep a suitable standard of cleanliness it is desirable for vehicles to be washed with either a steam cleaner or a high-pressure water washer at regular intervals and certainly immediately before they are taken to the test station.

If the maintenance of the vehicles is contracted out to a garage on a maintenance scheme, the operator should arrange either for the garage to clean the vehicle before inspection or for one of the many specialist vehicle cleaners to do it.

ENFORCEMENT OF MAINTENANCE STANDARDS

There has been increasing concern in recent times about the standards of vehicle maintenance. As a result of this concern, VOSA has stepped up the levels of checking on vehicles by enforcement staff particularly at night and at weekends. The purpose of these additional checks is to catch vehicles operating outside normal working hours – many legitimately but some possibly deliberately running the gauntlet.

19 Maintenance Records

There is a legal requirement under the Road Traffic Act 1988 for goods vehicle operators to keep records of maintenance work carried out on their vehicles. When completing form GV79, 'O' licence application, operators have to make the statutory 'declaration of intent' in which they promise to fulfil undertakings made at that time throughout the duration of the 'O' licence. A number of items in the declaration of intent relate to maintenance records. From this can be determined what records are needed by law.

In the declaration of intent the operator promises to ensure that the following records will be kept:

- Safety inspections.
- Routine maintenance.
- Repairs to vehicles.

A promise is also made that drivers will report 'safety faults' in their vehicles and the Traffic Commissioners insist that these reports should be in writing and therefore they become part of the vehicle record-keeping system. The operator promises to keep all these records for a minimum period of 15 months and to make them available on request by Vehicle and Operator Services Agency (VOSA) examiners or the Traffic Commissioner (TC).

Additionally, operators are frequently asked by vehicle examiners and the TCs to keep a wall chart showing vehicles in the fleet and when they are due for inspection, service and annual test. The VOSA Code of Practice on vehicle maintenance (see p 418) recommends annual flow charts showing vehicle inspection-due dates.

DRIVER REPORTS OF VEHICLE FAULTS

It is a specific requirement (and part of the declaration of intent on an 'O' licence application form) that arrangements must be made for drivers to have a proper means of reporting 'safety faults' (ie defects) in the vehicle they are driving as soon as possible. As already mentioned, the TCs expect these defect reports to be made in writing, not verbally. Ideally, the report should be made either on an individual form which is completed and handed in by the driver, or in a defect book reserved for recording defects found on a vehicle which is kept in a convenient place where all drivers have easy access to it, and where whoever is responsible for ensuring that repair work is carried out can also easily reach it. To reiterate, it has been made abundantly clear that verbal reporting of defects is not in itself a system acceptable to the TCs.

Whichever method is used it is important that the repair of the defects is recorded on the form or in the book by a note of the work done and the signature of the person who has done it (see also under repair records). It is also important for operators to ensure that whatever system of defect reporting is used, they make regular checks on drivers and repair staff to see that the procedure is being followed correctly. This requirement has been pointed out by the TCs on a number of occasions. Using separate pads of defect sheets (see Figure 19.1) is the best alternative and where these can be made out in duplicate they provide the driver with his own copy of the report for future reference.

Systems of defect reporting which rely on a centrally located defect book or on verbal reports by drivers are open to the risks of drivers forgetting to report defects when they return to base. If a driver's attention is distracted by his manager who wants to talk to him, for example, just as he is about to report a defect, the defect is not reported and another driver may take the vehicle out next day with a defect which could result in a prohibition notice being issued in a roadside check. A driver may also forget to report a defect when he returns late from a journey and is in a rush to get home or if he cannot find the defect book.

It is the operator's responsibility to make sure that the system used is infallible in all these circumstances, firstly because it is an offence to fail to cause the defect to be reported and secondly because the vehicle could be found on the road subsequently with a safety fault not reported and not repaired.

INSPECTION REPORTS

The Road Traffic Act 1988 requires that records of regular safety inspections to vehicles must be made. For this purpose the vehicle inspector (see p 427) should have a sheet on which are listed all the items to be inspected (preferably in accordance with the contents of *The Heavy Goods Vehicle Inspection Manual* – see pp 387–88).

The inspection sheet should identify the necessary items for inspection with a cross-check reference number to the *Inspection Manual* to enable, if necessary, full details of the method of inspection of that item – and the reason for its rejection as not being within acceptable limits – to be determined. The sheet should have provision for the inspector to mark against each item whether it is 'serviceable' or 'needs attention' and space to comment on defects for immediate rectification and other items for attention at a future date or on which a watch should be kept if attention is not required immediately. The form should contain space for the inspector to sign his name and add the date.

DEFECT REPAIR SHEETS

Besides ensuring that proper records are kept of vehicle safety fault reports made by drivers and of regular safety inspections, the operator must also keep a record showing that any defects reported or found on inspection are rectified in order to keep the vehicle in a fit, serviceable and safe condition. Records of such repairs may be added to the driver defect report or the inspection report to provide combined records, or a separate repair or job sheet may be used. The important points about repair records are, first, that they should show comprehensive details of the actual repair work carried out, identifying components which were repaired or replaced and new parts added, and, second, that there should be a matching repair sheet for every defect reported or found on inspection so that the vehicle examiners, when they visit to examine records, can see the report of the defect and then subsequently a report of the repair work carried out to rectify it. Reports of defects which do not have a corresponding repair record can arouse suspicion in the examiner's mind that perhaps the necessary repair has not been carried out and that the defect still exists. This is a good reason for him then to consider examining that vehicle, or perhaps the whole fleet.

SERVICE RECORDS

In addition to the records of defects and vehicle inspections which have to be kept, a record should also be kept of all other work carried out on the vehicle, whether it is repair or replacement of working parts or normal servicing (oil changes and greasing, etc).

RETENTION OF RECORDS

It is a legal requirement that records of maintenance must be retained by operators. The original inspection report, or a photocopy of it, with the inspector's comments, the date and the mileage at which the inspection was carried out, must be retained and kept available for inspection if required by the VOSA examiners for 15 months from the date of the inspection. Work sheets showing the repair work carried out following the inspection and all other work done on the vehicle, including repairs following defect reports by drivers, as well as defect reports themselves, must be kept for 15 months, available for inspection by VOSA examiners or the TC, if requested.

Repair Records from Garages

When vehicle safety inspections, servicing and repair work are carried out for the operator by repair garages, the operator should obtain from the garage comprehensive documentation to enable him to meet the legal requirements detailed above. In many cases VOSA examiners are quite happy if the garage retains the

DRIVER'S DAILY VECHICLE DEFECT REPORT

Date:	**Driver's name:**
Vehicle No:	**Trailer Fleet/Serial No:**

Note: Drivers are responsible for the safe condition of their vehicle and load. You are required by law to report defects to your employer.

Tick items on this checklist that are in order: put cross against defective items.

DAILY CHECK			**Tick or cross**
Fuel:		Lights:	
Oil:		Reflectors:	
Water:		Indicators:	
Battery:		Wipers:	
Tyres:		Washers:	
Brakes:		Horn:	
Steering:		Mirrors:	
Security of body:		Markers:	
Security of load:		Sheets/ropes/chains:	
Artics/Lorry & trailer combinations			
Brake hoses:		Electrical connections:	
Coupling secure:		Trailer No. plate:	

REPORT DEFECTS HERE

WRITE NONE HERE IF NO DEFECTS

Driver's signature:

Action Taken By:

Signature:

Figure 19.1 *A typical example of a driver's vehicle defect report available in pad form*

records of inspection and repair, so long as they can be made available for examination when required. The operator must be certain, in these instances, that the garage is keeping proper records (for a period of 15 months) which satisfy the legal requirements and that they are being kept available for inspection, not bundled away out of easy reach in a store with thousands of others.

In the event of failure of the garage to keep records as required, the operator's licence would be at risk but there would be no penalty imposed on the garage. On the GV79 declaration of intent (see p 429) the operator promised that he would 'make proper arrangements so that records are kept for (15 months) of all driver defect reports, safety inspections, routine maintenance and repairs to vehicles and trailers and these are made available on request'.

It is important to note that in this respect, invoices, or copies of invoices from garages, for repair work are not in themselves sufficient to satisfy the record-keeping requirement. It is the actual inspection sheet and repair sheets, or photocopies of them, which are needed because of the greater and more precise detail which they contain. Similarly, maintenance records in computer printout form are unlikely to satisfy the requirement of enforcement staff to examine actual records – they will still want to see the original inspection sheets. This is an important point to consider with the increasing application of computers to transport operations and vehicle maintenance functions.

Location of Records

Where companies hold operators' licences in a number of separate Traffic Areas the maintenance records for the vehicles under each licence should be kept in the area covered by the individual licence (preferably at the vehicle operating centre). With the sanction of the local TC, records may be kept centrally at a head office or central vehicle workshop, although the vehicle examiners may ask for them to be produced for inspection at the operating centre of the vehicles in the Traffic Area – probably giving three days' to one week's notice to enable the records to be obtained from the central files.

WALL PLANNING CHARTS

While it is not strictly a legal requirement, many vehicle examiners (and the current West Midland TC) like to see operators using wall planning charts (see note above about use of flow charts in accordance with the VOSA Code of Practice on vehicle maintenance) to provide a visual reminder of important dates such as:

- Vehicle/trailer due for inspection
- Vehicle/trailer due for service
- Vehicle/trailer due for annual test
- Excise duty due.

Such charts usually provide facilities for a whole year's recording of these items for the fleet (either shown by vehicle registration number or by fleet number).

VEHICLE HISTORY FILES

For efficiency in record keeping, a system of vehicle history files – one for each vehicle and trailer in the fleet – is most useful. This provides the facility for keeping all relevant records relating to individual vehicles and trailers together and in one place. Individual files can have all the important details of the vehicle/trailer on the front cover for easy reference, as follows:

- Registration number/fleet number
- Make/type
- Date of original registration
- Price new plus extras/options
- Annual test date
- Taxation (ie VED) date
- Base/location
- Model designation
- Wheelbase (in/mm)
- Body type
- Special equipment
- Chassis number
- Engine number
- Gearbox type/number
- Rear axle type
- Electrical system – 12V/24V
- Plated weights – gross/axle
- Supplier's name and address.

20 Safety – Vehicle, Loads and at Work

Transport operators, along with all other sectors of business and industry, are under constant pressure to become ever more conscious of the need for safety in their operations, in the provision of facilities for their employees, and in the way their employees work and conduct themselves on work premises. Predominantly, this is influenced by the demands of the Health and Safety at Work etc Act 1974 and the stringent requirements which it imposes on employers and employees alike. But in transport, the requirements of the Road Vehicles (Construction and Use) Regulations 1986 (as amended), regarding the safety of vehicles and loads, place additional legal burdens on operators and drivers. The problems of safe loading and avoidance of vehicle overloading are not new, but increased enforcement activity has accelerated concern in these areas. There is concern also on the wider front of safety in load handling and in the use of loading aids (forklift trucks for example) and with regard to vehicle manoeuvring in depots and works premises, which alone results in many workplace deaths and injuries annually.

In fact, it is worth remembering that, according to the Health and Safety Executive (HSE), some 400 people are killed annually and a further 25,000 are forced to leave work each year and are unable to work again due to neglect of proper health and safety management procedures. According to the HSE, being struck by a moving vehicle is one of the most common causes of fatal injury among workers: in 2000/01 the figure was 22 per cent of all fatalities.

C&U REQUIREMENTS

The Road Vehicles (Construction and Use) Regulations 1986 require that all vehicles and trailers, and all their parts and accessories, and the weight, distribution, packing and adjustment of their loads, shall be such that no danger is caused or likely to be caused to any person in or on the vehicle or trailer or on the road. Additionally, no motor vehicle or trailer must be used for any purpose for which it is so unsuited as to cause or be likely to cause danger or nuisance to any person in or on the vehicle or trailer or on the road.

Under the regulations, provisions relating particularly to bulk and loose loads make it an offence if a load causes a nuisance as well as a danger to other road users and such loads must be secured, if necessary by physical restraint, to stop them falling or being blown from a vehicle.

These regulations include two notable terms relating to load safety: one is the use of the term 'nuisance' in addition to the term 'danger' so that to commit an offence the operator does not have to go so far as causing danger; merely causing nuisance is sufficient to land him in trouble. The other term is 'physical restraint' which clearly implies the need for sheeting and roping any load, such as sand or grain, hay and straw and even builders' skips carrying rubble that may be blown from the vehicle.

THE SAFETY OF LOADS ON VEHICLES

A Code of Practice – The Safety of Loads on Vehicles (Third edition, available from The Stationery Office) sets out general requirements in regard to the legal aspects of safe loading, information on the forces involved in restraining loads, the strength requirements of restraining systems and load securing equipment, and then details specialized requirements for containers, pallets, engineering plant, general freight, timber, metal and loose bulk loads. It provides a list of do's and don'ts for drivers and others concerned with the loading of vehicles as shown below. The new 2003 version includes advice on anchor points and restraint methods.

For particular note is the basic principle on which the Code is based which is that:

> 'the combined strength of the load restraint system must be sufficient to withstand a force not less than the total weight of the load forward and half the weight of the load backwards and sideways'.

DO'S

1. Do make sure your vehicle's load space and the condition of its load platform are suitable for the type and size of your load.
2. Do make use of load anchorage points.
3. Do make sure you have enough lashings and that they are in good condition and strong enough to secure your load.
4. Do tighten up the lashings or other restraining devices.
5. Do make sure that the front of the load is abutted against the headboard, or other fixed restraint.
6. Do use wedges, scotches etc, so that your load cannot move.
7. Do make sure that loose bulk loads cannot fall or be blown off your vehicle.

DON'TS

1. Don't overload your vehicle or its individual axles.
2. Don't load your vehicle too high.
3. Don't use rope hooks to restrain heavy loads.
4. Don't forget that the size, nature and position of your load will affect the handling of your vehicle.
5. Don't forget to check your load:
   a. Before moving off;
   b. After you have travelled a few miles;
   c. If you remove or add items to your load during your journey.
6. Don't take risks.
7. Do make sure that loose bulk loads cannot fall or be blown off your vehicle.

SAFE LOADING

Your own life and the lives of others may depend upon the security of your load

Taken from the original Code of Practice on the Safety of Loads on Vehicles (DfT)

The new Code of Practice is available on the DfT website: www.dft.gov.uk

ROAD SAFETY

Government statistics show the following road casualties in Great Britain in 2004: 3,221 people killed, 34,351 people seriously injured and 243,268 people slightly injured, making a total of 280,840 casualties. The figures do not identify the numbers of goods vehicle driver or passenger casualties.

BRAKE Campaign

A vigorous campaign to improve vehicle mechanical standards (especially in regard to heavy vehicle braking) and reduce death on the road is being conducted by road safety lobby group BRAKE. Among its specific activities are to promote the Road Risk Forum (supported by leading truck and component manufacturers and a number of major national transport fleet operators) whose aim is to develop road safety initiatives and promote best practice in road safety. BRAKE also publishes a booklet called *Procedures Following a Death on the Road*, designed to help bereaved relatives and provide information on criminal prosecutions, claiming compensation and the way the police investigate a death on the road.

BRAKE can be contacted at PO Box 272, Dorking, Surrey RH4 4FR (tel: 01306 741113; fax: 01306 888221 or via e-mail at risk@brake campaign.demon.co.uk).

SAFETY REPORT

A report entitled *Transport Kills* (The Stationery Office) produced by the Health and Safety Executive (HSE) and based on a study of fatal accidents in industry indicates that motor vehicles are one of the biggest causes of industrial deaths. These motor vehicle-related deaths occur during vehicle loading, unloading, maintenance and, of course, movement and are mainly caused by poor management, failure to provide safe working systems and inadequate training.

Included in the report is a checklist which is intended to help transport operators and others to reduce unnecessary risks and dangers. Companies should use it to help examine their current practices and to institute new and safer procedures. The checklist (reproduced below with acknowledgement to the HSE) is only a general guide, and it is emphasized that safety requirements vary with different types of operation. *However, it is important to note that firms could face prosecution if they do not meet the minimum safety standards outlined in the list.*

This checklist is intended as a general guide only. It will not necessarily be comprehensive for every operation and all points will not be relevant for all work.

Organization, Systems and Training

- Have all health and safety aspects of the transport operation been assessed?
- Has an organization (and arrangements) for securing such safety been detailed in the safety policy?
- Has a person been appointed to be responsible for the transport safety?
- Have safe systems of work been set up?

- What monitoring is carried out to ensure that the systems are followed?
- Have all drivers been adequately trained and tested?
- Is there a satisfactory formal licensing or authorization system for drivers?
- Have all personnel been trained, informed and instructed about safe working practices where transport is involved?
- Is there sufficient supervision?

External Roadways and Manoeuvring Areas

- Are they of adequate dimensions?
- Are they of good construction?
- Are they well maintained?
- Are they well drained?
- Are they scarified when smooth?
- Are they gritted, sanded, etc, when slippery?
- Are they kept free of debris and obstructions?
- Are they well illuminated?
- Are there sufficient and suitable road markings?
- Are there sufficient and suitable warning signs?
- Are there speed limits?
- Is there a one-way system (as far as possible)?
- Is there provision for vehicles to reverse where necessary?
- Are there pedestrian walkways and crossings?
- Are there barriers by exit doors leading on to roadways?
- Is there a separate vehicle parking area?
- Is there any storage positioned close to vehicle ways?
- Is the yard suitable for internal works transport, eg smooth surface, hard ground, no slopes?

Internal Transport

- Are internal roadways demarcated and separated where possible from pedestrian routes with crossings and priority signs?
- Are there separate internal doors for trucks and pedestrians? Have these vision panels?
- Are blind corners catered for by mirrors, etc?
- Are trucks kept apart from personnel where possible?
- Do the trucks use a satisfactory warning system?

Vehicles

- Is there a maintenance programme for vehicles and mobile plant?
- Is there a fault reporting system?
- Are there regular checks to ensure that the vehicles are up to an acceptable standard?
- Are keys kept secure when vehicles and mobile plant are not in use?
- Are vehicles and mobile plant adequate and suitable for the work in hand?
- Is suitable access provided to elevated working places or vehicles?

- Are tractors and lift trucks equipped with protection to prevent the driver being hit by falling objects and from being thrown from his cab in the event of overturning?
- Are there any unfenced mechanical parts on vehicles, eg power take-offs?
- Are there fittings for earthing vehicles with highly flammable cargoes?
- Are loads correctly labelled (especially hazardous substances)?
- Are the vehicles suitable for use in all the areas they enter? Do they need to be to Division I or II standards, etc?
- If passengers ride on vehicles, do they have a safe riding position?

Loading and Unloading

- Do loading positions obstruct other traffic? Do pedestrian ways need diverting?
- Are there special hazards, eg flammable liquid discharge? Do pedestrians need to be kept clear?
- Is there a yard manager to supervise the traffic operation, to control vehicular movement and to act as a banksman during reversing?
- Has he received satisfactory training? Does he use recognized signals, and has he cover during his absences?
- Are there loading docks? Will the layout prevent trucks falling off or colliding with objects or each other?
- Are there any mechanical hazards caused by dock levellers, etc?
- Are methods of loading and unloading assessed? Are loads stable and secured?
- Are safe arrangements made for sheeting?
- Is there a pallet inspection scheme?

Motor Vehicle Repair

- Are appropriate arrangements made for tyre repair and inflation?
- Are arrangements made for draining and repair of fuel tanks?
- Is access available to elevated working positions?
- Are arrangements made to ensure brakes are applied and wheels checked?
- Is portable electrical equipment low voltage and properly earthed?
- Are moving vehicles in the workshop carefully controlled?
- Are vehicles supported on both jacks and axle stands where appropriate?
- Are engines only run with the brakes on and in neutral gear?
- Are raised bodies always propped?

HEALTH AND SAFETY AT WORK

Stringent UK legislation on health and safety at work implements the EU's 'Framework Directive', namely EC Directive 391/1989. The overall objective of this directive is to impose on employers a duty to encourage improvements in the health and safety of people at work.

The main legislation is contained in the Health and Safety at Work, etc Act 1974, plus the regulations listed below – the so-called 'six pack' – which were introduced to implement EC Directive 391/1989 (the 'Framework Directive'). Additionally, there are many other relevant provisions dealing with such matters as fire protection and safety signs. Collectively, this legislation has largely replaced both the Factories Act 1961 and the Offices, Shops and Railway Premises Act 1963 which formerly applied.

- Management of Health and Safety at Work Regulations 1999;
- The Workplace (Health, Safety and Welfare) Regulations 1992;
- Manual Handling Operations Regulations 1992;
- Health and Safety (Display Screen Equipment) Regulations 1992;
- Provision and Use of Work Equipment Regulations 1998;
- Personal Protective Equipment at Work Regulations 2002.

Additionally, there are the Lifting Operations and Lifting Equipment Regulations 1998.

The Health and Safety at Work etc Act 1974

Since the original introduction of the Health and Safety at Work etc Act 1974 employers have had to take positive steps to draw up policy statements regarding health and safety at work, appoint safety representatives and establish safety committees in addition to ensuring that workplaces meet all the necessary safety requirements of the law. These responsibilities apply equally to employers in transport, and it should be remembered that here the requirements of the law apply to the transport operator's premises (ie offices, workshops, warehouses and yard) and to the vehicles which constitute the workplace of drivers.

The Act replaced certain parts of the Factories Act and the Offices, Shops and Railway Premises Act, and added other provisions. There are four parts to the Act:

- Part I relates to health, safety and welfare at work;
- Part II relates to the Employment Medical Advisory Service;
- Part III amends the law regarding building regulations;
- Part IV covers a range of general and miscellaneous provisions.

The main effects of the Act are:

- To maintain and improve standards of health and safety for people at work.
- To protect people other than those at work against risks to their health or safety arising from the work activities of others.
- To control the storage and use of explosives, highly flammable or dangerous substances, and to prevent their unlawful acquisition, possession and use.
- To control the emission into the atmosphere of noxious or offensive fumes or substances from work premises.
- To set up the Health and Safety Commission and the Health and Safety Executive.

Duties of Employers

The Act prescribes the general duties of all employers towards their employees by obliging them to ensure their health, safety and welfare while at work. This duty requires that all plant (including vehicles) and methods of work provided are reasonably safe and without risks to health. A similar injunction relates to the use, handling, storage and transport of any articles or substances used in connection with the employer's work. Provisions contained in the Data Protection Act mean that employers are no longer permitted to design health questionnaires for completion by employees or to screen or interpret employees' medical data. Now, only qualified health professionals may carry out these functions.

Provision of Necessary Information

In order that employees are fully conversant with all health and safety matters, it is the duty of the employer to provide all necessary information and instruction by means of proper training and adequate supervision.

Employers must either: 1) display the 'Health and Safety Law' poster; or 2) distribute the 'Health and Safety Law' leaflet to employees. (Copies can be obtained from: HSE Books, PO Box 1999, Sudbury, Suffolk CO10 2WA).

Condition of Premises

Workplaces generally, if under the employer's control, must be maintained in such a condition that they are safe and without risks to health, have adequate means of entrance and exit (again this applies equally to vehicles as it does to 'premises') and must provide a working environment that has satisfactory facilities and arrangements for the welfare of everybody employed in the premises.

Statements of Safety Policy

It is necessary for an employer of five or more employees to draw up and bring to the notice of all his workforce *a written statement of company policy* regarding their health and safety at work with all current arrangements detailed for the implementation of such a policy. Stress is laid on the necessity of updating the 'statement' as the occasion arises and of communicating all alterations to the personnel employed.

Appointment of Safety Representatives

Involvement of all employees in health and safety activities is envisaged by the appointment (by a recognized trade union) or election of safety representatives from among the workforce. A safety representative should be a person who has been employed in the firm for at least *two years* or who has had two years' similar employment 'so far as is reasonably practicable'. The broad duties of safety representatives are concerned with the inspection of workplaces, investigating possible hazards and examining the cause of accidents, investigating employees' complaints regarding health and safety matters and making representations to their employer on health and safety at work matters.

Safety Committees

At the written request of at least two safety representatives employers must establish a safety committee to review health and safety at work matters. The establishment of a safety committee creates a joint responsibility with the employer for concern with all health and safety measures at the workplace, together with any other duties arising from regulations or codes of practice.

Duty to Public

Employers and self-employed persons are required also to ensure that their activities do not create any hazard to members of the general public. In certain circumstances, information must be made publicly available regarding the existence of possible hazards to health and safety.

Summary of Duties of Employers to their Employees

- It shall be the duty of every employer to ensure so far as is reasonably practicable the health, safety and welfare at work of all his employees. That duty includes, in particular:
  - The provision and maintenance of plant and systems of work that are so far as is reasonably practicable safe and without risks to health.
  - Arrangements for ensuring so far as is reasonably practicable safety and absence of risks to health in connection with the use, handling, storage and transport of articles and substances.
  - The provision of such information, instruction, training and supervision as is necessary to ensure so far as is reasonably practicable the health and safety at work of employees.
  - So far as is reasonably practicable the maintenance of any place of work that is under the employer's control in a condition that is safe and without risks to health, and the provision and maintenance of means of access to and egress from it that are safe and without such risks.
  - The provision and maintenance of a working environment for employees that is so far as is reasonably practicable safe and without risk to health and adequate as regards facilities and arrangements for their welfare at work.
- Except in such cases as may be prescribed it shall be the duty of all employers to prepare and as often as may be appropriate revise a written statement of their general policy with respect to the health and safety at work of their employees and the organization and arrangements for the time being in force for carrying out that policy, and to bring the statement and any revision of it to the notice of all employees.
- It shall be the duty of any person who erects or installs any article for use at work in any premises where the article is to be used by persons at work to ensure, so far as is reasonably practicable, that nothing about the way in which it is erected or installed makes it unsafe or a risk to health when properly used.
- No employer shall levy or permit to be levied on any employee any charge in respect of anything done or provided in pursuance of any specific requirement of the relevant statutory provisions.

As well as these statutory duties, employers have a common law duty of care. Where a breach of that care results in negligence (eg where careless conduct results in injury to a person), they may be liable to pay compensation.

Duties of Employees

The Act states in general terms the duty of employees to take reasonable care for their own safety and that of others and to co-operate with others in order to ensure that there is a compliance with statutory duties relating to health and safety at work. In this connection no person shall interfere with or misuse anything provided in the interests of health, safety or welfare either intentionally or recklessly.

Summary of Duties of Employees at Work

- It shall be the duty of all employees while at work:
  - to take reasonable care for their own health and safety and of other persons who may be affected by their acts or omissions at work, and
  - as regards any duty or requirement imposed on their employer or any other person, to co-operate with them so far as is necessary to enable that duty or requirement to be performed or complied with.
- No person shall intentionally or recklessly interfere with or misuse anything provided in the interests of health, safety or welfare.

General Duties of Employers and Self-Employed to Persons other than their Employees

- It shall be the duty of all employers and self-employed persons to conduct their undertaking in such a way as to ensure so far as is reasonably practicable that persons not in their employment are not thereby exposed to risks to their health and safety.
- It shall be the duty of all employers and self-employed persons to give to persons not in their employment, who may be affected, the prescribed information about such aspects of the way in which they conduct their undertaking as might affect such persons' health and safety.

Improvement and Prohibition Notices

Improvement Notice

An improvement notice may be served on a person by a health and safety inspector in cases where he believes (ie is of the opinion) that the person is contravening or has contravened, and is likely to continue so doing or will do so again, any of the relevant statutory provisions. Such a notice must give details of the inspector's reason for his belief and requires the person concerned to remedy the contravention within a stated period.

Prohibition Notice

A prohibition notice with immediate effect may be served on a person under whose control activities to which the relevant statutory provisions apply are being carried on, or are about to be carried on, by an inspector if he believes that such activities involve or could involve *a risk of serious personal injury*. A prohibition notice must specify those matters giving rise to such a risk, the reason why the inspector believes the statutory provisions are, or are likely to be, contravened, if indeed he believes that such is the case. The notice must direct that the activities in question shall not be carried on unless those matters giving rise to the risk of serious personal injury, and any contravention of the regulations, are rectified.

Remedial Measures

Both improvement and prohibition notices may include directions as to necessary remedial measures and these may be framed by reference to an approved Code of Practice and may offer a choice of the actions to be taken. Reference must be made by the inspector to the fire authority before serving a notice requiring, or likely to lead to, measures affecting means of escape in case of fire.

Withdrawal of Notices and Appeals

A notice, other than a prohibition notice with immediate effect, may be withdrawn before the end of the period specified in the notice, or an appeal against it made. Alternatively, the period specified for remedial action may be extended at any time provided an appeal against the notice is not pending.

Penalties for Health and Safety Offences

Conviction for an offence under the Health and Safety at Work, etc Act 1974 may lead to a fine of £20,000, but where the case is tried in the Crown Court the amount of fine which may be imposed is unlimited. However, it should be stressed that while generally the fine imposed is intended to reflect the gravity of the particular offence and the employer's attitude towards health and safety matters, the Court will also bear in mind the offender's resources and the effects of a heavy penalty on his business and may reduce the penalty accordingly.

Corporate Manslaughter

In a situation where a death is caused through the negligence of an employer, whether health and safety related or as a result of a road accident, the employer risks facing a charge of corporate manslaughter.

It has been ruled by the courts that for an indictment on such a charge to be sustained, three key issues must be established by the jury as follows:

- The defendant owed a duty of care to the person killed.
- The defendant had breached his duty of care and that breach of care led to the death.
- The defendant's negligence was so gross that a jury would consider it justified to bring in a criminal conviction.

The government is still committed to introducing tough new corporate manslaughter legislation. This is expected to come into force in 2004. When enacted, this law will carry unlimited fines, disqualification for company directors, and imprisonment for convicted offenders.

The Management of Health and Safety

The new 'management' regulations (p 439) are both wide ranging and general in nature, overlapping with many existing regulations. They have to be viewed as a 'catch-all' regulation, sitting astride other more specific health and safety provisions (eg the COSHH regulations – see page 452). The Health and Safety Executive, in its Approved Code of Practice, advises that where legal requirements in these management regulations overlap with other provisions, compliance with duties imposed by the specific regulations will be sufficient to comply with the corresponding duty in the management regulations. However, the HSE says, where duties in the Management of Health and Safety at Work Regulations 1999 go beyond those in the more specific regulations, additional measures will be needed to comply fully with the management regulations.

Specifically, these regulations cover:

- Requirements for employers (and self-employed persons) to make assessments of the risks to the health and safety of:
  - employees while they are at work;
  - other persons not in their employ, arising out of or in connection with their conduct or the conduct of their undertaking.
- Employers with five or more employees must record the significant findings of their risk assessment (ie make an effective statement of the hazards and risks which lead management to take relevant actions to protect health and safety).
- Employers and self-employed persons must make and give effect to such arrangements, as are appropriate to the nature of their activities and the size of their undertaking, for the effective planning, organization, control, monitoring and review of preventive and protective measures.
- Health surveillance must be provided for employees as appropriate to the risks to their health and safety as identified by the assessment.
- Employers must appoint one or more (competent) persons to assist them in undertaking the measures necessary to comply with the requirements and prohibitions of this legislation.

The Workplace Regulations

These new regulations (see second item in the list of new legislation on p 440) add further to existing legislation on workplaces. They apply to new workplaces as of 1 January 1993 and to any modifications or conversions to existing workplaces started after this date. Existing workplaces (ie which are unaltered) must comply from 1 January 1996.

Specifically the regulations impose requirements on the:

- maintenance of workplaces;
- ventilation of enclosed workplaces and temperatures of indoor workplaces and the provision of thermometers;
- lighting (including emergency lighting);
- cleanliness of the workplace, and of furniture, furnishings and fittings (also the ability to clean floors, walls and ceilings) and the accumulation of waste materials;

- room dimensions and unoccupied space;
- suitability of workstations (including those outside) and the provision of suitable seating;
- condition of floors, and the arrangement of routes for pedestrians or vehicles;
- protection from falling objects, and from persons falling from a height, or falling into dangerous substances;
- material of and protection of windows and other transparent or translucent walls, doors or gates, and to them being apparent;
- way in which windows, skylights or ventilators are opened and their position when left open and the ability to clean these items;
- construction of doors and gates (including the fitting of necessary safety devices), and escalators and moving walkways;
- provision of suitable sanitary conveniences, washing facilities and drinking water (including cups and drinking vessels);
- provision of suitable accommodation for clothing and for changing clothes, for rest and for eating meals.

Manual Handling

The Manual Handling Regulations 1992 are intended to reduce back and other injuries suffered through the manual handling of loads. Employers must reduce the risk of injury to employees by using more mechanical aids, providing training in load handling, and providing precise information about load weights, centres of gravity and the heaviest side of eccentrically loaded packages.

Employers must avoid having employees undertake manual handling operations which involve risk of injury. Where it is not possible to avoid this, they have a duty to assess potential injury risks and take steps to reduce those risks to the absolute minimum or, better still, find an alternative way of achieving the same objective which does not involve manual handling, for example by mechanization of certain handling tasks or by eliminating some activities altogether.

Where such activities must take place, the employer should ensure that employees:

- are well trained in good handling techniques;
- understand how operations have been designed to ensure their safety; and
- make proper use of systems of work provided.

In particular employees should understand:

- how potentially hazardous handling operations can be recognized;
- how to deal with unfamiliar operations;
- the proper use of handling aids;
- the proper use of personal protective equipment;
- features of the working environment that contribute to safety;
- the importance of good housekeeping;
- factors affecting individual capability; and
- good handling techniques.

Training in good handling techniques should be tailored to the particular handling operations likely to be undertaken, beginning with relatively simple examples and progressing to more specialized handling operations.

It is especially useful to train employees to:

- recognize loads whose weight as well as their shape and other features, and the circumstances in which they are handled, might cause injury;
- treat unfamiliar loads with caution and not assume that apparently empty drums or other closed containers are in fact empty;
- test loads first by attempting to raise one end; and
- apply force gradually when lifting or moving loads until either undue strain is felt, in which case an alternative method should be considered, or it is clear that the task can be accomplished without injury.

The following key safe-handling points are taken from the HSE's booklet mentioned below.

Step 1. Plan the lift. Decide where is the load going to be placed. Use appropriate handling aids if possible. Determine whether you need help with the load. Remove obstructions such as discarded wrapping materials. For a long lift such as floor to shoulder height, consider resting the load midway on a table or bench in order to change grip.

Step 2. Place the feet apart, giving a balanced and stable base for lifting (tight skirts and unsuitable footwear make this difficult). Place the leading leg as far forward as is comfortable.

Step 3. Adopt a good posture. Bend the knees so that the hands when grasping the load are as nearly level with the waist as possible. Do not kneel or overflex the knees. Keep the back straight (tucking in the chin). Lean forward a little over the load if necessary to get a good grip. Keep shoulders level and facing in the same direction as the hips.

Step 4. Get a firm grip. Try to keep the arms within the boundary formed by the legs. The optimum position and nature of the grip depends on the circumstances and individual preference, but it must be secure. A hook grip is less fatiguing than keeping the fingers straight. If it is necessary to vary the grip as the lift proceeds, do this as smoothly as possible.

Step 5. Don't jerk. Carry out the lifting movement smoothly, keeping control of the load.

Step 6. Move the feet. Don't twist the trunk when turning to the side.

Step 7. Stay close to the load. Keep the load close to the trunk for as long as possible with the heaviest side next to the trunk. If a close approach to the load is not possible try sliding it towards you before attempting to lift it.

Step 8. If precise positioning of the load is necessary, put it down first, then slide it into the desired position.

The HSE booklet *Manual Handling* (available from The Stationery Office, price £5) provides excellent guidance on the regulations and their application. Additionally, these notes will provide useful information to employers on many aspects of manual handling.

The Use of Display Screen Equipment

For the purposes of these regulations (fourth item in the list on p 440), a display screen is any form of alphanumeric or graphic display screen regardless of the process involved – generally termed Visual Display Units (ie VDUs). Under the regulations employers are required to protect VDU users (ie those who habitually use VDUs for a significant part of their normal work) from the risks associated with habitual VDU use – mainly visual fatigue, mental stress, backache and upper limb pain. This must be done by controlling the design of workstations and actual working conditions. Employers must make an assessment of the risks to which VDU users are exposed and reduce these to the lowest reasonably practicable level. Work at VDU screens must be periodically interrupted by breaks or a switch to other work and employers must provide eye and eyesight tests for relevant employees and, where necessary, also provide special corrective appliances (ie spectacles) where normal spectacles cannot be used. Both existing and new VDU users must be given health and safety information about, and training in the use of, the workstation.

Provision and Use of Work Equipment

The Provision and Use of Work Equipment Regulations 1998 require employers and the self-employed to ensure that equipment provided for use at work complies with the regulations. The law applies to owned, hired, leased and second-hand equipment and all machinery, appliances (including cranes, lift trucks and vehicle hoists), apparatus, tools or component assemblies so arranged as to function as a whole unit. Work equipment must be:

- suitable for its intended purpose;
- assessed as to any risks associated with the equipment;
- subject to a recorded inspection where such an inspection would assist in identifying health and safety risks; and
- maintained in efficient working order, in an efficient state and in good repair, and a maintenance log kept.

Personal Protective Equipment

Employers are required to provide suitable personal protective equipment (PPE) to employees where there are risks to their health and safety which cannot be adequately controlled by other means – it is a last-resort measure. Self-employed persons must provide their own protective equipment where necessary. PPE is not suitable unless:

- It is appropriate for the identified risks.
- Account has been taken of: the environment that it will be used in, ergonomic factors such as the nature of the job, the need for communication and the health of the wearer.
- It fits the wearer correctly and comfortably and is capable of being adjusted
- It is effective, so far as is practicable, against the risks it is intended to control.

PPE is described as being any equipment designed to be worn or held by persons to protect them from one or more risks, including against extreme temperatures, poor visibility and adverse weather. Work clothing and uniforms, sports equipment and road safety protective wear such as crash helmets are excluded.

A recent fatal accident to a lorry driver on a motorway hard shoulder has highlighted the need for drivers, and others, to wear high-visibility clothing when working in vulnerable situations. In this particular case the driver was criticized for being careless with his own safety in not wearing his reflective jacket.

Lifting Operations

The Lifting Operations and Lifting Equipment Regulations 1998 apply to all lifting equipment which is defined as 'work equipment for lifting and lowering loads and includes its attachments used for anchoring, fixing or supporting it'. These regulations obviously apply to forklift trucks and to automated goods storage and retrieval systems and front-end loaders on tractors.

It is important to note that these regulations also apply to tipping vehicles. In this context it is important that the operator ensures he receives from the manufacturer adequate information about safe operation and correct methods of use and that such information is passed on to the driver, who should also be properly trained as per all health and safety requirements.

The regulations make provisions regarding the:

- strength and stability of lifting equipment;
- safety of lifting equipment for lifting persons;
- the way lifting equipment is positioned and installed;
- marking of machinery and accessories for lifting, and lifting equipment which is designed for lifting persons or which might so be used in error;
- organization of lifting operations;
- thorough examination and inspection of lifting equipment in specified circumstances;
- evidence of examination to accompany it (ie the lifting equipment) outside the firm's premises;
- exceptions for winding apparatus at mines;
- making of reports of thorough examinations and records of inspections; and
- keeping of information in the reports and records.

Work-Related Stress

All employers have a duty in law to ensure that their employees are not made ill by their work. Work-related stress is becoming recognized as a major cause of employee illness and absenteeism. Transport employers, in particular, need to recognize that stress can make their employees ill because the work involved, especially driving, is renowned as a stressful occupation.

Stress is a person's natural reaction to excessive pressure and thus it is not a disease. However, excessive and long-term stress can lead to mental and physical ill health causing, for example, depression, a nervous breakdown and even heart disease. Stress in one person can lead to stress in other people who have to cover their work.

The costs of stress can be high: high staff turnover, increased absence through sickness, reduced work performance, increased lateness and more customer complaints. An employer who ignores signs of stress among employees or who fails to take the necessary action to reduce stress may face claims for compensation from employees who have suffered ill health from work-related stress. Half a million people in Britain are thought to be suffering from work-related stress, according to the HSE, and this is costing industry £3.75 billion in lost time.

Where stress has been caused in a workplace, or made worse by work, it is the employer's duty to assess the risks among employees, in particular:

- looking for pressures at work that could cause high and long-lasting levels of stress;
- judging which employees may be affected or harmed by stress;
- deciding whether sufficient measures are being taken to prevent harm from stress;
- determining what further measures can be taken to reduce the stress-inducing pressures of work.

Employers are not responsible for preventing ill health from stress caused by problems occurring outside work, but they should be able to recognize the effects of such stress, for example, people being unable to cope with their work, or performing below their normal standard. In such cases, adopting an understanding attitude would be helpful to the individual concerned and is in the employer's best interests.

ELECTRIC STORAGE BATTERIES

The Health and Safety Executive has issued guidance on the safe charging and use of electric storage batteries in motor vehicle repair and maintenance. This is to help reduce the number of injuries which occur annually when batteries explode through mishandling and improper use, generally causing acid burns to face, eyes and hands as well as other injuries.

In a freely available leaflet the HSE warns of the dangers of charging batteries, particularly those described as maintenance free but which still give off flammable hydrogen gas which on contact with a naked flame will burn and cause the battery to explode. The advice it gives is as follows:

General Precautions

- Always wear goggles or a visor when working on batteries.
- Wherever possible, always use a properly designated and well-ventilated area for battery charging.
- Remove any metallic objects from hands, wrists and around the neck (eg rings, chains and watches) before working on a battery.

Disconnecting and Reconnecting Batteries

- Turn off the vehicle ignition switch and all other switches or otherwise isolate the battery from the electrical circuit.
- Always disconnect the earthed terminal first (often the negative terminal, but not always – CHECK) and reconnect it last using insulated tools.
- Do not rest tools or metallic objects on top of a battery.

Battery Charging

- Always observe the manufacturer's instructions for charging batteries.
- Charge in a well-ventilated area. Do not smoke or bring naked flames into the charging area.
- Make sure the battery is topped up to the correct level.
- Make sure the charger is switched off or disconnected from the power supply before connecting the charging leads, which should be connected positive to positive, negative to negative.
- Vent plugs may need to be adjusted before charging. Carefully follow the manufacturer's instructions.
- Do not exceed the recommended rate of charging.
- When charging is complete, switch off the charger before disconnecting the charging leads.

Jump Starting

Preparation:

- Always ensure that both batteries have the same voltage rating.
- If starting by using a battery on another vehicle, check the earth polarity on both vehicles.
- Ensure the vehicles are not touching.
- Turn off the ignition of both vehicles.
- Always use purpose-made, colour-coded jump leads with insulated handles – red for the positive cable and black for the negative cable.

Connection for vehicles with the *same earth* polarity:

- First connect the non-earthed terminal of the good battery to the non-earthed terminal of the flat battery.
- Connect one end of the second lead to the earthed terminal of the good battery.
- Connect the other end of the second lead to a suitable, substantial, unpainted point on the chassis or engine of the other vehicle, away from the battery, carburettor, fuel lines or brake pipes.

Connection for vehicles with *different* earth polarity:

> *The HSE warns that in view of the potential for confusion this should be attempted only by skilled and experienced personnel.*

- First connect the earthed terminal of the good battery to the non-earthed terminal of the flat battery.
- Connect one end of the second lead to the non-earthed terminal of the good battery.
- Connect the other end of the second lead to a suitable, substantial, unpainted point on the chassis or engine of the other vehicle, away from the battery, carburettor, fuel lines or brake pipes.

Starting

- Ensure the leads are well clear of moving parts.
- Start the engine of the 'good' vehicle and allow to run for about one minute.
- Start the engine of the 'dead' vehicle and allow to run for about one minute.

Disconnection

- Stop the engine of the 'good' vehicle.
- Disconnect the leads in the reverse order to which they were connected.
- Take great care in handling jump leads; do not allow the exposed metal parts to touch each other or the vehicle body.

CONTROL OF SUBSTANCES HAZARDOUS TO HEALTH (COSHH)

There are many situations where the use of chemicals or other hazardous substances at work can put people's health at risk; consequently employers are legally required to control exposure to such hazardous substances to prevent ill health under the *Control of Substances Hazardous to Health Regulations 2002* (COSHH) (as amended). Eight basic measures are established that employers, and sometimes employees, must take as listed below. If an employer fails to control hazardous substances adequately, employees or other persons may become ill with anything from mild eye irritation to chronic lung disease or, occasionally, even death may occur. Employers must undertake an assessment of their work environment to determine the potential hazards and to take steps to minimize any such hazards. They must inform employees of any risks which exist and train them in safety procedures such as the handling of hazardous materials and the monitoring of the work environment. Failure to comply with the regulations will result in prosecution under these regulations and under the Health and Safety at Work Act.

The eight essential steps of COSHH as recommended by the HSE:

Step 1 Assess the risks.

Step 2 Decide what precautions are needed.

Step 3 Prevent or adequately control exposure.

Step 4 Ensure that control measures are used and maintained.

Step 5 Monitor the exposure.

Step 6 Carry out appropriate health surveillance.

Step 7 Prepare plans and procedures to deal with accidents, incidents and emergencies.

Step 8 Ensure employees are properly informed, trained and supervised.

Acknowledgement: The above list is reproduced from the Health and Safety Executive guide to the COSHH regulations and may be freely downloaded from the HSE website at: www.hse.gov.uk/pubns.

NOTIFICATION OF ACCIDENTS

The Reporting of Injuries, Diseases and Dangerous Occurrences Regulations 1995 (commonly referred to as RIDDOR) require employers to notify fatal accidents, major accidents and those causing more than three days' incapacity for work, work-related diseases, gas incidents and any dangerous occurrence whether or not anybody is injured.

The report of any notifiable accident or occurrence must be made to the appropriate authority (see below) by a 'responsible person'. Usually this is the employer in the case of employees, but if the accident or occurrence involves a member of the general public, the responsible person is the person who controls the premises.

In cases where an accident occurs to an employee away from base (as may be the case with lorry drivers and sales representatives, for example), although it remains the responsibility of the employer to report the accident or occurrence, the Health and Safety Executive suggests that it would be helpful if the owner or occupier of the premises where the event occurred were to advise the person's employer as soon as possible (although there is no legal obligation to do so). Where a dangerous occurrence involves the carriage of dangerous goods by road the person responsible is the 'O' licence holder.

Under new proposals being considered by the HSE, employers may be required to report all accidents involving their vehicles as part of the RIDDOR system. This would allow safety inspectors, as well as the police, to investigate on-road accidents and incidents and to determine whether employers have adequate on-road risk assessment systems in force.

Reporting Authorities

Reports of fatal and injury accidents and notifiable dangerous occurrences must be made to the 'enforcing authority' (ie the authority responsible for enforcing the Health and Safety at Work Act). Principally, the authority is the HSE but reports should be made to its individual Inspectorates as listed below, depending on the type of premises activity:

Premises by main activity	*To whom to report*
Factories and factory offices	HM Factory Inspector
Mines and quarries	HM Mines and Quarries Inspector
Farms (and associated activities), horticultural premises, and forestries	HM Agricultural Inspector
Civil engineering and construction sites	HM Factory Inspector
Statutory and non-statutory railways	Inspecting Officer of Railways
Shops, offices, separate catering services launderettes	District Council (or equivalent)
Hospitals, research and development services, water supply, postal services and telecommunications, entertainment and recreational services, local government services, educational services and road conveyance	HM Factory Inspector

Reporting an Accident

Reports should be made to the local office of the enforcing authority as soon as possible, preferably by telephone, of:

- any accident causing death or major injury to an employee;
- any accident which occurs on premises which are under a person's control causing death or major injury to a self-employed person or to a member of the public;
- any notifiable dangerous occurrence; incidents of these types affecting the work or equipment either of a firm or self-employed persons working on premises which are under a member of the firm's control should be reported even if nobody is injured.

'Major injury' is defined as follows:

- fracture of the skull, spine or pelvis;
- fracture of any bone:
  - in the arm other than a bone in the wrist or hand;
  - in the leg other than a bone in the ankle or foot;
- amputation of a hand or foot;
- the loss of sight of an eye; or
- any other injury which results in the person injured being admitted into hospital as an in-patient for more than 24 hours, unless that person was detained only for observation.

If the person responsible does not know whom to report to, he should tell the nearest office of the HSE who will pass on the report.

Details of the accident/dangerous occurrence should be entered in the record book (or other record system). Within seven days a written report must be sent, on form F2508 (available from HSE offices), to the enforcing authority. This form should be used to report on all accidents causing fatal or major injury and notifiable dangerous occurrences. In the case of a reportable disease form F2580A is used.

Other injuries to employees are also notifiable if they result in more than three days' absence from normal work, but all that is needed is to:

- enter details of the accident in the accident record book (or other record system);
- complete form B176 (relating to a claim for Industrial Injury Benefit) when requested to do so by the local DSS office.

Accident Books

Since 31 December 2003 firms should have been using the new-style Accident Book (Form BI 510) devised by the HSE. The reason for this change is that the old-type book (also using the same Form number, BI 510) does not comply with data protection legislation because anyone entering accident details is able to read personal information contained in previous entries. The new book also contains useful information on first aid and how to manage health and safety in the workplace so as to avoid accidents in the first place.

Dangerous Occurrences

The list of dangerous occurrences in the regulations is selective, the aim being to obtain information about incidents with a high potential for injury but with a low frequency of occurrence.

It is important to note that dangerous occurrences must be reported even though no injury was actually caused to any person.

Examples of incidents which constitute dangerous occurrences are as follows:

- Failure, collapse or overturning of lifts, hoists, cranes, excavators, tail-lifts, etc.
- Explosion of boiler or boiler tube.
- Electrical short circuits followed by fire or explosion.
- Explosion or fire which results in stoppage of work for more than 24 hours.
- Release of flammable liquid or gas (ie over one tonne).
- Collapse of scaffolding.
- Collapse or partial collapse of any building.
- Failure of a freight container while being lifted.
- A road tanker to which the Hazchem regulations apply either overturning or suffering serious damage to the tank while a hazardous substance is being carried.

Note 1: This list is abbreviated. Where appropriate the regulations themselves should be consulted.

Note 2: A useful leaflet on the subject is available free from local offices of the HSE.

ROAD RISK

Road risk is the new 'in' term for transport management to contend with. While knowledge of and compliance with the *Highway Code* is essential for both drivers and managers alike, the D*f*T's free guideline publication *Driving at Work – Managing Work-Related Road Safety* is now also essential reading for transport management and for anybody who drives on company business – ie managers and staff as well as commercial vehicle drivers, those who drive full-time and those who drive only occasionally and, of course, self-employed drivers. In fact, according to road safety guru Dr Will Murray, these two documents go together like 'bacon and eggs'. The guidelines show how legal responsibilities under health and safety at work provisions must be complied with in addition to existing road traffic and vehicle construction and use provisions. They describe how to manage work-related road safety, assessing risks on the road and how to evaluate those risks. Overall, the aim is to cut the number of road accidents involving people at work – estimated by the HSE to account for over 20 fatalities and 250 serious injuries *every week*.

FIRE SAFETY

All employers are required by law to carry out a fire risk assessment and introduce appropriate safety measures within their workplaces. This applies irrespective of whether the employer's premises are covered by a Fire Certificate or not, whereas previously this duty did not apply where a Fire Certificate was in force.

This requirement has been brought about by the Fire Precautions (Workplace) (Amendment) Regulations 1999, which amend the Fire Precautions (Workplace) Regulations 1997 made under the Fire Precautions Act 1971. Fire safety measures must be assessed under the following headings:

- Means of raising the alarm.
- Means of escape.
- Firefighting equipment.
- Fire prevention.

Records must be kept of the periodic risk assessments carried out, staff training, alarm tests and fire drills. More specialized rules apply in places where explosives are handled.

A proposed new Order on fire safety is currently awaiting Parliamentary Select Committee approval. The Regulatory Reform (Fire Safety) Order 2004 aims to reduce burdens on business caused by the existence of multiple and overlapping general fire safety regimes – and consequently overlap of the responsibilities of enforcing authorities. The proposed Order will consolidate and rationalize much existing fire safety legislation (currently scattered across a large number of statutes and secondary legislation) into one Order. In doing so, it will reduce the number of enforcing authorities dealing with general fire safety matters. The reform will maintain and enhance the protection afforded to users of premises (and others who might be affected by a fire on the premises) by the existing legislation and provide for a risk-based approach to fire safety, allowing more efficient, effective enforcement by the fire and rescue service and other enforcing authorities. In fact, the removal of multiple and overlapping fire safety provisions and their replacement with a single fire safety regime will reduce a significant burden, as will the removal of the need to apply for fire certificates and the costs associated with the issue of certificates.

FIRST AID

Regulations require employers to train members of their staff in first aid techniques and to provide first aid equipment and facilities under The Health and Safety (First Aid) Regulations 1981. An approved Code of Practice established by the Health and Safety Commission provides both employers and the self-employed with practical guidance on how they may meet the requirements of the regulations. Further, The Stationery Office has published a booklet in its health and safety series entitled *First Aid at Work* (HS(R) 11; price: £2.50).

The regulations state that 'an employer shall provide, or ensure that there are provided, such equipment and facilities as are adequate and appropriate in the circumstances for enabling first aid to be rendered to his employees if they are injured or become ill at work'.

Trained First-Aiders

The employer must provide suitable persons to administer first aid and these persons must have had specific (see Code of Practice) training or hold appropriate qualifications – occupational first-aiders are no longer considered to be 'suitable persons' for this purpose.

First Aid Boxes

First aid boxes must be provided. These should be properly identified as first aid containers, preferably with a white cross on a green background, and contain sufficient quantities of first aid material *and nothing else*. In particular the boxes should contain only material which a first-aider has been trained to use. The old-style of contents list, according to the numbers of persons covered, has been scrapped in favour of a single specification list as follows:

1 × guidance card;
20 × individually wrapped sterile, adhesive dressings of assorted sizes;
2 × sterile eye pads, with attachment;
6 × individually wrapped triangular bandages;
6 × safety pins;
6 × sterile, unmedicated wound dressings (approx 13 cm × 9 cm);
2 × sterile, unmedicated wound dressings (approx 28 cm × 17.5 cm);
One pair of disposable gloves (latex).

Soap and water and disposable drying materials, or suitable equivalents, should also be available. Where tap water is not available, sterile water or sterile 9 per cent saline, in sealed disposable containers (refillable containers are banned) each holding at least 300 ml, should be kept easily accessible, and near to the first aid box, for eye irrigation.

The contents of first aid boxes should be replenished as soon as possible after use and items which deteriorate will need to be replaced from time to time. Items should not be used after the indicated expiry date on the packet. For this reason boxes and kits should be examined frequently to make sure they are fully equipped.

It is unwise to store non-listed medical supplies (ie proprietary items such as pain killer tablets, etc) or indeed any other items in first aid boxes. This may contravene the COSHH or other regulations and could result in prosecution – as in one case when a firm was prosecuted because an HSE inspector discovered its staff had placed nail varnish and acetate varnish remover in the box 'for safety'!

Travelling First Aid Kits

An employer does not need to make first aid provisions for employees working away from his establishment. However, where the work involves travelling for long distances in remote areas, from which access to NHS accident and emergency facilities may be difficult, or where employees are using potentially dangerous tools or machinery, small travelling first aid kits should be provided.

The contents of such kits may need to vary according to the circumstances in which they are likely to be used. The regulations suggest that in general the following items should be sufficient:

1 × guidance card;
6 × sterile adhesive dressings;
1 × large sterile, unmedicated dressing;
2 × triangular bandages;

2 × safety pins;

a supply of individually wrapped, moist cleaning wipes.

First Aid Rooms

When the siting of a new first aid room is under consideration it should be borne in mind that there should, where possible, be toilets nearby and ready access to transport. Any corridors and lifts which lead to the first aid room may need to allow access for a stretcher, wheelchair, carrying chair or wheeled carriage. Consideration should also be given to the possibility of providing some form of emergency lighting. Such rooms should be cleaned every day and clearly identified.

The following facilities and equipment should be provided in first aid rooms:

- Sink with running hot and cold water always available.
- Drinking water when not available on tap.
- Paper towels.
- Smooth-topped working surfaces.
- An adequate supply of sterile dressings and other materials for wound treatment. These should be at least equivalent in range and standard to those listed above, and kept in the quantities recommended.
- Clinical thermometer.
- A couch with pillow and blankets (frequently cleaned).
- A suitable store for first aid materials.
- Soap and nail brush.
- Clean garments for use by first-aiders and occupation first-aiders.
- Suitable refuse container.

First Aid on the Road

Over 300,000 people are injured in road traffic accidents annually, with unfortunately some 3,500 of them dying as a result. Any help that a road user can give at the scene of an accident may just save a life, or may give much needed comfort to a traumatized casualty. The following guidelines have been established by the West Yorkshire Metropolitan Ambulance Service for dealing with casualties at the scene of a road accident:

1. **Assess the situation:**
   - What has happened?
   - Is there any further danger to you or the casualty? (Remember, roadsides are particularly dangerous places to treat a casualty.)
   - How many people are injured?
   - Is anybody else able to help?
2. **Assess the casualty:**
   - What is wrong with the casualty?
   - Is it necessary to move the casualty, or can he/she be left safely in his/her current position?
   - Remember, further injuries can be caused through unnecessary movement of the casualty.

3. **Do you need assistance?**
   - Do you need assistance from the emergency services?
   - Don't forget, other first aiders/bystanders can help you.
   - When you ask for help from the emergency services they will need to know:
     - the precise location of the incident;
     - what happened;
     - the injuries involved;
     - how many people are injured.
4. **Make a diagnosis**
   - What happened to the casualty? Did he or she fall, faint or have a bump on the head?
   - Look for signs such as bleeding, swelling or unaligned limbs.
   - If the casualty is conscious, ask him or her where he or she feels pain or if he or she went dizzy before the accident.
5. **Priorities**

   The priorities of first aid are usually referred to as the ABC of first aid:
   - *Airway*: check inside the mouth and remove any visible obstruction. Put your fingertips under the point of the casualty's chin. Lift the chin to open the airway.
   - *Breathing*: spend at least 10 seconds looking, listening and feeling for breathing. If the casualty is breathing, put him/her in the recovery position. Continue to monitor the patient's breathing.
   - *Circulation*: look, listen and feel for normal breathing, coughing or movement by the victim. Only if you have been trained to do so, check the carotid pulse.
6. **Priorities of general treatment (the three Bs)**
   - *Breathing* (see above).
   - *Bleeding.* Stop bleeding by raising the injured limb above the heart. Apply pressure to the wound with pad, bandage, clean handkerchief or towel.
   - *Bones.* If necessary immobilize a broken arm in a sling. If a broken leg is suspected, keep the casualty still.
7. **While you are waiting for the arrival of the emergency services**
   - Look for changes in the casualty's condition, monitoring the vital signs.
   - Check your treatment is adequate and successful.
   - Co-operate with the Emergency Services when they arrive.

SAFETY SIGNS

A safety sign is defined as one which combines geometrical shape, colour, and a pictorial symbol to provide specific health or safety information or an instruction as to whether or not any text is included on the sign. Regulations concerning the specification of safety signs in work premises have applied to all such signs since 1 January 1986. Specifications for various types of safety sign are given in BS 5378 (Part I). They are briefly described as follows:

- ***Prohibition Sign***: round in shape with a white background and a circular band and cross-bar in red. The symbol must be black and placed in the centre of the sign without obliterating the cross-bar. Typical examples of such signs are those prohibiting smoking, pedestrians or the use of water for drinking purposes.
- ***Warning Sign***: triangular in shape with a yellow background and black triangular band. The symbol or words must be black and placed in the centre of the sign. Typical examples of such signs are those warning of the danger of fire, explosion, toxic substances, corrosive substances, radiation, overhead loads, industrial trucks, electric shocks, proximity of laser beams, etc.
- ***Mandatory Sign***: round in shape with a blue background, with the symbol or words placed centrally and in white. Typical examples of such signs are those advising that certain pieces of protective clothing must be worn (eg goggles, hard hats, breathing masks, gloves, etc).
- ***Safety Sign***: square in shape with a green background and white symbol or words centrally placed. Typical examples of such signs are those for fire exits, first aid posts and rescue points.

VEHICLE REVERSING

Among the statistics of industrial accidents and road accidents, those resulting from vehicles reversing feature significantly. According to the HSE, nearly one-quarter of all deaths involving vehicles at work occur while vehicles are reversing. In a new booklet entitled *Reversing Vehicles* (available from the HSE) the HSE gives the following advice to those concerned:

- Identify all the risks and decide how to remove them.
- Remove the need for reversing.
- Exclude people from the area in which vehicles are permitted to reverse.
- Minimize the distance vehicles have to reverse.
- Make sure all staff are adequately trained.
- Use a properly trained banksman or guide (the booklet illustrates the signals such a person should use to ensure safety when vehicles are reversing).
- Decide how the driver is to make and keep contact with the banksman.
- Make sure all visiting drivers are briefed.
- Make sure all vehicle manoeuvres are properly supervised.

Additionally:

- Increase the area the driver can see.
- Fit a reversing alarm (see below).
- Use other safety devices (ie trip, sensing and scanning devices, and barriers to prevent vehicles overrunning steep edges).

It is in consequence of dangers of reversing that legislation permits the voluntary fitment of reversing bleepers on certain goods and passenger vehicles (see p 354 for full details). With the introduction of these voluntary provisions, there is the risk that any operator deciding not to fit such equipment on a voluntary basis, who then has one of his vehicles involved in a reversing accident, could face proceedings under the Health and Safety at Work Act for not taking sufficient care in safeguarding the health of others. A number of successful prosecutions on this account have been reported and in some cases very heavy fines were imposed.

SAFE TIPPING

Another area where concern has been expressed over safety measures involves the use of tipping vehicles (and vehicles with lorry-mounted cranes and such like) whereby elevated bodies (or crane jibs) come into contact with overhead power cables or are sufficiently close for arcing to occur in wet conditions. The specific danger lies in touching the vehicle body or tipping controls while in contact with the power cable. Drivers are safe when a cable is touched, provided they remain in the cab where the vehicle tyres prevent completion of an electrical circuit. A spokesperson for the electricity supply industry advises drivers to remain in their cabs and drive clear. If this is not possible they should jump from the cab and NOT touch any part of the vehicle, remaining well clear until an electricity engineer has been contacted and reports that it is safe to return to the vehicle.

Tipper operators should also be aware of the provisions of the Lifting Operations and Lifting Equipment Regulations 1998 described on p 449.

SHEETING OF LOADS

A variety of load types need to be sheeted for both legal and operational reasons; particularly loads comprising waste skips, bricks, sand, gravel, ash, grain and various types of agricultural produce. Other loads need to be sheeted for weather protection and security reasons.

Many aspects of health and safety legislation and various Codes of Practice are relevant and should be carefully observed to avoid any breach of health and safety law. Also, specifically, it is an offence under C&U regulations to allow loose loads of any type to cause nuisance or danger to other road users as a result of such material falling or being blown from the vehicle. These loads must be adequately restrained by sheeting or by covering with netting.

Sheeting and unsheeting of loads is a notoriously high-risk procedure. Those carrying out this activity may slip or fall while walking on top of the load or balancing on vehicle sides or as a result of pulling on ropes that break. Employers have a statutory duty to assess such risks and to put in place safe working practices.

Health and safety guidelines suggest that priority should be given to the use of closed vehicles or, where this is not feasible, to vehicles with mechanically operated sheeting systems or, at the very least, with mechanical systems that can be operated from ground level. Alternatively, consideration should be given to the provision of gantries from which operatives can sheet loads with minimal risk to their personal safety.

Useful guidance on safe sheeting methods is to be found in a number of official publications as follows:

- Health and Safety Executive publications:
  - *Workplace Transport Safety* – HS(G)136
  - *Sheeting and Un-sheeting of Tipper Lorries* – HS(G)148

- Department for Transport
  - *Code of Practice – Safety of Loads on Vehicles* (third edition, 2002).

SAFE PARKING

Many accidents occur with parked vehicles and trailers. In particular within transport depots there should be proper procedures for parking, especially in regard to parking areas and level standing for detached semi-trailers. A common failing is to ensure that semi-trailers are dropped on to hard and level ground, allowing the landing gear on one side to sink and the trailer to topple over to the side. Nose-diving is another common accident where nose-heavy semi-trailers are not supported with trestles when left detached. Ground sinkage can also be a cause of difficulty when recoupling tractive units.

Recently reported cases (in *Commercial Motor*) of unsafe parking practices resulting in driver deaths concern the braking of parked semi-trailers. When recoupling tractive units, drivers insert the red air-hose which releases the emergency brakes on the semi-trailer without checking that both the semi-trailer ratchet brake and the tractive-unit handbrake are fully applied. In such circumstances the vehicle combination can move while the driver is out of the cab. Other potential sources of accidents are when articulated combinations are coupled near to walls and loading bays, where there is always a danger of an unseen person walking behind the trailer. Again, if brakes are not properly set the semi-trailer is likely to shunt backwards as the tractive unit is driven under the coupling plate.

WORK AT HEIGHT

Another new set of regulations concerning health and safety matters, and very likely to have a significant impact on the road haulage industry, are *The Work at Height Regulations 2005* (SI 2005 No. 735), which came into effect from 6 April 2005. These regulations are intended to help prevent the many accidents involving falls from height. In 2003/04 these accounted for 67 fatal accidents at work and nearly 4000 major injuries; in fact they are the biggest single cause of workplace deaths and major injury. Replacing earlier regulations, they consolidate previous legislation and implement EC Directive 2001/45/EC.

The new rules apply to most 'work at height' situations where there is a risk of a fall likely to occasion person injury – as, for example, climbing into or out of a vehicle cab or the load area. The employer's duty is to ensure that current practices are safe and to take steps to prevent falls, that relevant staff are trained, that work is planned accordingly and that work is suspended when weather conditions endanger health or safety. Provision in these regulations also relates to falling objects, which should be safeguarded against. In the context of this legislation, 'at height' means a place from which a person falling could be injured – and it also includes below ground level. Working means moving around at a place of work.

FORKLIFT TRUCK SAFETY

There has been much concern in recent years about the high level of industrial accidents which are caused by or which result from forklift truck misuse. In consequence of this a system of forklift truck driver licensing is to be established in order to ensure a safe standard of operation.

The scheme is voluntary but a Code of Practice* will be established which employers will be bound to follow if they wish to avoid conflict with the Factory Inspectorate who have powers to order an employer to have drivers trained, to order that the use of forklift trucks must be stopped immediately if they believe that danger is being caused, and to take an employer to court if an untrained driver causes an accident with a forklift truck.

** NB: This Code of Practice has been published and copies are available from HSE Books, price £5.00, ISBN 0 7176 2455 2.*

FREIGHT CONTAINER SAFETY REGULATIONS

Owners and lessees and others in control of freight containers must ensure that they comply with the International Convention for Safe Containers – Geneva 1972. The Freight Containers (Safety Convention) Regulations 1984 apply to containers designed to facilitate the transport of goods by one or more modes of transport without intermediate reloading, designed to be secured or readily handled or both, having corner fittings for these purposes and which have top corner fittings and a bottom area of at least 7 square metres or, if they do not have top corner fittings, a bottom area of at least 14 square metres.

Containers must have a valid approval issued by the Health and Safety Executive or a body appointed by the HSE (or under the authority of a foreign government which has acceded to the Convention) for the purpose of confirming that they meet specified standards of design and construction and should be fitted with a safety approval plate to this effect. If they are marked with their gross weight such marking must be consistent with the maximum operating gross weight shown on the safety approval plate. Containers must be maintained in an efficient state, in efficient working order and in good repair. Details of the arrangements for the approval of containers in Great Britain are set out in a document *Arrangements in GB for the Approval of Containers* available from the HSE.

The safety approval plate (issued by the HSE) as described in the regulations must be permanently fitted to the container where it is clearly visible and not capable of being easily damaged and it must show the following information:

CSC SAFETY APPROVAL

- Date Manufactured
- Identification Number
- Maximum Gross Weight . . . kg . . . lb
- Allowable Stacking Weight for 1.8g . . . kg . . . lb
- Racking Test Load Value.

OPERATION OF LORRY LOADERS

The use of hydraulically operated lorry loaders or lorry-mounted cranes, as they are more commonly called (in fact the name Hiab is becoming a generic term for this equipment), 'has reduced the risk of accident from the arduous and potentially injurious manhandling of loads and they reserve the strength of the driver for safe conduct of the vehicle', according to the Association of Lorry Loader Manufacturers and Importers of Great Britain (ALLMI). But, the Association says, despite the inherent safety of a properly designed and installed lorry loader, accidents still occur through lack of knowledge and understanding. For this reason ALLMI has published an excellent booklet called *Code of Practice for the Safe Application and Operation of Lorry Loaders*. Copies are available (£20 for the new 1999 version) from the Association, at 14 Manor Close, Droitwich, Worcestershire WR9 8HG (tel: 01905 770892/451040).

See also Chapter 7 dealing with driver training for lorry-loader operatives.

SAFETY IN DOCK PREMISES

Under the Docks Regulations 1988 made under the Health and Safety at Work etc Act 1974, when goods vehicle drivers work in or visit docks premises including roll-on/roll-off ferry ports they must be provided with high-visibility clothing to be worn when they leave the vehicle cab. The clothing may take the form of fluorescent jackets, waistcoats, belts or sashes and must be worn at all times when out of the cab on such premises including when on the vehicle decks of the ferry. Protective headgear (hard hats) must be supplied and worn in such areas where there is likely to be danger of falling objects from above (eg where cranes are working).

Drivers must leave the vehicle cab when parked on a straddle-carrier grid or where containers are being lifted on to or off the vehicle.

The HSC has published an approved Code of Practice – *Safety in Docks* – which is available from The Stationery Office.

21 Loads – General, Livestock, Food, etc

In addition to the regulations referred to in Chapter 13 regarding the way in which vehicles are constructed and used, operators will find many more regulations imposed on them which depend on the types of load they carry. Some of these provisions are included in the Road Vehicles (Construction and Use) Regulations 1986, and its many amendments, but others are to be found elsewhere. This chapter deals with normal loads and also covers some of the special points applicable to carrying food, livestock, sand and ballast, solid fuel and containers.

DISTRIBUTION OF LOADS

When loading a vehicle, care must be taken to ensure that the load is evenly distributed to ensure stability of the vehicle and to conform to the vehicle's individual axle weights as well as the overall gross weight and is secure so it cannot move or transfer its weight during transit. It is important on multi-delivery work to make sure that when part of the load has been removed in the course of a delivery, none of the axles has become overloaded because of the transfer of weight. This can happen even though the gross vehicle weight is still within permissible maximum limits and in such cases it is necessary for the driver to attempt to correct the situation by shifting the load, or part of it.

All loads should be securely and safely fixed, roped and sheeted, and chained or lashed if necessary. It is an offence to have an insecure load or a load which causes danger to other road users. Furthermore, it is a legal requirement of the C&U

regulations that loads must not cause or be likely to cause a danger or nuisance to other road users and that they should be physically restrained (ie roped and sheeted) if necessary to avoid parts of the load falling or being blown from the vehicle. *NB: tipping vehicles (and others), which carry loose loads that could emit dust into the atmosphere, must be sheeted before travelling on the road under the separate provisions of the Environmental Protection Act.* It is an offence also for the securing ropes or other devices and sheets to flap and cause nuisance or danger to other road users. Heavy penalties, with maximum fines of up to £5,000, can be imposed on conviction for offences relating to these matters.

Axle Load Calculations

Imposed axle loads can be calculated to determine whether a vehicle is operating legally in particular circumstances by using the following formula:

- Determine the vehicle wheelbase.
- Determine the weight of the load (ie payload).

$$\text{Weight on rear axle} = \frac{\text{front loadbase}}{\text{wheelbase}} \times \text{payload}$$

$$\text{Weight on front axle} = \text{payload} - \text{weight on rear axle}$$

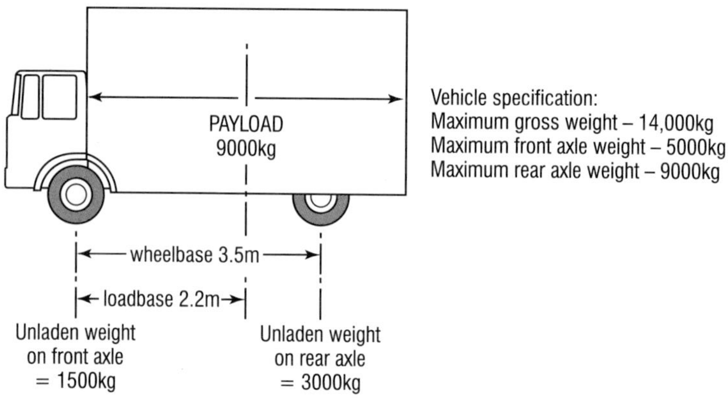

1. Rear axle weight calculation:

$$\frac{\text{front loadbase}}{\text{wheelbase}} \times \text{payload} = \text{payload weight on rear axle}$$

$$= \frac{2.2\text{m}}{3.5\text{m}} \times 9000\text{kg} = 5657\text{kg}$$

2. Front axle weight caculation:

Payload – rear axle weight = payload weight on front axle

9000kg – 5657kg = 3343kg

3. Total vehicle weight distribution:

Rear axle: payload weight + rear axle unladen weight
= 5657kg + 3000kg = 8657kg

Front axle: payload weight + front axle unladen weight
= 3343kg + 1500kg = 4843kg

- Calculate the front loadbase (ie centre line of front axle to centre of gravity of load).
- Apply the formula as above.

LENGTH AND WIDTH OF LOADS

Loads on normal goods vehicles (ie which come within the C&U regulations) in excess of the actual vehicle dimensions of length and width may be carried provided that certain special conditions are met, as indicated here. Details of the requirements relating to abnormal and projecting loads are given in Chapter 22.

Length

When moving a vehicle, which complies with C&U regulations, and its load which is more than 18.65 metres long, the police must be notified two clear days in advance and a statutory attendant must also be carried. If a long load is carried on an articulated vehicle which is specially designed to carry long loads but in all other respects complies with the C&U regulations, the 18.65 metre dimension is measured excluding the length of the tractive unit. Notice must also be given to the police and an attendant carried if the combination of a number of vehicles carrying one load is more than 25.9 metres long.

An overall limit of 27.4 metres is set for the length of a trailer and its load (the length of the towing vehicle is excluded from this dimension) above which movement can only be allowed by special order from the Secretary of State for Transport.

Further details of the requirements for markers, police notification and attendants on long loads is given in Chapter 22.

Width

The overall width of a normal load (ie not an indivisible load – see p 482 for definition) carried on a vehicle complying with the C&U regulations must not be more than 2.9 metres. The load itself must not project more than 305 millimetres on either side of the vehicle. There is an exception to this requirement when loose agricultural produce is carried.

If an indivisible load is carried on a normal vehicle which complies in all respects with the C&U regulations and the load is more than 2.9 metres wide, two days' notice must be given to the police of every district through which it is to pass. If such a load exceeds 3.5 metres width then the police must be notified as stated and an attendant must be carried.

If the load on a vehicle is more than 5 metres wide, two days' notice must be given to the police, a statutory attendant must be carried and a Special Order must be obtained from the Secretary of State for Transport on form VR1.

CARRIAGE OF LIVESTOCK

The Welfare of Animals (Transport) Order 1997, which became effective from 1 July 1997, signified an important step forward in strengthening animal welfare legislation in Great Britain. It revoked and re-enacted with modifications the Welfare of Animals during Transport Order 1994, as amended, and other British animal welfare legislation, which together with that Order implemented EU Council Directive 91/628/EEC on the protection of animals during transport. It also implemented the amending Council Directive 95/29/EEC.

There are two further items of relevant legislation. The Welfare of Animals (Staging Points) Order 1998 (SI 1998/2537) implements in the UK the provisions of EU Council Regulation 1255/97/EC on the provision of staging points for animals in transit. It specifies the approval authority and the powers of Veterinary Inspectors in relation to such premises and identifies the offences that may be committed for breach of the Order. The Welfare of Animals (Transport) (Amendment) Order 1999 (SI 1999/1622), which came into effect on 1 July 1999, implements in the UK the requirements of EU Council Regulation 411/98/EC on animal protection standards applicable to road vehicles used for the carriage of livestock on journeys exceeding eight hours, by creating an offence for any person to use a road vehicle for the transportation of animals in contravention of the regulation.

In August 1999, the Ministry of Agriculture, Fisheries and Food (MAFF) (now the Department of the Environment, Food and Rural Affairs (DEFRA)) issued a consultation document with the intention of reviewing (after two years in operation) current rules on the transport of animals.

EU Proposals

The EU plans to introduce new rules to help safeguard animal welfare, particularly while they are being transported. New regulation 1/2005/EC, which is due to come into force in 2007, identifies more effectively the chain of responsibility in animal transport, in particular defining 'who is responsible for what', so that more effective enforcement can be carried out. The regulations also introduce new and more efficient monitoring, such as checks on vehicles via satellite navigation systems, much stricter rules for journeys of more than eight hours' duration and a substantial upgrading of vehicle standards. Other measures proposed previously (as mentioned in the previous edition of this *Handbook*) but which have not yet been agreed upon – for example, revision of travelling times and animal densities on vehicles – are to be reviewed again no later than four years after this new regulation comes into force (ie in 2007). The new regulations will be fully described in next year's edition of the *Handbook*.

General Rules

As a general rule it is illegal to transport any animal in a way which is likely to cause injury or unnecessary suffering and which is in contravention of the new Order. The person in charge of animals in transit must ensure that they are not injured or do not suffer unnecessarily while waiting to be loaded or after being unloaded. On journeys over 50 kilometres animals must be given sufficient space having regard to their weight, size and physical condition, to weather conditions and the likely journey time. They must be fed and watered before and during a journey at specified intervals and must be transported to their destination without delay.

Fitness of Animals to Travel

It is illegal to transport any animal unless it is fit for the intended journey, and suitable provision has been made for its care during the journey and on arrival at its destination. Animals are not considered fit for an intended journey if they are ill, injured, infirm or fatigued, unless the intended journey is not likely to cause them unnecessary suffering.

Animals may be transported to the nearest available place for veterinary treatment or diagnosis, or to the nearest available place of slaughter, provided they are not likely to suffer unnecessarily. When transported under this provision they must not be dragged or pushed by any means, or lifted by a mechanical device, except under the supervision of a veterinary surgeon. Animals which fall ill or are injured during transport must be given first aid or veterinary treatment as soon as possible, or be slaughtered without unnecessary suffering.

Journey Times, Feeding and Watering

Normally, journey times for cattle, sheep, pigs, goats and horses must not exceed eight hours. However, where the transporting vehicle meets the following additional requirements, journey times may be extended:

- Where there is sufficient bedding on the floor of the vehicle.
- Provided appropriate feed is carried for the animals concerned and for the length of the journey.
- Where there is direct access to the animals.
- Where adequate ventilation is provided which can be adjusted depending on both inside and outside temperatures.
- Where movable panels are provided for creating separate compartments.
- If it is equipped for connection to a water supply during stops.
- If, when transporting pigs, sufficient liquid is carried for drinking during the journey.

Provided the above requirements are met, watering and feeding intervals, journey times and rest periods may be extended as follows:

- Unweaned calves, lambs, kids and foals which are still on a milk diet and unweaned piglets must be rested for at least one hour (ie sufficient for them to be given liquid and fed) after nine hours of travel. After this they may be transported for a further nine hours.
- Pigs may be transported for a maximum of 24 hours but they must have continuous access to liquid during the journey.
- Horses may be transported for a maximum of 24 hours but they must be given liquid during the journey and, if necessary, be fed every eight hours.
- All other cattle, sheep and goats must be rested for at least one hour (ie sufficient for them to be given liquid and fed) after 14 hours of travel. After this they may be transported for a further 14 hours.

When the maximum journey times specified have been reached, animals must be unloaded, fed and watered and be rested for at least 24 hours. Where it is solely in the interests of the animals, the maximum journey times mentioned above may be extended by two hours, depending on the proximity to the final destination (ie it would not be expected that journeys would be curtailed when within close proximity of their final destination).

Unweaned calves, lambs, kids and foals which are still on a milk diet and unweaned piglets may be transported for nine hours from a market if the journey to the market took not more than four hours (or nine hours if it was in a vehicle complying with items one to four listed above). Pigs or horses may be transported for nine hours from a market if the journey to market took not more than four hours (or eight hours if it was in a vehicle complying with items one to four listed above). All other cattle, sheep and goats may be transported for 14 hours from a market if the journey to market took not more than four hours (or 14 hours if it was in a vehicle complying with items one to four listed above).

Accompaniment by Competent Persons

Hauliers carrying animals on journeys of over 50 kilometres in length, or a person who accompanies the animals on the journey, must have had specific training or equivalent practical training to enable them to handle the animals properly and administer care where necessary.

A system for establishing competence in the handling and care for animals in transport is to be instituted (the Order sets out a framework for competence – see below). Those with practical experience (as opposed to specific training) must be assessed as to their ability, competence and knowledge and have a record to this effect which must be kept by the transporter while the person concerned is engaged on livestock journeys and for six months after such time. The record must be produced on request by an inspector.

The Order specifies the 'framework of competences', (ie the knowledge which people responsible for the welfare of animals during transport must have). They must have:

- an understanding of:
  - when to seek veterinary help and knowledge;
  - which body or organization to contact with general questions (eg about transport conditions);
  - matters of law or documentation;
- knowledge of the powers of enforcement authorities to inspect animals, documentation and vehicles before, during and after a journey;
- basic knowledge of authorization requirements for transporters and when such authorization is necessary;
- knowledge of how to plan a journey (taking account of such factors as maximum travelling times, required rest periods, and the time taken to load and unload), and the ability to anticipate changing conditions and make contingencies for unforeseen circumstances;
- understanding of when route plans or documentation are required and how to complete these documents;
- knowledge of vehicle construction and use requirements in current welfare legislation;
- the ability to load, operate and control a vehicle safely, efficiently and effectively so as to ensure the welfare of the animals;
- knowledge of the appropriate methods for handling animals during loading and unloading, including the use of visual fields and flight zones, lighting

and the appropriate use of such things as sticks, boards, blindfolds and electric goads – and know which handling methods are prohibited;
- knowledge of the specific requirements of the Order relating to different animal species for the provision of rest, feed and liquid;
- knowledge of stocking densities (ie the effects of overcrowding and understocking) and headroom and segregation requirements taking account of the species being transported, methods of transport, gender, condition, age, length of journey and ambient conditions;
- an understanding of the importance, for animal welfare, of temperatures both inside and outside the vehicle, including the effect on different species and the need for the adjustment of ventilation;
- an ability to clean and disinfect vehicles and a knowledge of when it is necessary to do so before and after the journey;
- elementary knowledge of the causes of stress in animals, ability to recognize the signs of stress and ill health, and basic knowledge of how to reduce such symptoms;
- an ability to care for animals which become unfit or injured during transport, including an understanding of when to seek veterinary advice;
- knowledge of the limited circumstances when it is permissible to transport unfit animals for veterinary treatment or for slaughter.

Tighter rules on competence for persons in charge of animals during transportation were introduced from 1 July 1998. Now persons must demonstrate that they are qualified to handle, transport, care for and safeguard the welfare of animals.

The Department of the Environment, Food and Rural Affairs (DEFRA) has produced a guidance booklet on the subject for employers called *Assessment of Practical Experience in the Handling, Transport and Care of Animals.*

Authorizations and Registration

Authorizations, issued in writing by the appropriate authority, are required by livestock transporters who carry animals on journeys over 50 kilometres.

Authorizations may be either 'general' or 'specific' and may be made subject to conditions.

Applicants for authorizations must be persons fit to transport animals (ie persons who have not committed any offences relating to animal welfare or contravened provisions of the Animal Welfare Act of 1981). Authorizations may be suspended or revoked for breach of the Order or of Council Directive 91/628/EEC whether this results in conviction or not.

Specific authorizations are issued by the Minister only to a named transporter and cover activities described in the authorization. General authorizations do not name individual transporters but cover individual transporters resident in Great Britain or transporter companies incorporated in Great Britain undertaking activities described in the authorization.

Since 1 October 1997 a specific authorization has been required for transporting cattle, sheep, pigs, goats and horses in a road vehicle on journeys exceeding eight hours' duration. For journeys of less than eight hours only a general authorization is needed.

Although a transporter may be covered by a general authorization for certain activities, this does not prevent him also holding a specific authorization for the same activities. However, where a specific authorization has been issued the general authorization no longer applies to animals covered by the specific authorization, and the transporter must not carry out any transport of those particular animals except under the terms of the specific authorization. Furthermore, if the specific authorization is revoked, or limited in any way (see below), then the transporter may not transport the animals concerned under the general authorization.

Both specific and general authorizations may be amended or revoked by the Minister, the former by notice in writing served on the transporter to whom the specific authorization was granted, and the latter by publication in a manner determined by the Minister. In the case of general authorizations the Minister may give notice in writing to a transporter, either excluding him from transport operations under a general authorization, or imposing on him additional conditions differing from the conditions in the general authorization.

Revocation or suspension of a specific authorization, or removal of a transporter from a general authorization, either temporarily or permanently, would occur if the transporter or any associate, employee or agent of the transporter:

- repeatedly infringes this Order in Great Britain (or any other Order implementing Council Directive 91/628/EEC);
- commits a single act resulting in serious suffering to animals (whether or not such an act leads to a criminal conviction);
- in the case of a journey taking place partly within and partly outside Great Britain, breaches the rules relating to route plans or repeatedly infringes national legislation implementing Council Directive 91/628/EEC, or a single such infringement involving serious suffering to animals (whether or not this leads to a criminal conviction).

The Minister must notify an offending transporter in writing that he is minded to revoke the authorization, or suspend it until a specified date or indefinitely.

Route Plans

Where horses, cattle, sheep, pigs and goats are traded between member states or exported to third countries on journeys exceeding eight hours, the transporter must draw up and sign a route plan. This route plan must be submitted along with an application for an export health certificate for approval by the Ministry. It will be returned with an authorization stamping and with the appropriate health certificates. Where more than one transporter is involved in such a journey the route plan requirements must be met by the person consigning the animals for the whole of the journey.

It is an offence to transport animals on such journeys unless an approved route plan is in force. The original of the route plan and the attached health certificates must accompany the consignment throughout the journey. All reasonable steps must be taken to comply with the route plan.

Feeding and watering times during the journey must be endorsed on the route plan (at the time they take place) by the person in charge of the animals.

When a route-plan journey is completed, the person who signed it when it was originally submitted must certify in writing on the route plan that it was complied with or, if not, must describe the actual journey together with the reasons for the route plan not being followed.

Within 15 days of completion of the journey the route plan must be returned to the issuing office and the sender must keep (for six months) proof that it was sent and of the date on which it was sent. The transporter must also keep a second copy of the route plan for six months from completion of the journey, and produce it on demand at the request of an inspector and allow copies to be taken.

The information to be provided on the route plan is as follows:

Section 1 – information to be completed before the journey:

- Name, full address, business name, telephone number and fax number of the transporter.
- Number and species of animals to be transported.
- Health certificate number(s).
- Number of the transporter's authorization (if any).
- Registration number of the vehicles to be used, and the trailers if different.
- Name of the person(s) in charge of the transport during the journey.
- Place where the animals are to be first loaded, and full address.
- Planned date and time of departure.
- Full itemized itinerary of the journey.
- Full address of the final destination.
- Estimated date and time of arrival at the final destination (local time).

Section 2 – information to be completed during the journey:

- Actual date and time of loading the first animal.
- Actual time of departure from the place of loading.
- Full itemized itinerary of the journey.
- Actual date and time of arrival at the final destination (local time).

Animal Transport Certificates

Where animals are transported other than under a route plan they must be accompanied by documentation which shows:

- The name and address of the transporter.
- The name and address of the owner of the animals.
- Where the animals were loaded, and their final destination.
- The date and time the first animal was loaded.
- The date and time of departure.
- The time and place where rest period requirements were met.

This documentary requirement does not apply where poultry and domestic birds are transported for distances of not more than 50 kilometres and where the number of such poultry and domestic birds is less than 50, or the entire journey is on land occupied by the owner of the poultry and domestic birds.

In the case of animals which are not cattle, sheep, pigs, goats, horses or poultry and domestic birds, the documentary requirement mentioned above does not apply to journeys of 50 kilometres or less.

Transporters are required to retain a copy of the documentation for a period of six months from the completion of the journey, and must produce it to an inspector on demand and allow a copy to be taken. Where such a copy is kept in electronic or magnetic form, the inspector may request its production in written form and for copies to be taken in writing.

Powers of Livestock Inspectors

Where they consider that animals are being transported, or are about to be transported, in a manner likely to cause injury or unnecessary suffering, or in any other way in contravention of the Order, inspectors appointed for the purposes of the Animal Health Act 1981 by the Minister or by a local authority* may serve notice on the person in charge of the animals requiring them to take any action necessary to ensure compliance with the Order and giving reasons for the requirements.

**NB: In practice these inspectors will be veterinary inspectors from the State Veterinary Service, a local veterinary inspector of the Department of Environment, Food and Rural Affairs (DEFRA) or an officer from the Trading Standards department of a local authority.*

In particular the inspector may:

- prohibit that movement of the animals, either indefinitely or for a specified period;
- specify conditions under which the animals may be transported;
- require the journey to be completed or the animals to be returned to their place of departure by the most direct route, provided this would not cause unnecessary suffering to the animals;
- require the animals to be held in suitable accommodation with appropriate care until the problem is solved;
- require the humane slaughter of the animals.

Inspectors may, where necessary for identification purposes, mark an animal. It is an offence for any person to remove, deface, obliterate or alter such marks.

Construction of Vehicles

The Order specifies general requirements for the construction and maintenance of vehicles and receptacles. In particular they must:

- be safe and not cause injury or suffering during transport, or while loading or unloading;
- be weatherproof;
- allow space for animals to lie down;
- have strong, non-slip floors which are free of protrusions;
- protect animals from the weather and from excessive humidity, heat and cold;
- be free of sharp edges;
- allow appropriate cleaning and disinfection;
- be escape proof;
- prevent animals from undue exposure to noise or vibration;

- provide sufficient natural or artificial light to enable animals to be properly cared for;
- have partitions if necessary which:
  - provide adequate support and prevent animals being thrown about during transport,
  - are of rigid construction and strong enough to bear the weight of any animal,
  - do not obstruct ventilation.

When transporting animals (ie all mammals and birds):

- they must not be subject to severe jolting or shaking;
- they must not be injured during loading or unloading;
- they must be driven without excessive use of prods and suchlike;
- the vehicle must carry means for emergency unloading;
- they must be segregated from goods carried on the same vehicle;
- carcasses must not be carried on the same vehicle except those of animals which die on the journey;
- the vehicle or receptacle must have been thoroughly cleaned and, where appropriate, disinfected;
- dead animals, soiled litter and droppings must be removed as soon as possible;
- vehicle/receptacle floors must be covered with sufficient litter, unless alternative arrangements are made or urine and droppings are regularly removed;
- receptacles must be marked/labelled to indicate:
  - they contain live animals of a named species,
  - the upright position;
- receptacles must be kept upright and secured during transport;
- at least one attendant must accompany the animals unless:
  - receptacles are secure,
  - adequately ventilated,
  - contain sufficient food and liquid, in dispensers which cannot tip over, for a journey of twice the anticipated time,
  - the transporter (driver) acts as the attendant,
  - the consignor has appointed an agent to care for the animals at stopping or transfer points.

Additional provisions relating to the construction and maintenance of vehicles and receptacles and for the transport of animals are as follows:

- There must be sufficient space for animals to stand normally.
- They must allow adequate ventilation and provide space above the animals to enable air to circulate properly.
- They must allow for the inspection of animals and their feeding and watering.
- Road vehicles must have a roof to protect against the weather.
- Have barriers (for horse-carrying vehicles, straps) to prevent animals falling out when doors are opened.
- Ramps on vehicles must (so as not to cause injury):
  - prevent slipping,
  - not be too steep for the animals being carried,

- not have a top or bottom step which is too high,
- not have gaps at the top or bottom which are too wide.

- Road vehicles must enable inspection of the animals from the outside (and must have suitable footholds).
- For animals which are normally tied, tying facilities must be provided and ties must be:
  - strong enough not to break during normal transport conditions,
  - designed to eliminate any danger of strangulation or injury,
  - long enough to allow animals to lie down and to eat and drink.

NB: Animals must not be tied by their horns or by a nose ring.

- On multi-deck vehicles, suitable ramps or lifting gear of sufficient strength must be provided.
- Certain animals must be segregated during transport as follows:
  - A cow with a suckling calf.
  - A sow with unweaned piglets.
  - A mare with a foal at foot.
  - A bull over 10 months old.
  - A breeding boar over six months old.
  - A stallion.

NB: Bulls, boars and stallions may be carried with others of the same species/gender if they have been raised in compatible groups or are accustomed to one another – other animals of mixed species may be carried together if separation from their companions would cause them distress.

- Animals must not be carried together if, due to differences in their age and size, injury or unnecessary suffering may be caused. Also, animals which are hostile to each other or fractious must not be carried together.
- Other animals which must not be carried together are:
  - uncastrated male adults with female animals (unless accustomed to each other),
  - horned cattle with unhorned cattle,
  - broken horses with unbroken horses (unless they are all secured).
- Animals must not be suspended by mechanical means, lifted or dragged by the head, horns, legs, tail or fleece.
- Excessive force must not be used to control animals including no use of:
  - electric shock instruments,
  - stick, goad or other instrument (to hit cattle under six months old),
  - stick, non-electric goad or other instrument (to hit or prod pigs).

NB: use of such items is allowed on the hindquarters of cattle over six months old and on adult pigs which are refusing to move forward when there is space to do so – but such action must be avoided as far as possible.

- Attendants must look after animals including, if necessary, feeding and watering them.
- Animals in milk must be milked at appropriate intervals – in the case of cows in milk this must be at intervals of about 12 hours but not exceeding 15 hours.

- Horses being transported in groups must wear halters unless they are unbroken and have their hind feet unshod.
- Horses must not be transported in vehicles with more than one deck in operation.

When poultry and domestic birds, and domestic rabbits, are carried further requirements are specified for the construction and maintenance of vehicles and receptacles:

- Adequate ventilation and air space must be provided.
- Receptacles must allow inspection of and care for the animals.
- Receptacles must be of such size as to protect the animals from injury or unnecessary suffering during transport.
- Receptacles for carrying birds must prevent the protrusion of heads, legs or wings.
- Birds must not be carried in a sack or bag.
- Birds must not be lifted or carried by the head, neck, wing or tail except:
  (a) ducks, which may be lifted or carried by the neck,
  (b) geese, which may be lifted or carried by the base of both wings.
- Birds must not be tied by the neck, leg or wing.
- Rabbits must not be carried in the same undivided pen, receptacle or road vehicle as an animal of any other species.
- Birds must be segregated according to sex and species except that:
  - female birds may be transported with their broods,
  - male and female chicks may be transported together,
  - male and female birds which are familiar with one another may be transported together,
  - chicks must be segregated from all other poultry except their mother or other chicks.
- Birds must not be transported next to any animal which is hostile to them or in the presence of any animal likely to cause them unnecessary suffering.

The term birds as used above means poultry and domestic birds.

Additional Standards for Road Vehicles

A new EU Council regulation, 411/98/EC, which came into force on 1 July 1999, requires additional standards to be applied to road vehicles used for the carriage of animals on journeys of over eight hours' duration. In particular, the new standards concern access to the vehicle, separation of animals by moveable partitions, detailed arrangements concerning feeding and watering, and the provision of adequate ventilation by either a forced system or a system that will ensure compliance with a prescribed temperature range.

FOOD

The Food Standards Agency established under provisions contained in the Food Standards Act 1999 controls and regulates food safety and standards in the UK, particularly including such issues as the transport of livestock, fresh meat and other foods. It will have power to intervene if hauliers jeopardize food safety.

Special regulations (the Food Safety (General Food Hygiene) Regulations 1995) apply to vehicles used for the carriage of food, excluding milk and drugs. All mobile shops and food delivery vehicles and the equipment carried by such vehicles must be constructed and maintained so that the food carried can be kept clean and fresh.

The driver of a food vehicle should wear clean overalls and if meat or bacon sides are carried, which the driver has to carry over his shoulder, he should wear a hat to prevent the meat touching his hair. He must not smoke while loading or unloading or serving the food (but he may do so in the cab of the vehicle if this is separate from the part of the vehicle in which the food is carried), and any cuts or abrasions on his hands must be covered with waterproof dressings.

If a driver or any other person concerned in the loading and unloading of food develops any infectious disease, his employer must notify the local authority health department immediately.

Food vehicles must have the name and address of the person carrying on the business shown on the nearside and the address at which the vehicle is garaged if this is a different address. If, however, a vehicle based in England or Wales has a fleet number clearly shown and is garaged at night on company premises, then the garage address is not required.

A wash hand basin and a supply of clean water must be provided on vehicles which carry uncovered food (except bread) unless the driver can wash his hands at both ends of his journey before he has to handle the food. When meat is carried, soap, clean towels and a nail brush must be provided on the vehicle. In Scotland all food-carrying vehicles must be provided with these items.

Mobile shops and food delivery vehicles must not be garaged with food still inside unless it can be kept clean.

An authorized officer of a council may enter and detain (but not stop while it is in motion) any food-carrying vehicle except those owned by a rail company or vehicles operated by haulage contractors.

Perishable Food

A whole range of legislation covers the carriage of food and particularly perishable food both in the UK and in Europe. When perishable foodstuffs are carried on international journeys to and through Austria, Belgium, Bulgaria, the CIS, the Czech Republic and Slovakia, Denmark, Finland, France, Germany, Italy, Luxembourg, Morocco, the Netherlands, Norway, Poland, Spain, Sweden and the former Yugoslavia the conditions of the Agreement on the International Carriage of Perishable Foodstuffs (known as the ATP agreement) must be observed.

The ATP agreement also applies in Britain (viz the International Carriage of Perishable Foodstuffs Act 1976) and under its provisions vehicles used to carry perishable food must be constructed and tested (at six-yearly intervals) to certain specified standards and must display an ATP approval plate to this effect. The principal requirement is that vehicles carrying certain specified perishable foodstuffs must comply with the body and temperature control equipment test standards and must be certified to this effect. The ATP agreement applies broadly to quick-frozen, deep-frozen, frozen and non-frozen foodstuffs but not fresh vegetables and soft fruit.

Information relating to food transport may be obtained from the Department of Health and Social Security which has published a number of Codes of Practice, and from the Department for Environment, Food and Rural Affairs.

Chilled Food Controls (ie Temperature-Controlled Food)

Following a series of food poisoning (salmonella and listeria) outbreaks in 1989 changes in legislation resulted in the introduction of the Food Hygiene (Amendment) Regulations 1990 which control the maximum temperature at which chilled food can be transported. The regulations apply to goods vehicles exceeding 7.5 tonnes gross weight carrying 'relevant food' products (which are listed in the schedules to the regulations). Such vehicles must have equipment capable of maintaining the temperature of the food at or below the specified temperature (1 April 1993, –5°C for certain foods as listed in the schedules and –8°C for other 'relevant' foods) or at or above 63°C as appropriate.

Vehicles up to 7.5 tonnes used for local deliveries must be capable of keeping relevant food at or below the specified temperature or at or above 63°C as appropriate. However, in a case where the regulations specify that food is to be kept at 5°C, provided it is not kept in the vehicle for more than 12 hours, it may be at a higher temperature, not exceeding 8°C.

There are exemptions to these requirements: where food is to be sold within two hours (ie if prepared at 63°C or over) or four hours (ie if prepared below 63°C). Other variations (ie not exceeding 2°C for a maximum of two hours) may be permitted in certain specified circumstances such as variations in processing; while equipment is defrosted; during temporary breakdown of equipment; while moving food from one place or vehicle to another; or any other unavoidable reason.

New regulations (ie the Food Safety (Temperature Control) Regulations 1995 and the Food Safety (General Food Hygiene) Regulations 1995) which came into effect in September 1995 implement various provisions of EU Directive 43/93/EEC on these matters and parts of the so-called 'Water Directive' (EU Directive 778/80/EEC) which relates to the use of water for food production purposes. These regulations only marginally affect transport operations and for that reason they are not detailed here.

Quick-Frozen Foodstuffs

The UK Quick-frozen Foodstuffs (QFF) Regulations 1990 are effective in setting general conditions for the quality and use of equipment for storing and transporting food labelled 'Quick-frozen' (but not ice cream). These regulations require certain temperatures to be maintained within a percentage range of –18°C, or colder, as follows:

- during transport (other than local distribution) a tolerance for brief periods of 3°C (but not warmer than –15°C).
- during local distribution 6°C (but not warmer than –12°C).

The Quick-frozen Foodstuffs (Amendment) Regulations 1994, which took effect from 1 September 1994, require refrigerated transport operators to fit temperature recorders to vehicles carrying frozen foodstuffs (but not chilled foods) and to keep the records for at least 12 months.

Waste Food

Vehicles used to collect unprocessed waste food, intended for feeding to livestock and poultry, must be drip-proof, covered and enclosed with material capable of being cleansed and disinfected.

Vehicles must be thoroughly cleansed and disinfected on the completion of unloading. No livestock or poultry or foodstuff or anything intended for use for any livestock or poultry may be carried in any vehicle which is carrying unprocessed waste food intended for feeding to livestock or poultry. Furthermore, processed and unprocessed waste food may not be carried in the same vehicle at the same time.

Grain Haulage

Grain hauliers are faced with new rules banning them from using grain trailers for the carriage of other loads (such as glass, toxins, waste, bonemeal and manure) which may contaminate the trailer. Auditors from UKASTA (the United Kingdom Agricultural Supply Trade Association) will monitor standards and check hauliers' records and invoices to ensure they have not broken the Association's code, which is based on provisions in the Food Safety Act 1990. A new style 'grain passport' must be carried on vehicles, otherwise loads may be rejected by consignees if they are not satisfied as to the origin or standard under which loads have been carried.

SAND AND BALLAST LOADS

The movement of sand and ballast comes to the attention of the Department of Trade and Industry and Trading Standards inspectors principally because under the Weights and Measures Act 1985 these materials must be sold in weighed quantities (normally by volume in metric measures – in multiples of 0.2 cubic metres) and carried in calibrated vehicles which display a stamp placed on the body by the Trading Standards department of the local authority.

When sand and ballast (including shingle, ashes, clinker, etc) loads are carried, the driver must have a signed note (ie conveyance note) from the supplier indicating the following facts:

- The name and address of the sellers.
- The name of the buyer and the address for delivery of the load.
- A description of the type of ballast.
- The quantity by net weight or by volume.
- Details of the vehicle.
- The date, time and place of loading the vehicle.

The document containing these details must be handed over to the buyer before unloading, or if he is not there it must be left at the delivery premises. Where a delivery is to be made to two or more buyers each must be given a separate document containing the details shown above. Similar requirements apply to the carriage of ready-mixed cement.

Quarry Tax

From April 2002, in an effort to increase the use of alternative recycled materials, a new 'quarry' tax of £1.60 per tonne applies to all primary aggregates such as sand, gravel and crushed rock taken from quarries. Recycled materials include wastes from construction, demolition, clay and coal extraction. Some of the revenue earned from this tax will go into an 'aggregates sustainability fund'.

SOLID FUEL LOADS

When solid fuel is carried a document giving similar details to those mentioned above for sand and ballast carrying must be held by the driver of the vehicle. When solid fuel is carried for sale in open sacks a notice on the vehicle in letters at least 60 mm high must contain the following words: 'All open sacks on this vehicle contain 25 kg or 50 kg'.

CONTAINER CARRYING

Container carrying has come to the fore in recent years, and while there are no specific regulations on this subject apart from the general safety provisions of the C&U regulations and the vehicle weight limitations, the authorities are concerned about the dangers arising from inadequate securing of containers. Containers, ideally, should be carried only on vehicles fitted with proper twistlocks or, failing this, should be secured by chains of sufficient strength with tensioners for adjustment and taking account of the recommended strength of restraint systems given in the DETR code of practice *Safety of Loads on Vehicles* (see p 436).

The use of ropes for securing containers should be avoided because the corner castings through which they are passed are rough and will cut through the rope. Containers where possible should be loaded against the headboard and directly on the platform of the vehicle, not on timber packing which is liable to move.

Under the Freight Containers (Safety Convention) Regulations 1984 owners and lessees and others in control of freight containers used or supplied must ensure that they comply with the conditions of use as stated in the International Convention for Safe Containers 1972 (see also Chapter 20).

FLY-TIPPING

The Control of Pollution (Amendment) Act 1989, implemented in 1991, tightened up existing measures to prevent the illegal tipping (fly-tipping) of waste, demolition rubble and such materials in unauthorized places. Among the specific measures legislated for under regulations called the Controlled Waste (Registration of Carriers and Seizure of Vehicles) Regulations 1991 are requirements for registration of waste transporters and tipper operators, the licensing of authorized operators, restriction of tipping to authorized sites and further powers to impound (and possibly sell) vehicles belonging to operators who dump loads illegally (such powers also exist under the Criminal Justice Act 1988).

Offences such as fly tipping can result in fines of up to £20,000 and three months' imprisonment for magistrates' court convictions under the Environment Protection Act 1990. Where a case goes before a Crown court, much higher penalties may be imposed.

NB: See Chapter 23 for more information on the carriage and disposal of waste materials.

22 Loads – Abnormal and Projecting

NEW SPECIAL TYPES ORDER

After a prolonged period of consultation, a new set of rules on the carriage of abnormal indivisible loads (AILs) on Special Types vehicles was made late in 2003 with the introduction of the Road Vehicles (Authorization of Special Types) (General) Order 2003 under Section 41 of the Road Traffic Act 1988. These new provisions are concerned primarily with four different types of vehicle, now referred to as Part 2 vehicles and vehicle combinations, as follows:

- abnormal indivisible load vehicles;
- mobile cranes;
- engineering plant;
- road recovery vehicles.

ABNORMAL INDIVISIBLE LOADS

For the purpose of regulations, abnormal indivisible loads (sometimes abbreviated to AILs) are loads which cannot, without undue expense or risk of damage, be divided into two or more loads for the purpose of carriage on the road and which cannot be carried on a vehicle operating within the limitations of the C&U regulations as described in Chapter 13.

Number of Abnormal Loads

While normally the carriage of only one abnormal load is permitted, two such abnormal loads may be carried on one vehicle within Category 1 or Category 2 (see below) provided the loads are from the same place and are destined for the same delivery address.

Engineering Plant

In the case of engineering plant, such plant and parts dismantled from it may be carried on the same vehicle (ie to constitute more than one or two loads) provided that the carriage of the parts does not cause the overall dimensions of the vehicle and the main load to be exceeded and that the parts are loaded and discharged at the same place as the main load.

SPECIAL TYPES VEHICLES

Dimensions

Width Special Types vehicles, locomotives and trailers and their loads are normally permitted to be up to 2.9 metres wide but, if necessary to ensure the safe carriage of large loads, they may be up to a maximum of 6.1 metres wide.

Length The overall length of a Special Types vehicle and its load must not exceed 27.4 metres which applies normally, but where the abnormal load is carried on a combination of vehicles and trailers or on a long articulated vehicle the dimension of 27.4 metres is measured excluding the drawing vehicle.

Weight The permissible maximum weight of a Special Types vehicle must not exceed 150,000 kg. There is a limit on the maximum weight which may be imposed on the road by any one wheel of the vehicle of 8250 kg and the maximum weight imposed by any one axle must not exceed 16,500 kg. These limits may be exceeded only if authorization by Special Order is obtained from the Secretary of State for Transport (D*f*T).

Vehicle Categories

The Special Types Order specifies three separate weight categories for abnormal load vehicles as follows:

Category 1 – up to 50 tonnes gcw.
Category 2 – up to 80 tonnes gcw.
Category 3 – up to 150 tonnes gcw.

Category 1 Vehicles

Vehicles within this category will normally fall within the C&U regulations in regard to permissible maximum weight, axle spacings and axle weights but where it is a five-axle articulated vehicle the weight may exceed 38 tonnes up to a maximum of 50,000 kg (ie 50 tonnes) provided the following minimum relevant axle spacings are observed:

Relevant axle spacing	*Maximum Weight*
At least 6.5 metres	40,000 kg
At least 7.0 metres	42,000 kg
At least 7.5 metres	44,000 kg
At least 8.0 metres	50,000 kg

Category 2 Vehicles

Vehicles within this category may operate up to a maximum weight of 80,000 kg (ie 80 tonnes) but they must have a minimum of six axles with a maximum weight of 50,000 kg on any group of axles and they must meet the minimum axle spacing requirements specified below.

Individual wheel and axle weight limits are as follows:

Distance between adjacent axles	*Maximum axle weight*	*Maximum wheel weight*
At least 1.1 metres	12,000 kg	6000 kg
At least 1.35 metres	12,500 kg	6250 kg

Minimum axle spacings and applicable maximum weights are as follows:

Distance between foremost and rearmost axles	*Maximum weight*
5.07 metres	38,000 kg
5.33 metres	40,000 kg
6.00 metres	45,000 kg
6.67 metres	50,000 kg
7.33 metres	55,000 kg
8.00 metres	60,000 kg
8.67 metres	65,000 kg
9.33 metres	70,000 kg
10.00 metres	75,000 kg
10.67 metres	80,000 kg

Category 3 Vehicles

A minimum of six axles is needed on Category 3 vehicles operating up to a permissible maximum weight of 150,000 kg (ie 150 tonnes) with a limit of 100,000 kg on any group of axles or 90,000 kg on any group of axles where the distance between adjacent axles is less than 1.35 metres. Vehicles in this category must meet the minimum axle spacing requirements specified below.

Individual wheel and axle weight limits are as follows:

Distance between adjacent axles	*Maximum axle weight*	*Maximum wheel weight*
At least 1.1 metres	15,000 kg	7,500 kg
At least 1.35 metres	16,500 kg	8,250 kg

Minimum axle spacings and applicable maximum weights are as follows:

Distance between foremost and rearmost axles	*Maximum weight*
5.77 metres	80,000 kg
6.23 metres	85,000 kg
6.68 metres	90,000 kg
7.14 metres	95,000 kg
7.59 metres	100,000 kg
8.05 metres	105,000 kg
8.50 metres	110,000 kg
8.95 metres	115,000 kg
9.41 metres	120,000 kg
9.86 metres	125,000 kg
10.32 metres	130,000 kg
10.77 metres	135,000 kg
11.23 metres	140,000 kg
11.68 metres	145,000 kg
12.14 metres	150,000 kg

VED for Special Types Vehicles

A separate VED taxation class applies to Special Types vehicles – see Chapter 8 for details.

Braking Standards

Vehicles operating within Category 1 must meet the C&U regulation braking standard requirements and those operating within Categories 2 and 3 from 1 October 1989 must meet the EU braking standards (ie EC 230/71, 132/74, 524/75, 489/79).

Identification Sign

Vehicles operating under the Special Types Order must display an identification sign at the front. This sign, on a plate at least 250 mm x 400 mm, must have white letters on a black background as follows:

STGO	letters 105 mm high
CAT ...	letters and figures 70 mm high

NB: A figure 1, 2 or 3 must follow the word 'CAT' as appropriate depending on the category of vehicle.

Special Types Plates

Vehicles falling within Category 2 and 3 which have been manufactured since 1 October 1988 must display Special Types plates (in a conspicuous and easily accessible position) showing the maximum operational weights recommended by the manufacturer when travelling on a road at varying speeds as follows: 12, 20, 25, 30, 35, 40 mph. The weights to be shown are the permissible maximum gross and train weights and the maximum weights for each individual axle. Plates on trailers (including semi-trailers) must show the permissible maximum weight for the trailer and the maximum weights for each individual axle. The plates must be marked with the words 'Special Types Use'.

Speed Limits

Maximum permitted speeds* are specified for Special Types vehicles as follows:

Vehicle category	*Motorways*	*Dual-carriageways*	*Other Roads*
1. Up to 46 tonnes	60 mph	50 mph	40 mph
2. Up to 80 tonnes	40 mph	35 mph	30 mph
3. Up to 150 tonnes	30 mph	25 mph	20 mph

Speeds for Wide Loads

Vehicles carrying loads over 4.3 metres but not over 6.1 metres wide are restricted to 30 mph on motorways, 25 mph on dual-carriageways and 20 mph on other roads.

**NB: It is important to note that tyre equipment on vehicles and trailers must be compatible with both the gross weight of the vehicle and the authorized maximum speed of operation.*

Attendants

An attendant must be carried on Special Types vehicles:

- when the vehicle or its load is more than 3.5 metres wide;
- if the overall length of the vehicle is more than 18.3 metres (not including the length of the tractive unit in the case of articulated vehicles);
- if the length of a vehicle and trailer exceeds 25.9 metres;
- if the load projects more than 1.83 metres beyond the front of the vehicle;
- if the load projects more than 3.05 metres beyond the rear of the vehicle.

If three or more vehicles carrying abnormal loads or other loads of dimensions that require statutory attendants to be carried travel in convoy, attendants need only be carried on the first and last vehicles in the convoy.

Police Notification

The police of every district through which a Special Types combination is to be moved must be given two clear days' notice (excluding Saturdays, Sundays and bank holidays in the case of notification under the STGO and excluding Sundays and bank holidays for notification under the C&U regulations) if:

- the vehicle and its load is more than 2.9 metres wide;
- the vehicle and its load (or trailer and load) is more than 18.3 metres long;
- a combination of vehicles and trailers carrying the load is more than 25.9 metres long;
- if the load projects more than 3.05 metres to the front or rear of the vehicle;
- the gross weight of the vehicle and the load is more than 80,000 kg.

The notice given must include details of the vehicle and the weight and dimensions of the load, the dates and times of the movement through the police district and the proposed route to be followed through that district.

In the case where notice has been given to the police of the movement of an abnormal load, they have the power to delay the vehicle during its journey if it is holding up other traffic or in the interests of road safety.

Under the provisions of the new STGO, which came into effect from 1 January 2004, responsibility for escorting abnormal loads has transferred to private escort companies. These companies, a list of which is available from the RHA (contact s.biddle@rha.net) must meet laid down criteria of standards and comply with a newly devised Code of Practice established jointly by the Highways Agency, the Association of Chief Police Officers (ACPO), ACPOS (Scotland) and the transport trade associations.

It should be noted that despite these new arrangements, notification about the movement of abnormal loads must still be made to the police, and that Chief Constables retain the right to escort vehicles 'where appropriate'.

Notification of Highway and Bridge Authorities

If a Special Types vehicle and its load weighs more than 80,000 kg, or the weight imposed on the road by the wheels of such a vehicle exceeds the maximum limit laid down in the C&U regulations (see pp 313–316), five clear days' notice must be given to the highway and bridge authorities for the areas through which the vehicle is to pass. Two days' notice is required when only the C&U axle weight limit is exceeded. The operator of such a vehicle is also required to indemnify the authorities against damage to any road or bridge over which it passes. These requirements also apply to vehicles which exceed the C&U gross weight limits and those which exceed their plated axle weight limits, plus mobile cranes and engineering plant which exceed the limits specified.

Stopping on Bridges

The driver of a vehicle carrying an abnormal load must ensure that no other such vehicle and load are on a bridge before he drives on to the bridge and once on the bridge he must not stop unless forced to do so.

Where a vehicle weighing more than 38,000 kg gross has to stop on a bridge for any reason it must be moved off the bridge as soon as possible. If it has broken down, the advice of the bridge authority (usually the Highways Department of the local authority) must be sought before the vehicle is jacked up on the bridge. In the event of damage being caused to a road or bridge by the movement of a heavy or large load over it the highway authority can take steps to recover from the vehicle operator the costs of repairing the damage.

Notification to Railtrack

All movements of abnormal loads over rail bridges must be notified to the BR Property Board by fax on 0990 143052.

Special Orders

Written approval has to be obtained from the Secretary of State (D*f*T) in cases where a Special Types vehicle and its load exceeds five metres in width. Application is made on form VR1 (ie the movement order) and a copy of this completed form must be carried on the vehicle. The route specified, and the date and timings for the journey notified in the application, must be adhered to, otherwise further approval will have to be sought.

Dump Trucks

When dump trucks (ie vehicles designed for moving excavated material) are used on the road the maximum permissible gross weight is limited to 50,800 kg and the maximum axle weight allowed for such vehicles is 22,860 kg. They must not exceed a speed of 12 mph on normal roads and an attendant must be carried if the width exceeds 3.5 metres. If a dump truck is more than 4.3 metres wide, permission in the form of a Special Order from the Secretary of State is required before it is moved. Where three or more such vehicles over 3.5 metres wide travel in convoy only the first and last vehicles need carry attendants.

Other Plant

The Special Types General Order also makes special provision for other items of plant such as grass-cutting machines and hedge trimmers, track-laying vehicles, pedestrian-controlled road maintenance vehicles, vehicles used for experimental trials, straddle carriers, land tractors used for harvesting, mechanically propelled hay and straw balers, vehicles fitted with moveable platforms and engineering plant. Any person proposing to move such items on public roads is advised to check that the appropriate legal requirements are met.

HIGH LOADS

While there are no general legal height restrictions on vehicles or loads (except those specific instances mentioned on p 306–07), clearly vehicles must be loaded so that they can pass under bridges on the routes to be used.

In particular it should be noted that motorway bridges are built to give a clearance of 16 ft to 16 ft 6 ins and overhead power cables crossing roads are set at a minimum height of 19 ft (5.8 metres) where the voltage carried does not exceed 33,000 volts and at 6.0 metres where this voltage is exceeded (see also note on p 442 about tipping vehicle dangers in regard to contact with overhead power cables). When planning to move loads above 19 ft it is a legal requirement to make contact with the National Grid Company or the appropriate regional electricity distribution company (ie previously the regional electricity boards) beforehand. It is also recommended that contact should be made with British Telecom regarding the presence of its overhead lines on the routes to be used. Some police forces now report that they will not action notification of abnormal load movements until both National Power and British Telecom have been notified.

PROJECTING LOADS

Projecting loads may be carried on normal C&U regulation vehicles as well as on Special Types vehicles as described above. A projecting load is one that projects beyond the foremost or rearmost points or the side of the vehicle and, depending on the length or width of the projection, certain conditions apply when such loads are carried on normal vehicles including requirements for lighting and marking.

Side Projections

Loads more than 4.3 metres wide cannot be moved legally under the C&U regulations although they can be carried on vehicles which comply with those regulations. In such cases the provisions of the Special Types General Order apply (see above).

The normal width limit for loads is 2.9 metres overall or 305 mm on either side of the vehicle except in the case of loose agricultural produce and indivisible loads. Where an indivisible wide load extends 305 mm or more on one or both sides of the vehicle or exceeds 2.9 metres overall, the police must be given two clear days' notice in advance and end marker boards must be displayed front and rear (fitted within 50 mm of the edge of the load), and if it exceeds 3.5 metres, police notification, end marker boards and an attendant are required.

Forward Projections

A load projecting more than 2.0 metres beyond the foremost part of the vehicle (on C&U vehicles) must be indicated by an approved side and end marker board (Figure 22.1) and an attendant must be carried. On Special Types vehicles only, where the load projects more than 1.83 metres to the front, an end marker must be displayed and an attendant carried.

If a load projects more than 3.05 metres beyond the front (on both C&U and Special Types vehicles), the police must be given two days' notice of its movement, both side and end approved marker boards must be displayed and an attendant must be carried.

If a load projects more than 4.5 metres beyond the front of the vehicle the provisions mentioned above must be observed and additional side marker boards must be carried within 2.5 metres of the first set of side markers.

Rearward Projections

Where a load projects more than 1.0 metre beyond the rear of a C&U vehicle (1.07 metres on Special Types vehicles) it must clearly be marked (the form which this must take is not specified, but a piece of rag tied to the end is usually sufficient). If the rear projection is more than 2.0 metres on C&U vehicles (1.83 metres on Special Types vehicles) an end marker board must be displayed.

Where the rearward projection is more than 3.05 metres the police must be notified, an attendant carried and approved side and end marker boards displayed. An end marker is not required if the projecting load is fitted with a rear reflective marker. If the rearward projection exceeds 5.0 metres additional side marker boards must be carried within 3.5 metres of the first set of side markers.

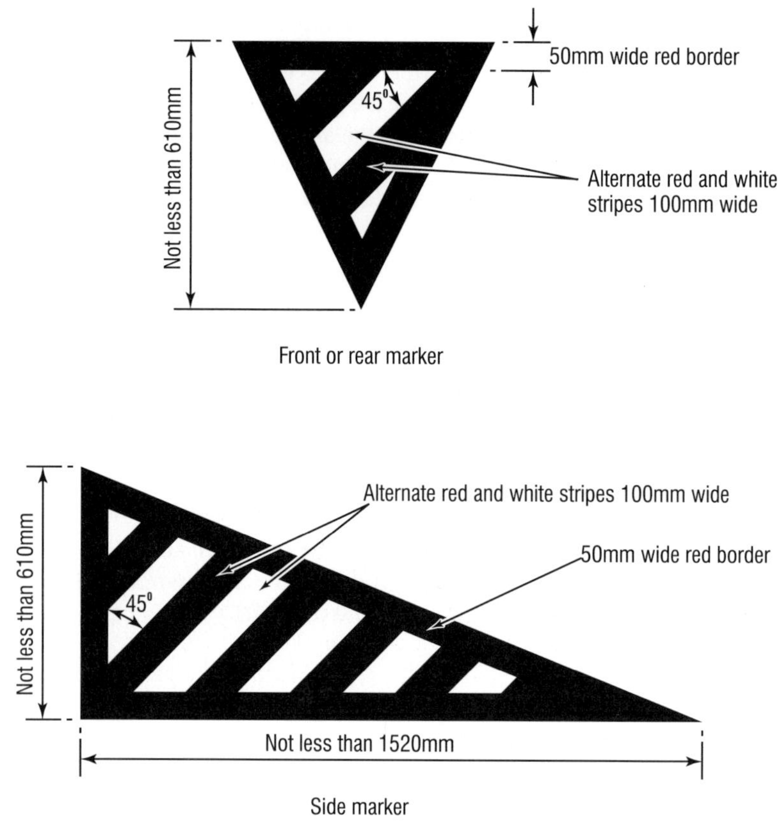

Figure 22.1 *The approved type marker boards which must be displayed when projecting loads are carried*

Marker Boards

Marker boards carried in accordance with the requirements described above must conform to the dimensions and colours shown in Figure 22.1 and they must be indirectly illuminated at night.

LIGHTING ON PROJECTING AND LONG LOADS

Rearward Projections

When carrying a load projecting more than 1.0 metre beyond the rear end of the vehicle an additional red rear position light must be carried within 1.0 metre of the end of the load or if the projecting load covers the rear lights and reflectors of the vehicle, additional lights and reflectors must be fixed to the load. This is usually best accomplished by having a complete lighting set and reflectors fitted to a board which can be fixed to the load.

Side Projections

When a load projects sideways more than 400 mm beyond the front and rear position lights of a vehicle, front position lights must be carried within 400 mm of the outer edges of the load and additional rear lights must also be carried within 400 mm of the outer edges of the load. White front and red rear reflectors must also be carried within 400 mm of the outer edges of the load.

Long Vehicles

A vehicle or combination of vehicles which, together with their load, are more than 18.3 metres long must, when on the road during the hours of darkness, carry side marker lights on each side positioned within 9.15 metres of the front of the vehicle or load and within 3.05 metres of the rear of the vehicle or load and other lights positioned between these at not more than 3.05 metre intervals. These requirements do not apply if approved illuminated marker boards are carried or if the combination is formed of a towing vehicle and a broken-down vehicle.

In the case of a combination of vehicles carrying a supported load (a load not resting on a vehicle except at each end) when the total length of the combination exceeds 12.2 metres but not 18.3 metres, side marker lights must be carried when on the road during the hours of darkness, positioned not more than 1.53 metres behind the rear of the drawing vehicle and if the load extends more than 9.15 metres beyond the drawing vehicle an additional side marker light must be carried not more than 1.53 metres behind the centre line of the load.

23 Loads – Dangerous, Explosive and Waste

According to the Health and Safety Executive (HSE), approximately 100 million tonnes of a wide range of dangerous goods are transported annually on British roads and railways. Legislation on the carriage of dangerous goods is both extensive and complex, and should be studied in detail by those responsible for such operations. Failure to comply with the strict provisions of the law can lead to harsh penalties. It could also result in risk to health and human life and cause serious environmental pollution.

THE LEGISLATION

UK national legislation on the carriage of dangerous goods by road is now aligned with the European Agreement Concerning the International Carriage of Dangerous Goods by Road (ADR) under Council Directive 94/55/EC, which harmonizes the law throughout the European Union. A so-called consolidated 'restructured' edition of ADR was published for 2005, taking effect from 1 January 2005 and being over 150 pages longer than the 2003 edition with many key changes, not least important new provisions concerning security measures to prevent the misuse of dangerous goods for terrorist actions (see p 516). The main changes are listed overleaf in summary form on a part-by-part basis.

THE ADR CONVENTION

ADR is published in two volumes and is consistent with the United Nation's Model Regulations on the Transport of Dangerous Goods (13th revised edition – 2003), the International Maritime Dangerous Goods Code, the International Civil Aviation's Technical Instructions for the Safe Carriage of Dangerous Goods by Air, and is fully harmonized with the Regulations concerning the International Carriage of Dangerous Goods by Rail (RID).

Volume 1 of this restructured edition contains the text of the actual ADR Agreement and the Protocol of Signature and Parts 1 to 3 of Annex A, which cover the general provisions and provisions concerning dangerous substances and articles as follows (ADR 2005 amendment items are shown in italics in each case below):

- Part 1 deals with the scope of the Agreement, definitions and units of measure, training of persons involved in the carriage of dangerous goods, safety obligations of the participants in the carriage of such goods, various derogations and transitional measures, the general requirements for the carriage of Class 7 goods (ie radioactive material), checks and other measures to ensure compliance with safety requirements, the rights of Contracting Parties to the Agreement to restrict or prohibit the carriage of certain dangerous goods for reasons other than safety, and it lists in an Appendix to Part 1 the Competent Authorities in each of the signatory states. In the UK this is the Department for Transport (DfT).
New provisions added dealing with security matters intended to prevent the misuse of dangerous goods for terrorist actions (see also p 516) and revised provisions for the renewal of Dangerous Goods Safety Adviser (DGSA) certificates.
- Part 2 deals with the general provisions concerning the classification of dangerous goods and lists the 9 main Classes and Sub-classes. It also describes the test methods to be used for the classification of dangerous goods and cross-references the procedures to the United Nation's *Manual of Tests and Criteria. Revised classifications for flammable aerosols, substances corrosive to metals, organo-metallic substances, infectious substances and pollutants to the aquatic environment.*
- Part 3 is taken up with the actual Dangerous Goods List in which all relevant substances are classified across 20 column headings under their UN Number and also by their names in alphabetical order.
Miscellaneous changes to packing instructions and tank instructions for dangerous goods already listed; 91 new UN entries to the Dangerous Goods List, many of them to differentiate chemicals carried as solids, liquids or solutions.

Volume 2 contains:

- Continuations of Annex A and Part 3 from volume 1;
- Part 4 dealing with packing and tank provisions;
revised provisions for testing the chemical compatibility of plastic packagings and IBCs, plus revised packing provisions for refrigerated gases, infectious substances (including medical diagnostic specimens) and used batteries, etc;

- Part 5 dealing with consignment procedures;
  new specifications for orange plates and revised transport documentation requirements for multimodal transport;
- Part 6 dealing with the construction and testing of packagings and IBCs;
  amended requirements for cryogenic receptacles and aerosol dispensers, for the testing and approval of gas receptacles, for packaging for radioactive material and miscellaneous changes to tanks as well as new requirements for bulk containers;
- Part 7 dealing with carriage by road, loading and handling of dangerous goods;
  revised provisions relating to the carriage of solid dangerous goods in bulk;
- Part 8 dealing with vehicle crews and equipment;
  new requirements relating to refresher training for the renewal of driver training certificates;
- Part 9 dealing with the construction and approval of dangerous goods carrying vehicles
  full revision of the procedures for type-approval of vehicles and new requirements for combustion (night) heaters.

Great Britain and Northern Ireland Legislation

A completely new, consolidated, set of regulations relating to the carriage of dangerous goods in Great Britain came into force from 10 May 2004. These completely replaced the previous conglomeration of more than a dozen sets of individual and separately published regulations which were revoked. The new regulations – The Carriage of Dangerous Goods and Use of Transportable Pressure Equipment Regulations 2004 (SI 2004/568) – implement in Great Britain three EC Directives; namely:

- Commission Directive adapting for the fourth time to technical progress Council Directive 94/55/EC on the approximation of the laws of the Member States with regard to the transport of dangerous goods by road (2003/28/EC);
- Commission Directive adapting for the fourth time to technical progress Council Directive 96/49/EC on the approximation of the laws of the Member States with regard to the transport of dangerous goods by rail 2003/29/EC; and
- 1999/36/EC.

The new regulations align GB domestic legislation directly with the ADR Agreement (concerning the transport of dangerous goods by road) and the RID Agreement (concerning the transport of dangerous goods by rail).

The new regulations bring about a number of significant changes, as follows:

- New limited quantity and load thresholds for transporting dangerous goods.
- Packaging requirements which are now based on ADR for road carriage and RID for rail carriage.
- New provisions bringing the carriage of diesel, gas oil and heating oil by road and/or rail within the regulations.
- New reporting requirements.
- Phasing out of the competent person periodic inspection regime for transportable pressure equipment by 1 July 2006.

- They take account of national derogations agreed with the European Commission in respect of certain transport operations involving local journeys or the movement of dangerous goods in small quantities.
- They also carry forward the mandatory Emergency Action Code system for placarding certain British registered vehicles on domestic journeys. Further details may be found on the HSE website at: www.hse.gov.uk/cdg.

Northern Ireland

Northern Ireland has its own independent regime of dangerous goods legislation which will be changed in due course, but for the moment (ie as at July 2004) the following pre-existing Northern Ireland regulations and statutory orders remain in force:

- Classification and Labelling of Explosives (Northern Ireland) Regulations 1991 (SR 1991/516).
- Radioactive Material (Road Transport) (Northern Ireland) Order 1992 (SI 1992/234).
- Packaging of Explosives for Carriage (Northern Ireland) Regulations 1993 (SR 1993/268).
- Carriage of Dangerous Goods (Classification, Packaging and Labelling) and Use of Transportable Pressure Receptacles Regulations (Northern Ireland) 1997 (SR 1997/247).
- Carriage of Dangerous Goods by Road Regulations (Northern Ireland) 1997 (SR 1997/248).
- Carriage of Dangerous Goods by Road (Driver Training) Regulations 1997 (SR 1997/249).
- Carriage of Dangerous Goods (Classification, Packaging and Labelling) and Use of Transportable Pressure Receptacles (Amendment) Regulations (Northern Ireland) 1997 (SR 1997/360).
- Carriage of Explosives by Road Regulations (Northern Ireland) 1997 (SR 1997/474).
- Carriage of Explosives by Road (Driver Training) Regulations (Northern Ireland) 1997 (SR 1997/475).
- Packaging, Labelling and Carriage of Radioactive Material by Rail Regulations (Northern Ireland) 1998 (SR 1998/132).
- Carriage of Dangerous Goods (Classification, Packaging and Labelling) and Use of Transportable Pressure Receptacles (Amendment) Regulations (Northern Ireland) 1998 (SR 1998/438).
- Carriage of Dangerous Goods (Amendment) Regulations (Northern Ireland) 1998 (SR 1998/448).

DfT Guide, *Working with ADR*

The Department for Transport (jointly with the HSE) published a guide in May 2004 to accompany the new regulations. This guide is a single document replacing the many separate, individual, sets of guidance notes and codes of practice produced to aid understanding of the previous complex legislative regime. The new guide: *Working with ADR – An introduction to the carriage of dangerous goods by road* is available free of charge from the following website: www.hse.gov.uk/cdg/guide.htm. A further set of notes highlighting and clarifying specific issues arising from the new carriage regulations and ADR are also available on the website as follows:

- Diesel
- Fire extinguishers
- Infectious substances
- Load thresholds Class 1
- Load thresholds Classes 2–9
- Limited quantities
- Carriage of black powder – this specifically relates to the carriage by private individuals of black powder and small arms ammunition, etc.

In Northern Ireland, it continues to be good practice to follow the pre-existing Codes of Practice and Guidelines issued by the HSE in Great Britain until such time as the new NI regulations and its related guidance has been published. At the time of writing, the following documents remain relevant in Northern Ireland only:

- Approved requirements and test methods for the classification and packaging of dangerous goods for carriage (HSE ref: L88) – Supplement 1999 (Supp 02) effective from 1 July 1999.
- Approved vehicle requirements 1999 (HSE ref: L89).
- Approved carriage list 1999 (Third edition) (HSE ref: L90). This list is also available on CD-ROM.
- Approved code of practice – Suitability of vehicles and containers and limits on quantities for the carriage of explosives (HSE ref: L91).
- Approved requirements for the construction of vehicles intended for the carriage of explosives by road 1999 (HSE ref: L92).
- Approved tank requirements 1996 (HSE ref: L93).

Additionally, the HSE publishes a range of Guidance Notes for road and rail which are also taken as good practice in Northern Ireland, as follows:

- *Are you involved in the carriage of dangerous goods by road or rail?* (Free HSE leaflet ref: IND(G)234L.)
- Guidance for consignors of dangerous goods by road and rail (classification, packaging and labelling and provision of information). (HSE ref: HS(G)160) – 1999 supplement SUPP 05.
- Guidance for road vehicle operators and others involved in the carriage of dangerous goods by road. (HSE ref: HS(G)161) – 1999 supplement SUPP 06.
- Guidance for operators, drivers and others involved in the carriage of explosives by road 1999 (HSE ref: HS(G)162).
- Guide to the packaging of explosives for carriage regulations 1991 (HSE ref: L.13).

DEFINITION AND CLASSIFICATION OF DANGEROUS GOODS

Definition

Dangerous goods are defined as being those substances and articles the carriage of which is prohibited by ADR (or RID in the case of carriage by rail), or authorized only under the conditions prescribed in these Agreements.

In general terms the following types of article and substance would be defined as dangerous goods:

- goods named in the ADR classification list as set out in ADR Part 2 (see also SI 2004/568, regulation 15);
- other goods which have one or more hazardous properties;
- explosives; or
- radioactive material.

Dangerous articles or substances falling within the scope of the legal requirements may range from:

- quite small quantities of individually packaged goods such as domestic cleaning products;
- tanker loads of acids and corrosive substances;
- explosives (both substances and explosive articles – eg detonators, etc);
- radioactive substances.

Identification

The purpose of, and legal necessity for, identifying dangerous goods is to determine their classification and identify their most hazardous properties and the dangers which may arise or be created when they are loaded, transported and unloaded. In many cases such goods have a number of hazards, not all of which may be immediately apparent.

It is for the manufacturer or consignor initially to determine the classification for dangerous goods intended to be consigned for carriage by deciding whether they are:

- explosives;
- radioactive material;
- named individually in the ADR Chapter 2;
- not named individually in the ADR Chapter 2.

Dangerous Goods Named Individually in the ADR Dangerous Goods List

Most of the commonly transported dangerous goods are named individually in ADR under their proper name, in alphabetical order and with their UN number. If the goods comprise a mixture or preparation of substances these too may be listed, as may dangerous goods that have a generic name. This 'dangerous goods list' is to be found in ADR Chapter 3.2.

Dangerous Goods not Named Individually in the ADR Dangerous Goods List

Where the dangerous goods are not 'named individually' the most appropriate generic entry (where the goods are of one chemical family) must be established or, alternatively, the relevant 'Not Otherwise Specified' (NOS) entry for those goods. This is done by first establishing the hazardous properties of the goods and then checking to see whether any alternative name can be found in the List that could be used to classify the goods.

Classification

It is important to note that the carriage of dangerous goods is prohibited unless their classification, packing group and any subsidiary hazards have been determined. An established classification system applies to the carriage of dangerous goods as set out in the United Nations so-called *Orange Book** and in ADR Part 2 as follows:

**The internationally recognized recommendations for the classification, package testing and marking requirements for the transport of dangerous goods. It is in two volumes comprising a set of model regulations (volume I) and tests and criteria (volume II) – currently in their tenth revised edition.*

The UN system classifies dangerous goods into:

- classes (of which there are nine);
- divisions;
- packing groups.

The nine classes cover the main types of dangerous goods (eg explosives, gases, flammable and oxidising substances, etc) – see full list below. Each of these classes is further sub-divided into a number of divisions which more specifically identify dangerous goods falling within each of the classes. A further degree of classification is achieved by assigning goods to a packing group according to the degree of danger they present.

Classes and Divisions

The following list shows the classes and divisions for dangerous goods:

- Class 1: Explosive substances and articles.
  - Division 1.1: Substances and articles which have a mass explosion hazard.
  - Division 1.2: Substances and articles which have a projection hazard, but not a mass explosion hazard.
  - Division 1.3: Substances and articles which have a fire hazard and either a minor blast hazard or a minor projection hazard or both, but not a mass explosion hazard.
  - Division 1.4: Substances and articles which present no significant hazard.
  - Division 1.5: Very insensitive substances that have a mass explosion hazard.
  - Division 1.6: Extremely insensitive articles that do not have a mass explosion hazard.

- Class 2: Gases.
  - Division 2.1: Flammable gases.
  - Division 2.2: Non-flammable, non-toxic gases.
  - Division 2.3: Toxic gases.
- Class 3: Flammable liquids.
- Class 4: Flammable solids, substances liable to spontaneous combustion and substances which, on contact with water, emit flammable gases.
  - Division 4.1: Flammable solids, self-reactive and related substances and desensitized explosives.
  - Division 4.2: Substances liable to spontaneous combustion.
  - Division 4.3: Substances which, in contact with water, emit flammable gases.
- Class 5: Oxidizing substances and organic peroxides.
  - Division 5.l: Oxidizing substance.
  - Division 5.2: Organic peroxides.
- Class 6: Toxic and infectious substances.
  - Division 6.1: Toxic substances.
  - Division 6.2: Infectious substances.
- Class 7: Radioactive material.
- Class 8: Corrosive substances.
- Class 9: Miscellaneous dangerous substances and articles.

UN Numbers and Proper Names

Besides the system of classification described above, dangerous goods are assigned a UN number comprising four digits which are used in conjunction with the letters 'UN' (eg UN 2426 which identifies ammonium nitrate liquid in a hot concentrated solution); and a proper shipping name according to their hazard classification and depending on their composition (eg as above).

UN Packing Groups

Dangerous substances are allocated UN packing groups as follows:

- Packing group I for goods presenting high danger;
- Packing group II for goods presenting medium danger;
- Packing group III for goods presenting low danger.

UN Table of Precedence

The UN *Orange Book* sets out a table of precedence to take account of situations where dangerous goods not specifically identified in the Dangerous Goods List (in ADR Chapter 3.2) have more than one risk, or where dangerous goods have multiple risks. In such cases the most stringent packing group (ie Packing group I) takes precedence over other packing groups.

The following tables indicate the relevant classifications for dangerous goods, along with their packing group number (ie I, II or III), their class number and any optional lettering to be shown on packages (Table 1) and the relevant danger signs for dangerous goods indicating the symbol to be used and the colouring of the lettering and background (Table 2).

Table 1

Classification	*Packing Group*	*Class No*	*Optional lettering*
Non-flammable, non-toxic gas	–	2.2	Compressed gas
Toxic gas	–	2.3	Toxic gas
Flammable gas	–	2.1	Flammable gas
Flammable liquid	I, II or III*	3	Flammable liquid
Flammable solid	I, II or III*	4.1	Flammable solid
Spontaneously combustible substance	I, II or III*	4.2	Spontaneously combustible
Substance which in contact with water emits flammable gas	I, II or III*	4.3	Dangerous when wet
Oxidizing substance	I, II or III*	5.1	Oxidizing agent
Organic peroxide	II	5.2	Organic peroxide
Toxic substance	I, II or III*	6.1	Toxic
Infectious substance	–	6.2	Infectious substance
Corrosive substance	I, II or III*	8	Corrosive
Miscellaneous dangerous goods	–	9	–

** NB: Depending on its relevant properties*

Table 2

Description of sign	*Symbol*	*Lettering*	*Background*
Non-flammable, non-toxic gas	Black gas cylinder	Black or white	Green
Toxic gas	Black skull & crossbones	Black	White
Flammable gas	Black flame	Black or white	Red
Flammable liquid	Black flame	Black or white	Red
Flammable solid	Black flame	Black	Vertical white/red stripes
Spontaneously combustible substance	Black flame	Black or white	White top/red bottom
Substance which in contact water emits flammable gas	Black flame	Black or white	Blue
Oxidizing substance	Black 'O' & flame	Black (5.1)	Yellow
Organic peroxide	Black 'O' & flame	Black (5.2)	Yellow
Toxic substance	Black skull & crossbones	Black	White
Infectious substances	Black symbol	Black	White
Corrosive substance	Black symbol	White	White top/black bottom
Miscellaneous			Vertical white/black stripes at top

RESPONSIBILITIES

Responsibility for Classification of Dangerous Goods

Responsibility for the classification of dangerous goods rests with:

- the appropriate competent authority (eg the United Nations) in publishing the classification system; and
- the manufacturer of the goods and the consignor (ie the sender) – these may be one and the same or separate entities who must:
  - identify their hazards;
  - determine the most appropriate description for them; and
  - establish their correct UN identification number (ie a four-digit number) by reference to the ADR Dangerous Goods List and their packing group.

Responsibility for Applying ADR

The competent authority applying ADR requirements in the UK rests with the Department for Transport (DfT) - (Dangerous Goods Branch).

Responsibility for Ensuring that Legal Requirements are Met

Responsibility for ensuring that relevant legal requirements are met (including the task of correctly identifying and classifying dangerous goods) rests with:

- the competent national authority, which in the UK is the DfT;
- consignors of dangerous goods; and
- those who load, unload and transport such goods, namely:
  - employers/carriers;
  - dangerous goods safety advisers (DGSAs);
  - drivers of dangerous goods carrying vehicles;
  - employees and other people.

Employer Responsibilities

Firms (described in the legislation as undertakings) or individuals involved in the loading, unloading or transport (ie carriage) of dangerous goods by road must comply with statutory requirements on the prevention of risks to the health and safety of persons, or to damage of property or the environment. They also need to appoint a qualified safety adviser (DGSA) (SI 2004 No 568, regulation [12]) and ADR 1.8.3). An 'undertaking' is any legal body (eg limited liability company), association or group (eg voluntary and charitable organizations), or any official body that transports, loads or unloads dangerous goods.

Employers are responsible for ensuring that:

- a qualified DGSA is appointed, where relevant; and
- dangerous goods vehicle drivers:
  - hold vocational training certificates; and
  - are trained in emergency action procedures.

Rules for Drivers

Drivers must not:

- carry unauthorized passengers on vehicles transporting dangerous goods (ADR 8.3.1);
- open any package containing dangerous goods unless authorised to do so (ADR 8.3.3);
- carry matches or lighters (or anything else capable of producing a flame or sparks) on vehicles carrying dangerous goods (except where the only goods on the vehicle are infectious substances), nor must the vehicle crew smoke during handling operations inside or in the vicinity of the vehicle (ADR 8.3.4 and 8.3.5).

Responsibility for Safety

Responsibility for compliance with the dangerous goods regulations, in particular in relation to safety issues, rests with the participants who are defined in ADR 1.4 as the consignor, the carrier and the consignee together with other personnel such as loaders, fillers, packers, etc (ADR 7.5.1 and SI 2004 No 568, regulation 23[5]). In particular these participants must take appropriate measures according to the nature and extent of foreseeable dangers, so as to avoid damage or injury and, if necessary, to minimize their effects. Where there is an immediate risk that public safety may be jeopardized, the participants must immediately notify the emergency services and make available to them the information they need to take appropriate action. Chapter 1.4 of ADR specifies the individual responsibilities of the three main participants as outlined below.

The Consignor

The consignor must ensure, when handing over dangerous goods for carriage, that consignments conform to the requirements of ADR, in particular by:

- ensuring that the goods are classified and authorized for carriage in accordance with ADR;
- giving the carrier relevant information, data and the required transport documents;
- using only packaging, intermediate bulk containers (IBCs) and tanks (ie tank vehicles, demountable tanks, battery vehicles, MEM, portable tanks and tank containers) approved for and suited to the carriage of the substances concerned and displaying the markings prescribed by ADR;
- complying with the requirements on the means of dispatch and on forwarding restrictions; and where he uses the services of other participants (packers, loaders, fillers, etc), he must still ensure that the consignment meets the requirements of ADR.

The Carrier

The carrier must ascertain that:

- the dangerous goods to be carried are authorized for carriage in accordance with ADR;
- the next due test date for tank vehicles, portable tanks and tank containers has not expired;
- vehicles are not overloaded;
- visually, vehicles and loads have no obvious defects such as leakages, cracks or missing equipment, etc;
- appropriate danger labels and prescribed markings have been affixed to the vehicle;
- prescribed documentation is being carried on board the vehicle; and
- the equipment prescribed in the driver's written instructions is on board the vehicle.

If, in carrying out the above checks, the carrier observes any infringement of the ADR requirements, he must not allow the consignment to be shipped until the matter has been rectified. Similarly, if during the journey, an infringement is detected which could jeopardize the safety of the operation; the consignment must be stopped as soon as possible bearing in mind the requirements of traffic safety, the safe immobilization of the consignment, and of public safety.

The Consignee

Consignees may not defer acceptance of a consignment of dangerous goods without compelling reasons. They must also, after unloading, verify that the requirements of ADR have been complied with.

Other Participants

Other participants may be involved with the carriage of dangerous goods consignments, such as employees described as tanker operators, loaders, packers, fillers, etc. Broadly, such employees have a duty to ensure that their actions and the loads they deal with comply in all relevant respects with the requirements of ADR.

Duty of Care

Employers have a general duty of care towards their employees under the Health and Safety at Work Act etc 1974, to ensure their health, safety and welfare. They also have a duty to ensure that their premises are safe and without risk to the health of those who are not employees but who use the premises as a place of work. Employees must take reasonable care for their own health while at work. Employers and self-employed persons must ensure that their work activities do not expose other persons (ie not in their employ) to risks to their health and safety through their work activities.

The relevant Northern Ireland legislation is the Health and Safety at Work (Northern Ireland) Order 1978.

EXEMPTIONS

A number of exemptions to ADR are specified in Chapter 1.1 (ie 1.1.3) and in SI 2004 No 568, regulation 7 and are set out in simplified form in *Working with ADR* (pages 5–7) as follows:

- the carriage of dangerous goods by private individuals where the goods are packaged for retail sale and are intended for their personal or domestic use or for their leisure or sporting activities;
- the carriage of machinery or equipment not specifically listed in the Dangerous Goods List and which contain dangerous goods in their internal or operational equipment;
- carriage undertaken by enterprises which is ancillary to their main activity;
- carriage by, or under the supervision of, the emergency services, in particular by breakdown vehicles carrying vehicles which have been involved in accidents or have broken down and contain dangerous goods;
- emergency transport intended to save human lives or protect the environment.

Special Provisions Exemptions

Certain special provisions of the ADR (ie Chapter 3.3) exempt, either partially or totally, the carriage of specific dangerous goods from the requirements of ADR, as identified in column 6 of the Dangerous Goods List (Chapter 3.2, Table A), against the appropriate entry for that article or substance.

Limited Quantities Exemptions

Exemptions also relate to dangerous goods packed in limited quantities where a 'limited quantity' code Q1–Q29 (see Table at ADR 3.4.6) appears in column 7 of the appropriate entry of the Dangerous Goods List, in which case that article or substance is exempted from the requirements of ADR, but only if it meets packaging and marking requirements; and the prescribed maximum quantities per inner packaging, and per package, are not exceeded.

When the code LQ0 appears in column 7 of the Dangerous Goods List (ADR 3.2), the substance or article is not exempted from any of the requirements of ADR (unless otherwise specified).

Transport Unit Exemptions

This class of exemptions is sometimes also referred to as being a 'limited quantities exemption', but it should not be confused with the limited quantities exemption described above. In this case, the exemption applies where:

- a specified maximum total quantity per transport unit, depending on its transport category (ADR 1.1.3.6) is not exceeded; or
- where the dangerous goods carried in the transport unit are in more than one transport category and the total quantity carried does not exceed the value calculated in accordance with ADR 1.1.3.6.4.

Transport categories and maximum quantities per transport unit

Transport category	*Max total quantity per transport unit**
0	0
1	20
1A	50
2	333
2A	500
3	1,000
4	unlimited

**NB: For articles, gross mass in kilograms (for articles of Class 1, net mass in kilograms of the explosive substance); for solids, liquefied gases, refrigerated liquefied gases and dissolved gases, net mass in kilograms; for liquids and compressed gases, nominal capacity of receptacles (see definition in ADR 1.2.1) in litres.*

If either of the two bullet point conditions (above) applies, the dangerous goods may be carried in packages in a single transport unit and the following requirements of ADR will not apply.

Other Exemptions

ADR	*Covering*
5.3	Placarding and marking of containers, etc
5.4.3	Documentation – instructions in writing
7.2 except for V5, V7 and V8 of 7.2.4	Carriage in packages
CV1 of 7.5.11	Prohibitions on loading and unloading in public places
Part 8, except for 8.1.4.2–8.1.4.5, 8.2.3, 8.3.4, 8.4 and S1(3), S1(6), S2(1), S4, S14–S21 of 8.5	Vehicle crews, equipment operation and documentation
Part 9	Construction and approval of vehicles

ADR also includes exemptions relating to:

- the carriage of gases;
- the carriage of liquid fuels;
- empty uncleaned packaging;
- carriage in a transport chain including maritime or air carriage;
- the use of portable tanks approved for maritime transport; and
- carriage other than by road.

CARRIAGE BY ROAD

The new dangerous goods regulations, namely The Carriage of Dangerous Goods and Use of Transportable Pressure Equipment Regulations 2004 **(SI 2004 No 568), generally referred to as** the Carriage Regulations, deal with the carriage by road and rail of dangerous goods (other than explosives and radioactive material) in packages or in bulk in any container, tank or vehicle. However, unlike former UK domestic dangerous goods legislation, these regulations relate specifically to ADR and cross-reference to the provisions set out in the Agreement.

Carriage in Bulk

Specific requirements relating to the carriage of dangerous goods in bulk are set out in SI 2004 No 568, regulation 23(2) and ADR 7.3. In particular, ADR states that goods may not be carried in bulk in either vehicles or containers unless a special provision, identified by the code 'VV', explicitly authorizing this mode of carriage is indicated in the ADR Dangerous Goods List (see ADR Chapter 3.2 column 17). The particular provisions that apply to carriage in bulk are set out in ADR 7.3 items VV1 to VV14.

Loading and Unloading

Requirements concerning the load and the unloading and handling of dangerous goods are specified in SI 2004 No 568, regulation 23(1) and detailed in ADR 7.5. Among the general provisions that apply are requirements that:

- on arrival at a loading or unloading site, the driver must ensure that he and his vehicle comply with the relevant legal provisions especially those concerning safety, cleanliness and satisfactory operation of the vehicle equipment used in loading and unloading;
- loading must not be commenced if examination of the documents and a visual inspection of the vehicle and its equipment show that the vehicle or driver do not comply with legal requirements;
- unloading must not be commenced if the above mentioned inspections reveal deficiencies that might affect the safety of the unloading operation;
- in the case of loading mixed packages bearing different danger labels into the same vehicle, this is permissible only in accordance with the tables shown at ADR 7.5.2.1 and 7.5.2.2 and the following provisions in 7.5.4 to 7.5.5;
- dangerous goods loads must be properly stowed on the vehicle or in the container and adequately secured to prevent them from being displaced;
- drivers or other crew members do not open any package containing dangerous goods;
- vehicles and containers must be properly cleaned before reloading;
- there shall be no smoking during handling operations in or in the vicinity of vehicles and containers;
- to avoid electrostatic charges a good earth connection between the vehicle and the ground must be made;
- the additional provisions shown in column 18 of the Dangerous Goods List (ADR 3.2) should also be observed.

Supervision and Parking of Vehicles

Vehicles carrying dangerous goods, when parked, must be supervised. Alternatively they may be parked in an isolated position, unsupervised, in a secure depot or factory premises. If such facilities are not available the vehicle, if properly secured, may be parked in a:

- vehicle park supervised by an attendant who knows both the nature of the load and the whereabouts of the driver;
- public or private vehicle park where it is not likely to suffer damage from other vehicles; or
- suitable open space separated from the public highway and from dwellings, where the public does not normally pass or assemble.

Packaging and labelling

Packaging

ADR specifies in Chapter 4.1 that:

- dangerous goods must be packed in good quality packaging, strong enough to withstand the shocks and loadings normally encountered in transit and during trans-shipment between vehicles and between vehicles and warehouses, as well as during any removal from a pallet or overpack for subsequent manual or mechanical handling;

- packaging must be constructed and closed to prevent any loss of contents when prepared for transport (ie caused by vibration, or by changes in temperature, humidity or pressure);
- no dangerous residue must be allowed to adhere to the outside of packaging during carriage.

The above provisions apply, as appropriate, to new, reused, reconditioned or remanufactured packaging.

Parts of packaging that are in direct contact with dangerous goods:

- must not be affected or significantly weakened by those dangerous goods;
- must not cause any dangerous effect (ie by reacting with the dangerous goods); and
- where necessary, must be provided with a suitable inner coating or treatment.

Unless provided elsewhere in ADR, packaging must conform to a design type successfully tested in accordance with ADR requirements.

When filling packaging with liquids, sufficient ullage (outage) must be left to ensure that neither leakage nor permanent distortion of the packaging occurs as a result of an expansion of the liquid caused by temperatures likely to occur during transport.

Labelling

Dangerous goods packages must be marked with:

- their designation;
- their UN number (preceded by the letters 'UN');
- the relevant danger sign; and
- any relevant subsidiary hazard signs. (ADR 5.2.1 and 5.2.2.)

Markings must be clear (ie indelibly marked on the package or printed on a label securely fixed to the package), in English, or the official language of another EU Member State if being supplied to that State, so they can be read easily, and so they stand out from the background to enable them to be readily noticeable*)*.

Northern Ireland

The relevant Northern Ireland legislation is the Carriage of Dangerous Goods (Classification, Packaging and Labelling) and Use of Transportable Pressure Receptacles (Amendment) Regulations (Northern Ireland) 1997 (SR 1997 No 247) and the Carriage of Dangerous Goods (Amendment) Regulations (Northern Ireland) 1998 (SR 1998 No 448).

Vehicles, Tanks and Pressure Receptacles

Vehicles and Tanks

In general terms, all containers, tanks and dangerous goods vehicles must be suitable for the purpose, in particular for the hazardous properties of the goods and journey to be undertaken, and must be adequately maintained to comply with the constructional and approval requirements of ADR Part 9 (9.1 to 9.7) and as specified in SI 2004 No 568, regulations 23 and 25.

ADR Certification of Vehicles and Tanks

Road tanker vehicles and demountable tanks (ie tanks contained within a strong metal frame and capable of being lifted on and off road vehicles and rail wagons) used for the carriage of dangerous goods by road under ADR must be examined annually and certified by the relevant authority in their country of registration (ie the DfT in the case of UK registered vehicles) under provisions contained in ADR 9.1.2.

In the UK, road tankers are examined by the Vehicle Operator and Services Agency (VOSA) at DfT Goods Vehicle Testing Stations, while examination of demountable tanks is carried out by a number of commercial testing organizations. On satisfactory completion of the examination an ADR certificate is issued (see ADR 9.1.2.1.5) and this must be carried on the vehicle when undertaking a dangerous goods journey under ADR.

Pressure Receptacles

Designers, manufacturers, importers or suppliers of transportable pressure receptacles* (ie gas receptacles, tank containers and other forms of gas storage vessel) must ensure that they are safe and suitable for their purpose and comply fully with ADR 6.2.

**NB: Receptacles are defined in the regulations as a vessel or the innermost layer of packaging which is in contact with any dangerous goods therein, and includes any closure or fastener.*

TRANSPORT INFORMATION AND DOCUMENTATION

Information to be Provided

Consignors of dangerous goods must provide the transport operator (ie carrier) with a transport document showing:

- the designation, classification code and UN number for the goods;
- any additional information needed to determine their transport category and their control and emergency temperatures;
- for packaged goods, the number and weight or volume of individual packages, or the total mass or volume in each transport category;
- for bulk loads, the weight or volume in each tank or container and the number of tanks or containers;
- the name and address of both consignor and consignee;
- any other information which the operator must give to the driver;
- a 'consignor's declaration' that the goods may be carried as presented, that they, their packaging and any container or tank in which they are contained is fit for carriage and is properly labelled.

It is an offence to provide false or misleading information and where one transport operator sub-contracts a dangerous goods consignment to another operator he must, by law, pass on the information provided by the consignor.

Documentation to be Carried during Carriage

Drivers of vehicles carrying dangerous goods must be provided with the following 'transport documentation', in writing:

- the information provided by the consignor (see above);
- details of the weight or volume of the load;
- the relevant hazard identification number or the emergency action code (where appropriate);
- emergency information comprising:
  - the dangers inherent in the goods and safety measures;
  - what to do and the treatment to be given should any person come into contact with the goods;
  - what to do in the event of fire and what fire-fighting appliances or equipment must not be used;
  - what to do in case of breakage or deterioration of packaging or of the goods, particularly where this results in a spillage on to the road;
  - what to do to avoid or minimize damage in the event of spillage of goods likely to pollute water supplies;
- any relevant additional information about the particular type of dangerous goods being carried.

It is an offence to provide false or misleading information to drivers about the particular type of dangerous goods being carried.

Drivers must keep the transport documentation readily available during dangerous goods journeys and produce it on request by the police or a VOSA goods vehicle examiner. Where a dangerous goods carrying trailer is detached from the towing vehicle the transport documentation (or an authenticated copy) must be given to the owner or manager of the premises where it is parked, or attached to the trailer in a readily visible position.

Documentation relating to dangerous goods no longer carried on a vehicle must be either removed completely, or placed in a securely closed container clearly marked to show that it does not relate to dangerous goods still on the vehicle.

Operators must keep a record of journey transport documentation for at least three months (SI 2004 No 568, regulation 54).

Transport Emergency Cards (Tremcards)

Transport Emergency Cards, commonly known as Tremcards, are a widely used system (devised by the European Council of Chemical Manufacturers' Federation – CEFIC) by which the emergency and other relevant information that a driver must carry, as detailed above, is provided in a convenient, ready-made format (covering single and multi-substance loads – for use where combinations of substances are loaded on the same vehicle). Tremcards are available in paper format for single journey use or laminated for repeated use with the relevant text in a single language, or four or eight languages from CEFIC's range of 27 language versions. Specifically, they:

- detail information on the substance/s to which they relate;
- specify any protective devices needed when such substances are handled;

- provide information on the action to be taken in the event of an emergency; and
- provide any other additional information provided by the manufacturer.

INFORMATION TO BE DISPLAYED ON CONTAINERS, TANKS AND VEHICLES

Containers, tanks and vehicles used for carrying dangerous goods must display information as described below. All panels and danger signs must be kept clean and free from obstruction.

It is an offence to display information when the container, tank or vehicle is not carrying dangerous goods, and to cause or permit the display of any information likely to confuse the emergency services.

Signs and panels relating to dangerous goods no longer being carried must be covered or removed. Where an orange-coloured panel is covered, the covering material must remain effective after 15 minutes' engulfment in fire. Danger signs, hazard warning panels, orange-coloured panels or subsidiary hazard signs need not be covered or removed if the mass or volume of dangerous goods in packages falls below the following limits:

Transport category	*Total mass/volume (kg/litres)*
0	0
1	20
2	200
3	500
4	unlimited

It is an offence to remove panels or signs from a container, tank or vehicle carrying dangerous goods (except for updating the information) and to falsify information on any panel or sign.

Danger Signs and Panels

A reflectorized orange-coloured, black-bordered panel (plain with no letters or figures) must be displayed at the front of vehicles carrying dangerous goods. A similar panel must be attached to the rear of vehicles carrying dangerous goods in packages.

Single Load Labelling

Where a *single load* of dangerous goods is carried in a container, tank or vehicle an orange-coloured panel showing the appropriate UN number and emergency action code (see Figure 23.1) must be displayed:

- one at the rear of the vehicle;
- one on each side of the vehicle, the container or the tank;
- and, in the case of a tank, one on each side of the frame of the tank, or on the vehicle positioned immediately below the tank.

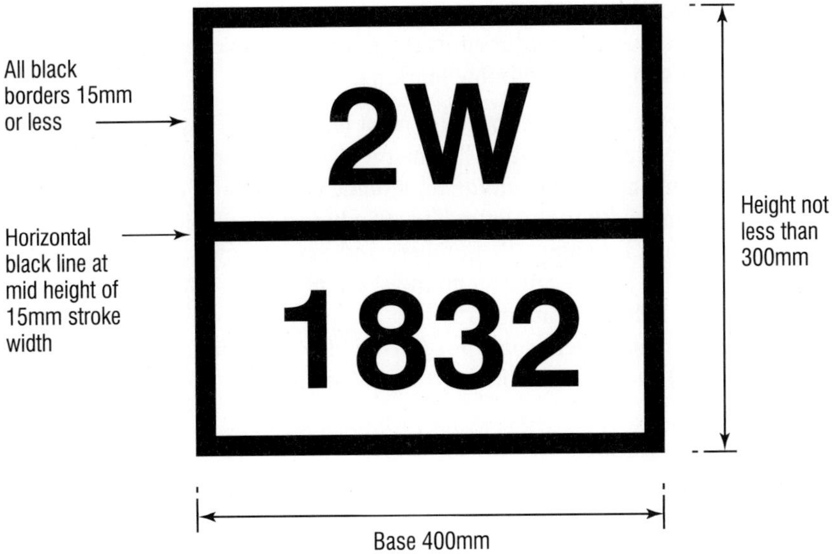

Figure 23.1 *Orange panel showing the emergency action code and the UN number identifying the dangerous substance being carried*

Multi-load Labelling

Where a vehicle is carrying a *multi-load* in tanks, or in bulk in separate compartments of the vehicle, or in separate containers, an orange-coloured panel showing the appropriate emergency action code only (see below) must be displayed at the rear of the vehicle.

Additionally, orange-coloured panels as follows are required on both sides of each tank (or, if it has multiple compartments, on each compartment), on each compartment of the vehicle or on each container on the vehicle:

- At least one on each side showing the appropriate UN number and emergency action code (Figure 23.1); and
- the remainder showing only the appropriate UN number.

Alternatively, for dangerous goods carried in a tank, the panels may be displayed on both sides of the frame of each tank, or on the vehicle positioned immediately below the tank or tank compartment concerned.

Where diesel fuel or gas oil or heating oil (UN 1202), petrol, motor spirit or gasoline (UN 1203), or kerosene (UN 1223) is carried in a multi-compartment road tanker it may be labelled as a single load only, showing the UN number and emergency action code for the most hazardous of the products carried.

Detail of Panels

The orange-coloured panels must be either:

- a rigid plate fitted as near vertical as possible; or
- in the case of a vehicle carrying dangerous goods in a tank container or in bulk in a container,

- orange-coloured self-adhesive sheets; or
- orange-coloured paint (or equivalent), provided the material is weather-resistant and ensures durable marking.

UN numbers and emergency action codes must be shown in black, at least 100 mm high and 15 mm wide – but where the emergency action code is white on a black background, it must appear as orange on a black rectangle at least 10 mm greater than the height and width of the letter. Except where panels comprise self-adhesive sheets or are applied by paint, UN numbers and emergency action codes must be indelible and remain legible after 15 minutes' engulfment in fire (not applicable to tanks constructed before 1 January 1999).

Where there is insufficient space for full-sized panels, these may be reduced to 300 mm wide by 120 mm high with a 10 mm black border.

Emergency Action Codes

The emergency action codes are as follows:

1. By numbers 1 to 4 indicating the equipment suitable for fire fighting and for dispersing spillages (ie 1 = water jets, 2 = water fog, 3 = foam, 4 = dry agent).
2. By letters indicating the appropriate precautions to take as follows:

Letter	*Danger of violent reaction*	*Protective clothing and breathing apparatus*	*Measures to be taken*
P	Yes	Full protective clothing	Dilute
R	No	Full protective clothing	Dilute
S	Yes	Breathing apparatus	Dilute
S*	Yes	Breathing apparatus for fire	Dilute
T	No	Breathing apparatus	Dilute
T*	No	Breathing apparatus for fire	Dilute
W	Yes	Full protective clothing	Contain
X	No	Full protective clothing	Contain
Y	Yes	Breathing apparatus	Contain
Y*	Yes	Breathing apparatus for fire	Contain
Z	No	Breathing apparatus	Contain
Z*	No	Breathing apparatus for fire	Contain

**These symbols are shown as orange (or can be white) letters reversed out of a black background.*

Where a letter 'E' is shown at the end of an emergency action code this means that consideration should be given to evacuating people from the neighbourhood of an incident.

Display of Telephone Number

A contact telephone number, comprising black digits at least 30 mm high on an orange background, must be shown on vehicles carrying single or multi-loads of dangerous goods in tanks, positioned:

- at the rear of the vehicle;
- on both sides of the tank (or each tank if more than one), the frame of each tank, or the vehicle; and
- located in the immediate vicinity of the orange-coloured panels.

Instead of a telephone number, the words 'consult local depot' or 'contact local depot' may be substituted, but only if:

- the name of the operator is clearly marked on the tank or the vehicle; and
- the fire chief for every area in which the vehicle will operate has been notified in writing of the address and telephone number of that local depot, and has confirmed in writing that he is satisfied with the arrangements.

Display of Danger Signs and Subsidiary Hazard Signs

Where a vehicle is carrying:

- packaged dangerous goods in a container:
  - any danger sign or subsidiary hazard sign required on the packages must also be displayed on at least one side of the container;
- dangerous goods in a tank container or in bulk in a container:
  - any danger sign or subsidiary hazard sign required on the packages containing such goods must be displayed on each side of the tank container or container, and where such signs are not visible from outside the carrying vehicle, the same signs must also be shown on each side of and at the rear of the vehicle;
- dangerous goods in a tank, other than a tank container, or in bulk in a vehicle, but not in bulk in a container on a vehicle:
  - any danger sign or subsidiary hazard sign required on the packages containing such goods must be displayed on each side of and at the rear of the vehicle.

Danger signs for a particular classification, or subsidiary hazard signs, need not be shown more than once on the sides or rear of any container, tank or vehicle.

Danger signs and subsidiary hazard signs must have sides at least 250 mm long; have a line the same colour as the symbol 12.5 mm inside the edge and running parallel to it; and be displayed adjacent to one another and in the same horizontal plane.

Display of Hazard Warning Panels

Despite the requirements described above for the display of orange-coloured panels the regulations permit, wherever such a panel is required on the sides or rear of a container, tank or vehicle, the alternative use of existing-type combined hazard warning panels (see Figure 23.2).

These are mainly orange with black borders and lettering – except for the white background where a reduced size (ie 200 mm sides) danger sign is located and any subsidiary hazard sign must be the same size and displayed adjacent to it and in the same horizontal plane.

All danger panels and signs required on the front or rear of a vehicle must be positioned at right angles across its width, and those on the sides of a container, tank or vehicle at right angles along its length. All signs must be clearly visible.

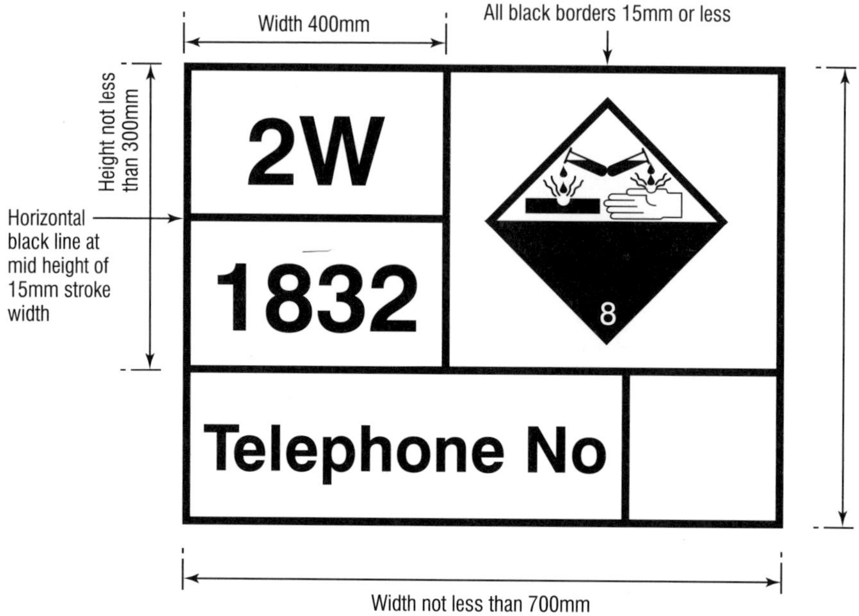

Figure 23.2 *Existing-type combined hazard warning panel which may be used on the sides and rear of a bulk chemical tanker vehicle or container*

EMERGENCY PROCEDURES

Vehicles carrying dangerous goods must be equipped so that the driver can take emergency measures, as follows:

- At least one portable fire extinguisher with a minimum capacity of 2 kg of dry powder (or other extinguishant with an equivalent test fire rating of at least 5A and 34B – defined in British Standard BSEN 3–1: 1996), suitable for fighting a fire in the engine (unless the vehicle has an automatic extinguisher system) or cab, and not likely to aggravate a fire in the load.

  NB: Such an extinguisher is not required on a detached trailer.

- At least one portable fire extinguisher with a minimum capacity of 6 kg of dry powder (or other extinguishant with an equivalent test fire rating of at least 21A and 183B – defined as above)*, suitable for fighting a tyre or brake fire, or a fire in the load, and not likely to aggravate a fire in the engine or cab. Fire extinguishers with a minimum total capacity of 12 kg are required.

  **NB: Where the vehicle has a gross weight under 3.5 tonnes, a 4 kg dry powder extinguisher (or another suitable extinguishant with a test fire rating of at least 5A and 34B) will suffice.*

Portable fire extinguishers must not be liable to release toxic gases into the driver's cab, or under the heat of a fire. They must be marked in compliance with a recognized standard; fitted with a seal verifying they have not been used; and, where manufactured after 31 December 1996, be inscribed with the date for their next inspection – it is an offence to carry an extinguisher with an overdue inspection date.

A fire extinguisher is not needed where only infectious substances are carried.

Where toxic gases are carried, vehicles must carry respiratory equipment to enable the crew to escape safely.

Accidents and Emergencies

In accident or emergency situations drivers must comply with the emergency information given to them and apply the emergency procedures for which they are trained and which are stated on the emergency action information given to them (eg on the Tremcard), including:

- notifying the relevant emergency services by the quickest practical means;
- taking steps to prevent chemical contamination of the environment (eg water courses);
- taking steps to prevent the risk of fire;
- keeping the public away from the scene;
- applying first aid where necessary.

Accident Investigation and Reports

All accidents and dangerous goods incidents must be reported and investigated to determine how and why they were caused and how any such event may be prevented in the future – possibly by changing procedures and the issue of fresh safety instructions.

Dangerous Goods Safety Advisors (DGSAs) have a statutory duty to prepare reports on serious accidents, incidents or legal infringements occurring during the loading, unloading and transport of dangerous goods and to implement appropriate measures to avoid recurrence of such events.

Reporting to the Health and Safety Executive

Accidents and serious incidents involving dangerous goods must be reported to the HSE as required under the Reporting of Injuries, Diseases and Dangerous Occurrences Regulations 1995 (**SI 1995/3163** – commonly known as RIDDOR). The report must be made to the HSE both immediately by telephone and in writing within 10 days (using HSE Form 2508) concerning any such occurrence causing:

- death or major injury to an employee;
- any injury occasioning an employee to be away from work for more than three days;
- any work-related disease.

Any dangerous occurrence (even if it does not result in personal injury to any person) must also be reported including:

- splitting or collapse of a tank or tank container;
- failure of pipework or valves resulting in spillage;
- overturning of a dangerous goods carrying vehicle;

- a dangerous goods carrying vehicle catching fire; or
- a fire in the loading or unloading premises.

In Northern Ireland the relevant legislation is the Reporting of Injuries Diseases and Dangerous Occurrences Regulations (Northern Ireland) 1997 (SR No. 455).

Road Traffic Accidents

The procedures to be followed in the event of a road traffic accident, apart from the special considerations relating to the carriage of dangerous goods as set out above, are those described on pages 255–56 of this *Handbook*.

NEW SECURITY PROVISIONS

From 22 July 2005, in response to new measures included in the revised ADR 2005 (referred to on p 493), the government has introduced new security provisions by means of the *Carriage of Dangerous Goods and Use of Transportable Pressure Equipment (Amendment) Regulations 2005* (SI 2005 No. 1732) relating to the carriage of all dangerous goods except nuclear materials. The new requirements fall into four groups as shown below; namely, general provisions, provisions relating to High Consequence Dangerous Goods, training requirements and the need to implement security plans.

General Provisions

These require that:

- all persons engaged in the carriage of dangerous goods must consider the security requirements for the carriage of dangerous goods commensurate with their responsibilities;
- dangerous goods may only be offered to carriers that have been appropriately identified;
- areas within temporary storage terminals, temporary storage sites, vehicle depots, berthing areas and marshalling yards used for temporary storage during carriage of dangerous goods must be properly secured, well lit and, where possible and appropriate, not accessible to the general public;
- each crew member of a vehicle carrying dangerous goods must carry with them means of photographic identification;
- safety inspections must cover appropriate security measures.

High Consequence Dangerous Goods

Dangerous goods that have the potential for misuse in a terrorist incident and which may, as a result, produce serious consequences such as mass casualties or mass destruction are now defined as High Consequence Dangerous Goods (see list below).

Training Requirements

Additional provisions relating to training require that up-to-date registers of all valid training certificates must be maintained by the competent authorities and that relevant training, and refresher training, must include elements of security awareness, in particular addressing the following issues:

- the nature of security risks;
- how to recognize security risks;
- how to address and reduce such risks;
- what actions to take in the event of a breach of security.

Security Planning

Another key new provision is that all carriers (eg road hauliers), consignors and other participants engaged in the carriage of High Consequence Dangerous Goods (as listed below) must formulate, implement and comply with a security plan that covers at least all the requirements specified below:

- The plan must specifically allocate responsibility for security to competent and qualified persons with appropriate authority to carry out their responsibilities.
- Records must be kept of the dangerous goods or types of dangerous goods involved in the carriage.
- A review must be undertaken of current operations and assessment of security risks, including at any stops necessary to the transport operation, the keeping of dangerous goods in the vehicle, tank or container before, during and after the journey and the temporary storage of dangerous goods during the course of intermodal transfer or transhipment between units.
- A clear statement must be made of the measures that are to be taken to reduce security risks, commensurate with the responsibilities and duties of the participant, including:
  - training matters;
  - security policies;
  - operating practices;
  - equipment and resource use to reduce security risks.
- Effective and up-to-date procedures must be established for reporting and dealing with security threats, breaches of security or security incidents.
- Procedures must be put in place for the evaluation and testing of security plans and actions.
- Measures must be taken to ensure the physical security of transport information contained in the security plan.
- Measures must be taken to ensure that the distribution of information relating to the transport operation contained in the security plan is limited to those who need to have it.
- A process needs to be established to cooperate in the exchange of security threat information, to apply appropriate security measures and to respond to security incidents.
- Steps must be taken to prevent the theft of vehicles carrying High Consequence Dangerous Goods or their cargo by means of suitable security devices, equipment or arrangements. These must be kept operational and effective at all times, but must not jeopardize any emergency response.

- Transport telemetry or other tracking methods or devices should be used to monitor the movement of High Consequence Dangerous Goods, when it is appropriate to do so and when such systems are already fitted.

High Consequence Dangerous Goods are those listed in the table below and carried in quantities greater than shown.

Class	Division	Substance or article	Quantity		
			Tank (ltrs)	Bulk (kg)	Packages (kg)
1	1.1	Explosives	a	a	0
	1.2	Explosives	a	a	0
	1.3	Compatibility group C explosives	a	a	0
	1.5	Explosives	0	a	0
2		Flammable gases (classification codes including only the letter F)	3000	a	b
		Toxic gases (classification codes containing the letters T, TF, TC, TO, TFC or TOC) excluding aerosols	0	a	0
3		Flammable liquids of packing groups I and II	3000	a	b
		Desensitized explosives	a	a	0
4	4.1	Desensitized explosives	a	a	0
	4.2	Packing group I substances	3000	a	b
	4.3	Packing group II substances	3000	a	b
5	5.1	Oxidizing liquids of packing group I	3000	a	b
		Percholates, ammonium nitrate and ammonium nitrate fertilizers	3000	3000	b
6	6.1	Toxic substances of packing group I	0	a	0
	6.2	Infectious substances of Cateogry A	a	a	0
7		Radioactive material	3000 A1 (special form) or 3000 A2, as applicable, in Type B or Type C packages		
8		Corrosive substances of packing group I	3000	a	b

ENFORCEMENT

In the interests of public safety, the police and authorized examiners of the Vehicle and Operator Services Agency (VOSA) are given wide-ranging powers to ensure compliance with the law on dangerous goods carriage. For example, they can demand production of the statutory information which drivers must carry when on a dangerous goods journey as well as production of the driver's own Vocational Training or ADR Certificate which he must carry at all times during such journeys.

VOSA is empowered to enforce certain aspects of dangerous goods law solely relating solely to vehicles (the HSE retains its responsibility for premises-based inspections). VOSA inspectors are authorized to serve Prohibition Notices and prosecute offenders for breaches of the law, but they *will* not serve Improvement Notices as these are not normally appropriate for on-the-road work.

NB: VOSA is an executive agency of the DfT responsible for operating the goods vehicle plating and testing scheme at test stations and enforcing goods vehicle safety on the road.

Offences and Penalties

Dangerous goods legislation is complex and extensive with a large number of potential offences which may be committed, either singly or in conjunction with other offences. Offenders can expect to be prosecuted and dealt with severely by the Courts on conviction.

On summary conviction in the Magistrate's Court (Sheriffs' Court in Scotland) for an offence under the HSWA 1974, currently a fine of up to £5,000 may be imposed, while in more serious cases indictment by the Crown Court for an offence may lead to an unlimited fine and/or a two-year prison sentence.

Obstruction of an inspector who is fulfilling his statutory duties is a serious offence carrying a fine of up to £5,000 on conviction while breach of any notice served by an inspector can result in a fine up to £20,000 on conviction or in either case an unlimited fine on indictment.

Defence

A person charged under these regulations can, in his own defence, prove that the offence was due to the act or default of another person (other than his own employees), and that he had tried to avoid committing the offence, but he must give the prosecutor, in writing, at least seven days before the court hearing and to the best of his knowledge, information identifying or assisting in the identification of that other person.

CARRIAGE OF EXPLOSIVES

Regulations which mainly came into force on 3 July 1989 control the movement of explosives by road. The Road Traffic (Carriage of Explosives) Regulations 1989 were made under the Health and Safety at Work etc Act 1974 and replace parts of the requirements of the Explosives Act 1875 as do the Packaging of Explosives for Carriage Regulations 1991 which require that explosives in packages being transported should be tested, certified and labelled in a specified manner.

For the purposes of the regulations the term 'explosives' means explosive articles or substances which have been classified under the Classification and Labelling of Explosives Regulations 1983 as being in Class 1 or those which are unclassified. An 'explosive article' is an article which contains one or more explosive substances and an 'explosive substance' is a solid or liquid substance or a mixture of solid or liquid substances, or both:

> which is capable by chemical reaction in itself of producing gas at such a temperature and pressure and at such speed as could cause damage to surroundings or which is designed to produce an effect by heat, light, sound, gas or smoke or a combination of these as a result of non-detonative self-sustaining exothermic chemical reactions.

The term 'Compatibility Group' also has a meaning assigned to it by the 1983 regulations referred to above. The term 'carriage' means from the commencement of loading explosives into a vehicle or trailer until they have all been unloaded, whether or not the vehicle is on a road at the time. However, the term carriage is not applied where explosives are loaded on an unattached trailer or semi-trailer; carriage begins and ends when the trailer is attached to and later is detached from the towing vehicle or when the explosives have been unloaded, whichever is the sooner.

The regulations prohibit the carriage of any explosive in Compatibility Group K in a vehicle and any unclassified explosive except where it is being carried in connection with an application for their classification and in accordance with conditions approved in writing by the Health and Safety Executive (or, in the case of military explosives, by the Secretary of State for Defence).

Explosives must not be carried in any vehicle being used to carry passengers for hire or reward except that a passenger in such a vehicle may carry explosives under the following conditions:

- the substance carried is an explosive listed in Schedule 1 to the regulations (eg certain cartridges, fireworks, distress-type and other signals, certain flares, fuses, igniters, primers and other pyrotechnic articles), gunpowder, smokeless powder or any mixture of them;
- the total quantity of such explosive carried does not exceed 2 kg;
- the explosives are kept by that person and are kept properly packed;
- all reasonable precautions are taken by the person to prevent accidents arising from the explosives.

The person carrying the explosives on the vehicle remains totally responsible for them and no responsibility for them is legally attached to the driver or the vehicle operator.

Suitability of Vehicles and Containers

It is the vehicle operator's duty under the regulations to ensure that any vehicle or any freight container used for the carriage of explosives is 'suitable' to ensure the safety and security of the explosives carried bearing in mind their type and quantity. The operator is also responsible for ensuring that the specified maximum quantities of any particular class of explosive carried on a vehicle or in a freight container are not exceeded and that no greater quantity of explosive is carried than that for which the vehicle or container is 'suitable'.

The limits on quantities of explosives permitted to be carried are shown in the following table:

Type of explosives		Maximum quantity
Division	*Compatibility Group*	
1.1	A	500 kg
1.1	B, F, G or I	5 tonnes
1.1	C, D, E or J	16 tonnes
1.2	Any	16 tonnes
1.3	Any	16 tonnes
Unclassified explosives carried solely in connection with an application for their classification		500 kg

The operator must ensure that explosives in different Compatibility Groups are not carried unless permitted as shown in Schedule 3 to the regulations.

Marking of Vehicles

Vehicles used for the carriage of explosives must be marked at the front and rear with a 400 mm x 300 mm rectangular reflectorized orange plate with black border (max 15 mm wide). Additionally a square placard set at an angle of 45° must be displayed on each side of the vehicle container or trailer containing the explosives (see Figure 23.3). The placard must conform to minimum dimensions as shown in the illustration overleaf and have an orange-coloured background with a black

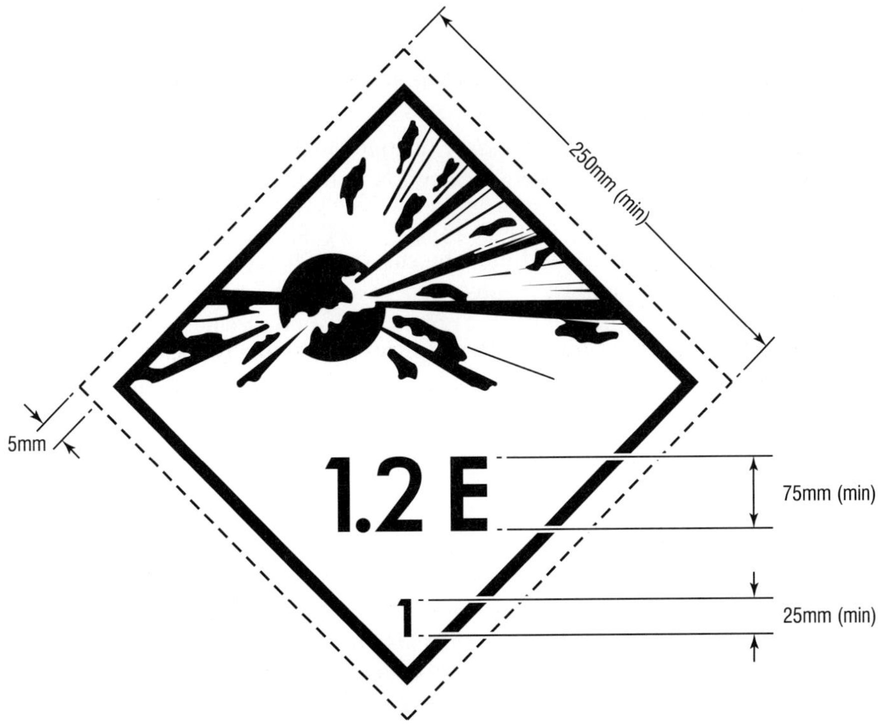

Figure 23.3 *Placard to be displayed on each side of vehicle or freight container carrying explosives*

border and with a 'bomb blast' pictograph and any figures or letters denoting classification and Compatibility Group shown in black.

Certain exemptions apply to the display of markings as described above where the quantities of explosives of particular categories carried are below limits set out in Schedule 4 to the regulations.

Markings on vehicles and containers must be clearly visible, be kept clean and free from obstruction and must be completely covered or completely removed when all explosives have been removed from the vehicle or container. Both the vehicle driver and operator are responsible for ensuring these marking provisions are complied with.

Duty to Obtain Information

Operators must obtain information in writing from consignors of explosives which enable them to comply with the regulations. The consignor's duty is to ensure the information given is both accurate and sufficient to allow the operator to comply.

Information to be Given to Drivers

The operator must give the driver or the vehicle attendant the following information in writing:

- The Division and Compatibility Group for classified explosives.
- The net mass (in tonnes or kg) of each type of explosive carried (or the gross mass if the net mass figure is not available).
- Whether the explosives carried are explosive articles or explosive substances (in the case of Group C, D or G explosives).
- The name and address of the consignor, the operator of the vehicle and the consignee.
- Such other information as necessary to enable the driver to know the dangers which may arise and the emergency action to be taken.

This information must be carried on the vehicle at all times during the carriage from the start of the journey and must be shown on request to a police officer or goods vehicle traffic examiner. It must also be shown to a fire brigade officer or inspector if required. Information must not be carried on a vehicle when the explosives it refers to are no longer on that vehicle. It must be removed, destroyed or placed in a securely closed container marked to show that the contents do not relate to explosives then being carried. If the necessary information is not available to the driver or vehicle attendant the explosives must not be carried.

Safe and Secure Carriage

The vehicle operator and the driver and any other person involved in the carriage of explosives must take all reasonable steps to prevent accidents and minimize the harmful effects of any accident. They must also prevent unauthorized access to, or removal of, all or part of the load. The operator and driver must ensure that a competent person is constantly in attendance with the vehicle whenever the driver is not present except during stops in a safe and secure place (as defined in the regulations – namely within a factory or magazine licensed under the Explosives Act

1875 or at a place with an exemption certificate granted under the Explosives Act 1875 (Exemptions) Regulations 1979) and when the vehicle is on a site where adequate security precautions are taken.

The operator and driver of a vehicle used to carry more than five tonnes of explosives in Division 1.1 must follow a route agreed with the chief officers of police for each area through which it is to pass.

Procedure in the Event of Accident

The driver or vehicle attendant must contact the police, fire brigade and vehicle operator as quickly as possible in the event of the following circumstances:

- spillage of explosives such as to constitute a safety risk;
- damage to the explosives or their packaging such as to constitute a safety risk;
- if the vehicle overturns;
- if a fire or explosion takes place on the vehicle.

When such circumstances arise the driver, vehicle attendant and the operator must take all proper precautions to ensure the security of the explosives and the safety of persons likely to be affected and the vehicle operator must immediately notify the Health and Safety Executive.

Duration of Carriage and Delivery

Both the vehicle operator and the driver are responsible for ensuring that the carriage of explosives is completed within a reasonable period of time having regard to the distance involved, that explosives are unloaded as soon as reasonably practicable on arrival, that the explosives are delivered to the consignee or his agent or to another person who accepts them in custody for onward despatch provided they are delivered to a safe and secure place or a designated parking area in an airport, a railway transhipment depot or siding, a harbour or harbour area. If they cannot be delivered as required they must be returned to the consignor or his agent. If loaded in a trailer, the trailer must not be detached except in a safe place or in an emergency.

Training of Drivers and Attendants

Vehicle operators must ensure that drivers and vehicle attendants have received adequate training and instruction to enable them to understand the nature of the dangers which may arise from the carriage of the explosives, the action to be taken in an emergency and their duties under these regulations and under the Health and Safety at Work etc Act 1974.

Vocational training certificates are required by drivers of explosives vehicles in accordance with the requirements of the Road Traffic (Training of Drivers of Vehicles Carrying Dangerous Goods) Regulations 1992 and EU legislation – see Chapter 7 for full details.

Minimum Ages

The regulations specify a minimum age of 18 years for those engaged in the carriage of explosives as a driver or vehicle attendant, for being made responsible for the security of explosives or for travelling in a vehicle carrying explosives unless in the presence of and supervised by a competent person over 18 years of age.

RADIOACTIVE SUBSTANCES

Complex legislation controls the carriage of radioactive substances. In particular, the Radioactive Material (Road Transport) Regulations 2002 made under the provisions of the Radioactive Material (Road Transport) Act 1991 apply to the carriage of such materials, while the Ionizing Radiations Regulations 1985 and the Ionizing Radiations (Outside Workers) Regulations 1993 cover controls on radiation doses received by persons working with such materials. Further, *A Code of Practice for the Carriage of Radioactive Materials by Road* is available (from TSO) which sets out the law and gives advice on all aspects of transporting these materials.

DRIVER TRAINING

Drivers of vehicles carrying dangerous goods must be instructed and trained so they understand the dangers of the particular goods being carried and the emergency action to be taken, as well as their duties under the Health and Safety at Work etc Act 1974, and current dangerous goods legislation. Operators must keep training records and give copies to the drivers concerned.

The driver training specified in ADR 8.2 applies (from the time of loading until the goods have been unloaded and, where appropriate, the compartment has been cleaned or purged) to drivers of the following dangerous goods vehicles:

- road tankers exceeding 1 m^3 capacity;
- those carrying tank containers exceeding 3 m^3 capacity;
- those exceeding 3.5 tonnes permissible maximum weight when carrying dangerous goods:
  - in bulk;
  - as a road tanker with a capacity not exceeding 1000 litres;
  - in a tank container with a capacity not exceeding 3000 litres;
  - where any of the goods are in transport category 0;
  - comprising more than 20 kg/litres* of category 1 goods in packages;
  - comprising more than 200 kg/litres* of category 2 goods in packages;
  - comprising more than 500 kg/litres* of category 3 goods in packages;
- those carrying explosives;
- those carrying radioactive materials.

* *NB: This refers to 'total mass or volume' (ie as measured in kg or litres).*

Vocational Training Certificates

Transport operators must ensure their drivers hold valid 'vocational training certificates' appropriate to the dangerous goods work they are employed in.

These certificates are granted to drivers on successful completion of an approved course of theoretical study and practical exercises and passing an approved examination. Certificates are valid for five years and may be extended for further five-yearly periods if in the 12 months preceding their expiry the holder has successfully completed a refresher course and passed the examination.

Current driver ADR certificates (if valid for the type of operation engaged in) are accepted as vocational training certificates for these purposes.

Certificates to be Carried and Produced to the Police

Drivers must carry their vocational training certificates on all dangerous goods journeys and produce them on request by the police or a goods vehicle examiner.

Minimum Training Requirements for Issue of Vocational Training Certificates

Approved training for drivers must cover at least:

- general requirements on dangerous goods carriage;
- main types of hazard;
- environmental protection in the control of the transfer of wastes;
- preventive and safety measures appropriate to various types of hazard;
- what to do after an accident (first aid, road safety, the use of protective equipment, etc);
- labelling and marking to indicate danger;
- what to do and not do when carrying dangerous goods;
- the purpose and operation of technical equipment on vehicles used for carrying dangerous goods;
- prohibitions on mixed loading in the same vehicle or container;
- precautions during loading and unloading of dangerous goods;
- civil liability;
- multi-modal transport operations.

For drivers of vehicles carrying packaged dangerous goods, training must also cover the handling and stowage of packages, and for road tanker or tank container drivers training must cover the behaviour of such vehicles on the road, including load movement during transit.

DANGEROUS GOODS SAFETY ADVISERS (DGSAS)

From 31 December 1999, firms which load, unload or carry dangerous goods by road (and by rail and inland waterway) must appoint a qualified Dangerous Goods Safety Adviser (DGSA) under provisions contained in EU Directive 96/35/EC. This also applies to self-employed persons, who must become qualified and appoint themselves. The person appointed must hold a valid Vocational Training Certificate confirming they have passed the official DGSA examination conducted by the

Scottish Qualifications Agency (SQA – appointed by the DfT to conduct all UK DGSA examinations). *Details of the examination system and a copy of the syllabus may be obtained from the SQA, Testing Services Section, at Hanover House, 24 Douglas Street, Glasgow G2 7NQ, or it may be downloaded from the SQA's website: www.sqa.org.uk/dgsa/*

To qualify as a DGSA it is necessary to study the subject material contained in the official syllabus, sit the relevant examinations and pass in at least three of the following subjects:

- the core examination, which is compulsory for all candidates;
- one modal paper covering either Road, Rail or Inland Waterways;
- at least one dangerous goods Class paper covering either:
  - Class 1;
  - Class 2;
  - Class 3 (specifically UN 1202, 1203 and 1223 – ie mineral oils);
  - Class 7;
  - General Chemical Classes (3, 4.1, 4.2, 4.3, 5.1, 5.2, 6.1, 6.2, 8 and 9);
  - all classes.

It is illegal to carry on a business in which dangerous goods are loaded, unloaded or transported after 31 December 1999 unless a qualified DGSA has been appointed and taken up his duties as outlined below.

Tasks and Functions of DGSAs

Appointed DGSAs must effectively carry out the legal duties and bear the responsibilities set out in law as follows:

- monitor compliance with the law on the transport of dangerous goods;
- advise their employer on the transport of dangerous goods;
- prepare an annual report to their employer on the firm's activities in transporting dangerous goods (to be kept for five years and made available to the authorities on request).

The DGSA must also monitor:

- procedures for identifying dangerous goods being transported;
- practices for taking account of any special requirements in connection with dangerous goods being transported;
- procedures for checking equipment used in the transport, loading or unloading of dangerous goods;
- employee training and maintenance of training records;
- emergency procedures to be taken in the event of accidents that may affect safety during the transport, loading or unloading of dangerous goods;
- investigation and preparation of reports on serious accidents or legal infringements during the transport, loading or unloading of dangerous goods;
- implementation of steps to avoid the recurrence of accidents, incidents or serious legal infringements;
- account taken of the legal requirements in the choice and use of sub-contractors;
- operational procedures and instructions that employees must follow;

- introduction of measures to increase awareness of the risks inherent in the transport, loading and unloading of dangerous goods;
- verification procedures to ensure that vehicles carry the documents and safety equipment required and that they comply with the law;
- verification procedures to ensure that the law on loading and unloading is complied with.

CONTROLLED AND HAZARDOUS WASTE

Many transport operators (and skip-hire firms) are concerned with the disposal of 'waste' and, as such, they are affected by newly imposed and strictly enforced legislation. For these purposes waste may be considered in two forms:

- controlled waste which comprises 'household, industrial and commercial waste or any such waste' (ie including waste paper, scrap metal and recyclable scrap);
- hazardous waste which is material defined as 'special waste' in the controlled waste legislation, or material which falls within the classification of dangerous substances for the purposes of the road tanker or packaged dangerous goods legislation described earlier in this chapter.

A range of legislation applies in this area of activity including the Disposal of Poisonous Waste Act 1972 (which makes it an offence to dispose of poisonous waste in an irresponsible way), the Criminal Justice Act 1988 (which provides powers for the authorities to impound vehicles engaged in illegal fly-tipping) and the Control of Pollution (Amendment) Act 1989, the Controlled Waste (Registration of Carriers and Seizure of Vehicles) Regulations 1991 and the Controlled Waste Regulations 1992. Two other sets of regulations are important, the Waste Management Licensing Regulations 1994 and the Special Waste Regulations 1996. Offences against the regulations can result in fines of £2,000 on conviction.

A new set of regulations dealing with hazardous waste was about to come into effect as this edition of the *Handbook* was closing for press. These are the *Hazardous Waste (England and Wales) Regulations 2005* (SI 2005 No 894). A more detailed description of the new provisions will be given in the next edition.

Controlled Waste

Controlled waste should not be disposed of, to, or transported away by, a person or firm not legally authorized for this purpose.

Registration of Operators

The law (see above) requires operators who transport controlled waste within Great Britain to register with the appropriate waste regulation authority (ie in the area in which they have their business). It is an offence to fail to register or to carry controlled waste when not registered. Besides fines following prosecution and conviction for such offences, legislation provides powers for the seizure and disposal of vehicles used for such illegal purposes.

Exemptions are provided for charities, voluntary organizations, domestic householders disposing of their own waste, waste collection authorities and producers of controlled waste. Builders and demolition companies are not exempt and must register and otherwise comply with the legislation.

Registration costs £95 initially and, currently, £65 at each three-yearly renewal period. Registration may be refused to operators (ie companies or individuals) convicted of relevant offences as set out in the regulations.

Duty of Care

Under the Environmental Protection Act 1990 firms and individuals that produce, import, store, treat, process, recycle, dispose of or transport controlled waste (see above for definition) have a statutory 'duty of care'. This places responsibility for the completion of paperwork (ie 'Waste Transfer Notes') and to take all reasonable steps to stop waste escaping and ensuring its safety and security.

Waste Transfer Notes

Waste transfer notes comprise written descriptions of waste handed over to other persons to transport and/or dispose of, and a transfer note signed by both parties (allowable as a single document) containing the following details:

- What the waste is and the quantity.
- The type of container in which it is carried.
- The time and date of transfer.
- The place where the transfer took place.
- The names and addresses of both parties (ie consignor and recipient).
- Detail as to which category each falls into (eg producer and registered waste carrier).
- A certificate number if either or both parties hold waste licences and the name of the authority from whom it/they was/were issued.
- Reasons for any exemption from registration or waste licensing.

Copies of documents (ie descriptions of waste and/or transfer notes) given and received must be retained for at least two years. Both or either party may be required to produce these and prove in court where particular consignments of controlled waste originated.

Seizure of Vehicles

The seizure of vehicles aspects of the legislation are a new method of penalty in the UK – and incidentally, are also being sought in regard to 'O' licence offenders. The law gives powers to waste regulation authorities to seize the vehicles of offenders, remove and separately store or dispose of loads as necessary, and dispose of (or destroy) vehicles following set procedures to publicize details of the seizure in local newspapers. Attempts will be made to seek out legitimate owners who may reclaim their vehicles on satisfactory proof of entitlement and identification.

Hazardous/Poisonous Waste Disposal

Broadly, the law on hazardous waste requires that waste which is poisonous, noxious or polluting is not deposited on land where its presence is liable to give rise to an environmental hazard. And it is necessary for anyone removing or depositing poisonous material to notify both the local authority and the river authority before doing so.

An environmental hazard is defined as waste that is deposited in a manner or in such quantity that it would subject persons or animals to material risk of death, injury or impairment of health or threaten the pollution or contamination of any water supply. A booklet, *Guidelines on the Responsible Disposal of Wastes,* is available from the Confederation of British Industry. The European Commission has also produced a booklet on this subject for local authorities in EU member states.

There are additional controls over the carriage and disposal of particularly hazardous waste. The Control of Pollution (Special Waste) Regulations 1980 were introduced in order to comply with EU directives:

The main requirements of these regulations are outlined below:

- Certain types of waste are to be regarded as special waste and subject to the additional controls. These are wastes that are regarded as dangerous to life as set out in the regulations.
- Waste producers have to give not less than three days' and not more than one month's prior notice to waste disposal authorities of their intention to dispose of a consignment of special waste.
- A set of consignment notes must be completed when special wastes are transported. This means that a consignment of special waste can be transported from the producer to the disposal site only if each person has signed for it and taken on responsibility for it. This is to ensure that waste disposal authorities know who is carrying the waste and they have to be informed within 24 hours of when it reaches the disposal site. Waste producers should take particular note of the requirement that all notices must be made on the statutory forms. Each form contains a unique reference number to assist an authority in making sure that waste is safely disposed of.
- A record of the location of the point of disposal on site of all special wastes must be kept in perpetuity. This is to ensure that proper arrangements can be made to bring the site back into use after the waste disposal operation has ceased.
- Proper registers of consignments must be kept by the producers, carriers and disposers.
- There will be a 'season ticket' arrangement for regular consignments of special wastes of similar composition disposed of at the same site. The Waste Disposal Authorities will decide which producers and disposers in their areas qualify.
- The Secretary of State will have emergency powers to direct receipt of special wastes at a particular site. This is likely to be rarely used.
- Radioactive waste which also has the characteristics of special waste will be subject to the new controls.

Failure to comply with any of the requirements of the regulations is an offence.

Advice on particular problems can be obtained from waste regulation authorities (usually the district or county council).

Waste Site Licensing

The site licensing provisions of the Control of Pollution Act 1974 require all commercial, industrial or domestic wastes to be disposed of at a site licensed for that purpose and nowhere else. The licences are issued by the waste disposal authorities (these are the county councils in England and the district councils in Wales and Scotland). These licences, which are a matter of public record, set conditions for the operation of each site including the types of waste which can be disposed of at that site, the manner of disposal, site supervision, boundary fences, notice boards, etc. Waste disposal authorities must inspect sites regularly and make sure the operators are following the conditions of the licence. Failure to comply with licence conditions is a criminal offence. In addition to prosecution, an authority can amend or even revoke the licence. See also section dealing with unauthorized tipping of waste (ie fly-tipping) on p 481.

PACKAGING WASTE

The Producer Responsibility Obligations (Packaging Waste) Regulations 1997, made under the Environment Act 1995, are designed to encourage businesses to recover value from products at the end of their life with the aims of:

- achieving a more sustainable approach to dealing with packaging waste;
- reducing the amount of packaging waste going to landfill;
- implementing the recovery and recycling targets in the EC Directive 94/621/EC on packaging and packaging waste.

Businesses affected by the regulations are those with an annual turnover in excess of £5 million (reducing to a threshold of £1 million annually since 1 January 2000) and which handle more than 50 tonnes of packaging or packaging material annually. If one or other of the thresholds are not met the business is excluded from the regulations.

The regulations impose three main obligations on relevant businesses, namely:

- Registration.
- Recovery and Recycling.
- Certifying.

Registration

Registration requires producers to register with the Environment Agency in England or Wales, or with the Scottish Environment Protection Agency (SEPA) if their principal place of business or registered office is in Scotland. A registration fee must be paid and packaging data must be provided on a data form (set out in Schedule 4 to the regulations) which will be provided by the Agency.

The deadline for businesses continuing their registration or registering for the first time is 1 April each year.

Recovery and Recycling

Recovery and recycling obligations started in 1998 and require producers to take reasonable steps to recover and recycle specific tonnages of packaging waste calculated on the basis of three factors:

- the tonnage of packaging handled by the producer in the previous year;
- the 'activities' that the producer performs and the percentage obligation attached to each activity; ie
  - manufacturing packaging raw materials
  - converting materials into packaging
  - packing and filling packaging
  - selling packaging to the final user;
- the national recovery and recycling targets (as set out in the regulations).

Certifying

Producers must write to the Agency by 31 January following the end of the calendar year in which it falls within scope of the regulations (ie by virtue of turnover and the tonnage of packaging waste handled) certifying that it has recovered and recycled the necessary tonnages of packaging waste. The person writing must be an 'approved person' (ie a director of a company, a partner in a partnership business or a sole trader).

24 Fleet Car and Light Vehicle Operations

Transport managers and other staff responsible for the operation of large goods vehicles within companies frequently have additional responsibilities for the operation of company-owned motor cars used by management, sales and service personnel, and light commercial vehicles which are outside the general scope of much of the legislation explained in this book. It is equally important that such vehicles should be operated strictly within the law. Most of the offences and penalties described in earlier chapters apply when operating such vehicles and the consequences of operating illegally can be serious. As already mentioned, for example, Traffic Commissioners will take account of failure to operate light vehicles safely and within the law when deciding whether an applicant is fit or is of sufficient good repute to hold an 'O' licence for larger vehicles.

For the purposes of this chapter, light goods vehicles are vehicles with a gross plated weight not exceeding 3.5 tonnes (ie below the 'O' licensing, LGV plating and testing and the EU drivers' hours and tachograph requirement thresholds).

Much of the legislation applicable to the use of private motor cars and light goods vehicles has already been dealt with under such headings as excise duty, insurance and traffic regulations. The same system as described previously in Chapter 8 applies to the registration and excise licensing of such vehicles and the legal requirements for insurance cover have been outlined in Chapter 9.

Most of the traffic restrictions, particularly with regard to parking and waiting, road signs, motorway lights, breath tests and zebra crossings, which apply equally to the private car and light vehicle driver and to the large goods vehicle driver, have been dealt with in Chapter 10. Nevertheless, some of these items of legislation are worth emphasizing here for the benefit of the light vehicle fleet manager along with other matters applicable to such operations within firms.

EXCISE DUTY

Private-type motor cars owned by firms and used for business purposes can be licensed at the private/light goods (PLG) rate of duty – currently £160 per year. Similarly, estate cars can be taxed at the £160 duty rate. See page 202 for details of graduated VED for smaller-engined, less polluting cars and light goods vehicles. If a private car is adapted for carrying goods, or an estate car is used for carrying goods, tools or samples, the appropriate goods rate of duty must be paid if the permissible maximum weight exceeds 3500 kg.

In determining whether an estate car or dual-purpose vehicle (see p 201 for definition) is subject to the goods vehicle rate of duty, if its permissible maximum weight exceeds 3500 kg then consideration has to be given to the nature of the goods or burden carried. No specific guidelines are laid down but local Vehicle Registration Offices take the view that if any goods or samples, service equipment or spare parts are carried in connection with a business from which profit may result, then the vehicle should be taxed as a goods vehicle. Samples which can be accommodated in a normal briefcase would not generally constitute goods for this purpose.

The tax (ie VED) disc must be displayed on the windscreen of the vehicle on the near side where it can be easily seen. See Chapter 8 for details of vehicle excise duty requirements.

INSURANCE

Private-type motor cars and light goods vehicles, like all other vehicles, are required by law to be covered for third-party insurance risks as a minimum, but clearly in the case of fleet cars additional cover would be taken out, and in most instances comprehensive cover is advisable. Most insurance policies contain a variety of option clauses, some of which are included in the basic premium and others are available at extra cost.

Correct Cover

Of importance to the fleet manager is the need to ensure that company cars are fully insured for business use by employees of the company. The difference in insurance classification for business cars used for commercial travelling and for those which are not used for this purpose is an important point for the fleet manager to consider. Most policies differentiate between such use, and in a car fleet where vehicles are used by both salesmen and other staff and management, it is essential to ensure that either the policy covers the whole of the car fleet for commercial travelling or, if it does not do so, the sales force should be restricted to driving only those cars which have this cover. If a salesman, or indeed any other staff member, uses a car for a purpose which can in any way be described as soliciting an order, then in insurance terms this use is classed as commercial travelling and the cover on the car must include this clause.

Dual-purpose vehicles registered and licensed as goods vehicles must be insured with cover permitting goods to be carried for business purposes.

Cover for Special Cars

There can be problems in a fleet with certain individual cars on insurance cover which may not be driven by certain employees. For example, if the managing director has a high-powered sports-type car it is likely that the insurers would impose severe restrictions on the driving of that car, for whatever purpose (chauffeuring, ferrying back and forth for service or repairs, for example), by relatively inexperienced drivers or by, say, young drivers under 25 years of age. Alternatively, such cars may be restricted to named drivers only.

Private Use of Company Cars

A further point which needs consideration is cover for the employee's use of a company car for private purposes. It is general practice for firms to allow their employees this concession both for commuting to work and for family motoring at the weekends and for holidays. The cover taken out for fleet cars should specifically include this provision if such use is permitted, otherwise the vehicle owner (the firm) may be guilty of permitting the use of an uninsured car. In these circumstances, if an accident and claim were to result, the consequences for the employee could be very serious in terms of meeting damage claims and in prosecution for using an uninsured car. The same provisions should be applied if the firm permits the employee's wife (or husband) or even his children (if they are qualified drivers) to drive the car at any time.

Employees' Use of Own Cars

If employees are ever required, or likely to be required, to use their own private car for business purposes, even for only the occasional errand, the fleet manager should ensure that they have adequate insurance cover on their own vehicle for such purpose. This usually means that employees' own policies must include provision for their car to be used in connection with the business of their employer.

The fleet manager, confronted with this type of situation, should ask to see evidence of the cover (ie a valid Certificate of Insurance or a temporary cover note showing the conditions for use covered by the policy) and not just rely on the word of the employee. Similarly, the fleet manager would be wise to inform all other persons in the firm, management and staff alike, that employees should not use, or be requested to use, their own car on company business without first having the insurance cover verified.

To avoid the dangers which could arise from an employee using his own private car for business purposes when it was not covered for such use, when the policy was not in force at all because the premium had not been paid, or because the policy was invalid as a result of the employee giving incorrect information at the time of completing the policy application, the company can take out a motor contingency policy which will indemnify it against any claims arising from an accident involving an employee undertaking company business in their own uninsured car.

This type of policy is very cheap to obtain and well worth the cost when considered against the risks. For example, if the employee had an accident causing serious injury to one or more third parties and his own insurance proved to be void, the third parties would look to the employer on whose business the employee was engaged at the time of the accident to meet their claims.

Dual Car Use

When an employee who is provided with a company car also owns a car which his wife uses, an anomalous situation on insurance cover can arise if his own car is insured in his, and not his wife's name. This problem occurs because the cover provided on his own car automatically provides him with third-party cover while driving any other car not belonging to him. Consequently, when driving the firm's car for pleasure purposes he has the double cover provided by both the firm's policy and his own policy. If he should then be involved in an accident resulting in a third-party claim, his own insurers could be held partially liable for the damages arising out of the claim.

In order to overcome this particular problem the major insurance companies have made an undertaking indicating that it is not their intention to take advantage of the cover provided by the driver's own personal insurance policy in such circumstances. However, not all insurance companies are party to this undertaking.

Payment by Passengers

Payment by passengers towards the cost of petrol consumed on a journey no longer infringes the 'hiring' exclusion clause on insurance cover. However, payment towards other motoring expenses such as parking fees or depreciation could still fall within the 'prohibition of hiring' clause in most insurance policies. In cases where this situation arises a check should be made with insurers to ascertain the current position.

COMPANY CARS AND INCOME TAX

Since 6 April 2002, there has been a fundamental change in the way company cars are taxed to help to protect the environment. Cleaner, more fuel-efficient cars will be rewarded by linking the tax and NIC (National Insurance Contributions) charge to the car's exhaust emissions, in particular its carbon dioxide (CO_2) emissions.

From the tax year commencing 6 April 2002, company cars are taxed on a percentage of their price, which, for the most part, varies according to the level of their carbon dioxide (CO_2) emissions, measured in grams per kilometre (g/km). A normal minimum charge of 15 per cent of the car's price applies to cars emitting CO_2 at or below a specified qualifying level, measured in g/km. The percentage charge builds up in 1 per cent steps for every additional full 5g/km over that level, up to a maximum charge of 35 per cent of the car's price. There is no longer an adjustment for business mileage or for older cars.

There is a supplement of 3 per cent on the list price for diesel cars, up to the maximum charge of 35 per cent. The supplement is waived for cars that meet a

European Community standard for cleaner cars, called Euro IV. The Euro standard is shown on the new Vehicle Registration Document for all cars first registered since 1 March 2001.

There are also discounts for certain cleaner, alternatively fuelled cars, which may reduce the minimum charge to less than 15 per cent. These diesel supplements and the discounts for cleaner alternatives apply only to cars first registered on or after 1 January 1998.

Cars without an approved figure of CO_2 emissions are taxed according to their engine size.

To calculate the car benefit, take the following seven steps:

1. Calculate the price of the car for tax purposes.
2. Find the approved CO_2 emissions figure for the car, if it has one.
3. The ready reckoner (see below) shows the qualifying levels for the minimum charge that will apply in the tax years 2002/3, 2003/4 and 2004/5, and the percentage charges that will apply for each CO_2 emissions band.
   - If the exact CO_2 emissions figure does not end in 0 or 5, it is rounded down to the nearest 5g/km.
   - If the emissions figure is below the minimum qualifying level shown for the relevant year, the percentage charge appropriate to the minimum level is used.
   - If the car has no CO_2 emissions figure, use the table on page 524 to find the percentage charge applicable to the engine size and age of the car.
4. The diesel supplement, or any discount for alternatively fuelled vehicles, is added to the percentage charge, where appropriate (see below).
5. The basic car benefit charge is calculated by multiplying the car's price by the percentage charge.
6. If the car has been available for part of the tax year only, the car benefit is reduced proportionately.
7. Any payments made by the employee for private use of the car are deducted.

The approved CO_2 emissions figure for car benefit purposes is recorded on the type approval certificate, which summarizes the results of the type approval testing procedure. The figure that applies at the date on which the car is first registered is set for the life of the car. For cars first registered in the UK from 1 March 2001, the CO_2 emissions figure will be shown on the Vehicle Registration Document (V5).

To calculate the car benefit charge it is necessary to know the price of the car for tax purposes and the approved CO_2 emissions figure for the particular vehicle.

Car Benefit Charges for Cars with an Approved CO_2 Emissions Figure

CO_2 emissions in grams per kilometre				*Percentage of car's price to be taxed 2002/3*
2002/3	*2003/4*	*2004/5*	*2005/6*	
165	155	145	140	15*
170	160	150	145	16*
175	165	155	150	17*
180	170	160	155	18*
185	175	165	160	19*
190	180	170	165	20*
195	185	175	170	21*
200	190	180	175	22*
205	195	185	180	23*
210	200	190	185	24*
215	205	195	190	25*
220	210	200	195	26*
225	215	205	200	27*
230	220	210	205	28*
235	225	215	210	29*
240	230	220	215	30*
245	235	225	220	31*
250	240	230	225	32*
255	245	235	230	33**
260	250	240	235	34***
265	255	245	240	35****

Diesel supplements:
* Add 3 per cent if car runs solely on diesel.
** Add 2 per cent if car runs solely on diesel.
*** Add 1 per cent if car runs solely on diesel.
**** Maximum charge, so no diesel supplement.

The basic car benefit charge will be the car's price multiplied by the percentage charge for the CO_2 emissions band in which the car falls, subject to:

- the addition of any diesel supplement, or
- discount for cars using alternative fuels and technologies.

Discounts for Cars that Run on Alternative Fuels and Technologies

Battery Electric Cars

Electric cars that run solely on electricity (normally by battery) do not emit CO_2. These vehicles attract the minimum percentage charge of 15 per cent of the price of the car, less a discount of 6 per cent, so tax is charged on 9 per cent of the price of the car.

Hybrid Electric Cars

For cars that run on a combination of petrol and electricity, the discount is 2 per cent of the price of the car, plus an additional 1 per cent of the price for each full 20g/km that the CO_2 emissions figure is below the qualifying level for the minimum charge for the year.

Car Benefit Charges for Cars with no Approved CO_2 Emissions Figure

One scale of charges applies for all cars registered before 1 January 1998, and another for the small number of cars registered from 1 January 1998 onwards.

	Car first registered before 1 January 1998	Car first registered on or after 1 January 1998
Engine size (cc)	*Percentage of car's price to be taxed*	
0–1400	15%	15%*
1401–2000	22%	25%*
2001+	32%	35%**
Cars without a cylinder capacity (eg rotary- engined petrol cars)	32%	35%

Diesel Supplements (cars registered on or after 1 January 1998 only):
* Add 3 per cent if car runs solely on diesel.
** Maximum charge, so no diesel supplement.

The basic car benefit charge is the car's price multiplied by the percentage charge appropriate to the car's age and engine size (including any diesel supplement).

Source: Inland Revenue website

Fuel Benefit Charge

Where an employee who has a company car is provided with free fuel for private use, a new fuel benefit charge came into effect from the tax year 2003/04 replacing the previous fuel scale charge. The change is in how the benefit is measured and, as with the company car benefit charge, it is linked directly to the CO_2 emissions of the company car.

Like car benefit, the fuel benefit charge will be based on the level of CO_2 emissions of the car, with the same 3 per cent supplement for diesels that do not meet Euro IV emission standards and reductions for alternatively fuelled cars. Also like car benefit, the percentages for petrol and diesel range from a minimum of 15 to a maximum of 35 per cent and the percentage for alternative fuels and hybrid cars can be below 15 per cent with reductions.

To calculate the benefit charge on free fuel the percentage figure will be multiplied against a set figure for the year. For the tax years 2003/04 and 2004/05 the set figure is the same; namely, £14,400.

Private Use of Vans

The Inland Revenue is reported to be checking and clamping down on van drivers earning more than £8,500 annually, who are permitted to use their vehicles for private purposes (including their use as transport between home and their place of work). Such private use should be declared on the relevant tax form (ie P11D) and the employee can be liable for tax based on 20 per cent of the vehicle's value. Safeguards against such tax liabilities include where the vehicle is taken home en route to a next-morning delivery, or where it is necessary for overnight security reasons, because the employee's own tools and personal work equipment are kept on board.

Since 6 April 1993, employees who have private use of a company van are taxed by reference to a scale charge – for 1997/98, £500, but only £350 if the van is more than four years old – which will include the provision of private fuel. Vehicles over 3.5 tonnes will be exempt from this arrangement, as will any pooled van which is not normally kept at or near an employee's home.

National Insurance on Private Fuel Usage

Where employees are provided with free fuel for private use there is a liability to pay National Insurance contributions on the value of the benefit to the employee. Where such value has not been taken into account by the employer there is a risk of claims by the DSS for retrospective payment of contributions going back one year.

CONSTRUCTION AND USE REGULATIONS

Apart from the sections of the C&U regulations which apply to private cars concerning their construction, lighting, noise, silencing and all the provisions which require the motor car to be maintained in accordance with the regulations, they are not involved in the special requirements relating to goods vehicles, unless the car has in any way been adapted to carry goods, for example, by removing seats or fitting racks on which goods can be carried. In this case the goods vehicle rate of excise duty applies and the vehicles become subject to certain aspects of the C&U regulations regarding goods vehicles (see Chapter 13).

It should be remembered that the overriding requirement of the C&U regulations for all vehicles to be maintained at all times in such a condition that they shall not cause danger to people carried on the vehicle and other road users applies equally to private cars and to goods vehicles irrespective of their size or weight.

New cars manufactured for sale in Britain must meet the braking system requirements of EC Directive 71/320 unless the system is of the dual-line type. This directive specifies maximum stopping distances equivalent to a braking efficiency of 27 per cent.

Also as a result of EU directives, reversing lamps and four-way hazard warning flashers are compulsory on all cars made in Britain. Regulations require rear fog lights to be fitted to new vehicles manufactured since 1 October 1979 and first used since 1 April 1980 (see p 365 for more details).

Vehicles for Unleaded Petrol

Petrol-engined vehicles first used on or after 1 April 1991 must be designed and constructed to run on unleaded petrol (it is illegal to reconvert such vehicles to run on leaded petrol only). Vehicles which fall into this category must not be capable of being filled with a petrol-pump nozzle greater than 23.6 mm in diameter (unless certain other conditions are met and the fuel filler is clearly marked with the word 'UNLEADED' or the symbol for unleaded fuel use).

Towing

From 1 August 1998, all tow-bars fitted to new cars must be EC Type Approved and marked accordingly. It is an offence to fit – or have fitted – a tow-bar which does not comply with Euro-legislation contained in Directive 94/20/EC.

Local Authority Emissions Testing

Seven local authorities, namely Birmingham, Bristol, Canterbury, Glasgow, Middlesbrough, Swansea and the City of Westminster, are currently operating a pilot scheme of roadside emissions testing. Drivers of vehicles which fail the test will be issued with a fixed penalty notice requiring payment of £60, rising to £90 if unpaid within 28 days. The test levels are the same as those that vehicles would be expected to meet at the annual test. Commercial vehicles that fail an emission test invariably have poorly maintained engines. The scheme covers all types of vehicles, including commercial vehicles, and if the tests prove successful they will be extended nationally.

SAFE USE OF MOBILE PHONES, ETC

Mobile telephones emit electrical energy when transmitting and should be switched off before entering a filling station. Receiving incoming calls presents the same hazards as when making outgoing calls. A number of serious fires have resulted from mobile phone use in filling stations.

The Highway Code (section 127) advises against the use of hand-held mobile phones or microphones while driving. It also says that using hands-free equipment may distract a driver's attention from the road. It is an offence to fail to exercise proper control of a vehicle at all times.

A similar injunction applies (section 128) in regard to in-vehicle (or on-board) systems such as those used for route guidance, navigation and congestion warning, and the use of laptop or hand-held PCs and multi-media, etc.

In all instances, when it is necessary to use such equipment, the driver should stop the vehicle in a safe place (ie not on the hard shoulder of a motorway except in an emergency).

Carrying Dangerous Goods in Cars

According to the National Chemical Emergency Centre (NCEC), the carrying of small quantities of dangerous goods, such as samples, in company cars can present both legal problems and safety risks.

The legal situation is determined by the type of goods carried, the size of the receptacles in which they are carried, and the total size of the load. The quantity thresholds for UK carriage are shown in the CDG Road regulations (schedule 1, Table 2) – see pp 505–08 for further details.

For the most part, it is unlikely that sufficient quantities of such goods in Class 3 (Flammable), Packing Group III (Transport Category 3) would be carried in a private car to require the full legal gamut of vehicle placarding, the provision of Tremcards and the need for driver training. Provided these goods are in receptacles with a volume of 25 litres or less it would be difficult to exceed the quantity threshold of 500 litres above which these legal requirements apply.

However, should the samples fall within the requirements for Packing Group I substances (Transport Category 1), goods must then be in a receptacle with a capacity of 1 litre or less, with a total volume of 20 litres or less, to avoid these requirements.

So far as the risk element is concerned, it is necessary to consider what would happen in the event of a road accident. With no Tremcards on the vehicle, there is no emergency information available for either the driver or the emergency services, nor, invariably, is there likely to be any emergency equipment such as the appropriate fire extinguishers, first aid kit, respirator, gloves, spill kit or emergency warning triangle to place on the road.

Besides all the dangers implicit in the carrying of dangerous samples, there is the question of the vehicle insurance to consider. Generally, it may be expected that an insurance company would repudiate any claim if they were not advised beforehand of the carriage of such goods, especially if legal requirements were not being met at the time.

DRIVERS' HOURS AND RECORDS

Light goods vehicle drivers (ie of vehicles not exceeding 3.5 tonnes gross weight) coming within the scope of this chapter and employees driving company-owned private-type motor cars are exempt from the EU driving hours regulations – consequently, only the relevant British provisions apply in certain cases.

The applicable limits for drivers of goods vehicles not exceeding 3.5 tonnes gvw are as follows:

Maximum daily driving time 10 hours
Maximum daily duty time 11 hours

There are no specified break or daily or weekly rest period requirements; no limits on continuous duty or weekly limits on duty or driving.

Drivers of light goods vehicles of not more than 3.5 tonnes gross plated weight and dual-purpose vehicles (see p 201 for definition) of any weight used for certain specialized duties are required to observe only a daily maximum driving time of 10 hours. This applies to light goods vehicles used:

- by doctors, dentists, nurses, midwives or vets;
- for any service of inspection, cleaning, maintenance, repair, installation or fitting;
- by a commercial traveller and carrying only goods used for soliciting orders;
- by an employee of the AA, the RAC or the RSAC;
- for business of cinematography or of radio or television broadcasting.

Record Keeping

Light goods vehicle drivers (ie vehicles under 3.5 tonnes gross weight) and drivers of dual-purpose vehicles are exempt from the requirements to keep records of their driving, duty and rest periods.

Employees driving vehicles in this category remain exempt from the requirements of the goods vehicle drivers' hours and record requirements, but only so long as they do not drive vehicles to which these requirements do apply. If at any time an employee who normally drives only light goods vehicles or motor cars also as part of his work needs to drive a goods vehicle to which the drivers' hours regulations do apply then any time spent driving a van or a private car (his own or the company's) for his firm's business must be counted within his total daily working time (ie duty time). If the goods vehicle exceeds 3.5 tonnes gross weight the driving of the car should be shown on his record sheet (ie tachograph chart) for the day on which he drives a goods vehicle as 'other work'.

Full details of the goods vehicle drivers' hours and records are given in Chapters 3 and 4.

Tachographs

Light goods vehicles of not more than 3.5 tonnes gross plated weight are exempt from the EU tachograph regulations requiring tachograph fitment and use, but if such a vehicle is coupled to a goods carrying trailer so that the total of the combined gross weights exceeds 3.5 tonnes then the vehicle will come within the scope of the EU tachograph regulations, unless it is otherwise exempt due to special use. This means that a fully calibrated tachograph must be fitted and must be used by the driver when the trailer is drawn (see Chapter 5 for full details of the tachograph regulations) and the EU drivers' hours rules must be followed (see Chapter 3 for details).

SPEED LIMITS

Private cars and dual-purpose vehicles not drawing trailers are restricted to maximum permitted speeds on certain roads in accordance with the restriction sign-posted on the section of road and to overall maximum speeds of 60 mph on single-carriageway roads and 70 mph on dual-carriageways and motorways.

Speed limits for cars and light goods vehicles are as follows:

	Motorways	Dual-Carriageways	Other Roads
• Cars and car-derived vans	70	70	60
• Cars and car-derived vans towing trailer	60	60	50
• Rigid goods vehicles not exceeding 7.5 tonnes	70	60	50
• Articulated vehicles and rigid goods vehicles not exceeding 7.5 tonnes drawing trailer	60	60*	50

**In Northern Ireland the limit for this category of vehicle is 50 mph.*

NB: In all cases of speed restrictions mentioned above, if specific lower limits are in force on any section of road, then it is the lower limit which must be observed.

SEAT BELTS

Compulsory fitment of seat belts applies in the case of the following light vehicles (see also p 257):

- Goods vehicles not exceeding 1525 kg unladen (first registered since April 1967).
- Goods vehicles not exceeding 3500 kg maximum gross weight (first registered since 1 April 1980).
- Dual-purpose vehicles first registered since 1 January 1965.
- Private cars first registered since 1 January 1965.

Vehicles to which the regulations apply first used since 1 April 1973 must be fitted with belts which can be secured and released with one hand only and must also be fitted with a device to enable the belts to be stowed in a position where they do not touch the floor. The belts must be maintained in a fit and serviceable condition and kept free from permanent or temporary obstruction which would prevent their being used by a person sitting in the seat for which the belt is provided. Failure to comply with these requirements is an offence and can lead to failure of the MoT test.

In cases where the regulations apply, as above, belts must be provided for the driver and one front seat passenger.

Wearing of Seat Belts

The wearing of front seat belts in vehicles fitted with them by law has been compulsory since 1 January 1983.

Since 1 September 1989 children up to 14 years of age riding in the back seats of motor cars fitted with rear seat belts or child restraints must wear those belts or restraints.

Since 1 July 1991 adult rear seat passengers travelling in cars or taxis fitted with rear seat belts must wear the belts provided, irrespective of the age of the vehicle.

Certain exemptions to the wearing of belts have been included in the regulations (see pp 258–59 for list of exemptions).

Heavy fines may be imposed on conviction for failing to wear a seat belt as required by law. The responsibility for seat belt wearing rests with the person sitting in the seat for which the belt is provided except that responsibility for ensuring that children wear seat belts as required by law either in the front or rear seats rests with the driver of the vehicle (ie not the parent or guardian who may be accompanying the child).

FUEL CONSUMPTION TESTS

Concern about energy conservation led to the Energy Act 1976 and the Passenger Car Fuel Consumption Order 1977 being enacted. Because of interest among readers of the Handbook in this subject, details of the legal requirements and the scheme are outlined here.

Since 1 April 1978 new cars on display in showrooms and on forecourts must carry a label showing official fuel consumption figures for that model of car. Every dealer must have details of officially approved fuel consumption tests for all cars listed in a booklet available in his showroom for buyers to consult on request. In addition, where reference is made in promotional literature, such as advertisements, technical specifications and sales brochures, to the petrol consumption of a new car the test results must be included. In all such cases the results of all tests carried out must be quoted (urban cycle, 90 km/h – and 120 km/h where appropriate) in both miles per gallon and litres per 100 km. These requirements do not apply in the Channel Islands or the Isle of Man.

Official Tests

The official tests are carried out in approved laboratories or on test tracks. They have been designed to be representative of real-life driving situations and the results achieved provide a guide to the models which are likely to be more economical in their fuel use.

The test results do not guarantee the fuel consumption of any particular car. Each new car has not itself been tested and there will inevitably be differences between cars of the same model. The driver's style, the loading of the car, road, weather and traffic conditions, the overall mileage of the car and its standard of maintenance will all affect its fuel consumption. For all these reasons the fuel consumption achieved on the road will not necessarily accord with the tests results.

The Standard Test

The tests follow an internationally agreed procedure and consist of two compulsory parts:

Part 1. A cycle simulating urban driving.

Part 2. A constant speed test at 56 mph (90 km/h).

Models Included in the Tests

Almost all types of new passenger cars are covered by the tests. However, certain types are excluded as follows:

- Cars manufactured before 1 January 1978.
- Second-hand cars.
- Cars adapted to carry more than eight passengers (excluding the driver).
- Three-wheelers.
- Invalid carriages.
- Van-derived passenger cars.
- Cars built specially for export.
- Cars operating on four-wheel drive only.
- Cars whose engines run on diesel, liquefied petroleum gas or other such fuels.

These vehicles will not, therefore, be labelled in the showrooms. Heavy goods vehicles, vans and motorcycles are also excluded from these tests. Also, a small number of manufacturers and importers have been granted exemption from testing because of the low volume of production involved. Consequently, these particular makes and models are not likely to be of significant interest or concern to the fleet user.

Urban Test Cycle

The urban test cycle is carried out in a laboratory where equipment simulates the loads experienced under normal driving conditions and the standard patterns of urban driving. The car is driven from a fully warmed-up start and is taken through a cycle of acceleration, deceleration and idling with a maximum speed not exceeding 31 mph (50 km/h).

Constant Speed Test

The constant speed test at 56 mph (90 km/h) is intended to be representative of open road driving. It may be carried out in the laboratory or on a test track (under strictly controlled road and weather conditions).

Optional Constant Speed Test

This test is carried out at 75 mph (120 km/h) in a laboratory or on a test track. Although it is recognized that this test exceeds the UK maximum speed limit, it is included to illustrate to car drivers the worsening fuel consumption at higher speeds. It may also be useful to manufacturers exporting to some parts of Europe where speed limits are higher.

Only one production car is tested as a representative of each model. It must have been run in and have been driven for at least 1800 miles (3000 km) before testing. In some cases several models, which do not differ significantly in certain technical characteristics thought to be important in determining fuel consumption, may be grouped together into a 'class'. Only one car in the class needs to be tested.

Testing

The responsibility for testing lies with the manufacturers and importers themselves. They must either carry out the tests themselves or arrange for them to be carried out on their behalf. Department of Transport officials have the right to inspect the test laboratories and to witness tests in progress to ensure that they are being carried out correctly.

Manufacturers must submit their fuel consumption test results to the Department of Transport who record the results in an official fuel economy certificate.

25 Rental, Hiring and Leasing of Vehicles

Rental, hiring and leasing of commercial vehicles is a major and very cost-effective alternative to fleet ownership. Transport operating companies have shown increasing interest in the advantages of these means of vehicle acquisition. In many cases operators have reduced their owned fleets to the bare minimum required to service basic and predictable delivery requirements, topping up with short-term rental vehicles to meet peak trading demands, or using this source for the replacement of vehicles off the road for service, repair or annual test preparation. At the other extreme, firms have disposed completely of their owned fleets in favour of contract hire arrangements where vehicles are provided by a third-party contractor and maintenance is taken care of as part of the contractual arrangement. Between these extremes are the firms who use a combination of owned vehicles, long-term hired vehicles and short-term rental, to ensure the most economical and efficient overall transport operation whatever the seasonal fluctuations or other trading exigencies of their business.

These alternative means of adding vehicles to the fleet have both legal and operational implications; hence the reason for including this basic outline of the subject in the *Handbook*.

THE VEHICLE USER

One of the most significant legal points in connection with the renting, hiring or leasing of vehicles concerns the status of the vehicle 'user'. If the person or company renting or hiring a vehicle provides the driver, then that person or company, as the employer of the driver and consequently as the user of the vehicle, carries the full weight of legal responsibility for both the safe mechanical condition and the safe operation of the vehicle when it is on the road.

It is the user's responsibility to ensure that the vehicle complies fully with the Construction and Use Regulations in all respects but especially with regard to safety items such as brakes, lights, steering, horn, tyres, speedometer/ tachograph and vehicle markings. The fact that the rental or hiring company which owns the vehicle *should* ensure that all these items are in order (and indeed usually proclaims that it does ensure they are in order) makes no difference to where the blame lies and where the prosecution will be aimed if they are found not to be in order when the vehicle is being used on the road.

Furthermore, it is the user's responsibility to ensure that operator's licence provisions where applicable are fully complied with and that the drivers' hours and tachograph requirements are observed where these are applicable.

RENTAL

Rental of vehicles on a short-term basis of a few days or a few weeks, which is the usual arrangement, does not impose onerous contractual obligations on the hirer.

'O' Licensing Provisions

However, it does involve other legal obligations in respect of the vehicle itself and its use. For a start, much depends on the gross weight of the vehicle. If it is over 3.5 tonnes permissible maximum weight and has been rented for use in connection with a trade or business, then the person or firm renting it must hold an 'O' licence and there must be a margin on that licence to cover the renting of one or more additional vehicles.

There is no need to advise the Licensing Authority (LA) of details of the vehicle unless it is to be retained on hire for more than 28 days, after which time the LA must be notified so an 'O' licence windscreen disc can be issued for the vehicle. If the vehicle is rented for a shorter period and then returned to the rental company to be replaced by another vehicle, the LA does not have to be notified if the combined total of the two rental periods exceeds 28 days unless both are part of the same rental agreement.

If the over-3.5 tonnes vehicle is rented by a firm for use in another traffic area different from the one in which the 'O' licence is held, then an 'O' licence must be obtained in that other traffic area before a vehicle is permitted to operate from a base there.

Where the gross weight of the rented vehicle does not exceed 3.5 tonnes there are no legal obligations in respect of 'O' licensing unless it is used to tow a goods-carrying trailer with an unladen weight over 1020 kg when the combined weights may take it over the 3.5 tonnes limit and into 'O' licensing.

Whether or not the rented vehicle comes within the scope of 'O' licensing, the person or firm renting it carries the user responsibility for its safe mechanical condition when it is on the road. Consequently, if vehicle faults result in prosecution the user will have to pay any fines imposed (not the rental company) and the user's 'O' licence will be put in jeopardy (even if the vehicle is not specified on his 'O' licence). Therefore careful selection of a reputable rental company with high maintenance standards is essential.

Charges and Payments

Charges for rented vehicles are usually on a time plus mileage basis in accordance with published scales so there is little scope for improvement in prices, although large users can sometimes negotiate discounts.

The particular advantage of rental is that payment for a vehicle is only made when the use of the vehicle is really required, and then the payment is out of revenue and not out of capital reserves. Further, because rental is normally for short periods only, it is easy to establish the total costs involved because there are no additional costs for the upkeep of the vehicle. Maintenance, tyre replacements, licences and most other costs apart from fuel and insurance are built into the rental price.

The disadvantage of rental is its high price if vehicles are taken for longer periods, because the rental companies try to recover their costs over a short time and the price covers the fact that only a certain number of hire days can be sold in a period.

HIRE AND CONTRACT HIRE

Hiring of vehicles (or more specifically contract hire), as opposed to rental, implies a longer term arrangement with a more rigid agreement as to the obligations of the parties involved. Hiring arrangements vary considerably since the vehicle provider and the customer draw up a contract to incorporate the services required. There are two principal forms of contract hire – vehicles supplied with drivers and vehicles supplied without drivers.

The important difference is that in the former case the contract hire company, as the employer of the driver, is the 'user' of the vehicles in law and therefore holds the 'O' licence and shoulders the legal responsibilities previously described, while the hirer merely operates the vehicles exclusively to suit his requirements. However, in the latter case the hirer is the 'user' and 'O' licence holder and, as with vehicles purchased and leased with his own employee drivers at the wheel, he carries the full legal responsibilities.

Between these two categories, a package is made up to suit individual company needs. A complete package normally includes full maintenance (ie safety inspections, service and repairs), fuel, licensing, insurance, parking areas, administration (ie checking of records, etc), replacement vehicles to cover downtime at no extra cost and free driver replacement to cover holidays and sickness.

Contract hire charges are normally made on a time and mileage basis (ie a fixed or standing charge and running charge). These charges are either increased annually to cover the hiring company's increased costs or linked to a published index. Sometimes such items as fuel surcharges are raised.

Advantages of Contract Hire

This method of vehicle acquisition offers a number of advantages. Principally, there is no investment of capital (generally not even an initial deposit to be found) and cash flow for transport services is predictable throughout the year, thus allowing easy budgeting. One regular monthly invoice covers all capital and operating costs. The hire charges are fully allowable against tax.

Overall, full contract hire with driver is advantageous to the operator, because it relieves him of the burdens of capital expenditure on an ancillary activity and of a welter of legal responsibilities and yet provides him with the right vehicles for his exclusive use to fulfil his delivery requirements as he wishes. He thus has the best of both worlds – all his transport needs met without the major burdens usually encountered by own-fleet operators.

A further financial advantage can arise for a firm operating its own fleet but wishing to switch to contract hire to gain the benefits outlined. Contract hire companies will usually purchase a whole existing fleet and then contract-hire it back to the operator, thus still giving him resources to meet his transport needs and yet providing him with an immediate refund of the capital tied up in vehicles. This proposition can be used to advantage in relieving cash flow pressure.

Transfer of Undertakings Provisions (TUPE)

In recent years there has been considerable activity in firms switching from their own, in-house transport operations to contract hire, and from one contract hire firm to another. This trading of contracts has raised a number of problems, not least the contractual obligations of transferring employees between one employer and another. At one time it was thought that a loophole had been found in the law but now it is clear that the stringent provisions of the Transfer of Undertakings (Protection of Employment) Regulations 1981 apply in most such cases. These regulations are notoriously complex but failure to comply with their provisions could lead firms into substantial financial penalties and compensation claims.

Broadly, the regulations require that a firm acquiring a contract by which a transport operation is transferred, is obliged to take on the existing staff on payment terms and conditions which match those of the previous employer, or pay redundancy on terms equal to those which the employees would have secured under their contracts of employment with the previous employer.

LEASING

Leasing is a totally different concept from outright purchase or hire purchase in that the operator (ie the lessee) never actually owns the vehicle but he has the full use of it as though it were his own. It is also a different concept from rental and hiring arrangements in that it is purely a financial means of acquiring vehicles. In other words those putting up the money are not transport or vehicle operators, they are finance houses.

Several different forms of leasing are available (basically divided by the assumption of risk with the lessee taking the risk with a pure finance lease and the lessor retaining the risk with an operating lease) and legislation governing leasing arrangements is subject to change. Also, the way in which the accountancy profession treats leasing is subject to variation, so it is important to discuss any proposed leasing arrangement with a professional accountant before commitment to an agreement.

The general concept of leasing is that a finance house (ie the lessor) purchases a vehicle, for which the operator has specified his requirements and negotiated the price and any available discount from the supplier, and then it spreads the capital cost, interest charges, overhead costs and its profit margin over a period of time to determine the amount of the periodic repayments.

The three basic types of finance lease are as follows:

- Full amortization lease which runs for an agreed fixed period (the primary lease) followed by an optional secondary period (if the vehicle is still required) when the lease is extended for a nominal (ie 'peppercorn') rental.
- Open-ended lease which allows the lessee to terminate the agreement on payment of a previously agreed settlement figure at any time after a fixed period (normally one year).
- Balloon lease which has one large payment and a number of relatively low rental payments. The balloon can be at the beginning with a large payment to start the lease or at the end of the lease term with a large precalculated final payment reflecting a forecast residual value for the vehicle.

Usually this full range of choice only applies in the case of smaller vehicles. For heavy vehicles it is customary to apply 'full pay-out' types of lease, whereby the vehicle is fully paid for in the rentals with no residual value.

Because no capital outlay is involved in a leasing agreement, beyond the initial lease payment which is sometimes a number of monthly rentals lumped together, the lessee is able to obtain the vehicles he needs and yet still invest his own capital in more profitable business avenues.

Changes in accounting practice brought into effect (from the start of accounting years beginning after 30 June 1987) by the introduction of the accounting profession's Statement of Accounting Practice 21 (SSAP21) mean that no longer is it a case where leased vehicles do not appear as assets in the lessee's balance sheet. Consequently leased vehicles and plant appear on the balance sheet as assets matched by outstanding lease payments showing on the other side as liabilities.

Leasing is a complex financial area and, as already mentioned, it is important that proper professional advice is obtained before signing any agreement, otherwise the promised tax and other benefits may not materialize.

Where leasing is purely a financial arrangement, the advantages and disadvantages from an operational viewpoint are the same as for outright purchase. Because in principle the lessee operates the vehicle as though he owns it and he employs the driver, the full weight of legal responsibility, as already outlined, applies to him so he needs to have a full transport back-up of administration and operational staff, maintenance facilities and policies for selection of the correct vehicles and for replacement at the most economic intervals.

26 Vehicle Fuel Efficiency

Next to wage costs, fuel is the most expensive goods vehicle operating cost item, accounting for up to one-third of road transport operating costs. It is a high-cost commodity which is a major budget feature of all vehicle fleet operations. It is also subject to occasional and dramatic shortages as a result of political unrest in some of the major oil-producing countries, as we have seen with the Middle East. Scientists predict total extinction as world supplies of crude oil are consumed ever more rapidly by developed nations which have become increasingly dependent on transportation systems powered by oil-based fuels. Even the once much-heralded finds of oil in offshore waters of the British Isles and Eire are now known to have limited life expectancy.

While the search for and research into acceptable alternative fuels and power units goes on, it is important to take steps to minimize consumption of our present fuel supplies. As well as being a problem for nations, this is a problem which concerns all fleet operators, whatever their size. Besides any conscience they may have about energy conservation they will readily appreciate that fuel consumption must be reduced in the campaign to keep vehicle operating costs down.

FUEL AND VEHICLE

Fuel consumption is substantially related to the type of vehicle, its power unit and drive line, its mechanical condition, the use to which it is put and how it is driven. In recent times manufacturers have offered fuel economy models within their ranges so the cost-conscious operator can choose between economy or outright performance.

The fuel-conscious operator who is in a position to buy new vehicles will undoubtedly choose fuel economy models where these are suited to his particular needs. However, for the most part, fleet operators have to stick with the vehicles they already have in their fleets and are faced with the need to consider how improved fuel economy can be achieved with existing vehicles. Three principal areas exist for improvement in the vehicle itself. These are as follows:

- Mechanical condition.
- Efficient use.
- Addition of fuel economy aids.

Mechanical Condition

A vehicle which is poorly maintained will inevitably consume more fuel. Particular attention should be paid to efficient maintenance of the following items:

- **Fuel system** (fuel tank, pipe lines, filters, pump and injectors). There should be no leaks and the vehicle should not emit black smoke. Both of these are causes or consequences of excessive consumption as well as matters which could result in test failure and prosecution. Fuel pumps and injectors should be properly serviced as recommended by the manufacturers.
- **Wheels and brakes.** Wheels should turn freely and without any brake binding. Front wheels should be correctly aligned. Brake binding and misalignment cause unnecessary friction which is only overcome by the use of more fuel. Axles on bogies should be correctly aligned because tyres running at slip angles have a high rolling resistance and therefore are a source of increased fuel consumption.
- **Driving controls.** Throttle cables, clutch and brake pedals should be correctly adjusted so the driver has efficient control over the vehicle. In particular, engine tickover should be accurately adjusted to save throttle 'blipping' to keep it running when the vehicle is stationary.

Efficient Use

Inefficient use of vehicles constitutes the greatest waste of fuel. The following activities should be avoided by careful route planning, scheduling and prior thought about the cost consequences:

- Vehicles running long distances when only partially loaded.
- Vehicles covering excessive distances to reach their destination.
- Large vehicles being used for running errands or making small-item deliveries which could be accomplished more efficiently, and certainly more economically, by other means.
- Vehicles running empty.

It is frequently argued that traffic office staff have little control over these matters, since customer demands for orders and the need to give drivers freedom to choose routes are dictates which overrule efficient planning. Nevertheless, attempts should be made to persuade those concerned of the need for restraint, in the quest for saving fuel and thereby reducing costs.

FUEL ECONOMY AIDS

The quest for fuel saving has led to a market for economy aids which can be added to existing vehicles. These aids fall into three general categories:

- Streamlining devices such as cab-top air deflectors, under-bumper air dams, front corner deflectors for high trailers and box vans, in-fill pieces for lorry and trailer combinations and shaped cones for addition to the front of van bodies.
- Road speed governors which restrict maximum speed – one of the greatest causes of excessive fuel consumption (speed limiters are now a mandatory requirement on certain heavy vehicles – see Chapter 13).
- Engine fans and radiator shutters which are designed to ensure that diesel engines are always operating at the correct temperature to give the most efficient performance and fuel economy.

All these types of device can be economically justified to a varying degree, but it is important to note that fitting streamlining devices in isolation only reduces fuel consumption if vehicle speeds are kept down. If the driver is able to use the few extra miles per hour which these devices provide – which he will do unless otherwise restricted – then there will be little fuel saving and the cost of fitting would not be wholly justified.

FUEL AND TYRES

The type and condition of tyres on a vehicle play a significant part in its fuel consumption. It is a proven fact that the lower rolling resistance inherent in radial ply tyres adds considerably to the fuel economy of the vehicle compared to the greater resistance of cross-ply tyres.

Improvements in fuel consumption of 5 to 10 per cent can be expected from the use of radial ply tyres. Low-profile tyres which offer a number of operational benefits over conventional radial tyres – such as reduced platform height and reduced overall height – also offer further possibilities for fuel saving.

The savings mentioned will only be achieved if the tyres are in good condition, are correctly inflated to the manufacturer's recommended pressures and are properly matched, especially when used in twin-wheel combinations. Neglect of tyre pressures is common in fleets and underinflation is one of the major causes of tyre failure. It is also a major contributor to excessive fuel consumption.

FUEL AND THE DRIVER

Driving techniques, above all else, influence the overall fuel consumption of vehicles. A driver with a heavy right foot will negate all fuel-saving measures and devices and destroy any expectations of acceptable fuel consumption. Poor driving which has these consequences falls into two categories:

- High-speed driving.
- Erratic, stop-go driving.

Fast driving consumes excessive fuel: this fact is beyond question but the extent of the extra consumption is difficult to assess accurately. Tests carried out some time ago (but still making a valid point) by the National Freight Company with a 32-ton articulated vehicle on motorway operation indicated that fuel consumption increased quite dramatically when the vehicle was travelling at over 40 mph. In the tests, at 40 mph the fuel consumption was 10.5 mpg, at 50 mph this reduced by 2.6 mpg to 7.9 mpg and at 60 mph a further reduction of 1.5 mpg was experienced, making a 3.75 mpg difference between 40 mph and 60 mph travelling speeds. This represents a 35.7 per cent increase in fuel consumption. The real significance of these figures will be fully appreciated when annual motorway travel is calculated and this is multiplied by the increase in consumption, by the number of vehicles in the fleet and by the cost per gallon of diesel fuel.

In round figures the consequences of results of the calibre shown above may be as follows:*

Vehicles travel 40,000 miles per year on motorways.
At 40 mph, consumption is 10.5 mpg = 3809.5 gallons @ £4.08 per gallon = £15,542.76 pa
At 50 mph, consumption is 7.9 mpg = 5063.3 gallons @ £4.08 per gallon = £20,658.26 pa
At 60 mph, consumption is 6.4 mpg = 6250 gallons @ £4.08 per gallon = £25,500.00 pa
Extra consumption: 40 to 50 mph = 1253.8 gallons @ £4.08 = £5115.50 pa
50 to 60 mph = 1186.7 gallons @ £4.08 = £4841.74 pa

If five vehicles are involved the extra cost of fuel for 60 mph travel amounts to nearly £25,000 per year.

**NB: In the calculation above a June 2005 UK average diesel price of 89.7 pence per litre was converted to an equivalent gallon price.*

The effects of erratic driving are more difficult to determine in quantitative terms, but it is sufficient to say that it results in abnormally high fuel consumption as well as causing excessive wear and tear on vehicle components. Impatience behind the wheel and an inability to assess in time what is happening on the road ahead leads the driver to see-saw between fierce acceleration to keep up with the traffic and violent braking to avoid running into the vehicle in front. Hence, the excessive use of fuel.

More economical driving is achieved by concentration on the road and traffic conditions ahead, anticipating well in advance how the traffic flow will move and what is happening in front, so that acceleration and braking can be more progressive and a smooth passage assured.

One other fuel-saving practice which the driver can adopt is to stop the engine while the vehicle is stationary rather than letting it tick over for unnecessarily long periods. If he feels this is necessary because his battery is in poor condition, then it is much cheaper to deal with the battery and charging problems than pay for the extra fuel to compensate.

FUEL AND FLEET MANAGEMENT

Fleet operators can take a number of steps to reduce fuel consumption besides ensuring that the measures already mentioned are implemented.

Bulk Supplies/Buying

Control over the buying of bulk supplies and issues and over the buying of supplies outside from filling stations at the higher pump prices are important aspects for management attention. So too is accurate record keeping without which it is impossible to compare the fuel consumption of vehicles or to see which vehicles are consuming excessive amounts of fuel. Without records to identify these problems, there is no hope of remedying high fuel costs.

Recording Issues

Overfilling of vehicle tanks, causing spillage, is a common occurrence which wastes fuel and prevents accurate consumption records being obtained, as well as causing a mess and a health hazard. Accurate recording of issues against individual vehicles is another area which demands close attention. Modern electronic and computerized fuel-dispensing systems are available which ensure security of bulk supplies by preventing access except with a known key or card. This stops unauthorized drawing of fuel and it monitors issues to identified vehicles or key holders. The cost of such systems can be quickly recouped through savings in missing or unaccounted fuel and through better record keeping which enables high vehicle consumption to be quickly identified and investigated.

Buying Away from Base

Where bulk supplies are available at base, drivers should be discouraged from buying supplies from outside sources at higher prices except when absolutely necessary and even then they should buy only sufficient to get them home. Commonly, drivers fill tanks to the top which is enough to cover the trip home and to do further journeys as well on fuel which cost much more than that which they could have drawn once they got back to base. It is important to keep watch on the purchase of outside fuel supplies especially for cash because of the incentives which are offered to drivers to fill to the top. If outside fuel drawings are necessary then recognized bunkering card systems should be established.

Long-Range Tanks

In the case of vehicles which operate regularly on long distances – on international work, for example – the fitting of long-range tanks is an economic proposition because of the savings achieved by using bulk-purchased supplies. A word of caution though: the extra fuel carried should not be at the expense of payload unless this can be justified, and on international work there is the cost of fuel levies or taxes to be borne in mind. These are chargeable when entering some countries and are based on the amount of fuel in the tank. British hauliers returning to the UK may also find they have to pay an excess fuel tax on supplies bought outside the UK. (See also p 212–13.)

Following a number of cases where road hauliers have been stopped for using over-large diesel fuel tanks on vehicles, the RHA has reminded its members of the official definition for 'standard' fuel tanks.

Under Article 8a of Council Directive 92/81/EEC, the diesel tanks fitted to new vehicles are defined as follows.

> The tanks permanently fixed by the manufacturer to all motor vehicles of the same type as the vehicle in question and whose permanent fitting enables fuel to be used directly, both for the purpose of propulsion and, where appropriate, for the operation, during transport, of refrigeration and other systems.

In the case of used vehicles, the RHA advice is that:

> any tank already fitted to such a vehicle when purchased by its current owner, would satisfy the 'standard tank' criteria. This means that should the original tank have been replaced by a new, larger tank prior to purchase by a subsequent owner, then that larger tank would be regarded as having been fitted by the manufacturer for the purpose of the legal definition.

Customs and Excise has warned that the focus of attention on fuel import issues is on the misuse of duty relief and has emphasized that fuel brought into the UK may only be used by the particular vehicle carrying it and should not be offloaded from that vehicle for use by other vehicles or for transfer to a bulk fuel installation.

Route Planning/Scheduling

Better planning of vehicle schedules and routes offers the prospect of quite considerable fuel savings, besides other savings which may also result. A reduction in the miles travelled to fulfil particular delivery schedules will inevitably result in fuel savings. If the schedules can be planned so that fewer vehicles are needed to carry out the operation then, besides the broader savings in vehicle costs, the fleet as a whole will use less fuel. Therefore, the elimination of unnecessary trips or trips where vehicles are only partly loaded is a major priority in the search for fuel cost savings.

Tachographs in Fuel Saving

The use of tachographs in vehicles and detailed analysis of tachograph charts are steps which have provided many fleet operators with fuel savings if nothing else. The value of chart recordings in this connection should not be overlooked.

Agency Cards

There are a number of agency and fuel cards available to fleet operators. Generally they can be categorized between those from the major oil companies (eg Shell, Esso, BP) and those from other commercial organizations (eg All-Star, Overdrive, Dial Card, Petrocheck, BRS Transcard).

Principally these systems offer the opportunity to buy fuel at pump prices, paying only when fuel is actually purchased. The differences in the various schemes is in the levels of service which they offer.

Shell Scheme

Shell claims to lead the field with its portfolio of five fuel cards, each designed to suit a particular business need. For example, the Shell Gold Card is intended mainly for the car and light vehicle operator with 20-plus vehicles, but for the heavy truck operator the convenient Shell agency card is more appropriate.

The Shell scheme offers a complete fleet management service enabling the authorized card holder to purchase all the legitimate items a vehicle user would need on the road (eg fuel, oil, tyres, batteries, windscreens, other parts, repair assistance). Prices charged are those which reflect maximum fleet discount no matter who or where the supplier. Thus the small fleet operator gets the same benefit as the large operator. Purchases are consolidated on to one VAT invoice and these invoices are submitted twice monthly. A 25-day credit period is given.

In addition to the invoice, Shell sends the operator three computer 'reports' as follows:

- Vehicle transaction report and analysis.
- Cost-centre summary.
- Fleet report.

Use of the Shell Gold Card automatically confers AA membership and this can be voluntarily extended to include AA's Relay and Home Start services. The balance of any existing AA membership at the time of joining the Shell scheme is refunded by the AA. The Shell European agency card, EuroShell, enables holders to obtain fuel and service at some 11,500 locations in 27 countries, of which very many have facilities to accommodate heavy goods vehicles, and pay tolls on certain European motorways (eg in France, Portugal and Austria) and at a number of tunnels.

BP Scheme

The BP Agency Card can be used at some 5000 filling stations in the UK, but only for fuel and lubricants. There is no charge for the card and fuel is paid for at the average BP UK pump price for the appropriate grade of fuel irrespective of where the fuel is purchased. The system provides the operator with an average of 27 days of credit – a useful saving device in these needy days of stringent economies.

An international version of the card enables holders to obtain fuel at over 6000 service stations in 14 European countries.

Total Scheme

Total's Eurotrafic card can be used to pay for all on-the-road expenses such as fuel, lubricants, accessories and tolls, in 9* partner networks with more than 11,000 sites spread out across 14 European countries (and Morocco). The scheme also offers VAT recovery abroad, a breakdown and recovery service and a fines payment system.

Detailed statements enable a check to be made of fuel consumption, expenditure, mileage and so on for every vehicle on the scheme. A 'TOTAL*Card*' is available for use solely in the UK.

**NB: The partner networks are TOTAL, FINA, ELF, CEPSA, ERG, LOMO, SVG, WESTFALEN and DCB.*

Internet address: www.eurotrafic.total.com and www.totalcard.co.uk

Automatic Bunkering

A number of automatic diesel fuel bunkering systems have been established recently to improve fuel supply services to heavy truck operators. Notable among these are Kuwait Petroleum (GB) Limited (selling under the Q8 brand name) which has introduced its International Diesel Service (IDS) in the UK, in addition to its Europe-wide network. The Kuwait scheme features fully automatic diesel sites, operating round the clock every day of the year, at which card-holding truck drivers can obtain fuel at an agreed price. The scheme is secure because the cards have a unique and secret four-digit PIN code which prevent unauthorized use. The latest-technology pumps enable drivers to refuel rapidly in their own currency and their own language. The operator receives a comprehensive invoice at regular intervals with fuel drawn charged at a competitive price, with VAT recorded and a facility to reclaim VAT where appropriate.

A basically similar scheme is operated by Mobil Oil with its Mobil Diesel Club (MDC). The company has about 100 outlets on key truck routes in 14 countries including the UK. A security-coded card is used as described above to obtain fuel from automatic high-speed pumps.

Both Phillips Petroleum (RouteMate) and Gulf Oil (GB) Limited (G-Stop) have agency systems providing bunkering networks with competitive fuel price structures. Other systems include independents such as Keyfuels and Fuelink (NFC Group).

FUEL ECONOMY CHECKLIST

Check

- Fuel systems free from leaks.
- Fuel pump and injectors serviced and correctly adjusted.
- Exhaust not emitting black smoke.
- Air cleaners not blocked.
- Engine operating at correct temperature.
- Wheels turning freely.
- Controls properly adjusted and lubricated.

Tyres

- Condition and inflation pressures.
- Possibility of changing to radials on all vehicles.

Drivers

- Speed limits not being exceeded.
- Driving methods smooth and gentle.
- Engines stopped when vehicle standing.

Management

- Control over supplies and issues.
- Avoidance of spillage, loss and unauthorized use.
- Purchases from outside suppliers kept to a minimum.

- Record systems accurate and up-to-date.
- Possibility of installing fuel issue and monitoring systems.
- Possibility of fitting fuel economy aids (deflectors, engine fans, etc).
- Possibility of using long-range fuel tanks on vehicles.
- Routeing and scheduling practices to reduce wasted journeys and unnecessary mileage.
- Fuel economy programme to ensure all possible steps being implemented efficiently and recorded accurately.

ENERGY EFFICIENCY BEST PRACTICE PROGRAMME

The Government has produced a comprehensive *Fuel Management Guide* for the road haulage industry, providing advice on fuel economy, under its Energy Efficiency Best Practice Programme (EEBPP). The loose-leaf guide covers many aspects of best practice in fuel efficiency, including easy methods of capturing and monitoring of data, vehicle specification and maintenance as well as driver training.

The Guide, which complements the *Save It: The Road to Fuel Efficiency* video, is available free to operators. Field trials with a range of operators using the techniques described in the guide and video have shown fuel consumption savings ranging from 7–15 per cent. Copies of the *Guide* and the video can be obtained by contacting the Environment and Energy Hotline on 0800 585 794 or the EEBPP on 01235 43285.

NB: The Energy Efficiency Best Practice Programme promotes energy efficiency through free impartial, authoritative information and advice for industry, buildings and transport. The website can be found on: www.energy-efficiency.gov.uk/transport

FUEL WASTAGE

According to a recent Government press release ('Recent survey reveals fleet managers' ignorance of fuel costs') from its Energy Efficiency Bureau (ETSU), of the 265 fleets surveyed only 30 per cent knew their total expenditure on fuel while only 20 per cent knew how much fuel their fleets actually consumed.

With fuel duty set to rise annually by 6 per cent above inflation, there's never been a greater need to control fuel costs and if fleet managers took advantage of the EEB's Energy Efficiency Best Practice Programme, they would not only save £220 million a year but would also cut CO2 emissions by over a million tonnes annually.

ETSU can be contacted at: Harwell, Didcot, Oxfordshire, OX11 0RA; (tel: 01235 436747; fax: 01235 433066; e-mail etsuenq@aeat.co.uk).

REDUCING FUEL BILLS

The University of Huddersfield has published a 20-point plan to help transport and distribution companies reduce their fuel bills following recent increases in fuel duty. It says the following aspects are key to improving fuel efficiency:

- Ensure that you have accurate fuel consumption figures.
- Get the best from your present fleet.
- Make fuel-efficient purchasing decisions when buying new or second-hand equipment.

It also suggests following the points listed below.

- Check that weekly or monthly averages are produced by total distance divided by total fuel used rather than average of the daily averages. Quarterly summaries (using average of the daily averages) have been found to be inaccurate by as much as half a mile per gallon (mpg).
- In most cases there is a seasonal pattern to mpg, which peaks in July and August and bottoms out in December and February. It is very important to know this if you are going to test products that claim to improve mpg.
- When you find a large discrepancy in daily mpg, investigate it. Don't just average it or ignore it. Take measures to prevent it from happening again.
- If you want to accurately determine the effect of different equipment under controlled conditions, then enter the Institute of Road Transport Engineers (IRTE) fuel trials that are run every year in June. It is a lot cheaper than hiring a facility on your own and you get the benefit of mixing with like-minded fleet managers and engineers.
- Driver training consistently achieves better miles per gallon (mpg), but it needs a reinforcement mechanism otherwise it will fail.
- Reinforcement mechanisms for fuel efficient driving can be:
  - simple feedback on a noticeboard;
  - individual letters to drivers; or
  - a fuel bonus; however, while an annual bonus can be based on the annual average mpg, shorter-term bonus systems (eg weekly or monthly) should not use the annual average mpg. (Remember too that bonuses must not endanger road safety.)
- The most senior person with an LGV licence should be trained first.
- Identify the most fuel-efficient vehicles and, if possible, bearing in mind other operational factors, place them on the operations or routes that use the most fuel.
- Think of aerodynamics. For example, ensure that the gap between the back of the cab and the front of the trailer is minimized to reduce aerodynamic resistance. Tippers with easy sheets should have them closed when empty to prevent the airflow hitting the inside of the tailboard.
- If adjustable air deflectors are fitted, get the drivers to adjust them for maximum effect. If the deflector is too low, you will see a tide mark on the front of the trailer.
- Specify the correct bodywork – it should be no higher or wider than the job requires.
- Monitor maintenance records – poor mpg and short brake-lining life are good indicators of a driving style that wastes fuel.
- When buying new vehicles, calculate which is best over the life of the vehicle. Is there a better residual at the end of the vehicle's life from a larger engine, or are there reduced fuel costs from a smaller engine that is just as capable of doing the job? Remember to include the effect of the 'fuel escalator' in the fuel costs.
- Specify trailers or bodywork with rounded leading edges (minimum 200 mm radius).
- Aerodynamic aids may not be cost-effective on vehicles that do not undertake long, high-speed journeys as part of their regular work.

- Beware of claims made for aerodynamic equipment tested at 56 mph and translating the saving to your vehicle(s). Aerodynamics is highly sensitive to speed. As a rough guide, calculate the average speed of your vehicle(s) and ask for test results conducted at that speed.
- When buying a new vehicle, consider specifying a tachograph with the fourth needle activated to record engine speed. While you can get more information from electronic systems, activating the fourth needle will be far cheaper. However, you must be prepared to use the information, otherwise you are wasting your money.
- Specify and activate an engine speed limiter. (In some vehicles these are a legal requirement.)
- When buying a second-hand vehicle, take it for a test drive and note the engine speed at 56 mph. If the vehicle is being purchased for medium or long-distance work, you do not want to purchase a vehicle that is geared for local work. If you get this wrong, you will end up cruising at too high an engine speed and subsequently wasting a lot of fuel.
- When purchasing a new vehicle, get the manufacturer to provide free driver training. Most do, now, so take advantage of it.

CITY PETROL

City petrol is a specially formulated unleaded petrol designed to reduce the emission of harmful substances. It contains 10 times less sulphur than standard unleaded petrol, as well as lower levels of other chemicals such as aromatics, benzene and olefins. This greatly reduces the emission of harmful substances like sulphur dioxide, carbon monoxide and producers of smog, thereby helping to improve air quality.

Any vehicle, whether fitted with a catalytic converter or not, can use this petrol and no alterations are needed to the vehicle or engine adjustments, allowing users to switch between ordinary unleaded and city petrol at any time.

GAS-POWERED HEAVY TRUCKS

Increasingly, environment-conscious transport operators are looking to natural gas for powering heavy vehicles. Among the key players is the Somerfield supermarket group, which has put 300 compressed natural gas-powered, Detroit-engined ERF vehicles into service. Besides the savings in fuel costs, these vehicles are much quieter than conventional diesel-powered vehicles; quieter than a saloon car, it is said, thus allowing night-time deliveries to stores without upsetting local residents.

There are three main types of gaseous fuels in use in the UK:

1. Liquefied petroleum gas (LPG).
2. Compressed Natural Gas (CNG).
3. Liquefied Natural Gas (LNG).

All these three types of gas are used by both van and truck operators due to their environmental benefits, particularly:

- clean exhaust emissions;
- reduced engine noise;

- favourable duty treatment (ie 29 per cent reduction in road fuel gas duty in the 1999 Budget).

The Government's National Air Quality Strategy requires the monitoring of airborne pollutants such as NOx (oxides of nitrogen) and particulates while the Air Quality Regulations 1997 oblige local authorities to assess air quality. As a result of this, many councils are now running gas-fuelled vehicles.

As well as providing an attractive rate of duty, the Government is encouraging the use of alternative fuels through its Powershift programme. It offers contributions of between 25 and 75 per cent of the premium an operator pays for a gas-powered van or truck, over and above the cost of a conventionally fuelled version.

Further details can be obtained from the Energy Saving Trust; tel: 020 7222 0101. See also the website: www.transportaction.org.uk

LOW-SULPHUR DIESEL

The International Road Transport Union (IRU) is calling for supplies of low-sulphur diesel fuel to be made readily available throughout Europe long before the EU's current proposal for universal supplies from 2009. With legislation requiring all new heavy vehicles from October 2005 to meet Euro-4 exhaust emission standards, this fuel is essential since these engines need fuel that does not exceed a sulphur content of 10 parts per million (ppm) whereas current diesel supplies have a sulphur content of up to 50 ppm. The IRU is concerned that without adequate supplies of the low-sulphur diesel operators will not be inclined to invest in new Euro-4 vehicles until they are absolutely forced to do so by compulsory legislation.

27 Mobile Communications and Information Technology in Transport

There has been a significant increase in interest in all types of mobile communications in recent years. No longer is the in-vehicle radio or radio telephone an executive toy to be fitted only in the chief executive's car. These days such equipment has proved and continues to prove itself to be a very cost-effective and efficient aid to a wide range of vehicle users from the company chairperson down through all levels of executive vehicle users, to the sales representative, the service engineer and the delivery driver. In fact, wherever it can be established that there is a need and justification for the vehicle driver to be in contact with his base, with colleagues, with customers and with others, there are the requirements for vehicle-based mobile communications.

Before looking at the systems of communication available it is useful to consider just what the benefits are to have vehicle drivers – whatever their status or purpose – in contact with others outside the vehicle. It is widely recognized that once a person gets into a vehicle and drives off he is totally cut off from his workplace and the people with whom he normally deals in the way of business. And until he arrives at a known destination where a message can be relayed to him, or unless he manages to find a roadside telephone which operates (it is commonly felt that few of them do so when needed), he remains out of contact and out of touch with what is going on in his business or the business of his employer.

Clearly, in these days of high costs, competitive marketplaces and a fast pace of business life, and taking account of technological developments, this is becoming an unacceptable penalty of having people travel by road, especially during working hours. The inability to contact a top executive could, in the extreme, result in missed opportunities in business deals. The inability to contact a sales representative or a service engineer could mean a lost order or at least an irate customer; the same applies with a delivery driver. In addition, the inability to contact drivers could mean wasted journeys because of cancellations or changed plans which arise after they have set out on their journey. It could mean a driver getting back to base and having to go back to a customer visited earlier, simply because he could not be contacted en route to be warned of late, or forgotten, orders or items.

It is a common experience in transport operations for considerable cost to be wasted through late, changed, redirected and cancelled orders and instructions. Try as they might, transport or fleet managers can rarely avoid their share of these annoying frustrations and there is nothing that they can do, usually because they cannot contact the driver while he is travelling. Vehicle-based mobile communications at today's level of sophistication can change all this and the wasted costs of the past can be turned into savings and even into profit quite simply by being able to contact the driver, relay details of the changed plans and generally divert vehicles to meet the current needs of the business. The savings in wasted time and miles, the avoidance of heavy vehicles returning home empty, because they can be directed to pick up return loads, and the response to last-minute customer demands are significant benefits which in themselves, or with other benefits, add up to offset the capital costs of buying and installing communications equipment and the ongoing costs of rentals and call charges.

CHOICE OF COMMUNICATIONS

Mobile communications can be reviewed under four broad headings as follows:

- CB radio;
- radio paging;
- private mobile radio (PMR);
- cellular telephone.

CB Radio

Citizens' Band (ie CB) radio was officially inaugurated in Britain in 1982 when the law made it permissible to operate such systems, which had hitherto been illegal and had caused difficulty because their use interfered with the radio links of the emergency services and other official networks. Initially, CB was used by enthusiasts as a means of 'friendly' communication, to chat to other users and generally communicate non-business information. In time, however, it proved capable of serving more important needs such as the reporting of accidents and other emergencies, breakdowns, road blockages, diversions and so on. Similarly, it proved to have some use in providing communication between vehicle drivers and their base – within limited range – for the purposes of passing information about loads, schedules and changed instructions, for example.

Despite its obvious use for these purposes, CB has significant disadvantages: the frequencies are cluttered with undisciplined, long, sometimes foul and frequently frivolous conversation and chatter which can be heard by all users due to lack of privacy. This in itself is another disadvantage, along with the general interference experienced and the congestion on channels. Furthermore, the equipment itself has its limitations in terms of power and range, and in some areas there is less than satisfactory reception.

Although CB use has declined over recent years from a peak of around 300,000 licences in 1980, some 33,000 licences are still current, mainly in use by truck drivers.

Licences

CB users (who must be over 14 years of age) must, by law, obtain an annual radio licence from the Radio Licensing Centre, Subscription Services, PO Box 884, Bristol BS99 5LF (tel: 0117 925 8333) at a cost of £15 per annum. The licence permits the licensee to operate using only apparatus which conforms in all respects to the Department of Trade and Industry (DTI) specifications MPT 1320 and/or MPT 1321. Suitable equipment may be recognized by the presence of one or other of the DTI approval marks.

Radio Pagers

Radio pagers (commonly called bleepers) are a portable means of contact but only on a one-way basis from the sender to the receiver. Pagers fall into two main categories, tone pagers and voice pagers. With either type a person can be alerted to the fact that he is required and with more sophisticated equipment can be advised by varying tones to follow specific predetermined instructions, for example, to ring one telephone number or another. Voice pagers or 'talking bleepers' convey a spoken message which is usually repeated twice, thereby alerting the user to follow specific courses of action. Some pagers are capable of alerting the user by vibration (silent pagers) so that outsiders are not aware of the sound or so that the user is not interrupted in mid-conversation by an audible 'bleep'. Others have the ability to display messages on a liquid crystal screen from a text memory of up to some 800 characters.

While the cost of pagers is relatively low (usually only a few pence per day) and the unit itself is small and unobtrusive to carry around in the pocket (some are no bigger than credit cards), there are disadvantages to their use. The first is the limited range (generally not more than about 15 miles) and the second is the need for the receiver to find a working telephone in order to make the call for which he has been alerted.

Private Mobile Radio (PMR)

Portable radios are used for private two-way communication usually between a base station and a number of mobiles (ie radio-equipped vehicles) with the added facility in some cases of the mobile units being able to talk to each other. Generally, mobile radio operates over a limited range of some 10 to 20 miles, depending on location and the height of the base station aerial. Much depends on the type of terrain between the base station and the mobile unit. In open country far greater range may be obtained than in a city with built-up areas and many tall buildings.

Many systems have the disadvantage that anybody in the vehicle or within earshot can hear the message being relayed and that all mobile units hear a message intended for one only. However, some equipment has a facility for selective calling so that only one unit need be contacted at a time.

In the main, mobile radio is restricted to a closed system but there are facilities whereby this can be extended by linking into one of the national relay networks which have many base stations throughout the country. The Securicor 'Relayfone' system is a good example. Developments are in hand to allow, in the future, interconnection of private mobile radio systems into the public switched telephone network (PSTN).

Typical of PMR systems are those used by local taxi services where the driver has a hand-held microphone with an 'ON/OFF' switch. This PTT (press to talk) switch must be depressed to enable the driver to talk to the base station and then released while he listens to the returning message. The disadvantage of this is the road safety risk created by a driver trying to control his vehicle and operate the microphone switch (the *Highway Code* warns against such practices, and the police are alert to this habit and will take action against offending drivers).

Mobile radio operates on a number of alternative radio frequencies as follows:

- **Low band** VHF (25–50 MHz) provides the greatest range but it suffers from high noise levels and heavy channel loadings (used by police and emergency services).
- **High band** VHF (150–174 MHz) provides good coverage in built-up areas with less noise but there is still heavy usage of the channels.
- **UHF** (450–512 MHz) has a much shorter range than VHF but there is much less congestion on the channels and it provides good penetration in urban areas where there are many buildings and tall structures.

Within these radio frequencies equipment may be obtained for AM or FM operation.

Band 3

Increased demand for mobile communication has led to the allocation by the Government of the VHF slot left vacant by the old 405-line black and white TV network for use in providing communications services. This is now called Band 3 PMR and is operated by the two official franchise holders GEC and Band Three Radio each with 200 channels.

This system will be of interest to those requiring only brief communication (ie not full conversation) between base station and vehicle (mobile unit), but with little need for communication outside, and who wish to avoid the relatively high call costs of cell-phones and the rather higher capital costs of the cellular telephone units themselves.

Mobile Phones

The advent of the mobile phone system has revolutionized mobile communications. Today it is possible to have a telephone interconnected to the national and international telephone networks from a vehicle-based (ie mobile) installation or from a set carried neatly in an executive briefcase or a jacket pocket. Such systems provide the user with the facility to dial direct to almost any telephone number in the UK or to reach such numbers via the British Telecom operator, and to make international calls and calls to any other cellular telephone. Similarly, any telephone user can dial direct to a mobile cell-phone number.

The principle of the Total Access Communications System (TACS) cellular telephone system is that instead of a connection by wire as with the conventional telephone, the link is made by radio airwaves in the 900 MHz (also 1800 MHz) radio band frequency using only 50 MHz bandwidth. The UK is now divided into a number of individual cells like a honeycomb. Each cell is anything from 2 kilometres to 30 kilometres across, with a transceiver which relays cellular calls to and from the normal public switched telephone network (PSTN), as well as from one cellular telephone to another.

A central computer (the brains of the system) monitors all traffic in the system and switches calls from one transceiver to another as the mobile cell-phone user travels from one cell into another. Calls go through both the cellular network and the British Telecom network, which is why call charges are higher than with the normal PSTN system and especially in the London area.

The Government has licensed four operators to provide the network, namely BTCellnet, Orange, One-2-One and Vodafone, and currently some 90 per cent of the UK population is covered by these systems.

Equipment for operation within the cellular telephone system can be divided into three groups:

- Mobile units (eg installed in vehicles).
- Transportable units (for use in vehicles or can be carried in a briefcase for example).
- Portable units (small units which are plugged in for use with hands-free installations).

Voice Activation

A relatively new development in vehicle-based mobile units is voice activation, to overcome the problems (and illegalities*) of answering the telephone and dialling numbers for outward calls while actually driving the vehicle. This equipment is programmed to recognize a voice signal (usually just a single word) spoken into the handset which sets off the dialling of a predetermined number (eg base, office, home). Other equipment is dashboard mounted so calls can be received and made 'hands off' to avoid the road safety risks and contravention of the advice given in the new *Highway Code* against using 'phones' while on the move.

**NB:* *While it is not strictly illegal to answer or dial out when driving, the police may prosecute for not exercising full control of the vehicle.*

Equipment and Charges

A wide range of equipment is currently available from many suppliers in the market, operating on either or both Cellnet and Vodafone systems. Equipment can be purchased outright, leased or rented for a short period. The costs and charges incurred fall under the following headings:

- Purchase/lease of equipment.
- Installation charges (ie fitting transceiver and aerial to vehicle).
- Connection charges (once only).
- Monthly subscription charge.
- Call charges.

Call charges are invoiced to the user on a monthly or quarterly basis, along with the subscription charge. These call charges are not governed by distance in the same way as normal BT call charges, thus a local call costs the same as a long distance call of the same duration. Outsiders making calls from a BT (ie PSTN) telephone to a cellular telephone number also incur higher costs at the 'M' rate (currently equivalent to a call from the UK to the Republic of Ireland).

Developments

Such is the rapid pace of development in this field that already a wide range of 'add on' facilities are, or soon will be, available to users. Links to data transmission equipment to allow communication between computers, the ability to send telexes and to interface with fax machines are all current possibilities which substantially extend the use of a mobile or transportable cellular telephone.

Choosing Cell-Phone Suppliers

Potential cell-phone users should contact a number of suppliers for details and demonstrations of the variety of equipment available before selecting alternatives and making final decisions. The RHA has negotiated a special deal with Racal Vodac for its members. In particular, attention should be given to the strength of signal obtained on each system (Cellnet or Vodafone) in certain areas, the efficiency of installation and the back-up services provided, as well as the equipment itself. There is a great deal of choice and competition in this field so keen prices should be obtained.

Warning

It is useful to caution users and potential users of cellular telephones about three particular matters as follows:

- Proper installation of equipment in vehicles is essential for both efficient operation and for safety reasons. Special care is needed in the case of installation in heavy vehicles with 24 volt electrical systems to avoid wiring faults and other electrical problems.

- Insurance of equipment in vehicles is important; such equipment is highly attractive to thieves. If stolen or lost in a vehicle accident or fire an insurer may decline to accept a claim if the installation of the equipment had not been notified beforehand. In general, motor insurance does not cover a cellular phone. It is relatively easy for the subscriber to prevent fraudulent use of a stolen set by notifying the air-time retailer. The retailer can disconnect the stolen unit remotely, thus preventing further use. Mobile cell phones are not easy to steal because the handset, transceiver and wiring loom have to be removed.
- Users should be aware that foreign customs officials may impound mobile communications handsets when entering certain countries because use of the equipment is not compatible with overseas telecommunications networks and might interfere with emergency services.

SAFETY

The *Highway Code* now includes a special section on the use of microphones and car telephones. It advises against using a hand-held microphone or telephone handset while driving, except in an emergency. It says the driver should only speak into a fixed, neckslung or clipped-on microphone when it would not distract attention from the road. Drivers should not stop on the hard shoulder of a motorway to answer or make a call, no matter how urgent.

SATELLITE-BASED COMMUNICATIONS SYSTEMS

Technological developments in satellite-based communication systems provide facilities for long-range telephone, fax and paging links between base and vehicle, as well as positive vehicle tracking systems. Such systems are now widely used in North America and are of increasing interest to UK and European transport fleet operators. In fact, the European Transport Council announced in 2002 the launch of the development phase of the European 'Galileo' global navigation satellite system (GNSS). Using the same basic technology that puts instantaneous live pictures from sporting events and news reports on our television screens, the precise location of vehicles can be pinpointed and messages passed, but on a one-way basis only from base to vehicle, not vice versa, so drivers cannot abuse the system by calling friends and relatives worldwide.

The Global Positioning System

The Global Positioning System (GPS), originally devised by the US military to assist navigation of both ground and air crew, incorporates over 24 satellites orbiting the earth in outer space. As the system developed, it has been utilized for civilian purposes, in particular the marine, leisure and commercial markets.

The nature of the satellite signal enables the GPS receiver, ranging in size from a standard mobile phone to a small television set, to track it accurately and efficiently. At least three satellites are in view of the receiver at any one time, and this decodes the information to display global position in longitude and latitude, and a number of different coordinate formats. Because it does not rely on radio signals GPS can be used anywhere in the world.

A relatively new use for GPS is for 'in-vehicle systems' which now are capable of incorporating CD, radio and GPS in a single unit. For navigation purposes, it is just a case of entering a destination (such as the name of a road) and within seconds the distance, route and travel time is calculated and displayed. A voice gives commands for turning at the correct junctions, while a map on the screen is a useful navigation aid.

The Blaupunkt system, for example, gives accurate information for navigation in the UK, but the system also incorporates CD ROMs that can be downloaded in seconds to allow navigation on almost any road in Europe. Norwich supplier of the Blaupunkt system, KTF, has had interest from the Ambulance Service which is planning to have a system of this type installed by 2002, and many systems have been sold to couriers and haulage firms.

For a free, no obligation, demonstration of the Blaupunkt system telephone KTF on 01603 486020.

Siemens VDO System

Siemens VDO Automotive has become the first company in the UK to incorporate the recently introduced Traffic Message Channel service (TNIC) into its after-market satellite navigation systems, enabling it to combine satellite navigation with real-time traffic information. TMC provides up-to-the minute information on current traffic conditions which is relayed to vehicles equipped with any VDO Dayton or Philips CARiN navigation system (the two brands are owned by Siemens VDO) using RDS TMC, a digital data stream carried by radio broadcaster Classic FM. Data is gathered from a wide variety of sources, including real-time information from vehicles actually out on the road. The system is also 'intelligent', being capable of differentiating between different types of delays, for example slow-moving traffic, accidents, fog or roadworks. TMC data can also be used by the navigation system to automatically plan alternative routes around incidents or congested areas and to predict arrival times. This form of real-time traffic information significantly adds to the usefulness of on-board navigation, turning it into a driving aid that can assist in reducing journey times, even for drivers travelling on routes they use regularly.

INFORMATION TECHNOLOGY IN TRANSPORT

Reports prepared by Management Consultants KPMG Peat Marwick McLintock for the Institute of Logistics indicate that the use of information technology in distribution is growing rapidly with expenditure on this element amounting to some 2 per cent of turnover.

Information technology (IT) is quite simply the means by which information is collected accurately, fully analysed, transmitted to all relevant functions within and without an organization, and disseminated by those charged with decision making. Many efficient transport and distribution operations these days are totally dependent on reliable IT systems (for example 'just-in-time' stocking principles hinge on rapid and efficient communications) for communication with customers, suppliers and contractors.

Within IT systems, one of the fastest-growing areas is that of Electronic Data Interchange (EDI) where transport and distributor firms receive orders, delivery documentation and invoices direct from their customers' computers into their own (compatible) systems via direct communications links, completely eliminating the delays, errors and other difficulties associated with the creating and movement of paper documents via postal and courier systems. EDI systems provide the benefits of rapid and accurate order passing and processing (at less direct cost) and with the potential for reducing stock levels.

KPMG noted in its reports to the Institute of Logistics the trend away from centralized computer systems for IT and a growth in the use of personal computers operated in linked systems or networked to a central mainframe computer.

ELECTRONIC DATA INTERCHANGE (EDI)

Widescale computerization in the road freight, shipping and export/import industries and by Customs and Excise departments has led to the introduction of modern systems for passing shipping and export/import data between traders, shippers and H.M. Customs. This is now done via direct or indirect computer links without the need for paper documents to pass or confirm information. This whole concept is known as EDI (Electronic Data Interchange), sometimes called paperless systems of trading, and its use results in automatic handling and actioning of data. To ensure uniformity in this method of trading the United Nations has set a standard for EDI known as EDIFACT (an acronym for Electronic Data Interchange for Administration, Commerce and Transport), under which there are standard procedures and messages for passing such matter as shipping instructions.

E-COMMERCE

More recently, the development of e-commerce (transactions via computer) has become a major talking point following on from the e-mail revolution, particularly in respect of the home shopping (also termed online or internet shopping) explosion. This latest trend for e-business has come about as a result of the combination of the latest technology in both telematics and computing power, and is likely to lead to a whole new logistics concept where shoppers order via the internet and expect their goods to be delivered direct to their home at a time to suit them, in effect replacing the usual shopping trip and weekly visit to the supermarket.

COMPUTERIZATION IN TRANSPORT

Computerization provides many facilities of value to the transport industry, especially vehicle routeing and load scheduling systems which determine the most cost-effective and load-efficient delivery routes and schedules.

Bar coding systems allow packages to be scanned for data both in the warehouse and on the vehicle, allowing data to be transferred back to base as soon as a consignment has been delivered to its destination. In the office, all manner of data can be held on file, readily accessible by staff, to give information about employees, vehicles, customers, loads, etc.

Computerized tachograph chart analysis systems can be installed for compliance with the legal requirement to check drivers' tachograph charts on a regular basis.

The current trend is towards the development of on-board systems to allow vehicle drivers to remain in constant communication via computer with their base for transmitting data and instructions. Precise vehicle location (ie to within a few metres irrespective of where the vehicle is) can be determined via the GPS (global positioning satellite) and other tracking systems, which is particularly useful for tracking high-value cargo vehicles and stolen vehicles.

Additionally a full range of data recording, information and analysis operations, including accounting, staff records and vehicle maintenance records, can be accessed by all relevant staff and management, both in the office and away from base on laptop or hand-held computers via an internet connection.

The use of e-mail and internet connection to the world wide web has led to enormous improvements in business efficiency and information, providing rapid person-to-person communication and access to unlimited data on just about every subject under the sun.

INTELLIGENT TRANSPORT SYSTEMS

Intelligent Transport Systems (ITS) that use telematics (ie the combination of information technology and telecommunications) provide online (electronic) information and control systems for all modes of transport, including vehicle-based systems. Typical of these systems used in road transport are those that operate:

- variable message signs (VMS) for use on, for example, motorways;
- on-board route and traffic information, such as Trafficmaster;
- traffic control and enforcement signs, such as the new-type digital speed cameras;
- electronic road charging;
- traffic monitoring;
- vehicle-to-base communications;
- electronic ticketing.

TELEROUTE

Teleroute is a leading European online matching service for freight and vehicle space, providing freight forwarders and road hauliers with opportunities to maximize efficiency and increase profits. Established in 1985, the system has over 56,000 freight offers posted and consulted online daily by more than 35,000 subscribers in 16 European countries. Each year more than 120 million tonnes of freight are efficiently transacted in 40 different countries.

Teleroute claims to provide:

- more efficient and strategic transport planning and reduced empty running;
- flexible reaction to customers' needs;
- better load combinations;
- new UK and international contacts;
- improved time management;
- immediate results with only data that really matters;
- increased profitability through optimized efficiency;
- lower communication costs.

THE DATA PROTECTION ACT 1998

This Act is designed to protect living and identifiable persons from the misuse or unauthorized disclosure of any information about them, including the expression of any personal opinion. Thus firms or individuals such as self-employed road hauliers who hold personal data on file, especially in electronic – ie computerized – format, about employees, job applicants and customers, fall within the scope of the Act. Exemptions apply when information is held:

- by persons for their own family or domestic purposes;
- for historical or statistical research;
- in certain instances relating to journalism, art and literature;
- for the purposes of national security.

In road haulage, for example, if customer names and addresses held on file include the name of an individual person (such as the managing director, transport manager, etc) then such data falls within scope of the Act. The Act defines data as information that is:

- being processed by means of automatic equipment;
- recorded to be processed by such equipment;
- kept as part of a filing system, or with that intention;
- relating to certain health and associated records.

Processing of personal data means obtaining, recording or holding information or data or carrying out any operation on it, including its:

- organization, adaptation or alteration;
- retrieval, consultation or use;
- disclosure by transmission, dissemination or otherwise making it available;
- alignment, combination, blocking, erasure or destruction.

Special rules apply to the processing of 'sensitive' data such as an individual's racial origin.

Any firm or individual holding personal data must: 1) register with the Data Protection Commissioner; 2) conform to the data protection principles set out in the Act.

It is a criminal offence (punishable by a fine of up to £5,000) for an unregistered person to hold or disclose personal data about any person, and no processing of data should take place until after registration under the Act and an entry has been made in the Data Protection Register.

It is the responsibility of data controllers (ie those who process and determine the use of personal data) to operate within the terms of their Register entries. They can be held liable to pay compensation for any damage or associated distress suffered as a result of holding or disclosing inaccurate personal data.

On registration, a data controller must notify the Data Protection Commissioner of:

- the name and address of their principal place of business; in the case of limited companies, the address of the registered office;
- the name and address of any nominated representative for the purposes of the Act, where this differs from that of the company's registered office;
- a description of the personal data to be or being processed, and the category or categories of data subject to which they relate;
- a description of the purpose or purposes for which the data are to be or are being processed;
- a description of the recipients to whom the data will be disclosed;
- the names of countries outside the European Economic Area (EEA) to which the data controller will directly or indirectly transfer the data.

It is illegal for a registered data user to:

- hold personal data of any description other than that specified in their entry in the Data Protection Register;
- hold or use any such data for purposes other than those described in their entry;
- obtain or hold data, or the information to be contained in such data, from any source that is not described in the register entry;
- disclose the data they hold to any person who is not described in the register entry, or directly or indirectly transfer data they hold to any country or territory outside the EEA which does not have similar data protection laws.

The Act sets out eight so-called 'data protection principles' under which registered data users must ensure that data is:

1. processed fairly and lawfully;
2. used only for specified and lawful purposes;
3. adequate, relevant and not excessive;
4. accurate and up to date;
5. not kept longer than necessary;
6. processed in accordance with the rights of data subjects under the Act;
7. protected against unauthorized or unlawful processing and against accidental loss, destruction and damage;
8. not transferred outside the European Economic Area (EEA), unless an adequate level of protection exists for the rights and freedoms of data subjects.

Individuals who know or believe that a firm or organization holds information about them have a statutory right to request details of the information held. Thus, for example, an employee may ask his employer about the information held about himself. This is a right under the Data Protection Act and under other employment legislation. In the case of health records, some information must be statutorily provided on request. However, there are certain areas where information about data held will not be given, for example, by the police about their records while 'enquiries' are in progress.

Warnings

Two disturbing factors have come to light recently regarding application of the Data Protection Act, as follows.

First, road hauliers could face fines of up to £5,000 for operating unregistered CCTV systems in their premises for security purposes. Such systems must be registered under the Act and a £35 registration fee paid to the Information Commissioner (formerly the Data Protection Registrar).

Second, demands (including official-looking red 'Final Demands') for a £95 payment plus VAT by an organization called The Data Protection Act Registration Agency, and a number of similar names, have been shown to be a scam. The registration fee, legally due under the Act, is payable direct to the Information Commissioner and is only £35.

28 Transport and the Environment

One of the most important and widely discussed issues in transport throughout the 1990s has been (and will continue to be in the new millennium) the impact which the industry as a whole, and heavy lorries in particular, have on the environment. So-called 'green' issues feature in every aspect of transport operation, from the siting of vehicle depots to the routeing of heavy goods traffic and the disposal of certain loads, especially waste.

This *Handbook* charts in earlier chapters the legal requirements regarding choice of vehicle operating centres to satisfy 'O' licensing requirements, and the need for licence applicants to advertise their proposals in this regard and defend their premises against potential environmental representations from local residents. In the technical section, legal requirements concerning the emission of noise, smoke, fumes and exhaust pollutants are detailed (together with new information on qualifying for lower rates of VED with 'Reduced Pollution' vehicles), and in Chapter 23 the subject of waste carriage is covered with descriptions of the legal requirements regarding hazardous waste and the more recent legislation on 'controlled' waste – which is basically everyday waste emanating from commercial and industrial premises and is now subject to stringent control as to its carriage and disposal.

Many of the UK's major transport groups (eg Exel Logistics – part of the NFC Group – the Transport Development Group and Wincanton Logistics) have developed clear environmental policies. These initiatives are concentrated mainly on fuel conservation at this stage which has a direct payback benefit by way of cost savings and an indirect benefit in the shape of an improved company 'image', as well as achieving actual reductions in the amount of carbon dioxide which their respective company vehicles discharge into the atmosphere. This in turn helps to reduce global warming which scientists have identified as being a problem of catastrophic proportions. However, there are many other ways in which these and other firms are contributing to the environmental effort, even down to the use of recycled paper for routine stationery needs, for example.

Major steps towards tackling the environment 'problem' have been taken with reports by the Royal Commission and more recently the Government's consultation documents on roads and on developing an integrated transport policy – both aiming to find ways of reducing traffic congestion and the adverse effects of air pollution from road traffic. These consultation exercises came to fruition in 1998 with a White Paper in which positive and sustainable solutions were put forward.

In the meantime, one step towards improving the environmental impact of transport and distribution operations has been taken by the Institute of Logistics with the publication of a three-part practical handbook covering three distinct aspects of the subject – namely, Volume 1 which deals with transport, Volume 2 which covers non-transport and specialized logistics operations and Volume 3 which concerns environmental management.

The Department for Transport (now the D*f*T) foresaw the need for initiative on the environmental front by publishing its booklet, *Transport and the Environment*, to coincide with the European Year of the Environment. This booklet was seen as a useful introduction for school pupils to the complex issues of environmental decision making and it provided a guide to environmentalists as to the Department's activities in this field. It considers the impact of the major transport modes (including road freighting) and examines the effects of the Channel Tunnel and the problems of transport in towns, as well as outlining the work of the Transport Research Laboratory (TRL).

Yet another publication devoted to the subject is produced by IBM UK Limited from a text researched and written by Dr Peter Davis of the National Materials Handling Centre, Cranfield, Bedford MK43 0AL, under the title *Gearing up for the Environment – A Guide for Managers in Distribution*. More recently, the Road Haulage Association (RHA) published an environmental *Code of Conduct* following publication of its *Care of the Environment* leaflet published in 1989.

Such is the concern for this topic that some of the trade associations now have full-time officers concentrating on environmental issues and many firms have linked responsibility in this area with those of safety, health matters and quality.

NEW DEVELOPMENTS

More recently, in 1999, an Environment and Energy Helpline for UK businesses was launched on 0800 585794. This initiative is jointly funded by the Department of Industry (DTI) and the D*f*T and provides free information and advice on such matters as:

- how to increase profits by reducing waste;
- how to reduce energy costs;
- how to cut water and effluent bills;
- how to control transport costs;
- how to improve environmental performance and image;
- legislation and how it affects businesses;
- cleaner and more efficient technologies.

Pollution Prevention and Control

The Pollution Prevention and Control Act 1999, based on the EU's Integrated Pollution Prevention Control (IPPC) Directive, brings in tougher environmental controls to be observed by individuals and business, particularly regarding the efficient use of raw materials and energy and the pollution of land, water and air. Clearly this legislation has significant implications for transport operators.

Oil Storage

New regulations, which came into effect from 1 September 2005 to protect against water pollution, will apply to road hauliers who store more than 200 litres of oil on any site that is located near to a river, waterway, bore hole or even close to the water table. These regulations, the *Control of Pollution (Oil Storage) Regulations 2001*, require the person with custody of the oil to carry out such works or to take such precautions or actions as are necessary to minimize the risk of oil-related water pollution. The regulations specify technical details relating to the storage facility, particularly its bund base and wall, which must be checked regularly, and the pipework, valves and gauges. It is essential that the tank or its pipework is not positioned where there is a risk from vehicle impact, especially when reversing.

IMPACT OF TRANSPORT

Transport impacts on the environment in a variety of ways; some are more distinctly controllable by fleet operators than others; some produce more tangible benefits both to the vehicle operator and to the community than others; but it has to be remembered that, inevitably, any form of transport, serving any and every need imaginable, has an adverse impact on the environment. There is no such thing as a totally environmentally acceptable form or means of transport. What there can be, however, are means and systems of transport which are more 'friendly' towards the environment, and which can be controlled and managed in such a way that the environmental impact is minimized.

So far as transport operations are concerned the following list identifies some of the main areas of environmental impact which can be, and should be, challenged by management:

Vehicle Depots

- siting;
- noise, fumes, vibration and light emitted;
- disposal of waste.

Vehicle Operations

- engine, exhaust, tyre, body and load noise;
- smoke, fumes, gases, spray emitted;
- fuel/oil consumption;
- visual impact;

- routes and schedules;
- load utilization.

In addition to these more environmentally controllable aspects of transport, there are other aspects which have a powerful impact, but over which the transport manager has virtually no control, namely vehicle design and manufacture, road planning and building, legislative controls which are not necessarily environmentally oriented and customer demand, which is influenced more by commercial pressure and financial consideration than by the vehicle operator's quest to, among other things, reduce fuel consumption, deliver during non-congested times, or combine loads to improve vehicle efficiency.

POSSIBLE SOLUTIONS

Transport managers and small fleet operators are undoubtedly limited in the steps they can individually take towards improving the environment, but this does not mean they should take no steps at all. Simple measures are available to them which will make a valuable contribution. The following list provides just a few examples:

In the Depot

- Examine the way that waste material is stored and disposed of.
- Ensure that controlled waste (see Chapter 23) is correctly and safely stored on site and then handed over to licensed disposal contractors.
- Avoid burning of waste which can cause pollution and lead to complaints.
- Ensure that recyclable material is identified and saved for proper disposal – including waste paper and packing materials from office and stores.
- Take steps to ensure that vehicle washing does not result in dirty (ie grease-laden) water draining on to neighbouring properties as well as into sewage systems.
- Ensure that oil and fuel spillages do not pollute drains.
- Consider the use of recycled products such as paper for administrative uses and packing.
- Undertake regular depot clean-up campaigns (in particular, ensuring that the outside appearance of the depot is 'environmentally friendly' to local residents, business visitors and others).

On the Vehicle

- Ensure that legal requirements regarding noise, smoke, exhaust emissions and spray suppression are fully complied with.
- Take steps to economize on fuel consumption*.
- Provide driver training to ensure courtesy and consideration on the road and the use of defensive driving methods* (which saves on wear costs).
- Fit speed limiters (now a legal requirement on certain vehicles – see Chapter 13).
- Ensure that drivers obey rules about parking and causing obstruction with their vehicles, and are aware of the problem of visual intrusion, noise and

vibration on domestic properties (contravention of these matters can jeopardize 'O' licences – see Chapter 1).
- Route vehicles and plan journeys to avoid congestion – ensure full utilization of vehicles to avoid extra or unnecessary journeys (which add to congestion, air pollution, and the operator's own costs).
- Consider the visual impact of vehicles in terms of their general appearance and livery (change aggressive liveries to present a 'softer' image).

**NB: These matters are discussed in more detail in Chapter 26.*

In the Community

- Consider the sponsorship of local community efforts to improve the environment and encourage staff to undertake environmental protection projects.

THE ENERGY SAVING POWERSHIFT PROGRAMME

The Powershift Programme has been established by the Energy Saving Trust (EST), an independent non-profit-making body, to:

- encourage the efficient use of all forms of energy;
- develop a market for Clean Fuel Vehicles (CFVs) in the UK; and
- create conditions in which these clean fuels can be made practically and economically viable.

The Powershift Programme offers financial help to vehicle operators seeking to purchase new vehicles that use the approved clean fuels, LPG and CNG (see definitions below), as their method of propulsion. Grants are also available for owners to convert vehicles from petrol/diesel use to LPG or CNG. Cars and light commercial vehicles are only eligible for funding if they are under one year old.

Funding is via the Powershift Register which contains the names of companies and vehicles that have gained the Programme's seal of approval. The register provides impartial advice allowing vehicle buyers to make informed choices about the different CFVs on offer. The level of funding is linked to the emissions and efficiency of the CFV types. Grants range from 25 to 75 per cent of the additional cost of the CFV compared to a equivalent conventionally fuelled vehicle.

Air pollution and global warming are the most damaging environmental effects of traffic growth, especially in urban areas. It is well known that traffic is the main cause of poor air quality in towns and cities and that vehicle emissions cause significant health problems. It is also widely recognized that the emission of carbon dioxide from fossil fuel-powered vehicles is a major contributor to global warming. For these reasons, the use of alternative fuels to replace diesel and petrol is playing an increasingly important role in reducing the impact of traffic on the environment.

Alternative fuels which are proven to offer an environmental benefit over traditional fuels include:

- Liquefied petroleum gas (LPG). Predominantly propane in liquid form at normal temperatures. Its advantage is that it produces fewer harmful gas emissions resulting in less damage to the environment.

- Natural gas, such as compressed natural gas (CNG). Predominantly methane, is substantially non-reactive in the urban environment and has a very low impact on air quality.
- Electric hybrid vehicles. Although still in their infancy and currently more suited to smaller vehicle operations, there are developments in heavier diesel/electric vehicles which run on battery power in urban areas and use the diesel engine in rural areas and for journey work.

NATURAL GAS VEHICLES

In an effort to make towns and cities cleaner and quieter places in which to live and work, the government is making progress in cutting vehicle emissions and noise by encouraging greater use of cleaner vehicles.

A UK Government Minister speaking at the Commercial Vehicle show at the NEC in Birmingham in May 2002 said that natural gas vehicles are part of the solution, offering transport operators an opportunity to contribute to a cleaner, quieter environment as well as save money on their fuel bills. The government is encouraging the wider availability of natural gas vehicles by offering demonstration grant funding through the Energy Saving Trust's 'TransportAction' initiative. This supports part of the cost of introducing new natural gas vehicles on to the market (see below).

Vehicles powered by natural gas – essentially the same product as used for domestic cooking and heating – can offer substantial environmental benefits, in particular compared with equivalent diesel vehicles. They are much quieter than diesel vehicles and emit lower levels of the two air pollutants of most concern in the UK, particulates and oxides of nitrogen.

ENERGY SAVING TRUST 'TRANSPORTACTION' INITIATIVE

The Government's Energy Saving Trust 'TransportAction' initiative, which is funded mainly by the D*f*T, aims to reduce the damaging effects of transport on the environment through its PowerShift and CleanUp programmes. PowerShift provides grants towards the additional cost of purchasing gas or electric powered vehicles, while the CleanUp campaign provides grants (of up to 75 per cent of the cost) towards the cost of fitting emissions reduction equipment to the most polluting diesel vehicles. The total budget for PowerShift and CleanUp for 2002/03 is around £20 million. Further information on these initiatives can be found at: www.transportaction.org.uk

THE ALTER EUROPE PROJECT

The 'Alter Europe Project', launched in 1998, aims to generate a demand via joint action by European cities for environmentally cleaner (ie low emission) vehicles. The idea is to encourage vehicle manufacturers to shift to volume production of such vehicles at lower cost.

The first European Alter Project Convention and Clean Vehicle Exhibition was held in Florence in October 1998.

More than 120 European cities have signed the Declaration of Florence, with support from 30 more. These cities have undertaken to:

- renew their own vehicle fleets on a low-emission basis;
- convert where possible to cleaner fuels and power systems;
- introduce and progressively extend urban low-emission or clean-air zones.

In return the Alter Europe Project undertakes to:

- organize conferences and workshops for the exchange of best practice and experience;
- enable producers to present information on the availability of cleaner fuels and vehicles;
- coordinate joint procurement of low-emission vehicles through the Alter Leasing and Procurement Consortium.

The 150 cities, towns and local authorities that support the Alter Europe Project are:

In the UK

Bath & North-East Somerset; Birmingham; Bradford; Bridgend; Bristol; Cardiff; Chester; Coventry; Derby; Edinburgh; Exeter; Hull; Leeds; Leicester; Luton; Liverpool; Merseyside Metropolitan Boroughs; Manchester; Middlesbrough; Newcastle; Nottingham; Oxford; Reading; Stratford-on-Avon; Swansea; Telford; Warwick; York; Cheshire; Derbyshire; Hampshire; Hertfordshire; Norfolk; Oxfordshire; Westminster; Camden; Southwark; Hammersmith; Fulham; Sutton; Kensington and Chelsea; Glasgow; Southampton; Plymouth; Brighton; Oldham; Wolverhampton and Northampton are expected.

In Austria

Vienna.

In Belgium

Ghent.

In Denmark

Aalborg.

In Finland

Helsinki is expected to join.

In France

Paris; Lyon; Clermont-Ferrand; Nantes; Nimes; Grenoble; La Rochelle; Strasbourg; Saint-Germain-en-Laye. The Alpes-Maritimes, Orne, Jura and Corsica regions are supporting the project.

In Germany

Stuttgart; Berlin; Augsburg; Darmstadt; Nurnberg; Osnabruck; Pinneberg. Expected to join are Bremen; Dresden; Freiburg; Forchheim; Hamburg; Heidelberg; Potsdam; Saarbucken; and Rheingau-Taunus-Kreis.

In Greece

Athens; Thessaloniki; Volos. Other Greek cities are expected to join.

In Ireland

South Dublin; Dunlaoghaire; Rathdown.

In Italy

Milan; Turin; Venice; Trieste; Genoa; Florence; Livorno; Bologna; Rome; Naples; Catania; Messina; Palermo; Verona; Cagliari; plus 20 other cities, and several provinces.

In The Netherlands

Amsterdam; Groningen; Hertogenbosch; North Brabant. The Hague is expected to join.

In Norway

Oslo is expected to join.

In Portugal

Lisbon; Braga; Oeiras; Funchal; Barreiro; Coimbra; Oporto. Sintra is expected to join.

In Spain

Barcelona; Bilbao, Catalonia; Castiglion; Seville; San Sebastian.

In Sweden

Stockholm; Göteborg; Malmö.

ROAD PRICING

Road pricing has been proclaimed as being one of the key solutions to the problems of congestion, carbon dioxide emissions and noise in cities and urban areas; problems that are increasing with the proliferation of motor vehicles (both private and commercial) in these areas. EU-financed research confirms that road pricing is an effective tool for reducing traffic problems. A number of research projects have been carried out under the EU's Fourth Framework Programme – some showing 10–20 per cent reductions in daily car travel as drivers switched to public transport.

Further information on these projects can be accessed via the EU's Research for Sustainable Mobility website: www.europa.eu.int/comm/transport/extra/home.html

29 Business Management in Transport

QUALITY MANAGEMENT

Increasingly, road hauliers are facing demands from customers, and from principal contractors, to meet recognized standards of quality assurance in the form of certification to the British Standard BS 5750, or alternative national standards in the United Kingdom – or to EN 29000 (Europe) or ISO 9000 (International). This chapter outlines the basic requirements for meeting quality assurance standards to achieve recognized certification to these standards.

A great deal has been written about the subject of 'quality', and many myths have been spread, but it is quite simply the concept of doing things right first time, and right every time. This saves having to repeat operations at extra cost, and annoyance to the customer, whether in production of goods or the provision of a service – road haulage for example. It means supplying customers with the service they need, not what suppliers think they can best provide. In haulage, it means, particularly, providing cost-effective deliveries – on time, to the right address with the load intact and undamaged and delivered by a courteous driver in a presentable vehicle. It does not mean a service which causes customer complaint.

The concept of 'quality' is not new. It has been applied to production for many years, especially by the Japanese, who have captured world markets for cars, motorcycles, electronics and cameras due to the inherent quality, reliability, durability and desirability of their products. Only more recently has the quality concept been applied to service industries, and particularly to road haulage. Achieving quality assurance (QA) certification involves complex steps, changed ideas, new thinking and acceptance that 'old ways' must be replaced by new methods, despite extra paperwork, form filling, writing of manuals, checking and rechecking of standards, and visits from inspectors. What are quality systems and quality management? A quality system is one where problems, queries, faults, and anything which could give rise to customer dissatisfaction or complaint, are identified and eliminated. Every aspect of operating procedure is critically examined to ensure that nothing unexpected (short of pure accident, and contingencies can even be established for these) can arise to jeopardize service to the customer. Quality management is the management of quality systems – a totally new way of doing business, hence the expression 'total quality management' (TQM).

QUALITY ASSESSMENT AND ACCREDITATION

Quality assessment and accreditation is the process by which a firm demonstrates to an accredited certification body that its services meet pre-established quality standards, followed by certification of this fact.

Accreditation

The British Standards Institution (BSI) standard for quality management systems is designated BS 5750. This incorporates the provisions of both the International and European standards and is fully accepted in the member states of the European Community and the 7 European Free Trade Association (EFTA) members – Norway, Sweden, Finland, Iceland, Austria, Switzerland and Liechtenstein. There are a number of separate constituent parts of BS 5750 and ISO 9000 but those which specifically relate to road haulage are BS 5750 part 2 and ISO 9002.

Firms whose quality systems meet specified standards may register with an approved body and, on satisfactory completion of the formalities, receive accreditation 'Registered Firm' status when they may use the accreditation body's symbol of approval on its company literature (ie letterheads, brochures) and on vehicles. A road haulier accredited under BS 5750 in the UK is additionally accepted as meeting both European and international quality standards, an essential ingredient for trading in the single European market.

Standards

The essence of a quality haulage service is that every step in fulfilling customer orders is undertaken in accordance with a documented standard – a set of rules governing the best way to operate and against which day-to-day operations are compared. Thus, performing to standard means performing as set out in the rules. BS 5750 is a standard for quality systems which identifies the basic disciplines and specifies the procedures and criteria to be applied to ensure that services are of a quality that will always meet specified customer requirements.

Assessment by Accreditation Certification Bodies

Assessment for accreditation involves a number of stages. Among these, one is the need to establish, document and maintain a quality system demonstrating a commitment to quality and to meeting customer service needs. Another is the selection of an appropriate accreditation body to which application for registration is made. This involves providing information about the business, its size and scope, the number of employees, how many locations are involved, the particular nature and manner of its operation and the services provided. Applicants must submit their documented quality system for examination and approval. The accreditation body checks this to ensure it covers all aspects of the quality system standard and then follows up with site visits to thoroughly review the operation in practice. These inspections – made by specialists in quality assessment – are a key element in quality assurance. Documented procedures are examined in detail and systems observed in operation to ensure that day-to-day procedures follow the documented quality system in every respect and comply with the laid-down standards of the accreditation body.

Monitoring

A process of continuous monitoring for compliance with standards is maintained through inspectors from the accreditation body making regular visits – up to four times each year. Any drop in performance will require renewed efforts to bring procedures back to standard. Continued failure to meet the standard will result in withdrawal of registration.

ESTABLISHING A QUALITY SYSTEM

To establish a quality system within a firm, it is necessary to identify key aspects of the business where quality principles are to be applied and record precisely how these should be carried out. This is achieved by translating individual tasks into descriptive text in manual form, and by making step-by-step checklists or by establishing Codes of Practice. The key aspects will include:

- Setting company policy and objectives for quality systems.
- Determining the structure of the organization and responsibilities.
- Preparing instructions and Codes of Practice for quality work standards.
- Monitoring sub-contracted supplies and services for quality.
- Establishing operational methods and controls to meet specified standards.
- Inspection and monitoring of the transport service.
- Controlling defective work.
- Determining corrective actions before problems arise.
- Measuring and recording quality performance.
- Determining training requirements.
- Establishing quality audit and review procedures.

Determining Customer Needs

Quality starts with the customer; the need to determine exactly what every customer wants. Never *assume* that one customer's service requirements are identical, or even remotely similar, to those of another. Each must be asked individually about his precise expectations when making enquiries and placing orders. They will differ widely: a customer may even have differing service requirements in differing circumstances. These requirements, whether few or many, whether standard in all circumstances or varying widely for individual consignments, must be clearly understood by the haulier, leaving no doubt whatsoever as to what is needed and expected in all circumstances. This information should be carefully recorded (and regularly updated) to form the basis for the quality system. It will become the standard against which all future responses to each customer's demands will be measured to ensure satisfaction.

The next step is to consider internal workings of the haulage operation to determine how these relate to the provision of a quality service. This task should be split into individual components or identifiable activities and for each there should be a documented set of procedures to be followed. The following are some examples of such activities:

- Receipt of customer orders.
- Planning daily work schedules.
- Allocation of work to vehicles and drivers.
- Preparation of collection/delivery note sets.
- Instruction of drivers.
- Preparation of vehicles.
- Confirmations to customers.
- Driver conduct on arrival.
- Driver debriefing on completion of deliveries.
- Recording/confirmation of work completed.
- Invoicing.

Documenting the System

The individual tasks mentioned above – and many more if appropriate – should be written down in the form of checklists to ensure that no matter who takes the order, who allocates the vehicle or driver, who prepares the delivery notes or who carries out any other of these functions, the established step-by-step procedure is followed and no key element is missed which could lead to customer dissatisfaction. Documenting the system fully, effectively means preparation of a 'Quality Manual'.

QUALITY PROCEDURES MANUAL

A quality procedures manual is a statement of the firm's commitment to quality, a constant reminder to management of their obligation to the firm's customers based on these documented procedures. It becomes the bible of operating practice for the staff, which has to be followed from when a customer first rings with an enquiry to when the job papers are finally filed away. It should detail the:

- Organizational structure of the company.
- Relationships between, and the responsibilities of, individual operating departments and functional managers.
- How the company's quality system is to work in practice:

- every day;
- every time an order is received;
- every time a customer rings up;
- every time a load is scheduled; and
- every time a driver/vehicle is allocated to a job.

No matter what process or function is carried out, the manual should define it, who does it, how he does it, when, where and in what sequence, what follow-up action is carried out, by whom, to whom everyone reports in the event of difficulties or potential problems, what records are kept and so on. It is the ultimate guidebook to the firm: every single company operation is described, and cross-referenced to every other related function.

Mistakes may still be made, but the procedures manual should take account of such possibilities by detailing the action to be taken when errors are discovered (or pointed out by customers), and by whom, as well as to whom in the firm such matters should be reported, what reports should be written and to whom they are to be submitted. These 'failure reports' should be acted upon promptly, and details of the corrective action taken should also be recorded for future reference. By documenting failures, they should progressively be eliminated from the system.

Manuals within Manuals

A simple manual would have an opening section containing basic reference material as follows:

- Firm's name.
- Description of the business in which it is engaged (eg haulage of aggregates and excavated materials, also contract haulier to . . .).
- Number of locations (eg one only or head office and vehicle base at . . .).
- General description of resources (eg 'x' number of vehicles/plant).
- Names/functions of key departments (eg traffic/administration/workshop).
- Names/titles of key managers/personnel.
- Job descriptions for key positions (eg operations manager/fleet engineer/ company accountant/marketing manager).
- Date(s) and name(s) of person(s) compiling manual.
- Names of persons issued with the quality manual and those responsible for keeping it up to date.
- Statement of firm's quality policy.
- Name(s) of person(s) with ultimate responsibility for compliance with quality policy.

From this point, the manual could divide into sections covering each functional department (eg traffic operations/accounts and administration/workshop). For each there could be a general statement of departmental responsibility; the names, positions and individual responsibilities of key personnel; and the lines of reporting and communication (ie upwards to the company's top management or board of directors, downwards to supervisors, shift leaders, etc, and sideways by liaison with other departmental heads).

To compile the procedures manual, the work of each department must be examined in detail to see what current practices exist, who does what job (routine ones, special ones, urgent ones, etc), what controls are imposed and what safety procedures, if any, are followed to avoid failures (eg loads missed, jobs not invoiced, vehicles missing out on services, etc). Every step should be categorized within a functional heading (eg receipt of customer orders, recording orders on daily work sheets, allocating orders to vehicles, planning vehicle loads and so on) and the correct procedure for carrying out each step needs to be recorded. It is necessary to describe how details relating to the job should be entered, in what form or sequence, and the safety measures necessary to ensure that no essential information (eg a customer's special instructions) are missed.

The manual must be capable of being read – and the procedures understood and followed – by any member of the firm and by the inspector from the quality accreditation organization which checks the procedures. If the wording is tortuous and the manual littered with technical terms and unexplained abbreviations its point will be lost. For this reason manual writing demands the use of clear and simple language. Jargon should be avoided and technical terms kept to a minimum. When they are essential they should be defined or explained. Essential legal obligations (such as goods vehicle drivers' hours, tachographs, safe loading, etc) must be explained or cross-referenced to a source of detailed legal information.

Manuals in a form other than full-size pages in ring binders (eg pocket-size handbooks) may be more appropriate for drivers. These should detail the rules to be followed on drivers' hours, the correct use of tachographs, safe loading, dangerous load procedures, routine daily vehicle inspections, what to do in an emergency or the event of an accident and many other instructions. Workshop staff could have their own handbooks concentrating on safety procedures, what to do in the event of injury or accident, use of tools and equipment, procedures for drawing spare parts and consumable supplies and so on. Fitters who are required to road test vehicles should also be issued with a copy of the driver's handbook.

OTHER DOCUMENTATION

Besides the main manual, other documentation systems are needed – a 'day book' in which to record customer orders, collection notes, delivery notes, receipt notes or combined consignment note sets, invoices, statements and such like. Many of the forms may relate to ongoing or periodic events such as the annual examination of employee driving licences; annual medical examinations and eyesight tests; inspections of safety equipment and safety signs, workshop equipment, fire fighting equipment and first aid kits. Other forms would record company meetings with staff, drivers, workshop staff, and training sessions on new procedures or new legislation. Examples of all forms used should be included in the procedures manual to show which form to complete in any particular set of circumstances, where supplies of the form can be found, how it should be completed and what to do with it following completion.

TRAINING TO ACHIEVE QUALITY STANDARDS

All staff within a firm seeking quality accreditation should be properly trained to operate in accordance with specified quality procedures (ie to perform to standard).

Management, staff and workers should be updated on the latest techniques and new developments. For example, heavy vehicle drivers may need refresher courses in driving skills to eradicate bad habits and training in efficient and economical driving techniques for new vehicles, or vehicle equipment such as new types of gearbox; they need reminders and updating on essential safety procedures and the use of safety equipment, on legal requirements (such as on drivers' hours rules and tachograph use) and on the use of mechanical loading aids.

At all levels from management to driver/operative, training improves job skills. Additionally, it contributes to the individual's motivation and job interest and provides an incentive to do that job better and take more of an interest in the firm's overall objectives – namely to provide a quality service. The following list shows examples of training from which various grades of management and staff may benefit:

Top management –
Management techniques.
Sales/marketing/public relations.
Control of people.
Motivational skills.
Employment legislation.
Financial controls.
Taxation matters.

Middle management –
Administrative controls.
Business systems.
Computer familiarization.
Management techniques.
Motivational skills.
Health and safety law.

Functional management –
Transport legislation.
Safety systems.
Engineering skills.
Computer techniques.
Accounting practices.

Goods vehicle drivers –
Economic driving.
Safety procedures.
Legal requirements.
Dealing with customers.
Dangerous loads requirements.

Loaders –
Safety matters.

Fork-lift truck drivers –
Driver training.
Safety procedures.
Truck maintenance.

Vehicle workshop staff –
Safety procedures.
Vehicle manufacturer training.
Component training.
First aid.

Administrative and secretarial staff –
Office procedures.
Computer/word processing systems.
Telephone techniques.
Use of office equipment (photocopiers/fax machines, etc).
First aid.

While the range of courses and training opportunities is endless, there is a need to identify activities which demand acquired skills; for example where special competence is legally required such as for goods vehicle vocational driving licences and dangerous goods training. These require priority over other training because failure could result in prosecution, or loss of operating licences.

Training Records

All training should be carefully recorded, whether a 30-minute in-company explanation of new legislation or the desired telephone answering techniques, or a longer external training course for management. This should show dates, who attended, who presented the training session, the duration, the location, the facilities/aids used, the objectives of the session, the results achieved and any other relevant data, including any follow-up action necessary. Similarly, where staff achieve other qualifying standards as a result of company-sponsored incentives (eg attendance at evening or day-release classes or by home-study learning), or on a voluntary basis, these should be recorded.

MONITORING OF QUALITY STANDARDS

Successful operation of quality management systems involves a continuous monitoring process to ensure quality procedures are maintained and do not slip back into old inferior methods. A variety of monitoring methods may be employed. The establishment of Quality Circles is one, with groups or teams of volunteer employees meeting to discuss the application of quality, particularly in relationship to their own departments or work sections. They identify quality 'problem' areas and put forward suggestions as to possible solutions.

In the small firms, where the cost structure does not warrant full-time monitoring, two methods in particular will ensure that standards are maintained. First, regular monitoring of customer perception of the quality of service they are receiving; customers should be asked to point out any deficiencies and their answers noted, analysed and corrective action taken. Second, maintaining full and clear communication in the firm will ensure that everybody, from top to bottom, is on the same quality wavelength and working towards the same goals. In this way the whole firm will feel united in the quest for total quality management.

FURTHER INFORMATION

Further information on the establishment of quality systems in road haulage may be obtained from the following:

Road Haulage Association (RHA)
Tel: 01932 841515

British Standards Institution (BSI)
Business Development Advice
Tel: 01908 220908

Department of Trade and Industry (DTI)
Tel: 020 7215 5000

DEBT COLLECTION

The Late Payment of Commercial Debts (Interest) Act 1998, which came into force on 1 November 1998, allows small businesses to claim interest on commercial debts that relate to the supply of goods and services that are paid late. It does not apply to consumer debt. In effect, interest becomes chargeable from the day after the agreed date for payment. Where no specific date has been agreed for payment, interest becomes payable after 30 days from either: 1) the date on which the contract was formed; or 2) the date of invoice, or notice of the amount of the debt (whichever is the later).

For these purposes, small businesses are limited liability companies with an annual turnover not exceeding £2.8 million or a balance sheet value not exceeding £1.4 million, or if they are not an incorporated limited company, they should have no more than 50 employees. This obviously includes many road haulage businesses and some own-account vehicle operators.

The rate of interest chargeable is the bank base rate plus 8 per cent (ie 11.5 per cent at the time of preparing this edition of the *Handbook* (August 2003)).

The Act is being implemented in stages, the first part having been applied since 1 November 1998 to small businesses as mentioned above; the second part having been applied since November 2000, allowing small businesses to claim against all businesses and the public sector. The final part will be implemented in November 2002 when all businesses and the public sector will be able to claim from all other businesses.

INSOLVENCY

People often refer to a company being insolvent, but what does that actually mean? It is important at the outset to establish exactly what form of insolvency you are dealing with. Each type of insolvency has its own specific office holder, who has differing obligations, duties and powers. This will undoubtedly affect how you act, and whether or not you have any chance of recovery of a debt from that company.

Administration

The holder of office is called 'the administrator' and is appointed by the Court following a petition. His or her aim is either to achieve the survival of the company and the realization of assets and liquidation, or to enable a Company Voluntary Arrangement. The effect is a moratorium preventing any creditor action. While the administrator owes a duty to creditors in general, he or she will not agree creditors' claims, nor has he or she the power to distribute funds to the creditors.

Administrative Receivership

The holder of office is called 'the receiver' and is appointed by someone holding a debenture over (substantially) all the assets and undertaking of the company. The receiver's aim is to continue trade with a view to sale of the business as a going concern. The directors remain in office, albeit the receiver has day-to-day control of the business. There is no moratorium preventing creditor action, but the administrative receiver's duty is solely to get in and realize the assets of the company, sufficient to repay the indebtedness of the debenture holder.

Liquidation

There are three types of liquidation. These are:

- compulsory liquidation;
- creditors' voluntary liquidation; and
- members' voluntary liquidation.

In all three cases the office holder is called 'the liquidator', but his or her appointment is different, depending on what type of liquidation it is.

- In the case of compulsory liquidation the company is placed in liquidation by the court following the making of a winding-up order.
- In the case of creditors' voluntary liquidation, the liquidator is appointed by the directors/shareholders where the company is insolvent and unable to pay its debts.
- In the case of members' voluntary liquidation, the liquidator is appointed by the directors/shareholders where the directors have sworn a statutory declaration of insolvency. The aim of the liquidation is to realize assets, shut down the company and distribute the assets in accordance with a strict order of priority.

With company voluntary arrangement (CVA), the office holder is called 'the supervisor' and his or her rights and powers are set out in a document entitled *the proposal*, which is put to and voted upon by the company's creditors. The company can, and invariably will, trade under the auspices of a CVA, which is often used to effect a reconstruction or reorganization of the company's trading enterprise. The proposal that leads to the making of the CVA forms a private contract between the company and its creditors. If the proposal is approved, creditors are bound to accept the terms in full and final settlement of their indebtedness.

THE COMPETITION ACT

The Competition Act 1998, which came into force on 1 March 2000, is designed to ensure that UK businesses remain competitive. It applies to all businesses irrespective of the trade in which they are engaged and their size. It even applies to sole traders such as owner-driver road hauliers.

Principally, the Act prohibits anti-competitive agreements and abuse of a dominant position in the market.

Anti-Competitive Agreements

The prohibition on anti-competitive agreements covers agreements and practices that prevent, restrict or distort competition, or are intended to do so. These can be formal or informal, written or not. In particular, they include agreements to:

- fix purchase or selling prices or other trading conditions;
- limit or control production, markets, technical development or investment;
- share markets or sources of supply;
- apply different trading conditions to equivalent transactions, which will put some parties at a competitive disadvantage;
- make contracts subject to unrelated supplementary conditions.

The effect of an agreement must be 'appreciable' to be caught by the Act. There are certain exemptions and exclusions.

Abuse of a Dominant Position

Whether a business is in a dominant position will depend on how much it can act independently of its competitors and customers. As a general rule, a dominant position is unlikely if its market share is less than 40 per cent. This will, however, depend on the structure of the market. Examples of abuse are:

- imposing unfair purchase or selling prices or other unfair trading conditions;
- limiting production, markets or technical development to the prejudice of consumers;
- applying different trading conditions to equivalent transactions, which will put some parties at a competitive disadvantage;
- making contracts subject to unrelated supplementary conditions.

The Director-General of Fair Trading has wide-ranging powers to investigate suspected breaches of the law. His officials can enter premises and demand relevant documents, and may even get a warrant to make a search. They can order offending agreements or conduct to be terminated. Businesses can be fined up to

10 per cent of their UK turnover; prohibited agreements will be made void and unenforceable, and any third party harmed by an unlawful agreement or conduct may be able to sue for damages.

If Your Business is a Victim of Unfair Competition

A complaint may be made to the Office of Fair Trading (OFT) by anyone who suspects that the competition rules are being broken. This may be in writing or by telephone and the complainant's identity may be protected although evidence will be needed to back up any such complaint. The OFT may launch a formal investigation if there are reasonable grounds for suspecting that an offence is being or has been committed.

Where a business finds that it is part of an illegal cartel and provides information voluntarily to the OFT, it may receive a significant reduction in any financial penalty imposed or, if it is the first to notify the OFT of the situation, it may gain complete immunity from financial penalty.

Further information can be obtained from:

The Office of Fair Trading
Field House Room 107
15–25 Bream's Buildings
London EC4A 1PR
e-mail: enquiries.competitionact@oft.gov.uk

30 UK Road Network Developments

Britain's roads have come in for scathing comment in the recent past. Reports suggest that Britain has 5 million potholes and a backlog of £7.4 billion of vital repair work outstanding, but councils cannot carry out these repairs because, it is alleged, they are having to spend as much as £1.2 billion annually on meeting compensation claims for injury and damage (source: *Mail on Sunday* 5 and 12 May 2002).

DEVELOPMENTS IN 2004

In July 2004 Secretary of State for Transport Alistair Darling made a key statement on roads policy designed firstly, to encourage better use of our roads, and secondly to outline proposals for future development of the M6 motorway between Birmingham and Manchester, as follows:

> Decades of under investment and a growing economy have put our roads under severe pressure. Two steps are necessary if we are to reduce congestion – especially at peak times. *First* we need to make better use of our existing roads and *second* provide additional capacity on key strategic routes. I have two new proposals today [July 8 2004] both to tackle congestion and provide for more reliable journey times. I'll start with a scheme to make better use of the existing capacity on the road network by encouraging people to share vehicles at peak times.

Car pool lanes have been used to great effect both in America and Australia. And if it works in other parts of the world – there is no reason in principle why it should not work here. Car pool lanes could play a key role in reducing congestion on key commuter routes. So I have asked the Highways Agency to carry out a feasibility study of four sites on the motorway network for trialling High Occupancy Vehicle or car pool lanes. These could be created by widening the road to create an additional lane or in some cases by using the hard shoulder as a running lane, and they would be reserved for the use of vehicles carrying two or more people, offering a faster more reliable journey.

The sites for the Highways Agency to look at are:

- M62 Junction 25 to 27 (Brighouse to Leeds);
- M3 Junction 3 to 2 (Bagshot to Thorpe);
- M1 Junction 13 to 7 (Milton Keynes South to St Albans); and
- M61 Junctions 6 to 3 (north west of Manchester).

In general the Government favours providing extra capacity for car pool lanes, but in this particular case it may be possible to use one of the existing lanes and I have asked the Highways Agency to look at this. An assessment of each site will be carried out over the coming months which will consider operational suitability as well as safety and environmental implications. This proposal extends the measures we are already putting in place to make better use of our existing roads – including piloting hard shoulder running on the M42 and since April introducing Traffic Officers on our motorways to help clear up incidents and get traffic flowing as quickly as possible.

I now turn to the need for additional capacity on one of the most heavily used road arteries in the country – the M6. We need more capacity on road and rail. That's why we're investing over £7 billion on the West Coast Main Line – which when the first stage is complete will cut half an hour from the Manchester to London train times. But we also need more road capacity on this key route.

The new M6 Toll motorway opened in December last year [2003]. Today [8 July 2004] the Highways Agency is publishing analysis of the first three-month traffic on the M6 Toll and the existing M6. This shows the M6 Toll regularly carries around one-fifth of all traffic flowing through the West Midlands conurbation each day. With vehicles saving up to half an hour on a journey that can take over an hour during peak times on the existing M6. And drivers who choose not to pay the toll are benefiting from reduced congestion. Since the M6 Toll opened, traffic on the existing non-tolled M6, has reduced by 10 per cent. Friday afternoons have seen a significant improvement in journey times. And there have also been improvements to roads as far afield as Stoke on Trent.

The M6 links two of the country's most important economic centres, Birmingham and Manchester. It is one of the busiest roads in the country, and drivers regularly suffer heavy congestion. That's why on 10 December 2002, I told the House [of Commons] that I would support the widening of a number of key strategic roads. Including widening the M6 between Junction 11a near Cannock and Junction 19 near Knutsford to four lanes in each direction.

Given the success of the M6 Toll in improving congestion, I believe that it is now right to look at extending the tolled motorway and building a new expressway to run parallel with the M6 between Birmingham and Manchester, as an alternative to widening the existing road. This 51 mile stretch of the M6 between Junctions 11a near Cannock and Junction 19 near Knutsford is being considered as a tolled expressway because of its importance to business; its link to the M6 Toll and because a significant percentage of its traffic travels over 20 miles or more.

There are a number of advantages to an Expressway. *First,* it would provide motorists and businesses with a choice of using either the existing M6 or paying for a faster more reliable journey on the new road. An Expressway could be designed to suit long distance journeys with fewer junctions. And could also improve conditions on the original road, freeing up space there. *Second,* an Expressway would provide double the extra capacity at a lower cost than widening the existing road by one lane. *Third,* construction of an Expressway would not cause disruption to road users whilst it was being built. In comparison, widening the existing M6 obviously would need to be done in a phased approach. And would cause disruption over 5 to 6 years.

We have also to carefully consider the wider social and environmental impacts – including the effect on landscape, biodiversity, heritage, as well as air quality and climate change. We will need to assess what impact it would have on traffic levels generally and the scope for high quality measures to combat any adverse effects. Sometimes new roads can have a positive impact on the environment. A new route, with good environmental mitigation plus environmental improvements to the existing M6, might overall be better in environmental terms than simply widening the existing M6. This will need to be looked at carefully.

A consultation paper is to be published on the proposals for an expressway to run parallel with the M6 between Birmingham and Manchester, as an alternative to widening the existing road.

£5.5 Billion Package of Road Improvements

In December 2002 Alistair Darling announced a £5.5 billion package of major national and local transport measures to tackle congestion, improve safety and reliability, and increase the quality of life. These measures included significant improvements to the M1, M4, M5 and M6, plus 19 major new local transport schemes. The green light was given to the following schemes:

- Widening the M6 between Manchester and Birmingham from three to four lanes between junctions 19 and 11a (but see more recent proposal above).
- Widening the M1 in the East Midlands generally to four lanes (between junctions 21 and 30).
- Providing climbing (crawler) lanes for HGVs and junction improvements on sections of the M4 and M5 near Bristol.
- Improving the A453 from the M1 to Nottingham.
- Junction improvements on the A419 Commonhead junction near Swindon.

- 1,600 traffic management schemes and over 400 new or improved road junctions.
- Completion of the dualling of the A1 between Morpeth and Alnwick.
- Dualling most of the remaining single carriageway on the A303 and improving sections of the A30 west of Exeter.

For more information on the Highways Agency's *Business Plan 2002/03*, visit the website at: www.highways.gov.uk

The M6 Toll Road – Midland Expressway

The 27-mile, six-lane Birmingham Northern Relief Road, variously known also as the M6 Toll Road or the Midland Expressway, which opened January 2004, provides a motorway link around the northern and eastern edge of the UK's West Midlands conurbation, offering a genuine alternative to the busy M6 motorway and relief to the local road network.

The M6 motorway, where it passes through the West Midlands, is a lifeline of both national and local strategic importance. It is also one of the country's busiest roads, with some sections carrying 100 per cent more traffic than they were ever designed for. At peak times the 25-minute journey between junctions 11 and 4 is said to take an hour or more. In fact, congestion and traffic delays are now so common on the M6 that they no longer even make local news headlines. Alternative routes for through traffic – the A5, A38, A446 and A452 – are also very busy, especially at peak times.

The M6 Toll motorway has these principal aims:

- to provide a free-flowing alternative to the heavily congested M6 between Laney Green (junction 11) and Coleshill (junction 4), one of Britain's busiest stretches of motorway.
- To provide a distributor to the north and east of the West Midlands region, improving communications to conurbations such as Cannock, Lichfield and Tamworth.

Although a privately built and managed toll road (the first ever in Britain), it remains ultimately the responsibility of the Secretary of State for Transport, and is as much a public highway as any conventional road.

Toll stations are sited at:

- Great Wyrley;
- Burntwood;
- Shenstone (Lichfield junction);
- Weeford (Tamworth junction);
- Weeford Park;
- Langley Mill (Sutton Coldfield Junction) – 2 toll stations, southbound entry and northbound exit.

With the exception of the southbound entry at Langley Mill, tolls are payable on exit from the road. Users pass a barrier once only, when they pay the toll. Toll collection at all stations is both manual and automatic. There are no toll stations on the section of the widened M42 or the M6 Toll south of Belfry.

Tolls

The tolls are as follows:

	Day 06.00–23.00	*Night* 23.00–06.00
Car	£3.00	£2.00
Van	£6.00	£5.00
HGV	£11.00	£10.00

The first 10 million users will benefit from a £1.00 reduction on all standard tolls.

NB: In June 2005, a discounted toll rate for HGVs using the M6 Toll motorway was increased from £6.00 (ie £5.11 plus VAT) to £7.00.

Midland Expressway Ltd (MEL) has operated the road since its completion in 2004 and is responsible for the tolling arrangements. Further information can be obtained from the interesting website at: www.m6toll.co.uk

BRITAIN'S EXISTING ROAD NETWORK

At the end of 1999, Britain had a network of 371,598 kilometres of roads. Since 1989, the network has grown by 17,000 km, but 16,000 km of this growth has been on unclassified roads, mainly in housing developments. Except for 77 km of local motorways, all other motorways and trunk roads are the direct responsibility of the Government, making up about 4.7 per cent of the road system. However, this small proportion carries 33 per cent of all traffic and 57 per cent of heavy goods traffic.

Total roads by type and percentage of the UK network are shown in the following table:

Type of Road	*Length (km)*	*% of network*
Motorways	3,303	0.9
Trunk roads (excl motorways)	12,228	3.3
Principal roads (excl motorways)	36,002	9.7
Classified (non-principal)	113,430	30.5
Unclassified	206,635	55.6
Total	371,598	100.0

Traffic flows, taken across all roads, grew by 33 per cent between 1985 and 1995, with growth on motorways exceeding 69 per cent (sections of the M25 and M6 regularly carry more than 200,000 vehicles per day against a normal design capacity of only 79,000 vehicles per day).

NB: Information and table above courtesy of the British Road Federation, Road Fact 2000. (This organization is now defunct.)

HIGHWAYS AGENCY INFORMATION LINE

The original Cones Hotline scheme has been relaunched as the 'Highways Agency Information Line' using the same telephone number – 0845 750 4030 – but offering a broader range of information and services.

TRANS EUROPEAN NETWORK PROGRAMME

The United Kingdom was to receive 41.5 million euros (approximately £28 million at the current rate of exchange) from Europe for the development of Trans European Network (TEN) schemes and studies in 1998. Eighty-five per cent of the funds were to go to the following TEN priority projects:

- £16.75 m to the Channel Tunnel Rail Link;
- £6.7 m to the West Coast Main Rail Line;
- £1.1 m for studies in England along the Ireland–UK–Benelux Road Link.

Other schemes to benefit include:

- £0.34 m to the Phase 1 (1998/99) works at Manchester Airport;
- £0.9 m for Cardiff International Airport access road study;
- £0.7 m to the Stranraer Area Integrated Transport Package;
- £0.7 m for Telematics and Traffic Management on the TEN in the UK;
- £0.34 m towards access improvements and interchange facilities at Harwich International Port.

The balance of the funding will go towards studies into rail access at five other UK ports, upgrade evaluations of vessel tracking systems in the Humber and Southampton, and a 'People Mover' study (to link with the new railway station for London Luton Airport).

31 Services

This chapter in the *Handbook* provides an opportunity to include information on Legal Services, Breakdown Services and Vehicle Recovery, Return Loading, Franchising, and Motorway Services. In future years further suitable and appropriate subjects may be included.

LEGAL SERVICES

As the growth of this *Handbook* and the expansion of its main subject material clearly demonstrate, the transport industry in general, and the operation of heavy goods vehicles in particular, is fraught with a burden of legislation verging on leviathan proportions. But while the volume of the legislation is one thing, understanding and complying with its strictures is quite another, and it is here that transport managers and fleet operators alike find problems. While this book, and others like it, attempt to explain the law in simple terms, none of them is much help when, having failed to understand the law on a particular point, or even failed to realize that it even applied, the operator faces prosecution and a court appearance.

Deciding how to plead, guilty or not guilty; knowing what defences can and should be put forward; knowing that, when an absolute offence has been committed that putting forward feeble excuses will add nothing to the defence; knowing what mitigating circumstances may be put forward in given circumstances; and knowing how to plead for leniency in the penalty to be imposed when convicted, these are all things that books are not good at explaining. This is when a good solicitor is necessary to weigh up the case, determine whether the police and Crown Prosecution Service (CPS) have got the charges right to suit the alleged offence (it is a fact that they do not always do so), and decide what plea the accused should make.

It is clear that many firms facing prosecution for transport or vehicle-related infringements prefer to plead guilty and pay the fine rather than incur the extra costs of legal representation, and the possible attendant (ie adverse) publicity by pleading not guilty and making a case of it. It is clear also that in so doing, some of those are convicted on the wrong grounds or for offences where they could have raised a legitimate defence (eg by being exempt from a particular requirement – which the enforcement authorities are not always good at recognizing).

What it is important to remember, for big firms as well as small fleet operators, is that convictions for most of the offences relevant to transport operations will bring, besides fines, the probable jeopardy of 'O' licences, and the driving licences of proprietors and partners.

For these reasons, in most cases, the retention of a solicitor is strongly advised where cases are to be dealt with by a court (in the case of fixed-penalty offences, taking the simple line of accepting that an offence has been committed and paying the penalty can be the easiest and cheapest way out). Importantly though, such is the complexity of transport, vehicle-related and road traffic law, that any solicitor chosen should be one who is recognized as having expertise and a reputation in these areas of law. Any fear that a known 'name' may cost extra (which is probably not the case) should be weighed against his ability to present an experienced case. The legal reports columns of the transport press regularly highlight cases where good advocates have charges against their clients quashed or cases dismissed because they really do know the ins and outs and technicalities of the law.

The following is a list of well-known transport lawyers who will provide suitable advice or defend cases as necessary.

Jonathan S Lawton (Solicitor/advocate)	Hill Dickinson, 34 Cuppin Street, Chester CH1 2BN (tel: 01244 896601; fax: 01244 896600).
John Backhouse (Backhouse Jones)	23 Wellington Street, St Johns, Blackburn, Lancs BB1 8DE (tel: 01254 828300; www.backhouses.co.uk).
Michael Carless (Carless, Davies & Co)	140 Stourbridge Road, Halesowen, Worcs (tel: 0121 550 2181/4429).
Stephen Kirkbright (Ford & Warren)	Westgate Point, Westgate, Leeds (tel: 0113 243 6601).
Malcolm Partridge (Eversheds)	Paston House, Princes Street, Norwich, Norfolk NR3 1BD (tel: 01603 660241).
Ian Rothera **Jane George** (Rothera Dawson)	2 Kayes Walk, Stoney Street, The Lace Market, Nottingham NG1 1PZ (tel: 0115 910 0600).

Other transport lawyers can be found via the Association of Road Transport Lawyers (AORTL) – contact Ian Rothera or Jane George as shown above.

BREAKDOWN AND RECOVERY SERVICES

There has been a proliferation of heavy vehicle breakdown and recovery services in recent years providing roadside services to help the driver with a broken-down truck, tyre problems or accident-damaged vehicle. Many such services have a freephone contact arrangement. These services are operated by independent firms (many are members of AVRO – see below) and by the vehicle manufacturers. Examples of the former are AA Truck Rescue (previously AA-BRS Rescue), RAC

Commercial Assistance (previously RAC-Octagon) and National Breakdown* and of the latter include DAF Aid, Action Service Volvo, Fodensure and MAN Rescue.

The Association of Vehicle Recovery Operators (AVRO) is a major player in this field and as such has developed a code of practice in conjunction with the Retail Motor Industry Federation (RMI), while the Road Rescue Recovery Association (RRRA) has its own code. Quite separately, and in conjunction with the British Standards Institution, the RAC too has its own code of practice.

The AVRO Directory lists the following members:

Northern region	
Alpha Auto Services	Cramlington, Northumberland. Tel: 0191 250 0009
Tebay Vehicle Repair Ltd	Tebay, Cumbria. Tel: 0158 74 241
Lynch Motors Ltd	Parkside, South Shields, Tyne & Wear. Tel: 0191 456 4665
Ron Perry Test & Tune Ltd	Hartlepool, Cleveland. 01740 644223
South Midlands region	
Brooks & Stratton Ltd	Welwyn Garden City, Herts. Tel: 01707 330678
3B's Rescue Banbury	Oxfordshire. Tel: 01295 750236
Eastern region	
John Canham Ltd	Clacton-on-Sea, Essex. Tel: 01255 432888
J S Holmes Ltd	Wisbech, Cambridgeshire. Tel: 01945 81243
J & A Recovery	Brandon, Suffolk. Tel: 01842 810146
Greater London	
Arcade Motors Ltd	Tottenham, London. Tel: 020 8363 2323
Kenfield Motors Ltd	Hayes, Middlesex. Tel: 020 8569 2323
Queens Motors Ltd	Penge, London. Tel: 020 8778 6666
MV Recovery	Croydon, Surrey. Tel: 020 8686 1883
J Winfield Motors	Hayes, Middlesex. Tel: 020 8848 7421
Southern region	
McAllisters Recovery	Aldershot, Hampshire. Tel: 01252 22289
Brighton Recovery	Brighton, East Sussex. Tel: 01273 430420
Dawes of Swanley	Swanley, Kent. Tel: 01322 62211
Langley Vale Recovery	Epsom, Surrey. Tel: 01372 277021
Thames Valley Motors	Newbury, Berkshire. Tel: 01635 48772
Western region	
Avon Commercial Recovery	Severn Beach, Bristol. Tel: 01454 52331
Walls Garage Services	Severn Bridge, Bristol. Tel: 01454 53472
Bristol Omnibus Co	Bristol, Avon. Tel: 0117 955 8211
Lamb Hill Recovery	Cullompton, Devon. Tel: 01884 38572
P G Hayes	Minehead, Somerset. Tel: 01643 705363
J & P Motors Ltd	South Petherton, Somerset. Tel: 01460 40553

Wales

Walls Truck Services Ltd Newport, Gwent. Tel: 01633 246622

**Information on National Breakdown's Trucklink and Trailerlink services can be obtained from: National Breakdown, FREEPOST, Leeds, Yorkshire LS99 2NB or by telephoning 0113 239 3666 (fax: 0113 257 3111).*

RETURN LOAD SERVICES

Goods vehicles running needlessly empty are a costly inefficiency on the part of the operator and wasteful of valuable natural resources (eg fuel), as well as contributing to the environmental problems outlined in Chapter 28. While it is not always possible, or indeed desirable for operational reasons, to seek and carry return loads, nevertheless many more vehicles travel the roads with empty space behind than is necessary or could be justified by an efficiency and cost-conscious industry.

It is recognized that return loading presents many problems of strategy as well as the uncertainties about securing payment (and at an acceptable price) for such work. However, there are now more reliable return load systems which allow hauliers to secure loads from reputable sources where the haulage rate has not been creamed in the way established by many disreputable clearing houses in the past.

A number of computerized load-matching systems are now operating of which the French Lamy Teleroute is probably the best known. This operates throughout Europe on a 24-hour/365-day basis and only requires the haulier to have a Videotex terminal or personal computer with a telephone line to connect to Teleroute's international network.

Other hi-tech return load services are provided by Returnline (tel: 01455 233998) and CargoFile (tel: 01277 363756).

FRANCHISING

Franchising is a business activity in which an existing (usually large and successful) operator contracts with newcomers or smaller firms in the business to provide a service under its well-established name and to its established standards (eg of quality and reliability, etc).

The benefit to the franchisor (ie the existing operator) is that it expands his business and spreads his name and reputation without significant capital investment on his part. The disadvantage is that if the franchisee proves to be unsatisfactory in the way he conducts business the franchisor's good name (and that of his products or services) will suffer.

The benefit to the franchisee is that he can start a new business (or convert his existing business) with the backing of an established name (ie a recognized corporate identity) and reputation, and with proven systems of operating, marketing and administration already in place. The disadvantage is that once committed to the franchise contract, the franchisee will have little freedom or flexibility to exercise his own entrepreneurial flair, management skills or administrative know-how. He will have to conform to the rigidly laid-down system of his franchisor, although this can be of significant advantage in imposing the rigid disciplines necessary to establish a sound and successful business.

Common examples of well-known franchising operations are those of McDonald's, to be found in the main streets of most of the world's major cities, and Colonel Sanders Kentucky Fried Chicken. But this type of business is not confined to fast-food chains. It covers many other fields as well, including Athena retail shops (selling pictures, prints cards, etc), and Ryman the Stationer retail outlets. In transport and related industries, franchised express parcels operations (such as Amtrak Express Parcels Limited and Interlink, part of the Geopost network) and workshop tools and equipment supply (eg Snap-on Tools) are two of the most popular franchise operations.

The Amtrak Franchise

Amtrak provides potential franchisees with a helpful information pack explaining the benefits of franchising, mainly as outlined above but including such other advantages as the franchisee not having to handle billing, or cash transactions (and therefore not incurring bad debts), these being a central, computerized function and the responsibility of the franchisor. Franchisees in this case are paid promptly, on a monthly basis, for the work they have done in the form of commission for every single collection and delivery made (and irrespective of whether of not the customer has actually paid his account).

Franchisees are allocated an exclusive territory (of which there are about 120 in mainland UK) in which they are licensed to trade and their role is to collect and deliver express parcels in these territories which are linked into the company's central sorting hub in the Midlands by a fleet of heavy trunk vehicles (owned by the company, not the franchisee).

Amtrak provides its franchisees with back-up support by means of a team of area managers who will assist with sales and marketing strategies and help the franchisee to operate within the company's guidelines. And, before starting, the candidate attends a training course to familiarize himself with the nature of the parcels business, the operational procedures to be followed and sales and marketing techniques.

Legal Considerations

While franchising as a means of starting in business provides many advantages (and possibly less risk of loss of initial capital – although there are no guarantees on this count), there are legal considerations which must be taken into account. Mainly these fall into two categories: one is the employment relationship between the franchisor and franchisee and the other concerns goods vehicle operator licensing where vehicles over 3.5 tonnes maximum gross weight are involved.

Employment Contracts

Normally, one would expect a franchised operation to involve a contract for the provision of service by an independent (possibly self-employed) contractor. In this case the franchisee is not an employee of the franchisor and is therefore responsible for (among other things) making his own National Insurance contributions, paying income tax and registering for VAT (subject to turnover being above the statutory limit at which registration is mandatory). However, should the contract be one of service (as opposed to one for the provision of) then this could be held to be an employment contract where the franchisor would be liable for such matters as National Insurance (ie deducting the employee's contribution and paying his own employer's contribution) and deducting income tax from the franchisee's payments.

These are not the only legal considerations under this heading, but they are the main ones – it still leaves open the question as to liability for redundancy payments and suchlike. For this reason, the inexperienced franchising candidate should seek advice either from a solicitor or an accountant before signing (irrevocably) any documents.

Operators' Licensing

This subject is covered extensively in Chapter 1 of this book, but it should be said here that should the vehicle which the franchisee plans to operate exceed 3.5 tonnes maximum gross weight, then he will need to obtain an 'O' licence before commencing operations. In fact, he should do this before even committing himself to the franchise contact because, irrespective of what the franchisor may say, there is no guarantee of getting an 'O' licence. A franchisor cannot buy one for the franchisee or allow him, as a self-employed contractor, to operate under his own 'O' licence; this would be illegal.

Further Information

Further information on franchising can be obtained from the following organizations:

British Franchise Association

Thames View, Newton Road, Henley-on-Thames, Oxon RG9 1HG (tel: 01491 578049).

National Federation of Self-Employed and Small Business

32 St Anne's Road West, Lytham St Annes, Lancs FY8 1NY (tel: 01253 720911).

The following high street banks have departments dealing specifically with franchise operations:

Barclays Bank Plc

Franchise Unit, PO Box 120, Longwood Close, Westwood Business Park, Coventry CV4 8JN (tel: 024 7653 2451).

Lloyds Bank Plc

Franchise Department, Commercial Banking, PO Box 112, Canons Way, Bristol BS99 7LB (tel: 0117 943 3136).

National Westminster Bank Plc

Franchise Section, Commercial Banking Service, 4th Floor, National House, 14 Moorgate, London EC2R 6BS (tel: 020 7728 1684).

Midland Enterprise
Midland Bank Plc, PO Box 2, 41 Silver Street, Sheffield S1 3GG
(tel: 0114 252 9037).

Advice on the use of franchising consultants is available from:

Franchise Consultants Association
James House, 37 Nottingham Road, London SW17 7EA
(tel: 020 8767 1371).

Franchise World is published bi-monthly by the Franchise Consultants Association (see above for address) and a directory (*Franchising World*) is also available from the same address.

MOTORWAY SERVICES IN THE UK

Motorway Service Areas (MSAs) in the UK are controlled by the Highways Agency (an Agency of the D*f*T). While the D*f*T, through the Highways Agency, guarantees and maintains minimum standards of operation and service, responsibility for identifying potential sites and obtaining planning consent rests with the private sector dealing directly with the relevant local authorities.

The following are the major service area operators who welcome heavy vehicle drivers and provide suitable services:

Guide to Truck Cafés and Truckstops

Richard Barker has published the seventh edition of *A Truck Driver's Guide to Cafés & Truckstops*. The new edition (2004/05) includes more cafés and truckstops than previous editions as well as a number of new lorry parks. Copies of the guide are available direct from the publisher at 49 Noon Close, Stanley, Wakefield WF3 4PT, price £3.80 (+ 50p postage and packing for UK orders) total £4.30 for one copy – cheques payable to R. A. Barker. For orders of six or more copies, postage and packing is free. E-mail enquiries and feedback to info@driversguide.plus.com

Moto

Moto is the new name in motorway services, having taken over the Granada sites from July 2001. It is part of the Compass Group and operates 47 sites in the UK.

- Kinross M90 J6
- Stirling M9/M80 J9
- Southwaite M6 J41/42
- Washington A1(M) J64
- Scotch Corner A1/A64
- Burton-in-Kendall M6 J35/36
- Lancaster M6 J32/33
- Ferrybridge M62/A1(M) J33
- Birch M62 J18/19
- Woolley Edge M1 J38/39

• Blyth	A1(M)/A614
• Doncaster	M18 J5
• Knutsford	M6 J18/19
• Grantham North	A1
• Trowell	M1 J25/26
• Stafford North	M6 J14/15
• Donnington Park	M1 J23a
• Hilton Park	M6 J10a/11
• Tamworth	M42/A5 J10
• Frankley	M5 J3/4
• Swansea	M4 J47
• Cardiff West	M4 J33
• Leigh Delamere	M4 J17/18
• Exeter	M5 J30
• Severn View	M48 J1
• Cherwell Valley	M40 J10
• Toddington	M1 J11/12
• Chieveley	M4/A34 J13
• Reading	M4 J11/12
• Heston	M4 J2/3
• Thurrock	M25/30/31
• Pease Pottage	A23
• Medway	M2 J4/5

Open:	24 hours/365 days
Parking:	Varies approx 6–120 heavy vehicles (parking fee includes free shower and restaurant/shop vouchers)
Accommodation:	Rooms with bathroom, colour TV, tea/coffee maker
Services:	24-hour diesel supply, breakdown recovery, fax, photocopier and shop

For further details tel: 01525 873881

Drivers are advised that free parking within most service areas lasts for two hours only, after which time a parking ticket must be purchased. Failure to do so will result in a £60 penalty charge.

Roadchef

- Clacket Lane (Westerham) – on M25
- Taunton Deane (Somerset) – on M5
- Sedgemoor (Somerset) – on M5
- Rownhams (Hants) – on M27
- Pont Abraham (Dyfed) – on M4
- Sandbach (Cheshire) – on M6
- Killington Lake (Cumbria) – on M6
- Hamilton (Lanarks) – on M74
- Bothwell (Lanarks) – on M74
- Harthill (Lanarks) – on M8.

Open:	24 hours/365 days
Parking:	For approx 30–100 heavy vehicles (parking fee includes free shower and restaurant voucher)
Accommodation:	Rooms with bath/shower, tea/coffee-maker, hairdryer
Services:	24-hour diesel supply, shop and cash dispensers

For further details tel: 01452 623300

The Strategic Lorry Park

- Beckton Roundabout (London Dockland) – on A13

Open:	24 hours/365 days
Parking:	For approx 120 heavy vehicles (with security surveillance, etc; showers available)
Accommodation:	Rooms with showers
Services:	24-hour diesel supply, fax, photocopier and shop

For further details tel: 020 8594 8730

Pavilion Services (formerly Rank Organisation)

- Medway Pavilion (formerly Farthing Corner) – on M2 between junctions 4 and 5
- Severn View (formerly Aust) – on M4 at junction 21
- Cardiff West – on M4 at junction 33
- Swansea – on M4 at junction 47
- Hilton Park (Wolverhampton) – on M6 between junctions 10A and 11
- Knutsford – on M6 between junctions 18 and 19
- Forton (Lancaster) – on M6 between junctions 32 and 33
- Rivington (Bolton) – on M61 between junctions 6 and 8
- Newark – at junction of A1/A46/A17
- Bangor – at junction of A5/A55.

Open:	24 hours/365 days
Parking:	For approx 30–60 heavy vehicles (with security patrol) (parking fee includes free shower and food voucher)
Accommodation:	Rooms with showers, TV, hairdryer, trouser press (food voucher)
Services:	24-hour diesel supply, breakdown recovery, fax, photocopier, shop and cash dispensers

For further details contact: Roadside Services, Baker's House, Baker's Road, Uxbridge, Middx UB8 1RG

Welcome Break

• London Gateway	M1 J2/4
• Newport Pagnell	M1 J14/15
• Leicester Forest East	M1 J21/22
• Woodall	M1 J30/31
• Fleet	M3 J4/5
• Membury	M4 J14/15
• Michaelwood	M5 J13/14
• Gordano	M5 J19
• Sedgemoor (Northbound)	M5 J21/22
• Corley	M6 J3/4
• Keele	M6 J15/16
• Charnock Richard	M6 J27/28
• Birchanger Green	M11 J8
• South Mimms	M25 J23
• Oxford	M40 J8
• Warwick	M40 J12/13
• Hopwood Park	M42 J2
• Burtonwood	M62 J8/9
• Hartshead Moor	M62 J25/6

For further details tel: 01908 698700

Truckers Rest

Located at West Bromwich, just off junction 1 of the M5.

Services	Serves variety of foods, often with a special 'national' flavour (eg Italian, Irish, Scottish, Spanish, etc) on a weekly rotation

For further details tel: 0121 500 5040

32 International Haulage

Since the removal of barriers to inter-Community trading from January 1993, UK road hauliers have been largely free to operate, uninhibited, across the whole of the EU which, since 1 May 2004, comprises the 15 pre-existing member states plus the 10 new member states* which were admitted to the EU, increasing the consumer market to almost half a billion people.

**NB: Cyprus, Czech Republic, Estonia, Hungary, Latvia, Lithuania, Malta, Poland, Slovakia, Slovenia. The countries of Bulgaria, Croatia, Romania and Turkey are expected to join later.*

TRANSPORT IMPLICATIONS

The transport implications of the SEM are very significant and whole new horizons have been opened up. In the past, road freighting across Europe was impeded by restrictive Customs procedures and road haulage permit requirements resulting in excessive administrative burdens, frustrating delays and inhibiting costs. In the wider arena of distribution past inhibition has been rather more to do with the inability to trade across frontiers. Now, with all these constraints swept away, road freighting into Europe and the setting up of Euro-wide distribution networks is basically no more complex and no more fraught with bureaucracy than operating in the domestic marketplace. Our concepts of what is 'local' and what is 'regional' in terms of distribution are undergoing considerable change.

NB: In 2003, almost 3 million goods vehicles travelled from the United Kingdom to mainland Europe.

ENTRY TO THE INTERNATIONAL ROAD HAULAGE MARKET

Access to the European haulage market is governed solely by a system of quality licences much on the lines of the present UK scheme of operators' licensing – which was amended in 1978 to align with the original EU requirement for establishment of a professional competence qualification for those wanting access to the road haulage industry. Entrants to the haulage business must meet standards of good repute, financial standing and professional competence and the granting of licences is dependent on satisfactory proof that these standards are met. Past convictions, mainly, but not exclusively, for transport and vehicle-related offences will result in licence penalty – even revocation – and inability to sustain adequate finances will have a similar effect. Licence holders have to prove resources or provide financial guarantees (this topic is dealt with in detail in Chapter 1).

COMMUNITY AUTHORIZATION

A system of 'Community Authorizations' was implemented from 1 January 1993 enabling EU road hauliers to operate freely (ie to undertake as many journeys as they wish) *between* member states* – not to be confused with the quite separate cabotage authorizations which are needed by hauliers wishing to collect and deliver goods *within* EU member states other than their own (see below). International hauliers operating within the EU (apart from when operating domestically within their own state) must hold a Community Authorization issued by the transport authority in their own member state.

**NB: Valid Community Authorizations also permit cross-border operations into three European Economic Area (EEA) states – namely, Liechtenstein, Norway and Iceland.*

It is important to note the continuing requirement for bilateral and ECMT permits for road haulage journeys *outside* the EU (and the three EEA states mentioned above) and for international furniture removals, and the particular requirements relating to journeys to and through Austria which are described later in this chapter.

Regulation 881/92/EEC

The system of Community Authorizations for intra-EU road haulage operations is established under Council Regulation (EEC) 881/92 and implemented in the UK by The Goods Vehicles (Community Authorisations) Regulations 1992 (SI 1992 No 3077). Regulation 881/92 amended earlier legislation (ie Council Regulation (EEC) 3164/76 as amended by Council Regulations (EEC) 1841/88 on Access to the Market in the International Carriage of Goods by Road) by effectively introducing qualitative criteria in place of the previous system of quantitative restriction. The qualitative criteria are as specified in Council Regulation (EEC) 561/74 as amended by Council Regulation (EEC) 438/89, namely a requirement that the road haulage operator be of good repute, of adequate financial standing and professionally competent in road haulage operations.

Issue of Community Authorizations

Community Authorizations are issued in the UK by the Traffic Commissioners (TCs) on an automatic basis to all standard international operator ('O') licences holders (ie there is no need for UK operators to make separate application for these authorizations). It should be noted that such 'O' licences are granted only to those applicants who fully satisfy the qualitative standards of the EU, in other words the legal requirements for good repute, adequate financial standing and professional competence in road haulage operations.

The authorization comprises an original document to be retained safely at the licence holder's main place of business, and a number of certified true copies equalling the total number of vehicles authorized on the operator's licence. One of the certified true copies of the authorization must be carried in each vehicle undertaking international journeys within the EU. Community Authorizations and the certified true copies carried on vehicles must be produced for inspection on request. Failure to do so, and to carry the certified true copy on a vehicle while on an international journey within the EU, is an offence.

Penalties for Infringement of the Law

UK-based international hauliers who jeopardize their 'O' licences by failing to meet the requirements of good repute, financial standing or professional competence also jeopardize their Community Authorization. In other words, where circumstances arise which, as a result of infringement of the law or failure to meet the qualitative requirements of good repute, financial standing and professional competence, require the TC to suspend, curtail or revoke an 'O' licence, the Community Authorization will also be automatically suspended, curtailed or withdrawn (ie revoked). The precise action taken by the TC will depend on the seriousness of the offence or offences. Serious or repeated minor infringement of carriage regulations (ie the Community Authorization regulation itself) will result in temporary or partial suspension of the certified true copies of the authorization.

In member states where 'O' licences of the type issued in the UK or its equivalent are not used, failure by international hauliers based in those states to meet (or maintain) the standards of good repute, financial standing and professional competence required under the EU regulation will, nevertheless, result in jeopardy of the Community Authorization.

It is a specific requirement of the regulation that where one member state becomes aware of infringement of Community Authorization legislation by a haulier from another member state, it shall inform the authorities in that member state and may ask that state to impose sanctions on the haulier in accordance with the regulations (ie for temporary or partial suspension of certified copies or withdrawal of the Community Authorization).

Validity and Duration of Authorizations

Community Authorizations are made out in the original licence holder's name and are not transferable to any third party and remain valid while the 'O' licence is in force unless otherwise revoked. Certified copies, as mentioned above, must be carried on the relevant vehicle when on an international journey and must be produced by the driver for examination whenever he is required to do so by an authorized inspecting officer.

On the expiry of a Community Authorization after five years, it is a requirement that the issuing authority (ie in the UK, the TC) must verify whether the operator still satisfies the legal conditions for its issue. Since these conditions are identical to those on which renewal of the haulier's 'O' licence depends, namely good repute, financial standing and professional competence, in the UK at least, operators whose 'O' licences are renewed can rest assured that their Community Authorization will be automatically renewed at the same time.

Community Authorization Documents

Annex I to EU Regulation (EEC) 881/92 specifies a model for the Community Authorization, the front page of which contains details of the haulier (ie name and full address), the date from which it is valid and the name of the authority by whom it is issued and the date of issue; on the rear are printed the general provisions for the use of such authorizations, in particular that while within the territory of any member state the holder (ie both the road haulage operator and the vehicle driver) must comply with the 'laws, regulations and administrative provisions in force in that state', especially in regard to transport and traffic.

Exemptions from Community Authorization Procedure

Certain transport operations are specifically exempt from the requirement for Community Authorizations in accordance with Annex II to the EU regulation as follows:

- Carriage of mail as a public service.
- Carriage of vehicles which have suffered damage or breakdown.
- Carriage of goods in vehicles with a permissible laden weight (including that of any trailer drawn) which does not exceed 6 tonnes or the maximum permitted payload of which does not exceed 3.5 tonnes.
- Carriage of goods* in vehicles owned (including hired) by an own-account firm solely for its own purposes and where the transport is no more than ancillary to its overall activities and where the vehicle is driven only by an employee of the firm.

  ** NB: The goods concerned may be the property of the firm, or have been sold, bought, let out on hire or hired, produced, extracted, processed or repaired by the firm.*
- Carriage of medicinal products, appliances, equipment and other articles required for medicinal care in emergency relief, in particular for natural disasters.

ROAD HAULAGE CABOTAGE

Cabotage operation is provided for under the Treaty of Rome. A UK Transport Minister was reported as saying that 'the liberalization of cabotage is essential to the creation of a true single market in road haulage', and furthermore, 'it would help to reduce the wasteful costs associated with empty running, bringing both economic and environmental benefits, and would open up exciting new markets for hauliers'.

Cabotage is quite simply internal haulage by foreign transport operators – the collection and delivery of goods by road within a country by a road haulier whose business is established in another country. The significance of cabotage, of course, is that it protects internal haulage markets against incursion – or in this case the abstraction of domestic traffics – by outsiders (see below). Hence the reason why, hitherto, it has always been an illegal practice, but now with the liberalization policies of the SEM in force, such restrictive practices have been swept away and road freight cabotage within EU member states is permitted by regulation.

Cabotage by EU own-account road transport operators is permitted, but only on the same basis as defined above (ie the fourth item under exemptions from Community Authorization requirements).

Distortion of Domestic Haulage Markets

Provision is included in the EU regulation for safeguard measures to be implemented – on the authority of the Commission of the EU – where cabotage operations cause or lead to serious disturbance of the national transport market in a given geographical area.

In practice, the operation of road haulage cabotage in the early days of the liberalization process appears to have had negligible impact on domestic haulage markets – no more than 0.25 per cent, according to former EU Transport Commissioner, Neil Kinnock – and most of that probably within Germany by Dutch and Belgian hauliers.

The Permanent Cabotage Regime

Council Regulation (EEC) 3118/93 (of 25 October 1993) initially limited access to cabotage within the EU on the basis of a quota system only, with specified annually increasing numbers of cabotage permits being available for member states. These permits were obtained by road hauliers who wished to operate internally within other member states so long as they held a valid Community Authorization permitting international road haulage operations. The total allocation of cabotage permits throughout the EU increased by some 30 per cent or more annually until the quota system was legally abolished and replaced by the so-called 'permanent cabotage regime' (ie total liberalization of the European Union road haulage market) from 1 July 1998.

With the permanent cabotage regime in place since 1 July 1998, all international road hauliers holding Community Authorizations are entitled to operate temporary* road haulage services in member states other than their own without any restriction as to quantitative limits or any requirement for a registered office or any other establishment in that state. Since this date there is no longer any requirement for transport operators to obtain cabotage authorization or for goods vehicles to carry cabotage permits. However, they have to comply with the laws, regulations and administrative provisions in force in the 'host' country including those concerning:

- rates and conditions incorporated in haulage contracts;
- weights and dimensions of road vehicles – which may, in fact, exceed those of the home country, but must not in any case exceed the vehicle's design standards;
- dangerous goods, perishable foodstuffs and live animals;
- goods vehicle drivers' hours and rest periods;
- VAT on transport services (see also below).

** The word temporary, used in the context above, means that cabotage permit holders may enter a member state (ie temporarily) and carry out internal road haulage journeys as required. It does not mean that they have the right to establish a permanent haulage operation in that country. If a haulier is established on a permanent basis within a member state, or wishes to be so, so that he can operate domestic haulage, then he must conform to the relevant national legislation of that state relating to internal haulage.*

VAT on Cabotage Operations

Internal transport operations under cabotage authorization requires operators to comply with national VAT regulations. For this purpose, operators may need to register in the member states in which they are operating or appoint a suitable VAT agent or fiscal representative to handle these matters on their behalf.

Prohibited Operations, Offences and Penalties

Cabotage by hauliers and own-account operators in non-EU states is prohibited.

Non-resident hauliers who infringe either the cabotage rules when operating in a state other than that in which their business is established, or who otherwise offend against Community or national transport legislation while in such states, may be penalized by the host nation, on a non-discriminatory basis. Penalties may comprise an official warning or, in the case of more serious or repeated infringements, a temporary ban on cabotage. Where falsified cabotage documents are found these will be confiscated immediately and returned to the appropriate authority in the haulier's own country.

Member states are required to co-operate in applying the cabotage rules and may ask another member state to impose penalties on its own hauliers who are found to have breached these rules – even to the point of withdrawing an offending haulier's right to operate (ie in the case of a UK haulier this could mean loss of his 'O' licence). Additionally, the haulier may be prosecuted for relevant offences and brought before a court in his home country for offences committed in another EU member state.

BILATERAL ROAD HAULAGE PERMITS

Certain road haulage operations from the UK and other EU member states to non-EU member states still require the issue of a bilateral road haulage permit (see table below). At the present time road haulage journeys to or through Austria, Belarus, Morocco, Russia, Turkey, Tunisia and Ukraine require such permits for specified transport operations (see table below). Third-country permits are required for journeys from either Germany or Romania.

Bilateral road haulage permits are not required for transport operations within the EU. However, hauliers on transit journeys across Community territory to non-EU destinations must be in possession of a Community Authorization.

Validity of Permits

Where bilateral road haulage permits are required as described above (see also table below), such permits are available covering single journeys only, allowing just *one* return journey to be undertaken between the dates shown on the permit. Outside of these dates the permit is invalid and it would be illegal to commence or continue the journey.

In the case of Turkey, single journey permits as described above are available as well as multiple journey permits authorizing four journeys.

For Austria, normal termination permits are available for journeys destined for that country, but for transit traffic the Eco-points system applies whereby the haulier has to obtain stamps to affix to an Eco card (see below).

Single journey permits are valid only between the dates shown, as mentioned above. Permits for Austria are valid for two months from the date of issue while those for Turkey, Russia, Ukraine and Belarus are valid from the date of issue until 31 December.

Table of Permit Requirements

Type of operation
1. Own-account carriage
2. Unaccompanied trailer/semi-trailer
3. Unladen in transit
4. Unladen relief vehicle
5. Unladen entry to collect goods
6. Airports – re-routed goods
7. Airports – carriage of luggage
8. Carriage of broken-down vehicles
9. Funeral transport
10. Works of art for fairs/exhibitions
11. Works of art for commercial purposes
12. Carriage of antiques
13. Goods for publicity or information purposes
14. Sports/theatre/media
15. Fairs and exhibitions
16. Animal carcasses *not* for human consumption
17. Animal carcasses *for* human consumption
18. Household removals
19. Carriage of mails
20. Refuse and sewage
21. Bees and fish fry

22. Valuable goods
23. Medical emergencies
24. Vehicles with plw* not over 3.5 tonnes
25. Vehicles with ulw* not over 6 tonnes
26. Payload not over 3.5 tonnes
27. Abnormal loads
28. Spare parts for sea-going vessels
29. Ships provisions
30. Transit traffic

* plw = permissible laden weight; ulw = unladen weight

	EU States	Austria	Belarus*	Morocco	Russia	Tunisia	Turkey	Ukraine
1.	–	0	P	P	P	P	P	–
2.	C	–	P	P	P	P	P	P
3.	C	E	P	P	P	P	P	P
4.	C	–	P	P	P	P	P	P
5.	C	P	P	P	P	P	P	P
6.	C	–	P	–	P	P	–	P
7.	C	–	P	P	P	–	–	P
8.	–	–	–	P	–	–	–	–
9.	C	–	–	P	–	P	–	P
10.	C	–	–	–	–	–	P	–
11.	C	–	P	P	P	–	P	–
12.	C	–	P	P	P	–	P	P
13.	C	–	P	P	P	P	P	–
14.	C	–	–	P	–	–	–	–
15.	C	–	–	P	–	–	–	–
16.	C	–	P	P	P	P	–	P
17.	C	–	P	P	P	P	P	P
18.	C	–	–	P	P	–	P	P
19.	–	–	–	P	–	P	–	–
20.	C	–	P	P	P	P	–	P
21.	C	–	P	P	P	P	–	P
22.	C	–	P	P	P	P	–	P
23.	–	–	–	–	P	–	–	–
24.	–	–	–	P	P	–	P	–
25.	–	–	–	P	P	P	P	–
26.	–	–	–	P	P	P	P	–
27.	S	S	S	S	S	S	S	S
28.	C	–	P	P	P	–	P	P
29.	C	–	P	P	P	P	P	P
30.	C	E	P	P	P	P	P	P

– = no permit needed; C = Community Authorization; E = Eco-point system; 0 = own-account document; P = bilateral permit required; S = special permit required; * bilateral agreement not signed yet (informal arrangement at present)

Source: A Guide to Taking Your Lorry Abroad, International Road Freight Office, Cambridge

Third-Country Traffic

Third-country traffic, which is the carriage of goods between two countries other than the country in which the vehicle is registered, is permissible in certain cases (ie between any EU country) but not in others. For UK hauliers, journeys are permissible between any two EU countries.

NB: Permits valid for third-country traffic are available and full details can be obtained from the IRFO – now part of VOSA (see p 646).

Issue of Permits

Road haulage permits where necessary as described above are issued by the relevant authority in each member state. Normally, this involves completion of application forms, advance payment of the relevant fee and submission by the applicant of a copy of his authority to operate (eg his Community Authorization).

Return of Used Permits

Used and expired permits must be returned to the issuing authority not later than 15 days after the relevant journey has been completed or the permit expiry date, whichever is earlier.

Journey record sheets issued with period permits (eg the four-journey permit for Turkey) must be returned within the same timescale.

Lost or Stolen Permits

Road haulage permits are valuable transit documents and as such should be treated with care and appropriate security. They are not transferable to another operator and such misuse is illegal throughout the Community, with harsh penalties imposed on offenders (see also below). Replacement of lost or stolen permits is not normally automatic, and in any case a full written explanation of the circumstances surrounding the loss or theft is required, together with a copy of the police report.

Journeys to or Through Non-Agreement Countries

If vehicles are to travel to or through a country with which an EU member state has no agreement, permission to operate in that country has to be sought direct from its transport authority. Application should be made well before the journey is due and full details of the vehicle, the load and the route should be given.

ECMT Permits for Non-EU Journeys

A number of ECMT (European Conference of Ministers of Transport) permits are allocated to the UK each for haulage journeys between ECMT member countries (ie all EU member states plus Bulgaria, Bosnia-Herzegovina, Croatia, Moldova, Romania, Switzerland and Turkey). However, the validity of some permits is limited in certain countries, particularly Austria.

These ECMT permits allow journeys between member countries, including laden or empty transit journeys and third-country journeys to other ECMT countries, which are prohibited by certain bilateral agreements. However, they cannot be used for transit of ECMT countries on journeys to non-ECMT states or for cabotage. They are for hire or reward journeys only and may not be used by unaccompanied trailers or semi-trailers. They are valid for one calendar year and allow an unlimited number of journeys within that period, but they may be used with only one vehicle at a time. The quota for their issue is limited, so these permits are allocated before the beginning of the year in which they are issued. Usually no further supplies are available during the course of the year, but should the quota be increased an announcement is made in the trade press.

ECMT Removals Permits

These permits are quota-free and can be used for international removals between, or crossing, ECMT member countries. They are available only to firms employing the specialized equipment and staff needed to undertake such operations and are valid for one year from the date of issue.

PERMIT CHECKS

As a result of the exposure of a number of cases of permit frauds, stringent regulations exist to prevent vehicles on international journeys travelling without valid permits (where relevant – see above) and checks are made on vehicles to ensure that these regulations are complied with. A vehicle will be prevented from continuing its journey if it does not carry a valid permit. In the UK it is an offence to forge or alter permits, to make a false statement to obtain a permit or to allow one to be used by another person.

ENVIRONMENTAL PLAQUES ON HEAVY LORRIES

The European Conference of Ministers of Transport (ECMT) scheme (see above), for the issue of multilateral quota road haulage permits, allows for an additional quota of permits for those countries participating in the so-called 'green lorry' scheme. In order to gain the additional quota the vehicles used on ECMT multilateral permit journeys must comply with specified minimum standards for noise emissions and chemical exhaust emissions for diesel engines as well as certain other safety requirements. Compliance with these standards is confirmed by the issue of an environmental certificate of conformity, which must be carried on the vehicles during relevant journeys, and by the display of an environmental plaque – usually stuck in a prominent place on the vehicle front bumper or cab front. These plaques and certificates indicate to the relevant authorities, while in cross-border and internal transit, that the vehicle in question meets ECMT standards for either 'green' or 'greener and safe' lorries; the former with Euro-1 specification engines and the latter with Euro-2 spec engines. Vehicles with Euro-3 engines are required to display a plaque showing the number '3'.

These requirements are set out in the ECMT's document *Guide for Government Officials and Carriers on the Use of the ECMT Multilateral Quota*. This document can be viewed and/or downloaded from the ECMT's website at: www1.oecd.org/cem/pub/pubpdf/MQguide.pdf.

Also accompanying the *Guide* is a document called: *Documents to be on board a Tractor Vehicle.*

The plaques (referred to in the *Guide* as stickers) fall into the following categories:

- *For 'green' lorries*: a green circular plaque of 200 mm diameter with a 20 mm-wide white border and displaying:
  - a white letter 'E' (Environment) or white letter 'U' (Umwelt) denoting vehicle with a Euro-1 engine.
- *For 'greener and safe' lorries*: a green circular plaque of 200 mm diameter with a 20 mm-wide white border and displaying either:

- a white letter 'S' (Sicher/safe) denoting a vehicle with a Euro-2 engine and safety features*; or
- a white figure '3' denoting a vehicle with a Euro-3 engine and safety features*.

**NB: These safety features include: rear underrun protective devices, lateral protection, rear view mirrors, lighting and light signalling devices, tachograph, speed limiters, rear marking plates, anti-lock braking systems and steering, all of which must comply with relevant EC/EEC/ECE/UN Directives and Regulations.*

Austria and Germany have their own plaques to indicate vehicles that meet low noise standards as follows:

- 'L' – denoting Lärmarm Fahrzeug (ie low noise vehicle): Austrian version of the symbol for 'quiet' diesel vehicles, which are:
  - small and medium-sized lorries of no more than 150 kW producing noise levels no greater than 78dB; and
  - heavy lorries exceeding 150 kW with noise levels no greater than 80dB.
- 'G' – denoting Geräuscharm Fahrzeug (ie noiseless/quiet vehicle): German version of the Austrian 'L' plate for use only on diesel vehicles meeting German low-noise requirements – similar to those for Austria as above.

Emissions Standards

ECMT-specified noise emission and chemical exhaust emission requirement standards for so-called 'green' vehicles are as follows.

Noise Emissions Standards for 'Green', 'Greener and Safe' and 'Euro-3 Safe' Lorries

- For vehicles with maximum power output up to 150 kW = 78 dB.
- For vehicles with maximum power output over 150 kW = 80 dB.

Chemical Exhaust Emissions for Diesel Engines

	'Green lorries'	*'Greener and safe lorries'*	*'Euro-3 safe lorries'*
CO	4.9 g/kWh	4.0 g/kWh	2.1 g/kWh
HC	1.23 g/kWh	1.1 g/kWh	0.66 g/kWh
NO_x	9.0 g/kWh	7.0 g/kWh	5.0 g/kWh
Part.	0.4 g/kWh^2	0.15 g/kWh^2	0.10 g/kWh^2

Other Vehicle Markers

In some European countries, particularly Germany and Austria for example, vehicles may also be seen displaying other plaques such as the following:

- 'A' – required in Germany on all vehicles carrying waste;
- 'K' – required in Germany for vehicles making use of exceptions to the Motor Tax Law.

In Italy the following plaques (displayed on the rear and on both sides of the vehicle) are used to indicate vehicles carrying special products and which are therefore exempt from weekend driving bans as follows:

- 'a' – for vehicles carrying easily perishable foodstuffs;
- 'd' – for vehicles carrying milk.

In Austria, three plaques are used to indicate vehicles that have had their maximum weight modified under certain conditions:

- 'R' – for vehicles authorised for use only on special routes;
- 'H' – for vehicles that have had their maximum permissible weight increased;
- 'E' – for vehicles that have had their maximum permissible weight reduced.

AUSTRIAN 'MAUT' SYSTEM

The former Austrian Eco-points system described in previous editions of this *Handbook* has been completely abolished and replaced by a new 'Maut' (toll) 'per mile' system for trucks over 3.5 tonnes gross weight similar in concept to the German LKV-Maut system described on p 247. The previous Austrian 'Vignette' sticker system has also been replaced. The system requires relevant vehicles to be fitted with an easy-to-install, DSRC-TAG electronic (microwave) transponder, colloquially called a 'GO-Box', and payment is via the Internet or at point-of-sale terminals, located mainly at roadside fuel filling stations.

All 2,000 kilometres of Austria's motorways are wired for DSRC (short-range electronic tolling) to operate the nationwide toll system, which is achieved with the aid of an electronic transponder or by off-road payment, normally by internet or at one of 200-odd point-of-sale terminals, mostly located at fuel filling stations. When paying the toll the truck driver enters his registration number, which is recorded on a list of payers. When the vehicle is seen on the road by cameras mounted on some 400 overhead gantries, it is matched against the so-called 'white' payment list.

The truck toll currently is €0.13/km, €0.18/km and €0.27/km for two-, three- and four-axle vehicles respectively. A higher toll rate is charged for use of a number of special road sections on the A9, A10, A11, A13 (the Brenner Pass) and A16 motorways.

OWN-ACCOUNT TRANSPORT OPERATIONS

Own-account transport operations within Community territory (including cabotage as defined above) are now free from all bilateral permit requirements (under the provisions of EU Regulation 881/92 Annex II) provided that goods are carried solely in connection with the trade or business of the vehicle user and are not carried for hire or reward, and that the following conditions are also met:

- The goods carried must be the property of the business (of the vehicle user) or must have been sold, bought, let out or hired, produced, extracted, processed or repaired by the business.
- The purpose of the journey must be to carry the goods to or from the business or to move them, either within the business or outside for its own needs.

- Motor vehicles used for the carriage must be driven by employees of the business.
- The vehicles carrying the goods must be owned by the business, bought on deferred terms or hired by it (this does not apply where a replacement vehicle is used during a short breakdown of the vehicle normally used).
- Road haulage must not be the major activity of the business.

Own-account operations between the UK, Austria, Cyprus and Hungary are free from permit requirements but in the case of such journeys, drivers should carry on the vehicle a document containing the following information to confirm that the operation is solely for own-account purposes:

- The name and address of the vehicle operator (ie user).
- The nature of the operator's trade or business.
- The nature of the goods being carried.
- The location of the loading and unloading points.
- The registration number of the vehicle on which the goods are carried.
- Details of the route to be followed.

In all cases, own-account vehicle operators (and their drivers) should be aware that they may be asked to provide satisfactory evidence to help the authorities to determine the ownership of the goods and that they are being carried solely for own-account purposes.

INTERNATIONAL CARRIAGE OF GOODS BY ROAD – CMR

Transport operators – whether just a one-man haulage business (ie owner-driver) or a large firm – carrying goods for reward on international road haulage journeys must comply* with the Convention on the Contract for the International Carriage of Goods by Road 1956 (*Convention Relative au Contrat de Transport International de Marchandises par Route* – commonly referred to as the CMR Convention). This Convention is applied in the UK under English law by the provisions of the Carriage of Goods by Road Act 1965. The Convention defines the carriers' liability and the documents to be carried on vehicles engaged in the international movement of goods between different countries, of which at least one is a party to the CMR Convention.

** NB: Since, in effect, CMR applies automatically to an international road haulage journey, the haulier has no choice in the matter – but may be carrying out the operation in ignorance of this fact.*

The following countries are party to the CMR convention: Austria, Belgium, Bulgaria, Czech Republic and Slovakia, Denmark, Finland, France (including its overseas territories), Germany, Gibraltar, Greece, Hungary, Italy, Luxembourg, the Netherlands, Norway, Poland, Portugal, Romania, Spain, Sweden, Switzerland, United Kingdom (including Northern Ireland), and the former Yugoslavia.

Applicable Law

International conventions such as CMR override the relevant provisions of national law but only insofar as they cover the point at issue. However, not all points of issue that may arise by way of dispute or claim out of a contract for international carriage are covered by the CMR Convention, in which case relevant national law would apply (basically this means the law of the country in which the dispute or claim arose and to whose courts the matter is referred). In deciding the outcome of any dispute or claim, the courts may refer to either the French or English language texts of the Convention although the French text is not reproduced in the schedule to the Carriage of Goods by Road Act 1965. When referring to this Act, a court is still free to consider the French language version of the Convention for clarification should difficulty be found in determining the precise meaning of any aspect of the English text.

Non-CMR Operations and Journeys

For the CMR Convention to apply there must be clear evidence of a contract for the international carriage of goods for reward. In other words the carriage of goods on an international journey at no charge, and therefore outside of a contract for carriage as specified in the Convention, would not be covered by CMR. Furthermore, the carriage must be of *goods* to allow the Convention to apply.

Besides these exclusions arising from defining the precise terms of applicability, the Convention has a number of more specific exemptions. Namely, its terms do not apply to:

- own-account operations which involve international journeys;
- furniture removals;
- funeral consignments which are transported abroad;
- carriage under an International Postal Convention.

The Convention is also *not* applicable in respect of international haulage operations between the United Kingdom (including Northern Ireland) and the Republic of Ireland, or to contracts for the carriage of goods between the UK mainland and the Channel Islands (ie Guernsey, Jersey, Alderney, Sark and Herm); such operations not being legally classed as international journeys. Cabotage journeys (ie internal journeys within a country by a road haulier from another country) are also outside the provisions of the CMR.

Basic Requirements of CMR

The CMR Convention automatically applies to every contract for the international carriage of goods by road in vehicles for reward, even when the vehicle containing the goods is carried over part of its journey by sea, rail or inland waterway, although other conventions may also apply and take precedence over CMR. A CMR-type consignment note must be completed for the journey. There is no escape from the Convention's provisions and road hauliers may not opt out (by agreement with consignors or otherwise) from its legal (ie liability) requirements. Thus, for example, if a road haulier was induced to carry goods for reward on an international journey knowing nothing of the CMR and its provisions and his consequential liability responsibilities, and even without a CMR consignment note in force, the provisions of the Convention would still apply. Should a legal dispute or claim subsequently arise he could be liable to pay substantial compensation not covered by his domestic Goods in Transit insurance policy.

Even in circumstances where a road haulier may not be aware that a load being carried on one of his vehicles is destined to continue on, or has previously been moved on, an international journey, and where he is not aware of the conditions and implications of CMR, the Convention's provisions still apply. This is particularly relevant where loaded articulated semi-trailers are collected or delivered – it is the load and that part of the vehicle which are together at the time of crossing national boundaries which determines that a particular journey is legally an international journey to which CMR applies. The Convention is not relevant where containers are carried domestically after having been transferred from the vehicle or rail wagon which crossed national boundaries; or where loads are transhipped between the vehicle which undertook the international part of the journey (to which CMR did apply), and local delivery vehicles (to which CMR does not apply).

CMR Conditions for International Road Haulage Journeys

An outline of the principal conditions of the Convention is given here:

1. The Convention applies to every contract for carriage of goods, whether wholly by road or partly by road and partly by rail, sea or inland waterway, as long as the goods remain in the original vehicle, on a journey from one country to another, one of which is a contracting party to the Convention (with the exception of UK–Eire and UK mainland–Channel Islands journeys which are ruled not to be international journeys for this purpose). Exemptions to CMR apply to carriage under international postal conventions, funeral consignments and furniture removals.
2. The carrier (ie road haulier) is responsible under the Convention for the actions and omissions of his agents and any other persons whose services are used in carrying out the movement. Even if the original road haulier contracted to undertake the movement sub-contracts the whole of the operation to another road haulier (whose name appears on the CMR consignment note) the first (original) haulier remains fully liable under CMR should a dispute or claim arise. This can present problems in a case where, for example, the first (original) haulier operates only in domestic transport and thus is covered only by, say, Road Haulage Association Conditions of Carriage and is insured accordingly – not being insured to the much higher CMR level of liability.
3. A contract for the international carriage of goods for reward is confirmed by making out a CMR consignment note in three original copies which should be signed by the sender and carrier. Each keeps a copy and the third copy travels on the vehicle with the goods. While a CMR consignment note confirms that a contract exists, the absence of, or failure to raise such a note does not invalidate the contract or dis-apply the terms of the Convention.
4. If the goods are carried under a single contract in different vehicles or are divided owing to their different nature, the carrier or the sender can specify that a separate consignment note should be made out for each vehicle or each load of goods.
5. The consignment note must contain certain specified details (see pp 631–32) and may also contain additional information of use to the parties to the

contract. It must state that the carriage is subject to CMR. Although not a 'title' to the goods, the consignment note is evidence of the facts it contains (ie the details shown are presumed to be correct), such as the number of packages, etc, and any claim which disputes such facts would have to be backed by substantial independent evidence to the contrary. Normally a standard note such as that available from the International Road Transport Union (IRU) is used (or in the UK from the Road Haulage Association or the Freight Transport Association – both IRU members).

6. The sender is responsible for all expenses, loss and damage sustained by the carrier as a result of inaccuracies in completion of the consignment note in relation to information supplied by him – even if the road haulier completes the note from information given to him by the sender.
7. On receipt of the goods, the carrier must check the accuracy of the details shown in the consignment note particularly, for example, as to the number of packages, the apparent condition of the goods, their packaging and how they are marked. Any discrepancies or comments about other relevant matters such as the condition of the goods or packages should be noted by a 'reservation' on the note. Should the sender request the carrier to check the contents of packages or to have the consignment weighed he must reimburse the carrier any costs incurred in doing so.
8. The sender is liable to the carrier for damage and expenses due to defective packing of the goods unless the defects were known to the carrier when taking over the goods and he indicated this fact by way of a 'reservation' on the note. The absence of such a reservation means that the carrier, if he was aware of the damage, accepted any likely risks of subsequent claims.
9. The sender must attach to the consignment note or make available to the carrier the necessary documents to complete Customs formalities. The sender is liable to the carrier for any damage caused by the absence, inadequacy or irregularity of such documents.
10. The sender has the right of disposal of the goods and may stop transit of the goods or change the delivery address up to the time of delivery to the consignee unless he has stated on the consignment note that the consignee has this right. Once the goods are delivered to the address on the consignment note, the consignee has the right of disposal.
11. A carrier who fails to follow the instructions on the consignment note or who has followed them without requesting the first copy of the consignment note to be produced is liable for loss or damage caused by such failure.
12. The carrier must provide the consignee with a second copy of the consignment note at the time of delivering the goods.
13. If the carrier cannot follow the instructions on the consignment note for any reason, he must ask the sender or the consignee, depending on who has the right of disposal (see item 10 above), for further instructions.
14. The carrier is liable for the total or partial loss of the goods and for any damage to them occurring between the time when he takes over the goods and the time of their delivery unless the loss, damage or delay was caused by a wrongful act or neglect of the claimant. The burden of proof in this case rests with the carrier.

15. Failure to deliver goods within 30 days of a specified time limit, or within 60 days from the time when the first carrier took them over if there is no time limit for delivery, results in the goods being considered to be lost.
16. When goods of a dangerous nature are consigned, the carrier must be informed of the nature of the danger and the precautions to be taken.
17. Calculation of compensation in the event of loss or damage is related to the value of the goods at the place and time they were accepted for carriage but will not exceed a set value (ie related to SDR – see p 630 for explanation).
18. Carriage charges, Customs duties and other charges in respect of the carriage are refunded in the case of total loss of the goods and proportionately in the case of partial loss.
19. Higher levels of compensation may be claimed where the value or a special interest in delivery has been declared or where a surcharge has been paid in respect of a declared value exceeding the limit mentioned in item 17 above.
20. In the case of damage the carrier is liable for the amount by which the value of the goods has diminished.
21. The claimant may demand interest in respect of the amount of any claim at 5 per cent per annum from the date on which the claim was sent to the carrier.
22. A carrier cannot avail himself of exclusions or limiting clauses if damage to goods was caused by his wilful misconduct or default which constitutes wilful misconduct.
23. The consignee is considered to have accepted the goods in a satisfactory condition if he does not indicate his reservations at the time of delivery or within seven days (excluding Sundays and public holidays).
24. In legal proceedings, the plaintiff may bring an action in any court or tribunal of a contracting (ie CMR contracting) country, or of a country in which the defendant is normally resident or has his principal place of business, or of a country where the goods were taken over by the carrier or where they were designated for delivery, and in no other courts or tribunals.
25. The period of limitation for an action under the Convention is one year, or three years in the case of wilful misconduct.
26. Where successive road carriers are involved in a contract under the Convention, each one of them is responsible for the whole operation as a party to the contract. Each successive carrier must give the previous carrier a dated receipt and must enter his name and address on the second copy of the consignment note.
27. A carrier who has paid compensation arising from a claim may recover the compensation plus interest, costs and expenses from other carriers who were parties to the contract subject to:
    (a) the carrier responsible for the loss or damage paying the compensation;
    (b) each carrier responsible for loss or damage jointly caused shall be liable to pay proportionate compensation, or compensation proportionate to their share of the carriage charges, if responsibility cannot be apportioned.
28. If a carrier who is due to pay compensation is insolvent, his share must be paid by the other carriers who are party to the contract.

In particular, road hauliers should note that the terms of the carriage contract require the carrier taking over the goods to check the accuracy of the statements in the consignment note as to the number of packages, their marks and numbers, the apparent condition of the goods and their packaging – and they should obviously do so for their own protection in the event of later disputes or claims. Under the Convention the carrier is responsible for loss, damage or delay from the time of taking over the goods until the time of their delivery.

Furthermore, where goods are handled by a number of carriers on an international journey (eg by transferring a loaded articulated semi-trailer from one to another), provisions are contained in CMR to apportion the liability for loss or damage between all the carriers (based on the relative proportions of the total carriage cost charged by each one of them). This is because of the difficulties which may arise in pinpointing the exact time and place when the damage occurred (unless specific responsibility can be determined), but should one or more of the carriers in this situation default (ie through insolvency) in meeting their share of any claim for damage to or loss of the goods then the remaining carrier or carriers will have to meet the share of those defaulting.

CMR Liability

As stated above, in international road haulage operations (but not own-account road transport operations), the carriage automatically comes within the terms of the CMR Convention under which the carrier's liability for claims resulting from loss of or damage to the goods carried is determined by comparison with a measure known as 'Special Drawing Rights' (SDRs), whereby compensation must not exceed 8.33 units of account per kilogram of gross weight short (gws). Special Drawing Rights are defined by the International Monetary Fund (IMF) as being a unit for converting currency values based on a 'basket' of the currencies of the key member states of the IMF and are converted to the national currency of the country in which any claim is dealt with in court, and is assessed as to value on the date of the judgement, or on a date agreed to by the parties. A treasury certificate stating the value for that day is taken to be conclusive proof of that fact.

NB: The value or exchange rate of SDRs on the date of judgement or agreement referred to above must not be confused with the date of calculation of the value of the goods which are subject to the claim as referred to in item 17 on p 629).

SDR Conversion Rate

The daily conversion rate for SDR to national currencies can be found in the financial newspapers. With changing values it is essential that the current value should be established at any particular point in time, and adequate insurance cover to at least this level of liability should be carried. To give an approximate idea of what this value represents, on 3 September 2001, a conversion of SDR to sterling at a rate of £1.00 = 1.131900* per kilogram represented a value of £7,359.30 per tonne (ie calculated as £1.00/1.131900 = 0.8834702 p × 8.33 × 1000 kg).

* *NB: It is important to note that the SDR rate may change daily.*

CMR Consignment Notes for International Haulage Journeys

Road hauliers carrying goods for hire and reward on international journeys under the provisions of the Convention on the Contract for the International Carriage of Goods by Road (CMR) as described above must complete special CMR consignment notes to be carried on the vehicle. These consignment notes confirm that the carriage is being conducted under a contract subject to the terms of the CMR Convention, but even in the absence of a CMR note, the carriage will still be subject, under international law, to the terms of the Convention although the carrier may not have been aware of this fact.

The consignment note is made out in three original copies all of which should be signed by both the carrier and the consignor of the goods. One copy of the note (with red lines) is retained by the consignor, the second copy (with blue lines) is for the consignee, and the third copy (with green lines) is for the carrier and must travel forward with the vehicle and remain with it while the goods are on board – a fourth copy (with black lines) may be retained on file by the originator of the document.

Where a consignment is divided to travel by different vehicles or by separate means, separate CMR consignment notes can be made out for each individual part of the consignment.

The following details must be entered on CMR consignment notes:

Box 1 Sender (name, address, country)
Box 2 Customs reference/status
Box 3 Sender's/agent's reference
Box 4 Consignee (name, address, country)
Box 5 Carrier (name, address, country)
Box 6 Place and date of taking over the goods
Box 7 Successive carriers
Box 8 Place designated for delivery of goods
Box 9 Marks and numbers; number and kind of packages; description of goods*
Box 10 Gross weight (kg)
Box 11 Volume (m^3)
Box 12 Carriage charges
Box 13 Sender's instructions for Customs
Box 14 Reservations
Box 15 Documents attached
Box 16 Special agreements
Box 17 Goods received
Box 18 Signature of carrier
Box 19 Company completing the note
Box 20 Place, date, signature
Box 21 Copies to:
(i) Sender
(ii) Consignee
(iii) Carrier

**For dangerous goods indicate:*

(i) Correct technical name (ie proper shipping name);
(ii) Hazchem class;

(iii) UN number;
(iv) Flashpoint (°C), if applicable.

Where applicable, the consignment note must also contain the following particulars:

- A statement that transhipment to another vehicle is not allowed.
- The charges which the sender undertakes to pay.
- The amount of 'cash on delivery' charges.
- A declaration of the value of the goods and the amount representing special interest in delivery.
- The sender's instructions to the carrier regarding insurance of the goods.
- The agreed time limit within which the carriage is to be carried out.
- A list of documents handed to the carrier.

The consignor or consignee can also add to the consignment note any other particulars which may be useful to the road haulier.

CONSIGNMENT NOTES FOR OWN-ACCOUNT CARRIAGE BY ROAD

Own-account operators are not required to use the CMR consignment note for international journeys. For most journeys a simple consignment note is all that is necessary to prove that the journey is on own-account (ie not for hire or reward). It should contain details of the following:

- The vehicle operator.
- His trade or business.
- The goods being carried.
- Their loading and delivery points.
- The vehicle being used (ie by registration number, etc).
- The route to be followed.

In the case of own-account traffic to Germany a more detailed document is required containing the following particulars:

- The place at which the document was made out and the date it was made out.
- The name and address of the own-account operator and an accurate description of the nature of his business.
- If the goods are to be accepted from, or delivered to, any other person, the name and address of that person and an accurate description of the nature of his or her business.
- Details of the loading point or points.
- Details of the place or places at which the vehicle is to deliver.
- Details of the nature of the load (ie a description of the goods carried).
- The gross weight or other indication of the quantity of the load.
- The carrying capacity of the vehicle by weight.
- The index mark and registration number of the vehicle or, if these do not exist, the chassis number.
- The distance of the loaded journey in Germany in kilometres.
- The point or points at which the frontier is to be crossed.
- A signature of the operator or his authorized representative.

CUSTOMS PROCEDURES AND DOCUMENTATION

Many changes have taken place in Customs procedures in recent years, both Euro-wide and domestically within the UK; not least, in the latter case, a merger and re-branding of the government departments that deal with the issues of tax and Customs duties. From April 2005 a new, so-called 'super-agency' has been created known as Her Majesty's Revenue and Customs (HMRC).

Community Transit

There have been other changes in Customs procedures, most notably the removal of barriers to intra-Community trade with the opening of the Single European Market (SEM) from 1 January 1993. Many former complicated document systems have been replaced with simplified paperwork systems and the term 'export' no longer applies to shipments within the territories of the EU. The EU's key procedures for dealing with this trade, known as Community Transit and Common Transit in their current manifestation (as described in the *HM Customs and Excise Guide to Exporting and Importing* – version 5 of December 2004), are Customs procedures that allow respectively:

- goods that are not in free circulation, and in certain cases, EC goods to move between two points in the EC, including two points in a single Member State, with the duties and other charges suspended. The procedure is also used to control the movement of goods to and from Andorra, San Marino and the 'special territories' of the Community (such as the Channel Islands) – Community Transit; and
- the movement of both Community and 'non free circulation' goods to, from and through European Free Trade Association (EFTA) countries such as Iceland, Norway, Liechtenstein – Common Transit.

For Community Transit (CT) purposes goods are said to have a particular customs 'status' which is based on their liability to customs duty and other charges (eg excise duty and VAT) or claims to Common Agricultural Policy (CAP) refunds. Goods are divided into two distinct categories as follows, either:

1. goods which have Community status (ie Community goods). These are goods which:
    - originate in the Community;
    - have been imported from a non-EC country and have been put into free circulation (see below) in the Community; or
    - have been manufactured in the Community from materials or parts imported from a non-Community country provided the imported materials or parts are in free circulation.

    The term 'free circulation' is used to describe imported goods on which all import formalities have been complied with and any customs duties or equivalent charges which are payable have been paid and not repaid in whole or in part. Goods originating in the Community are also in free circulation unless a CAP export refund or other refund has been claimed on them. Consequently, as a general rule, Community goods can move within the EC without any customs controls.

or

2. goods without Community status (ie 'non-Community goods').

The Tariff

All the information and advice needed to help with importing and exporting procedures is to be found in 'The Tariff' (ie *The Integrated Tariff of the United Kingdom*). This document, published in three volumes and to the same format in all EU Member States, contains all the relevant commodity codes (more than 65,000 of them); duty rates; procedures, including a box-by-box completion guide for C88 documents (see below); as well as references to the legal provisions.

Customs Documentation

An eight-part Customs document (C88), known as the 'Single Administrative Document' (SAD), is used throughout the Community and EFTA countries for the purposes of import, export and transit controls. A Community Transit declaration comprises completed copies of pages 1, 4 and 5 of the SAD. In certain circumstances a Community status document (T2L) may be used where required to provide evidence to Customs of Community status. A T2L can be either a copy 4 of the SAD, or a commercial document such as an invoice or a transport document or a manifest.

The SAD (form C88) comprises an eight-part document set, but not all of it is used since the opening of the SEM. Only the following copies are now required:

Copy 1 Copy for the Customs office of departure.
Copy 3 Consignor/exporter's copy.
Copy 4 For Customs office of destination, or Community status (T2L) declaration.
Copy 5 Return copy from Customs office of destination to prove that the goods arrived intact.
Copy 7 Statistical copy for Customs in the country of destination.

Copies 1 and 3 remain in the country of origination (ie export) and copies 4, 5 and 7 travel forward with the goods, copy 5 eventually being returned to the office of departure.

Community Transit movements

There are two types of CT procedure:

1. The external CT procedure. This is mainly used to control the movement of non-Community goods and to control the movement of Community goods which are subject to a Community measure involving their export to a third country, eg CAP (Common Agricultural Policy) goods which are liable to an export refund. This procedure is also known as the 'T1' procedure.
2. The internal CT procedure. This procedure controls the movement of free circulation goods to or from the special territories, to or via the EFTA countries and between the EFTA countries. This procedure is also known as the 'T2' procedure (or the 'T2F' procedure for goods moving to or from the special territories).

Control Procedures

All transit movements are the responsibility of a 'Principal' who is the person or company who undertakes to ensure that the goods are delivered to the office of destination within the prescribed time limit. The Principal must put up a guarantee to secure the relevant duties and other charges in case the goods do not arrive intact at the office of destination. The guarantee can be cash (lodged at the office of departure), vouchers (purchased from a guaranteeing association) or a guarantee undertaking (provided by a guarantor).

To start a CT movement the completed declaration must be presented to Customs at an 'office of departure' where it will be authenticated and a time limit for completing the movement will be set. The goods may also be sealed and an itinerary set. Copies 4 and 5 of the CT declaration (ie the SAD) must travel together with the goods until they reach the EU Member State or EFTA country where the movement is to end. Here the declaration and goods must be presented to the appropriate Customs at the 'office of destination'. This office will then stamp copy 5 of the SAD and return it to the office of departure so that the movement can be discharged. If copy 5 is not received within two months the Principal is contacted and asked to provide proof that the procedure has ended correctly. If proof is not provided within four months, the enquiry procedure is initiated to recover the potential debt.

There is a range of transit simplifications available for use by compliant and reliable traders. These include the use of comprehensive and reduced guarantees or waivers, becoming Authorized Consignors and Consignees, using trader seals on vehicles, special loading lists, having an exemption from prescribed itineraries, goods moved by pipeline, use of other simplified procedures and being able to use commercial documentation instead of the SAD in the case of air, rail and sea environments. All of these simplifications are subject to authorization by the Customs authorities.

Computerization of CT

At the time of writing this text in mid-2005, the Community and Common Transit systems are undergoing a planned computerization programme. The aim of this is to replace the paper CT declaration with electronic messages. This will enable Customs offices throughout the Community and Common Transit countries to exchange information electronically about goods despatched and goods received, thus speeding and tightening the whole process of control. The new system will be able to carry out certain validation and verification checks, thus reducing the workload for Customs staff. Additionally, authorized CT operators will be able to access the system directly, enabling them to reduce administrative and other costs. All UK Customs transit offices are now connected to what is known as NCTS (ie the New Computerized Transit System) and traders can submit electronic transit declarations to Customs and take advantage of the facilities offered by NCTS. In fact, all countries that joined the European Union with effect from 1 May 2004 are now required to make electronic transit declarations.

NCTS

The New Computerized Transit System (NCTS) is a European-wide system, based on the concept of electronic declaration and processing, and is designed to replace, in due course, the existing, paper-based Community Transit system and to provide better management and control of CT. It involves all EU Member States and the EFTA countries, 29 countries in all. Each country develops its own NCTS processing system, according to centrally defined architecture, and these systems are connected, through a central domain in Brussels, to all of the other countries – some 3000 European Customs offices.

The UK, in common with many other participating countries, is using the Minimal Common Core (MCC) software developed by the European Commission, which provides all of the basic data capture and messaging functionality for effective connection to the European network. It is important to note that UK NCTS is a completely separate system from CHIEF (see below) and is not interconnected to it in any way, and separate declarations for transit will continue to be required.

Customs Handling of Import and Export Freight

CHIEF is one of the largest and most advanced Customs declaration processing systems in the world, providing a sound technological platform for Customs & Excise and international trade. Its sophisticated computer software controls and records the UK's international trade movements, whether by land, sea or air, and links with several thousand businesses. It uses the latest technology and is designed to cope with continuous growth in international trade. While it enhances Customs & Excise control of imports and exports, it is also claimed to improve the service provided to the international trade community and reduce overall administrative costs.

Further details of the extensive CT procedures can be found in *The Transit Manual* on the European Commission website at http://europa.eu.int/comm/taxation_customs/resources/documents/customs/procedural_aspects/transit/common_community/transit_manual_en.pdf

Information on UK Customs procedures can be found at www.hmrc.gov.uk.

'DE SUIVI' DOCUMENT

Drivers of UK haulage and own-account vehicles entering France must be able to produce a 'de suivi' document in addition to the normal CMR or own-account consignment note.

There is no official or formal layout for this additional document, it being left to operators to determine how they present the relevant information which can be added to the existing CMR document if space permits. The relevant information required is as follows:

- Date the consignment was drawn up.
- Transport operator's name and address.
- Transport operator's VAT number.
- Date when goods were taken into charge.
- Nature, weight and volume of goods.

- Name of the consignor.
- Address of loading point.
- Name of consignee.
- Address of unloading point.

Space must be provided on the note for entering of the following information:

- Time of vehicle arrival at loading point.
- Time of departure of loaded vehicle.
- Required time of arrival at destination.

A separate document must also be made out to show:

- Date and time of arrival at destination.
- Date and time of departure of unloaded vehicle.

On international journeys the driver must also carry a further document showing the relationship between the transport operator and the driver (ie whether he is a direct employee, an owner-driver or an employee of a sub-contractor).

CARNETS DE PASSAGE

Most countries to which vehicles are likely to travel permit the temporary importation of foreign vehicles and containers (not to be confused with the loads they carry) free of duty or deposit and without guaranteed Customs documents.

However, a Customs document known as a *Carnet de Passage en Douane* is required for the following:

- Vehicles and trailers entering Iran, Iraq, Jordan, Kuwait, Saudi Arabia, Syria, Turkey and other Middle Eastern countries.
- Vehicles remaining in Italy for more than three months.
- Vehicles remaining in Greece for more than 10 days.
- Vehicles remaining in Pakistan for up to three months.

A triptyque is required for vehicles carrying spare parts into Portugal.

Where *Carnets de Passage* are needed for travel they can be obtained from the Automobile Association (AA), the Royal Automobile Club (RAC) and the Royal Scottish Automobile Club (RSAC).

ATA CARNETS

Goods that are being imported only temporarily into certain countries,* such as samples, professional equipment and items for display at exhibitions and fairs, and which, eventually, will be returned to the UK can be moved under an international Customs clearance document known as an ATA Carnet. These documents, valid for 12 months from the date of issue, are issued by chambers of commerce to members without the need for, payments of, or deposits against, duty although the Carnet fees of £94.00 to chamber of commerce members and £170.38 to non-members must be paid. It should be noted, however, that holding an ATA Carnet does not relieve the operator from observing Customs requirements in each individual country.

* *NB: The countries to which this is applicable include all EU member states plus Bulgaria, Gibraltar, Israel, Romania,Switzerland and Turkey.*

THE TIR CONVENTION

The TIR Convention system applies to road journeys to all countries (except member states of the EU) which are party to the Convention but only where the haulier elects to conform to the Convention's requirements.

Under the Customs Convention on the International Transport of Goods by Road (TIR Carnets) 1959, to which the UK is party, goods in Customs-sealed vehicles or containers may travel through intermediate countries en route to their final destination with the minimum of Customs formalities provided a TIR Carnet has been issued in respect of the journey (UK vehicles may operate on international haulage journeys outside the EU without the protection of TIR but in this case they will be subject to the full weight of Customs formality and bureaucracy (and delay) at each border crossing and on arrival at destination – see also below).

The Carnet is a recognized international Customs document intended purely to simplify Customs procedures; it is not a substitute for other documents, nor is it mandatory for any operator to use it; it does not give any operator the right to run vehicles in any European country. Use of a Carnet frees the operator from the need to place a deposit of duty in respect of the load he is carrying in each country through which the vehicle is to pass.

The issuing authorities for the Carnets (in this country the FTA and the RHA – see Appendix I for addresses) act as guarantors on behalf of the IRU (International Road Transport Union – the international guarantor), and for this reason Carnets are issued only to bona fide members of these two associations.

Goods may only be carried under a TIR Carnet provided the vehicle in which they are carried has been specifically approved by the Vehicle and Operator Services Agency (VOSA). This means complying with constructional requirements so that the load-carrying space can be sealed by Customs, after which it must not be possible for any goods to be removed from or added to the load without the seals being broken, and there must be no concealed spaces where goods may be hidden.

Detailed requirements are laid down concerning the structure of the body, particularly regarding the manner in which it is assembled, so that there is no possibility of panels being removed by releasing nuts and bolts and so on. The manner in which doors and roller shutters are secured must also meet stringent specifications. Sheeted vehicles or containers may be used provided conditions relating to the construction of the sheet are observed and as long as when the closing device has been secured it becomes impossible to gain access to the load without leaving obvious traces.

VOSA examines vehicles (by appointment at Goods Vehicle Test Stations) to ensure that they meet the technical requirements for operation under the TIR Convention and issues a certificate of approval, which must be renewed every two years and must be carried on the vehicle when it is operating under a TIR Carnet. This point is particularly important as Customs authorities carry out checks on vehicles leaving the UK to ensure that this certificate is being carried where necessary.

Application for TIR Certification

Application for the examination of vehicles or containers must be made to the Clerk to the Traffic Commissioner for the traffic area in which they are available for inspection. These offices provide the application form GV 62, and a leaflet setting out the technical conditions which have to be met. If a TIR-approved vehicle is sold to another operator, the TIR certificate (form GV 60) is not transferable and the new owner must have the vehicle recertified if he wishes to use it for TIR operations

The fee for examination of a vehicle for TIR purposes is £77, unless it is not of a previously approved TIR design type, in which case the fee is £473. The fee for a TIR certificate is £11. For a reinspection of a failed vehicle, the fee is £51.

TIR Plates

When a vehicle has been approved it must display at the front and the rear a plate showing the letters 'TIR' in white on a blue background. Such plates are obtainable from the FTA and RHA. They should be removed or covered when the vehicle is no longer operating under TIR.

TIR Carnets

TIR Carnets are internationally recognized Customs documents. Carnets are in pairs and have counterfoils in a bound cover. They are in four parts and contain 6, 14 or 20 pages (ie *volets* in French). A 6-page Carnet is valid only for a journey between the UK and one other country. Journeys to more than one other country require 14- or 20-page Carnets, which are valid for two months and three months respectively. A Carnet covers only one load and if a return load is to be collected, a separate Carnet is needed (each individual voucher covers one frontier crossing) and the driver should take this with him on the outward journey. At each Customs point *en route* a voucher is detached and the counterfoil is stamped. Careful attention must be paid to the completion of the Carnet if delays and difficulties are to be avoided during the journey.

Carnets are valid for limited periods only (see above), and if not used they must be returned to the issuing authority for cancellation. Those which are used and which bear all the official stampings acquired *en route* must also be returned within 10 days of the vehicle's return.

Strict instructions regarding the use of Carnets are supplied by the issuing authorities both for the operator and the driver. For example, the driver should never leave the Carnet with any Customs authority without first obtaining a signed, stamped and dated declaration quoting the Carnet number and certifying that the goods on the vehicle conform with the details contained in the Carnet. Drivers should also ensure that the Customs officials at each departure office, transit office and arrival office take out a voucher from the Carnet and stamp and sign the counterfoil accordingly.

The four parts of the Carnet comprise the following:

1. Details of the issuing authority, the Carnet holder, the country of departure, the country of destination, the vehicle, the weight and the value of the goods as shown in the manifest (see 3 below).

2. A declaration that the goods specified have been loaded for the country stated, that they will be carried to their destination with the Customs seals intact and that the Customs regulations of the countries through which the goods are to be carried will be observed.
3. A goods manifest giving precise details of the goods, the way in which they are packed (the number of parcels or cartons) and their value.
4. Vouchers which Customs officials at frontier posts will remove, stamping the counterfoil section which remains in the Carnet.

Before obtaining a Carnet the applicant must sign a form of contract with the issuing authority, agreeing to abide by all the necessary legal and administrative requirements. A financial guarantee is required to ensure that any claims which may be made against the applicant will be met.

Seal Breakage

If a Customs seal on a TIR vehicle is broken during transit as a result of an accident or for any other reason, Customs or the police must be contacted immediately to endorse the Carnet to this effect.

Parties to the TIR Convention

Besides all EU member states being signatories to the Convention, the following non-EU member countries are party to the TIR Convention:

Afghanistan	Iran	Moldova	Switzerland
Albania	Israel	Morocco	Tunisia
Belarus	Jordan	Norway	Turkey
Bulgaria	Kazakhstan	Romania	Ukraine
Georgia	Kuwait	Russia	Uzbekistan

Non-TIR Journeys

Goods may be sent abroad in vehicles without TIR cover and no Carnet is required. In this case, however, it is necessary to comply with the individual Customs requirements of each country through which the vehicle passes. A guarantee in lieu of import duty, or a deposit against such duty, will have to be paid before the vehicle is allowed to enter the country to which it is travelling, or any country through which it needs to pass, and the vehicle will be subject to stringent Customs scrutiny not only at the port of exit from the UK and the port of entry to the Continent, but at all further frontier crossings during the journey.

THE ADR AGREEMENT

When dangerous goods are transported on international road journeys the provisions apply of the European Agreement concerning the International Carriage of Dangerous Goods (the ADR Convention). The current edition applicable from 2003 is called the *Restructured ADR*. This is a long and complex document contained in two large volumes. However, for most general purposes, if the reader refers to the provisions of the UK dangerous goods legislation in Chapter 23 of this *Handbook* he will find all the information he needs, since our rules are based on the international ADR requirements and will be even more so when the current (mid-2003) consultation process is completed and new regulations are made in 2004.

Eurovignettes

Motorway charges (ie tax) must be paid (in accordance with EC Directive 1999/62 EC – the so-called Eurovignette Directive) for goods vehicles of 12 tonnes maximum weight and over, including those towing trailers where the combined maximum weight is 12 tonnes or more, when travelling in or through The Netherlands, Belgium, Luxembourg, Denmark and Sweden. The tax is charged in Euros (€), payable in advance, and is based on a sliding scale according to the pollution emission rating of the vehicle engine (ie Euro 0, Euro 1 or Euro 2/3 – see Chapter 12), the number of axles and the amount of time during which the motorways will be used (eg day, week, month etc), as per the following table. Payment covers jouneys in all five countries.

	Vehicles with up to 3 axles			*Vehicles with 4 or more axles*		
	Euro 0	*Euro 1*	*Euro 2/3*	*Euro 0*	*Euro 1*	*Euro 2/3*
Daily	8	8	8	8	8	8
Weekly	26	23	20	41	37	33
Monthly	96	85	75	155	140	125
Annually	960	850	750	1550	1400	1250

Source: FTA International Road Transport Guide

NB: The currency exchange rate at the time of writing (September 2003) is around €1.38 to the pound sterling meaning that the price range for Eurovignettes ranges from approximately £5.90, this being the daily rate for a two-axle vehicle, to £1,123.19, being the annual rate for a non-Euro-engined, four-axle vehicle.

When the tax is paid, a certificate/receipt called a Eurovignette is issued, showing the vehicle registration number, the date and period of validity, and the amount paid. The vignette must be carried on the vehicle at all times as proof of payment. Failure to obtain a vignette or to be able to produce it on request can result in a fine of up to £5,000.

Eurovignettes can be purchased at petrol stations, motorway service areas and near motorway access points. In England they are obtainable from P&O at Dover, the Ferry Terminal at Ramsgate, the Port Services Office at Harwich and from Eurotunnel at Folkstone.

The Czech Republic introduced a one-day Eurovignette on 1 January 2001. No further details were available at the time of preparing this edition of the *Handbook*.

German Road Toll System

Germany has a scheme for charging heavy lorries over 12 tonnes gross weight for use of the country's 12,000 kilometres of highways (autobahns). After a couple of false starts the scheme finally took off from January 2005. It applies to trucks over

12 tonnes gross weight, which must be equipped with 'on-board' units (OBUs) tracked by global positioning satellite (GPS) from outer space. Road users must register with the toll operator, Toll Collect GmbH, in advance of road usage and fit the vehicle OBU, which calculates the toll due based on the road being used and the distance travelled. A manual back-up system is available which ensures that non-registered users and non-OBU-equipped trucks still pay the toll. The current toll charge is €0.125 per kilometre.

Trailer Certification

UK international road hauliers travelling in Germany and Italy are required by these countries to carry legal documentation relating to the trailer/semi-trailer. Principally, this means a Certificate of Ownership (available from VOSA at a cost of £20) and from the second year of service, a copy of the plating certificate which shows the trailer's roadworthiness.

Driver Contracts for France

Drivers travelling to or through France must carry either:

- a copy of their contract of employment with their current employer; or
- a letter of engagement from their employer; or
- a valid recent (ie their last) payslip.

This requirement is enforced in connection with the campaign to ensure that only EU-based drivers are employed in EU member states.

NB: The employment of non-EU nationals by EU-based companies is not permitted.

Impounding in Italy

The Italian authorities are strict on law compliance and increasingly use vehicle impounding as a means of penalty for infringements – sometimes holding vehicles for as long as three months even for relatively minor infringements, according to FTA reports.

Speed Limits in France

Speed-limit offenders in France face instant loss of their driving licence if they exceed statutory limits by more than 40 km/h (ie approximately 25 mph). Heavy vehicles in France are restricted to 90 km/h (ie approx 56 mph) on motorways, 80 km/h (ie approx 50 mph) on dual carriageways, 80 km/h on country roads (only 60 km/h (ie approx 37 mph) for vehicles drawing trailers) and 50 km/h (ie approx 31 mph) in built-up areas.

Euro Currency

The euro officially came into use from 1 January 2002. The exchange rate for the euro against the pound sterling is published daily in newspapers and on the internet.

Appendices

Appendix I

THE TRAFFIC AREA NETWORK (TAN)

National number for all queries: The telephone contact number for all Traffic Area Offices is 0870 60 60 440. This is the VOSA national telephone number, which is manned by staff trained to deal with all general enquiries between the hours of 07.30–17.30 Monday to Friday, and 08.00–14.30 Saturdays.

Traffic Area	Areas Covered
North-Eastern Hillcrest House, 386 Harehills Lane, Leeds LS9 6NF Tel 0870 60 60 440 Fax 0113 248 9607 *Traffic Commissioner:* *Tom Macartney*	The Metropolitan Boroughs within South Yorkshire, Tyne and Wear and West Yorkshire; the Counties of Durham, Northumberland, Nottinghamshire and North Yorkshire; the Districts of Darlington, East Riding of Yorkshire, Hartlepool, Kingston upon Hull, Middlesbrough, North Lincolnshire, North East Lincolnshire, Nottingham, Redcar and Cleveland, Stockton on Tees and York.
North-Western Hillcrest House, 386 Harehills Lane, Leeds LS9 6NF Tel 0870 60 60 440 Fax 0113 248 9607 *Traffic Commissioner:* *Beverley Bell*	The Metropolitan Boroughs within Greater Manchester and Merseyside; the Counties of Cheshire, Cumbria, Derbyshire and Lancashire; the Districts of Blackburn with Darwen, Blackpool, Derby City, Halton and Warrington.

West Midland 38 George Road, Edgbaston, Birmingham B15 1PL Tel 0870 60 60 440 Fax 0121 608 1001 *Traffic Commissioner:* *David Dixon*	The Metropolitan Boroughs within the West Midlands; the Counties of Shropshire, Staffordshire, Warwickshire and Worcestershire; the Districts of Herefordshire, Stoke on Trent and Telford and Wrekin.
Eastern City House, 126–130 Hills Road, Cambridge CB2 1NP Tel 0870 60 60 440 Fax 01223 309684 *Traffic Commissioner:* *Geoffrey Simms*	The Counties of Bedfordshire, Buckinghamshire, Cambridgeshire, Hertfordshire, Essex, Leicestershire, Lincolnshire (except the Districts of North Lincolnshire and North East Lincolnshire), Norfolk, Northamptonshire and Suffolk; the Districts of Leicester, Luton, Milton Keynes, Peterborough, Rutland, Southend on Sea and Thurrock.
Wales 38 George Road, Edgbaston, Birmingham B15 1PL Tel 0870 60 60 440 Fax 0121 608 1001 *Traffic Commissioner:* *David Dixon*	Counties of Clwyd, Dyfed, Gwent, Gwynedd, Mid Glamorgan, Powys, South Glamorgan and West Glamorgan.
Western 2 Rivergate, Temple Quay, Bristol BS1 6EH Tel 0870 60 60 440 Fax 0117 929 8352 *Traffic Commissioner:* *Philip Brown*	The Counties of Cornwall, Devon, Dorset, Gloucestershire, Hampshire, Oxfordshire, Somerset and Wiltshire; the Districts of Bath and North East Somerset, Bournemouth, Bracknell Forest, Bristol, Isle of Wight, North Somerset, Plymouth, Poole, Portsmouth, Reading, Slough, Southampton, South Gloucestershire, Swindon, Torbay, West Berkshire, Windsor and Maidenhead and Wokingham.
South-Eastern and Metropolitan Ivy House, 3 Ivy Terrace, Eastbourne BN21 4QT Tel 0870 60 60 440 Fax 01323 721057 *Traffic Commissioner:* *Christopher Heaps*	Greater London; the Counties of Kent, Surrey, East Sussex and West Sussex; the Districts of Brighton and Hove, The Medway Towns.

Scottish
'J' Floor, Argyle House,
3 Lady Lawson Street,
Edinburgh EH3 9SE
Tel 0870 60 60 440
Fax 0131 529 8501
Traffic Commissioner:
Joan Aitken

Scotland.

Former International Road Freight Office

The issue of authorizations and permits for international road freight transport has transferred to VOSA in Cambridge at:
City House,
126–30 Hills Road,
Cambridge CB2 1NP
Tel 0870 60 60 440
Fax 01223 309684

Appendix II

TRANSPORT TRADE ASSOCIATIONS AND PROFESSIONAL BODIES

Freight Transport Association Regional Offices

Head Office:
Hermes House,
157 St John's Road,
Tunbridge Wells,
Kent TN4 9UZ
Tel 01892 526171/Fax 01892 534989

Midlands, West and Wales:
Hermes House,
20 Coventry Road
Cubbington
Leamington Spa
Warwickshire CU32 7JN
Tel 01926 450020/Fax 01926 452765

Northern:
Springwood House,
Low Lane,
Horsforth,
Leeds LS18 5NU
Tel 0113 258 9861/Fax 0113 258 6501

Scottish (including Northern Ireland):
Hermes House,
Melville Terrace,
Stirling FK8 2ND
Tel 01786 457500/Fax 07186 450412
(Belfast: Tel 028 9046 6699/
Fax 028 9046 6690)

London and South-eastern:
Hermes House,
157 St John's Road,
Tunbridge Wells,
Kent TN4 9UZ
Tel 01892 526171/Fax 01892 552352

Cardiff:
Regus House,
Falcon Drive,
Cardiff Bay CF10 4RU
Tel 029 20 504070

Enquiries regarding management training should be made to Training and Personnel Services,
Hermes House,
Tunbridge Wells (see Head Office).

Road Haulage Association District Offices

Head Office:
Roadway House,
35 Monument Hill,
Weybridge,
Surrey KT13 8RN
Tel 01932 841515/Fax 01932 852516

Scotland and Northern Ireland:
Roadway House,
The Rural Centre,
Ingliston
Newbridge EH28 8NZ
Tel 0131 472 4180/Fax 0131 472 4179

Northern:
Roadway House,
Little Wood Drive,
West 26 Industrial Estate,
Cleckheaton,
West Yorks BD19 4TQ
Tel 01274 863100/Fax 01274 865855

Southern and Eastern:
Roadway House,
Bretton Way,
Bretton,
Peterborough PE3 8DD
Tel 01733 261131/Fax 01733 332349

Midlands and Western:
Roadway House,
Cribbs Causeway,
Bristol BS10 7TU
Tel 01179 503600/Fax 01179 505647

Other Trade Associations

Local Government Association
Local Government House
Smith Square
London SW1P 3HZ
Tel 020 7664 3131

Association of Lorry Loader Manufacturers and Importers (ALLMI)
Unit 6,
Wincombe Trading Estate,
Albert Road,
St Phillips,
Bristol BS2 0XW
Tel 0117 977 4655 or 0117 972 3980

Association of Load Restraint Equipment Manufacturers (ALREM)
Tel 01538 382312

Association of Vehicle Recovery Operators (AVRO)
1 Bath Street,
Rugby CV21 3JP
Tel & Fax 01788 572850

British Association of Removers (BAR)
3 Churchill Court,
58 Station Road,
North Harrow HA2 7SA
Tel 020 8861 3331/Fax 020 8861 3332

British International Freight Association (BIFA)
Redfern House,
Browells Lane,
Feltham,
Middlesex TW13 7EP
Tel 020 8844 2266/Fax 020 8890 5546

British Industrial Truck Association
Scammell House,
High Street,
Ascot,
Berkshire SL5 7JF
Tel 01344 623800/Fax 01344 291197

British Vehicle Rental and Leasing Association (BVRLA)
River Lodge,
Badminton Court,
Amersham
Bucks HP7 0DD
Tel 01494 434747/Fax 01494 434499

Chemical Industries Association
Kings Buildings,
Smith Square,
London SW1P 3JJ
Tel 020 7834 3399/Fax 020 7834 4469

Electric Vehicle Association (EVA)
Aberdeen House,
Headley Road,
Grayshott,
Hindhead,
Surrey GU26 6LA
Tel 01428 735536

UK Warehousing Association (formerly the National Association of Warehouse Keepers)
Walter House,
418–422 Strand,
London WC2R 0PT
Tel 020 7836 5522

National Association of Waste Disposal Contractors (NAWDC)
Mountbarrow House
6–20 Elizabeth Street
London SW1W 9RB
Tel 020 7824 8882/Fax 020 7824 8753

National Tyre Distributors Association (NTDA)
Broadway House,
The Broadway,
Wimbledon,
London SW19 1RL
Tel 0870 900 0600

Retail Motor Industry Federation (RMI)
201 Great Portland Street,
London W1N 6AB
Tel 020 7580 9122/Fax 020 7580 6376

Road Haulage and Distribution Training Council (RHDTC)
14 Warren Yard,
Warren Farm Office Village,
Stratford Road,
Milton Keynes MK12 5NW
Tel 01908 313360/Fax 01908 313006

Road Operators Safety Council
395 Cowley Road
Oxford O4 2DJ
Tel 01865 775552/Fax 01865 711745

Society of Motor Manufacturers and Traders (SMMT)
Forbes House,
Halkin Street,
London SW1X 7DS
Tel 020 7235 7000/Fax 020 7235 7112

Tachograph Analysis Association
Merseyside Innovation Centre,
131 Mount Pleasant,
Liverpool L3 5TF
Tel 0151 708 0123/Fax 0151 709 4707

Transfrigoroute (UK) Ltd
39a Meneage Street,
Helston,
Cornwall TR13 8RB
Tel 01326 569657
www.transfrigoroute.co.uk

Transport Association
c/o Peter Acton Associates,
Aqua House,
30–32 High Street,
Epsom KT19 8AH
Tel 01372 846482/Fax 01372 727130

Vehicle Builders and Repairers Association (VBRA)
Belmont House,
Finkle Lane,
Gildersome,
Leeds LS27 7TW
Tel 0113 253 8333/Fax 0113 238 0496

Professional Bodies in Transport

Institute of Advanced Motorists
IAM House,
510 Chiswick High Road,
London W4 5RG
Tel 020 8996 9600/Fax 020 8996 9601

Institute of Grocery Distribution
Grange Lane,
Letchmore Heath,
Watford,
Hertfordshire WD25 8GD
Tel 01932 857141/Fax 01923 852531

Chartered Institute of Logistics and Transport (UK)
Earlstrees Court,
Earlstrees Road,
PO Box 5787
Corby,
Northamptonshire NN17 4AX
Tel 01536 740100/Fax 01536 740101

London Office
11/12 Buckingham Gate
London SW1 6LB
Tel 020 7592 3112

Institute of the Motor Industry (IMI)
'Fanshaws',
Brickendon,
Hertford SG13 8PQ
Tel 01992 511521/Fax 01992 511548

Institution of Mechanical Engineers (IME)
1 Birdcage Walk,
London SW1H 9JJ
Tel 020 7222 7899/Fax 020 7222 4557

Institute of the Moving Industry
3 Churchill Court,
58 Station Road,
North Harrow HA2 7SA
Tel 020 8861 3331/Fax 020 8861 3332

Institute of Transport Administration (IoTA)
IoTA House,
7B St Leonards Road,
Horsham
West Sussex RH12 2NW
Tel 01403 242412/Fax 01403 242413

Society of Operations Engineers
22 Greencoat Place,
London SW1P 1PR
Tel 020 7630 1111/Fax 020 7630 6677

Appendix III

OTHER ORGANIZATIONS CONNECTED WITH TRANSPORT AND TRANSPORT JOURNALS

Automobile Association (AA)
Carr Ellison House,
William Armstrong Drive,
Newcastle upon Tyne NEA 7YA
Tel 0870 600 0371/Fax 0191 235 5111
www.theaa.com

BRAKE (for safe road transport)
PO Box 272,
Dorking,
Surrey RH4 4FR
Tel 01306 741113/Fax 01306 888221
e-mail risk@brake-campaign.demon.co.uk

British Standards Institution (BSI)
Linford Wood,
Milton Keynes MK14 6LE
Tel 01908 220022/Fax 01908 320856

Department for Transport,
Eland House,
Bressenden Place,
London SW1E 5DU
Tel 020 7890 3333

International Road Freight Office
See page 646

International Road Transport Union (IRU)
Centre International,
3 rue de Varembe,
PB 44,
1211-CH Geneve 20,
Switzerland
Tel (from UK) 00 + 41 22 734 13 30/
Fax 00 + 41 22 733 06 60

Green Flag National Breakdown
Cote Lane,
Pudsey,
Leeds LS28 5GF
Tel 0113 390 4000/Fax 0113 257 3111

Road Haulage and Distribution Training Council (RHDTC)
14 Warren Yard,
Warren Farm Office Village,
Stratford Road,
Milton Keynes MK12 5NM
Tel 01908 313360/Fax 1908 313806
e-mail: rhdtc@dial.pipex.com

Royal Automobile Club (RAC)
RAC House,
M1 Cross,
Brent Terrace,
London NW2 1LT
Tel 020 8452 8000/Fax 020 8208 0679

OCR
Westwood Way,
Westwood Business Park,
Coventry CV4 8JQ
Tel 02476 470033/Fax 02476 421944

Recruitment and Employment Federation
36–38 Mortimer Street
London W1W 7RG
Tel 020 7462 3260/Fax 020 7255 2878

Royal Society for the Prevention of Accidents (RoSPA)
Cannon House,
Priory Queensway,
Birmingham B4 6BS
Tel 0121 248 2000

Vehicle Inspectorate: Executive Agency
Goods Vehicle Centre,
Welcombe House,
91–92 The Strand,
Swansea,
Glamorgan SA1 2DH
Tel 01792 458888

Transport Journals

Commercial Motor
Quadrant House,
The Quadrant,
Sutton,
Surrey SM2 5AS
Tel 020 8652 3673/3302

Distribution Business
Cygnet House,
12–14 Sydenham Road,
Croydon CR9 2ET
Tel 020 8680 7474/Fax 020 8649 9747

Export and Freight
139 Thomas Street,
Portadown BT62 3BE,
Northern Ireland
Tel 01762 334272/Fax 01762 351046

Logistics and Transport Focus (Journal of Institute of Logistics and Transport)
Supply-Chain Centre,
PO Box 5787,
Corby,
Northants NN17 4XQ
Tel 01536 740100/Fax 01536 740103

Freight (Journal of FTA)
Hermes House,
St John's Road,
Tunbridge Wells TN4 9UZ
Tel 01892 26171

International Freighting Weekly (IFW)
Informa Freight Publishing
69-77 Paul Street,
London EC2A 4LQ
Tel 020 7505 3551

Logistics Europe
Cygnet House,
12–14 Sydenham Road,
Croydon CR9 2ET
Tel 020 8680 7474/Fax: 020 8649 9747

Materials Handling News
Quadrant House,
The Quadrant,
Sutton,
Surrey SM2 5AS
Tel 020 8661 3500

Motor Transport
Quadrant House,
The Quadrant,
Sutton,
Surrey SM2 5AS
Tel 020 8661 3500

Removals and Storage (Journal of British Association of Removers)
3 Churchill Court,
58 Station Road,
North Harrow HA2 7SA
Tel 020 8861 3331

Roadway (Journal of RHA)
Roadway House,
35 Monument Hill,
Weybridge,
Surrey KT13 8RN
Tel 01932 841515/Fax 01932 852516

Transport Engineer
22 Greencote Place,
London SW1P 1PR
Tel 020 7630 1111/Fax 020 7630 6677

Trucking
Messenger House,
33-35 St Michael's Square,
Gloucester GL1 1HX
Tel 01452 307181/Fax 01452 307170

Appendix IV

VEHICLE AND OPERATOR SERVICES AGENCY (VOSA) GOODS VEHICLE TESTING STATIONS

Highlands and Islands
Regional office:
Arran Road,
North Muirton Industrial Estate,
Perth PH1 3DZ
Tel 01738 632037/8

Inverness
Seafield Road,
Longman Industrial Estate,
Inverness IV1 1RG
Tel 01463 235505/Fax 01463 715229

Fort William
Highland Bus & Coach Depot,
Fort William,
Inverness-shire PH33 6PP
Tel 01463 235505

Lairg
County Council Roads Depot,
Laundry Road,
Lairg,
Sutherland IV27 4DE
Tel 01463 235505

Wick
Airport Industrial Estate,
Wick,
Caithness KW1 4QS
Tel 01463 235505

Kirkwall
BT Workshop,
Hatson Industrial Estate,
Kirkwall,
Orkney KW16 1RE
Tel 01463 235505

Portree
Industrial Estate, Dunvegan Road,
Portree,
Isle of Skye IV51 9HF
Tel 01463 235505

Stornaway
Building No. 29,
Airport,
Isle of Lewis PA86 0BN
Tel 01463 235505

Aberdeen
Cloverhill Road,
Bridge of Don Industrial Estate,
Aberdeen AB23 8EE
Tel 01224 702357/Fax 01224 703774
Opening times: 08:00–17:00 Mon–Thurs, 08:00–16:30 Fri, 08:30–12:30 Sat (Weds eves on demand).

Keith
Banff Road,
Keith,
Banffshire AB55 3ET
Tel 01542 882819
Opening times: 09:30–17:00 Mon, 08:00–16:30 Tues–Thurs, 08:00–12:30 Fri, no Sats.

Lerwick
Anderson Base,
Gremista,
Lerwick,
Shetland ZE1 0PX
Tel 01595 694900
Opening times: Mon–Fri every third week, 13:30–17:00 Mon, 08:00–16:30 Tue–Thurs, 08:00–11:00 Fri (additional testing on demand).

Perth
Arran Road,
North Muirton Industrial Estate,
Perth PH1 3DZ
Tel 01738 632037/8/Fax 01738 440519
Opening times: 08:00–17:00 Mon–Fri and 08:30–12:00 Sats (when open (no class 5 or 7)).

Kirkcaldy
Park Road,
Gallatown,
Kirkcaldy,
Fife KY1 3ER
Tel 01592 651233/Fax 01592 651238

Montrose
Broomfield Industrial Estate,
Montrose,
Angus DD10 9BB
Tel 01674 673760/Fax 01674 673760

Lowlands and Borders
Regional Office:
216 Western Road,
Kilmarnock KA3 1LP
Tel 01563 524312/Fax 01563 539858

Bishopbriggs
Crosshill Road,
Bishopbriggs,
Glasgow G64 2QA
Tel 0141 7726321/Fax 0141 7623290
Opening times: 07:00–17:00 Mon–Thurs, 08:00–16:30 Fri and 08:30–12:00 Sat (supplement charged). Out of hours testing on demand (supplement charged).

Kilmarnock
216 Western Road,
Kilmarnock KA3 1LP
Tel 01563 524312/Fax 01563 539858
Opening times: 08:30–16:30 Tues–Thurs.

Lochgilphead
Unit 12, Kilmory Industrial Estate,
Lochgilphead,
Argyll PA31 8PR
Tel 01546 603206

Berwick upon Tweed
Tweedside Trading Estate,
Berwick upon Tweed,
Northumberland TD15 2XF
Tel 01289 306004/Fax 01289 306004
Opening times: 07:30–17:00 Mon–Thurs, 07:30–16:30 Fri. Out of hours testing 17:00– 20:00 Tues & Thurs, 07:30–12:00 Sat (supplement charged).

Livingston
Grange Road,
Houston Industrial Estate,
West Lothian, EH54 5DD
Tel 01506 430053/Fax 01506 437425
Opening times: 07:45–17:00 Mon–Thurs, 07:45–16:30 Fri, 07:45–12:00 Sat (on demand, supplement charged).

Charlesfield
Charlesfield,
St Boswells,
Melrose,
Roxburghshire TD6 0HH
Tel 01835 823701/Fax 01835 823701
Opening times: 08:30–16:30 Mon–Fri.

East Fortune
East Fortune Airfield,
Drem,
North Berwick,
East Lothian EH39 5LF
Tel 01620 880350/Fax 01620 880350
Opening times: 08:30–16:30 Weds.

Carlisle
Brunthill Road,
Kingstown Industrial Estate,
Carlisle,
Cumbria CA3 0EH
Tel 01228 528106/Fax 01228 592720
Opening times: 08:00–17:00 Mon–Thurs, 08:00–16:30 Fri. Out of hours testing on demand (supplement charged).

Workington
Pittwood Road,
Lillyhall Industrial Estate,
Workington,
Cumbria CA14 4JP
Tel 01900 64456/Fax 01900 64456
Opening times: 08:00–16:30 Tues–Fri. Out of hours testing on demand (supplement charged).

Newton Stewart
Industrial Estate,
Wigtown Road,
Newton Stewart,
Wigtownshire DG8 6JZ
Tel/Fax 01671 402516

Dumfries
Heathall Industrial Estate,
Locharbriggs,
Dumfries, DG1 3PH
Tel 01387 261141/Fax 01387 261141

North East

Regional Office:
Patrick Green,
Woodlesford,
Leeds LS26 8HE
Tel 0113 282 5060/Fax 0113 282 7942

Scarborough
Cayton,
Scarborough,
North Yorkshire YO11 3BY
Tel 01723 582695/Fax 01723 581332

Darlington
Banks Road,
McMullen Road,
Darlington,
Co Durham DL1 1YE
Tel 01325 460547/Fax 01325 360354
Opening times: 08:00–17:00 Mon–Thurs, 08:00–16:30 Fri, 08:00–12:00 Sat (supplement charged).

Newcastle
Sandy Lane,
Gosforth,
Newcastle upon Tyne NE3 5HB
Tel 0191 2365011/Fax 0191 2170144

Walton
Wighill Lane,
Walton,
Wetherby,
West Yorkshire LS23 7DU
Tel 01937 844560/Fax 01937 842346
Opening times: 08:00–16:30 Mon–Thurs, 08:00–16:00 Fri, 08:00–12:00 Sat, Sunday and out of hours testing on demand (supplement charged).

Leeds
Patrick Green,
Woodlesford,
Leeds LS26 8HE
Tel 0113 2825060/Fax 0113 2827942

Beverley
Oldbeck Road,
Off Grovehill Road,
Beverley,
North Humberside HU17 0JG
Tel 01482 881522 /881629/Fax 01482 883365
Opening times: 08:00–16:30 Mon–Fri (lunch 12:00–13:00), 08:00–12:00 Sat (usually the last one of the month, others on demand).

Doncaster
Wellsdyke Road,
Adwick-le-Street,
Doncaster DN6 7DU
Tel 01302 724404/Fax 01302 337461
Opening times: 08:00–21:00 Mon–Fri (lunch 12:00–13:00), 08:00–12:00 Sat (usually the last one of the month, others on demand).

Grimsby
Estate no. 1,
South Humberside Industrial Estate,
Pyewipe,
Grimsby,
South Humberside DN31 2TB
Tel 01472 353703/Fax 01472 345310
Opening times: 08:00–17:00 Mon, 08:00–17:00 Tues–Fri, 08:00–12:00 Sat (supplement charged).

North West

Regional Office:
Stopgate Lane,
Simonswood,
Kirkby,
Liverpool L33 4YA
Tel 0151 5474445/Fax 0151 5473494

Steeton
Steeton Grove,
Steeton,
West Yorkshire BD20 6RW
Tel 01535 653433/Fax 01535 653073
Opening times: 08:00–17:00 Mon–Thurs (lunch 12:00–12:30), 08:00–16:30 Fri (lunch 12:00–12:30), 07:30–13:00 Sat. Out of hours testing on demand (supplement charged).

Chadderton
Broadway Business Park
Broadgate
Chadderton
Oldham OL9 9XA
Tel 0161 947 1000/Fax 0161 681 7818

Kirkham
Freckleton Road,
Kirkham,
Preston,
Lancs PR4 2RA
Tel 01772 684809/683785/Fax 01772 672123
Opening times: 08:00–17:00 Mon–Thurs (lunch 12:00–12:30), 08:00–16:30 Fri. Out of hours testing on demand (supplement charged).

Milnthorpe
Dock Road South,
Bromborough,
Wirral,
Merseyside L62 4SH
Tel 01539 563751/Fax 01539 564502

Bromborough
Dock Road South,
Bromborough,
Wirral,
Merseyside L62 4SH
Tel 0151 6431013/Fax 0151 6431025
Opening times: 08:00–17:00 Mon–Thurs, 08:00–16:30 Fri (lunch 12:00–13:00), 08:00–16:00 Sat (depending on demand) (supplement charged).

Bredbury
Lingard Lane,
Bredbury,
Stockport,
Cheshire SK6 2QX
Tel 0161 4305160/Fax 0161 4940600
Opening times: 08:00–17:00 Mon–Thurs, 08:00–16:00 Fri, 06:00–13:00 Sat. Out of hours testing on demand (supplement charged).

Liverpool
Stopgate Lane,
Simonswood,
Kirkby,
Liverpool L33 4YA
Tel 0151 5474445/Fax 0151 5473494
Opening times: 08:00–17:00 Mon–Fri, 17:00–19:30 Tues & Thurs, 08:00–12:00 Sat (supplement charged).

Wrexham
Llay Road,
Llay,
Wrexham,
Clwyd LL12 0TL
Tel 01978 852422/Fax 01978 856610

Caernarvon
Cibyn Industrial Estate,
Caernarfon,
Gwynedd LL55 2BD
Tel 01286 672567/Fax 01286 672567

Mid East
Regional Office:
Orgreave Way,
Handsworth,
Sheffield S13 9LT
Tel 01142 693691/693729

Kidderminster
Worcester Road,
Kidderminster,
Worcs DY11 7RD
Tel 01562 745857/Fax 01562 753049

Birmingham
Garretts Green Industrial Estate,
Birmingham B33 0SS
Tel 0121 783 6560/1/Fax 0121 628 9090

Nottingham
Main Road,
Watnall,
Nottingham NG16 1JF
Tel 01159 382591/2/Fax 01159 383469
Opening times: 08:00–17:00 Mon–Thurs, 08:00–16:30 Fri, 08:15–12:00 Sat (supplement charged).

Derby
Raynesway,
Alvaston,
Derby DE21 7AY
Tel 01332 571961/Fax 01332 756143
Opening times: 07:30–17:00 Mon–Fri, 08:00–12:00 Sat (supplement charged).

Leicester
40 Cannock Street,
Barkby,
Thorpe Road,
Leicester LE4 9HT
Tel 0116 276 0144/276 7405/Fax 0116 246 0674
Opening times: 08:00–17:00 Mon–Thurs (lunch 12:30–13:30), 08:00–16:00 Fri, 08:00–12:00 Sat (first or last one of the month, supplement charged).

Sheffield
Orgreave Way,
Handsworth,
Sheffield S13 9LT
Tel 01142 693691/Fax 01142 548242
Opening times: 08:00–16:30 Mon–Thurs (lunch 12:00–13:00), 08:00–16:00 Fri (lunch 12:00–13:00), 08:00–12:00 Sat.

Grantham
Spitalgate Airfield,
Blue Harbour,
Grantham,
Lincs NG31 7TX
Tel 01476 562012/Fax 01476 751234
Opening times: 08:00–17:00 Mon–Thurs, 08:00 –16:30 Fri, 08:00–12:00 Sat (last one of the month, early morning tests available).

Crimplesham
Bexwell Airfield,
Crimplesham,
Kings Lynn,
Norfolk PE33
Tel 01366 382481/382866/386891/
Fax 01366 383821
Opening times: 08:00–17:00 Mon–Thurs, 08:00–16:30 Fri, 08:00–12:00 Sat.

Norwich
Jupiter Road,
Hellesdon,
Norwich NR6 6SS
Tel 01603 408128/Fax 01603 482573
Opening times: 08:00–16:30 Mon–Thurs (lunch 12:00–13:00) 08:00–16:00 Fri (lunch 12:00–13:00) 08:00–12:00 Sat.

Peterborough
Saville Road,
Westwood,
Peterborough PE3 6TL
Tel: 01733 263399/263423/
Fax: 01733 262273
Opening times: 08:00–17:00 Mon–Thurs, 08:00–16:30 Fri, 08:00–12:00 Sat (normally last Saturday of the month).

Mid West
Regional Office:
Ennerdale Road,
Harlescott,
Shrewsbury SY1 3LF
Tel 01743 462530/Fax 01743 450107

Shrewsbury
Ennerdale Road,
Harlescott,
Shrewsbury SY1 3LF
Tel 01743 462530/Fax 01743 450107
Opening times: 08:00–17:00 Mon–Thurs, 08:00–16:30 Fri and Sat (last one of each month, additional on demand).

Wolverhampton
Cat & Kittens Lane,
Featherstone,
Wolverhampton WV10 7JD
Tel 01902 397722

Stoke
Station Road,
Cold Meece,
Stone,
Staffs ST15 0QP
Tel 01785 760213/Fax 01785 760912
Opening times: 08:00–17:00 Mon–Thurs, 08:00–16:30 Fri (lunch 12:30–13:30) and Sat (last one of each month, additional on demand, supplement charged).

Swansea
Tirydail Lane,
Ammanford,
Dyfed SA18 3AR
Tel 01269 592875/Fax 01269 591627

Aberystwyth
Llanrystyd,
Dyfed SA18 3AR
Tel 01974 202447/Fax 01974 202447

Haverfordwest
Withybush,
Haverfordwest,
Dyfed SA62 4BN
Tel 01437 764402/Fax 01437 764402

Hereford
Faraday Road,
Westfield Trading Estate,
Hereford HR4 9NS
Tel 01432 267956/Fax 01432 354013

Cardiff
School Road,
Miskin,
Pontyclun,
Mid Glamorgan CF72 8YR
Tel 01443 224701/Fax 01443 237873
Opening times: 07:30–17:00 Mon–Thurs, 07:30 –16:30 Fri and 07:00–12:30 Sat (lunch 12:30–13:30).

Pontypool
Polo Ground Industrial Estate,
New Inn,
Pontypool,
Gwent NP4 0YN
Tel 01495 756001/3/Fax 01495 759865
Opening times: 08:00–17:00 Mon–Thurs, 08:00–16:30 Fri, 07:00–12:00 Sat (normally 2 per month or on demand (HGVs only)).

Llandrindod Wells
Gun Park,
Waterloo Road,
Llandrindod Wells,
Powys LD1 6DH
Tel 01597 822788/Fax 01597 822788

South Central
Regional Office:
Hambridge Lane,
Newbury,
Berks RG14 5TZ
Tel 01635 47649/Fax 01635 552770

Bicester
Launton Road,
Bicester,
Oxon OX6 0JG
Tel 01869 243416/242562/Fax 01869 324051

Northampton
Cavalry Hill Industrial Park,
Weedon,
Northampton NN7 4PP
Tel 01327 340697/Fax 01327 349049
Opening times: 08:00–17:00 Mon–Fri (lunch 12:30–13:30), 08:00–12:00 Sat (third one of the month, supplement charged).

Royston
Orchard Road,
Royston,
Herts SG8 5HA
Tel 01763 242697/244822/249639/
Fax 01763 249864
Opening times: 06:30–16:30 Mon–Thurs, 06:30–16:00 Fri, 06:30–12:00 Sat (usually 3 per month).

Leighton Buzzard
Stanbridge Road,
Leighton Buzzard,
Beds LU7 8QG
Tel 01525 373074/Fax 01525 378383
Opening times: 06:30–17:00 Mon–Fri (lunch 12:45–13:45), 08:00–12:00 Sat (usually 3rd in the month, supplement charged).

Southampton
Hillson Road,
Bottings Trading Estate,
Botley,
Southampton SO3 2DY
Tel 01489 785522/Fax 01489 798384
Opening times: 08:00–17:00 Mon–Thurs, 08:00–16:30 Fri, 08:00–12:00 Sat (supplement charged for out of hours testing).

Newbury
Hambridge Lane,
Newbury,
Berks RG14 5TZ
Tel 01635 47649/Fax 01635 552770

Cowes
Prospect Road,
Cowes,
Isle of Wight PO31 7AD
Tel 01983 293171/Fax 01983 281766

Salisbury
Brunel Road,
Churchfields Industrial Estate,
Salisbury,
Wilts SP2 7PU
Tel 01722 326733/322898/
Fax 01722 410605

Poole
Hanwell Road,
Nuffield Industrial Estate,
Poole, Dorset BH17 0GE
Tel 01202 672844/Fax 01202 673731
Opening times: 08:00–17:00 Mon–Thurs, 08:00–16:30 Fri, out of hours testing available on demand (supplement charged).

Mitcham
Redhouse Road,
Croydon,
Surrey CR0 3AQ
Tel 0208 684 1499/Fax 0208 665 1108
Opening times: 08:00–12:15/13:30–16:00 Mon–Thurs, 08:00–12:15/13:30–15:30 Fri, 07:00–12:30 Sat.

Lancing
Chartwell Road,
Churchill Industrial Estate,
Lancing,
West Sussex BN15 8TU
Tel 01903 753305/754276/
Fax 01903 851434

Guildford
Moorfield Road,
Slyfield Industrial Estate,
Guildford,
Surrey GU1 1SA
Tel 01483 565151/Fax 01483 451489

South West
Regional Office:
Ashton Vale Road,
Ashton Gate,
Bristol BS3 2JE
Tel 0117 963 2673/Fax 0117 963 7600

Calne
Port Marsh Road,
Calne,
Wiltshire SN11 9BW
Tel 01249 812351/Fax 01249 812132
Opening times: 08:00–17:00 Mon–Thurs, 08:00–16:30 Fri (early mornings 7:15 supplement). Last Sat of every month if required.

Bristol
Ashton Vale Road,
Ashton Gate,
Bristol BS3 2JE
Tel 0117 966 1472/Fax 0117 963 7600

Gloucester
Ashville Road,
Gloucester GL2 6ET
Tel 01452 529749/520401/Fax 01452 529380
Opening times: 08:00–12:30/13:30–16:30 Mon–Thurs, 08:00–12:30 Fri, no Sats (out of hours testing on demand).

Taunton
Taunton Trading Estate,
Norton Fitzwarren,
Taunton,
Somerset TA2 6RX
Tel 01823 282525/Fax 01823 325329
Opening times: 08:00–17:00 Mon–Thurs, 08:00–1630 Fri, 08:30–12:00 Sat (once a month, supplement charged). Early morning tests available (supplement charged.)

Exeter
Grace Road,
Marsh Barton Trading Estate,
Exeter,
Devon EX2 8PH
Tel 01392 278267/Fax 01392 493628
Opening times: 08:00–17:00 Mon–Thurs, 08:00–16:30 Fri (early morning tests available).

South Molton
Station Road,
Pathfield Industrial Estate,
South Molton,
Devon EX36 3LL
Tel 01769 572248/Fax 01769 572248
Opening times: 08:00–17:00 Mon–Thurs, 08:00–16:30 Fri and Sunday opening (once a month, supplement charged). Early morning tests available (supplement charged).

Plymouth
Agaton Fort,
Budshead Road,
Ernesettle,
Plymouth PLS 2QY
Tel 01752 362294/Fax 01752 351946
Opening times: 08:00–17:00 Mon–Thurs, 08:00–16:30 Fri (lunch 12:30–13:30).

Redruth/Camborne
Wilson Way,
Redruth,
Cornwall TR15 3RP
Tel 01209 216251
Opening times: 08:00–12:30/13:30–16:30 Mon–Thurs, 08:00–12:30/13:30–16:00 Fri, 08:00–13:00 Sat (when required).

St Austell
Par Moor Road,
Par,
Cornwall PL24 2SQ
Tel 01726 812218/Fax 01726 812218
Opening times: 08:00–12:30/13:30–16:30 Mon–Thurs, 08:00–12:30/13:30–16:00 Fri.

South East

Regional Office:
Tank Hill Road,
Purfleet,
Essex RM16 1SX
Tel 01708 866651/Fax 01708 863378

Chelmsford
Widford Industrial Estate,
Chelmsford,
Essex CM1 3AE
Tel 01245 259341/Fax 01245 346585
Opening times: 07:30–17:00 Mon–Thurs, 07:30–16:30 Fri, 08:00–12:00 Sat. Out of hours testing from 17:00 onwards (supplement charged).

Ipswich
Holbrook Road/Landseer Road,
Ipswich,
Suffolk IP3 0DF
Tel 01473 259061/Fax 01473 221166
Opening times: 08:00–17:00 Mon–Thurs, 08:00–16:30 Fri, 08:00–12 noon Sat (excluding Christmas, New Year, Easter).

Yeading
Willow Tree Lane,
Yeading,
Hayes,
Middlesex UB4 9BS
Tel 0208 841 9205/Fax 0208 845 8133
Opening times: 08:00–17:00 Mon–Thurs and 08:00–16:30 Fri. Supplement charged for all out of hours testing.

Purfleet
Tank Hill Road,
Purfleet,
Essex RM16 1SX
Tel 01708 866651/Fax 01708 863378

Edmonton
Anthony Wharf,
Lea Valley Trading Estate,
Edmonton,
London, N18 3JR
Tel 0208 803 7733/Fax 0208 887 0392

Canterbury
Hersden,
Canterbury,
Kent CT3 4HB
Tel 01227 710010/710852/
Fax 01227 713259

Gillingham
Ambley Road,
Gillingham,
Kent ME8 0SJ
Tel 01634 232541/232754/
Fax 01634 261623

Hastings
Ivy House Lane,
Hastings,
East Sussex TN35 4NN
Tel 01424 430248/Fax 01424 713151

Appendix V

DRIVING STANDARDS AGENCY* (DSA) AREA OFFICES AND LGV DRIVING TEST CENTRES

Head Office
Stanley House,
56 Talbot Street,
Nottingham NG1 5GU
Tel 0115 901 2500
Fax 0115 901 2510

Area Offices	**LGV Driving Test Centres**
Northern DSA PO Box 280, Newcastle upon Tyne NE99 1FP Tel 0191 201 4000 Fax 0191 201 4010 Recorded message 0191 201 4100	Berwick, Bredbury (Manchester), Carlisle, Darlington, Grimsby, Heywood (Manchester), Kirkham, Newcastle, Patrick Green, Sheffield, Simonswood, Steeton, Upton, Walton (York).
Midlands and Eastern DSA PO Box 287 Newcastle upon Tyne Tel 0121 697 6700 Fax 0121 697 6750 Recorded message 0121 697 6730	Birmingham, Cambridge, Chelmsford, Derby, Featherstone, Harlescott, Ipswich, Leicester, Leighton Buzzard, Norwich, Nottingham, Peterborough, Swynnerton, Weedon.

Wales and Western DSA PO Box 286 Newcastle upon Tyne NE99 1WA Tel 0122 258 1000 Fax 0122 258 1050 Recorded message 0122 258 1030	Bristol, Caernarfon, Camborne, Exeter, Gloucester, Llantrisant, Neath, Plymouth, Pontypool, Poole, Taunton, Swindon, Withybush, Wrexham.
London and the South-East DSA PO Box 289 Newcastle upon Tyne NE99 1WE Tel 020 7957 0957 Fax 020 7468 4550 Recorded message 020 7468 4530	Croydon, Enfield, Guildford, Purfleet, Yeading.
Scotland DSA PO Box 288 Newcastle upon Tyne NE99 1WD Tel 0131 529 8580 Fax 0131 529 8589 Recorded message 0131 529 8592	Aberdeen, Bishopbriggs (Glasgow), Connel, Galashiels, Inverness, Kirkwall, Lerwick, Livingston (Edinburgh), Locharbriggs (Dumfrieshire), Machrihanish (Kintyre), Oban, Perth, Port Ellen, Stornoway, Wick.

NB: Tests are conducted only occasionally at some centres – test applicants and candidates are advised to check with their local DSA Area Office for address details, telephone numbers and opening times.

*The Driving Standards Agency is an Executive Agency of the Department for Transport. Contact the DSA on the internet: www.dsa.gov.uk

Appendix VI

TACHOGRAPH MANUFACTURERS AND APPROVED CENTRES

This appendix lists tachograph manufacturers, independent tachograph repairers and the tachograph centres approved by the Secretary of State for the Department for Transport (D*f*T) in the UK (ie 'approved workshops' as referred to in EU legislation), as listed in August 2002 on the Department's website as follows: www.roads.dft.gov.uk/roadsafety/tachograph/centres02.htm

NB: This Crown copyright material is published under the terms set out on the Department's website and acknowledgement for its use is hereby recorded.

Tachograph Manufacturers/Suppliers

VDO Kienzle UK Limited (now Siemens VDO)
36 Gravelly Industrial Park,
Birmingham B24 8TA
Tel 0121 326 1200/Fax 0121 326 1299

TVI Europe Ltd (now owned by Stoneridge Electronics)
Kilspindie Road,
Dundee DD2 3QJ
Tel 01382 83303/Fax 01382 832382

Time Instrument Manufacturers Ltd (a division of ACCTIM)
5 Alston Drive,
Bradwell Abbey,
Milton Keynes MK13 9HA
Tel 01908 220020/Fax 01908 220145

Volvo Truck (Great Britain) Ltd
Wedgnock Lane,
Warwick,
CV34 5YA
Tel 01926 401777/Fax 01926 490991

Approved Tachograph Repairers

VDO Kienzle UK Limited GBH01R
36 Gravelly Industrial Park,
Birmingham B24 8TA
Tel 0121 328 5533 /Fax 0121 327 4864
Can repair tachographs of Family One (mechanical), Family Two (electro-mechanical) and Family Three (microprocessor based).

TVI Europe Limited GBM01R
Kilspindie Road,
Dundee DD2 3QJ
Tel 01382 83303 /Fax 01382 832382
Can repair tachographs of Family One (mechanical), Family Two (electro-mechanical) and Family Three (microprocessor based).

Instrument Repair Services GBE01R
35 Redcliffe Road,
West Bridgford,
Nottingham NG2 SFF
Tel 01602 819988/818109/
Fax 01602 455358
Can repair tachographs of Family One (mechanical), Family Two (electro-mechanical) and Family Three (microprocessor based).

Tachodisc Limited GBD01R
19 Kingsland Grange,
Woodstan,
Cheshire WA1 4RW
Tel 01925 283341 /Fax 01925 852203
Can repair tachographs of Family Two (electro-mechanical) and Family Three (microprocessor based).

Gifford Tachograph Services Limited GBN01R
Assured House,
Perry Road, Chequers Lane,
Dagenham Dock,
Dagenham RM9 6QD
Tel 0181 593 1550
Can repair tachographs of Family Three (microprocessor based).

Tacho-Serve GBA01R
Unit 3,
Aston Cross Industrial Estate,
51 Lichfield Road,
Aston,
Birmingham B6 5RW
Tel 0121 328 5235/Fax 0121 328 8091
Can repair tachographs of Family One (mechanical), Family Two (electro-mechanical) and Family Three (microprocessor based).

Approved Tachograph Calibration Centres

Area 1

AEW Ltd GBL218
t/a Lucas Service UK,
Units 3&4,
Girdleness Road,
Torry, Aberdeen,
Grampian AB1 4DQ
Tel 01224 895050

Grampian Transport Ltd GBL107
395 King Street,
Aberdeen AB24 5RP
Tel 01224 650000

Harper Motor Company Ltd GBL204
218 AuchMill Road,
Bucksburn,
Aberdeen AB21 9NB
Tel 01224 714741

Norscot Truck Centre GBL314
Services (Aberdeen) Ltd,
The Parkway,
Bridge Of Don,
Aberdeen AB23 8JZ
Tel 01224 824444

Reliable Vehicles GBL315
Altens Industrial Estate,
Aberdeen AB12 3LH
Tel 01224 896312

Volvo Truck & Bus Scotland Ltd GBL313
Barclayhill Place,
Portlethen,
Aberdeen AB12 4PF
Tel 01224 781782

AEW GBL219
t/a Lucas Service UK,
165 Brook Street,
Dundee,
Tayside DD15DE
Tel 01382 227122

G Mutch Mechanical GBL110
Riverside Distribution Centre,
Riverside Avenue,
Dundee DD2 1UD
Tel 01382 640776

Wincanton Ltd GBL316
t/a PullMan Fleet Services,
Smeaton Road,
Western Gourdie Estate,
Dundee DD2 4UT
Tel 01382 623623

Baillie Bros (Truck Services) GBL217
Linkwood Industrial Estate,
Elgin IV30 1XB
Tel 01343 555312

Elgin Truck & Van Centre Ltd GBM125
Grampian Road,
Elgin IV30 1XN
Tel 01343 542171

A M Phillips Ltd GBL214
Muiryfaulds,
Forfar DD8 1XP
Tel 01307 474000

J & G Riddell GBL304
Woodside Garage,
Knock,
Huntly AB5 5LJ
Tel 01466 771245

Norscot Truck & Van Ltd GBL216
52 Seafield Road,
LongMan Industrial Estate,
Inverness IV1 1SG
Tel 01463 712000

W M Dunnet & Co Ltd GBL312
Masons Lane,
Thurso,
Caithness KW14 8EW
Tel 01847 893101

Frews Cars Ltd GBL500
Riggs Road,
Perth PH2 0NT
Tel 01738 625121

G Mutch Mechanical Services GBL307
Shore Road,
Perth PH2 8BH
Tel 01738 626688

G BannerMan (Tain) Ltd GBL103
Shore Road,
Tain IV19 1EH
Tel 01862 892480

Area 2

Central Motors (Calderbank) Ltd GBM213
105 Carlisle Road,
Airdrie ML6 8RD
Tel 01236 762881

B Mundell Ltd GBM229
Bardaravine,
Tarbert PA29 6YF
Tel 01880 820223

Caledonian Trucks Ltd GBM228
Whitfield Drive,
Heathfield Industrial Estate,
Ayr KA8 9RX
Tel 01292 610888

Man Central Scotland GBM216
Clark Way,
Motherwell Food Park,
Bellshill ML4 3NX
Tel 01698 327328

Reliable Vehicles Ltd GBM227
Melford Road,
Righead Industrial Estate,
Bellshill ML4 3LF
Tel 01698 841994

Lex Commercials Ltd GBM301
8 WardPark Court,
WardPark South,
Cumbernauld G67 3HE
Tel 01236 727771

Kerr & Smith (Cumnock) Ltd GBM103
Riverside Garage,
Ayr Road,
Cumnock KA18 1BJ
Tel 01290 422440

Carlisle Commercials Ltd GBM325
New BWest,
Lockerbie Road,
Dumfries DG1 3PG
Tel 01387 267646

Reliable Vehicles Ltd GBM224
Heathhall Industrial Estate,
Dumfries DG1 3PH
Tel 01387 250502

Barr Ltd GBM316
Braehead,
Barrhill,
Giran KA26 0QR
Tel 01465 821300

ERF Strathclyde Ltd GBM226
30 Clydesmill Drive,
Glasgow G32 8RJ
Tel 0141 641 6172

Kerr & Smith GBM231
3000 Springhill Parkway,
Glasgow Business Park,
Glasgow G69 6GA
Tel 0141 4296363

Leaseway Services Ltd GBM324
Blochairn Distribution Centre,
Blochairn Road,
Glasgow G21 2XQ
Tel 0141 552 8989

Lex Commercials GBM303
79 Hardgate Road,
Goan,
Glasgow G51 4SX
Tel 0141 445 3917

Lex Commercials GBM115
131 Bogmoor Road,
Shieldhall,
Goan,
Glasgow G51 4TH
Tel 0141 425 1530

Lucas Services UK Ltd GBM204
200–210 Garscube Road,
Glasgow G4 9RR
Tel 0141 332 6591

Reliable Vehicles Ltd GBM218
Clyde Street,
Renfrew,
Glasgow G76 7TZ
Tel 0141 886 5633

Renault Trucks Commercials Ltd GBM122
Penilee Road,
Hillington Industrial Estate,
Glasgow G52 4UW
Tel 0141 882 3304

Vardy Continental Ltd GBM502
North Canalbank Street,
Port Dundas,
Glasgow G4 9XT
Tel 0141 334 8155

Volvo Truck & Bus Ltd GBM230
101 Kelburn Street,
Barrhead,
Glasgow G78 1LB
Tel 0141 881 5851

Volvo Truck & Bus (Scotland) Ltd GBM401
2 Whistleberry Industrial Park,
Blantyre,
Hamilton ML3 0ED
Tel 01698 823300

James Mckinnon Jnr (Truck Services) Ltd GBM221
West Hillhead,
Western Road,
Kilmarnock KA3 1PH
Tel 01563 544888

Johnstone & Dryan GBM313
1 Fullarton Street,
Kilmarnock KA1 2RB
Tel 01563 523008

W H Malcolm Ltd GBM123
Burnbrae Drive,
Linwood,
Paisley PA3 3BU
Tel 01505 327123

E Reid & Son GBM505
8A Underwood Road,
Paisley PA3 1TD
Tel 0141 889 1263

Area 3

Central Scotland Maintenance GBM129
Bathville Business Park,
Lower Bathville,
Armadale EH48 2JS
Tel 01501 733667

Volvo Truck & Bus Scotland Ltd GBM222
Drovers Road,
East Mains Industrial Estate,
Broxburn EH52 5ND
Tel 01506 856892

J E Douglas & Sons GBM320
Station Road,
Duns TD11 3HS
Tel 01361 883411

Fulton Auto Electric Ltd GBM318
14 West Bowling Green Street,
Leith,
Edinburgh EH6 5PQ
Tel 0131 5550396

A M Phillips Ltd GBM326
Simpson Road,
East Mains Industrial Estate,
Broxburn,
West Lothian
Tel 01506 865 000/Fax 01506 85 005

I S Fairbairn GBM207
Netherdale Garage,
Netherdale Industrial Estate,
Galashiels TD1 3EY
Tel 01896 756598

Leaseway Services Ltd GBL111
1 Abbotsinch Road,
Abbotsinch Industrial Estate,
Grangemouth FK3 9YE
Tel 01324 666974

Drummond Motor Co Ltd GBL109
Ferrard Road,
Kirkcaldy KY2 5RY
Tel 01592 201555

Caledonian Commercials Ltd GBM232
Naysmith Square,
Houston Industrial Estate,
Livingston EH54 5EG
Tel 01506 430000

Lothian Leyland Daf Ltd GBM220
Pentland Industrial Estate,
Loanhead,
Edinburgh EH20 9QH
Tel 0131 4404100

Reliable Vehicles Ltd GBM317
Newbridge Industrial Estate,
Newbridge EH28 8PJ
Tel 0131 333 2362

Area 4

Gilbraith Commercials Ltd GBC220
Market Street,
Church,
Accrington BB5 0DN
Tel 01254 31431

Lynch Truck Services Ltd GBC124
Barnfield Way,
Altham Business Park,
Altham BB5 5YT
Tel 01282 773377

JDS Trucks GBC305
Navigation Garage,
Forrest Street,
Blackburn BB1 3BB
Tel 01254 675111

Carlisle Commercials GBA303
Kingstown Broad Way,
Kingstown Industrial Estate,
Carlisle CA3 0HA
Tel 01228 529262

C G Trucks GBA304
Wakefield Road,
Kingstown Industrial Estate,
Carlisle CA3 0HE
Tel 01228 517290

Duncans Tachograph Centre GBA102
Unit 17–19, Kingstown Industrial Estate,
Gearshill Road,
Carlisle CA3 0EP
Tel 01228 515284

Solway Leyland Daf Ltd GBC235
Kingstown Broad Way,
Kingstown Industrial Estate,
Carlisle CA3 0HD
Tel 01228 39394

Thomas Hardie Chorley GBC403
Ackhurst Road,
Chorley PR7 3EH
Tel 01257 269651

Steadplan Ltd GBC229
Salthill Industrial Estate,
Clitheroe B7 1QL
Tel 01200 427415

Thomas Armstrong GBA206
(Transport Services) Ltd,
Workington Road,
Flimby,
Maryport,
Cumbria CA15 8RY
Tel 01900 68114

Lakeland Commercials Ltd GBC504
Mintsfleet Industrial Estate,
Kendal LA9 6LZ
Tel 01539 723956

PYE Motors Ltd GBC317
Oangle Road,
Lancaster LA3 3PF
Tel 01524 63553

Carlisle Commercials Ltd GBC147
White Lund Industrial Estate,
Lancaster LA3 3PT
Tel 01524 62866

Cabus Garage GBC128
Lancaster New Road,
Cabus,
Preston PR3 1AD
Tel 01524 791417

Lancashire Leyland Daf GBC328
223–224 Walton Summit Centre,
Bamber Bridge,
Preston PR5 8AL
Tel 01772 338111

Leyland Auto Electrical & Diesel Ltd GBC242
Unit 232,
Walton Summit Industrial Estate,
Bamber Bridge,
Preston PR5 8AL
Tel 01772 695000

Ribblesdale Auto Electrics Ltd GBC303
Marsh Lane,
Preston PR1 8YN
Tel 01772 555011

Area 5

Hartshorne (E Midlands) Ltd GBC405
Unit C,
Cotes Park West Industrial Estate,
Alfreton,
Derby DE55 4QR
Tel 01773 529400

Sherwood Commercial Vehicles Ltd GBE226
Berristow Lane,
Blackwell, Alfreton,
Derby DE55 5HP
Tel 01773 863311

Ciceley Commercials Ltd GBC251
Weston Street,
Bolton BL3 2BZ
Tel 01204 370377

H & J Quick Ltd GBC339
t/a Quicks Trucks,
Gladstone Road,
Farnworth,
Bolton BL4 7EA
Tel 01204 700111

Jeffries Vehicle Services Ltd GBC239
Swadlincote Road,
Woodville,
Swadlincote DE11 8DD
Tel 01283 214326

Hazlecroft Garage Ltd GBC343
Fairfield,
Buxton SK17 7HP
Tel 01298 79671

F B Atkins & Son Ltd GBC243
Burton Road,
Findern,
Derby DE6 6BG
Tel 01332 516151

Lex Commercials Ltd GBC337
Ashbourne Road,
Mackworth,
Derby DE22 4NB
Tel 01332 824371

Bridge Mills Service Centre GBC134
Unit 5, Bridge Mills,
Rochdale Road,
Edenfield BL0 0RE
Tel 01706 826344

D Hulme Ltd GBC240
Broad Way Industrial Estate,
Dunkinfield Road,
Hyde SK14 4QY
Tel 0161 366 9400

Chatfields of Manchester GBC247
40–46 Ashton Old Road,
Ardwick,
Manchester M12 6NA
Tel 0161 273 7351

ERF Manchester Ltd GBC138
Trafford Park Road,
Trafford Park,
Manchester M17 1NJ
Tel 0161 848 8331

Man Truck & Bus UK Ltd GBC149
Redclyffe Road,
Trafford Park,
Urmston M41 7BY
Tel 0161 7470530

Thomas Hardie Commercials GBC342
Bredbury Parkway,
Bredbury,
Stockport SK6 2SN
Tel 0161 9356121

Thomas Hardie Commercials GBC404
5th Avenue Village,
Trafford Park M17 1TR
Tel 0161 9354121

Sanderson Ford GBC309
Ashton Old Road,
Ardwick,
Manchester M12 6JD
Tel 0161 272 7000

West Pennine Trucks Ltd GBC324
Stakehill Industrial Estate,
Middleton,
Manchester M24 2RW
Tel 0161 653 9700

J A Leach Transport Ltd GBC246
t/a Leach Commercials,
Chichester Street,
Rochdale OL16 2AU
Tel 01706 868668

Tameside Tachograph Centre GBC506
Bayfreight Ltd,
Premier Mill,
Tame Street,
Stalybridge SK15 1ST
Tel 0161 338 8794

Les Botham Commercials GBC145
12 Stephenson Road,
Staveley S43 3JN

Roy Braidwood & Sons GBC119
Worsley Trading Estate,
Lester Road,
Little Hulton,
Worsley M28 6PT
Tel 0161 799 3801

Area 6

Transis GBC224
Dinas Depot,
Llanwnda,
Caernarvon LL54 5UD
Tel 01248 374133

Quick Trucks Ltd GBC108
The Truck Centre,
Bretton,
Nr Chester CH4 0DS
Tel 01244 660681

K J Ford GBC143
Mochdre Business Park,
Conwy Road,
Colwyn Bay LL13 8BJ
Tel 01492 546756

Chamberlains Transport Ltd GBC116
t/a Crewe Tachograph Centre,
Western Gate North,
Crewe CW1 6NB
Tel 01270 502800

Thomas Hardie Commercial NRT/W
GBC326
23 Fourth Ave,
Deeside Industrial Park,
Deeside CH5 2NR
Tel 01244 281004

Tachograph Chester Ltd GBC225
Rossfield Road,
Rossmore Trading Estate,
Ellesmere Port L65 3AW
Tel 0151 3552101

Haydock Commercial Vehicles Ltd GBC325
Yew Tree Trading Estate,
Kilbuck Lane,
Haydock WA11 9XW
Tel 01942 714103

Holyhead Truck Services GBC510
The Garage,
Llanfaethlu,
Holyhead LL65 4NW
Tel 01407 730759

Lucas Services UK Ltd GBC205
Andries Street,
Liverpool L3 7BJ
Tel 0151 2367063

North West Trucks GBC331
Unit 22 Huyton Industrial Estate,
Wilson Road,
Huyton,
Liverpool L36 6AJ
Tel 0151 4800098

Peoples (Liverpool) Ltd GBC502
Hawthorne Road,
Bootle,
Liverpool L20 6AF
Tel 0151 9228481

Perris & Kearon Ltd GBC215
173–175 Crown Street,
Liverpool L7 3LZ
Tel 0151 7094262

Steve Gray Motor England Ltd GBC338
Brookfield Drive,
Aintree,
Liverpool L9 7AJ
Tel 0151 5233393

Thomas Hardie Commercial GBC314
Newstet Road,
Knowsley Industrial Park North,
Liverpool L33 7TJ
Tel 0151 549 3003/Fax 0151 549 3030

A M Bell (Garage) Ltd GBC141
Hawkshead Quarry,
Leek Old Road,
Sutton,
Macclesfield SK11 0JB
Tel 01260 253232

Beechs Garage (1983) Ltd GBC136
Brooks Lane,
Middlewich CW10 0JH
Tel 01606 832930

Deeside Truck Services GBC509
Pinfold Lane,
Alltami,
Mold CH7 6NY
Tel 01244 547202

North West Trucks GBC302
Griffiths Road,
Lostock Gralam,
Northwich CW9 7NU
Tel 01606 48611

Needhams of Oldham GBC245
Fields New Road,
Chadderton,
Oldham OL9 8NH
Tel 0161 6287788

Roberts Motors GBC113
t/a St Helens Ford,
City Road,
St Helens WA10 6NZ
Tel 01744 26381

Woodwards Services Ltd GBC105
Merton Street,
Merton Bank Road,
St Helens WA9 1HU
Tel 01744 20266

Manchester Truck Centre GBC336
(Sandbach Branch),
Moss Lane,
Elworth,
Sandbach CW11 3WZ
Tel 01270 759974

Ryland North West Ltd GBC330
John Street,
Warrington WA2 7UD
Tel 01925 633271

Polar Trucks GBC503
Winwick Road,
Dallas Lane,
Warrington WA2 7NY
Tel 01925 651111

P & O Ferrymasters Ltd GBC507
Leacroft Road,
Risley,
Warrington WA3 6NN
Tel 01925 810000

Man Truck & Bus UK Ltd GBC150
t/a Man NorthWest,
Gorsey Lane,
Widnes,
Cheshire WA8 0SH
Tel 0151 4205111

Thomas Hardie Commercials GBC253
Lockett Lane,
Bryn,
Wigan WN4 8DE
Tel 01942 505100

Lex Commercials GBC231
Wrexham Road,
Rhostyllen,
Wrexham LL14 4DP
Tel 01978 291915

Border Tachograph Services Ltd GBC252
Penybont Works,
Pentre,
Chirk,
Wrexham LL14 5AW
Tel 01978 823434

Area 7

Gerard Mann GBD506
2 Lichfield Road,
Aston,
Birmingham B6 5SV
Tel 0121 3274411

Guest Motors Ltd GBD133
31 Shefford Street,
Aston,
Birmingham B6 4PQ
Tel 0121 3595888

AEW Ltd t/a Lucas Service UK GBD339
171 Lichfield Road,
Aston,
Birmingham B6 5SN
Tel 0121 327 1525

Excel Logistics GBD331
Derby Road,
Stretton Business Park,
Burton-on-Trent DE13 0BB
Tel 01283 516570

Hartshorne (E Midlands) Ltd GBD317
Derby Street,
Burton-on-Trent DE14 2LN
Tel 01283 515777

Marley Building Materials GBD126
Lichfield Road,
Branston,
Burton-on-Trent DE14 3HD
Tel 01283 722510

Cannock Tachograph Centre GBD124
Unit 13, Cannock Industrial Centre,
WalkMill Lane Bridgetown,
Cannock WS11 3LN
Tel 01543 574489

Hartshorne Motor Company GBD336
Hammond Road,
Parkhouse Industrial Estate,
Newcastle under Lyme ST5 7EF
Tel 01782 568600

Greenman (Motor Services) Ltd GBD132
Unit 14, Springfield Industrial Estate,
Oldbury,
Warley B69 4HH
Tel 0121 5440080

AEW Ltd t/a Lucas Service UK GBD229
Lancaster Road,
Harlescott,
Shrewsbury SY1 3JN
Tel 01743 465061

Furrows Commercial Vehicles Ltd GBD116
Ennerdale Road,
Harlescott Industrial Estate
Shrewsbury SY1 3NP
Tel 01743 447971

Hartshorne (Shrewsbury) Ltd GBD326
Ainsdale Drive,
Harlescroft,
Shrewsbury SY1 3TL
Tel 01743 444555

Stan Robinson (Stafford) Ltd GBD224
Ladford Fields,
Seighford,
Nr Stafford ST18 9QE
Tel 01785 282501

Beechs Garage (1983) Ltd GBD101
Shelton New Road,
Cliffe Vale,
Stoke-on-Trent ST4 7DL
Tel 01782 848485

Excel Logistics GBD219
Vernon Road,
Stoke-on-Trent ST4 2QF
Tel 01782 848281

Lex Commercials GBD315
Leek New Road,
Cobridge,
Stoke-on-Trent ST6 2DE
Tel 01782 264121

West Pennine Trucks Ltd GBD227
Cross Street,
Off Chemical Lane,
Longport,
Stoke-on-Trent ST6 4PU
Tel 01782 577955

Furrows Commercial Vehicles GBD325
Kemberton Road,
Halesfield,
Telford TF7 4QS
Tel 01952 684433

Renault Trucks Birmingham GBD334
Power Way,
Black Country New Road,
Tipton DY4 0PW
Tel 0121 505032

Brownhills Tachograph Centre GBD322
Lindon Road,
Brownhills,
Walsall WS8 7BW
Tel 01543 372528

Hartshone Motor Services Ltd GBD305
Bentley Mill Close,
Walsall WS2 0BN
Tel 01922 704600

S Jones Truck Centre Ltd GBD226
Westgate,
Aldridge,
Walsall WS9 8EZ
Tel 01922 743783

Guest Motors Ltd GBD503
Kenrick Way,
West Bromwich B70 6BY
Tel 0121 5532737

Keltruck Ltd GBD328
Kenrick Way,
West Bromwich B71 4JW
Tel 0121 524 1800/Fax 0121 524 1858
E-mail info@keltruck.co.uk
Website http: //www.keltruck.co.uk

A F Glaze Ltd GBD316
Dixon Street,
Wolverhampton WV2 2BT
Tel 01902 455434

Guest Motors Ltd GBD127
55 Wilenhall Road,
Wolverhampton WV1 2HL
Tel 01902 352888

Greenhous Leyland Daf Ltd GBD223
Neachells Lane,
Willenhall,
Wolverhampton WV13 3SF
Tel 01902 305090

Area 8

Oldhams of Barford GBD117
Wellesbourne Road,
Barford,
Warwick CV35 8DS
Tel 01926 624333

Excel Engineering GBD333
Landor Street,
Birmingham B8 1AE
Tel 0121 3590961

Lex Transfleet GBD324
Bannerly Road,
Garrets Green Industrial Estate,
Birmingham B33 0SA
Tel 0121 7844000

Lucas Kienzle Instruments Ltd GBD228
36 Gravelly Industrial Park,
Birmingham B24 8TA
Tel 0121 3261234

Carwood Motor Units Ltd GBD221
Herald Way,
Binley,
Coventry CV2 2RQ
Tel 01203 449531

TLS Plc GBD337
Kingswood Close,
Holbrooks,
Coventry CV6 4BJ

Volvo Truck & Bus (South) Ltd GBD319
Siskin ParkWay West,
Middlemarch Business Park,
Coventry CV6 5HE
Tel 01203 576100

Central Tachograph Services GBD128
Top Level 1,
Cochrane House,
Pedmore Road,
Dudley DY2 0RL
Tel 01384 482002

Lex Commercials Ltd GBD209
Park Road,
Halesowen B63 2RL
Tel 01384 424500

AEW GBD230
t/a Lucas Service UK,
Mortimer Road,
Hereford HR49SP
Tel 01432 265571

Burgoyne Bros Ltd GBD330
t/a Hereford Tachograph Centre,
Perseverence Road,
Hereford HR4 9SN
Tel 01432 344377

Noden Truck Centre GBD123
3 Avon Industrial Estate,
Butlers Leap,
Rugby CV21 3UY
Tel 01788 579535

Allelys Ltd GBD130
Benavon,
The Slough,
Studley B80 7EN
Tel 01527 857090

Keltruck Ltd GBD335
Watling Street,
Dordon,
Tamworth B78 1TS
Tel 01827 330100

Worcester Tachograph Centre GBD225
Unit 16A, Blackpole Trading Estate,
Worcester WR3 8SG
Tel 01905 755889

Note: please also refer to Area 7 entries for certain Central & South West Midlands Tachograph Centres not listed in this area, eg Aston, Walsall, West Bromwich and Wolverhampton.

Area 9

Williams Motors (Cwmdu) Ltd GBG207
The Walton,
Rich Way,
Brecon LD3 7EH
Tel 01874 622223

Euro Commercials (S Wales) Ltd GBG216
10 Millers Avenue,
Brynmenyn Industrial Estate,
Brynmenyn,
Bridgend CF32 9TD
Tel 01656 723279

Smiths Mechanical SVS Ltd GBG313
Tower Works,
Tower Close,
Bridgend Industrial Estate
Bridgend CF31 3TH
Tel 0165 6647370

Cardiff Truck Centre Ltd GBG307
Whittle Road,
Leckworth Industrial Estate,
Cardiff CF1 8AT
Tel 01222 308595

Euro Commercials Ltd GBG217
Ipswich Road,
Cardiff CF3 7AQ
Tel 01222 310310

Lucas Services (UK) Ltd GBG,
Hadfield Road,
Cardiff CF1 8TR
Tel 01222 228361

South West & Wales GBG114
Newlands Road,
Wentloog Corporate Park,
Cardiff
Tel 01222 361133

Tanner Electrics GBG109
Whittle Road,
Leckwith Industrial Estate,
Cardiff CF1 8AT
Tel 01222 225580

Merlin Motor Co Ltd GBG105
Fishguard Road Industrial Estate,
HaverfordWest SA62 4BT
Tel 01437 762468

Smiths Mechanical Services Ltd GBG312
Heol Alltycnap,
Johnstown,
Carmarthen,
Dyfed SA31 3NE
Tel 01267 230650

Knighton Trucks Co GBG308
Station Road,
Knighton LD7 1DT
Tel 01547 528600

Mansel Davies + Son (Garage) Ltd GBG206
Taffvale Garage,
Station Yard,
Llanfyrnach SA35 0BZ
Tel 01239 831631

Lewis's Coaches GBG212
Brynceithin,
Llanrhystyd SY23 5DN
Tel 01974 202495

Commercial Motors (Newport) Ltd GBG111
Frederick Street,
Newport NP9 2DR
Tel 01633 251576

Euro Commercials (S Wales) Ltd GBG215
164 Malpas Road,
Newport NP9 5PP
Tel 01633 821721

Griffin Mill Ltd GBG401
Spytty Road,
Leeway Industrial Estate,
Newport NP9 0QU
Tel 01633 290929

S A Trucks (Newport) Ltd GOG309
10 Newquay Road,
Stephenson St Industrial Estate,
Newport NP9 0PL
Tel 01633 290360

Grooms Industries GBC216
Pool Road,
Newtown,
Powys SY16 1DL
Tel 01686 626731

Bryan Brothers Ltd GBG112
Treforest Industrial Estate,
Pontypridd CF37 5YG
Tel 01443 844777

Griffin Mill Truck & Bus GBG314
Upper Boat,
Treforest,
Near Pontypridd
Tel 01443 841168

City Electro Diesel Services GBG108
Site F3, Nantyffin Road South,
Llansamlet,
Swansea SA7 9RG
Tel 01792 792010

M R Fussell Ltd GBG310
Millbrook Bean Group,
18/20 Morfa Road,
Swansea SA1 2EN
Tel 01792 650646

Griffin Mill Garages Ltd GBG113
21 Viking Way,
Winch Wen Industrial Estate,
Swansea SA1 7DA
Tel 01792 795462

Shorts Auto Electrical Services Ltd GBG209
43–51 Station Road,
Landore,
Swansea SA1 2JE
Tel 01792 469595

Area 10

Westward Commercials Ltd GBH402
Gurney Slade,
Nr Bath BA3 4TQ
Tel 01749 840777

Hickley Valtone Ltd GBH129
Unit 1, East Quay,
Bridgewater TA6 4DB
Tel 01278 423570

AEW Ltd GBH140
t/a Lucas Service Ltd,
Short Street,
St Phillips,
Bristol BS2 0SW
Tel 0117 972 4007

Bryan Bros Trucks Ltd GBH202
Albert Crescent,
St Phillips,
Bristol BS2 0UD
Tel 01179 772671

Lancaster Truck & South West GBH134
Days Road,
Barton Hill,
Bristol BS5 0JA
Tel 0117 955 1571

Lex Commercials GBH222
Days Road,
St Phillips,
Bristol BS2 0QP
Tel 01179 557755

Truck and Bus UK Ltd GBH139
Nova Way,
Avonmouth,
Bristol BS11 9DJ
Tel 0117 938 4100

Renault Trucks Bristol GBH132
c/o Excel Logistics,
Smoke Lane,
Avonmouth,
Bristol BS11 0YA
Tel 01179 821325

S A Trucks Ltd GBH216
Third Way,
Avonmouth,
Bristol BS11 9YL
Tel 01179 821241

Westward Commercials Vehicles Ltd
GBH400
Burcott Road,
Severnside Trading Estate,
Avonmouth,
Bristol BS11 8AP
Tel 01179 823741

Westward Commercials GBH403
Clark Avenue,
Porte Marsh Industrial Estate,
Calne SN11 9PZ
Tel 01249 817345

Norman Webb Transport Services GBH138
Unit 6, Bumpers Farm Industrial Estate,
Chippenham,
Wiltshire SN14 6NQ
Tel 01243 443523

Southway Scania Ltd GBH329
Avonmouth Way,
Bristol BS11 8DB
Tel 01179 379800

J R Harding & Sons (Frome) Ltd GBH115
17 Handlemaker Road,
Marston Trading Estate,
Frome BA11 4RW
Tel 01373 454477

Bristol St Commercials GBH502
Bristol Road,
Hempsted,
Gloucester GL2 5YB
Tel 01452 521581

Joseph Rice Truck Services GBH327
26a Hempsted Lane,
Hempsted,
Gloucester G12 5FH
Tel 01452 522563

Lucas Ltd GBH227
Sparrows Wharf,
Bristol Road,
Gloucester GL2 6DH
Tel 01452 524951

Watts Truck Centre Ltd GBH233
Mercia Road,
Gloucester GL1 2SQ
Tel 01452 525721

Richard Read (Transport) Ltd Longhope GBH321
Gloucester GL17 0QG
Tel 01452 830456

Mendip Truck & Bus Services Ltd GBH237
Crowne Trading Estate,
Shepton Mallett BA4 5QU
Tel 01749 343963

Exel Logistics GBH312
Radway Road,
Britannia Trade Park,
Swindon SN3 4ND
Tel 01793 831515

Rygor Commercial Ltd GBH230
Hunts Rise,
South Marston Park,
Swindon SN3 4TB
Tel 01793 821820

Mudie Bond Ltd GBH126
Newtown Trading Estate,
Tewkesbury,
Gloucester GL20 8JD
Tel 01684 295090

Rygor Commercials Ltd GBH228
The Broad Way,
West Wilts Trading Estate,
Westbury BA13 4JX
Tel 01373 855555

Coombs Travel GBH323
Searle Crescent,
Winterstoke Commercial Centre,
Weston Super Mare
Tel 01934 632612

Area 11

First Red Bus Ltd GBH241
Coney Avenue,
Barnstaple,
North Devon EX32 8QJ

AEW Ltd GBK136
t/a Lucas Services UK,
Elliot Road,
West Howe Industrial Estate,
Bournemouth BH11 8LN
Tel 01202 570507

Adams Morey Ltd GBH318
Yeomans Way,
Yeomans Industrial Park,
Bournemouth BH8 0BJ
Tel 01202 524422

West Dorset Coaches Ltd GBH141
t/a Dorchester Coachways,
Grove Trading Estate,
Dorchester DT1 1ST
Tel 01305 262992

MC Truck & Bus Ltd GBH404
Higher Shaftsbury Road,
Blandford,
Dorset DT11 7FB
Tel 01258 480404

AEW Ltd GBH239
t/a Lucas Services UK,
Grace Road,
Marsh Barton,
Exeter EX2 8PE
Tel 01392 270235

Frank Tucker (Commercials) Ltd GBH213
Peamore Truck Centre,
Alphington,
Exeter EX2 9SL
Tel 01392 833600

People 2000 GBH324
t/a Western Truck & Van Grace Road,
Marsh Barton,
Exeter EX2 8QB
Tel 01392 276561

Stuarts Commercials GBH401
Hillbarton Business Park,
Clyst St Mary,
Nr Exeter EX5 1DR
Tel 01395 232800

Westrucks Ltd GBH325
Heron Road,
Sowton Industrial Estate,
Exeter EX2 7LL
Tel 01392 444660

Cawsey Commercials Ltd GBH319
Unit 11, Newport Industrial Estate,
Launceston PL15 8EX
Tel 01566 772805

Pannell Commercials GBH235
Pennygillam Industrial Estate,
Launceston PL15 7ED
Tel 01566 773896

Wessex Leyland Daf Ltd GBH133
1 Roundhead Road,
Heathfield,
Newton Abbot TQ12 6EU
Tel 01626 833737

AEW Ltd GBH240
t/a Lucas Services Ltd,
Plymouth City Bus,
Milehouse,
Plymouth
Tel 01752 226450

Hendy Trucks GBH234
6 Witney Road,
Nuffield Industrial Estate,
Poole BH17 0GH
Tel 01202 494700

Saltash Truck & Bus Ltd GBH232
Stanley Way,
Cardrew Industrial Estate,
Redruth TR15 1SP
Tel 01209 314496

Cornish Ford GBH106
Slades Road,
St Austell PL25 4HP
Tel 01726 672333

Lancaster Truck & Van South West
GBH238
Bucklers Lane,
Holmbush,
St. Austell,
Cornwall PL25

Hickley Valtone Ltd GBH203
Castle Street,
Tangier,
Taunton TA1 4AH
Tel 01823 276041

Taunton Trucks Ltd GBH328
148 Priorswood Industrial Estate,
Taunton TA2 8DW
Tel 01823 331275

Whites of Taunton Ltd GBH107
Commercial Division,
South Street,
Taunton TA1 3HR
Tel 01823 335481

Unigate Fleet Services Ltd GBH236
Dart Mills,
Babbage Road,
Totnes TQ9 5JA
Tel 01803 869995

Marsh Road Garages GBH500
Marsh Road,
Weymouth DT4 8JD
Tel 01305 776116

Abbey Hill Truck Centre Ltd GBH309
The Parade Ground,
Boundary Way,
Lufton,
Yeovil BA22 8HZ
Tel 01935 432399

Western Truck & Van GBH137
9 Oxford Road,
PenMill Trading Estate,
Yeovil BA21 5HR
Tel 01395 426526

Area 12

PDE (Farnham) Ltd GBK220
Pavilion Road,
Aldershot GU11 3NX
Tel 01295 267528

Trimtruk GBK319
286 Weyhill Road,
Andover SP10 3NP
Tel 01264 334600

Banbury Truck & Transport Centre GBE118
Unit 1,
Station Approach,
Banbury OX16 8AB
Tel 01295 881581

Exel Logistics GBK331
Wildmere Industrial Estate,
Banbury,
Oxford OX16 7JU
Tel 01295 276819

Jacksons (Basingstoke) Ltd GBK113
Roentgen Road,
Daneshill,
Basingstoke RG24 8NT
Tel 01256 461656

Volvo Truck & Bus (South) GBK317
Houndmills Road,
Basingstoke RG21 6XL
Tel 01256 840402

John Lewis Tachograph Centre GBK307
John Lewis plc,
Doncastle Road,
Bracknell RG12 8YB
Tel 01344 824429

Oxfordshire Truck Centre GBK226
Collett,
Southmead Park,
Didcot OX11 7ET
Tel 01235 511115

Hendy Lennox Trucks GBK100
Bournemouth Road,
Chandlers Ford,
Eastleigh SO53 3ZG
Tel 01703 271271

Pentagon Ltd GBK128
Standard Way,
Fareham Industrial Park,
Fareham PO16 8XL
Tel 01329 286224

Southway Scania Ltd GBK316
26 Brunel Way,
Segensworth East,
Fareham PO15 5SD
Tel 01489 579880

Southern Vectis Comm Ltd GBK400
Nelson Road,
Newport,
Isle of Wight PO30 1RD
Tel 01983 821135

Volvo Truck & Bus Ltd GBK324
Stafferton Way,
Maidenhead SL6 1AY
Tel 01628 771787

Volvo Truck & Bus South GBK325
Hanesworth Industrial Estate,
Didcot OX11 7EN
Tel 01235 832217

Berkshire Commercial Repairs GBK229
Daytona Drive,
Colthrop Lane,
Thatcham,
Newbury RG13 4NT
Tel 01635 873746

Adams Morey Ltd GBK132
Riverway Industrial Estate,
Newport,
Isle of Wight PO30 5UZ
Tel 01983 522552

Mudie Bond Ltd GBK230
t/a Evenlode Truck Centre,
Eynsham Road,
Cassingham,
Witney OX81DD
Tel 01865 881581

R P Cherry & Son Ltd GBK330
Thrupp Lane,
Radley,
Abingdon,
Oxford OX14 3NG
Tel 01235 529756

AEW Ltd GBK234
Lucas Service UK,
Airport Services Road,
Portsmouth PO3 5PY
Tel 01705 650776

The Tacho Centre GBK135
Unit 9, Harnham Trading Estate,
Salisbury SP2 8NW
Tel 01722 322004

Adams Morey Ltd GBK323
Burrfields Road,
Copnor,
Portsmouth PO3 5NN
Tel 01705 691122

Hendy Lennox Trucks Ltd GBK121
t/a Hendy Truck Cosham,
Southampton Road,
Cosham,
Portsmouth PO6 4RW
Tel 01705 322900

AEW Ltd GBK235
Lucas Service UK,
16–20 Long Barn Lane,
Reading RG2 7SZ
Tel 01734 861202

Barnes of Reading GBK508
Station Road,
Theale,
Reading RG7 4AG
Tel 0118 9304111

Reg Vardy Trucks GBK506
20 Commercial Road,
Reading RG2 0RN
Tel 01734 868511

Renault Trucks Reading GBK328
Bennet Road,
Reading RG2 0QX
Tel 01189 752355

Terranova Lifting Ltd GBK133
Terranova House,
Bennet Road,
Reading RG2 0QX
Tel 01189 312345

AEW Ltd GBK233
Lucas Services UK,
Oakley Road,
Shirley,
Southampton SO16 4LL
Tel 01703 777111

Adams Morey Ltd GBK201
The Causeway,
Redbridge,
Southampton SO15 0DR
Tel 01703 663000

Hendy Lennox Trucks Ltd GBK304
2nd Avenue,
Mill Brook,
Southampton SO15 0LP
Tel 01703 701500

MC Truck & Bus Ltd GBK333
Test Lane,
Nursling Industrial Estate,
Southampton SO1 9JX

Area 13

Rossetts Commercials GBK137
Unit R, Rudford Industrial Estate,
Ford,
Arundel BN18 0BD
Tel 01903 722444

Morgan Elliott Ltd GBN348
87 Beddington Lane,
Croydon CR0 4TD
Tel 020 8689 4414

Summit Motor Investments UK Ltd GBN241
t/a Dees of Croydon,
2 Imperial Way,
South Croydon CR2 0BP
Tel 020 8680 4466

Volvo Truck & Bus (South) GBK322
Beddington Farm Road,
Croydon CR0 4XB
Tel 020 8665 5775

Cummings + Foster Ltd GBN500
t/a Benhill Motors,
London Road,
Ewell Bypass,
Epsom KT17 2PT
Tel 020 8393 2247

Barnes of Guildford GBN207
Slyfield Industrial Estate,
Woking Road,
Guildford GU1 1RT
Tel 01483 537731

Grays Truck Centre Ltd GBN307
Slyfield Industrial Estate,
Woking Road,
Guildford GU1 1RY
Tel 01483 571012

Horley Services Ltd GBN107
Salfords Industrial Estate,
Nr Redhill RH1 5ES
Tel 01293 771481

Evans Halshaw (Sussex) Ltd GBK205
78 Billingshurst Road,
Broad Bridge Heath,
Horsham RH12 3LP
Tel 01403 256464

Hancock Ford GBN502
53–55 Bishopric,
Horsham RH12 1QJ
Tel 01403 254331

Barnes of Shoreham GBK504
44 Dolphin Road,
Shoreham-by-Sea BN43 6PB
Tel 01273 454887

Vales Truck Centre Ltd GBN116
t/a Heathrow Truck Centre,
Lakeside Industrial Estate,
Colnbrook Bypass,
Colnbrook,
Slough SL3 0ED
Tel 01753 681818

Southway Scania Ltd GBN347
Bedfont Way,
Stanwell,
Middlesex TW19 7LZ
Tel 01784 240777

Rossetts Commercials GBK129
Meadow Road Industrial Estate,
Worthing BN11 2RU
Tel 01903 204127

Area 14

Channel Commercials plc GBK332
Brunswick Road,
Cobbs Wood Estate,
Ashford TN23 1EH
Tel 01233 629272

Thanet Commercials GBK116
Unit 12 Hornet Close,
Pysons Road Industrial Estate,
Broadstairs CT10 2YD
Tel 01843 602194

AEW Ltd GBK231
t/a Lucas Service UK,
Mainard Road,
Wincheep Industrial Estate,
Canterbury CT1 3RH

Invicta Trucks GBK102
73–74 Ellingham Way,
Ashford,
Kent TN23 6JU
Tel 01233 665200

Rossetts Commercials GBK326
7 Birch Road,
Eastbourne BN23 6PD
Tel 01323 410192

EuroTrucks Ltd GBK211
Eastbourne Road,
Westham,
Pevensey,
Nr Eastbourne BN24 5NH
Tel 01323 767626

Kent Truck Services Ltd GBK329
East Kent International Freight Terminal,
Hernhill,
Faversham ME13 9EN
Tel 01227 771111

MGC Ltd GBK134
Royal Oak Garage,
London Road,
Flimwell,
East Sussex

M C Truck & Bus Ltd GBK320
Unit A1, Lympne Industrial Estate
Hythe CT21 4LR
Tel 01303 266864

M C Truck & Bus Ltd GBK303
Beddow Way,
Forstal Road,
Aylesford North,
Maidstone ME20 7BT
Tel 01622 710811

South Eastern Auto Electrical Services Ltd
GBK222
Wharf Road,
Tovil,
Maidstone ME15 6RR
Tel 01622 690004

Hancock Group GBK500
t/a Stormont Trucks,
Ellen Street,
Portslade,
Brighton BN41 1DY
Tel 01273 430828

Lucas Service UK GBK232
Unit 8, Victoria Road Trading Estate,
Portslade,
Brighton BN41 1XO
Tel 01273 439955

Channel Commercials plc GBK228
Whitewall Road,
Medway City Estate,
Strood,
Rochester ME2 4DZ
Tel 01634 296686

ERF Medway Ltd GBK315
Arnold Close,
Sir Thomas Longley Road,
Medway City Estate,
Rochester ME2 4QW
Tel 01634 711144

Lavers Repairs Ltd GBK126
North Downs Business Park,
Pilgrims Way,
Dunton Green,
Sevenoaks TN13 2TL
Tel 01732 462320

Sparshatts of Kent Ltd GBK212
Unit 10, Eurolink Industrial Estate,
Murston,
Sittingbourne ME10 3RN
Tel 01795 479571

Scantruck Ltd GBK327
Unit 15a, Eurolink Industrial Estate,
Sittingbourne ME10 3RN
Tel 01795 430304

Stormont Truck GBK124
London Road,
Hildenborough,
Tonbridge TN11 8NN
Tel 01732 833005

Kent & Sussex Truck Centre GBK501
Longfield Road,
North Farm Industrial Estate,
Tunbridge Wells TN2 3EY
Tel 01892 515333

Lucas Service UK GBK507
North Farm Road,
High Brooms Industrial Estate,
Tunbridge Wells TN2 3EA
Tel 01892 510444

Area 15

Dagenham Motors GBN506
51 River Road,
Barking IG11 0SW
Tel 0181 4774027

Acorn Truck Sales Ltd GBN335
Acorn Industrial Park,
Crayford Road,
Crayford DA1 4AL
Tel 01322 556415

Gifford Tachograph Services + Commercial Ltd GBN322
Assured House,
Chequer Lane,
Dagenham RM9 6QD
Tel 0181 5931550

Duffields of East Anglia Ltd GBN346
Mollison Avenue,
Brimsdown,
Enfield EN3 7NJ
Tel 0181 8059911

Hunter Vehicles Ltd GBN100
Crown Works,
Southbury Road,
Enfield EN1 1UD
Tel 0181 8051016

Norfolk Trucks Ltd GBN242
Mollison Avenue,
Brimsdown,
Enfield EN3 7NE
Tel 0181 4432540

South Eastern Auto Electrical Services Ltd GBN226
Unit 26, Manford Industrial Estate,
Erith DA8 2AB
Tel 01322 342277

Foden Heathrow GBN119
The Griffin Centre,
Staines Road,
Feltham TW14 0JL
Tel 0181 8932270

Volvo Truck & Bus Ltd GBN223
t/a Volvo Truck & Bus (South),
5 Bulls Bridge Centre,
North Hyde Road,
Hayes UB3 4QR
Tel 0181 624 515

M C Truck & Bus Ltd GBN400
Waterglade Industrial Park,
Barclay Way,
West Thurrock RM20 3FB
Tel 01708 868956

The Londoners Ltd GBN350
1a Brabourn Grove,
Peckham,
London SE15 2BS
Tel 0171 539 1212

County Truck Services Ltd GBN118
Auriol Drive,
Oldfield Lane North,
Greenford UB6 0AY
Tel 0181 5785588

Dagenham Motors (Hayes) Ltd GBN240
Dawley Road,
Hayes,
Middlesex UB3 1EH
Tel 020 8561 4888

Comprehensive Commercial Services Ltd GBN331
Railway Terrace,
Kings Langley WD4 8JA
Tel 01923 262199

Capital Vehicle Maintenance GBN238
207–209 Worton Road,
Isleworth,
Middlesex TW7 6DS
Tel 020 8758 0888

A23 Tacho Centre GBK225
146/156 Brixton Hill,
London SW2 1SD
Tel 020 8671 7781

Transfleet Services Ltd GBN507
17 Western Road,
Park Royal,
London NW10 7LT
Tel 020 8961 5225

Martin Endersby GBN345
Truck + Trailer Services,
91 Cowley Road,
Uxbridge UB8 2AG
Tel 01895 239834

Alperton Trucks GBN508
West Links,
Alperton Lane,
Alperton,
Harrow HA0 1ER
Tel 020 8810 7976

Area 16

Scantruck Ltd GBF330
Unit B, Stanstead Distribution Centre,
Start Hill,
Great Hallingbury,
Bishops Stortford CM22 7DG
Tel 01279 758088

Boyton Cross Motors GBF121
Boyton Cross,
Roxwell,
Chelmsford CM1 4LL
Tel 01245 248658

Lucas Services UK Ltd GBN339
3 Montrose Road,
Dukes Park Industrial Estate,
Chelmsford CM2 6TE
Tel 01245 466166

Colchester Fuel Injections Ltd GBF223
Haven Road,
Colchester CO2 8HT
Tel 01206 862049

Dovercourt Ford GBF509
114 Ipswich Road,
Colchester,
Essex CO4 4AA
Tel 01206 791171

Man Eastern GBF124
West Side Centre,
London Road,
Stanway CO3 5PB

Harris Commercial Repairs Ltd GBN329
601 London Road,
West Thurrock,
Grays RM20 4AU
Tel 01708 864426

Lancaster Europa Ltd GBN120
Lakeside Estate,
Heron Way,
West Thurrock,
Grays RM20 3WJ
Tel 01708 861321

Renault Trucks Essex GBN117
Weston Avenue,
Waterglade Industrial Park,
West Thurrock,
Grays RM20 3FZ
Tel 01708 866643

S+B Commercials plc GBN217
Travellers Lane,
Welham Green,
Hatfield AL9 7JL
Tel 01707 261111

Valley Trucks Ltd GBF123
Unit 1, Bingley Road,
Hoddesdon EN11 0NX
Tel 01992 441551

Scantruck Ltd GBN206
Arterial Road,
Purfleet RM19 1PB
Tel 01708 864915

Scantruck Ltd GBN340
Bignells Corner,
St Alans Road
South Mimms EN6 3NG
Tel 01707 649955

Waljohn Vehicle Services Ltd GBN239
Upminster Trading Park,
Warley Street,
Upminster RM14 3PJ
Tel 01708 250485

Harris Commercials GBF326
Station Approach,
Waltham Cross EN8 7NA
Tel 01992 651166

Vales Truck & Van Centre GBN219
Tolpits Lane,
Watford WD1 8QP
Tel 01923 713820

RO-Truck Ltd GBF329
Unit 5 Moss Road,
Witham Industrial Estate,
Witham CM8 3UQ
Tel 01376 503003

Area 17

Brand (Motor) Engineers Ltd GBF224
Common Lane North,
Beccles NR34 9BN
Tel 01502 718800

MRCT Ltd GBF118
Unit 1, Vinces Road,
Diss IP22 3HQ
Tel 01379 640790

Roy Humphrey (Car + Commercial) GBF324
A140 Ipswich Road,
Brome,
Eye IP23 8AW
Tel 01379 870666

Jack Richards + Son Ltd GBF119
2 Garold Drive Industrial Estate,
Fakenham NR21 8NL
Tel 01328 863111

RO-Trucks Ltd GBF331
6 Hodgkinson Road,
Trimley Industrial Estate,
Felixstowe,
Suffolk IP11 8QT
Tel 01394 676625

Duffields of East Anglia Ltd GBF401
Gapton Hall Industrial Estate,
Gapton Hall Road,
Great Yarmouth NR31 0NL
Tel 01493 443001

L G Perfect (Engineering) Ltd GBF306
Jubilee Works,
Hafreys Road,
Great Yarmouth NR31 0JL
Tel 01493 657131

AEW Ltd GBF226
t/a Lucas Service UK,
Hadleigh Road Industrial Estate,
Arkwright Road,
Ipswich IP2 0HB
Tel 01473 215932

Duffields of East Anglia Ltd GBF320
Foxtail Road,
Ransomes EuroPark,
Ipswich IP3 9RT
Tel 01473 718223

Norfolk Trucks Ltd GBF117
Lodge Lane,
Great Blakenham,
Ipswich IP6 0LB
Tel 01473 834200

GDM Transport Engineering Ltd GBF112
Saddle Bowest,
Staple Road,
Kings Lynn PE34 3AH
Tel 01553 761112

Hamlin Way GBF333
Hardwick Narrows,
Kings Lynn,
Norfolk PE30 4NG

AEW Ltd GBF227
Weston Road North,
Norwich NR3 3TL
Tel 01603 410301

Bussey Ltd GBF313
95 Whiffler Road,
Boundary Road,
Norwich NR3 2EW
Tel 01603 424022

Duffields of East Anglia Ltd GBF402
24–28 Frensham Road,
Sweet Briar Industrial Park,
Norwich NR3 2BT
Tel 01603 785100

Ford & Slater Norwich Ltd GBF116
Unit 8, Kerrison Road Industrial Estate
Kerrison Road,
Norwich NR1 1JA
Tel 01603 621415

Norfolk Trucks Ltd GBF325
School Lane,
Sprowston,
Norwich NR7 8TL
Tel 01603 253300

RO-Truck GBF316
Violet Hill Road,
Stowmarket IP14 1NN
Tel 01449 613553

Almalco Motors Ltd GBF309
Caxton Way,
Thetford IP24 3RY
Tel 01842 752457

Duffields of East Anglia Ltd GBF400
34 Howlett Way,
Thetford IP24 1HZ
Tel 01842 763830

A G Potter (Framlingham) Ltd GBF508
Station Road,
Framlingham,
Woodbridge IP13 9EE
Tel 01728 723215

Area 18

Chambers Engineering Ltd GBK127
Warmstone Lane,
Waddesdon,
Bedford HP18 0NF
Tel 01296 651380

Pentagon Ltd GBE232
Bicester Road,
Aylesbury HP19 3BL
Tel 01296 481641

Banks Trucks Centre Ltd GBF317
3 Brunel Road,
Barkers Lane Industrial Estate,
Corby MK41 9TL
Tel 01234 211241

K Watson Commercials Ltd GBE233
6 Kenneth Way,
Wilstead Industrial Park,
Aylesbury MK45 3PD
Tel 01234 742266

Progress Motor Company GBF105
t/a Polar Trucks,
Hudson Road,
Bedford MK41 0HR
Tel 01234 340041

Volvo Truck & Bus (South) GBE403
Woburn Road Industrial Estate,
Wolseley Road,
Kempston,
Bedford MK42 7WR
Tel 01234 853877

R C S Commercials Ltd GBE127
Crucible Road,
Phoenix Parkway Industrial Estate,
Corby
Tel 01536 203370

Hartwell Truck GBN222
Skimpot Road,
Dunstable LU5 4JX
Tel 01582 597575

Renault Trucks Chiltern GBF323
Luton Road,
Dunstable LU5 4QF
Tel 01582 476446

Biffa Tachograph Centre GBN306
London Road,
Loudwater,
High Wycombe HP10 9TD
Tel 01494 427257

Paynes Garage Ltd GBE210
Watling Street,
Hinckley LE10 3ED
Tel 01455 238911

Derek Jones Commercials Ltd GBE327
Kettering Road,
Islip,
Kettering,
Northants NN14 3JW

Stormont Truck GBE137
Bartley Drive,
Telford Way Industrial Estate,
Kettering NN16 8JN
Tel 01536 517079

Volvo Truck & Bus (South) Ltd GBE131
Pytchley Road Industrial Estate,
Kettering NN15 6JJ
Tel 01536 516233

A B Butt Ltd GBE217
Frog Island,
Leicester LE3 5AZ
Tel 0116 25133344

Alltruck Services GBE130
Scudamore Road,
Braunstone Industrial Estate,
Leicester LE3 1UR
Tel 0990 168777

Arriva Motor Retailing plc GBE300
t/a Arriva Vauxhall,
Freemans Common Road,
Aylestone Road,
Leicester LE2 7SL
Tel 0116 2557567

East Midlands Commercials Ltd GBE323
Unit 2, Midlands Distribution Centre,
Markfield Road,
Groby,
Leicester LE6 0FS
Tel 01530 243133

Ford + Slater of Leicester GBE305
Hazel Drive,
Narborough Road South,
Leicester LE3 2JG
Tel 0116 2632900

Sanderson Truck GBE125
302 Melton Road,
Leicester LE4 7SL
Tel 0116 2667721

Volvo Truck & Bus (South) GBE404
Station Road,
Stoney Station,
Leicestershire LE9 6LJ
Tel 01455 27326

Chassis Developments Ltd GBF217
Grovebury Road,
Leighton Buzzard LU7 8SL
Tel 01525 374151

Luton Truck + Van Centre GBE325
166 Camford Way,
Sunden Park,
Luton LU3 3AN
Tel 01582 505464

Brian Currie Milton Keynes Ltd GBN113
Chesney Wold,
Bleak Hall,
Milton Keynes MK6 1LH
Tel 01908 663991

City Truck Sales Ltd GBF218
10 Northfield Drive,
Northfield,
Milton Keynes MK15 0DE
Tel 01908 665152

Polar Ford Commercials GBE128
Northfield Drive,
Northfield,
Milton Keynes MK15 0DA
Tel 01908 672400

The Truck Centre GBE236
Colts Holm Road,
Old Wolverton MK12 5QD

Volvo Truck & Bus (South) GBN342
Delaware Drive,
Tongwell,
Milton Keynes MK15 8JH
Tel 01908 210525

Arlington of Northampton GBE318
Bedford Road,
Northampton NN1 5NS
Tel 01604 250151

Brian Currie Ltd GBE124
Milton Trading Estate,
Gayton Road,
Milton Malsor,
Northampton NN7 3AB
Tel 01604 858810

Derek Jones Truck & Van Centre GBE133
Stewarts Road,
Finedon Road Industrial Estate,
Wellingborough,
Northampton NN84RJ

Northampton Diesel + Electrical Services Ltd GBE211
Holloway Industrial Estate,
25–29 Gambrel Road,
Northampton NN5 5DG
Tel 01604 755321

Stormont Trucks GBE328
Jackdaw Close,
Crow Lane,
Northampton NN3 9ER

The Truck Centre Ltd GBE132
t/a The Truck Centre,
37 Crow Lane,
Great Billing,
Northampton NN3 4BZ
Tel 01604 401405

Cummins Diesel GBE229
Rutherford Drive,
Park Farm South,
Wellingborough NN10 0DL
Tel 01933 672200

Area 19

C F Parkinson Ltd GBE231
Fydell Crescent Workshop,
Fydell Crescent,
George Street,
Boston PE21 8XQ
Tel 01205 363008

AEW Ltd GBF228
t/a Lucas Service UK,
442 Newmarket Road,
Cambridge CB5 8JU
Tel 01223 315931

City Ford (UK) Ltd GBF506
350 Newmarket Road,
Cambridge CB5 8JT
Tel 01223 323105

Marshall Commercial Vehicles GBF100
The New Airport Garage,
699 Newmarket Road,
Cambridge CB5 8SQ
Tel 01223 377900

D H M Truck & Tacho Centre GBF332
Unit 1, Sanfield House,
Alconbury Hill,
Alconbury Weston,
Huntingdon PE17 5JL
Tel 01480 433300

C F Parkinson Ltd GBE235
Outer Circle Road,
Lincoln LN2 4UH
Tel 01522 530176

Ford & Slater (Leyland Daf Ltd) GBE224
Sleaford Road,
Bracebridge Heath,
Lincoln LN4 2NQ
Tel 01522 522231

G A Smalley Ltd GBE401
Freeman Road Industrial Estate,
North Hykeham,
Lincoln LN6 9AP
Tel 01522 684496

John Longden Ltd GBE102
PO Box 19,
Crofton Road,
Allenby Road Industrial Estate,
Lincoln LN3 4NW
Tel 01522 538811

Fengate Commercials Services GBF315
Bretton Way,
Bretton,
Peterborough PE3 8YO
Tel 01733 333232

Ford & Slater GBF111
316 Padholme Road,
Eastern Industrial Estate,
Peterborough PE1 5XL
Tel 01733 347100

Sellars and Batty GBF225
Fengate,
Peterborough PE1 5XG
Tel 01733 560591

T C Harrison Group GBF215
Oxney Road,
Peterborough PE1 5YN
Tel 01733 558111

Foulgers Garage GBF328
Melda Farm,
Bury Lane,
Melbourn,
Royston SG8 6DK
Tel 01763 248332

Ford + Slater Ltd GBF321
58 Station Road,
Donnington,
Spalding PE11 4UJ
Tel 01775 820777

R C Edmondson GBF308
St Johns Road,
Spalding PE11 1JA
Tel 01775 767651

Duffield of East Anglia Ltd GBF322
Boleness Road,
Wisbech PE13 2RE
Tel 01945 463355

Area 20

Excel Logistics GBB112
Shawfield Road,
Barnsley S71 3HS
Tel 01226 298321

North Way Scania Ltd GBB502
Barrow Field Road,
Platts Common Trading Estate,
Hoyland,
Barnsley S74 9SF
Tel 01226 742654

Mercedes Benz UK Ltd GBB226
Wentworth Way,
Tankersley,
Barnsley S75 3DH

The Polar Truck Centre GBB331
Wombwell Lane,
Stairfoot,
Barnsley S70 3NX
Tel 01226 732732

Paul Barton Tachographs GBE116
63 High Road,
Chilwell,
Beeston NG14 6NX
Tel 0115 922 1441

AEW Ltd GBB228
t/a Lucas Service UK,
Unit 9, Tenpound Walk,
Doncaster DN4 5HX
Tel 01302 342794

E & G Charlesworth Ltd GBB503
Truck Services Dept,
Decoy Bank North,
Carr Grange Industrial Estate,
Doncaster DN4 5JE
Tel 01302 327111

Trailer Supermarket (Bawtry) Ltd GBB114
Doncaster Road,
Bawtry,
Doncaster DN10 6NX
Tel 01302 710711

K+M (Haulage) Ltd GBE107
The Aerodrome,
Watnall Road,
Hucknall NG15 6EN
Tel 01159 630630

Charnwood Truck Services GBE234
Hillside,
Gotham Road,
Kingston-on-Soar NG11 0DF
Tel 0115 983 0093

C F Parkinson (Notts) Ltd GBE227
Brunel Drive,
Northern Road Industrial Estate,
Newark NG24 2EG
Tel 01636 672631

Lex Commercials Ltd GBE319
Boulevard Industrial Park,
Beacon Road,
Beeston,
Nottingham
NG9 2JR

R H Commercials Ltd GBE228
Lenton Lane,
Nottingham NG7 2NR
Tel 0115 943 8000

Sherwood Truck & Van GBE121
522 Derby Road,
Lenton,
Nottingham NG7 2GX
Tel 0115 978 7274

AMB Commercials GBE129
Crookford Hill,
Elkesley,
Retford DN22 8BT
Tel 01777 838380

CrossRoads Commercials Ltd GBB402
Canklow Meadows Industrial Estate,
West Bawtry Road,
Rotherham S60 2XL
Tel 01709 365566

Sheffield Tachograph and Tailift Services
GBB312
Unit 4, Thornhill Industrial Estate,
Hope Street,
Rotherham S60 1LH
Tel 01709 839000

Unitec GBB330
Denby Way,
Hellaby,
Rotherham S66 8HR
Tel 01709 535101

AEW GBB227
t/a Lucas Service UK,
300 Savile Street,
Attercliffe,
Sheffield S4 7UD
Tel 0114 275 2522

Chatfields of Sheffield GBB142
7 Orgreave Drive,
Handsworth,
Sheffield S13 9NR
Tel 0114 254 8854

Plaxton Parts & Services GBB211
Ryton Road,
South Anston,
Sheffield S31 7ES
Tel 01909 551155

Sherwood Commercial Vehicles GBB316
Highfield Garage,
Highfield Lane,
Sheffield S13 9DB
Tel 0114 2693230

South Yorkshire Trucks Ltd GBB133
5–6 Parkhouse Lane,
Tinsley,
Sheffield S9 1XA
Tel 0114 2610055

T C Harrison Group Ltd GBB201
Sheffield Truck Operations,
Shepcote Lane,
Sheffield S9 1TX
Tel 0114 2012424

Trent Trucks GBE209
Sandcliffe Motor Group,
Nottingham Road,
Stapleford NG9 8AU
Tel 01159 395000

Scania Bus and Coach UK Ltd GBE502
Claylands Avenue,
Worksop S81 7DJ
Tel 01909 500822

Area 21

Direct Commercial Vehicle Systems Ltd
GBB140
Island Carr Industrial Estate,
Brigg DN20 8PD
Tel 01652 656885

H&L Garages Ltd GBB225
Junction 38 M62,
Newport,
Brough HU15 2RD
Tel 01430 422297

Jayserv Ltd GBB321
Mariner Street,
Goole DN14 5BW
Tel 01405 767977

E T Commercials Ltd GBB141
Estate Road No 5,
South Humberside Industrial Estate,
Grimsby DN31 2TG
Tel 01472 347244

Scanlink Ltd GBB315
Estate Road One,
South Humberside Industrial Estate,
Grimsby DN31 2TA
Tel 01472 346913

Crossroad Commercials Ltd GBB404
Valletta Street,
Kingston-upon-Hull HU9 5NP
Tel 01482 781831

Lex Commercials GBB213
Hedon Rd,
Kingston upon Hull HU9 5PJ
Tel 01482 795111

Scanlink Ltd GBB221
Central Orbital Trading Estate,
Waverley Street,
Kingston-upon-Hull HU1 2SH
Tel 01482 225591

Thompson of Beverley Ltd GBB109
t/a Thompson Commercials,
Salvesen Way,
Clive Sullivan Way,
Kingston upon Hull HU3 4UQ
Tel 01482 322331

Torridon Commercial Vehicles GBB332
Littlefair Road,
Hedon Road,
Kingston upon Hull HU9 M5LP

Lex Commercials Ltd GBE302
Midland Industrial Estate,
Kettering Road,
Scunthorpe DN16 1UW
Tel 01724 282444

H & L Garages Ltd GBB218
Grange Lane North,
Scunthorpe DN16 1BT
Tel 01724 856655

H+L Garage Ltd GBE219
Humber Road,
South Killingholme DN40 3DL
Tel 01469 571666

Fishlake Commercial Motors Ltd GBB224
Jubilee Bridge Works,
Selby Road,
Thorne,
Nr. Doncaster,
South Yorkshire DN8 5JD
Tel 01405 740086

John Hebb (Motors) Eng Ltd GBB400
Cherry Lane,
Wootton,
Ulceby DN39 6RJ
Tel 01469 588431

Area 22

Crossroads Commercials Ltd GBB405
t/a The Crossroads Group,
Pheasant Drive,
Birstall,
Batley WF17 9LR
Tel 01924 425000

Lucas Service UK Ltd GBB210
232 Bradford Rd,
Batley WF17 6LF
Tel 01924 472415

Boroughbridge Motors Ltd GBB117
Barr Lane,
Roecliffe,
Boroughbridge,
York YO5 9NN
Tel 01423 322741

Northside Truck Centre GBB125
Legrams Lane,
Bradford BD7 2HR
Tel 01274 57731

Northern Commercials Mirfield GBB324
Armitage Road,
Brighouse HD6 1PG
Tel 01484 380111

Reliance Commercials GBB507
Wakefield Road,
Brighouse HD6 1QQ
Tel 01484 712611

Sanderson Truck GBB329
Derwent Valley Industrial Estate,
Common Lane,
Dunnington,
York YO1 5PD
Tel 01904 488005

Arriva Bus & Coach Ltd GBB137
Lodge Garage,
Whitehall Road West,
Gomersal,
Cleckheaton BD19 4BJ
Tel 01378 999311

Doves Leyland Daf Ltd GBB223
Bradley Junction Industrial Estate,
Leeds Road,
Huddersfield HD2 1UR
Tel 01484 300500

Knottingley Trucks Ltd GBB130
Common Lane,
Knottingley WF11 8BG
Tel 01977 672217

Linpac Leyland Daf GBB222
Moor Lane Trading Estate,
Sherburn-in-Elmet,
Leeds LS25 6ES
Tel 01977 683133

Northway Scania Ltd GBB139
Royds Farm Road,
Beeston Royds Industrial Estate,
Beeston,
Leeds LS12 6DX
Tel 01132 311411

Transfleet Services Ltd GBB127
Parkside Lane,
Dewsbury Road,
Leeds LS11 5TD
Tel 01132 477378

Pelican Engineering Co Ltd GBB146
Wakefield Road,
Rothwell Haigh,
Leeds LS26 0RU
Tel 0113 2822181

Polar Trucks GBB102
Albert Road,
Morley,
Leeds LS27 8TT
Tel 01132 383110

Renault Trucks Ltd GBB327
Howley Park Road,
Morley LS27 0BN
Tel 0113 252 7333

Sewell of Leeds GBB202
49 Marshall Street,
Leeds LS11 9SU
Tel 01132 435101

WASS (Leeds) Ltd GBB104
123 Hunslet Road,
Leeds LS10 1LD
Tel 01132 439911

Slaters Transport Ltd GBB212
Kirby Misperton,
Malton YO17 0UE
Tel 01653 668275

Northway Scania GBB328
Ripley Drive,
Normanton Industrial Estate,
Normanton WF6 1QT
Tel 01924 891254

Firth Truck + Van Services Ltd GBB144
Wakefield Road,
Flushdyke,
Ossett WF5 9JU
Tel 01924 281575

Chatfields of Leeds GBB501
Grangefield Industrial Estate,
Richard Shaw Lane,
Pudsey LS28 6SD
Tel 01132 571701

Crossroads Commercials Ltd GBB403
Stockton Road,
Thirsk YO7 1AX
Tel 01845 522057

Calderford Motor Co Ltd GBB118
Barnsley Road,
Wakefield WF1 5JS
Tel 01924 290290

York Auto Electrics Ltd GBB207
ta Yortec,
58 Layerthorpe,
York YO3 7YN
Tel 01904 654513

Area 23

Cochranes Garage (Berwick) Ltd GBA308
Tweedside Trading Estate,
Berwick upon Tweed TD15 2XF
Tel 01289 305585

North East Truck + Van Ltd GBB326
Cowpen Bewley Road,
Haverton Hill,
Billingham TS23 4EX
Tel 01642 370021

Renault Trucks NE GBB132
Nuffield Road,
Cowpen Lane Industrial Estate,
Billingham TS23 4DA
Tel 01642 370370

North East Truck & Van Ltd GBB318
Chainbridge Road,
Blaydon NE21 5TR
Tel 0191 4143333

Lex Transfleet GBA309
Penshaw Way,
Portabello Trading Estate,
Birtley,
Chester-le-Street DH3 2SA
Tel 0191 4104437

Tyne Tees Leyland Daf Ltd GBA113
Drum Road,
Barleymow Birtley,
Chester-le-Street DH3 2AF
Tel 0191 4921155

Victoria Garage GBB131
Black Friar Street,
Shotton Colliery,
Co. Durham DH6 2NX
Tel 0191 5170577

Darlington Commercials GBA312
Lingfield Way,
Yarm Road Industrial Estate,
Darlington DL1 4PY
Tel 01325 355161

Tyne Tees Leyland Daf Ltd GBB325
Bridge House,
Middleton St George,
Darlington DL2 1HR
Tel 01325 332941

Albany Motors GBB323
Saltmeadows Road,
Gateshead NE8 3AH
Tel 0191 4770501

Highway Truck Rental Ltd GBA114
Green Lane House,
Saltmeadows Road,
Gateshead NE8 3AH
Tel 0191 4777788

Lucas Ltd GBA200
Earlsway,
Team Valley Trading,
Gateshead NE4 0RQ
Tel 0191 4911720

Bell Truck Sales Ltd GBB322
Bellway Industrial Estate,
Whitley Road,
Long Benton,
Newcastle-upon-Tyne NE12 9SW
Tel 0191 2150060

Patterson Ford GBB505
Melbourne Street,
Newcastle upon Tyne NE1 2ER
Tel 0191 2611471

Union Trucks Ltd GBB500
Mylord Crescent,
Killingworth,
Newcastle-upon-Tyne NE12 0UW
Tel 0191 2683141

Renault Trucks North East GBA110
Unit B6, Third Avenue,
Tyne Tunnel Trading Estate,
North Shields NE29 7SP
Tel 0191 2962848

G Abbots + Son GBA112
Aumans House,
Leeming,
Northallerton DL7 9RZ
Tel 01677 422858

Heathline Commercials Ltd GBB217
Stakeford Lane,
Choppington,
Stakeford NE62 5QJ
Tel 01670 824006

Auto Electrics(Teeside) Ltd GBA210
Thornaby House,
Thornaby Place,
Stockton on Tees TS17 6BN
Tel 01642 607901

Electro Diesel North East Ltd GBA208
Portrack Grange Road,
Stockton on Tees TS18 2PH
Tel 01642 679741

Hargreaves Vehicle Distributors Ltd
GBA100
Bowesfield Lane,
Stockton-on-Tees TS18 3HF
Tel 01642 614121

Arriva Ford Commercials GBA307
North Hylton Road,
Southwick,
Sunderland SR5 3HQ
Tel 0191 5491111

Darlington Commercials Ltd GBB401
Crowther Road,
Washington NE38 0AZ
Tel 0191 4151111

Appendix VII

VEHICLE AND OPERATOR SERVICES AGENCY (VOSA) WEIGHBRIDGES

This list shows the geographical location of VOSA dynamic axle weighbridges within Traffic Areas. Locations marked * provide a self-weighing facility.

North-Western

A5, Holyhead, Anglesey, North Wales.

Wallasey Tunnel (East), Liverpool, Merseyside.

Wallasey Tunnel (West), Liverpool, Merseyside.

M62, Junction 20, Thornham, Greater Manchester.

A59, Samlesbury, Preston, Lancashire.

A556, Rostherne, Cheshire.

A5117/M56, Dunkirk, Cheshire.

A74, Harker, Carlisle, Cumbria.

A494, Ewloe, Clwyd, North Wales.

North-Eastern

A1, Scotch Corner, Middleton Tyas, North Yorkshire.

A19, Wellfield, County Durham.

A1/A659, Boston Spa, Wetherby, West Yorkshire.

A63, South Cave, North Humberside.

*Nepshaw Lane, Gildersome, Morley, Leeds, Yorkshire.

A1, Robin Hoods Well, Skellow, Doncaster, South Yorkshire.

Blackley New Road, Ainley Top, Huddersfield, West Yorkshire.

A61, Barnsley Road, Tankersley, Barnsley, South Yorkshire.

*King George V Docks, Kingston upon Hull, Humberside.

*A15, Approach Road, Humber Bridge, Hessle, South Humberside.

A46/52, Saxondale, The Old Nottingham Road, Bingham, Nottinghamshire.

West Midlands

A449, Link Road, Warndon, Nottinghamshire.

A5, Wall Island, Lichfield, Staffordshire.

M6, Doxey, Nr Stafford, Staffordshire.

A45/M45, Thurleston Island, Dunchurch, Rugby, Warwickshire.

M5, Junction 3, Quinton, Nr Birmingham, West Midlands.

*A40, Three Crosses, Ross-on-Wye, Gloucestershire.

Eastern

A43, Towcester, Northamptonshire.

A17/A15, Holdingham, Sleaford, Lincolnshire.

A428/M1, Junction 18, Crick, Northamptonshire.

*Stone Grove Road, The Dock, Felixstowe, Suffolk.

Old A45 Road, Risby, Suffolk.

A1, Southbound, Connington Fen, Sawtry, Cambridgeshire.

A1, Northbound, Lower Caldecote, Sandy, Bedfordshire.

A414/M11, Junction 7, Harlow, Essex.

M1, Junction 14, Newport Pagnell, Buckinghamshire.

Southlands Road, Denham, Buckinghamshire.

South Wales

A40, Penblewin, Narberth, Dyfed.

M4, Coldra, Newport, Gwent.

Western

*Millbay Docks, Plymouth, Devon.

A303, Wylye, Wiltshire.

Off A361, Sampford Peverell, Tiverton, Devon.

*Continental Freight Ferry Terminal, New Harbour Road, Poole, Dorset.

A38, Anchor Inn Lay-by Kennford, Devon.

A35, Lay-by, Puddletown, Dorchester, Dorset.

*Portsmouth Docks, Portsmouth, Hampshire.

A46, Tormarton, Avon.

A34/A415, Marsham Road, Abingdon, Oxfordshire.

South-Eastern and Metropolitan

A30, Staines Bypass, Surrey.

M25, Junction 9, Leatherhead interchange, Surrey.

A13, Barking, Essex.

A3, Burpham, Nr Guildford, Surrey.

A1, Holloway Road, London (North).

*Dover, East Dock, Dover, Kent.

*Dover, West Dock, Dover, Kent.

A23, Handcross, West Sussex.

Stockers Hill, Boughton, Kent.

A27, Beddingham, East Sussex.

A26/A22, Millpond, Maresfield, East Sussex.

*A27, Withy Patch, Lancing, West Sussex.

Sheerness Docks, Sheerness, Kent.

Scottish

A75, Castle Kennedy.

*A90, Cramond.

*A92, Findon, Nr Aberdeen.

A74, Beattock Summit.

M9, Craigforth, Stirling.

*Tayside Truckstop, Smeaton Road, Dundee.

For a fully detailed list on currently available weighbridges consult the *Commercial Motor* list or visit www.chrishodgetrucks.co.uk and click on weighbridges.

Appendix VIII

POLICE FORCES IN GREAT BRITAIN (FOR NOTIFICATION OF ABNORMAL LOAD MOVEMENTS)

In most cases, Police Forces require fax notifications of abnormal load movements to be clearly identified as such in the heading or prefix to the message. Communications should be addressed to the Chief Constable, unless a specific department is indicated.

England

Constabulary/Force	*Address/Tel/Fax Nos for Notification*
Avon and Somerset	Abnormal Loads Dept, PO Box 37, Valley Road, Portishead, Bristol BS20 8QJ. Tel 01275 816885 Fax 01275 816884
Bedfordshire Police	Abnormal Loads, 4th Information Room, Woburn Road, Kempston, Bedford MK43 9AX. Tel 01234 842249 Fax 01234 842051
Cambridgeshire	Hinchingbrooke Park, Huntingdon PE18 8NP. Tel 01480 456111 Fax 01480 414448/412848
Cheshire	Castle Esplanade, Chester CH1 2PP. Tel 01244 612215 Fax 01244 612268
City of London Police	Abnormal Loads Dept, 37 Wood Street, London EC2P 2NQ. Tel 020 7601 2190 Fax 020 7601 2760 *(For City of London movements only – not Metropolitan area)*
Cleveland	Abnormal Loads, Road Traffic Division, Cannon Park, Middlesbrough, Cleveland TS1 5RY. Tel 01642 301578 Fax 01642 301579
Cumbria	Carleton Hall, Penrith, Cumbria CA10 2AU. Tel 01768 891991 Fax 01768 217455
Derbyshire	Operations Room, Butterley Hall, Ripley, Derby DE5 3RS. Tel 01773 570100 Fax 01773 572225

Devon and Cornwall	Abnormal Loads Office, Middlemoor, Exeter EX2 7HQ. Tel 01392 452268 Fax 01392 452426
Dorset Police	Winfrith, Dorchester, Dorset DT2 8DZ. Tel 01305 223731 Fax 01202 223678
Durham	Operational Support, Aykley Heads, Durham DH1 5TT. Tel 0191 386 4929 Fax 0191 375 2160
Essex Police	PO Box 2, Springfield, Chelmsford CM2 6DA. Tel 01245 491491 Fax 01245 452424
Gloucestershire	Force Control Room, Holland House, Lansdown Road, Cheltenham GL51 6QH. Tel 01242 276037 Fax 01242 276194
Greater Manchester Police	Traffic Management Section, PO Box 47, Chester House, Boyer Street, Manchester M16 0SD. Tel 0161 856 1678/9 Fax 0161 856 1676
Hampshire	Romsey Road, Winchester, Hants SO22 5DB. Tel 0845 045 4545 Fax 01983 538650 For Isle of Wight movements: Tel 01983 528000
Hertfordshire	Stanborough Road, Welwyn Garden City, Herts AL8 6XF. Tel 01707 354571 Fax 01707 354572
Humberside Police	Hessle High Road, Hull, Humberside HU4 7BA. Tel 01482 597683 Fax 01482 597645
Kent County	Police HQ Ops Centre, Sutton Road, Maidstone, Kent ME15 9BZ. Tel 01622 690690 Fax 01622 654169
Lancashire	PO Box 77, Hutton, Nr Preston PR4 5SB. Tel 01772 209786 Fax 01772 410788
Leicestershire	Abnormal Loads Dept, St Johns, Narborough, Leicester LE9 5BX. Tel 0116 222 2222 Fax 0116 248 2327
Lincolnshire Police	PO Box 999, Lincoln LN5 7PH. Tel 01522 558125 Fax 01522 558098
Merseyside Police	Traffic Dept, Smithdown Lane, Liverpool L7 3PR. Tel 0151 777 5759 Fax 0151 777 5760
Metropolitan Police	Room 154, New Scotland Yard, Broadway, London SW1H 0BG. Tel 020 7230 1212 Fax 020 7230 2133 Covers Greater London but not City of London.
Norfolk	Martineau Lane, Norwich NR1 2DJ. Tel 01603 276022 Fax 01603 276087
Northamptonshire Police	Abnormal Loads Office, Mereway, Northampton NN4 8BH. Tel 01604 700700 Fax 01604 703028
Northumbria Police	Area Operations Room, Ponteland, Newcastle upon Tyne. Tel 01661 869754 Fax 01661 869788
North Yorkshire Police	Force Communications Centre, Racecourse Lane, Northallerton DL7 8RB. Tel 01609 783131 Fax 01609 789213
Nottinghamshire Police	Force Control Room, Sherwood Lodge, Arnold, Nottingham NG5 8PP. Tel 0115 967 0111 Fax 0115 967 2145
South Yorkshire Police	Abnormal Loads Officer, Snig Hill, Sheffield S3 8LY. Tel 0114 252 3251 Fax 0114 252 3250
Staffordshire Police	Abnormal Loads Section, Weston Road, Stafford ST18 0YY. Tel 01785 232674 Fax 01785 232673
Suffolk	Martlesham Heath, Ipswich IP5 3QS. Tel 01473 613500 Fax 01473 613737

Surrey	Mount Browne, Sandy Lane, Guildford, Surrey GU31HG. Tel 01483 571212 Fax 01483 300279 NB: Certain parts of the county of Surrey fall within the jurisdiction of the Metropolitan Police.
Sussex Police	Malling House, Lewes, East Sussex BN7 2DZ. Tel 01273 520379 Fax 01273 520369
Thames Valley Police	Headquarters, Kidlington, Oxford OX5 2NX Tel 01865 846422/3 Fax 01865 846726 *Includes movements in Berkshire, Buckinghamshire and Oxfordshire*
Warwickshire	Greys Mallory Traffic Base, Europa Way, Warwick CV34 6RN Tel 01926 415260 Fax 01926 415388
West Mercia	Abnormal Loads Officer, Hindlip Hall, Hindlip, Worcester WR3 8SP. Tel 01905 723000 Fax 01905 756161
West Midlands Police	Central Motorway Police Group, Thornbridge Avenue, Birmingham B42 2AG Tel 0121 322 6018 Fax 0121 322 6039
West Yorkshire Police	Operations Division PO Box 9, Wakefield WF1 3QP Tel 01924 375222 Fax 01924 292182
Wiltshire	London Road, Devizes SN10 2DN. Tel 01380 722341 Fax 01380 734135

Wales

Constabulary/Force	*Address/Tel/Fax Nos for Notification*
Dyfed-Powys Police	PO Box 99, Llangunnor, Carmarthen, Carmarthenshire SA31 2PF. Tel 01267 222020 Fax 01267 226128
Gwent	Headquarters, Croesyceiliog, Cwmbran, Gwent NP44 2XJ Tel 01633 838111 Fax 01633 642312
North Wales Police	Central Control Room, Glan-y-Don, Colwyn Bay LL29 8AW. Tel 01492 511968 Fax 01492 511055
South Wales	Communications Dept., Bridgend, Mid Glamorgan CF31 3SU. Tel 01656 869379 Fax 01656 869397

Scotland

Constabulary/Force	*Address/Tel/Fax Nos for Notification*
Central Scotland Police	Randolphfield, Stirling FK8 2HD. Tel 01786 456506 Fax 01786 462681
Dumfries and Galloway	Cornwall Mount, Dumfries DG1 1PZ. Tel 01387 252112 Fax 01387 250763
Fife	Detroit Road, Glenrothes KY6 2RJ. Tel 01592 418888 Fax 01592 418444
Grampian Police	Traffic Dept, Nelson Street, Aberdeen AB2 3EQ. Tel 01224 386744/386015 Fax 01224 625735
Lothian and Borders Police	Fettes Avenue, Edinburgh EH4 1RB. Tel 0131 311 3418 Fax 0131 311 3038
Northern	Old Perth Road, Inverness IV2 3SY. Tel 01463 715555 Fax 01463 230800 For Highland Region, Orkney, Shetland and Western Isles.
Strathclyde Police	Traffic Operations, Meiklewood Road, Glasgow G51 4EU. Tel 0141 532 6400 Fax 0141 532 6430
Tayside Police	PO Box 59, West Bell Street, Dundee DD1 9JU. Tel 01382 596433 Fax 01382 596439

Appendix IX

LOCAL AND OTHER AUTHORITIES (FOR NOTIFICATION OF ABNORMAL LOAD MOVEMENTS)

Local Authorities (responsible for highways and bridges, or their agents for abnormal load notifications)

Further information may be obtained by accessing the Local Government website: www.lga.gov.uk

ENGLAND

Counties

County	*Address*
Bedfordshire	Engineering Policy and Planning, Bedfordshire County Council, County Hall, Bedford MK42 9AP Tel: 01234 363222
Buckinghamshire	Head of Transportation, Buckinghamshire County Council, County Hall, Walton Street, Aylesbury, Bucks HP20 I UY Tel: 01296 382420 Fax: 01296 383990
Cambridgeshire	Assistant Director (Highways and Engineering) Department of Environment and Transport, Cambridgeshire County Council, Castle Court, Shire Hall, Castle Hill, Cambridge CB3 OAP Tel: 01223 717715

Cheshire	Director of Environment Services, Cheshire County Council, County Hall, Chester CHI ISF Tel: 01244 603770 Fax: 01244 603608
Cornwall	Director of Planning, Transportation and Estates, Cornwall County Council, County Hall, Truro TRI 3AY Tel: 01872 322000
Cumbria	Director of Community Economy and Environment, Cumbria County Council, Citadel Chambers, Citadel Road, Carlisle CA3 8SG Tel: 01228 606060 Fax: 01228 606755
Derbyshire	Director of Environmental Services, Traffic Section, Derbyshire County Council, County Offices, Matlock DE4 3AG Tel: 01629 580000 extn 7591 Fax: 01629 585255
Devon	County Environment Directorate, Devon County Council, Lucombe House, County Hall, Topsharn Road, Exeter EX2 4QW Tel: 0 1392 383249 Fax: 01392 382135
Dorset	Head of Highways and Transportation, Dorset County Council, Colliton Park, County Hall, Dorchester, Dorset DTI IXJ Tel: 01305 251000
Durham	Director of Environment and Technical Services, Durham County Council, County Hall, Durham DHI 5UQ Tel: 0191 386 4411
East Sussex	Director of Transport and Environment, East Sussex County Council, County Hall, St Anne's Crescent, Lewes, East Sussex BN7 I UE Tel: 01273 481000
Essex	Head of Transportation and Operational Services, Essex County Council, County Hall, Chelmsford CM1 1QH Tel: 01245 437065
Gloucestershire	Executive Director: Environment, Gloucestershire County Council, Shire Hall, Gloucester GLI 2TH Tel: 01452 425523 Fax: 01452 541305
Hampshire	County Surveyor, Hampshire County Council, The Castle, Winchester S023 SUD Tel: 01962 841841 Fax: 01962 847055
Hertfordshire	Director of Environment, Hertfordshire County Council, County Hall, Hertford SG13 8DE Tel: 01992 555201 Fax: 01992 588900
Kent	Director of Strategic Planning, Kent County Council, Sessions House, County Hall, Maidstone, Kent ME14 IXQ Tel: 01622 221612
Lancashire	Environment Director, Lancashire County Council, Guild House, Cross Street, Preston PRI 8RD Tel: 01772 264468 Fax: 01772 262154
Leicestershire	Director of Planning and Transportation, Leicestershire County Council, County Hall, Glenfield, Leicester LE3 8RJ Tel: 0116 265 7000 Fax: 0116 265 7962
Lincolnshire	Director of Highways and Planning, Lincolnshire County Council, City Hall (4th Floor), Lincoln LN1 1DN Tel: 01522 553053 Fax: 01522 553149

Norfolk	Director of Planning and Transportation, Norfolk County Council, County Hall, Martineau Lane, Norwich NR I 2SG Tel: 01603 222143 Fax: 01603 223219
Northamptonshire	Highways Service, Northamptonshire County Council, PO Box 221, John Dryden House, 8-10 The Lakes, Northampton NN4 7DE Tel: 01604 237155 Fax: 01604 236393
Northumberland	Director of Environment, Northumberland County Council, County Hall, Morpeth, Northumberland NE61 2EF Tel: 01670 533000 Fax: 01670 533086
North Yorkshire	Director of Environmental Services, North Yorkshire County Council, County Hall, Northallerton, North Yorkshire DL7 8AH Tel: 01609 780780
Nottinghamshire	Director of Environment, Nottinghamshire County Council, Trent Bridge House, Fox Road, West Bridgford, Nottingham NG2 6BJ Tel: 0115 977 4500 Fax: 0115 945 4531
Oxfordshire	Director of Environmental Services, Oxfordshire County Council, Speedwell House, Speedwell Street, Oxford OX1 1NE Tel: 01865 815700
Shropshire	Corporate Director, Community and Environment Directorate, Shropshire County Council, The Shirehall, Abbey Foregate, Shrewsbury SY2 6ND Tel: 01743 253021 Fax: 01743 252505
Somerset	Corporate Director Environment and Property, Somerset County Council, County Hall, Taunton, Somerset TA1 4DY Tel: 01823 355600 Fax: 01823 356107
Staffordshire	Development Services Department, Staffordshire County Council, Riverway, Stafford ST16 3TJ Tel: 01785 276633
Suffolk	Director of Environment and Transport, Suffolk County Council, St Edmund House, County Hall, Ipswich 1P4 1LZ Tel: 01473 583305 Fax: 01473 230078
Surrey	Head of Engineering, Surrey County Council, County Hall, Kingston-upon-Thames KTI 2DY Tel: 020 8541 9375 Fax: 020 8541 9447
Warwickshire	Director of Planning, Transport and Economic Strategy, Warwickshire County Council, PO Box 43, Shire Hall, Warwick CV34 4SX Tel: 01926 410410 Fax: 01926 412903
West Sussex	Transport Planning Service, Environment and Development Group, West Sussex County Council, The Grange, Tower Street, Chichester, West Sussex P019 1RH Tel: 01243 777100
Wiltshire	Director of Environmental Services, Wiltshire County Council, County Hall, Trowbridge, Wilts BA14 8JD Tel: 01225 713000

Greater London and Metropolitan County Areas

Greater London

Corporation of London	Director of Technical Services, Corporation of London, PO Box 270, Guildhall, London EC2P 2EJ Tel: 020 7606 3030
Barking and Dagenham	Controller of Development and Technical Services, London Borough of Barking and Dagenham, Town Hall, Barking, Essex IG11 7LU Tel: 020 8592 4500
Barnet	Chief Highways Officer, London Borough of Barnet, Barnet House, 1255 High Road, Whetstone, London N20 0EJ Tel: 020 8359 4481 Fax: 020 8359 4116
Bexley	Chief Engineer Directorate of Environmental Services, Bexley Civic Offices, Broadway, Bexleyheath, Kent DA6 7LB Tel: 020 8303 7777 extn 2929
Brent	Director of Transportation, London Borough of Brent, Brent House, High Road, Wembley HA9 6BZ Tel: 020 8937 5128 Fax: 020 8937 5125
Bromley	Director of Environmental Services, London Borough of Bromley, Bromley Civic Centre, Stockwell Close, Bromley BR1 3UH Tel: 020 8313 4443
Camden	Director of Environment, London Borough of Camden, Argyle Street, London WC1H 8EQ Tel: 020 7974 2605 Fax: 020 7974 5556
Croydon	Planning and Transportation, London Borough of Croydon, Taberner House, Park Lane, Croydon CR9 3JS Tel: 020 8686 4433
Ealing	Executive Director: Environment, London Borough of Ealing, Perceval House, 14–16 Uxbridge Road, Ealing W5 2HL Tel: 020 8825 6859 Fax: 020 8579 5453
Enfield	Director of Environmental Services, London Borough of Enfield, PO Box 52, Civic Centre, Silver Street, Enfield EN1 3XD Tel: 020 8379 3503
Greenwich	Strategic Planning, London Borough of Greenwich, Peggy Middleton House, 50 Woolwich New Road, Woolwich SE18 6HQ Tel: 020 8854 8888 extn 5488
Hackney	Borough Services, London Borough of Hackney, Maurice Bishop House, Reading Lane, Hackney, London E8 1HH Tel: 020 8356 3000 Fax: 020 8356 8142
Hammersmith and Fulham	Environment Department, London Borough of Hammersmith and Fulham, Town Hall, King Street, London W6 9JU Tel: 020 8748 3020 extn 3004
Haringey	Director of Environmental Services, Haringey Council, Tottenham Town Hall, Town Hall Approach Road, London N15 4RY Tel: 020 8489 4537 Fax: 020 8489 4596

Harrow	Head of Environment and Transportation, London Borough of Harrow, PO Box 37, Civic Centre, Harrow, Middlesex HA1 2UY Tel: 020 8863 5611 Fax: 020 8424 1541
Havering	Transportation Planning Manager, Building and Technical Services, London Borough of Havering, The Whitworth Centre, Noak Hill Road, Harold Hill, Romford RM3 7YA Tel: 01708 432804 Fax: 01708 433719
Hillingdon	Head of Transportation, London Borough of Hillingdon, Civic Centre, High Street, Uxbridge, London UB8 IUW Tel: 01895 250111
Hounslow	Head of Traffic and Parking, Environmental Services Department, Borough of Hounslow, Civic Centre, Lampton Road, Hounslow, Middlesex TW3 4DN Tel: 020 8583 4961 Fax: 020 8583 5350
Islington	Head of Planning Transport, London Borough of Islington, PO Box 3333, 222 Upper Street, London N1 1YA Tel: 020 7527 2000
Kensington and Chelsea	Mr Michael Stroud, Executive Director of Environmental Services Royal Borough of Kensington and Chelsea, Council Offices, Pembroke Road, London W8 6PW Tel: 020 7341 5101 Fax: 020 7341 5155
Kingston upon Thames	Director of Environmental Services, Royal Borough of Kingston upon Thames, Guildhall, Kingston upon Thames KT1 1EU Tel: 020 8546 2121
Lambeth	Head of Transportation and Highways, London Borough of Lambeth, 3rd Floor, Blue Star House, 234–244 Stockwell Road, London SW9 9SP Tel: 020 7926 1236 Fax: 020 7926 2486
Lewisham	Directorate of Regeneration, London Borough of Lewisham, Laurence House, Catford Road, Catford SE6 4RU Tel: 020 8314 7167
Merton	Director of Environmental Services, London Borough of Merton, Civic Centre, London Road, Morden, Surrey SM4 5DX Tel: 020 8543 2222 Fax: 020 8545 4105
Newham	Director of Environment, London Borough of Newham, 25 Nelson Street, East Ham, London E6 6EH Tel: 020 8430 2000 extn 22301 Fax: 020 8430 3051
Redbridge	Highways Services, London Borough of Redbridge, Lynton House, 255/259 High Road, Ilford, Essex IG1 1NY Tel: 020 8478 3020. Fax: 020 8708 3970
Richmond upon Thames	Transport Planning Manager, London Borough of Richmond upon Thames, 44 York Street, Twickenham TW1 3BZ Tel: 020 8891 7310
Southwark	Director of Regeneration, London Borough of Southwark, Chiltern House, Portland Street, London SE17 2ES Tel: 020 7525 5000 Fax: 020 7525 5484
Sutton	Head of Transportation, London Borough of Sutton, 24 Denmark Road, Carshalton SM5 2JG Tel: 020 8770 5000

Tower Hamlets	Highways Department, London Borough of Tower Hamlets, 60 Southern Grove, London E3 4PN Tel: 020 7364 5000
Waltham Forest	Head of Street Services, Low Hall Depot, Argyll Avenue, Leyton E10 7AS Tel: 020 8496 2510 Fax: 020 8496 2512
Wandsworth	Director of Technical Services, Wandsworth Borough Council, Town Hall, Wandsworth High Street, London SW18 2PU Tel: 020 8871 6547 Fax: 020 8871 0218
Westminster	Director of Planning and Transportation, Westminster City Council, PO Box 240, Westminster City Hall, 64 Victoria Street, London SW1 E 6QP Tel: 020 7641 6000

Metropolitan Borough Councils

Greater Manchester Area

Bolton	Director of Environment, Bolton Metropolitan Borough Council, The Wellsprings, Civic Centre, Bolton BL1 1US Tel: 01204 336470
Bury	Borough Engineer, Development and Environmental Services Department, Bury Metropolitan Borough Council, Lester House, Broad Street, Bury BL9 OAW Tel: 0161 253 5804 Fax: 0161 253 5329
Manchester	Engineering Services Division, Chief Executives Department, Manchester City Council, Town Hall, Manchester M60 2JT Tel: 0161 455 2258 Fax: 0161 455 2151
Oldham	Executive Director, Environment and Transportation, Oldham Metropolitan Borough Council, Henshaw House, Cheapside, Oldham OL1 1NY Tel: 0161 911 4300 Fax: 0161 911 3411
Rochdale	Director of Technical Services, Rochdale Metropolitan Borough Council, PO Box 20, Electric House, Smith Street, Rochdale OL16 1YP Tel: 01706 647474
City of Salford	Director of Development Services, City of Salford, Civic Centre, Chorley Road, Swinton, Salford M27 5BW Tel: 0161 793 3800 Fax: 0161 793 3848
Stockport	Director of Environment and Economic Development, Stockport Metropolitan Borough Council, Stopford House, Piccadilly, Stockport SK1 3XE Tel: 0161 474 4934
Tameside	Head of Engineering Services, Tameside Metropolitan Borough Council, Council Offices, Wellington Road, Ashton-under-Lyne, Tameside OL6 6BL Tel: 0161 342 8355 Fax: 0161 342 3911
Trafford	Executive Director of Economic and Community Regeneration, Trafford Metropolitan Borough Council, Trafford Town Hall, Talbot Road, Stretford, Trafford, Greater Manchester M32 0YX Tel: 0161 912 4453 Fax: 0161 912 4407

Wigan	Director of Engineering Services, Wigan Metropolitan Borough Council, Civic Buildings, New Market Street, Wigan WN1 1RP Tel: 01942 244991 Fax: 01942 404210

Merseyside Area

Knowsley	Head of Transportation and Engineering Services, Department of Planning and Development, Knowsley Metropolitan Borough, PO Box 26, Yorkon Building, Archway Road, Huyton, Knowsley, Merseyside L36 9FB Tel: 0151 443 2235
Liverpool	Executive Director of Regeneration, Liverpool City Council, 2nd Floor, Kingsway House, Hatton Garden, Liverpool L69 2RA Tel: 0151 233 3000 Fax: 0151 225 2924
St Helens	Transport Department, St Helens Metropolitan Borough Council, Hardshaw Brook Depot, Warrington New Road, St Helens, Merseyside WA9 1JR Tel: 01744 456000
Sefton	Technical Services Director, Sefton Metropolitan Borough Council, Balliol House, Balliol Road, Bootle, Liverpool L20 3NJ Tel: 0151 922 4040
Wirral	Borough Engineer, Metropolitan Borough of Wirral, Town Hall, Civic Way, Bebington CH63 7PT Tel: 0151 643 9000

South Yorkshire Area

Barnsley	Transportation Manager, Planning and Transportation, Barnsley Metropolitan Borough Council, Central Offices, Kendry Street, Barnsley S70 2TN Tel: 01226 772629 Fax: 01226 772199
Doncaster	Infrastructure Services, Directorate of Development and Transport Engineering, Scarborough House, Chequer Road, Doncaster DNI 213B Tel 01302 735113 Fax: 01302 735028
Rotherham	Transportation Unit Manager, Regeneration and Planning Service, Bailey House, Rawmarsh Road, Rotherharn S60 1QT Tel: 01709 822953 Fax: 01709 822139
Sheffield	Head of Service, Planning, Transport and Highways, Sheffield City Council, Howden House, 1 Union Street, Sheffield S1 2SH Tel: 0114 273 5499 Fax: 0114 273 5002

Tyne and Wear Area

Gateshead	Head of Transport Strategy, Gateshead Council, Civic Centre, Regent Street, Gateshead NE8 1HH Tel: 0191 433 3000 Fax: 0191 477 8422
Newcastle upon Tyne	Planning and Transportation, Newcastle City Council, Civic Centre, Newcastle upon Tyne NE1 8PD Tel: 0191 232 8520 Fax: 0191 211 4984
North Tyneside	Environmental Services, North Tyneside Council, PO Box 113, Station Road, Killingworth, Newcastle upon Tyne NE12 6WJ Tel: 0191 219 2198 Fax: 0191 219 2534

South Tyneside	Director of Development Services, South Tyneside Metropolitan Borough Council, Town Hall and Civic Offices, Westoe Road, South Shields NE33 2RL Tel: 0 191427 1717 Fax: 0191 427 717 1
Sunderland	Director of Environment, City of Sunderland, Civic Centre, Sunderland SR2 7DN Tel: 0191 553 1000 Fax: 0191 553 1460

West Midlands Area

Birmingham	Director of Transportation, Birmingham City Council, 1 Lancaster Circus, Queensway, Birmingham B4 7BQ Tel: 0121 303 7400 Fax: 0121 303 1352
Coventry	Director of City Development, Coventry City Council, Tower Block, Much Park Street, Coventry CV1 2PY Tel: 024 7683 2037 Fax: 024 7683 1324
Dudley	Environment, Engineering and Transportation Department, Council House, Mary Stevens Park, Stourbridge, West Midlands DY8 2AA Tel: 01384 814005
Sandwell	Highways Direct, Sandwell Metropolitan Borough Council, PO Box 42, Lombard Street, West Bromwich, West Midlands B70 8RU Tel: 0121 569 4141 Fax: 0121 569 4072
Solihull	Director of Environment Services, PO Box 19, Council House, Solihull B91 3QT Tel: 0121 704 6000 Fax: 0121 704 6404
Walsall	Engineering and Transportation Services, Walsall Metropolitan Borough Council, Civic Centre, Walsall WS1 1DG Tel: 01922 652558 Fax: 01922 649036
Wolverhampton	Head of Transportation, Wolverhampton City Council, Salop Street, Wolverhampton WV3 0SQ Tel: 01902 555700 Fax: 01902 555656

West Yorkshire Area

Bradford	Head of Transportation and Planning, City of Bradford Council, 3rd Floor, Jacobs Well, Bradford BDI 5RW Tel: 01274 754605 Fax: 01274 754583
Calderdale	Calderdale Metropolitan Borough Council, Northgate House, Northgate, Halifax HX1 1UN Tel: 01422 357257 Fax: 01422 392195
Kirklees	Head of Highways Services, Flint Street, Fartown, Huddersfield HD1 6LG Freephone: 0800 731 8765
Leeds	Director of Highways and Transportation, Selectapost 6, 2 Rossington Streets, Leeds LS2 8HB Tel: 0113 247 7500
Wakefield	Head of Engineering, Engineering Division Development Department, PO Box 92, Newton Bar, Wakefield, West Yorkshire WFl lXS Tel: 01924 306009 Fax: 01924 306052

Unitary Authorities

Bath and North East Somerset	Head of Transportation, Access and Waste Management, Bath and North East Somerset Council, Riverside, Temple Street, Keynsham, Bristol BS31 1LA Tel: 01225 394128 Fax: 01225 394335
Blackburn with Darwen	Director of Technical Services, Blackburn with Darwen Borough Council, Town Hall, Blackburn, Lancs BB1 7DY Tel: 01254 841240
Blackpool	Head of Technical Services, Business Services Dept, Blackpool Borough Council, PO Box 117, Westgate House, Squires Gate Lane, Blackpool FY4 2TS Tel: 01253 476111 Fax: 01253 476112
Bournemouth	Director of Development Services, Bournemouth Borough Council, Town Hall Annexe, St Stephens Road, Bournemouth BH2 6EA Tel: 01202 451451 Fax: 01202 451006
Bracknell Forest	Planning and Transportation Department, Bracknell Forest Borough Council, Time Square, Market Street, Bracknell, Berkshire RG12 1JD Tel: 01344 424642 Fax: 01344 3 51260
Brighton and Hove	Chief Executive, Brighton and Hove City Council, King's House, Grand Avenue, Hove BN3 2LS Tel: 01273 290000 Fax: 01273 291003
Bristol City	Director of Environment, Transport and Leisure, Bristol City Council, Colston House, Colston Street, Bristol BS1 5AQ Tel: 0117 922 2000
Darlington	Chief Executive, Darlington Borough Council, Town Hall, Darlington DL1 5QT Tel: 01325 380651 Fax: 01325 382032
Derby City	Transport Policy Officer, Development and Cultural Services, Derby City Council, Roman House, Friar Gate, Derby DE1 1XB Tel: 01332 715034 Fax: 01332 715040
East Riding of Yorkshire	Director of Operational Services, East Riding Council of Yorkshire, County Hall, Beverley HU17 9BA Tel: 01482 887700 Fax: 01482 884944
Halton	Executive Director, Environment and Development, Halton Borough Council, Grosvenor House, Halton Lea, Runcorn WA7 2GW Tel: 0151 424 2061 Fax: 0151 471 7304
Hartlepool	Head of Highways and Transportation, Hartlepool Council, Bryan Hanson House, Hanson Square, Hartlepool TS24 7BT Tel: 01429 523241 Fax: 01429 523541
Herefordshire	County Offices, Herefordshire Council, Engineering Services, PO Box 234, Hereford HR1 2HQ Tel: 01432 260779 Fax: 01432 261983
Isle of Wight	Directorate of Environment Services, Isle of Wight Council, County Hall, Newport, Isle of Wight PO30 1UD Tel: 01983 821000
Kingston upon Hull City	Director of Technical Services, Kingston upon Hull City Council, Kingston House, Bond Street, Kingston upon Hull HU1 3ER Tel: 01482 300300

Leicester City	Director of Environment and Development, Leicester City Council, New Walk Centre, Welford Place, Leicester LE1 6ZG Tel: 0116 254 9922 Fax: 0116 254 5531
Luton	Luton Borough Council, Department of Environment, Regeneration, Engineering and Transportation Division, Town Hall, Luton LU1 2BQ Tel: 01582 546000
Medway	Planning and Development, Front Line Task Force Division, Compass Centre, Chatham Maritime, Chatham, Kent ME4 4YH Tel: 01634 331145 Fax: 01634 331613
Middlesbrough	Head of Transport and Design Services, Middlesbrough Council, PO Box 99A, Town Hall, Middlesbrough TS1 2QQ Tel: 01642 264002
Milton Keynes	Strategic Director Environment, Milton Keynes Council, PO Box 108, Civic Offices, 1 Saxon Gate East, Milton Keynes MK9 3ZL Tel: 01908 691691
North East Lincolnshire	Head of Engineering, North East Lincolnshire Council, Civic Offices, Knoll Street, Cleethorpes DN35 8LN Tel: 01472 313131 Fax: 01472 324302
North Lincolnshire	Assistant Director Highways and Transportation, North Lincolnshire Council, PO Box 42, Church Square House, Scunthorpe DN15 6XQ Tel: 01724 297440
North Somerset	Highway Services, North Somerset Council, Kings Court, 87/89 High Street, Nailsea, Somerset BS48 lAW Tel: 01934 882001 Fax: 01934 882002
Nottingham City	Director of Development and Environmental Services, Nottingham City Council, Lawrence House, Talbot Street, Nottingham NG1 5NT Tel: 0115 915 6225 Fax: 0115 915 6150
Peterborough City	Head of Transport and Engineering Services, Peterborough City Council, Bridge House, Town Bridge, Peterborough PE1 1XG Tel: 01733 453500 Fax: 01733 453444
Plymouth City	Head of Transport, Plymouth City Council, Civic Centre, Plymouth PL1 2EW Tel: 01752 66800
Poole	Chief Executive, Borough of Poole, The Civic Centre, Poole, Dorset BH15 2RU Tel: 01202 633633 Fax: 01202 633706
Portsmouth City	Chief Executive, Portsmouth City Council, Civic Offices, Guildhall Square, Portsmouth POI 2AL Tel: 0239 282 2251 Fax: 0239 282 8441
Reading	Transport Strategy Manager, Reading Borough Council, Civic Centre, Reading RG1 7TD Tel: 0118 939 0813 Fax: 0118 939 0633
Redcar and Cleveland	Director of Technical Services, Redcar and Cleveland Council, Cargo Fleet Offices, Middlesbrough Road, PO Box South Bank 20, Cleveland TS6 6EL Tel: 01642 444000 Fax: 01642 444600
Rutland	Head of Highways and Transportation, Rutland County Council, Council Offices, Catmose, Oakham LE15 6HP Tel: 01572 722577 Fax: 01572 771616

Slough	Head of Engineering, Slough Borough Council, Wellington House, PO Box 570, Slough SL1 1FA Tel: 01753 552288 Fax: 01753 875296
South Gloucestershire	Director of Planning, Transportation and Strategic Environment, South Gloucestershire Council, The Council Offices Castle Street, Thornbury BS35 lHF Tel: 01454 863400 Fax: 01454 863642
Southampton City	Chief Executive, Southampton City Council, Civic Centre, Southampton S014 7LY Tel: 023 8022 3855 Fax: 023 8083 3232
Southend-on-Sea	Technical Services Department, Southend-on-Sea Borough Council, Civic Centre, Southend-on-Sea SS2 6ER Tel: 01702 215731
Stockton-on-Tees	Director of Environment and Technical Services, Stockton-on-Tees Borough Council, Municipal Buildings, Church Road, Stockton-on-Tees TS18 1LD Tel: 01642 391230 Fax: 01642 391316
Stoke-on-Trent City	City Services, Stoke-on-Trent City Council, Cromer Road, Northwood, Stoke-on-Trent ST1 6XB Tel: 01782 234567 Fax: 01782 236257
Swindon	Director of Environmental Services, Swindon Borough Council, Premier House, Station Road, Swindon SH1 1TZ Tel: 01793 466360 Fax: 01793 466165
Telford and Wrekin	Corporate Director, Environment and Economy, Telford and Wrekin Council, Dowby House, Telford TF3 4LB Tel: 01952 202367 Fax: 01952 200764
Thurrock	Head of Strategic Planning, Thurrock Borough Council, Civic Offices, New Road, Grays RM17 6SL Tel: 01375 652207 Fax: 01375 652780
Torbay	Director of Environment Services, Torbay Council, Roebuck House, Abbey Road, Torquay TQ2 5TF Tel: 01803 207701 Fax: 01803 208856
Warrington	Director of Environment and Regeneration, Warrington Borough Council, New Town House, Buttermarket Street, Warrington WA1 2NH Tel: 01925 442700 Fax: 01925 442704
West Berkshire	Highways and Engineering, West Berkshire District Council, Council Offices, Faraday Road, Newbury RG14 2AF Tel: 01635 42400 Fax: 01635 519865
Windsor and Maidenhead	Director of Planning and Environment, Royal Borough of Windsor and Maidenhead, Town Hall, St Ives Road, Maidenhead, Berkshire SL6 1RF Tel: 01628 798888 Fax: 01628 796408
Wokingham	Chief Executive, Wokingham District Council, Council Offices, Shute End, Wokingham RG40 1WQ Tel: 0118 974 6000 Fax: 0118 978 9078
Worcestershire	Director of Environmental Services, PO Box 373, County Hall, Spetchley Road, Worcester WR5 2XG Tel: 01905 763763 Fax: 01905 766899
York	Director of Environment and Development Services, City of York Council, 9 St Leonards Place, York YO1 7ET Tel: 01904 613161 Fax: 01904 551340

WALES

Counties

Cardiff	Head of Highways and Parks Services, Cardiff County Council, County Hall, Atlantic Wharf, Cardiff CF1 5UW Tel: 029 2087 2000 Fax: 029 2087 3653
Carmarthenshire	Director of Environment and Economic Development, Carmarthenshire County Council, Llanstephan Road, Johnstown, Carmarthen SA31 3LZ Tel: 01267 234567
Ceredigion	Director of Highways, Property and Works, Ceredigion County Council, County Hall, Market Street, Aberaeron SA46 0HT Tel: 01545 570881 Fax: 01545 571089
Swansea	Director of Technical Services, City and County of Swansea, County Hall, Swansea SA1 3SN Tel: 01792 636149 Fax: 01792 652712
Denbighshire	Head of Highways and Transportation, Denbighshire County Council, Caledfryn, Smithfield Road, Denbigh LL16 3RJ Tel: 01824 706801 Fax: 01824 706970
Flintshire	Director of Highways, Transportation and Engineering, Flintshire County Council, County Hall, Mold CH7 6NB Tel: 01352 704500 Fax: 01352 703222
Gwynedd	The Surveyor, Highways and Engineering, Gwynedd Council, Council Offices, Caernarfon, Gwynedd LL55 1SH Tel: 01286 679444 Fax: 01286 675961
Isle of Anglesey	Corporate Director of Highways, Transportation and Property, Isle of Anglesey County Council, Council Offices, Swyddfa'r Sir, Llangefni, Anglesey LL77 7TW Tel: 01248 752301 Fax: 01248 724839
Monmouthshire	Head of Highways & Waste Management, Monmouthshire County Council, County Hall, Cwmbran NP44 2XH Tel: 01633 644707 Fax: 01633 644701
Pembrokeshire	Director of Transport, Pembrokeshire County Council, County Hall, Haverfordwest SA61 1TP Tel: 01437 775828 Fax: 01437 775008
Powys	Director of Highways, Transportation and Property, Powys County Council, County Hall, Llandrindod Wells LD1 5LG Tel: 01597 826630 Fax: 01597 826269

County Boroughs

Blaenau Gwent	Director of Community Services, Central Depot, Barleyfield Industrial Estate, Brynmawr NP23 4YF Tel: 01495 355600 Fax: 01495 312357
Bridgend	Director of Environmental and Planning Services, Bridgend County Borough Council, Transportation and Engineering Department, Morien House, Bennett Street, Bridgend Industrial Estate, Bridgend CF31 3SH Tel: 01656 642501 Fax: 01656 642581

Caerphilly	Chief Engineer, Caerphilly County Borough Council, Council Offices, Pontllanfraith, Blackwood NP12 2YW Tel: 01495 235319 Fax: 01495 235012
Conwy	Chief Highways Officer, Conwy County Borough Council, The Heath, Penmaenmawr Road, Llanfairfechan LL33 0PF Tel: 01492 574001 Fax: 01248 681881
Merthyr Tydfil	Head of Engineering, Merthyr Tydfil County Borough Council, Ty Kier Hardie, Riverside Court, Avenue de Clichy, Merthyr Tydfil CF47 8XF Tel: 01685 725000 Fax: 01685 722146
Neath Port Talbot	Director of Technical Services, Neath Port Talbot County Borough Council, Civic Centre, Penllergaer SA4 1GH Tel: 01792 512703 Fax: 01792 512666
Newport	Director of Development and Transportation, Newport County Borough Council, Civic Centre, Newport, South Wales NP20 4UR Tel: 01633 232707 Fax: 01633 244721
Rhondda Cynon Taff	Divisional Director of Transportation, Rhondda Cynon Taff County Borough Council, Sardis House, Sardis Road, Pontypridd CF37 1DU Tel: 01443 494700 Fax: 01443 494888
Torfaen	Director for the Environment, Torfaen County Borough Council, 4th Floor, County Hall, Cwmbran NP44 2WN Tel: 01633 648601
Vale of Glamorgan	Director of Environmental and Economic Regeneration, Vale of Glamorgan Council, Dock Office, Barry Docks, Barry CF63 4RT Tel: 029 206 73000
Wrexham	Chief Transport and Engineering Officer, Wrexham County Borough Council, Crown Buildings, PO Box 1293, Wrexham LL11 1WQ Tel: 01978 292000 Fax: 01978 297103

SCOTLAND

Unitary Councils

Aberdeen	Director of Environment and Property, Aberdeen City Council, St Nicholas House, Aberdeen AB10 1EZ Tel: 01224 523037 Fax: 01224 626301
Aberdeenshire	Director of Transportation and Infrastructure, Aberdeenshire Council, Woodhill House, Westburn Road, Aberdeen AB16 5GB Tel: 01224 665520 Fax: 01224 662005
Angus	Director of Planning and Transport, Angus Council, St James House, St James Road, Forfar DD8 2ZP Tel: 01307 461460 Fax: 01307 461895
Argyll and Bute	Director of Transportation and Property, Argyll and Bute Council, Manse Brae, Lochgilphead PA31 8RD Tel: 01546 602233 Fax: 01546 606618

City of Edinburgh	Director of City Development, City of Edinburgh Council, 1 Cockburn Street, Edinburgh EH1 1ZJ Tel: 0131 529 3595 Fax: 0131 529 3498
City of Glasgow	Director of Land Services, Glasgow City Council, Richmond Exchange, 20 Cadogan Street, Glasgow G2 7AD Tel: 0141 287 9000 Fax: 0141 287 9013
Clackmannanshire	Roads and Transport Manager, Development and Environmental Services, Clackmannanshire Council, Lime Tree House, Alloa FK10 1EX Tel: 01259 452532 Fax: 01259 452530
Dumfries and Galloway	Head of Strategy and Transportation, Dumfries and Galloway Council, Council Offices, English Street, Dumfries DG1 2HR Tel: 01387 260115 Fax: 01387 260111
Dundee City	Director of Planning and Transportation, Dundee City Council, Tayside House, 28 Crichton Street, Dundee DD1 3R13 Tel: 01382 434000 Fax: 01382 433013
East Ayrshire	Head of Roads and Transport, East Ayrshire Council, Greenholm Street, Kilmarnock KA1 4DJ Tel: 01563 576310 Fax: 01563 576312
East Dumbartonshire	Strategic Director, Development and Environment, East Dumbartonshire Council, Tom Johnston House, Civic Way, Kirkintilloch, Glasgow G66 4TJ Tel: 0141 578 8000 Fax: 0141 578 8333
East Lothian	Head of Transportation, East Lothian Council, John Muir House, Haddington, East Lothian EH41 3HA Tel: 01620 827658 Fax: 01620 827923
East Renfrewshire	Head of Roads and Transportation, East Renfrewshire Council, Refuge House, 17/19 Gaule Street, Paisley PA1 1ES Tel: 0141 577 3417
Falkirk	Head of Planning and Transportation, Falkirk Council, Abbotsford House, David's Loan, Falkirk FK2 7YZ Tel: 01324 504820 Fax: 01324 504914
Fife	Head of Transportation, Fife Council, Fife House, North Street, Glenrothes KY7 5LT Tel: 01592 414141 Fax: 01592 416100
Highland	Director of Roads and Transport, The Highland Council, Glenurquhart Road, Inverness IV3 5NX Tel: 01463 702601 Fax: 01463 702606
Inverclyde	Head of Transportation and Roads Service, Inverclyde Council, 71 East Hamilton Street, Greenock PA15 2UA Tel: 01475 714800 Fax: 01475 714825
Midlothian	Head of Transportation, Midlothian Council, Fairfield House, 8 Lothian Road, Dalkeith EH22 3ZN Tel: 0131 270 7500 Fax: 0131 271 3537
Moray	Head of Direct Services, Moray Council, Roads Service, Academy Street, Elgin IV30 1LL Tel: 01343 562500 Fax: 01343 545628
North Ayrshire	Head of Roads, North Ayrshire Council, Perceton House, Irvine KA11 2AL Tel: 01294 225200 Fax: 01294 225244
North Lanarkshire	Director of Planning and Environment, North Lanarkshire Council, Fleming House, Cumbernauld G67 1JW Tel: 01236 616201 Fax: 01236 616206

Perth and Kinross	Director of Roads, Transport and Architectural Services, Perth and Kinross Council, Puller House, 35 Kinnoull Street, Perth PH1 5GD Tel: 01738 475650 Fax: 01738 476510
Renfrewshire	Head of Roads, Department of Planning and Transport, Renfrewshire Council, South Buildings, Cotton Street, Paisley PA1 1LL Tel: 0141 842 5741 Fax: 0141 842 5040
Scottish Borders	Director of Transport and Environmental Standards, Scottish Borders Council, Council Headquarters, Newtown St. Boswells, Melrose TD6 0SA Tel: 01835 825075 Fax: 01835 825071
South Ayrshire	Head of Strategy and Design, South Ayrshire Council, Macadam House, 34 Charlotte Street, Ayr KA7 1EA Tel: 01292 612301 Fax: 01292 612342
South Lanarkshire	Head of Roads and Transportation, South Lanarkshire Council, Montrose House, 154 Montrose Crescent, Hamilton ML3 6LL Tel: 01698 454484
Stirling	Director of Technical Services, Stirling Council, Viewforth, Stirling FK8 2ET Tel: 01786 442692 Fax: 01786 443053
West Dumbartonshire	Head of Roads, Transport and Environmental Services, West Dumbartonshire Council, Council Offices, Garshake Road, Dumbarton G82 3PU Tel: 01389 737598 Fax: 01389 737637
West Lothian	Strategic Services, West Lothian Council, West Lothian House, Almondvale Boulevard, Livingston EH54 6QG Tel: 01506 777000 Fax: 01506 777102

Island Councils

Orkney Islands	Director of Technical Services, Orkney Islands Council, Council Offices, Kirkwall, Orkney KW15 1NY Tel: 01856 873535 Fax: 01856 876094
Shetland Islands	Executive Director Infrastructure Services, Shetland Islands Council, Grantfield, Lerwick, Shetland ZE1 0NT Tel: 01595 744800 Fax: 01595 695887
Western Isles	Director of Technical Services, Comliairle nan Eilean Siar (Western Isles Council), Council Offices, Sandwick Road, Stornoway, Isle of Lewis HS1 2BW Tel: 01851 703773 Fax: 01851 709482

NORTHERN IRELAND

Headquarters	Clarence Court, 10–18 Adelaide Street, Belfast BT2 8GB Tel: 028 9054 0540
Belfast	Roads Service, Eastern Division, Hydebank, 4 Hospital Road, Belfast BT8 8X Tel: 028 9025 3000
Coleraine	Roads Service, Northern Division, County Hall, Castlerock Road, Coleraine, Co. Londonderry BT51 3HS Tel: 028 7034 1300

Craigavon	Roads Service, Southern Division, Marlborough House, Central Way, Craigavon, Co. Armagh BT64 IAD Tel: 028 3834 1144
Omagh	Roads Service, Western Division, County Hall, Drumragh Avenue, Omagh, Co. Tyrone BT79 7AF Tel: 028 8225 4111

Tunnel and Bridge Authorities

Blackwall and Rotherhithe	Tunnels Manager, Greenwich Borough Council, Blackwall Tunnel Office, Naval Row, London E14 9PS. Tel 020 7987 4208/3601 Fax 020 7987 6571
Dartford Tunnel	Dartford River Crossing Ltd., Tunnel Operating Division, South Orbital Way, Dartford, Kent DA1 5PR. Tel 01322 221603 Fax 01322 294224
Mersey Tunnels	General Manager, The Mersey Tunnels,Georges Dock Building, Georges Dock Way, Pier Head, Liverpool, Merseyside L3 1DD. Tel 0151 236 8602 extn 200 Fax 0151 255 0610
Tyne Tunnel	Tyne and Wear PTA, Howdon, Wallsend, Tyne and Wear NE28 OPD. Tel 0191 262 4451 Fax 0191 262 1031
Erskine Bridge	The Bridge Manager, Erskine Bridge, Bishopston, Renfrewshire, Strathclyde PA7 5QD Tel 0141 812 7022 Fax 0141 812 6625
Forth Road Bridge	The General Manager, Forth Road Bridge Joint Board, South Queensferry, West Lothian EH30 9SF. Tel 0131 319 1699 Fax 0131 319 1903
Humber Bridge	The Bridgemaster, Humber Bridge Board, PO Box 6, Ferriby Road, Hessle, Humberside HU13 0JG. Tel 01482 647161 Fax 01482 640838
Severn Bridges	Severn River Crossing, Aust Services, Pilning, Bristol BS12 3BE. Tel 01454 633522 (Toll office 01454 632436)
Skye Road Bridge	Skye Bridge Limited, Kyle of Lochalsh. Tel 01599 534880
Tamar Road Bridge	Tamar Bridge Authority, Plymouth, Devon. Tel 01752 361577
Tay Bridge	The Bridge Manager, Tay Road Bridge Joint Board, Marine Parade, Dundee, Tayside DD1 3JB Tel 01382 21881/2 Fax 01382 201529

Electricity Authorities/Companies

(for notification of high loads – ie above 5.8 metres)

Area	*Board/Company Address*
Eastern	Engineering Director, Eastern Electricity, Wherstead Park, PO Box 40, Wherstead, Ipswich, Suffolk IP9 2AQ. Tel 01449 772378 Fax 01449 772396
East Midlands	Power Systems Manager, East Midlands Electricity, PO Box 4, North PDO, 398 Coppice Road, Arnold, Nottingham NG5 7HX. Tel 0115 926 9711 Fax 0115 935 8585

London	Transport Manager, London Electricity, 54 Bengeworth Road, London SE5 9AJ. Tel 020 7733 5611 Fax 020 7326 6278
Merseyside and North Wales	Chief Engineer, Merseyside and North Wales Electricity Board Plc, Head Office, Sealand Road, Chester CH1 4LR. Tel 01244 652234 Fax 01244 652353
Midlands	Chief Engineer, Midlands Electricity, PO Box 8 Mucklow Hill, Halesowen, West Midlands B62 8BP. Tel 0121 423 2345 Fax 0121 422 7977
North-Western	Chief Engineer, Norweb Plc, Talbot Road, Manchester M16 0HQ. Tel 0161 875 7111 Fax 0161 875 7456
Northern	Director of Engineering, Northern Electric Plc, Carliol House, Market Street, Newcastle upon Tyne NE1 6NE. Tel 0191 221 2000 Fax 0191 235 2211
Scotland	Distribution Operations Manager, Scottish Hydro-Electric Plc, South Inch Business Centre, Shore Road, Perth PH2 8BW. Tel 01738 455050 Fax 01738 455055
South-Eastern	Network Operations, Seeboard Plc, Wealdon House, Lewes Road, East Grinstead RH19 3TB. Tel 01342 413252 Fax 01342 315620
South of Scotland	General Manager (Operations), Scottish Power Plc, Catheart House, Spean Street, Glasgow G44 4BE. Tel 0141 637 7177 Fax 0141 568 2016
South Wales	Chief Engineer, South Wales Electricity Plc, St Mellons, Cardiff CF3 9XW. Tel 029 2079 2111 Fax 029 2077 7759
South-Western	Operations Manager, Engineering, South-Western Electricity Plc, Sowton Industrial Estate, Osprey Road, Exeter EX2 7HZ. Tel 01392 444111 Fax 01392 444827
Southern	Operations Administration, Southern Electricity Plc, Westacott Way, Littlewick Green, Maidenhead, Berkshire SL6 3QB. Tel 01628 822166 Fax 01628 584403
Yorkshire	Systems Operations Manager, Yorkshire Electricity, 161 Gelderd Road, Leeds LS1 1QZ. Tel 0113 241 5000 Fax 0113 241 5057

Rail and Water Authorities

(for notification of matters concerning rail crossings, bridges, eg abnormal load movements and damage to bridges caused by goods vehicles)

British Waterways Board	Technical Services Department, Wellington Park House, Thirsk Row, Leeds LS1 4DD. Tel 0113 245 0711 Fax 0113 246 0493
Railtrack	Eastern Region – Room B004, Hudson House, York YO1 1HP. Tel 01904 522351 Fax 01904 522802

	Western Region – 1 Gloucester Street, Swindon, Wilts SN1 1GW. Tel 01793 515898 Fax 01793 515723
	Network SouthEast (South) – Wellesley Grove, Croydon, Surrey CR9 1DY. Tel 020 8666 6789 Fax 020 8666 6519
	(North) – Grosvenor House, Room 305, 112 Prince of Wales Road, Norwich NR1 1NZ. Tel 01603 285567 Fax 01603 762791
	Regional Railways – 1st Floor West Wing, Stanier House, 10 Holliday Street, Birmingham B1 1TE. Tel 0121 654 4251 Fax 0121 654 4528
	ScotRail – ScotRail House, 58 Port Dundas Road, Glasgow G4 0LQ. Tel 0141 335 3102 Fax 0141 335 2862
London Underground	4th Floor, 30 The South Colonnade, Canary Wharf, London E14 5EU. Tel 020 7308 2571 Fax 020 7308 2700

NB: See note on p 487 about notifications to British Rail Property Board.

Appendix X

TRAINING FACILITIES FOR DANGEROUS GOODS DRIVERS

The following establishments, approved by the City and Guilds of London Institute on behalf of the Department for Transport, provide training in the handling of various classes of dangerous goods for tanker vehicle drivers and for those drivers whose vehicles carry goods in packages – see Chapter 7 and 23 for full details of the relevant requirements.

NB: The list indicates the courses held (ie T = tanker, P = packaged goods) and the classes of dangerous goods covered as at 24 July 2003.

ENGLAND

Avon

Bristol

Chemfreight Training Ltd, The Avon Lodge, Third Way
Avonmouth BS11 9YP
Tel 01928 580505
(Courses T + P covering all classes).

Training Force, Lane Group plc, Regional Freight Centre,
Portbury, Bristol BS20 9XX
Tel 01275 374401
(Courses T + P covering all classes).

Berkshire

Slough

Fullers Transport Training Services, 724 Dundee Road,
Slough SL1 4JL
Tel 01753 430829
(Courses T + P covering classes 2, 3, 4, 5, 6, 8 and 9).

Buckinghamshire

Bletchley	Shanks Waste Solutions Ltd, Dunedin House, Aukland Park, Mount Farm, Bletchley MK1 1BU Tel 01908 650720 *(Contact for details of courses).*

Cambridgeshire

Peterborough	Apex Training Services, Unit 3, Oxney Road Industrial Estate, Peterborough PE1 5YW Tel 01733 567478 *(Courses T + P covering classes 1, 2, 3, 4, 5, 6, 8 and 9).*

Cheshire

Great Sutton	Prismatic Training Ltd, 15 Wakefield Road, Great Sutton, CH66 2HW Tel 01513 393669 *(Courses T + P covering classes 2, 3, 4, 5, 6, 8 and 9).*
Runcorn	Chemfreight Training Ltd, Cormorant Drive, Picow Farm Road, Runcorn WA7 4UD Tel 01928 580505 *(Courses T + P covering all classes).*
Warrington	North Cheshire Training Ltd, 11/4 Pallatine Industrial Estate, Causeway Avenue, Warrington WA4 6QQ Tel 01952 658342 *(Courses T + P covering classes 2, 3, 4, 5, 6, 8 and 9).*
Widnes	Chemtrain Ltd, Stanley House, Ditton Road, Widnes WA8 0NE Tel 0151 257 9820 *(Courses T + P covering classes 1, 2, 3, 4, 5, 6, 8 and 9).*
	PST Management Services, Mathieson Road, West Bank Dock Estate, Widnes WA8 0NX Tel 01514 953114 Fax 01514 953394 *(Courses T + P covering classes 2, 3, 4, 5, 6, 8 and 9).*

Cumbria

Carlisle	System Driver Training, 62–66 Lowther Street, Carlisle CA3 8DP Tel 01228 574010 *(Courses T + P covering all classes).*

Devon

Exeter	Transport Mastertrain, Exeter Livestock Centre, Matford Park Road, Exeter EX2 8FD Tel 01392 426242 *(Courses T + P covering classes 2, 3, 4, 5, 6, 8 and 9).*

	RMT Consultancy, The Beeches, Dunchideock, Exeter EX2 9TZ Tel 01392 832888 *(Contact for details of courses).*
Modbury	Training for Transport, 3 Modbury Court, Church Street, Modbury PL21 0GR Tel 01548 831383 Fax 01548 831384 *(Courses T + P covering all classes).*
Dorset	
Bournemouth	P&O GBE Logistics, Elliot Road, Bournemouth BH11 8JR Tel 01202 278104 Fax 01202 278111 *(Courses T + P covering classes 2, 3, 4, 5, 6, 8 and 9).*
Poole	Wessex Transport Training Ltd, 59 Old Wareham Road, Parkestone, Poole BH12 4QN Tel 01202 717881 *(Courses T + P covering all classes).*
Essex	
Battlesbridge	Stevand Training Services, Unit 4, Mayphill Industrial Estate, Hawk Hill, Battlesbridge SS11 7RJ Tel 01268 571777 *(Courses T + P covering classes 1, 2, 3, 4, 5, 6, 8 and 9).*
Braintree	TDT Management Services Ltd, 7 Broomhills Industrial Estate, Podsbrook Road, Braintree CM7 2RG Tel 01245 425725 *(Courses T + P covering classes 1, 2, 3, 4, 5, 6, 8 and 9).*
Stanford-le-Hope	Sigmas Studios ADR Training, 18 Avondale Gardens, Stanford-le-Hope SS17 8DB Tel 01375 673254 *(Courses T + P covering classes 1, 2, 3, 4, 5, 6, 8 and 9).*
West Thurrock	Wright Training Services, c/o Truckworld Thurrock, Oliver Road, West Thurrock, Grays RM20 3ED Tel 01708 867564 *(Courses T + P covering classes 2, 3, 4, 5, 6, 8 and 9).*
	RTA Training Services, Unit 9, Lyndale Court, London Road, West Thurrock, Grays RM20 3BJ Tel 01708 861339 *(Courses T + P covering classes 2, 3, 4, 5, 6, 8 and 9).*
Gloucestershire	
Whitchurch	Proforce (Southwest) Ltd, Unit 8 Axis, Hawkfield Business Park, Whitchurch BS14 0BY Tel 01179 399988 Fax 01179 399994 *(Courses T + P covering classes 2, 3, 4, 5, 6, 8 and 9).*

Hampshire

Southampton — Southampton Training Services Ltd, Nutley Lane, Colmore Industrial Estate, Totton, Southampton SO40 3ND
Tel 023 8086 0608
(Courses T + P covering classes 1, 2, 3, 4, 5, 6, 8 and 9).

Hertfordshire

St Albans — West Herts Training Ltd, VER House, Frogmore, St Albans AL2 2WH
Tel 01727 873878 Fax 01727 872322
(Courses T + P covering classes 2, 3, 4, 5, 6, 8 and 9).

Kent

Ashford — Occupational Safety Logistics Ltd, 10 Little Hempen, Ashford TN23 4YS
Tel 01233 643017
(Courses T + P covering classes 2, 3, 4, 5, 6, 8 and 9).

Dartford — Transed Europe Ltd, 10 Masterhead, Capstan Court, Crossways Business Park, Dartford DA2 6QG
Tel 01322 287954
(Courses T + P covering classes 2, 3, 4, 5, 6, 8 and 9).

Hythe — Hythe Transport Training, Unit G6, Lympne Industrial Park, Hythe CT21 4LR
Tel 01732 780830
(Courses T + P covering classes 2, 3, 4, 5, 6, 8 and 9).

Sevenoaks — Hythe Transport Training, Wirral, Huntsman Lane, Wrotham Heath, Sevenoaks TN15 7SS
Tel 01732 780830 Fax 01732 780611
(Courses T + P covering classes 2, 3, 4, 5, 6, 8 and 9).

Sittingbourne — Mainstream Training, St Georges House, St Georges Business Park, Castle Road, Sittingbourne ME10 3TD
Tel 01795 432562 Fax 01795 432732
(Courses T + P covering classes 2, 3, 4, 5, 6, 8 and 9).

Tunbridge Wells — Freight Transport Association, Hermes House, St John's Road, Tunbridge Wells TU4 9UZ
Tel 01892 786000
(Courses T + P covering all classes).

Lancashire

Accrington — Specialist Training & Consultancy Services Ltd, Suite 12, Technology Management Centre, The Globe Centre, Accrington BB5 0GE
Tel 01254 230444 Fax 01254 235444
(Courses T + P covering classes 1, 2, 3, 4, 5, 6, 8 and 9).

Burnley	Nightfreight Training Services, PO Box 6, Sandringham Court, Network 65 Business Park, Burnley BB11 5GG Tel 01282 420358 *(Courses T + P covering all classes).*
	Cameon Ltd, 1 Mill Hill Lane, Hapton, Burnley BB11 9QU Tel 01744 886000 *(Courses T + P covering all classes).*
Lancaster	Lancaster Training Services Ltd, Training Centre, St Georges Quay, Lancaster LA1 5QJ Tel 01524 87285 *(Courses T + P covering all classes).*
Oldham	Hazchem Training Services, 19 Helmshore Avenue, Moorside, Oldham OL4 2QB Tel 0161 628 8402 Fax 0161 628 8505 *(Courses T + P covering classes 2, 3, 4, 5, 6, 8 and 9).*
Leicestershire	
Birstall	North Lindsay Training Services, 19 Fairfield Avenue, Birstall, Leicester LE4 4DS Tel 01472 354264 *(Courses T + P covering classes 1, 2, 3, 4, 5, 6, 8 and 9).*
Earl Shilton	Driver Training and Management Services Ltd, West Street, Earl Shilton, Leicester LE9 7EJ Tel 01455 848578 *(Courses T + P covering classes 1, 2, 3, 4, 5, 6, 8 and 9).*
Hinckley	Automotive Transport Training, Jacknell Road, Dodwells Bridge Industrial Estate, Hinckley LE10 8BS Tel 01455 251516 *(Courses T + P covering classes 1, 2, 3, 4, 5, 6, 8 and 9).*
Leicester	Securicor Omega Express, Training Centre, 23 Ealing Road, Freemans Common, Leicester LE2 7SZ Tel 0116 255 9923 *(Courses P only covering classes 2, 3, 4, 5, 6, 8 and 9).*
	J Coates (HGV Services) Ltd, 46–50 Great Central Street, Leicester LE1 4NF *(Courses T + P covering classes 2, 3, 4, 5, 6, 8 and 9).*
Lincolnshire	
Brigg	North Lindsey Training Services, Bishop Burton College, North Lincolnshire Centre, Brigg Road, Broughton, Brigg DN20 0JW Tel 01472 354264 *(Courses T + P covering classes 1, 2, 3, 4, 5, 6, 8 and 9).*

	Crownship Developments, Charlton House Mews, Bridge Street, Brigg DN20 8NQ Tel 01652 658151 *(Courses T + P covering classes 2, 3, 4, 5, 6, 8 and 9).*
Gainsborough	JMW Driver Training Centre, Water Tower Complex, Caenby Corner Estate, Hemswell Cliff, Gainsborough Tel 01427 668424 *(Courses T + P covering classes 2, 3, 4, 5, 6, 8 and 9).*
Grimsby	North Lindsey Training, 20 Great Coates Road, Grimsby DN34 4NE Tel 01472 354264 *(Courses T + P covering classes 1, 2, 3, 4, 5, 6, 8 and 9).*
Lincoln	Focus Transport Training, Odder Farmhouse, Saxilby Road, Burton, Lincoln LN1 2BB Tel 01522 704615 Fax 01522 704800 *(Courses T + P covering classes 2, 3, 4, 5, 6, 8 and 9).*
Scunthorpe	Project ASC Group Ltd, 3 Park Square, Scunthorpe DN15 6JH Tel 01724 278331 Fax 01724 850860 *(Courses T + P covering classes 2, 3, 4, 5, 6, 8 and 9).*
Stallingborough	Transafe (Immingham) Ltd, Osborne Road, Kiln Lane, Stallingborough DN37 8DG Tel 01469 574433 *(Courses T + P covering classes 1, 2, 3, 4, 5, 6, 8 and 9).*
Greater London	
Wembley	Kamillionex Communications Ltd, Heron House, 109–115 Wembley Hill Road, London HA9 8DA Tel 0208 900 9038 *(Courses T + P covering classes 1, 2, 3, 4, 5, 6, 8 and 9).*
Greater Manchester	
Manchester	Manchester Training, Greengate, Middleton, Manchester M24 1RU Tel 0161 653 5767 *(Courses T + P covering classes 1, 2, 3, 4, 5, 6, 8 and 9).*
	Trafford Training Ltd, Bailey Road, Ashburton, Manchester M17 1SA Tel 0161 876 0006 *(Courses T + P covering classes 1, 2, 3, 4, 5, 6, 8 and 9).*
Radcliffe	Red Rose Training, Europa Way, Stoneclough, Radcliffe, Manchester M26 1GG Tel 01204 862999 *(Courses T + P covering classes 2, 3, 4, 5, 6, 8 and 9).*

Norfolk

Norwich	Norfolk Training Services, Harford Centre, Hall Road, Norwich NR4 6DG Tel 01603 259900 *(Courses T + P covering classes 2, 3, 4, 5, 6, 8 and 9).*

Northamptonshire

Daventry	TGB Training, 2 Brunel Close, Drayton Fields Industrial Estate, Daventry NN11 5RB Tel 01327 79547 *(Courses T + P covering classes 1, 2, 3, 4, 5, 6, 8 and 9).*
Northampton	Christian Salvesen Industrial Division, Caswell Road, Brickmills Industrial Estate, Northampton NN4 7PW Tel 01604 662800 Fax 01604 662844 *(Courses T + P covering classes 2, 3, 4, 5, 6, 8 and 9).*

Nottinghamshire

Kimberley	Trent Transport Training, The Watson Centre, Artic Way, Kimberley, Nottingham NG16 2HS Tel 0115 938 4982 *(Courses P only covering classes 1, 2, 3, 4, 5, 6, 8 and 9).*
Kirkby-in-Ashfield	Joseph Merritt & Son plc, Byron Avenue, Lowmoor Industrial Estate, Kirkby-in-Ashfield NG17 7LA Tel 01623 759737 *(Courses P only covering classes 1, 2, 3, 4, 5, 6, 8 and 9).*
Mansfield	Risk & Safety Management Services Ltd, Pleaseley Vale Business Park, The Old Mill, Outgang Lane, Mansfield NG19 9DJ Tel 01623 810000 Fax 01623 811999 *(Courses T + P covering classes 1, 2, 3, 4, 5, 6, 8 and 9).*
Worksop	City Centre Training Ltd, Claylands Avenue, Dukeries Industrial Estate, Worksop S81 7DJ Tel 01623 744199 *(Courses T + P covering classes 2, 3, 4, 5, 6, 8 and 9).*
	North Notts Create Ltd, North Nottinghamshire College, Carlton Road, Worksop S81 7HP Tel 01909 473561 Fax 01909 512540 *(Courses T + P covering classes 2, 3, 4, 5, 6, 8 and 9).*

Oxfordshire

Banbury	JOB Training and Consultants Ltd, 48 West Bar, Banbury OX6 9RZ Tel 01295 255811 *(Courses T + P covering classes 2, 3, 4, 5, 6, 8 and 9).*

Enstone	EP Training Services Ltd, Rollright School of Transport, Enstone Airfield Complex, Enstone OX7 4NP Tel 01608 677618 *(Courses T + P covering classes 2, 3, 4, 5, 6, 8 and 9).*
Somerset	
Shepton Mallet	Skilltrain Associates, 17 Town Street, Shepton Mallet BA4 5BE Tel 01749 345681 *(Courses P only covering classes 2, 3, 4, 5, 6, 8 and 9).*
Taunton	Friendberry ltd, Stogumber, Taunton TA4 3TP Tel 01984 656310 *(Courses T + P covering all classes).*
South Yorkshire	
Doncaster	Doncaster, Rotherham & District MTG Training Association, Rands Lane Industrial Estate, Rands Lane, Armthorpe, Doncaster DN3 3DY Tel 01302 832831 *(Courses T + P covering classes 2, 3, 4, 5, 6, 8 and 9).*
Staffordshire	
Leek	Transed Associates, Springbuck House, Leekbrook Industrial Estate, Cheadle Road, Leek ST13 7AP Tel 01538 381313 *(Courses T + P covering classes 2, 3, 4, 5, 6, 8 and 9).*
Stafford	Stan Robinson (Stafford) Ltd, Ladfords Fields, Seighford, Nr Stafford ST18 9QE Tel 01785 282501 *(Courses P only covering classes 3, 4, 5, 6, 8 and 9).*
Tamworth	John Terry Training & Business Services, 4 Avill, Hockley, Tamworth B77 5QE Tel 01827 260626 *(Contact for details of courses).*
Suffolk	
Felixstowe	P&O Transport Training Ltd, Sub Station Road, Felixstowe IP11 8SE Tel 01394 674139 *(Courses T + P covering classes 1, 2, 3, 4, 5, 6, 8 and 9).*
Ipswich	RTT Training Services Ltd, Gippeswyk Hall, Gippeswyk Avenue, Ipswich IP2 9AF Tel 01473 602424 *(Courses T + P covering classes 1, 2, 3, 4, 5, 6, 8 and 9).*
Stanton	Driver Transport Training Ltd, Unit 2, Shepherds Grove Industrial Estate, Stanton IP31 2BG Tel 01359 253535 Fax 01359 251933 *(Courses T + P covering classes 1, 2, 3, 4, 5, 6, 8 and 9).*

Surrey

Egham	Training Link Europe Ltd, Unit 2 Beta Way, Thorpe Industrial Estate, Egham TW20 8RE Tel 01784 479700 Fax 01784 477321 *(Courses T + P covering classes 1, 2, 3, 4, 5, 6, 8 and 9).*
Great Bookham	EP Training Services Ltd, The Old Library, Lower Shott, Great Bookham, Leatherhead KT23 4LR Tel 01372 490800 *(Courses T + P covering classes 1, 2, 3, 4, 5, 6, 8 and 9).*

Sussex

Arundel	The Page Group, Robert House, Rudfords Industrial Estate, Ford, Arundel BN18 0BD Tel 01903 736300 *(Courses T + P covering classes 2, 3, 4, 5, 6, 8 and 9).*

Teesside

Middlesbrough	Teesside Training Enterprise, PO Box 36, Wilton Training Centre, Wilton Site, Middlesbrough TS6 8YX Tel 01642 433295 *(Courses T + P covering classes 2, 3, 4, 5, 6, 8 and 9).*
Thornaby	NETA Training Group, The Mandela Centre, Wilson Street, Bon Lea Estate, Thornaby-on-Tees TS17 7AR Tel 01642 616936 *(Contact for details of courses).*

Tyne & Wear

Gateshead	Van Hee Transport Ltd, William Street, Felling, Gateshead NE10 0SP Tel 0191 438 2512 *(Courses T + P covering classes 2, 3, 4, 5, 6, 8 and 9).*
	SITA, Merlin House, Princes Park, Princesway, Team Valley, Gateshead NE11 0NF Tel 0191 419 6700 Fax 0191 419 6702 *(Courses T + P covering classes 2, 3, 4, 5, 6, 8 and 9).*
Jarrow	TB Training Services, Unit 134, TEDCO Business Centre, Viking Industrial Park, Jarrow NE23 3DT Tel 0191 428 3422 *(Contact for details of courses).*
Newcastle upon Tyne	Tyneside Training Services Ltd, Airport Industrial Estate, Kingston Park, Kenton, Newcastle upon Tyne NE3 2EF Tel 0191 286 2919 *(Courses T + P covering all classes).*

North Shields	Minister Technical & Training Services Ltd, Unit C20, Tromso Close, Tyne Tunnel Trading Estate, North Shields NE29 7XH Tel 0191 249 6616 *(Courses T + P covering classes 2, 3, 4, 5, 6, 8 and 9).*
Warwickshire	
Atherstone	TNT Chambers Ltd, 102 Long Street, Atherstone CV9 1BS Tel 01826 715371 Fax 01826 714925 *(Courses P only covering classes 2, 3, 4, 5, 6, 8 and 9).*
West Midlands	
Aldridge	Westgate Training, Aldridge Manor House, Little Aston Road, Aldridge, Walsall WS9 8NJ Tel 01922 455514 *(Courses T + P covering classes 2, 3, 4, 5, 6, 8 and 9).*
Kingswinford	Kingswinford Industrial Training Services, Logistics & Warehousing Services Units A + B, Ham Lane, Kingswinford DY6 7JS Tel 0184 344700 Fax 01384 344701 *(Courses T + P covering classes 1, 2, 3, 4, 5, 6, 8 and 9).*
Oldbury	Targeted Training Services, 26–28 Birmingham Street, Oldbury, B69 4DS Tel 0121 543 5234 Fax 0121 543 5222 *(Courses T + P covering classes 1, 2, 3, 4, 5, 6, 8 and 9).*
	Benchmark Logistics Training, Union Road, Oldbury B69 3EX Tel 0121 552 0825 Fax 0121 522 0601 *(Courses T + P covering classes 2, 3, 4, 5, 6, 8 and 9).*
Stourbridge	Premier Training (UK) Ltd, Unit 3, Lye Business Centre, Enterprise Drive, Hayes Lane, Lye, Stourbridge DY9 8QH Tel 01384 898798 *(Courses T + P covering classes 2, 3, 4, 5, 6, 8 and 9).*
Wednesbury	MDM Training, 48 Church Hill, Wednesbury WS10 9DG Tel 0121 556 6147 *(Courses T + P covering classes 1, 2, 3, 4, 5, 6, 8 and 9).*
Wolverhampton	SDH Training, Unit 6a, Malborough Industrial Estate, Cockshutts Lane, Wolverhampton WV2 3NP Tel 01902 455677 *(Courses T + P covering classes 2, 3, 4, 5, 6, 8 and 9).*
Wiltshire	
Trowbridge	RLR Services, Rode House, 73 Fore Street, Trowbridge BA14 8HQ Tel 01225 754347 Fax 01225 774139 *(Courses T + P covering all classes).*

Wirral

Birkenhead	Wirral Bottled Gas, 117 New Chester Road, Birkenhead L41 9BW Tel 0151 666 2177 Fax 0151 647 3951 *(Courses T + P covering classes 2, 3, 4, 5, 6, 8 and 9).*

Yorkshire

Batley	TDG Nexus, Nab Lane, Birstall, Batley WF17 9NH Tel 01924 478235 *(Courses T + P covering classes 2, 3, 4, 5, 6, 8 and 9).*
Beverley	Welwyn LGV Driver Training, Plaxton Bridge, Woodmansey, Beverley HU17 0RT Tel 01482 867125 *(Courses T + P covering classes 2, 3, 4, 5, 6, 8 and 9).*
	Les Ritchings Training, 9 East Close, Molescroft, Beverley HU17 7JN Tel 01482 872521 Fax 01482 872521 *(Courses T + P covering classes 1, 2, 3, 4, 5, 6, 8 and 9).*
Bradford	Driver Hire Bradford Training, Puma House, Preston Street, Bradford, West Yorks BD7 1JE Tel 01274 778322 Fax 01274 778433 *(Courses T + P covering classes 1, 2, 3, 4, 5, 6, 8 and 9).*
Brough	Amoudi Ltd, 93 Welton Road, Brough, West Yorks HU15 1BJ Tel 01525 385565 Fax 01525 385564 *(Courses T + P covering classes 1, 2, 3, 4, 5, 6, 8 and 9).*
Goole	Norman Lewis Tankers Ltd, Hensall Garage, Weeland Road, Hensall, Nr Goole, DN14 0QE Tel 01977 661023 Fax 01977 662995 *(Courses T + P covering classes 2, 3, 4, 5, 6, 8 and 9).*
Hull	Training for Industry (Humberside) Ltd, Willow House, Clay Street, Chamberlain Road, Hull HU8 8HA Tel 01482 223520 *(Courses T + P covering classes 2, 3, 4, 5, 6, 8 and 9).*
Leeds	Hargreaves Training Services, Unit 3, Parkside Industrial Estate, Glover Way, Leeds LS11 5JP Tel 0113 270 1188 *(Courses T + P covering classes 2, 3, 4, 5, 6, 8 and 9).*
	Yorkshire Environmental, PO Box 63, Leeds LS9 0XH Tel 0113 2972491 Fax 0113 2972496 *(Courses T + P covering classes 2, 3, 4, 5, 6, 8 and 9).*

Oxenhope	Dangerous Goods Training & Advisory Services, The Croft, Hillhouse Lane, Oxenhope, West Yorks BD22 9JH Tel 01535 643546 *(Courses T + P covering classes 2, 3, 4, 5, 6, 8 and 9).*
Richmond	Roberts Training & Consultancy Services Unit 5, Racecourse Court, Gallowfields Trading Estate, Richmond, North Yorks DL10 4SP *(Courses T + P covering classes 1, 2, 3, 4, 5, 6, 8 and 9).*
Wakefield	Trident Training, 80 Lime Pit Lane, Stanley, Wakefield WF3 4DF Tel 01924 824200 *(Courses T + P covering classes 1, 2, 3, 4, 5, 6, 8 and 9).*
	Utilities Training (Northern) Ltd, Thornes Moor Road, Wakefield WF2 8PT Tel 019924 377379 *(Courses T + P covering classes 2, 3, 4, 5, 6, 8 and 9).*
Wiltshire	
Devizes	Wiltshire Transport Training Ltd, The College, Hopton Industrial Estate, London Road, Devizes SN10 2EX Tel 01380 723712 *(Courses T + P covering classes 2, 3, 4, 5, 6, 8 and 9).*

SCOTLAND

Aberdeen	J Gilbert Transport Training, Cloverhill Road, Bridge of Don Industrial Estate, Aberdeen AB2 8EE Tel 1224 825644 *(Courses T + P covering all classes).*
	Shore Porters Society, 1 Baltic Place, Aberdeen AB11 5EW Tel 01224 369500 Fax 01224 569515 *(Courses T + P covering classes 1, 2, 3, 4, 5, 6, 8 and 9).*
Aberlour	McPherson Ltd, Fisherton Garage, Aberlour Banffshire AB38 9LB Tel 01340 871401 Fax 01340 871721 *(Courses T + P covering classes 2, 3, 4, 5, 6, 8 and 9).*
Cove	Babcock Naval Services, Building 41, Training Centre, RNAD Caulport, PO Box 1 Cove, Helensburgh G84 0PD Tel 01436 674321 *(Courses T + P covering all classes).*

Dyce	Dangerous Goods Management, Unit 61, South Wellshead Centre, Wellshead Crescent, Dyce, Aberdeen AB21 7GA Tel 01224 773776 *(Courses T + P covering all classes).*
Coatbridge	LAGTA Group Training, 7 Palacecraig Street, Shawhead, Coatbridge ML5 4RY Tel 01236 425171 *(Courses T + P covering classes 1, 2, 3, 4, 5, 6, 8 and 9).*
Glasgow	Glasgow Training Group Ltd, 120 Crownhill Road, Bishopbriggs, Glasgow G64 1RP Tel 01417 621461 *(Courses T + P covering classes 1, 2, 3, 4, 5, 6, 8 and 9).*
	Ritchies HGV Training Centre, Hobden Street, Glasgow G21 4AQ Tel 01415 572212 *(Courses P only covering classes 2, 3, 4, 5, 6, 8 and 9).*
Heathhall	Nithcree Training Services, Unit 264, Heathhall Industrial Estate, Heathhall, Dumfries DG1 3PH Tel 01387 256528 Fax 01387 256528 *(Courses T + P covering classes 2, 3, 4, 5, 6, 8 and 9).*
Langbank	Hazardous Goods Management Ltd, 8 Douglas Avenue, Langbank, Renfrewshire PA14 6PE *(Courses T + P covering classes 1, 2, 3, 4, 5, 6, 8 and 9).*

WALES

Anglesey	
Holyhead	BRAMA Training & Development Centre, Royal Air Force Valley, Holyhead, Anglesey LL65 3NY Tel 01407 762241 Fax 01407 742812 *(Courses T + P covering classes 1, 2, 3, 4, 5, 6, 8 and 9).*
Dyfed	
Pembroke Dock	Mainport Training, Pembroke Enterprise Centre, Kingswood Estate, Pembroke Dock SA72 4RS Tel 01646 684315 *(Courses T + P covering classes 2, 3, 4, 5, 6, 8 and 9).*
Flintshire	
Deeside	Deeside College, Kelsterton Road, Connahs Quay, Deeside CN5 4BR Tel 01244 834454 Fax 01244 834555 *(Courses T + P covering classes 1, 2, 3, 4, 5, 6, 8 and 9).*

Mid Glamorgan

Bridgend	DLT Training Ltd, Ogmore House, 7 Ogmore Crescent, Bridgend Industrial Estate, Bridgend CF31 3TE Tel 01656 656382 Fax 01656 649889 *(Courses T + P covering classes 1, 2, 3, 4, 5, 6, 8 and 9).*
Treharris	Pass Go Ltd, 32/34 Cardiff Road, Quakers Yard, Treharris CF46 5DU Tel 01443 413324 *(Courses T + P covering classes 1, 2, 3, 4, 5, 6, 8 and 9).*

South Glamorgan

Barry	Barry Training Services, Holt Buildings, Powell Duffryn Way, No 1 Dock, Barry CF62 5QS Tel 01446 739457 *(Courses T + P covering classes 2, 3, 4, 5, 6, 8 and 9).*
Cardiff	Wales & South West Training, 'Calon', 210 Marshfield Road, Marshfield, Nr Cardiff CF3 8TU Tel 01633 680710 *(Courses T + P covering classes 1, 2, 3, 4, 5, 6, 8 and 9).*
Swansea	Powerhouse Transport Solutions, The Feudrod Office Centre, Feudrod Way, Swansea Enterprise Park, Swansea SA7 9DG Tel 01792 793555 Fax 01792 793388 *(Courses T + P covering classes 2, 3, 4, 5, 6, 8 and 9).*
Wrexham	Gatewen Training Ltd, Gatewen, New Broughton, Wrexham LL11 8YA Tel 01978 720907 *(Courses T + P covering classes 2, 3, 4, 5, 6, 8 and 9).*

NORTHERN IRELAND

Coleraine	Bush Training Centre, 57 Ballylagan Road, Cloyfin, Coleraine BT52 2PQ Tel 01265 52653 *(Courses T + P covering classes 2, 3, 4, 5, 6, 8 and 9).*
Crumlin	Transport Training Services Ltd, 15 Dundrod Road, Nutts Corner, Crumlin, Co. Antrim Tel 02890 825653 *(Courses T + P covering classes 1, 2, 3, 4, 5, 6, 8 and 9).*
Limavady	Sandy Arthur Training Services, Unit 3, Aghanloo Industrial Estate, Limavady, Co. Londonderry BT49 0HE Tel 02877 722221 *(Courses T + P covering classes 2, 3, 4, 5, 6, 8 and 9).*

Appendix XI

DVLA LOCAL OFFICES AND NEW VEHICLE REGISTRATION MARKS

Since 1 January 2001 the DVLA's Vehicle Registration Offices have been called Local Offices to reflect the inclusion of driver licensing work and the potential for other Agency business to be undertaken. Telephone numbers are centralized – tel: 0870 850 0007.

Aberdeen
Greyfriars House,
Gallowgate,
Aberdeen AB10 1WG

Bangor
Penrhos Road,
Penrhosgarnedd,
Bangor LL57 2JF

Beverley
Crosskill House,
Mill Lane,
Beverley HU17 9JB

Birmingham
2nd Floor,
Edward House,
Edward Street,
Birmingham B1 2RF

Bournemouth
Bourne Gate,
25 Bourne Valley Road,
Poole BH12 1DX

Brighton
4th Floor, Mocatta House,
Trafalgar Place,
Brighton BN1 4UE

Bristol
Northleigh House,
Lime Kiln Close,
Stoke Gifford,
Bristol BS34 8SR

Cardiff
Archway House,
77 Ty Glas Avenue,
Llanishen,
Cardiff CF14 5DX

Carlisle
Ground Floor,
3 Merchants Drive,
Parkhouse,
Carlisle CA3 0JW

Chelmsford
2nd Floor,
Parkway House,
49 Baddow Road,
Chelmsford CM2 0XJ

Chester
Norroy House,
Nuns Road,
Chester CH1 2ND

Dundee
Caledonian House,
Greenmarket,
Dundee DD1 4QP

Edinburgh
Department of Transport,
Saughton House,
Broomhouse Drive,
Edinburgh EH11 3XE

Exeter
Hanover House,
Manaton Close,
Matford Business Park,
Marsh Barton,
Exeter EX2 8EF

Glasgow
46 West Campbell Street,
Glasgow G2 6TT

Inverness
Longman House,
28 Longman Road,
Inverness IV1 1SF

Ipswich
Podium Level,
St Clare House,
Greyfriars,
Ipswich IP1 1UT

Leeds
1st Floor, 42 Eastgate,
Leeds LS2 7DQ

Lincoln
Firth Court,
Firth Road,
Lincoln LN5 7WD

Luton
2 Dunstable Road,
Luton LU1 1EB

Maidstone
Coronet House,
11 Queen Anne Road,
Maidstone ME14 1XB

Manchester
Trafford House,
Chester Road,
Manchester M32 0SL

Newcastle upon Tyne
Eagle Star House,
Regent Farm Road,
Newcastle upon Tyne NE3 3QF

Northampton
Wootton Hall Park,
Northampton NN4 0GA

Norwich
11 Prince of Wales Road,
Norwich NR1 1UP

Nottingham
Block 6, Government Buildings,
Chalfont Drive,
Nottingham NG8 3RA

Oxford
Ground Floor, 3 Cambridge Terrace,
Oxford OX1 1RW

Peterborough
88 Lincoln Road,
Peterborough PE1 2ST

Portsmouth
5th Floor, Baltic House,
Kingston Crescent,
Portsmouth PO2 8AH

Preston
Fulwood Park,
Caxton Road,
Fulwood PR2 9NZ

Reading
77–81 Basingstoke Road,
Reading RG2 0ER

Sidcup
12/18 Station Road,
Sidcup DA15 7EQ

Sheffield
Cedar House, Hallamshire Court,
63 Napier Street,
Sheffield S11 8HA

Shrewsbury
Whitehall, Monkmoor Road,
Shrewsbury SY2 5DR

Stanmore
Government Building,
Canon Park,
Honeypot Lane,
Stanmore,
Middlesex HA7 1BD

Stockton
St Marks House,
St Marks Court,
Thornaby,
Stockton on Tees TS17 6QR

Swansea
Heol Pentre Felen,
Swansea SA6 7HG

Truro
Pydar House,
Pydar Street,
Truro TR1 2TG

Wimbledon
Ground Floor, Connect House,
133–137 Alexandra Road,
Wimbledon SW19 7JY

Worcester
Clerkenleap Barn,
Broomhall,
Worcester WR5 3HR

See table of registration marks overleaf.

NB: In the event of one office receiving an exceptionally high demand that depletes its stock of registration marks, marks may be transferred between DVLA Local Offices.

Source: Reproduced from DVLA website www.dvla.gov.uk, last updated 20 July 2003.

Local memory tag	*DVLA office*	*Local identifier*
A	Peterborough Norwich Ipswich	A B C D E F G H J K L M N O P R S T U V W X Y
B	Birmingham	A Y
C	Cardiff Swansea Bangor	A B C D E F G H J K L M N O P R S T U V W X Y
D	Chester Shrewsbury	A B C D E F G H J K L M N O P R S T U V W X Y
E	Chelmsford	A Y
F	Nottingham Lincoln	A B C D E F G H J K L M N P R S T V W X Y
G	Maidstone Brighton	A B C D E F G H J K L M N O P R S T U V W X Y
H	Bournemouth Portsmouth	A B C D E F G H J K L M N O P R S T U V W X Y
		HW Reserved for the Isle of Wight
K	Luton Northampton	A B C D E F G H J K L M N O P R S T U V W X Y
L	Wimbledon Stanmore Sidcup	A B C D E F G H J K L M N O P R S T U V W X Y
M	Manchester	A Y
N	Newcastle Stockton	A B C D E F G H J K M N O P R S T U V W X Y
O	Oxford	A Y
P	Preston Carlisle	A B C D E F G H J K M N O P R S T U V W X Y
R	Reading	A Y
S	Glasgow Edinburgh Dundee Aberdeen Inverness	A B C D E F G H J K L M N O P R S T U V W X Y
V	Worcester	A Y
W	Exeter Truro Bristol	A B C D E F G H J K L M N O P R S T U V W X Y
Y	Leeds Sheffield Beverley	A B C D E F G H J K L M N O P R S T U V W X Y

Appendix XII

UK BRIDGE AND TUNNEL TOLLS

Erskine Bridge

Tolls suspended for legislative reasons

Dartford River Crossing

The Dartford–Thurrock Crossing is now operated by a French consortium called Le Cross Company Ltd.

Vehicle	*Single journey*	*10 pm–6 am*
Motor car, including motor car with three wheels or taxi	£1.00	£1.00
Light or medium goods vehicle, motor coach, omnibus or tractor, having no more than two axles	£1.80	£1.00
Heavy goods vehicle, motor coach or omnibus having more than two axles	£2.90	£1.00
Special types vehicle	£2.90	£1.00

Trailer: the same as for the vehicle towing it

Certain invalid carriages: no charge if exempt from payment of Vehicle Excise Duty.

Dart-Tag

Discount charges apply to holder of Dart-Tags as follows:

Car	£0.93
2-axle goods vehicles	£1.67
Multi-axle goods vehicles	£2.69

Proportionate tolls apply where a trailer is drawn (ie double the normal rate).

Charges for Large Vehicles and/or Loads

Passage through the tunnels		
Passage between 6.00 am and 10.00 pm	Wider than 2.75 m Longer than 18.75 m	£42 + toll
Passage between 10.00 pm and 6.00 am	Wider than 2.75 m Longer than 18.75 m	£21 + toll
Passage at any time of day or night	Wider than 6.10 m Longer than 27.40 m Over 150,000 kg	£190 + toll
Height limit:	West tunnel	15 ft 9 in (4.8768 m)
	East tunnel	16 ft 6 in (5.0292 m)
Passage over the bridge		
Passage between 6.00 am and 10.00 pm	Wider than 3.65 m	£42 + toll
Passage between 10.00 pm and 6.00 am	Wider than 3.65 m	£21 + toll
Passage at any time of day or night	Wider than 6.10 m Longer than 27.40 m Over 150,000 kg	£190 + toll

Breakdown Recovery

Breakdown recovery of vehicles causing an obstruction on the carriageway within the Crossing area will be undertaken free of charge to the nearest place of safety at the Crossing.

Forth Road Bridge

NB: One-way tolling system: Pay (northbound); Nil (southbound).

Vehicle	
Motor cycles	Nil
Cars; goods vehicles and tractors having maximum weight not exceeding 3.5 tonnes; buses constructed or adapted to carry not more than 16 passengers:	
• when the vehicle is the subject of a valid composition agreement with the joint Board	72p
• otherwise	80p
Buses constructed or adapted to carry more than 16 passengers	£1.40
Goods vehicles and tractors having a maximum weight exceeding 3.5 tonnes:	
• when the vehicle is the subject of a valid composition agreement with the joint Board	£1.30
• otherwise	£2
Vehicles (notwithstanding that they fall within any of the classes herein described) displaying a badge issued by a local authority in terms of its powers under section 21 of the Chronically Sick and Disabled Persons Act 1970 and conveying either as driver or passenger the disabled person in respect of whom such badge was issued	Nil

Vehicles (notwithstanding that they fall within any of the classes herein described) which, in the opinion of the bridgemaster, require an escort provided by him to ensure the safety of the bridge and/or other road users	£26

Discount vouchers can be purchased online from www.feta.gov.uk/webpages/payonline.php providing savings of 10 per cent on cars and light goods vehicles and 35 per cent on heavy goods vehicles.

Humber Bridge

	Single crossing	*Discounted price per book of 20 tickets*
Cars and goods vehicles having a maximum weight not exceeding 3.5 tonnes Motor caravans	£2.50	£45.00
Goods vehicles having a maximum gross vehicle weight exceeding 3.5 tonnes but not exceeding 7.5 tonnes Vehicles within Class 2 above, with trailers Small buses (with seating for 9–16 passengers)	£4.50	£81.00
Goods vehicles having a maximum gross vehicle weight exceeding 7.5 tonnes with 2 axles Large buses (with seating for 17 or more passengers)	£10.00	£180.00
Goods vehicles having a maximum gross vehicle weight exceeding 7.5 tonnes with 3 axles	£13.40	£241.20
Goods vehicles having a maximum gross vehicle weight exceeding 7.5 tonnes with 4 or more axles	£16.70	£300.60

Abnormal Loads

An additional scale of charges applies to abnormal load movements across the bridge. See www.humberbridge.co.uk

Mersey Tunnel

	Standard	*Fast Tag*
Private/light goods vehicle up to 3.5 tonnes (ie under 1.525 tonnes unladen)	£1.30	£1.15
Private/light goods vehicle up to 3.5 tonnes, with trailer	£1.30	£1.15
Heavy goods vehicle over 3.5 tonnes, with two axles	£3.90	£3.45
Heavy goods vehicle over 3.5 tonnes, with three axles	£3.90	£3.45
Heavy goods vehicle over 3.5 tonnes, with four or more axles	£3.90	£3.45

Severn Bridges

Tolls

Cars and motor caravans	£4.60
Small goods vehicles and small buses	£9.30
Heavy goods vehicles and buses	£13.90

Tag Charges

	Monthly	*Quarterly*
Cars and motor caravans	£80.96	£242.88
Small goods vehicles and small buses	£163.68	£491.04
Heavy goods vehicles and buses	£275.22	£825.66

Breakdown Fees

Small vehicles	£40 for first 30 minutes £20 for each subsequent 15 minutes or part thereof
Goods vehicles over 3.5 tonnes gvw	£100 for the first 30 minutes £50 for each subsequent 15 minutes or part thereof

Skye Bridge

Vehicle	*Toll category*	*Toll*	*Book of discount tickets*
Car (and transit vans < 5.5 m)	High season	£5.70	£26.80 for 20 tickets
	Low season	£4.70	
LGV (< 7.5 tonnes gross)	High and low season	£10.80	£78.40 for 10 tickets
HGV 1 (2 or 3 axles)	High and low season	£14.00	£101.30 for 10 tickets
HGV 2 (4 axles or more)	High and low season	£27.90	£202.60 for 10 tickets

Tamar Bridge

	Cash toll	*Discount toll*	*Voucher book*
Vehicles up to 3.5 tonnes mgw and up to 8* seats Cars, vans, motorcycle combinations	£1	50p	£10 (10 vouchers)
Vehicles over 3.5 tonnes mgw and vehicles with more than 8* seats according to axle count:			
2 axles:	£2.50	£1.25	£50 (40 vouchers)
3 axles:	£4	£2	£80 (40 vouchers)
4+ axles:	£5.50	£2.75	£110 (40 vouchers
Vehicles towing trailers	charge is doubled		

Special toll rates apply to abnormal loads. See www.tamarbridge.org.uk

Tyne Tunnel

	Standard	*Permit holder*
Car or light van (up to 3,500 kg gross weight)	£1.00	£0.90
Goods vehicle or passenger-carrying vehicle (over 3,500 kg gross weight)	£1.20	£1.08

Recovery Charges

Motorcycles:	£5
Cars:	28 per hr/min £14
LGV:	28 per hr/min £14
Breakdown due to fuel shortage:	£17.29 per hr/min £8.30 + recovery charge
Breakdown parking:	£2 per day after 48hrs

Index

NB: page numbers in *italic* indicate figures and tables

INDEX OF ADVERTISERS